PHENOMENOLOGY OF PARTICLES AT HIGH ENERGIES

1 Dave Salisbury
2 Bob Shaw
3 Bill Leithead
4 Thom Curtright
5 Giorgio Immirzi
6 Umberto Trevisan
7 Georges Girardi
8 Bellisario Esposito
9 Gavin Finnie
10 Shimon Yankielowicz
11 Rex Fidler
12 Eliezer Rabinovici
13 Simon Ruysenaars
14 Ferenc Csikor
15 Spyros Vlassopulos
16 Jim Elliott
17 Hans-Jurgen Willutzki
18 Piet van der Stok
19 Krishnappa Subbarao
20 Ray Hughes
21 Christian Gilain
22 David Scott
23 Kenneth Evans
24 Alec Fitton
25 Haim Harari
26 Peter Osborne
27 Quentin Jones
28 Vijai Dixit
29 Peter Croft
30 Andrew Davies
31 Ken Wilson
32 Ken Bowler
33 Volker Hepp
34 Arne Rognebakke
35 Lindsay Tassie
36 Nigel Parsons
37 Marcel Margulies
38 Dietmar Menze
39 Sverker Fredriksson
40 Moshe Moshe
41 Harold Kirk
42 David Jackson
43 Ian Hughes
44 Dick Roberts
45 Robert Wang
46 Ron Crawford
47 Norman White
48 Bob Jennings
49 Uzi Bar-Gadda
50 Erwin Hilger
51 Fred Gilman
52 Julian Uschersohn
53 Namik Pak
54 Roger Cashmore
55 Flavio Constantini
56 Marjorie Stewart
57 Yoav Eylon
58 Ethylin Wang
59 Edward McGaffigan
60 Frederique Lalanne
61 Guy Plaut
62 Dominique Lalanne
63 David Dorren
64 Helmut Ackermann
65 Edmund Crouch
66 Norisuke Sakai
67 Sunanda Banerjee
68 Michael Thorne
69 Ken Peach
70 Nicholas Kemmer
71 Claude Gilain
72 Susan Thorne
73 Imogen Thorne
74 Isobel Hughes
75 Barbara Jackson
76 Ray Chester
77 Ayeleth Harari
78 Jo Ann Curtright
79 Aviva Harari
80 Sharon Harari
81 Hanna Moshe
82 Kinga Eskreys
83 Tamara Eisenschitz

PHENOMENOLOGY OF PARTICLES AT HIGH ENERGIES

Proceedings of the Fourteenth Scottish Universities
Summer School in Physics, 1973.

A NATO Advanced Study Institute

Edited by

R. L. CRAWFORD

and

R. JENNINGS

University of Glasgow, Scotland

1974

ACADEMIC PRESS · LONDON AND NEW YORK

A Subsidiary of Harcourt Brace Jovanovich, Publishers

ACADEMIC PRESS INC. (LONDON) LTD.
24/28 Oval Road,
London NW1

United States Edition published by
ACADEMIC PRESS INC.
111 Fifth Avenue
New York, New York 10003

Library of Congress Catalog Card Number: 74 5666
ISBN: 0 12 197150 3

PRINTED IN GREAT BRITAIN BY PAGE BROS (NORWICH) LTD, NORWICH

List of Participants

ORGANIZING COMMITTEE

HUGHES, PROFESSOR and MRS. I. S. *University of Glasgow, Scotland (Director).*
KEMMER, PROFESSOR N. *University of Edinburgh, Scotland (Director).*
BOWLER, DR. K. C. *University of Edinburgh, Scotland (Secretary).*
CRAWFORD, DR. R. L. *University of Glasgow, Scotland (Editor).*
DAVIES, DR. A. T. *University of Glasgow, Scotland (Treasurer).*
JENNINGS, DR. R. *University of Glasgow, Scotland (Editor).*
OSBORNE, DR. P. *University of Edinburgh, Scotland (Steward).*
PEACH, DR. K. J. *University of Edinburgh, Scotland (Steward).*

LECTURERS

CASHMORE, DR. R. J. *SLAC, Stanford, U.S.A.*
GILMAN, PROFESSOR F. J. *SLAC, Stanford, U.S.A.*
HARARI, PROFESSOR H. *Weizmann Institute, Israel.*
HIGGS, DR. P. W. *University of Edinburgh, Scotland.*
JACKSON, PROFESSOR J. D. *University of California, Berkeley, U.S.A.*
LLEWELLYN SMITH, DR. C. H. *CERN, Switzerland.*
ROBERTS, DR. R. G. *RHEL, Chilton, England.*
SENS, PROFESSOR J. C. *CERN, Switzerland.*
TOUNSI, DR. A. *University of Paris, France.*
WILSON, PROFESSOR K. G. *Cornell University, U.S.A.*

STUDENT PARTICIPANTS

ACKERMANN, MR. H. *DESY, W. Germany.*
AMENDOLIA, MR. S. R. *University of Pisa, Italy.*
BANERJEE, MR. S. *Imperial College, London, England.*
BAR-GADDA, MR. U. *SLAC, Stanford, U.S.A.*
BARTOLI, DR. B. *Frascati National Laboratory, Italy.*
CONSTANTINI, DR. F. *University of Pisa, Italy.*
CROUCH, MR. E. A. C. *Cavendish Laboratory, Cambridge, England.*
CSIKOR, DR. F. *Eötvös University, Budapest, Hungary.*
CURTWRIGHT, MR. T. *Caltech., Pasadena, U.S.A.*

DIXIT, DR. V. V. *Purdue University, U.S.A.*
DORREN, DR. J. D. *ETH Zurich, Switzerland.*
ELLIOTT, MR. J. R. *Duke University, U.S.A.*
ESKREYS, DR. K. *Institute of Nuclear Physics, Krakow, Poland.*
ESPOSITO, DR. B. *Frascati National Laboratory, Italy.*
EVANS, MR. K. J. *DAMTP, Cambridge, England.*
EYLON, MR. Y. *Weizmann Institute, Israel.*
FIDLER, MR. R. *DAMTP, Cambridge, England.*
FINNIE, MR. G. *University of Glasgow, Scotland.*
FITTON, MR. A. *University of Durham, England.*
FREDRIKSSON, MR. S. *Royal Institute of Technology, Stockholm, Sweden.*
GILAIN, MRS. C. *Saclay, France.*
GILES, MR. R. C. *SLAC, Stanford, U.S.A.*
GIRARDI, MR. G. *Saclay, France.*
GORCZYCA, DR. BARBARA, *DESY, W. Germany.*
GRIFFITH, DR. R. W. *University of Bonn, W. Germany.*
HEPP, DR. V. *University of Heidelberg, W. Germany.*
HILGER, DR. E. *University of Bonn, W. Germany.*
HUGHES, MR. R. P. *DAMTP, Cambridge, England.*
IMMIRZI, DR. G. *University of Lecce, Italy.*
KARLINER, MISS INGA, *SLAC, Stanford, U.S.A.*
KIRK, DR. H. G. *Physikalisches Institut, Aachen, W. Germany.*
LALANNE, DR. D. *Orsay, France.*
McGAFFIGAN, MR. E. *Caltech., Pasadena, U.S.A.*
MARGULIES, MR. M. *Colombia University, U.S.A.*
MENZE, DR. D. *University of Bonn, W. Germany.*
MEUNIER, DR. J.-L. *University of Nice, France.*
MOSHE, MR. M. *Caltech, Pasadena, U.S.A.*
PAK, DR. N. *Hacettepe University, Turkey.*
PARSONS, MR. N. *University of Glasgow, Scotland.*
PLAUT, DR. G. *University of Nice, France.*
RABINOVICI, MR. E. *Weizmann Institute, Israel.*
RODRIGUES, DR. W. A. *University of Campinas, Brasil.*
ROGNEBAKKE, MR. A. *University of Bergen, Norway.*
RUYSENAARS, MR. S. N. M. *Institut Lorentz, Leiden, Holland.*
SAKAI, DR. N. *Max Planck Institute, W. Germany.*
SALISBURY, MR. D. *University of Sussex, England.*
SCOTT, MR. D. M. *DAMTP, Cambridge, England.*
SHAW, MR. K. R. *Cavendish Laboratory, Cambridge, England.*
SHKLOVSKAYA, MRS. ALLA Y. *Bulgarian Academy of Sciences, Sofia, Bulgaria.*
STOK, MR. P. D. V. VAN DER, *Zeeman Laboratory, Amsterdam, Holland.*
SUBBARAO, DR. K. *Cornell University, U.S.A.*
TASSIE, DR. L. J. *Australian National University, Canberra, Australia.*
THORNE, MR. M. C. *University of Sheffield, England.*
TREVISAN, DR. U. *University of Genoa, Italy.*

USCHERSOHN, MR. J. *University of Lund, Sweden.*
VLASSOPULOS, MR. S. D. P. *University of Durham, England.*
WANG, MR. R. *Caltech., Pasadena, U.S.A.*
WHITE, DR. J. N. J. *Max Planck Institute, W. Germany.*
WILLUTZKI, MR. H.-J. *DESY, W. Germany.*
YANKIELOWICZ, MR. S. *Weizmann Institute, Israel.*
YU, DR. W-Y, *ICTP, Trieste, Italy.*

Directors' Preface

The 1973 Scottish Universities Summer School was held at Middleton Hall near Edinburgh from 29 July to 18 August supported by grants from the Scottish Universities and the NATO Scientific Affairs Division programme for Advanced Study Institutes. The subject of the school was Phenomenology of Particles at High Energies. This was chosen as being particularly timely in view of the new and exciting data flowing from the CERN Intersecting Storage Rings and NAL, and in view of the very considerable developments in phenomenological theory for high energy processes. We were fortunate in being able to recruit a group of excellent lecturers who covered nearly all aspects of the subject in some depth as can be gauged from the contents of this volume. We also had a number of seminars from lecturers and students which were an important contribution to the school, although for reasons of availability of space we have been unable to include them in this book.

The school had its traditional world wide spread of students and of teachers and on this occasion was deliberately aimed at a mixed group of experimental and theoretical physicists. We believe that this resulted in benefits for both sets of students particularly in the field in question where the interaction between theory and experiment is presently so stimulating in the best classical tradition.

The task of the directors of this school was a particularly pleasant one due on the one hand to the collaboration of lecturers and students, and on the other to the careful and unsparing work of the officials of the organizing committee, the stewards Dr. Osborne and Dr. Peach, the treasurer Dr. Davies and most of all the secretary Dr. Bowler.

November 1973

I. S. Hughes
N. Kemmer
Directors

Editors' Note

These articles are based on the talks given at the School and have been written by the lecturers.

R. L. Crawford
R. Jennings

Contents

1. INTRODUCTION TO HADRONIC INTERACTIONS AT HIGH ENERGIES

J. D. JACKSON

Department of Physics and Lawrence Berkeley Laboratory, University of California, Berkeley, California 94720

I. Introduction

The subject of the phenomenology of hadronic interactions at high energies is extremely broad. In these lectures it is necessary to limit the discussion to only a few topics and merely to scratch the surface of most of these. Since my charge is to provide an introduction on which concurrent and subsequent lectures can build, I stick to basics. Furthermore, my approach is down-to-earth in the extreme. Elegance and rigor take a back seat to "Anschaulichkeit" and intuitive understanding. The lectures are thus aimed at plain and simple folk. Theorists and other sophisticates may, while wincing, also benefit, but they are not the intended audience.

In strong interactions at high energies for the past year and a half the scene has been dominated by the great extension of the energy range for controlled experimentation made available by the Intersecting Storage Rings (ISR) at CERN, Geneva, Switzerland, and the proton synchrotron at the National Accelerator Laboratory, Batavia, Illinois (NAL). Prior to the start-up of these facilities the highest available energy was at Serpukhov in the U.S.S.R. where a 70 GeV proton beam provided a c.m.s. energy $W \simeq 11{\cdot}5$ GeV in collisions with a stationary nucleon. Now we have available c.m.s. energies of $W = 20$–60 GeV at the ISR in proton–proton collisions and $W \sim 20$–27 GeV at NAL from 200–400 GeV protons or mesons on a stationary nucleon target. For the first time we have been able to look in detail at truly high energy phenomena. At these energies the symbol, "$\gg$", in the statement $W \gg m_N$ almost takes on its rigorous mathematical meaning, not merely the physicist's interpretation of somewhat larger than or at least slightly greater than! Of course, many of the phenomena observed at the ISR and NAL have been known, in outline at least, for many years from experiments with cosmic rays, as the experts in that field hasten to point out (e.g. Feinberg, 1972). Nevertheless, the wealth of detail possible with controlled experimental conditions and intense beams of particles has meant that gross features seen in cosmic rays are now firmly established and finer details, energy dependences, or rare processes are being explored for the first time.

Since the time of Rutherford elastic scattering has occupied an honorable position among phenomena designed to elucidate the forces between particles and their structures. Inelastic scattering, too, has played its role in the study of structures and excitations since the days of Franck and Hertz. In particle physics both elastic and inelastic scattering have been pursued continuously over the years, but with the availability of meson and baryon beams with energies of several GeV and the attendant production of particles emphasis shifted to the detailed study of the complete final state of three, four, or more particles. Quasi-two-body processes with one or two resonances in the final state were studied, decay correlations and energy dependences observed. A tremendous amount of information was and still is being accumulated on these so-called *exclusive* processes and a semi-quantitative understanding and theoretical framework was gained. Some aspects of this are summarized in Chapter 2. As higher energies became available the most probable occurrence in a hadronic collision was the production of many particles. Two-body or quasi-two-body channels were relatively improbable. Attempts were made to study resonance formation among the final state particles and to interpret the various invariant mass plots within some theoretical framework (e.g. n-point Veneziano amplitudes or Van Hove longitudinal phase space plots). At energies like Serpukhov or higher, however, the average number of produced particles is so large that little can be learned from exclusive experiments—the number of degrees of freedom is just too great. It is necessary to fall back on simpler things—total cross section measurements, topological cross sections, single particle production spectra (with elastic and inelastic scattering of the incident particle as special cases), two-particle correlations and perhaps slightly more complicated situations. We speak then of *inclusive* experiments or processes. The total cross section σ_{ab} is the zero-particle inclusive process—$a + b \rightarrow$ anything. The reaction $a + b \rightarrow c +$ anything, where the type of particle (c), its momentum and perhaps spin, is all that is observed, is called a single-particle inclusive process, and so on. Over the past three or four years inclusive processes have become an industry at least as large as the quasi-two-body industry once was. Counter experimenters have found their beam surveys upgraded in theoretical respectability and bubble chamber physicists have been able to get publishable results from the tremendous number of previously useless unfitted events. At first glance it might seem surprising that much of interest could come out of single-particle spectra, but we will see in Section III that their simplicity allows application of ideas closely related to two-body phenomenology.

The most important single feature of hadronic interactions to be discovered with the extension of the available energy range is the rising total cross section for proton–proton collisions [again, this was anticipated somewhat from cosmic ray evidence (Yodh, Pal, and Trefil, 1972)]. This raises the question of bounds on total and differential cross sections and other aspects. These topics are discussed in Section II after a review of the basic experimental facts and the general theoretical framework. The Froissart bound on total cross sections, a different treatment of the ratio of real to imaginary part of the forward scattering amplitude, discussion of the partial wave (impact parameter) distribution for p–p scattering, the MacDowell–Martin bound and the connection between the energy dependence of amplitudes and their J-plane structure completes this chapter. An introduction to inclusive processes is given in Section III. Again the basic facts are presented, followed by a discussion of the main theoretical ideas via the Feynman–Wilson fluid analogy. The relation between the fluid analogy and the Mueller–Regge description is outlined briefly, as is the "two-component" model of prong cross sections. A series of appendices summarize notation and some details that would burden and disturb the flow of argument in the text proper. A major omission is the discussion of processes involving large transverse momenta. Reliable data are just beginning to emerge and there are some fascinating theoretical speculations, but that could take a whole lecture series in itself.

References are cited in the text by authors and year of publication and are given in full in the bibliography, alphabetically by first author. Papers from conference or "summer" school proceedings are cited in the text in the same manner, but are listed in the bibliography by the conference location. The full citations for the conferences are given at the beginning of the bibliography. In such a rapidly developing field as high-energy physics the best sources of background information and leads to more detail are the conference proceedings, summer school notes, and the review literature (*Physics Reports, Reviews of Modern Physics*). Two recent books are noteworthy—D. Horn and F. Zachariasen, "Hadron Physics at Very High Energies", W. A. Benjamin, Reading, Mass. (1973) and R. P. Feynman, "Photon-Hadron Interactions", W. A. Benjamin, Reading, Mass. (1972).

II. Total Cross Sections, Elastic Scattering and Two-body Processes

1. Basic Facts and Samples of Data

The most recent reviews of total cross sections and elastic scattering are those of Diddens (1972), Giacomelli (1972), and Amaldi (1973), and on two-body and quasi-two-body inelastic processes those by Chiu (1972), Michael (1972), Phillips and Ringland (1972), Barloutaud (1973), and Fox and Quigg (1973). I shall lean heavily on these and other reviews both in the topics I discuss and for excuse on the topics I omit.

While the emphasis in this chapter is more on total cross sections and elastic scattering at high energies than on processes with nontrivial quantum number exchanges, it is expedient to summarize the gross empirical facts and main theoretical concepts for all two-body processes:

"(i) There exist $SU(3)$ singlets and octets of mesons, and singlets, octets, and decimets of baryons, of a variety of different spins and parities. Some of these mesons and baryons are stable, apart from electromagnetic or weak decays. Others appear as resonant states in scattering or production experiments.

(ii) The quantum numbers of the observed meson and baryon multiplets can be generated by the mnemonic of the quark model, with $(\bar{q}q)$ for the mesons and (qqq) for the baryons. (This particular empirical fact will need modification as soon as any "exotic" resonance is firmly established.)

(iii) Two-body and quasi-two-body processes are peripheral, showing peaking at forward directions (small t) and/or backward directions (small u).

(iv) Integrated cross sections, or differential cross sections at fixed momentum transfer, show approximate power-law behaviour in the energy. In particular, total cross sections seem to become constant asymptotically and obey Pomeranchuck's theorem.

(v) Virtually all occurrences or nonoccurrences of peripherality in a given process (iii) can be understood in terms of the exchanges of the internal quantum numbers of the known $SU(3)$ multiplets of mesons and baryons (i).

(iv) A modest amount of analyticity in the kinematic invariants, plus crossing symmetry, relates the phase of an amplitude at high energies to its power-law behaviour (iv). This connection is the same as, but more general than, that given by Regge pole theory.

(vii) The known mesonic and baryonic states (i) can plausibly be placed on Regge trajectories, and the trajectories are approximately linear in the square of the masses. This gives great impetus to the use of Regge exchanges to unify items (iv), (v), and (vi) into an aesthetically pleasing whole."

The above seven points were written down four years ago (Jackson, 1970) and are subject to only slight modification. On item (ii) there is increasing evidence, though not yet overwhelming, of the existence of exotic baryonic resonances (see Lovelace, 1972). In point (iv) the statement that "total cross sections seem to become constant asymptotically" should be omitted. Total cross sections may become constant asymptotically or they may not. As we shall discuss subsequently in detail, at the highest available energies total cross sections show energy dependence. If they become constant ultimately, it occurs at very much higher energies.

(a) *Total cross sections*

The high-energy behaviours of total cross sections of $\bar{p}$, p, π^-, π^+, K^-, and K^+ on protons are shown in Fig. 1(a), taken from Denisov *et al.* (1971). With the exception of the K^+p total cross section, which shows a very slight rise, all the cross sections fall smoothly from 5 to 60 GeV/c incident momentum. The total cross section differences, $\Delta\sigma = \sigma_t(\bar{x}p) - \sigma_t(xp)$ with $x = p$,

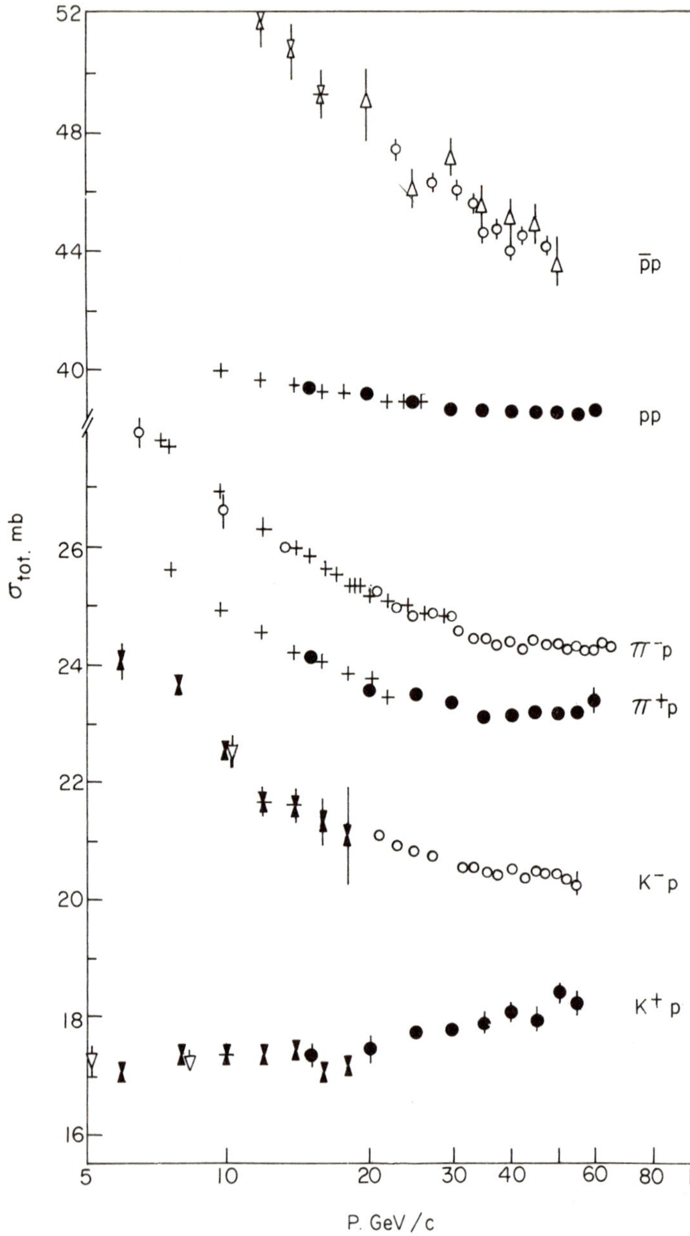

FIG. 1(a). Total cross sections in millibarns for $\bar{p}$, p, π^-, π^+, K^-, K^+ on protons versus incident laboratory momentum in GeV/c (from Denisov *et al.*, 1971).

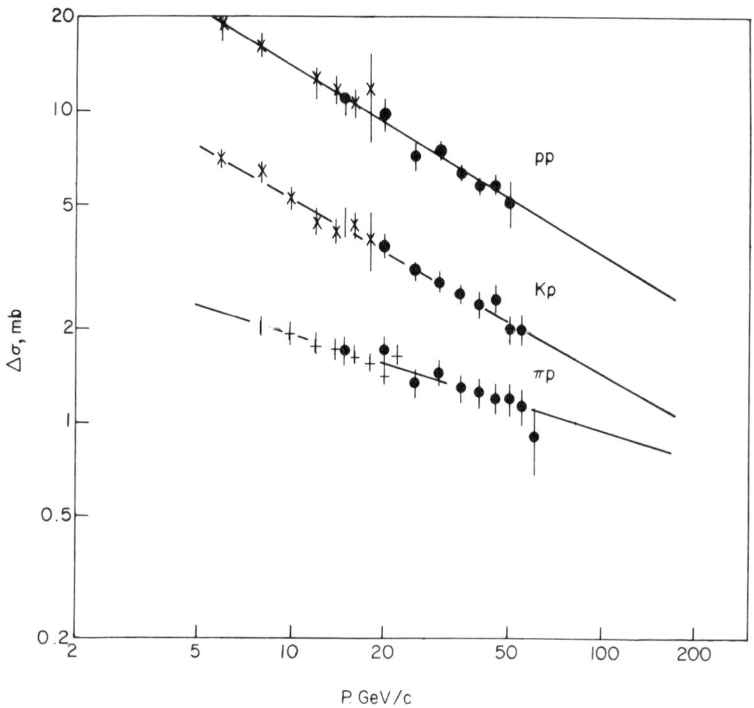

FIG. 1(b). Cross section differences in millibarns versus incident laboratory momentum in GeV/c on a log–log plot (from Denisov *et al.*, 1971).

π^+, K^+, are displayed in Fig. 1(b) on a log–log plot. The differences can be fitted by a power-law form $\Delta\sigma_i = A_i/P^{n_i}_{\text{lab}}$, with $n = 0{\cdot}64 \pm 0{\cdot}02$, $0{\cdot}54 \pm 0{\cdot}02$, and $0{\cdot}32 \pm 0{\cdot}02$ for $(\bar{p}, p)$, (K^-, K^+), and (π^-, π^+) differences, respectively (Table 4 of Giacomelli, 1972). This power-law behaviour supports the first part of statement (iv) above, and Fig. 1(a) the now discredited second part.

The constancy of total cross sections at high energies, so nicely indicated in Fig. 1(a), received a jolt with the commencement of operation of the ISR at energies equivalent to 300 to 2000 GeV incident in the laboratory. Right from the beginning there were rumours of large cross sections (45–50 mb). Furthermore, an analysis of cosmic ray data on the very high energy proton flux at an atmospheric depth of 550 gm/cm^2 on Mt. Chacaltya in Bolivia, compared with the flux at the top of the atmosphere, gave evidence that the nucleon–nucleon total cross section increased with energy significantly at laboratory energies above 500 GeV (Yodh, Pal, and Trefil, 1972). Data from the ISR were published early in 1973 (Amaldi *et al.*, 1973a, b; Amendolia *et al.*, 1973; Braccini, 1973). These and other results on the proton–proton total cross section are displayed in Fig. 2. The dashed curve is a lower bound deduced from analysis of the cosmic ray data. The data show that the asymptotic constancy inferred from the left-hand side of Fig. 2 (and all of Fig. 1a) is only a local minimum at $E \sim 100$ GeV and that the cross section rises from this minimum of 38·5 mb to 43^+ mb at the highest ISR energy of $W \simeq 60$ GeV.

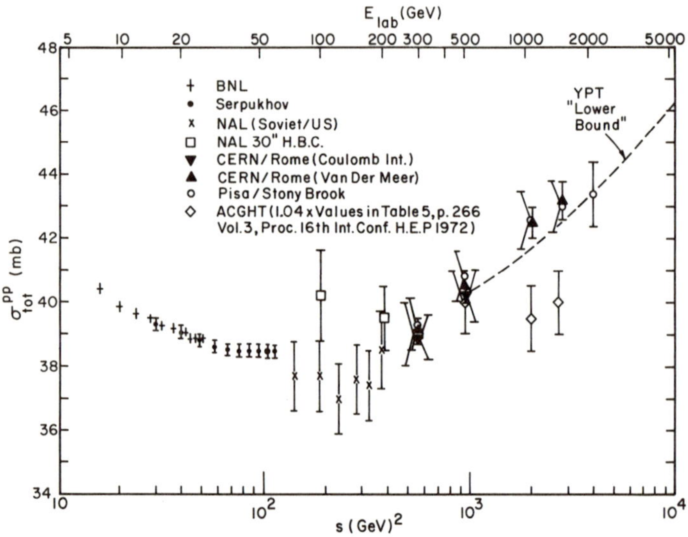

FIG. 2. Total cross section in millibarns for *pp* collisions versus $s = W^2$ in GeV^2 (bottom scale) and incident laboratory energy in GeV (top scale). The dashed curve is a lower bound estimated from cosmic ray data (Yodh, Pal, and Trefil, 1972).

For incident particles other than protons, data at energies higher than 70 GeV are almost nonexistent. While the next year will bring many results from NAL, at present the only very high-energy datum is $\sigma_t = 24{\cdot}0 \pm 0{\cdot}5$ mb at 205 GeV incident energy for π^- on protons in the NAL 30″ hydrogen bubble chamber (Huson, 1973). From Fig. 1(a) it can be seen that this result throws no light on the question of constancy versus rise of the $\pi^- p$ cross section.

(b) *Differential cross sections for elastic scattering*

The well-known peripheral nature of elastic scattering at high energies is

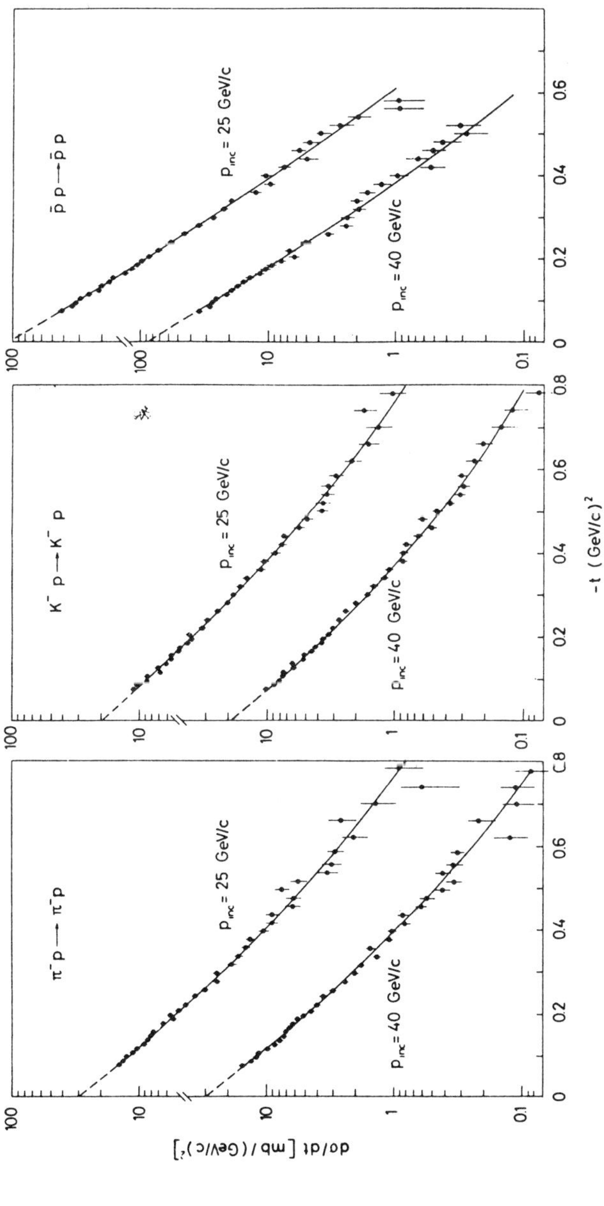

FIG. 3. Differential cross sections $d\sigma/dt$ for elastic scattering of $\pi^- p$, $K^- p$, and $\bar{p}p$ at 25 and 40 GeV/c incident momentum and $0 < |t| < 0{\cdot}8$ (GeV/c)^2 (Serpukhov data, Fig. 17 of Giacomelli, 1972).

illustrated for $\pi^- p$, $K^- p$, and $\bar{p}p$ scattering in Fig. 3. For small momentum transfers the cross sections are fitted roughly by $\exp(Bt)$ where $t = -q^2$ is the invariant momentum transfer variable (see Appendix A) and $B \simeq 7{\cdot}8, 8{\cdot}7$, and $11{\cdot}5\,(\text{GeV}/c)^{-2}$ for $K^- p$, $\pi^- p$, and $\bar{p}p$, respectively. In naive geometrical terms these "slope" parameters correspond to an extended scattering region with root mean square impact parameter $\langle b^2\rangle^{\frac{1}{2}} = (2B)^{\frac{1}{2}} = \sqrt{\{0{\cdot}0389[2B(\text{GeV}/c)^{-2}]\}}\,\text{fm} \simeq 0{\cdot}78\text{–}0{\cdot}95\,\text{fm}$.

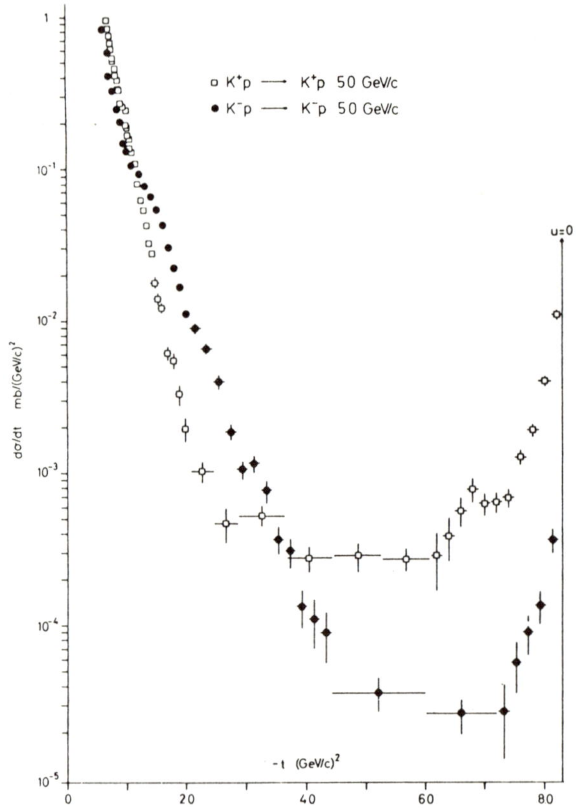

FIG. 4(a). K^+p and K^-p elastic differential cross sections over the entire momentum transfer range at 5 GeV/c (from Chabaud *et al.*, 1972).

That there is some structure within the forward peak and also backward peaks of various sizes is indicated by the data shown in Fig. 4. The K^-p differential cross section snakes back and forth around the smooth and featureless K^+p cross section. In the very forward direction ($|t| < 0{\cdot}2(\text{GeV}/c)^2$, not shown in Fig. 4a) the K^-p cross section is larger and falls off more rapidly than the K^+p cross section. For $0{\cdot}2 < |t| < 1{\cdot}0\,(\text{GeV}/c)^2$ the K^-p cross section is smaller than the K^+p, but at $|t| \sim 1{\cdot}0$ it crosses over and lies above the K^+p. Only beyond $|t| \sim 3{\cdot}5$ does it fall below and stay much below the K^+p cross section, each having a backward peak. The hints of diffraction maxima and minima in the K^-p cross section are more than hints in the $\bar{p}p$

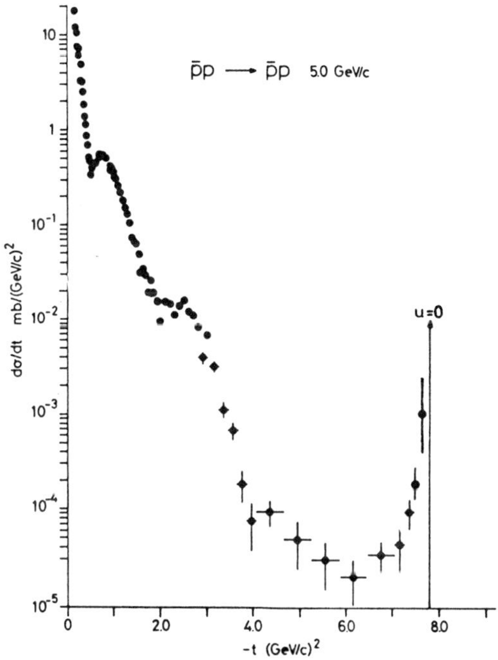

FIG. 4(b). $\bar{p}p$ elastic differential cross section over the entire momentum transfer range at 5 GeV/c (from Chabaud *et al*., 1972).

cross section shown in Fig. 4(b). By contrast, the pp differential cross section at the same momentum is extremely smooth, as can be seen in the compilation of Fig. 5. The evidence of Figs 1, 4, and 5 indicates that in geometrical terms $\bar{p}p$ and K^-p interactions correspond to larger absorbing regions, with more sharply defined edges, than pp and K^+p interactions.

The differential cross section for proton–proton elastic scattering at various energies is summarized in Fig. 5. The energy dependence is quite striking. The smooth behaviour at "low" energies gradually evolves into structure at $|t| \sim 1\text{–}2\ (\text{GeV}/c)^2$ at ISR energies as the cross section "shrinks" (becomes compressed to smaller and smaller $|t|$ values). The shrinkage of the very small $|t|$ region is best described by the energy dependence of the "slope parameter" B, defined by

$$B(s, t) = \frac{\mathrm{d}}{\mathrm{d}t} \ln\left[\frac{\mathrm{d}\sigma}{\mathrm{d}t}(s, t)\right]. \tag{1}$$

To the extent that B changes slowly with t this is equivalent to writing the differential cross section as

$$\frac{\mathrm{d}\sigma}{\mathrm{d}t}(s, t) = \frac{\mathrm{d}\sigma}{\mathrm{d}t}(s, 0) \exp\left[B(s, t)t\right]. \tag{2}$$

Furthermore, when we speak of *the* slope parameter $B(s)$ we mean $B(s, 0)$, or more commonly, some sort of an average value obtained by fitting $\mathrm{d}\sigma/\mathrm{d}t$ with an exponential in $|t|$ at small $|t|$. For p–p scattering the slope parameters $B(s, t)$ for two different small $|t|$ ranges are shown as functions of energy in Fig. 6. From 5 GeV/c to 2000 GeV/c laboratory momentum B increases by about 50 percent. At high energies simple Regge theory would predict $B(s, t) = A_1(t) \ln s + A_2(t)$, corresponding to a straight line on Fig. 6. For $P_{\text{lab}} > 10$ GeV/c the data are consistent with such a variation, but at ISR energies it is possible that the shrinkage has stopped at least momentarily. More will be said on this question in Section 5 below. At larger $|t|$ values, too, data from the ISR (the $W = 53$ GeV results are shown in Fig. 5) are consistent with little energy dependence from $W = 30$ to 53 GeV (Strolin, 1973).

(c) *Power law behaviour*

Part of the lore of high-energy phenomenology is that differential cross sections at fixed t, or cross sections integrated over the forward (or backward) peaks, show power law behaviour in energy and that the power depends on the reaction mechanism (Morrison, 1970). Simple Regge theory predicts, for the exchange of a single Regge pole in the t-channel,

$$\frac{\mathrm{d}\sigma}{\mathrm{d}t}(s, t) \simeq \beta(t) s^{2\alpha(t)-2}, \tag{3}$$

where $\alpha(t)$ is the trajectory function of the pole.

Figure 7 shows six examples of integrated cross sections for inelastic processes with nontrivial quantum number exchange in the t-channel (graphs from HERA reports, Bracci *et al.*, 1972a, b). All these processes show

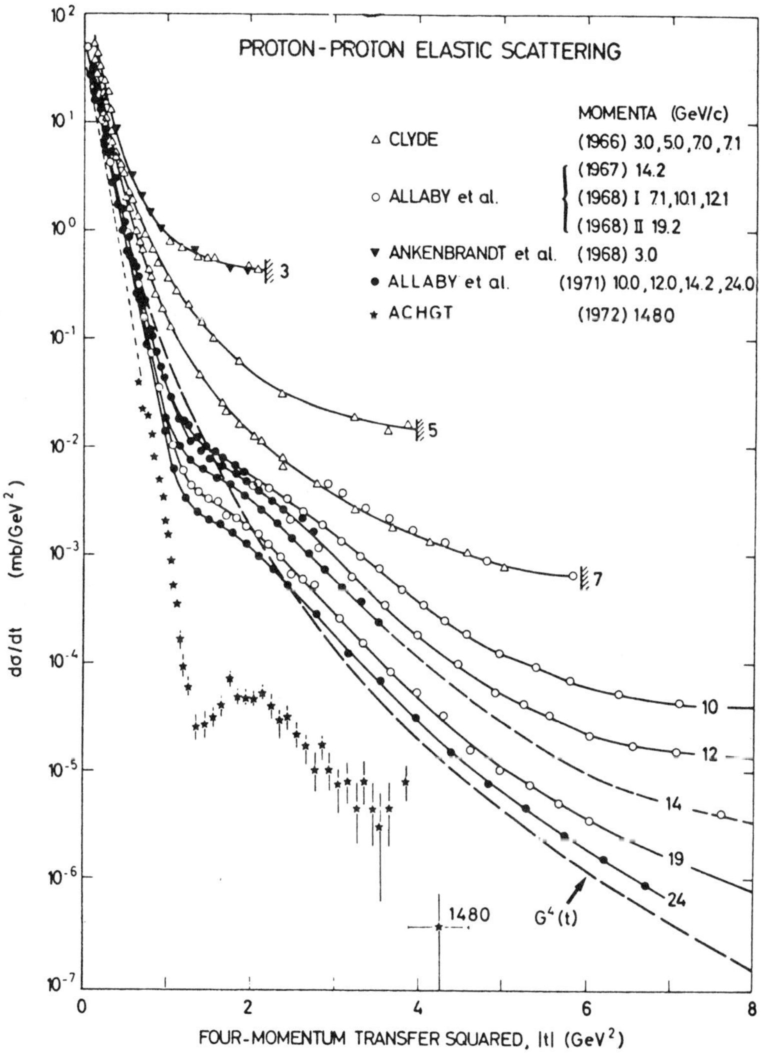

FIG. 5. Compilation of differential cross sections for p–p elastic scattering at various energies. The incident laboratory momentum is indicated at the right-hand end of each curve (Fig. 15 of Giacomelli, 1972).

a narrow forward peak in t; the integrated cross sections should, according to (3), show a power law behaviour with an exponent corresponding to a value of $\alpha(t)$ at some small negative value of t. The compilers have in each case fitted a power law in P_{lab} ($\propto s$) to the higher energy data. The six reactions,

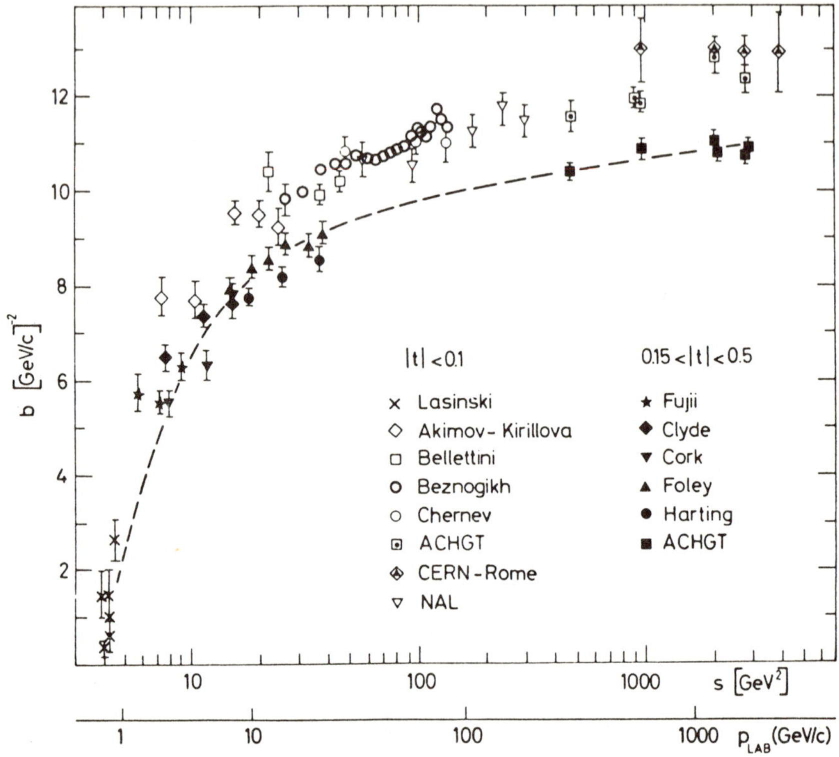

FIG. 6. The slope parameter $B(s, t)$ in $(\text{GeV}/c)^{-2}$ for p–p elastic scattering versus s in $(\text{GeV})^2$ for $0 < |t| < 0{\cdot}1$ (open points) and $0{\cdot}15 < |t| < 0{\cdot}5$ $(\text{GeV}/c)^2$ (solid points) (Fig. 14 of Giacomelli, 1972).

the anticipated t-channel exchanges, and the effective values of α deduced from the exponents, are

Reaction	*Exchanges*	*Effective* α
$\pi^- p \to \pi^0 n$	ρ	$0{\cdot}41 \pm 0{\cdot}07$
$\pi^+ p \to \pi^0 \Delta^{++}$	ρ	$0{\cdot}48 \pm 0{\cdot}09$
$\pi^- p \to \eta^0 n$	A_2	$0{\cdot}24 \pm 0{\cdot}09$
$K^- p \to \bar{K}^0 n$	ρ, A_2	$0{\cdot}28 \pm 0{\cdot}12$
$\pi^- p \to \rho^0 n$	π, A_2	$0{\cdot}06 \pm 0{\cdot}05$
$\pi^+ p \to \rho^+ p$	π, ω, A_2	$0{\cdot}04 \pm 0{\cdot}09$.

The first two reactions are classic ρ-exchange processes. The value $\alpha \simeq 0{\cdot}4$–$0{\cdot}5$ is in excellent agreement with a linear Regge trajectory $\alpha \simeq \frac{1}{2} + t$ that

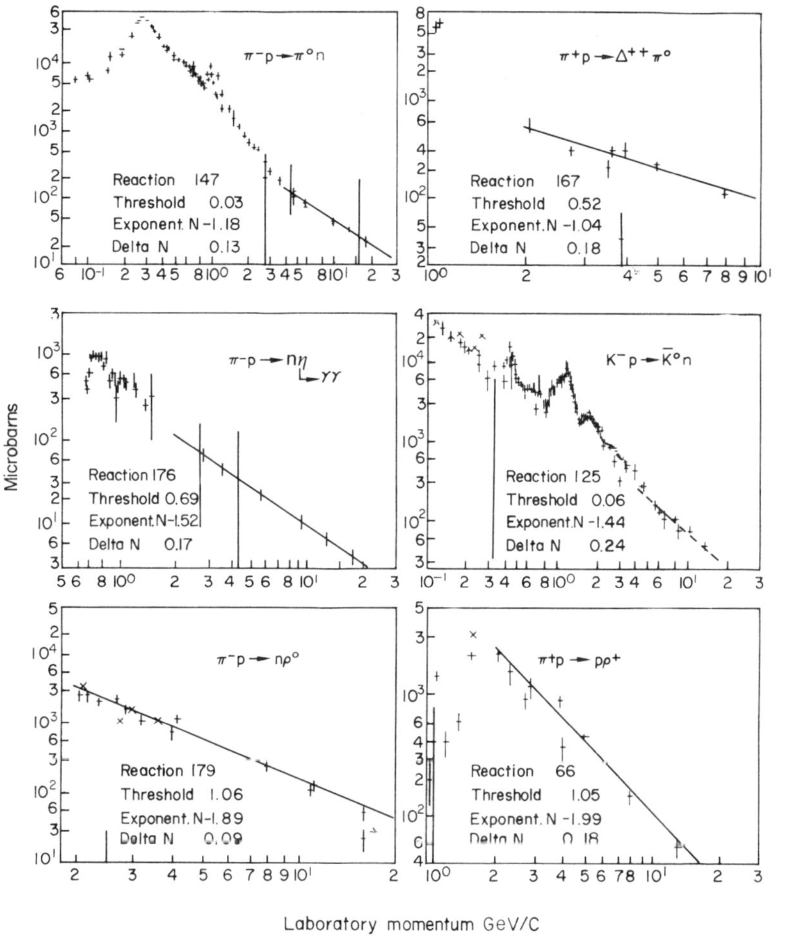

FIG. 7. Examples of the energy dependences of cross sections for inelastic processes with the exchange of mesonic quantum numbers in the t-channel. The data above a few GeV/c incident momentum are fitted with a power law, P^n_{lab}. The exponents are given on each graph (Bracci *et al.*, 1972a, b).

passes through the ρ ($J = 1^-$, $m_\rho^2 = 0{\cdot}58 \pm 0{\cdot}10\ \text{GeV}^2$) and the g ($J = 3^-$, $m_g^2 = 2{\cdot}82 \pm 0{\cdot}27\ \text{GeV}^2$). The third and fourth processes involve A_2 or A_2 and ρ exchange and seem to have a smaller effective α. Part of this may be a result of broader differential cross sections which sample more negative t values and so smaller $\alpha(t)$, but there is probably a residue that is evidence for the breaking of exact exchange degeneracy (EXD) of the even and odd signatured Regge trajectories. The final pair of reactions, ρ-meson production with and without charge transfer in the t-channel, show a much faster falloff with energy ($\alpha \simeq 0^+$). This behaviour is consistent with pion exchange as the dominant mechanism, at least in the forward peak, and is supported by the characteristic population of the zero helicity state of the ρ-meson.

Other examples of power law behaviour are less easy to understand, but the general trends are consistent and satisfying from a Regge exchange point of view. Barloutaud (1973) cites several more examples, including hypercharge (K^*, K^{**}) exchange reactions.

A final observation on power law behaviour is that it has been traditional to parametrize the total cross sections shown in Fig. 1(a) with the form,

$$\sigma_i = a_i + b_i P_{\text{lab}}^{-n_i} \tag{4}$$

with $n_i \simeq \frac{1}{2}$, consistent with the intercept $\alpha(0) \simeq 0{\cdot}5$ of the high lying Regge trajectories (ρ, ω, P', A_2). The range of exponents associated with the $\Delta\sigma$'s of Fig. 1(b) and more particularly the rising cross section shown in Fig. 2 show that (4) is at best a rough parametrization over a limited energy interval.

2. CROSSING SYMMETRY, SIGNATURE, POWER LAW BEHAVIOUR AND PHASE, POMERANCHUK THEOREMS

In order to discuss the phenomenology of total and elastic cross sections there are a few basic ideas that must be mentioned. These follow from the substitution law of field theory and from a modest amount of analyticity in the kinematic variables.

(a) *Substitution law*

Consider the process,

$$A + B \to C + D \tag{5}$$

where the 4-momenta of the particles are a_μ, b_μ, c_μ, d_μ, respectively. For simplicity suppose that the particles are spinless. Then there is one invariant amplitude $\mathscr{M}(s, t, u)$ describing the scattering, where according to Appendix A.5 s, t, and u are the standard kinematic variables, $s = (a + b)^2$ being the square of the total energy in the c.m.s., and $t = (a - c)^2$, $u = (b - c)^2$ being momentum transfer variables. From field theory it is known, and it is generally accepted as having wider applicability, that the amplitudes for other related processes can be obtained from $\mathscr{M}(s, t, u)$ by substitution according to the *substitution law*. For example, if we leave the 4-momenta of B and D unchanged but substitute $a_\mu \to -\bar{a}_\mu$ and $c_\mu \to -\bar{c}_\mu$, then $s \to s' = (\bar{a} - b)^2$, $t \to t' = (\bar{a} - \bar{c})^2$, $u \to u' = (\bar{c} + b)^2$, and

$$\mathscr{M}(s, t, u) \to \mathscr{M}' = \mathscr{M}(s', t', u')$$

with the amplitude $\mathcal{M}'$, which is just the old amplitude $\mathcal{M}$ evaluated at a different point in the (s, t, u) space, describing the process,

$$\bar{C} + B \to \bar{A} + D, \tag{6}$$

where $\bar{A}$ and $\bar{C}$ are the antiparticles of A and C, respectively. Process (5) is called the s-channel process, process (6) the u-channel process, because s plays the role of the energy variable in (5) while u plays that role for (6). The substitution $(a_\mu \to -\bar{a}_\mu, c_\mu \to -\bar{c}_\mu)$ is called *crossing* or *line reversal*, or more specifically $s - u$ crossing because of the interchange of the roles of s and u.

(b) *Crossing symmetry, analyticity, and signature*

The idea that a single amplitude $\mathcal{M}(s, t, u)$ can, depending on the range of the variables, describes several processes is a very important concept. With $s - u$ crossing in mind, we note that the constraint $s + t + u = m_a^2 + m_b^2 + m_c^2 + m_d^2$ indicates that for fixed t positive s requires generally a negative value of u and vice versa.† Thus positive u can be equally interpreted as negative s and the $s - u$ crossing can be viewed as a transformation from the positive s region to the negative s region. It is useful to introduce a new variable,

$$\nu = \frac{1}{4m_b}(s - u) \tag{7}$$

which together with t can be used as a kinematic variable.

For elastic scattering in the forward direction $(t = 0)$ ν has the simple interpretation of the total laboratory energy of A (or the negative of the lab energy of $\bar{C}$).

The invariant amplitude $\mathcal{M}(\nu, t)$ satisfies a dispersion relation in ν at fixed t. The dispersion relation follows from the analyticity of $\mathcal{M}(\nu, t)$ in the cut ν plane, as shown schematically in Fig. 8. The cut structure along the positive and negative real axes stems from unitarity in a familiar way, the s-channel thresholds opening up on the right and the u-channel on the left. The physical amplitude for the s-channel process $A + B \to C + D$ is obtained by letting ν approach the positive real axis from above, indicated by $\nu + i\varepsilon$. For the u-channel process $\bar{C} + B \to \bar{A} + D$ the physical region is just below the cut for ν negative.

The physical amplitudes for the s- and u-channel processes can be written

$$\begin{aligned} \mathcal{M}_s(\nu, t) &= \mathcal{M}(\nu + i\varepsilon, t) \\ \mathcal{M}_u(\nu, t) &= \mathcal{M}(-\nu - i\varepsilon, t). \end{aligned} \tag{8}$$

† This can be seen most easily at high energies where masses can be neglected. Then the constraint is $s + t + u \simeq 0$ and (A.16) and (A.18) show that for large positive s, the range of t is $-s \leqslant t \leqslant 0$.

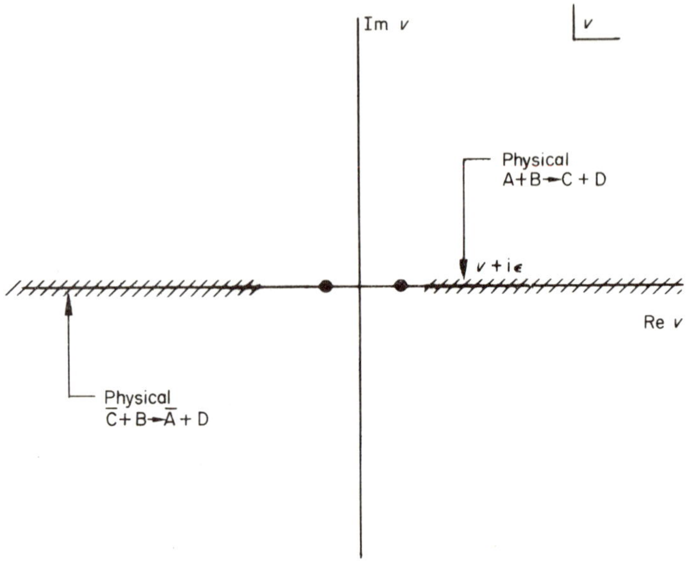

FIG. 8. The complex ν plane showing schematically the branch cut and pole structure of the scattering amplitude. For $A + B \to C + D$ the physical region is just above the right-hand cut, for $\bar{C} + B \to \bar{A} + D$ it is just below the left-hand cut.

It is useful to consider instead of $\mathcal{M}_s$ and $\mathcal{M}_u$ amplitudes that are even and odd in ν. We thus define for complex as well as real ν the even and odd amplitudes:

$$\mathcal{M}^{(\pm)}(\nu, t) = \tfrac{1}{2}[\mathcal{M}_s(\nu, t) \pm \mathcal{M}_u(\nu, t)]. \tag{9}$$

(Sometimes these are called crossing-even or crossing-odd amplitudes.) These amplitudes satisfy dispersion relations of the form,

$$\mathcal{M}^{(\pm)}(\nu, t) = \frac{1}{\pi}\int_0^\infty \mathrm{d}\nu' \,\mathrm{Im}\, \mathcal{M}^{(\pm)}(\nu', t)\left[\frac{1}{\nu' - \nu} \pm \frac{1}{\nu' + \nu}\right]. \tag{10}$$

In (10) the pole terms are implicit in the integral and the necessity of subtractions has been ignored.

The distinction between crossing-even and crossing-odd amplitudes has physical meaning when the amplitudes are considered in the t-channel ($A\bar{C} \to \bar{B}D$). Here t is the energy variable and s and u are momentum transfer variables related to the scattering angle θ_t. From eqn (A.20) we see that for elastic scattering at least (actually $m_a = m_c$ *or* $m_b = m_d$ is sufficient)

$$\nu = \frac{p_t p_t'}{m_b}\cos\theta_t. \tag{11}$$

Since p_t and p'_t are just functions of t, ν is equivalent in the t-channel to $\cos\theta_t$. Now in discussing the dynamics of Regge exchanges in the t-channel one first considers a partial wave expansion. Then because of the possibility of Majorana exchange forces (see Blatt and Weisskopf, 1952, p. 136, for this ancient terminology) one considers separately the forces occurring in the even partial waves and the odd partial waves. The Regge poles that arise from these two sets of forces are different in general. The poles coming from the even (odd) partial waves and having physical particles with even (odd) J values are called even (odd) *signature* Regge poles.

Because of the connection (11) between ν and $\cos\theta_t$ it is evident that for t-channel exchanges even-signature Regge poles contribute only to crossing-even amplitudes and odd-signature to crossing-odd amplitudes. Examples of even-signature Regge trajectories are the $P' \equiv f$ with the f^0 meson ($I = 0$, $J^P = 2^+$, $m^2 = 1{\cdot}60\ \mathrm{GeV}^2$) as an observed physical state, the A_2 with the A_2 meson ($I = 1$, $J^P = 2^+$, $m^2 = 1{\cdot}72$) and the K^{**} with the K^{**}-meson ($I = \frac{1}{2}$, $J^P = 2^+$, $m^2 = 2{\cdot}02$). Some odd-signature Regge trajectories are the ρ with the ρ-meson ($I = 1$, $J^P = 1^-$, $m^2 = 0{\cdot}58$) and the g-meson ($I = 1$, $J^P = 3^-$, $m^2 = 2{\cdot}82$) as physical states, the ω with the ω-meson ($I = 0$, $J^P = 1^-$, $m^2 = 0{\cdot}61$), and the K^* with the K^*-meson ($I = \frac{1}{2}$, $J^P = 1^-$, $m^2 = 0{\cdot}79$) as particles.

(c) *Power law behaviour and associated phase, Pomeranchuk theorems*

The evidence presented in Fig. 7 shows that high-energy reaction amplitudes exhibit power law behaviour in the energy, at least approximately. Such behaviour has important consequences for the phase of the amplitudes. The operative theorem of complex variables [for we do need to assume analyticity of the type displayed in Fig. 8 or eqn (10)] is the Phragmén–Lindelöf theorem (Titchmarsh, 1950, p. 183; Eden, 1967, p. 194). The application of the theorem is straightforward. We only state the results relevant for our purposes.

Let us assume the following properties for $\mathcal{M}(\nu, t)$ at fixed t:

(i) $\mathcal{M}(\nu, t)$ is analytic in the upper half ν plane,
(ii) $\mathcal{M}(\nu, t)$ does not increase exponentially for $|\nu| \to \infty$,
(iii) $\mathcal{M}(\nu, t) \to c(t)\nu^{\alpha(t)}(\ln \nu)^{\beta(t)}$ as $\nu \to \infty$ along the positive real axis, where $\alpha(t)$ and $\beta(t)$ are real functions of t,
(iv) $\mathcal{M}(\nu, t) \to \bar{c}(t)(-\nu)^{\alpha(t)}(\ln(-\nu))^{\beta(t)}$ as $\nu \to -\infty$ along the negative real axis.

Note that $\alpha(t)$ and $\beta(t)$ are the same in both limits, but $c(t)$ and $\bar{c}(t)$ are in principle different complex functions of t.

Application of the Phragmén–Lindelöf theorem establishes that

$$c(t) = \bar{c}(t)^* \, \mathrm{e}^{-i\pi\alpha(t)}. \qquad (12)$$

Suppose that $\mathcal{M}(\nu, t)$ is crossing-even. Then $\bar{c}(t) = c(t) = c^{(+)}(t)$ and (12) yields

$$c^{(+)}(t) = \pm |c(t)| \exp[-i\pi\alpha^{(+)}(t)/2] = \gamma^{(+)}(t)\left(i - \cot\frac{\pi\alpha^{(+)}(t)}{2}\right) \quad (13)$$

where $\gamma^{(+)}(t)$ is real. If $\mathcal{M}(\nu, t)$ is odd under crossing, $\bar{c}(t) = -c(t) = -c^{(-)}(t)$ and (12) gives

$$c^{(-)}(t) = \pm i |c(t)| \exp[-i\pi\alpha^{(-)}(t)/2] = \gamma^{(-)}(t)\left(i + \tan\frac{\pi\alpha^{(-)}(t)}{2}\right) \quad (14)$$

and $\gamma^{(-)}(t)$ is real. The phases given by (13) and (14) are the same as those that occur for even and odd signature Regge poles. The Regge amplitudes are of the form,

$$R^{(\pm)} \propto \frac{P_{\alpha^{(\pm)}(t)}(-\cos\theta_t) \pm P_{\alpha^{(\pm)}(t)}(\cos\theta_t)}{\sin \pi\alpha^{(\pm)}(t)}.$$

Using (11) and assuming that ν is large we find

$$R^{(\pm)} \propto \frac{(-\nu)^{\alpha^{(\pm)}(t)} \pm (\nu)^{\alpha^{(\pm)}(t)}}{\sin \pi\alpha^{(\pm)}(t)}$$

$$\propto \nu^{\alpha^{(\pm)}(t)} \begin{Bmatrix} i - \cot \pi\alpha^{(+)}(t)/2 \\ i + \tan \pi\alpha^{(-)}(t)/2 \end{Bmatrix} \quad (15)$$

The results (13) and (14) are more general, however, since they follow even if there is a $(\ln \nu)^{\beta(t)}$ variation of the amplitude times the power law behaviour.

The result (12) allows some other conclusions:

(i) Equality of elastic differential cross sections for particle and antiparticle scattering:

$$\lim_{\nu\to\infty}\left[\frac{d\sigma}{dt}(\bar{A}B \to \bar{A}B)\bigg/\frac{d\sigma}{dt}(AB \to AB)\right] = 1. \quad (16)$$

(ii) Pomeranchuk theorem of equality of particle and anti-particle total cross sections:

$$\lim_{\nu\to\infty}[\sigma_t(\bar{A}B)/\sigma_t(AB)] = 1, \quad (17)$$

provided $\alpha(0) = 1$. Pomeranchuk's original proof, based only on dispersion relations like (10), assumed $\sigma_t \to$ constant and required the weak condition, $\operatorname{Re}\mathcal{M}(\nu, 0)/\operatorname{Im}\mathcal{M}(\nu, 0) \ln \nu \to 0$. The present proof permits logarithmic variation of σ_t, but has the stricter assumption $\operatorname{Re}\mathcal{M}(\nu, 0)/\operatorname{Im}\mathcal{M}(\nu, 0) \to$ constant (perhaps zero). Needless to say, with the results of Fig. 2 known, there are more general proofs (e.g. Grunberg and Truong,

1973). Similarly, eqn (16) has been put on a firm and general footing for 2-body inelastic as well as elastic scattering by Cornille and Martin (1972).

The extent of the testing of (16) and (17) can be judged by inspection of Fig. 1 and 2 for total cross sections and Fig. 9 for the differential cross sections where the slope parameter defined in eqn (1) is shown for π^+p, π^-p, K^+p, K^-p, pp, and $\bar{p}p$ elastic scattering. Certainly the trends of the data in these figures support the asymptotic validity of (16) and (17), but as the data of Fig. 2 show, asymptopia may be far away. Experiments with meson and antiproton beams at NAL are eagerly awaited.

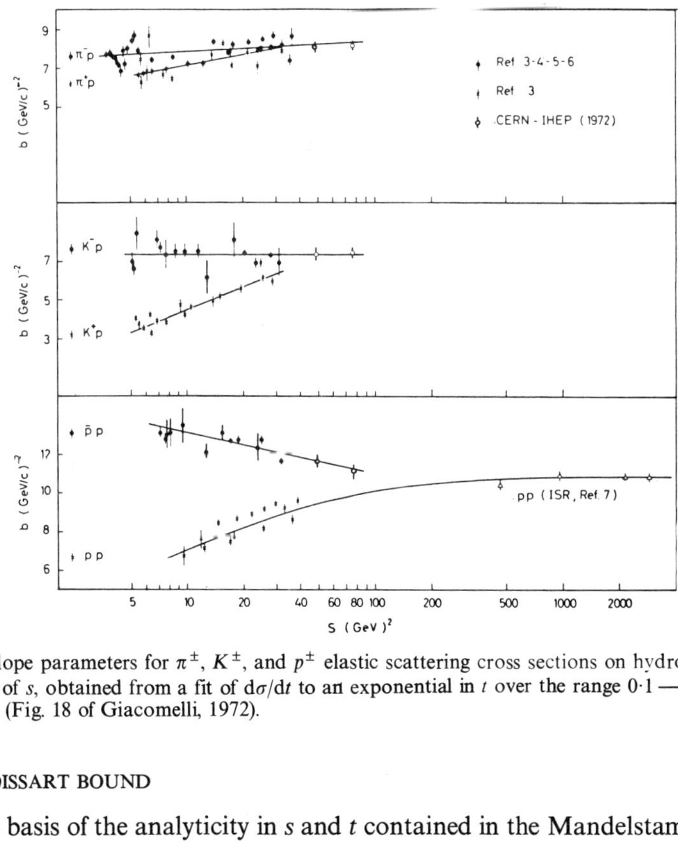

FIG. 9. Slope parameters for $\pi^{\pm}$, $K^{\pm}$, and $p^{\pm}$ elastic scattering cross sections on hydrogen as a function of s, obtained from a fit of $d\sigma/dt$ to an exponential in t over the range $0{\cdot}1 — |t| — 0{\cdot}4$ $(\text{GeV}/c)^2$ (Fig. 18 of Giacomelli, 1972).

3. FROISSART BOUND

On the basis of the analyticity in s and t contained in the Mandelstam representation Froissart (1961) proved that the total cross section is bounded from above according to

$$\sigma_t < C(\ln s)^2 \tag{18}$$

as $s \to \infty$. The right-hand side of (18) is called the *Froissart bound.* Its derivation has been generalized, simplified, made more rigorous, made plausible by many (not the same!) authors (e.g. Martin, 1963, 1966; Eden, 1967; Eden, 1971; Roy, 1972; Horn and Zachariasen, 1973). We will therefore not discuss the careful proofs, but confine our attention to the physical intuitive aspects.

Suppose that the interaction between two spinless particles is mediated by the exchange of a particle of spin J and mass μ in the t-channel. Then the lowest order amplitude will be real and at high energies of the form,†

$$F_B(s, t) = g^2 \frac{s^J}{\mu^2 - t}. \tag{19}$$

If we use the eikonal approximation (discussed in Appendix C) as an approximation to the full scattering amplitude, we obtain via eqn (C.14) an eikonal phase,

$$\delta_{\text{eikonal}}(s, b) = g^2 s^{J-1} K_0(\mu b). \tag{20}$$

The behaviour of δ as a function of b is indicated in the top part of Fig. 10. Asymptotically the modified Bessel function falls off as $\exp(-\mu b)/(\mu b)^{\frac{1}{2}}$. Hence the phase shift is small compared to unity provided

$$\mu b \gg \mu b_c = \ln (g^2 s^{J-1}). \tag{21}$$

The square of the partial wave scattering amplitude is sketched in the bottom part of Fig. 10. It is small compared with unity for $\mu b \gg \mu b_c$, rises to unity for $\mu b \lesssim \mu b_c$ and with the example of a real phase shift (20) oscillates between zero and unity for smaller values of μb. The integral (B.25) defining the elastic cross section (equal to the total here) can be estimated to be

$$\sigma_{\text{el}} = \sigma_t \simeq 4\pi \int_0^{b_c} \mathrm{d}(b^2) \times \tfrac{1}{2} = 2\pi b_c^2. \tag{22}$$

With the critical impact parameter given by (21) we obtain at high energies the estimate,

$$\sigma_t \sim \frac{2\pi}{\mu^2} (J - 1)^2 (\ln s)^2. \tag{23}$$

Note that (23) only has meaning for $J > 1$. For $J = 1$ the phase (20) is independent of s. This would lead to a constant cross section. For $J < 1$, μb_c decreases with increasing s; the phase is small at all impact parameters and the cross section falls with s.

† In the t-channel the single partial wave $l = J$ gives rise to a numerator proportional to barrier penetration factors $(p_t p_t')^J$ times $P_J(\cos \theta_t)$. At high energies in the s-channel, (11) leads to the stated form.

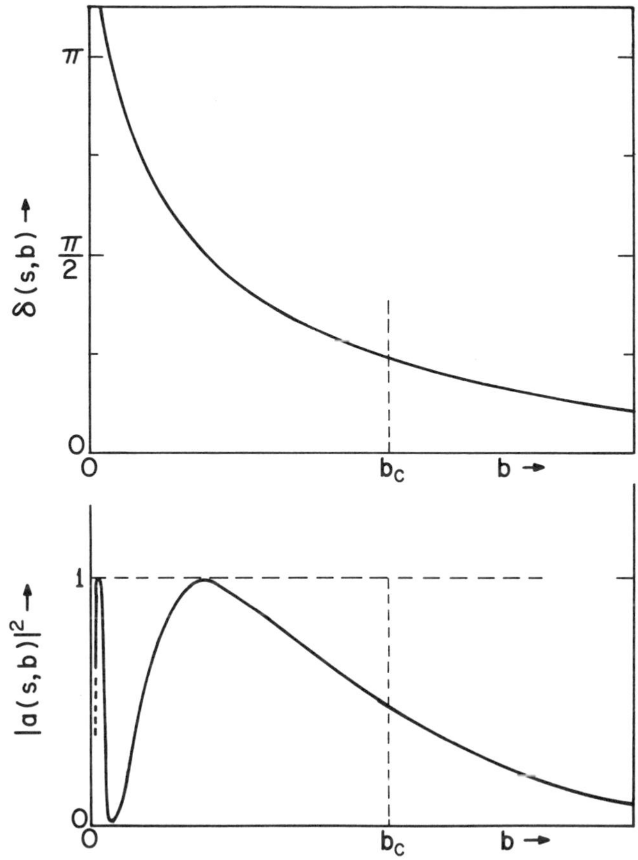

FIG. 10. Phase shift $\delta(s, b)$ as a function of impact parameter (top) and absolute square $|a(s, b)|^2$ of the partial wave amplitude as a function of impact parameter (bottom).

While the above example is somewhat unrealistic it does show how unitarization via the eikonal partial wave representation imposes the Froissart bound on the full amplitude even though the lowest order approximation (the t-channel exchange of something) may increase as a high power of s and violate the Froissart bound itself. This is, of course, related very intimately to Froissart's original proof.

A more realistic example would have had the phase shift becoming complex for $b < b_c$. Then the partial wave amplitude (B.26) would rapidly approach $i/2$ for $b < b_c$. Using (B.25) for σ_t and the above method of estimation we would still arrive at (22) for the total cross section, but would find $\sigma_{el} \simeq \sigma_t/2$.

The exchange of a Regge pole as the lowest order amplitude affords an instructive example of a complex phase shift and some subtleties in impact parameter space. For definiteness, consider the exchange of an even-signature Regge pole with amplitude,

$$F_B(s,t) = -\beta \exp(\gamma t) \exp\left[-i\pi\alpha(t)/2\right]$$

where β and γ are real, $\beta > 0$, and the exponential residue is chosen for convenience. With a linear Regge trajectory, $\alpha(t) = \alpha(0) + \alpha'(0)t$, this can be written

$$\left.\begin{aligned} F_B(s,t) &= -\beta(-is)^{\alpha(0)} \exp\left[B(s)t/2\right] \\ \text{where} \quad & \\ \tfrac{1}{2}B(s) &= \gamma + \alpha'(0)\ln(-is) \end{aligned}\right\}. \tag{24}$$

From (C.16) we deduce that the eikonal phase shift is

$$\delta_{\text{eikonal}}(s,b) = i\beta \frac{(-is)^{\alpha(0)-1}}{B(s)} \exp\left[-b^2/2B(s)\right]. \tag{25}$$

At high energies $B(s)$ is predominantly real with a small negative imaginary part. Thus the phase of δ is determined almost entirely by the factor, $i^2 \exp\left[-i(\pi/2)\alpha(0)\right]$. For $0 < \alpha(0) < 2$, this factor has a positive imaginary part.† For fixed impact parameter the power law increase in the magnitude of δ implies that for $1 < \alpha(0) < 2$ the phase shift will develop a large positive imaginary part and $e^{2i\delta} \to 0$ rapidly as $s \to \infty$. Another way to look at it is that $e^{2i\delta} \simeq 0$ for all values of b less than a critical value b_c that grows with energy. From (B.25) or (B.27) the total cross section will be given roughly by (22).

To estimate b_c from (25) we rewrite it as

$$\delta_{\text{eikonal}}(s,b) = \frac{i\beta}{B(s)} \exp\left[(\alpha(0) - 1)\ln(-is) - b^2/2B(s)\right]. \tag{26}$$

Evidently whatever the value of β, the imaginary part of δ will be very large until the second term in the exponent overcomes the first. This defines the critical impact parameter,

$$b_c^2 = 2B(s)\left[\alpha(0) - 1\right]\ln s \tag{27}$$

Note that if $B(s)$ were in fact independent of s, as occurs if the Regge trajectory has zero slope, the total cross section would grow only as $\ln s$, not as $(\ln s)^2$. This is just what we would expect with a phase shift that was a Gaussian of fixed shape rather than an exponential in impact parameter as was our

† The intercept $\alpha(0)$ is expected to be smaller than 2 on the basis of a theorem by Jin and Martin (1964) concerning the number of subtractions necessary in fixed-t dispersion relations.

elementary particle exchange. Assuming $\alpha'(0) \neq 0$, however, we obtain at high enough energies the estimate,

$$b_c^2 \simeq 4\alpha'(0)[\alpha(0) - 1](\ln s)^2. \tag{28}$$

The Yukawa (exponential) force and the Regge (Gaussian) force thus both give the same s-dependence. The Regge exchange does it in a sneaky way, however, by having the mean square radius of the Gaussian grow as $\ln s$, as well as having its magnitude increase as a power of s.

As already implied by the earlier discussion, if the full amplitude is described at high energies by a single Regge pole then the Froissart bound is violated if $\alpha(0) > 1$. Since total cross sections do not seem to decrease with energy it was natural within a Regge *pole* framework to assume that the leading Regge pole had $\alpha(0) = 1$. The pole, with the internal quantum numbers of the vacuum ($Q = 0, I = 0, Y = 0, B = 0, \ldots$), is known as the Pomeranchuk pole or pomeron. It occupies a unique position—it is the highest lying Regge trajectory; furthermore, it does not seem to have any particles associated with it and also seems to have an abnormally small slope (cf. the shrinkage of the forward peak in $d\sigma/dt$ for pp scattering, Fig. 6). The last two points have inevitably raised doubts in many minds as to whether diffraction scattering is properly described by a Regge *pole*. We now tend to speak of a Pomeranchuk *singularity*, leaving deliberately vague the type of singularity in the angular momentum plane. At some point, of course, the J-plane description may become so complicated as to be uneconomical and therefore inappropriate.

An alternative description of diffractive scattering is advocated by Cheng and Wu on the basis of a long study of massive quantum electrodynamics at very high energies (see Cheng and Wu, 1970, and the references cited there). They show that the leading behaviour as $s \to \infty$ at small t is given by the "one-tower" diagrams shown in Fig. 11. We call the sum of leading contributions of all the one-tower diagrams a single-tower exchange. The general form of this amplitude is

$$F_B(s, t) \sim \frac{is^{1+a}}{(\ln s)^n} f(t) \tag{29}$$

where a is real and positive ($a = 11\pi\alpha^2/32$, where α is the fine structure constant) and $n = 2$ in spinor QED. In Regge language this amplitude corresponds to the exchange of a fixed singularity (a cut) in the J-plane with $J > 1$. Cheng and Wu postulate, with plausible theoretical and physical arguments, that this single-tower exchange amplitude is a Born approximation whose two-dimensional Fourier transform gives the eikonal phase. The real function $f(t)$ is such that at large impact parameters $\delta(s, b) \propto e^{-\mu b}$ where $\mu < 2m_V$, m_V being the mass of the "photon" (vector meson). Apart

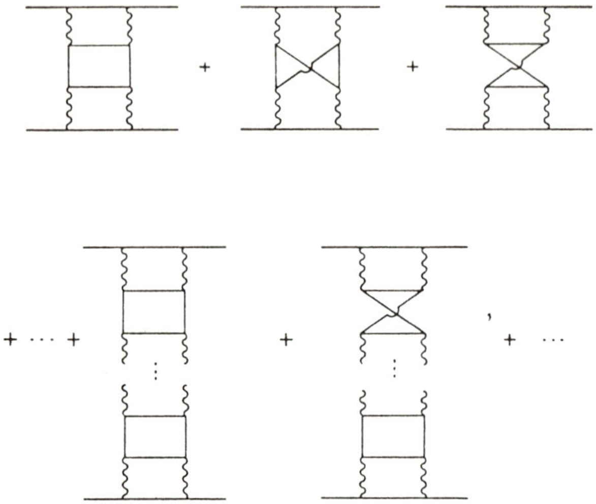

FIG. 11. One-tower diagrams in QED. These diagrams give the leading s dependence at high energies (Cheng and Wu, 1970).

from the logarithmic factors in (29) the Cheng and Wu eikonal phase shift is thus qualitatively similar to the elementary particle exchange (20) with $J = 1 + \varepsilon$. In the phenomenological fits (Cheng, Walker, and Wu, 1973a, b) the complicated and not totally explicit phase shift is approximated by

$$\delta_{cww}(s, b) = \frac{i}{2} f_j \frac{(-iE)^c}{[\ln(-iE)]^n} \exp\left[-\lambda\sqrt{(b^2 + b_{0j}^2)}\right]. \tag{30}$$

Here E is the lab energy of the incident particle, c and λ are fundamental parameters that are the same for all processes, while f_j and b_{0j} are different constants for πp, Kp, and pp scattering, but the same for particle-proton and antiparticle–proton scattering. Since (30) represents only the diffractive scattering contribution, fits to total cross section data at energies of the order of 2–30 GeV/c require an additional term, taken to be $A_j E^{-\frac{1}{2}}$, with A_j different for all six processes. With 14 parameters an adequate fit is obtained to all the total cross section data shown in Figs 1 and 2, obviously including the rising pp cross section at ISR energies. With $n = 0$ in (30), the parameter $c \approx 0{\cdot}083$; with $n = 1$, $c = 0{\cdot}20$. The fixed J-plane singularity is only slightly above unity. The estimate analogous to (23) is

$$\sigma_t \sim \frac{2\pi}{\mu^2}\left[\ln\left(\frac{s^c}{(\ln s)^n}\right)\right]^2.$$

Ultimately this gives a $(\ln s)^2$ behaviour with $\sigma_{el}/\sigma_t = \frac{1}{2}$, characteristic of a totally absorbing disc with a logarithmically growing radius, but the smallness of c and the presence of the $(\ln s)^n$ factor in the denominator makes asymptopia very far away.

4. RATIO OF REAL TO IMAGINARY PART OF THE FORWARD SCATTERING AMPLITUDE

Variation in energy of the real part of the forward scattering amplitude has traditionally been associated with structure in the total cross section, the classical optical dispersion of the index of refraction being the most familiar example. Experts in dispersion relations continue to polish the data and the equations in the resonance region and extend them to higher and higher energies. But in the classical age (denoted, I suppose, by B.I.S.R.) when 30 or 70 GeV was considered high energy the interest in real parts was not widespread. Above the resonance region ($P_{\text{lab}} > 5\,\text{GeV}/c$) the ratios of real to imaginary parts, usually denoted in the literature by $\rho(s)$ or $\alpha(s)$, are small and appear to decrease in magnitude smoothly with increasing energy. The ratios tend to be negative (only K^-p is positive; $\bar{p}p$ is nearly zero), with the pp and K^+p values largest in absolute value, of order $-0{\cdot}3$ or $-0{\cdot}4$ at $5\,\text{GeV}/c$ and $-0{\cdot}2$ at $30\,\text{GeV}/c$.

The rising total cross section for proton–proton interactions arouses interest in the real parts again. The monotonic decrease in magnitude of the ratio of real to imaginary part is cast in doubt. There is a fancy theorem (Khuri and Kinoshita, 1965) that states that if the total cross section continues to increase with energy then eventually $\rho(s)$ must become positive and stay positive; any approach to zero must be from above. On the experimental side there are recent ISR data (Amaldi *et al.*, 1973a) giving slightly positive or zero values, implying a change of sign of $\rho(s)$ for p–p scattering at ISR energies or below.

We discuss here a simple "do-it-yourself" way of understanding and calculating $\rho(s)$ from data on the total cross section. The method is not efficient (though applicable in principle) when cross sections vary rapidly with energy, but works admirably above the resonance region. Like dispersion relations it is based on analyticity of forward (or fixed t) amplitudes in energy. No integration is necessary, however; only differentiation! First a small amount of elementary complex variable theory. Let $f(z)$ be analytic inside some region of the complex z-plane. For points within that region, $f(z + \lambda)$ can be represented by a Taylor series expansion of $f(z)$:

$$f(z + \lambda) = f(z) + \lambda f'(z) + \frac{\lambda^2}{2!} f''(z) + \dots .$$

The series converges and represents $f(z + \lambda)$ uniquely provided the point $z + \lambda$ lies inside the circle of convergence defined by the distance from the point z to the nearest singularity of $f(z)$. This Taylor series can be represented compactly by the formal operator statement,

$$f(z+\lambda) = \exp\left(\lambda \frac{\mathrm{d}}{\mathrm{d}z}\right) f(z). \tag{31}$$

That is all the complex variable theory we will need.

Now consider the scattering amplitude $F(\nu, t)$ for the s-channel process, $ab \to ab$, where $\nu = (s-u)/4m_b$ and t is the momentum transfer variable. The u-channel process, $\bar{a}b \to \bar{a}b$, is described by the amplitude $\bar{F}(\nu, t)$. The domain of analyticity in ν of $F(\nu, t)$ is shown schematically in the top part of Fig. 12. From the substitution law we know that $\bar{F}(\nu, t) = F(-\nu, t)$, as indicated in Fig. 12 by the relationship of the points A and C. It is convenient to define symmetric and antisymmetric combinations of amplitudes,

$$F_{\pm}(\nu, t) = \tfrac{1}{2}[F(\nu, t) \pm F(-\nu, t)] \tag{32}$$

F_+ is said to be even under crossing ($\nu \to -\nu$), and F_- odd. The s-channel process then has an amplitude,

$$\left.\begin{aligned} &F_s(\nu, t) = F_+(\nu, t) + F_-(\nu, t), \\ &\text{while the } u\text{-channel amplitude is} \\ &F_u(\nu, t) = F_+(\nu, t) - F_-(\nu, t) \end{aligned}\right\}. \tag{33}$$

We wish to compare amplitudes at the points A and B of Fig. 12. We thus use the Schwarz reflection law for real analytic functions to obtain

$$F_{\pm}(\nu\,\mathrm{e}^{i\pi^-}) = \pm F_{\pm}^*(\nu). \tag{34}$$

In eqn. (34) and subsequent equations I suppress the fixed argument t for brevity.

Suppose now that we choose to use the variable $\xi = \ln \nu$ instead of ν to describe the energy variation of our amplitudes. The analytic structure in the ξ plane is sketched in the bottom part of Fig. 12. We see that now the points A and B are related by a displacement of $i\pi^-$, i.e. $\nu \to \nu\,\mathrm{e}^{i\pi^-}$ is equivalent to $\xi \to \xi + i\pi^-$. We wish to use the Taylor series representation (31) to express the left-hand side of (34), $F_\pm$ at the point B, in terms of the function and its derivatives at A. Because of the branch cuts this cannot be done without further consideration. The two branch cuts shown in the bottom diagram of Fig. 12 consist of very many superposed branch cuts, a new one running off the right from each new kinematic threshold. From a rigorous point of view we are thus stuck with the branch points and cuts where they are. In reality, however, the existence of all these higher branch points corresponding to the opening of more and more inelastic channels is of negligible importance for the total cross section and the analytic

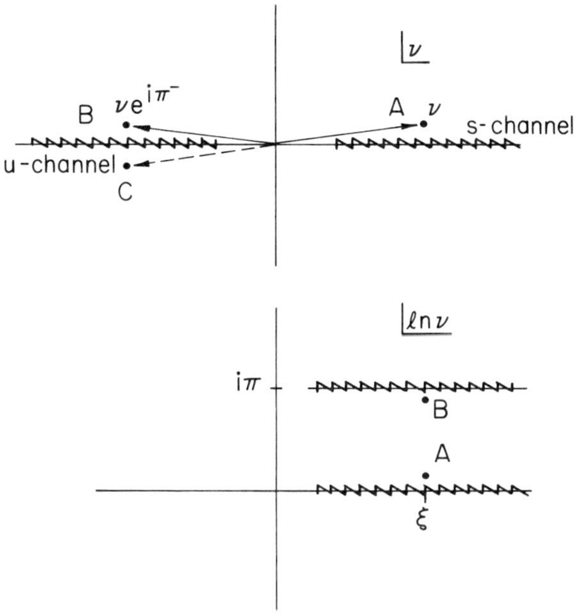

FIG. 12.

properties of the forward scattering amplitude at high energies. One need only recall the difficulty of seeing the cusp at the $K\Sigma$ threshold in the reaction $\pi N \to K\Lambda$ and the smoothness of the total cross sections above 2 or 3 GeV/c to appreciate this empirical fact.

If the existence of new thresholds beyond some finite energy is ignored as far as the elastic amplitude is concerned, then the branch cuts in the $\xi = \ln v$ plane can be bent back in such a way that the Taylor series expansion (31) can be used straightforwardly to express $F_\pm(\xi + i\pi)$ in terms of $F_\pm(\xi)$. When this is inserted in the left side of (34) we obtain:

$$\exp\left(i\pi\frac{\mathrm{d}}{\mathrm{d}\xi}\right)F_\pm(\xi) = \pm F_\pm^*(\xi). \tag{35}$$

Equation (35) can be put in a more symmetric form by operating on both sides with $\exp[(-i(\pi/2)(\mathrm{d}/\mathrm{d}\xi)]$:

$$\exp\left(i\frac{\pi}{2}\frac{\mathrm{d}}{\mathrm{d}\xi}\right)F_\pm(\xi) = \pm\left[\exp\left(i\frac{\pi}{2}\frac{\mathrm{d}}{\mathrm{d}\xi}\right)F_\pm(\xi)\right]^*. \tag{36}$$

Writing out the real and imaginary parts of both sides we find for *even* amplitudes the relation,

$$\operatorname{Im} F_+(\xi) = -\tan\left(\frac{\pi}{2}\frac{\mathrm{d}}{\mathrm{d}\xi}\right)\operatorname{Re} F_+(\xi) \tag{37}$$

and for *odd* amplitudes,

$$\operatorname{Re} F_-(\xi) = \tan\left(\frac{\pi}{2}\frac{\mathrm{d}}{\mathrm{d}\xi}\right)\operatorname{Im} F_-(\xi). \tag{38}$$

These formal equations were noted by John Bronzan in a talk at Argonne National Laboratory in March (Bronzan, 1973). Their application to the real world was impressed upon me by Gordon Kane. (They may be well known to André Martin and others.)

Before using (37) and (38) for p–p scattering I make a few observations. Firstly, for amplitudes having power-law behaviour in ν these relations yield immediately the standard Regge phase of the amplitude, a result usually deduced from the Phragmén–Lindelöf theorem (see Sect. II. 2(c) above). The reader can check that amplitudes varying as $\nu^{\alpha}(\ln \nu)^{\beta}$ have the standard phase with corrections of order $(\ln \nu)^{-1}$, as expected. The second observation is that if the functions on the right-hand sides of (37) and (38) are approximated by *finite* polynomials in ξ the infinitely many differentiations implied by

$$\tan\left(\frac{\pi}{2}\frac{\mathrm{d}}{\mathrm{d}\xi}\right) = \frac{\pi}{2}\frac{\mathrm{d}}{\mathrm{d}\xi} + \frac{1}{3}\left(\frac{\pi}{2}\right)^3\frac{\mathrm{d}^3}{\mathrm{d}\xi^3} + \ldots$$

terminate in a finite number. Thus the nonlocality inherent in the equations (and equivalent to a dispersion relation) is replaced in practice by a semi-local connection. Note that with a quadratic approximation in ξ only the first derivative contributes. Since cross section data above the resonance region can invariably be approximated, at least locally, by a quadratic form in ξ or a power law in ν, eqns (37) and (38) yield simple connections between real and imaginary parts.† A corollary is that the behaviour of $\rho(\xi)$ at any energy is determined almost completely by the energy variation of σ_t in the immediate neighbourhood. Nothing can be learned about asymptopia from the magnitude or energy dependence of $\rho(\xi)$ at finite energies.

† The use of the approximate relations

$$\operatorname{Im} F_+ \simeq -\frac{\pi}{2}\frac{\mathrm{d}}{\mathrm{d}\xi}\operatorname{Re} F_+ \quad \text{and} \quad \operatorname{Re} F_- \simeq \frac{\pi}{2}\frac{\mathrm{d}}{\mathrm{d}\xi}\operatorname{Im} F_-$$

resulting from keeping only the first term in the expansion of the tangent have been employed in the past, but not based on (37) and (38) and not with a clear explanation of what approximation was involved, at least to my knowledge.

For applications to scattering data it is convenient to introduce amplitudes $f_\pm$ that differ from the customary invariant amplitudes by one power of v and are normalized so that

$$\sigma_t = \operatorname{Im} f. \tag{39}$$

For proton–proton and antiproton–proton scattering we define $\sigma_\pm = \frac{1}{2}(\sigma_{\bar{p}p} \pm \sigma_{pp})$ so that $\sigma_{pp} = \sigma_+ - \sigma_-$ and $\sigma_{\bar{p}p} = \sigma_+ + \sigma_-$. The corresponding amplitudes are denoted by $f_\pm(\xi)$. For the imaginary part of the odd amplitude (actually even under crossing because of our division by v), we take the parametrization of Denisov *et al.* (1971):

$$\sigma_- = \operatorname{Im} f_- = A[P_{\text{lab}}(\text{GeV}/c)]^{-n}$$

with $A = 28{\cdot}4 \pm 2{\cdot}7$ mb and $n = 0{\cdot}61 \pm 0{\cdot}03$. Since $P_{\text{lab}} \simeq v$ at Serpukhov energies this power-law behaviour yields

$$\operatorname{Re} f_-(\xi) = \cot\left(\frac{\pi}{2} n\right) \sigma_-(\xi). \tag{40}$$

For the amplitude $f_+(\xi)$ we use (38) to find†

$$\operatorname{Re} f_+(\xi) = \tan\left(\frac{\pi}{2}\frac{\mathrm{d}}{\mathrm{d}\xi}\right) \sigma_+(\xi). \tag{41}$$

With a quadratic form for σ_+,

$$\left.\begin{aligned} &\sigma_+(\xi) \simeq a + b\xi + c\xi^2, \\ &\text{we have} \\ &\operatorname{Re} f_+(\xi) = \frac{\pi}{2}\frac{\mathrm{d}}{\mathrm{d}\xi} \sigma_+(\xi) = \frac{\pi}{2}(b + 2c\xi) \end{aligned}\right\}. \tag{42}$$

Then we have the ratio of real to imaginary parts for proton–proton scattering given by

$$\rho_{pp}(\xi) = \frac{\frac{1}{2}\pi(b + 2c\xi) - \cot(\pi n/2)\,\sigma_-(\xi)}{\sigma_+(\xi) - \sigma_-(\xi)}. \tag{43}$$

† There is a subtlety here. The usual amplitude $F_+(\zeta)$ satisfies (37). It is obvious that an additive real constant to F_+ will not affect the imaginary part calculated from (37). Correspondingly, the real part computed from (41) is uncertain by a term (c/v). The ambiguity is equivalent to an unknown subtraction constant in a dispersion relation or in a parameterized form satisfying analyticity and crossing requirements (Bourrely and Fischer, 1973). Such a contribution vanishes exponentially (in ξ). Since we are concerned only with the high energy behaviour we shall omit it.

The corresponding quantity for antiproton–proton scattering is

$$\rho_{\bar{p}p}(\xi) = \frac{\frac{1}{2}\pi(b + 2c\xi) + \cot(\pi n/2)\,\sigma_-(\xi)}{\sigma_+(\xi) + \sigma_-(\xi)}. \tag{44}$$

The same general forms for $\rho(\xi)$ hold for π^+p and π^-p and K^+p and K^-p scattering as for pp and $\bar{p}p$. The behaviour of the various $\rho(\xi)$ in the range from 5 to 70 GeV/c can now be understood. The "even" amplitude $f_+(\xi)$ is decreasing from the resonance region towards higher energies. The real part computed from (41) is therefore negative. For pp (and π^+p and K^+p) the two terms in the numerator in (43) are both negative, giving a negative $\rho(\xi)$ of appreciable magnitude. For $\bar{p}p$ (and π^-p and K^-p) on the other hand, the terms in the numerator tend to cancel, yielding a less negative (and perhaps even positive) value for $\rho(\xi)$.

The results of a calculation using (42) and (43), with the Denisov parametrization for $\sigma_-(\xi)$, are shown in Figs. 13 and 14. Two quadratic forms in ξ for $\sigma_+(\xi)$ were fitted to a smoothed $\sigma_{pp}(\xi)$ plus $\sigma_-(\xi)$. The ratio $\rho(\xi)$ was then calculated using (43). The solid and dashed curves in Figs. 13 and 14 represent the two parametrizations. The dotted curve is a modification illustrating a total cross section that flattens out at higher energies. The available data for $\rho(\xi)$ above 10 GeV/c are shown in Fig. 14. The agreement

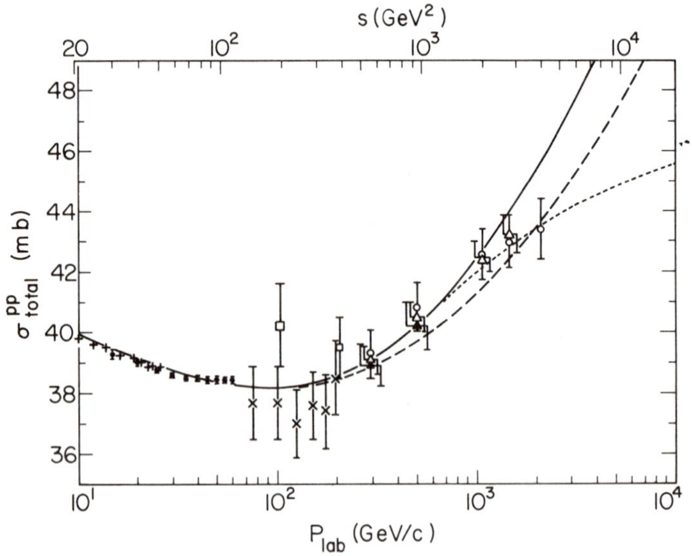

FIG. 13. Total cross section data for p–p scattering from Fig. 2 replotted to show two quadratic parametrizations in $\zeta = \ln \nu$. The dotted curve is yet another smooth behaviour at higher energies.

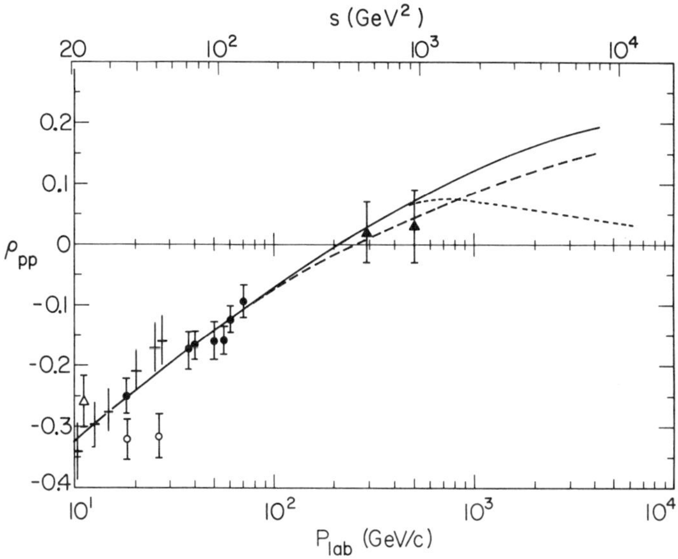

FIG. 14. The ratio ρ of real to imaginary part of the forward nonflip amplitude for p–p scattering calculated by differentiation from the total cross section of Fig. 13. The solid, dashed and dotted curves here correspond to the solid, dashed and dotted curves in Fig. 13.

between the curves and the data is quite satisfactory, showing the efficacy of our method. The whole calculation was an afternoon's work with an HP-35. The curves also agree in general trend with a recent dispersion relation calculation (Kroll, 1973) and the use of a parametrized analytic form (Bourrely and Fischer, 1973). Shortly there should be results from the US–Soviet collaboration using a gas jet target at NAL. This experiment will span the gap between Serpukhov and the ISR. It should locate the cross-over point precisely (near 200–250 GeV, I hope!). [Note added in proof (March, 1974): The experiment has been completed and published by Bartenev *et al.*, *Phys. Rev. Letters* **31**, 1367 (1973). The results lie more or less along the curves in Fig. 14. The crossover is quoted at 280 $\pm$ 60 GeV. The lowest energy and two highest energy points are $P = 51$ GeV, $\rho = -0{\cdot}157 \pm 0{\cdot}012$, $P = 349$ GeV, $\rho = 0{\cdot}025 \pm 0{\cdot}015$, $P = 393$ GeV, $\rho = 0{\cdot}039 \pm 0{\cdot}012$.]

Since (37) and (38) are unfamiliar it is perhaps worthwhile to show explicitly the connection with dispersion relations. The odd amplitude $F_-(\nu)$, for example, satisfies a dispersion relation of the form of (10):

$$\operatorname{Re} F_-(\nu) = \frac{2}{\pi}\,\nu P \int_0^\infty d\nu' \frac{\operatorname{Im} F_-(\nu')}{\nu'^2 - \nu^2}.$$

If the variables are changed, with $\nu = \nu_0\, e^{\xi}$, $\nu' = \nu_0\, e^{\xi+\eta}$, this becomes

$$\mathrm{Re}\, F_{-}(\xi) = \frac{1}{\pi} P \int_{-\infty}^{\infty} \frac{\mathrm{Im}\, F_{-}(\xi+\eta)\, \mathrm{d}\eta}{\sinh \eta}.$$

Unless $\mathrm{Im}\, F_{-}$ grows exponentially in η, as $e^{\lambda|\eta|}$ with $\lambda \geqslant 1$, the integral converges very rapidly away from $\eta = 0$. Excluding this circumstance, a Taylor series expansion of $\mathrm{Im}\, F_{-}(\xi+\eta)$ around $\eta = 0$ gives

$$\mathrm{Re}\, F_{-}(\xi) = \frac{2}{\pi} \sum_{n\,\mathrm{odd}} \mathrm{Im}\, F_{-}^{(n)}(\xi) \frac{1}{n!} \int_{0}^{\infty} \frac{\eta^n\, \mathrm{d}\eta}{\sinh \eta},$$

an expression that can be shown (exercise for the reader!) to be equivalent to (38).

5. PARTIAL WAVE DISTRIBUTION FOR PROTON–PROTON SCATTERING, MACDOWELL–MARTIN BOUND ON $B(s)$

In the previous two sections we have focused on the p–p elastic amplitude at $t = 0$. Now we consider the question of the shape of the differential cross section. The discussion will be naive and schematic, neglecting spin completely. The ISR data at $W \simeq 53$ GeV are shown in Fig. 5 as the stars with $E_{\mathrm{lab}} = 1480$ GeV indicated. At this energy the c.m.s. wave number is $k = W/2\,\mathrm{GeV} = 5{\cdot}07\ W/2\,\mathrm{fm}^{-1} \simeq 134\,\mathrm{fm}^{-1}$. A slope parameter $B \simeq 10\,\mathrm{GeV}^{-2}$ (see Fig. 6) implies a mean square extent of $\sim 0{\cdot}9$ fm. It can therefore be expected that of the order of 100 partial waves will be significant. The continuous impact parameter representation is quite appropriate.

(a) *Partial wave distribution for pp scattering at ISR energies*

The data of the ACGHT collaboration (Strolin, 1973) are replotted on a somewhat compressed vertical scale in Fig. 15. The data of this same group at very small $|t|$ values (Barbiellini *et al.*, 1972) are not shown, but for $0{\cdot}2\ (\mathrm{GeV}/c)^2 < |t| < 0{\cdot}5\ (\mathrm{GeV}/c)^2$ they are consistent with an exponential in t with slope parameter $B \sim 10$–$11\ \mathrm{GeV}^{-2}$. At $|t| < 0{\cdot}15\ (\mathrm{GeV}/c)^2$ the data show a steeper slope of order $B \sim 12$–$13\ \mathrm{GeV}^{-2}$, but this detail could not be seen in Fig. 15.

The dashed straight lines show that the data at small $|t|$ and for $2 < |t| < 4\ (\mathrm{GeV}/c)^2$ can be represented by exponentials in $|t|$. The dip at $|t| \sim 1{\cdot}3\ (\mathrm{GeV}/c)^2$ implies an interference between two contributions to the amplitude. Since we know from Fig. 14 that at $t = 0$ the real part of the nonspinflip amplitude is extremely small, we make the simplifying assumption that the amplitude is purely imaginary at all t values of interest. We are neglecting all other helicity amplitudes. Thus we fit the cross section with the expression,

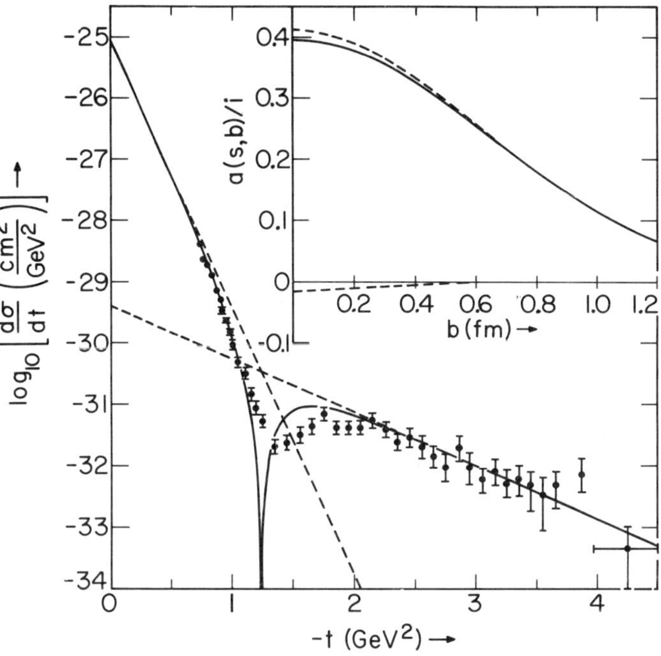

FIG. 15. Differential cross section for p–p elastic scattering at c.m.s. energy $W = 53$ GeV. Preliminary data of the ACGHT collaboration (Strolin, 1973). The dashed lines are proportional to exp $(10t)$ and exp $(2t)$. The solid curve is the cross section given by a purely imaginary nonflip amplitude whose partial wave profile is given by the solid curve in the inset at upper right.

$$\frac{d\sigma}{dt} = \frac{\sigma_t^2}{16\pi}\left|(1 + \lambda)\exp(B_1 t/2) - \lambda \exp(B_2 t/2)\right|^2. \tag{45}$$

The coefficient $\sigma_t^2/16\pi$ is the optical theorem value of $d\sigma/dt$ at $t = 0$. The solid cross section curve in Fig. 15 is (45) with $\sigma_t = 40$ mb, $B_1 = 10\,\text{GeV}^{-2}$, $B_2 = 2\,\text{GeV}^{-2}$, and $\lambda = 7 \times 10^{-3}$. The numbers were chosen for their simplicity, rather than in any attempt to give a least squares fit. The integrated elastic cross section with these parameters is 8·0 mb, in rough agreement with the value of 7·6 ± 0·3 mb quoted by Amaldi *et al.* (1973b).

The partial wave (impact parameter) amplitude corresponding to the scattering amplitude in (45) can be obtained by the methods of Appendices B and C. The scattering amplitude $F(s, t)$ is

$$F(s, t) = is\frac{\sigma_t}{8\pi}\left[(1 + \lambda)\exp(B_1 t/2) - \lambda \exp(B_2 t/2)\right] \tag{46}$$

and its partial wave projection, according to (B.24) and (C.16), is

$$a(s, b) = \frac{i\sigma_t}{8\pi}\left[\frac{(1 + \lambda)}{B_1} \exp(-b^2/2B_1) - \frac{\lambda}{B_2} \exp(-b^2/2B_2)\right]. \quad (47)$$

This partial wave profile (divided by i) is plotted as a function of impact parameter in fm in the inset of Fig. 15. The solid curve is the sum of the two terms in (47), while the dashed curves are the separate contributions. The most remarkable thing is that the distribution need be only slightly flatter than a Gaussian in order to introduce the dip and secondary maximum. [A counter argument might be that the secondary maximum is only $\sim 10^{-6}$ times the forward cross section and hence should be generated by a change of the order of only 10^{-3} times a Gaussian and so should be within the thickness of the lines on the figure!] Similarly, the mentioned steeper slope of the cross section at $|t| < 0{\cdot}15\,(\text{GeV}/c)^2$ can be incorporated by a third term in (45) or (47) that will cause the partial wave profile to extend slightly farther out in the region beyond 1 fm. The calculation of this is left as an exercise for the reader.†

The simple description contained in Fig. 15 applies at one energy. It is important to ask about energy dependence. It is clear from Fig. 5 that from 20 GeV to 2000 GeV there is significant energy variation in the cross section at fixed t. It is less clear over the ISR range (500 to 2000 GeV lab equivalent). Any model based on the eikonal approximation (or something like it) and with a largely imaginary phase shift will have destructive interference between successive terms, as in our simple description (45), but different predictions about the energy dependences of the several contributing terms (Barger, Phillips, and Geer, 1973). In the Chou-Yang model‡ the eikonal phase shift is energy independent and so is $d\sigma/dt$. In the Regge eikonal model of Frautschi and Margolis the eikonal phase from the pomeron pole is (25) with $\alpha(0) = 1$. There is thus s-dependence of the phase through $B(s)$ provided $\alpha'(0) \neq 0$. According to (C.16) the nth power of the phase shift leads to a term in the scattering amplitude proportional to $B^{1-n} \times \exp[Bt/2n]$. Thus the whole amplitude shrinks logarithmically, but successive terms involve relative powers of B^{-1} and so cause an energy dependence in the *shape* of the differential cross section, even when plotted versus

† It is amusing to note that the sharper peak at very small $|t|$ can be generated by assuming $a(s, b)$ is purely imaginary, using (B.5) to solve for $\text{Im}\, a = \frac{1}{2}(1 - \sqrt{(1 - H)})$, and assuming that the overlap function H is a Gaussian in b (Heckman and Henzi, 1972). There is no *a priori* reason to favour a Gaussian for H, of course. See Barger, Phillips, and Geer (1972) for an example of a peripheral addition to the basic Gaussian for $a(s, b)$ and de Groot and Miettinen (1973) for a more elaborate analysis with the overlap function.

‡ See Jackson (1970) for a description of this and the other models and also the appropriate references. Zachariasen (1971) also discusses the various models for diffraction scattering with emphasis on the J-plane structure.

$B(s)t$. There are other models, like the hybrid model of Chiu and Finkelstein, with features intermediate between these two and still others, like that of Cheng and Wu, with more drastic energy dependence at least asymptotically. Careful measurements at NAL energies and at the ISR should aid enormously in discriminating among models.

(b) *MacDowell–Martin lower bound on the slope parameter B*

While on the subject of partial wave distributions it is worthwhile to consider the following question: Given the total cross section and the integrated elastic cross section, can anything be said about the slope parameter B of the forward diffraction peak? Intuitively we expect a correlation. The larger the absorptive diffracting object, the narrower in angle its diffraction pattern. This has already been remarked on in correlating the total cross sections of Fig. 1 with the differential cross sections of Figs. 4 and 5 (see also Fig. 9). We are thus not surprised to learn that the answer to the question is yes (MacDowell and Martin, 1964). We give a slightly simplified derivation of the MacDowell–Martin bound using the impact parameter description and also discuss its limitations as a tool for learning about the partial wave content of the scattering amplitude.

The bound is on the logarithmic derivative with respect to t of the absorptive part of the forward scattering amplitude ($F = D + iA$),

$$A(s, t) = \frac{s}{2}\int_0^\infty \mathrm{d}b^2 \operatorname{Im} a(s, b)\, J_0(qb). \tag{48}$$

The derivative of $A(s, t)$ with respect to $t = -q^2$ evaluated at $t = 0$ is

$$\left[\frac{\mathrm{d}A(s, t)}{\mathrm{d}t}\right]_{t=0} = \frac{s}{8}\int_0^\infty \mathrm{d}b^2 b^2 \operatorname{Im} a(s, b). \tag{49}$$

The logarithmic derivative is thus

$$\left[\frac{1}{A}\frac{\mathrm{d}A}{\mathrm{d}t}\right]_{t=0} = \frac{\pi \int_0^\infty \mathrm{d}b^2\, b^2 \operatorname{Im} a(s, b)}{4\pi \int_0^\infty \mathrm{d}b^2 \operatorname{Im} a(s, b)}. \tag{50}$$

Because of the optical theorem (B.25) the denominator in (50) is just the total cross section.

The existence of a lower bound on (50) stems from the unitarity requirement that $\operatorname{Im} a(s, b) \geqslant 0$. In fact, the imaginary part of $a(s, b)$ is constrained to the range,

$$0 \leqslant \operatorname{Im} a(s, b) \leqslant 1. \tag{51}$$

We are thus invited to consider a variational problem subject to some constraints. We define $b^2 = x$, $\operatorname{Im} a(s, b) = f(x)$, and introduce the absorptive contribution to the elastic cross section,

$$\sigma_{\text{el abs}} = 4\pi \int_0^\infty \mathrm{d}b^2 |\mathrm{Im}\, a(s, b)|^2. \tag{52}$$

Then we minimize (50) subject to the constraints,

$$\left.\begin{aligned} \sigma_t &= 4\pi \int_0^\infty f(x)\,\mathrm{d}x \\ \sigma_{\text{el abs}} &= 4\pi \int_0^\infty [f(x)]^2\,\mathrm{d}x \\ 0 &\leqslant f(x) \leqslant 1 \end{aligned}\right\}. \tag{53}$$

This is a simple variational problem with Lagrange multipliers. The result is that (50) is minimized if

$$f(x) = \alpha\left(1 - \frac{x}{R^2}\right) \tag{54}$$

for $x < R^2$ such that $f(x) \leqslant 1$ and $f(x) = 1$ for smaller x. For all hadronic scattering processes at high energies $\sigma_{\text{el}}/\sigma_t < 0{\cdot}5$. From (54) we have

$$\frac{\sigma_{\text{el abs}}}{\sigma_t} = \frac{2}{3}\begin{cases} \alpha & \text{for} \quad 0 \leqslant \alpha \leqslant 1 \\ \dfrac{3\alpha - 2}{2\alpha - 1} & \text{for} \quad 1 < \alpha. \end{cases}$$

Thus the range of interest is $\alpha < 1$ and the partial wave distribution (54) is linear in x or parabolic in b. For $\alpha \leqslant 1$, $\sigma_t = 2\pi R^2\alpha$, $\sigma_{\text{el abs}} = 4\pi R^2\alpha^2/3$, and $(A^{-1}\,\mathrm{d}A/\mathrm{d}t)_{t=0} = R^2/12$. The parameters R^2 and α can be eliminated to yield the bound,

$$\left(A^{-1}\frac{\mathrm{d}A}{\mathrm{d}t}\right)_{t=0} \geqslant \frac{\sigma_t^2}{36\pi\sigma_{\text{el abs}}} \geqslant \frac{\sigma_t^2}{36\pi\sigma_{\text{el}}}. \tag{55}$$

This is the MacDowell–Martin bound, apart from an insignificant and totally justified simplification.

With the knowledge that the forward amplitude is largely imaginary at high energies we can equate the logarithmic derivative of the absorptive part to one half of the slope parameter defined by (1). In this régime the MacDowell–Martin bound reads

$$B(s, 0) \geqslant \frac{\sigma_t^2}{18\pi\sigma_{\text{el}}}. \tag{56}$$

Equation (56) is a nice bound, very solidly grounded in unitarity and nothing much else, but is it useful? Skeptics argue that no bound, even Froissart's, has ever had real practical use and the less assumptions needed to prove it, the less likely it is to be even vaguely useful. Certainly some bounds fall into this category (see Roy, 1972, for the most recent and detailed compendium of all kinds). On the other hand, a bound is a definite statement and should at least be given a chance to prove itself. We test (56) against the data on p–p elastic scattering. In Fig. 16 we show a compilation of data on

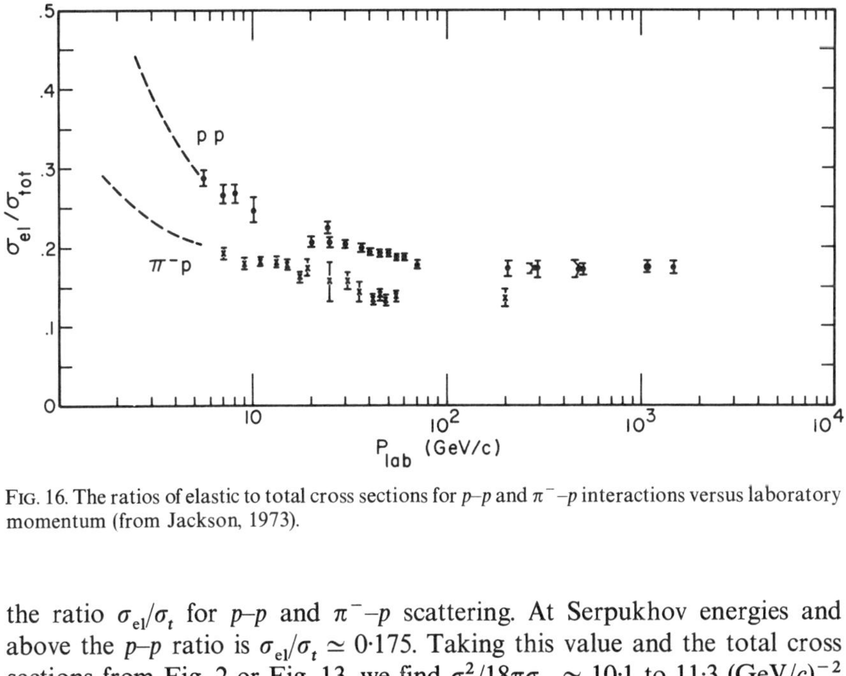

FIG. 16. The ratios of elastic to total cross sections for p–p and π^-–p interactions versus laboratory momentum (from Jackson, 1973).

the ratio σ_{el}/σ_t for p–p and π^-–p scattering. At Serpukhov energies and above the p–p ratio is $\sigma_{el}/\sigma_t \simeq 0{\cdot}175$. Taking this value and the total cross sections from Fig. 2 or Fig. 13, we find $\sigma_t^2/18\pi\sigma_{el} \simeq 10{\cdot}1$ to $11{\cdot}3$ (GeV/c)$^{-2}$ from the bottom to the top of the ISR energy range. These lower bounds are to be compared with the experimental values of $B(s, 0)$ from $11{\cdot}5 \pm 0{\cdot}6$ to $12{\cdot}6 \pm 0{\cdot}8$ over the same range. The experimental results are only slightly (10–15 %) greater than the lower bound.

The closeness of experiment to the bound can be put in perspective by considering an example of an exponential fit for all t. The differential cross section is then

$$\frac{d\sigma}{dt}(s, t) = \frac{\sigma_t^2}{16\pi}(1 + \rho^2)\exp\left[B_{\text{eff}}(s)t\right] \tag{57}$$

where we have explicitly exhibited the value at $t = 0$ via the optical theorem and the definition of ρ. Integration of (57) over t gives the relation,

$$B_{\text{eff}}(s) = \frac{\sigma_t^2(1 + \rho^2)}{16\pi\sigma_{\text{el}}}. \tag{58}$$

Comparison of (58) with (56) shows that if ρ^2 is negligible then $B_{\text{eff}} = (9/8)\,B_{\text{min}}$. Since the differential cross section at small $|t|$ *is* fitted well by an exponential (or two exponentials of slightly different slopes) it is quite reasonable that the experimental slope parameter is just slightly larger than the theoretical lower bound. In fact, to the extent that the cross section is exponential in t over the range contributing significantly to the integral anything else is impossible.

At this point we are a little disappointed in the significance of the MacDowell–Martin bound. The closeness of the experimental slope to the bound is merely a consequence of an approximately Gaussian shape in impact parameter of $a(s, b)$. Of course, the fact that it is roughly Gaussian and not rectangular or some other strange shape is progress, isn't it? Yes, it is progress, but not because of the bound. One might think, as did MacDowell and Martin (apparently because of an arithmetic slip), that comparison of experiment with their bound could distinguish between such grossly different partial wave distributions as a Gaussian and a rectangle. Not so. From (50) it is apparent that the slope parameter at $t = 0$ is determined by the average value of $x = b^2$. Requiring the first moment of a function to have a definite value constrains the function slightly, but still leaves almost unlimited freedom. To drive home this point I have constructed three different partial wave distributions, all having the following properties in common,

$$\sigma_t = 38{\cdot}9 \text{ mb}$$

$$\sigma_{\text{el}} = 6{\cdot}8 \text{ mb}$$

$$B(s, 0) = 11{\cdot}4 \text{ GeV}^{-2}$$

and

$$\frac{B(s, 0)}{B_{\text{min}}} = 1{\cdot}125.$$

The cross section and slope values are appropriate to 300 GeV laboratory energy of $W \simeq 24$ GeV at the ISR. The choice $B/B_{\text{min}} = 9/8$ is consistent with experiment and contains the Gaussian and rectangular partial wave distributions as examples. The cross section for a Gaussian is (57); for the rectangle it is

$$\frac{d\sigma}{dt} = \frac{\sigma_t^2}{16\pi}\left|\frac{2J_1(qR)}{qR}\right|^2 \tag{59}$$

with $B(0) = R^2/4$. The third partial wave distribution is a "peripheral" one. The cross section is chosen to be of the form

$$\frac{d\sigma}{dt} = \frac{\sigma_t^2}{16\pi} J_0^2(qR) \exp(-\lambda q^2). \tag{60}$$

The ratio $B(0)/B_{\text{min}}$ depends on the ratio of R^2/λ. For $R^2/\lambda = 7{\cdot}0$ we have $B(0)/B_{\text{min}} = 9/8$. The purely imaginary partial wave distribution corresponding to (60) is

$$a(b) = i\frac{\sigma_t}{8\pi\lambda} I_0(Rb/\lambda) \exp[-(b^2 + R^2)/2\lambda] \tag{61}$$

where $I_0(z)$ is a modified Bessel function of the first kind and order zero.

The three partial wave distributions are shown in Fig. 17 and the differential cross sections in Fig. 18. The distributions in b are quite different

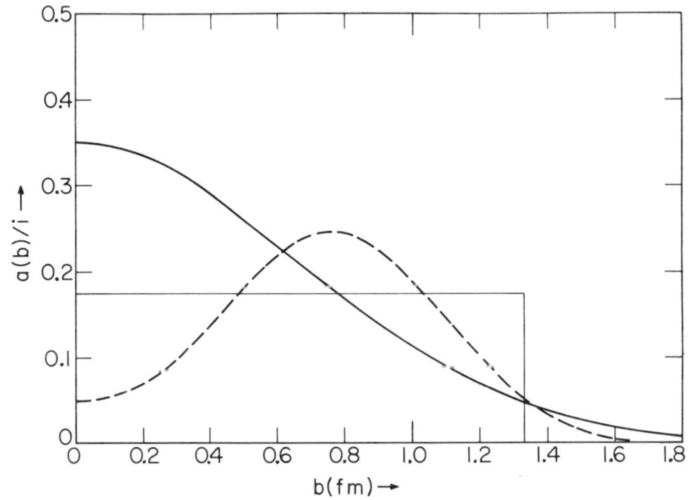

FIG. 17. Partial wave (impact parameter) distributions for p–p scattering. All three distributions give $\sigma_t = 38{\cdot}9$ mb, $\sigma_{\text{el}} = 6{\cdot}8$ mb, $B(0) = 11{\cdot}4$ GeV^{-2} ($B(0)/B_{\text{min}} = 9/8$).

in detail even though having the same $\langle x \rangle$ and the resulting cross sections are very different for $|t| > 0{\cdot}1$, too! The message is, I hope, clear—nearness of the slope parameter $B(0)$ to the lower bound (56) establishes little about the partial wave distribution. The shape of $d\sigma/dt$ at $t \neq 0$ can, of course, furnish much information, as has been illustrated already in Fig. 15.

Lest I leave the impression of scorning bounds like the MacDowell–Martin bound let me remark that (56) is useful in correlating various asymp-

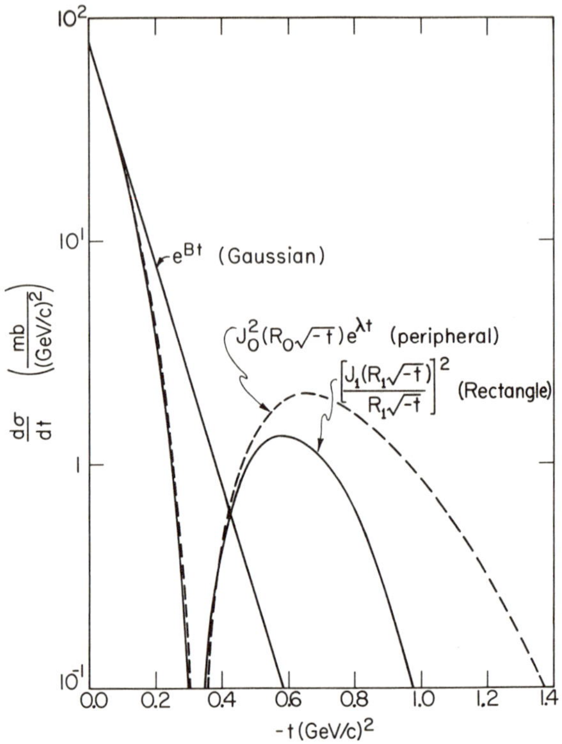

FIG. 18. Differential scattering cross-sections from the partial wave distributions of Fig. 17. All three cross-sections have the same slope at $t = 0$ and extrapolate to the same optical theorem point. For small $|t|$ they fit the p–p elastic data at ~ 300 GeV laboratory energy.

totic behaviours. If, for example, $\sigma_{el}/\sigma_t \to$ constant and $\sigma_t \to C(\ln s)^2$, (56) shows that the diffraction peak must exhibit rapid shrinkage with $B \sim (\ln s)^2$ asymptotically. More on this in the next section.

6. ASYMPTOTICS AND J-PLANE STRUCTURE

Although various remarks have already been made about energy dependences of total and differential cross sections in both theory and experiment we summarize here and also discuss briefly the J-plane structure of some models. By J-plane structure we mean the singularity structure of the t-channel partial wave amplitude analytically continued to complex angular momentum j. The dependence of an amplitude on s for fixed t is related to the singularity structure in the J-plane through the Watson–Sommerfeld transformations of the t-channel partial wave series and the connection (11) between ν and $\cos\theta_t$ (see, for example, Collins and Squires, 1968). At high energies it is possible to replace the Froissart–Gribov formula for the analytically continued t-channel partial wave amplitude by the simpler Mellin transform formula,

$$\left.\begin{aligned} F(t,j) &= \int_1^{\infty} ds\, s^{-j-1} A(s,t) \\ &\text{with its inverse,} \\ A(s,t) &= \frac{1}{2\pi i}\int_{c-i\infty}^{c+i\infty} dj\, s^j F(t,j) \end{aligned}\right\}. \tag{62}$$

In (62) c is any real number such that the vertical contour lies to the right of all the singularities of $F(t, j)$ and $A(s, t)$ is the s-channel absorptive part of the scattering amplitude (see Horn and Zachariasen, 1973, Appendix D, for the derivation and such details as signature). To gain faith in (62), assume that $F(t, j)$ has a pole at $j = \alpha(t)$. The second relation then yields $A(s, t) \propto s^{\alpha(t)}$, as expected for a Regge pole. A second order pole gives $A(s, t) \propto (\ln s)\, s^{\alpha(t)}$ and so on. It is left as an exercise for the reader to deduce the following examples:

$F(t, j)$	$A(s, t)$
$(j-\alpha)^{-1}$	s^{α}
$(j-\alpha)^{-2}$	$-s^{\alpha}\ln s$
$(j-\alpha)^{n}\ln(j-\alpha)$	$(-1)^n n!\, s^{\alpha}(\ln s)^{-n-1}$
$(j-\alpha)^{\nu}$	$-s^{\alpha}(\ln s)^{-\nu-1}/\Gamma(-\nu)$.

Here n is zero or a positive integer while ν is not an integer. The examples indicate some types of J-plane singularities and their associated s dependences. We saw that the successive terms in the Frautschi–Margolis (Regge eikonal) model had s-dependence $s^{\alpha}/(\ln s)^{n-1}$, $n = 1, 2, \ldots$. The first term corresponds to a pole in the J-plane, while higher terms evidently correspond to logarithmic singularities with softer and softer discontinuities at the tip of the branch cut. There are, of course, considerably more complicated singularities possible in the J-plane.

Two examples with increasing total cross sections can be mentioned. One is the self-consistent solution of a multiperipheral model for diffractive scattering (Ball and Zachariasen, 1972). The scattering amplitude is

$$F(s,t) = i\beta R_0^2 s \frac{J_1(qR_0 \ln(s/s_0))}{qR_0} \tag{63}$$

where $q^2 = -t$. The total cross section is $\sigma_t = 4\pi\beta R_0^2 \ln(s/s_0)$ so that the optical theorem point $d\sigma(s, 0)/dt$ is proportional to $(\ln s)^2$. On the other hand, the diffraction pattern shrinks as $(\ln s)^2$ so that the integrated *elastic* scattering cross section is constant in energy. The J-plane structure is given by

$$F(t, j) = \frac{i\beta}{t}\left[\frac{j-1}{\sqrt{((j-1)^2 - tR_0^2)}} - 1\right]. \tag{64}$$

There are, for $t < 0$, complex conjugate branch points at $\alpha_c(t) = 1 \pm iqR_0$. For $t \to 0$ these coalesce to give a second order pole at $j = 1$, yielding $\sigma_t \propto \ln s$ according to our examples quoted above. In terms of s-channel partial waves the amplitude (63) has a rectangular distribution in impact parameter out to $b_{max} = R_0 \ln(s/s_0)$, with a magnitude that decreases as $(\ln s)^{-1}$. It corresponds classically to an absorbing disc with a logarithmically growing radius, but with a decreasing opacity.

The other example is an amplitude appropriate to any model that saturates the Froissart bound, e.g. the model of Cheng and Wu (1970). At sufficiently high energy the scattering amplitude is

$$F(s, t) = i\frac{R_0^2}{2} s(\ln s)^2 \frac{J_1(qR_0 \ln s)}{qR_0 \ln s}. \tag{65}$$

The total cross section is $\sigma_t = 2\pi R_0^2(\ln s)^2$; $\sigma_{el}/\sigma_t = \frac{1}{2}$; the shrinkage of the diffraction peak is as $(\ln s)^2$. From (62) we find the J-plane projection to be

$$F(t, j) = \frac{R_0^2}{2}[(j-1)^2 - tR_0^2]^{-\frac{3}{2}}, \tag{66}$$

with complex conjugate branch points $\alpha_c(t) = 1 \pm iqR_0$ for $t < 0$. The singularities become a third order pole at $t = 0$ and give $\sigma_t \propto (\ln s)^2$. The physical interpretation in the s-channel has already been discussed.

We summarize the asymptotic behaviour of the various models in the following table:

Asymptotic Energy Dependence of Various Models

Model	σ_t	σ_{el}/σ_t	B_{eff}	ρ
Pomeron pole + *secondary poles*	$a + bs^{-\frac{1}{2}}$	$(\ln s)^{-1}$	$\ln s$	$(-)\, s^{-\frac{1}{2}}$
Pomeron eikonal	$a - b(\ln s)^{-1}$	$(\ln s)^{-1}$	$\ln s$	$(+)(\ln s)^{-1}$
Ball–Zachariasen (second order pole at $t = 0$)	$\ln s$	$(\ln s)^{-1}$	$(\ln s)^2$	$(+)(\ln s)^{-1}$
Saturation of Froissart bound (third order pole at $t = 0$)	$(\ln s)^2$	$\frac{1}{2}$	$(\ln s)^2$	$(+)(\ln s)^{-1}$
Chou-Yang (fixed pole at $j = 1$)	constant	constant	constant	(0)

Comparison of the predictions of this table with the p–p data of Fig. 2 (σ_t), Fig. 16 (σ_{el}/σ_t), Fig. 6 (B_{eff}), and Fig. 14 (ρ) shows several things. First of all, if one accepts the rising cross sections of Fig. 2 the simple Regge pole model and the Chou–Yang model are excluded. The other three models (and surely others) can accommodate the energy dependence of σ_t. Nothing can be said about saturation of the Froissart bound unless we include the cosmic ray evidence (Yodh, Pal, and Trefil, 1972). Figure 16 is on the face of it peculiar. For both pp and $\pi^- p$ interactions the ratio σ_{el}/σ_t seems to become energy independent at high energies and be quite small ($\sim$0·175 for pp, $\sim$0·135 for $\pi^- p$). For any of the models in the table, except Chou–Yang, we are asked to believe that this constancy is a transitional effect which will disappear at still higher energies. This is perhaps plausible for Cheng and Wu (although 0·5 is a long way off!), but less so for the Regge models. The third quantity, B_{eff}, displayed in Fig. 6 is also apparently in a transitional stage, at least for any model that saturates the Froissart bound. A steady growth with ln s is consistent with the Regge pole and Regge eikonal models, although the inferred slope of the pomeron trajectory ($\alpha'(0) \simeq 0·3$) is quite small. The evidence from Fig. 14 on $\rho(s)$ indicates that it is very far from its asymptotic behaviour at ISR energies and so cannot be sensibly compared with the expectations of the table. Its crossover to positive values does, of course, support some models.

There once was a time when theorists stated that asymptotic behaviour would occur at 5 or 10 GeV incident energy. From the evidence available today one might venture to say 5 or 10 TeV, but even that might be too low!

III. Inclusive Processes

1. PREAMBLE

As already mentioned in the Introduction, inclusive processes of the type

$$a + b \rightarrow c + \text{anything}$$

or

$$a + b \rightarrow c + d + \text{anything}$$

(and in principle more complicated processes) have become an important aspect of high-energy experiment and theory. Partly this is by default—very many particle states, often with several unseen neutrals, are difficult if not impossible to study in complete detail. Partly, however, it is by design. We have learned that in some senses inclusive reactions are simple and amenable to theoretical analysis.

Though the basic concepts and ideas have been known for 10 years or more from the work of Amati, Stanghellini, and Fubini (1962), Fubini (1963), and from Wilson's Schladming lectures (1963), it is only in the past four years that intensive theoretical and experimental work has been done. The renewed interest on the theoretical side was prompted mainly by work of Feynman (1969a, b) and of Yang and collaborators (Benecke *et al.*, 1969) with their ideas of scaling and of limiting fragmentation. The reasons for the experimental interest have already been discussed. By now hundreds of papers have been published; numerous conferences have been held; summaries of theory and of experiment exist in review journals and conference proceedings. Since these lectures are elementary and introductory I list a sampling of the reviews and conference reports where the hungry and/or dissatisfied reader can go for more or better information:

(a) Mostly experiment:
Wroblewski, 1970
Lander, 1971
Morrison, 1972, 1973

and the summary of contributed papers in Vol. 1 of the Proceedings of the Chicago–Batavia Conference (1972).

(b) Theory and experiment:
Quigg, 1971
Van Hove, 1971
Frazer *et al.*, 1972
Horn, 1972
Jacob, 1972
Mueller, 1972.

For entire conferences devoted to multiparticle production processes, see the proceedings of the Helsinki (1971) and the Zakopane (1972) meetings. Other lecturers here will carry on the list. In particular, the content of Wilson's hitherto unpublished Cornell report (Wilson, 1970) appears in these proceedings.

In this chapter we discuss the essential experimental facts and the general theoretical framework for inclusive processes. Most of the treatment is without bias towards any particular theoretical model, although some concepts, e.g. finite range correlations in rapidity, will be accepted as true without serious questioning. This facilitates establishing all the essential ideas and can be changed in the light of contrary facts without undue harm. The notation and kinematics of inclusive processes are described in Appendix D. Definitions of distributions, correlation functions, multiplicity moments, and sum rules are given in Appendix E.

2. BASIC FACTS AND SAMPLES OF DATA

(a) *Prong cross sections as functions of incident energy*

As mentioned in the Introduction, increasing energy brings production of more and more particles, mostly pions but with some heavier particles as well. A measure of the particle production is afforded by the values of the *topological* or *prong cross sections*. These are the cross sections σ_n for a specific number n of *charged* particles in the final state (whose ionization produces prongs or tracks in a bubble chamber or emulsion), independent of how many neutrals are produced. At a given incident energy the prong cross sections are expected to be given by something like a Poisson distribution, with events having half as many or twice as many as the average number of prongs being fairly frequent.† Figure 19 shows a typical set of prong cross sections (Charlton *et al.*, 1972). They happen to be from 205 GeV protons incident on the NAL 30″ hydrogen bubble chamber; the results from 50 and 69 GeV at Serpukhov (Ammosov *et al.*, 1972), 102 GeV at NAL (Chapman *et al.*, 1972), and 303 GeV at NAL (Dao *et al.*, 1972) are qualitatively similar. Later we will discuss the shape of the prong distributions in more detail, but now we turn to the energy dependence.

The prong cross sections for pp interactions at various energies are summarized in Fig. 20. The increasing numbers of particles produced as the energy increases is very evident. Not so dramatic but still evident is the peaking of a given cross section at some energy and then its decrease. The 2-, 4-, 6-, and even the 8-prong cross sections are decreasing at the highest energies. Whether these low prong number cross sections continue to decrease at higher energies or reach constant values is a point of considerable interest for "two-component" models of particle production. More later on this topic.

For pions and K-mesons incident the data are not available at as high energies, only up to 34 GeV for K and 50 GeV plus the one set of data at 200 GeV (Huson, 1973) for pions, but the same trends and features are visible. Figures 21 and 22 show the presently existing results (without the 200 GeV π^- data).

(b) *Average number of charged particles versus energy*

The average number of charged particles per inelastic collision $\langle n_{\text{ch}} \rangle$ is defined by

$$\langle n_{\text{ch}} \rangle \, \sigma_{\text{inel}} = \sum_n{}' \, n \, \sigma_n \tag{67}$$

† Completely independent emission of particles would lead to a Poisson distribution in the number of charged prongs. Crude imposition of charge conservation by the assumption of pair production of positively and negatively charged particles leads to a Poisson distribution in the number of negative prongs (Wang, 1969).

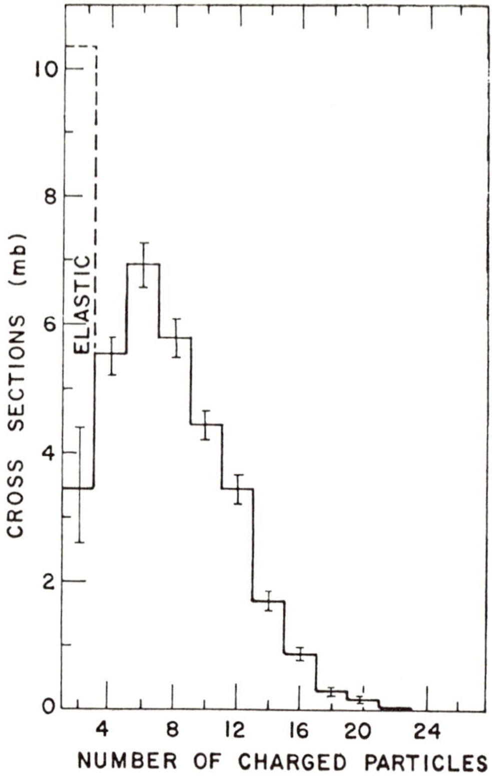

FIG. 19. Prong cross sections in millibarns versus prong number for 205 GeV p–p interactions. The first and second moments are $\langle n_{ch}\rangle = 7{\cdot}65 \pm 0{\cdot}17$, $\langle n_{ch}^2\rangle = 73{\cdot}6 \pm 2{\cdot}2$ (from Charlton *et al.*, 1972).

where σ_n is the n-prong cross section and the prime on the sum means that the elastic scattering contribution (to the 2-prong cross section usually) is omitted. This quantity and higher moments defined analogously are a useful way of characterizing the prong distribution. It is obvious from Figs. 20–22 that $\langle n_{ch}\rangle$ is an increasing function of energy. It is popular to plot the data on $\langle n_{ch}\rangle$ versus s in the manner of Fig. 2 with a linear ordinate and a logarithmic abscissa. The data then show a roughly linear rise, at least at high energies, corresponding to

$$\langle n_{ch}\rangle \simeq a_0 + a_1 \ln s \tag{68}$$

with $a_0 \simeq -3{\cdot}4$ and $a_1 \simeq 1{\cdot}94$ for p–p interactions and similar values for π^-p and K^-p collisions. A logarithmic increase is expected on the basis of elementary considerations (see Section 3(c) below), but for the sake of

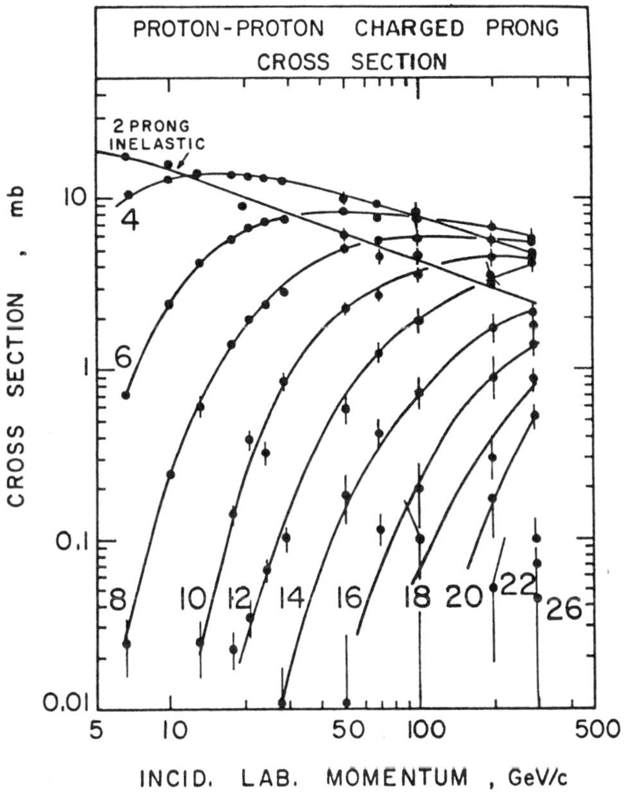

FIG. 20. Prong cross sections in millibarns versus incident laboratory momentum in GeV/c for p–p interactions (from Morrison, 1973).

perversity Fig. 23 displays the data for p–p interactions as a function of $Q^{\frac{1}{2}} = (W - 2m_p)^{\frac{1}{2}}$. At high energies $Q^{\frac{1}{2}} \rightarrow s^{\frac{1}{4}}$. The data lie on a reasonable straight line, $\langle n_{\text{ch}} \rangle \simeq 1{\cdot}85\, Q^{\frac{1}{2}}$. This is an example of something known among my friends as Jackson's theorem (see Fig. 18 of Jackson, 1970), the point being that over a limited range a ln s variation can be approximated by a power of s. Some models of multiparticle production, for example the hydrodynamic model of Landau, predict the $s^{\frac{1}{4}}$ variation of multiplicity.

(c) *Average multiplicity of different types of particles as a function of energy*

Figure 23 shows that the average number of charged particles per inelastic p–p collision is of the order of 10 to 13 at ISR energies. We expect that these are predominantly pions, but it is obviously of interest to learn the composi-

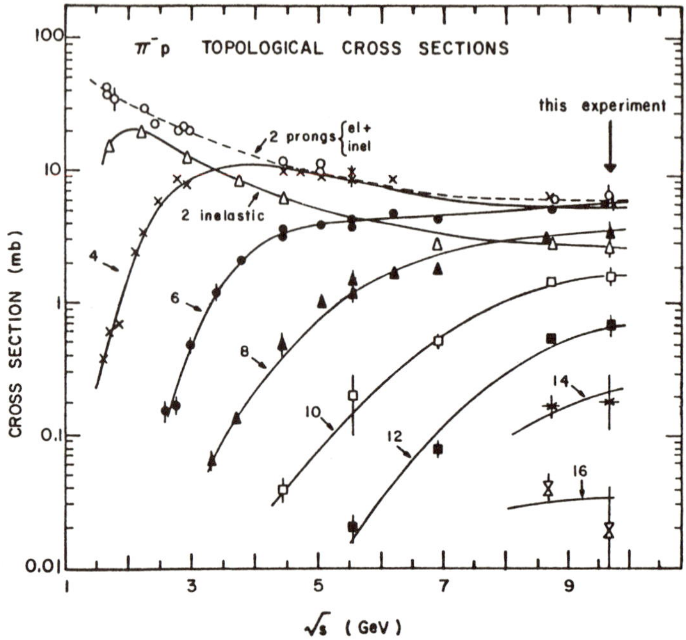

FIG. 21. Prong cross sections in millibarns versus centre of mass energy $W = \sqrt{s}$ in GeV for $\pi^- -p$ interactions. The highest energy ($W \simeq 9{\cdot}7$) corresponds to 50 GeV/c laboratory momentum (from Ammosov *et al.*, 1973).

tion in detail. For example, are the proportions of K-mesons and/or antiprotons relative to pions constant in energy, even though the total number of charged particles increases? Or is some (all?) of the increase above some energy accounted for by an increased production of $K^{\pm}$ and $\bar{p}$? A summary of available data is shown in Fig. 24. Several features are worthy of note. At low energies the charged multiplicity is built up with protons and to a lesser extent positive and then negative pions. Soon, however, the pions take over the bulk of the multiplicity and rise in proportion to the total. The average number of protons decreases slightly, from slightly less than 2 at low energies to $\sim 1{\cdot}3$ at ISR energies. The K^+, K^-, and $\bar{p}$ average multiplicities are quite small at low energies but have a steeper energy dependence than the pions. Nevertheless, even at ISR energies ($s \sim 10^3 - 3 \times 10^3$ GeV^2), their average numbers per collision are still small:

$$\langle n_{K^+} \rangle \simeq 0{\cdot}4\text{–}0{\cdot}5, \qquad \langle n_{K^-} \rangle \simeq 0{\cdot}3\text{–}0{\cdot}4, \qquad \langle n_{\bar{p}} \rangle \simeq 0{\cdot}10\text{–}0{\cdot}15.$$

There is some indication that the relative proportions of K^+ and K^- may

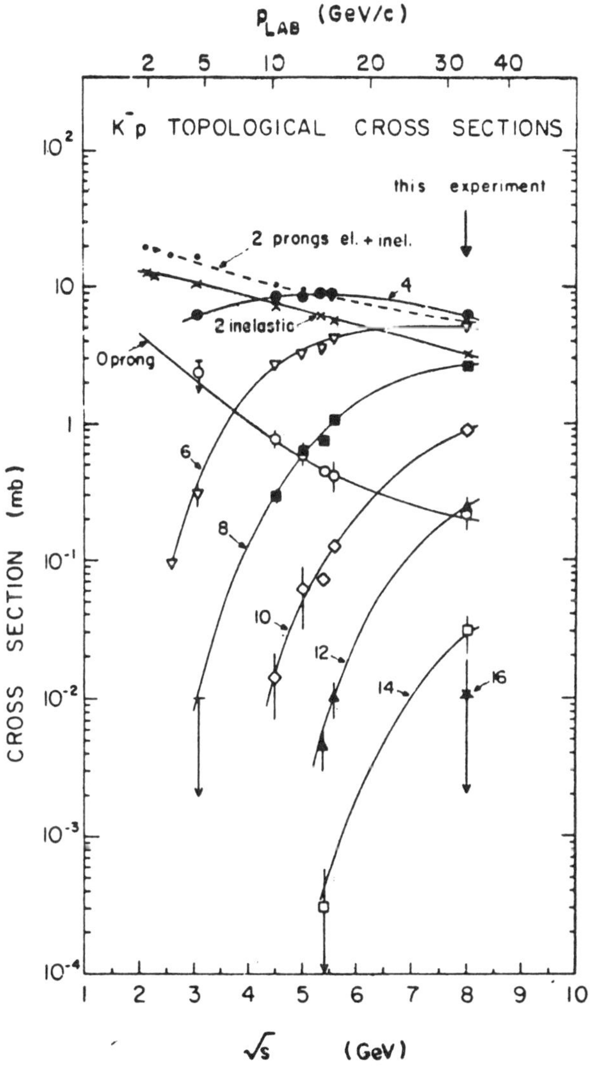

FIG. 22. Prong cross sections in millibarns versus centre of mass energy $W = \sqrt{s}$ in GeV for K^-p interactions. The highest energy ($W \simeq 8$ GeV) corresponds to 34 GeV/c laboratory momentum (from Ammosov *et al.*, 1973).

be becoming s-independent at the highest energies, but the average number of $\bar{p}$ is still growing relative to $\langle n_{ch} \rangle$.

All of this makes most reasonable sense on the naïve grounds of energetics, with the more massive objects being produced with greater difficulty. One

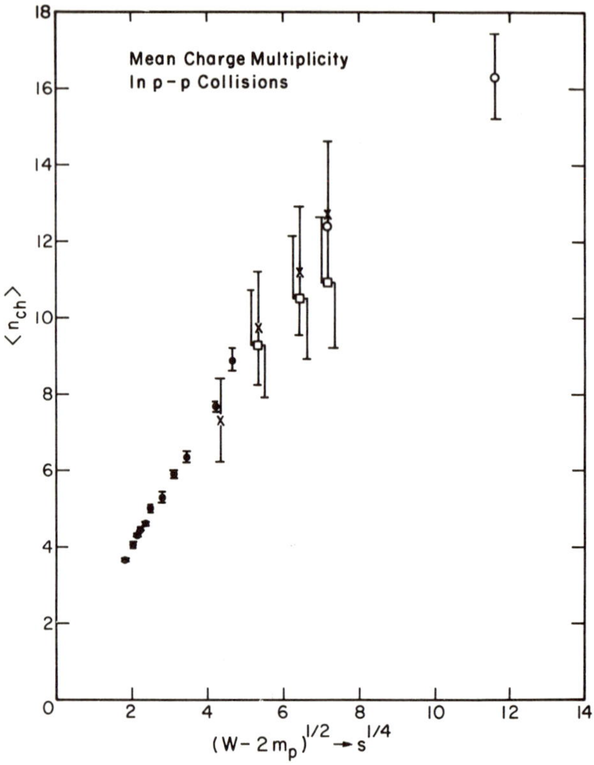

FIG. 23. Average charged particle multiplicity per inelastic p–p collision versus $(W - 2m_p)^{\frac{1}{2}} \to s^{\frac{1}{4}}$ in $(\mathrm{GeV})^{\frac{1}{2}}$ (from Jackson, 1973). The highest point is an estimate from cosmic rays.

can ask, of course, whether the range of s shown in Fig. 24 is plausible, or whether one might have expected a more rapid rise from threshold and earlier development of an asymptotic behaviour. After all, at $s \simeq 3000$ GeV it is energetically possible to make 28 nucleon–antinucleon pairs! (or ~ 380 pions!). The relatively expanded scale in Fig. 24 over which the multiplicities rise can be explained in part at least by the small inelasticity of the collisions. The collision partners, perhaps with their charges changed, carry off an appreciable fraction of the available energy. This "leading particle" effect is exhibited in more detail in item (e) below.

(d) *Limited transverse momentum*

One of the most striking features of multiparticle reactions at high energies is the limited extent of the transverse momenta, i.e. the magnitude of the component perpendicular to the beam direction. On a Peyrou plot of

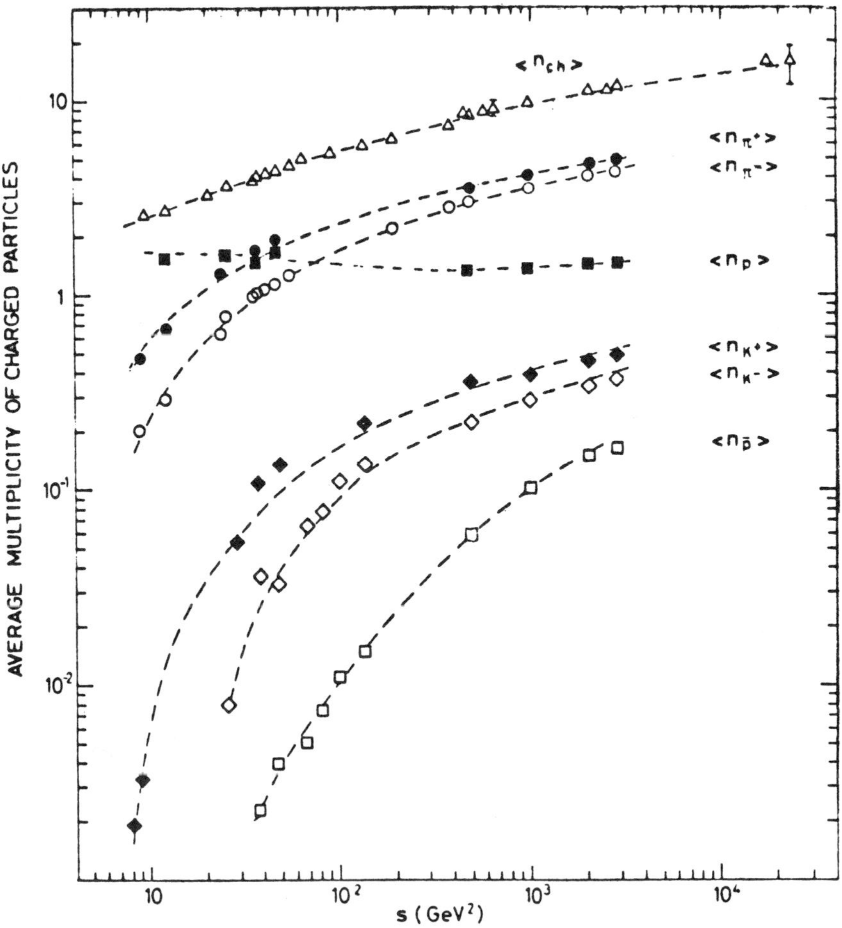

FIG. 24. Average multiplicity of various charged particles per inelastic p–p collision versus s in $(\mathrm{GeV})^2$ on a log–log plot. The top points and curve are for all charged particles. In order on the right from the top are π^+, π^-, p, K^+, K^-, $\bar{p}$ multiplicities. (Figure from Antinucci *et al.*, 1973)

$p_{\parallel}$ vs $p_{\perp}$, with the kinematic boundary a circle with radius, $p \simeq W/2$, events cluster along the x-axis. This behaviour has been known for a long time in cosmic rays (see, for example, Feinberg, 1972, and earlier references cited there). Two examples from recent experiments at accelerators are given in Figs. 25 and 26. The data shown in Fig. 25 are inclusive distributions in $p_{\perp}^2$ for π, K^0 and Λ from K^-–p interactions at 13 GeV/c (Barletta *et al.*, 1973). The results in Fig. 26 are from the ISR and show the inclusive invariant cross sections for π^+, π^-, K^+, K^-, p, and $\bar{p}$ at fixed $p_{\parallel}$ ($x = 0{\cdot}16$) versus $p_{\perp}$

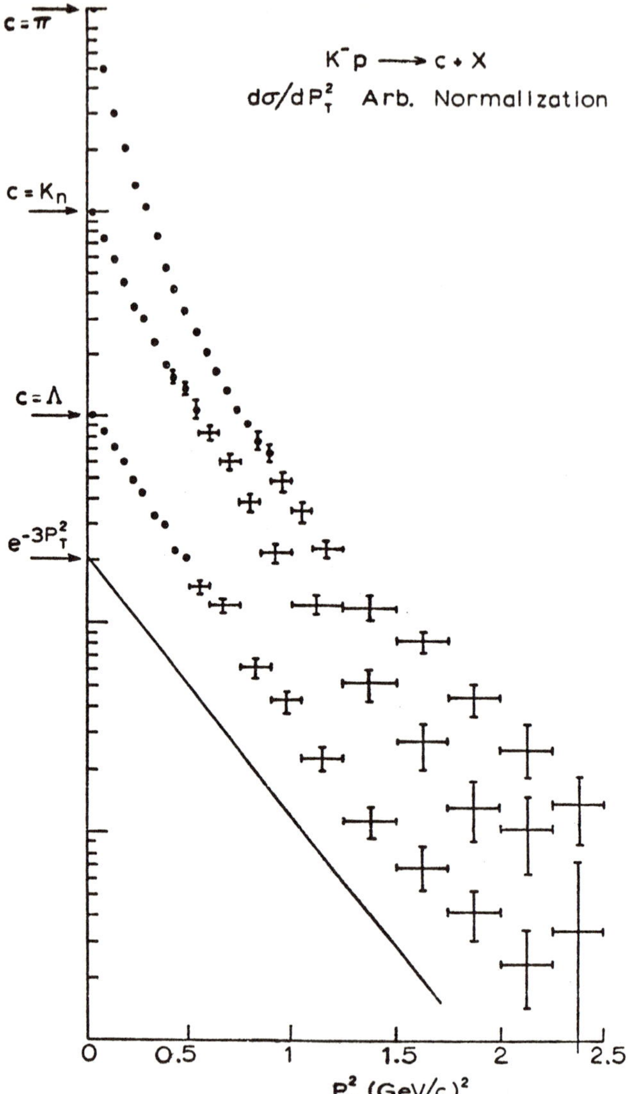

FIG. 25. Distributions in $p_\perp^2$ for pions, neutral K-mesons and Λ hyperons from K^-p interactions at 13 GeV/c (from Barletta *et al.*, 1973). $(p_\perp^2)_{max} \simeq 6{\cdot}4$ (GeV/c)^2.

(Bertin *et al.*, 1972). Independently of whether one fits with an exponential in $p_\perp$ or $p_\perp^2$ one finds mean values of transverse momenta of the order of $\langle p_\perp \rangle \simeq 0{\cdot}33$ GeV/c for pions and $\langle p_\perp \rangle \simeq 0{\cdot}4$–$0{\cdot}5$ GeV/c for $K^\pm$ or $p^\pm$. There is very little variation in $\langle p_\perp \rangle$ with bombarding energy from $\sim$10 GeV

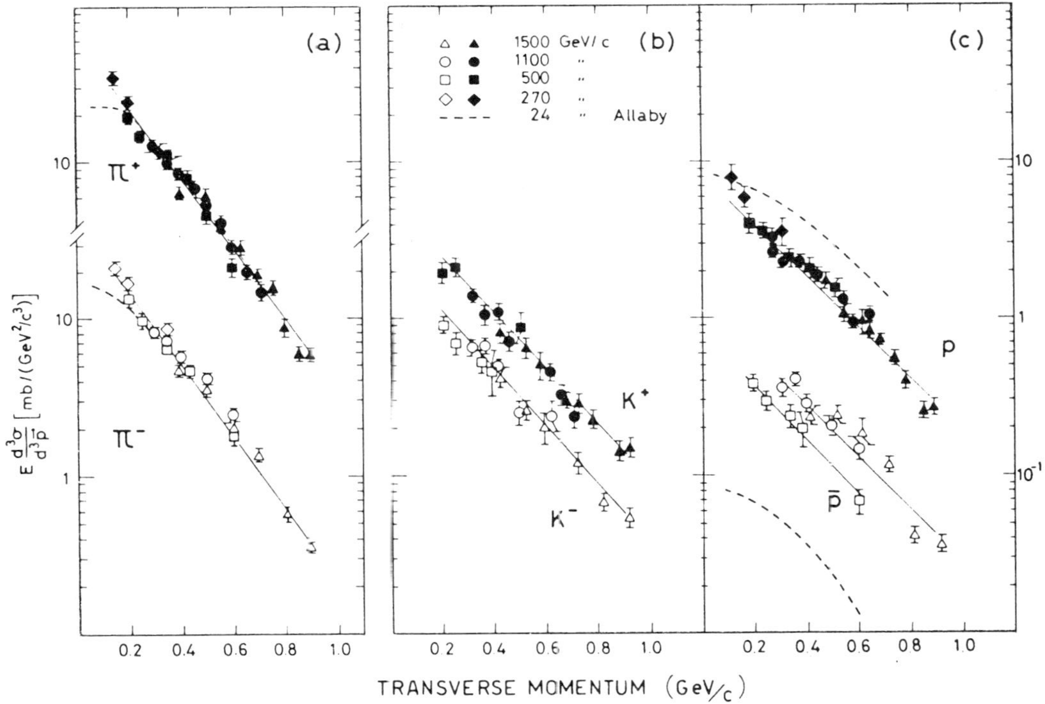

FIG. 26. Invariant cross sections $E(d^3\sigma/dp^3)$ for $\pi^+ . \pi^-$, K^+, K^-, p, and $\bar{p}$ produced in p–p collisions at ISR energies as functions of $p_\perp$ in GeV/c at $x = 0{\cdot}16$. The solid lines are exponential fits to the ISR data. The dashed lines show the trends of 24 GeV/c data from CERN. (Fig. from Bertin *et al.*, 1972).

to cosmic ray energies well above the ISR. There is some variation of $\langle p_\perp \rangle$ with $p_\parallel$ at a given energy. This can be understood in terms of the constraints imposed by the kinematic boundary.†

The smallness of $\langle p_\perp \rangle$ is another manifestation of a property of hadronic interactions already seen in the collimated character of elastic scattering where $\langle -t \rangle \simeq B^{-1} \sim 0{\cdot}1\ (\mathrm{GeV}/c)^2$. Hadrons are extended objects and their interactions are peripheral with the exchange of "soft" quanta as the dominant mechanism. Multiparticle production apparently proceeds in the same way. The multiperipheral model (Amati, Stanghellini, and Fubini, 1962, and hundreds of subsequent papers by others) is one explicit realization of this.

(e) *Longitudinal behaviour (in $p_\parallel$, x or y) of inclusive distributions*

The other kinematic dimension to be considered is the longitudinal momentum $p_\parallel$ or the equivalent variables x or y [see Appendix D, eqns (D.4) and (D.8)]. While in the transverse direction all types of particles tend to be limited in $p_\perp$, in the beam direction there are significant differences depending on particle type and the relation to the incident collision partners. Figure 27 shows some typical inclusive distributions in x for fixed $p_\perp = 0{\cdot}8$ GeV/c at the ISR (Albrow *et al.*, 1973). The data span the range of $0{\cdot}2 < x < 1{\cdot}0$. This may appear to be nearly the whole range of x, but the region $0 \leqslant x \leqslant 0{\cdot}2$ is more important than it seems because of the phase space $\mathrm{d}x/x$ [see (D.32)]. Rapidity phase space does not have any weighting factor; rapidity thus shows more clearly the fraction of available phase space covered. From (D.27) we deduce that at $p_\perp = 0{\cdot}8$ GeV/c, $x = 0{\cdot}2$ corresponds to $\Delta y \simeq 1{\cdot}5, 1{\cdot}6$, and $1{\cdot}9$ for π, K, and p respectively. Comparison with $\Delta y = Y/2 \simeq 3{\cdot}8$ for the range $0 < x < 1$ shows that the particles in Fig. 27 come from regions of phase space close to, or at least not remote from, the incident proton. They are all in the "fragmentation region" of Benecke *et al.* (1969).

The shapes of the spectra in Fig. 27 are characteristic of inclusive distributions at any energy. The protons are relatively flat (on a logarithmic scale) and persist up to $x = 1$. In fact the elastic peak and its tail of diffractively produced resonances and continuum in missing mass M^2 are not visible because of a broadened resolution and elimination of elastic events. The region near $x = 1$ for the protons is discussed in detail by Sens in these proceedings. The pion and K-meson spectra are typical of particles different from the incident ones. The distributions peak near $x = 0$ and fall more or less exponentially in $|x|$ away from that point, with negligible yields at

† There was for a time interest in something called the "seagull effect", namely a dip in $\langle p_\perp \rangle$ as a function of x at $x = 0$. If one evaluates $\langle p_\perp \rangle$ from the invariant cross section $\mathrm{d}^3\sigma/\mathrm{d}y\,\mathrm{d}^2p_\perp$ at fixed y instead, then $\langle p_\perp \rangle$ versus y shows a more or less monotonic behaviour away from $y^* = 0$. See, for example, Figs 12 and 13 of Bosetti *et al.* (1973).

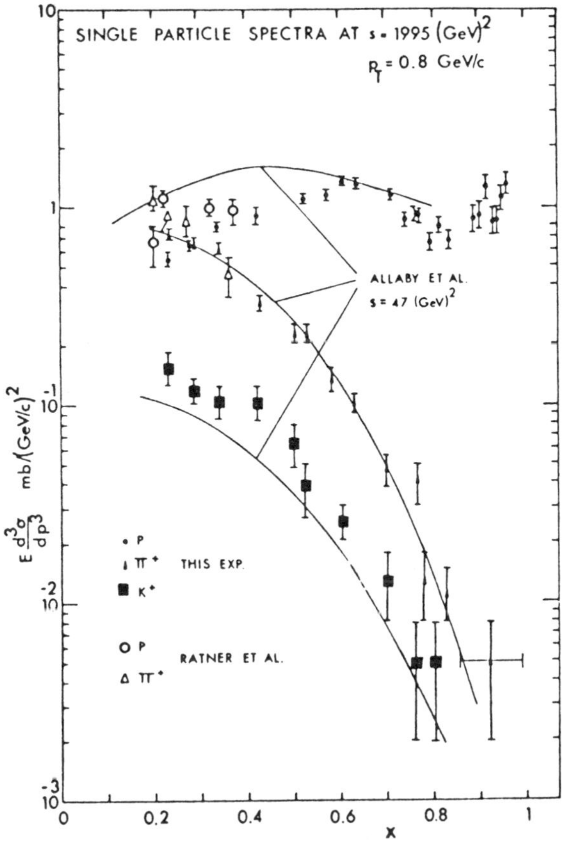

FIG. 27. Inclusive invariant cross sections $Ed^3\sigma/dp^3$ for $p + p \to c$ + anything where $c = p$, π^+, K^+ at $s = 1995\,(\text{GeV})^2$ and $p_\perp = 0{\cdot}8$ GeV/c versus $x = 2p_\parallel^*/\sqrt{s}$. The curves show the trend of CERN data at 24 GeV/c incident laboratory momentum [$s = 47\,(\text{GeV})^2$]. (Figure from Albrow *et al.*, 1973.)

the kinematic boundary ($x = 1$). Corresponding features for π^+ interactions at 8 and 16 GeV/c, this time integrated over all $p_\perp^2$, are shown in Fig. 28. The "leading particle" effect is visible near $x \simeq 1$ in the data on the left, $\pi^+ p \to \pi^+ X$. For negative x both distributions fall rapidly as $x \to -1$. This is expected because $x \to -1$ is the proton end of the scale. The inclusive proton spectra (not shown) peak at $x \simeq -1$ and fall monotonically with increasing x, being $\sim 0{\cdot}1$ in relative size at $x \simeq 0$ and still smaller for $x > 0$.

(f) *Scaling*

Feynman (1969a, b) gave a description of hadronic interactions that leads to the conclusion that as $s \to \infty$ the inclusive cross section, expressed in terms of $p_\perp^2$ and $x = 2p_\parallel^*/\sqrt{s}$, should be independent of s. This is called *Feynman scaling*. At more or less the same time Yang and collaborators (Benecke *et al.*, 1969) from a rather different point of view suggested the hypothesis of *limiting fragmentation* whereby at high enough energies the inclusive cross section for the production of a particle c from a target (or projectile) should be independent of the incident energy and type of the other collision partner provided the momentum of c is finite in the rest frame of the target (or projectile). For x away from $x = 0$, these two kinds of scaling are equivalent, as will be shown below.

Tests of scaling or the approach to scaling abound in the literature (see the references cited at the beginning of this chapter). We refer only to Figs 27 and 28 for an indication. The solid curves in Fig. 27 are representations of 24 GeV/c ($s = 47$ GeV^2) data from CERN while the points are ISR data at $s \simeq 2000$ GeV^2. For the π^+ data scaling is satisfied to within 10–20% accuracy. For the protons and K^+ scaling over such a wide range in s does not occur. The K^+ cross sections are rising from $s \simeq 50$ to $s \simeq 2000$, while the proton inclusive cross section falls slightly. Over the ISR range $s \simeq 500$ to $s \simeq 3000$ GeV^2 the data on $p + p \rightarrow c +$ anything, $c = \pi^+, \pi^-, K^+, K^-$, show scaling in the fragmentation region ($|x| > 0{\cdot}2$) to an accuracy of 10–15% (see Prof. Sen's lectures for examples). Only the $p + p \rightarrow \bar{p} +$ anything fails to scale at ISR energies. This is consistent with the behaviour of $\langle n_{\bar{p}} \rangle$ shown in Fig. 24.

Figure 28 shows $\pi^+ \ p \rightarrow \pi^\pm +$ anything at 8 and 16 GeV/c laboratory momentum. These energies are very low compared to the ISR energies, but the approach to scaling is visible. Details of the s dependence of the different regions of x need not concern us. These energies are sufficiently low that kinematic effects (e.g., in $\pi^+ p \rightarrow \pi^-$ only events with four or more charged prongs can contribute) can still have undue influence.

(g) *Quantum number transfer*

An interesting aspect of multiparticle production is the extent of transfer of additive quantum numbers such as charge, hypercharge, or baryon number from the region of phase space occupied by the initial particles to other regions in the final state. If we think of the colliding hadrons as extended bodies of hadronic matter making rather peripheral collisions with a relatively small fraction of the total energy going into particle production, we might expect that the "leading particles" would largely preserve their quantum numbers. In the multiperipheral model this naïve expectation occurs because of the correlation between the size of possible rapidity gaps and the Regge intercepts of the links. Crudely speaking the higher the intercept the larger the gap. Empirically (and dynamically in the models) the highest trajectories carry the fewest nonvacuum quantum numbers. Thus baryon number, for example, has difficulty migrating very far down the rapidity axis in such models.

Let Q be one of the additive quantum numbers such as charge. Then we define the differential distribution in rapidity dQ/dy by

$$\sigma_{\text{inel}} \frac{dQ}{dy} = \sum_c{}' Q_c \int \frac{d\sigma_{ab}^c}{dy\, d^2p_\perp} d^2p_\perp \tag{69}$$

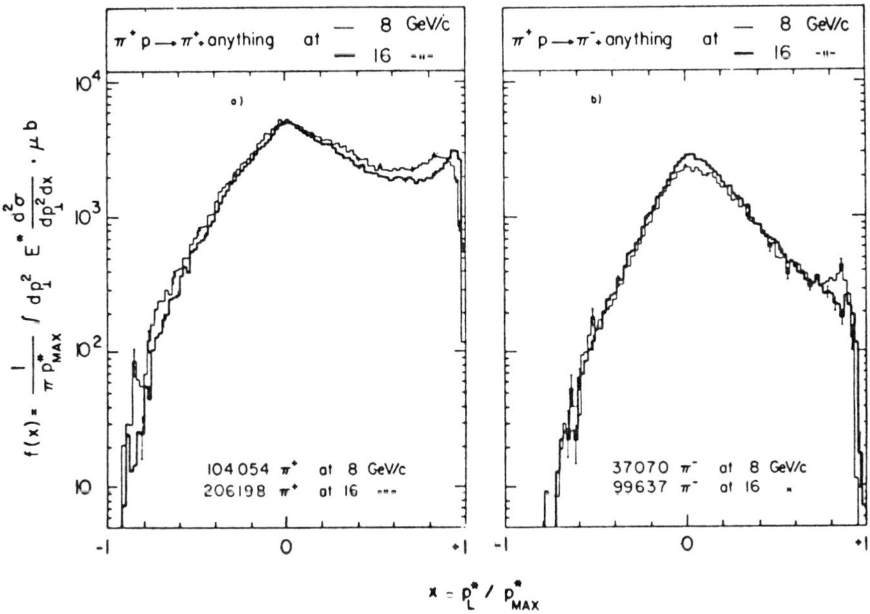

FIG. 28. Invariant inclusive cross sections, integrated over $p_\perp^2$ for $\pi^+ p \to \pi^{\pm} X$ at 8 and 16 GeV/c versus x. Left-hand figure, π^+; right-hand figure, π^- (from Bosetti *et al.*, 1973).

Figs 29 and 30 show differential distributions in rapidity for electric charge. The data of Fig. 29 are from $\pi^+ p$ and $\pi^- p$ interactions at 16 GeV/c. These data show no narrow spikes at the extremities of the plot with zero or very small values between. Rather, there is a gradual change from one end to the other, with the negative charge (belonging initially to the incident pion) being spread out somewhat more than the proton's positive charge. This can be accounted for by the lighter mass of the pion and its greater mobility in rapidity. The data of Fig. 30 are perhaps more revealing. These are from pp interactions at various energies and show the anticipated tendency as the rapidity interval widens to have the charge cling to its initial part of the

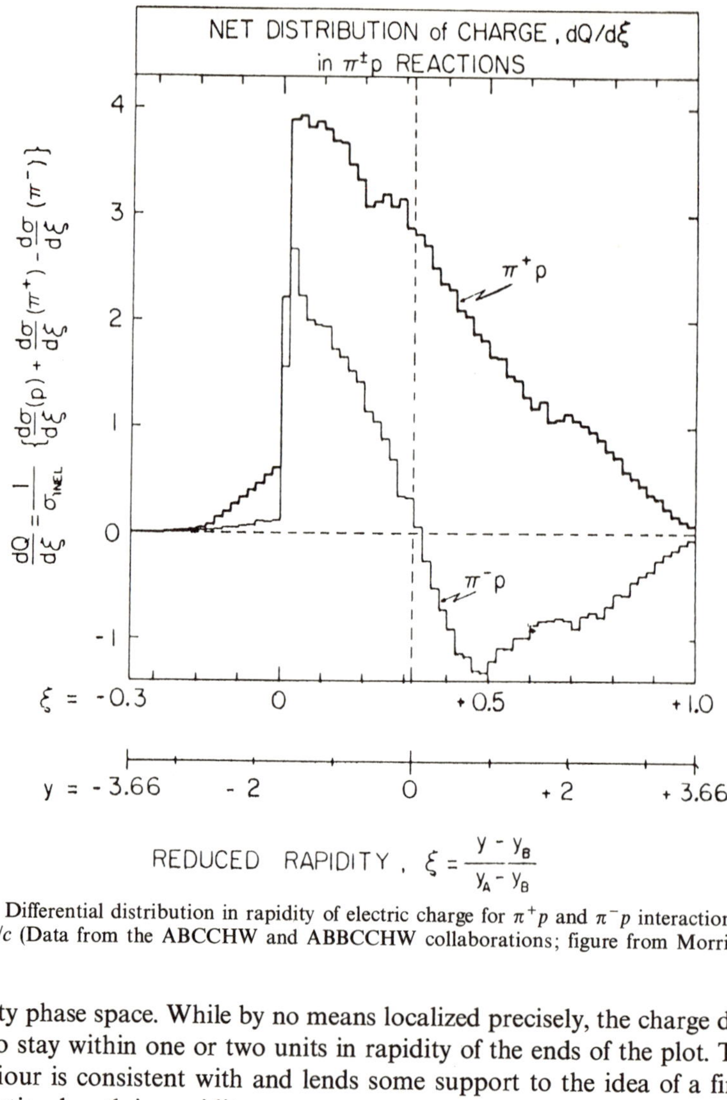

FIG. 29. Differential distribution in rapidity of electric charge for π^+p and π^-p interactions at 16 GeV/c (Data from the ABCCHW and ABBCCHW collaborations; figure from Morrison, 1972).

rapidity phase space. While by no means localized precisely, the charge does tend to stay within one or two units in rapidity of the ends of the plot. This behaviour is consistent with and lends some support to the idea of a finite correlation length in rapidity.

3. THEORETICAL FRAMEWORK ASSUMING A FINITE CORRELATION LENGTH: FEYNMAN–WILSON "GAS"

A very useful conceptual framework for inclusive processes is the fluid analogy, often called the Feynman–Wilson gas. The idea is that, since phase-space can be written $dy\, d^2p_\perp$ and the kinematics limits the possible range of

y, while the dynamics effectively limits the range of $p_\perp$, the particles in a multiparticle production process can be viewed as a fluid confined in a bottle in the $y - p_\perp$ space. As $s \to \infty$, the bottle becomes very long and the "motion"

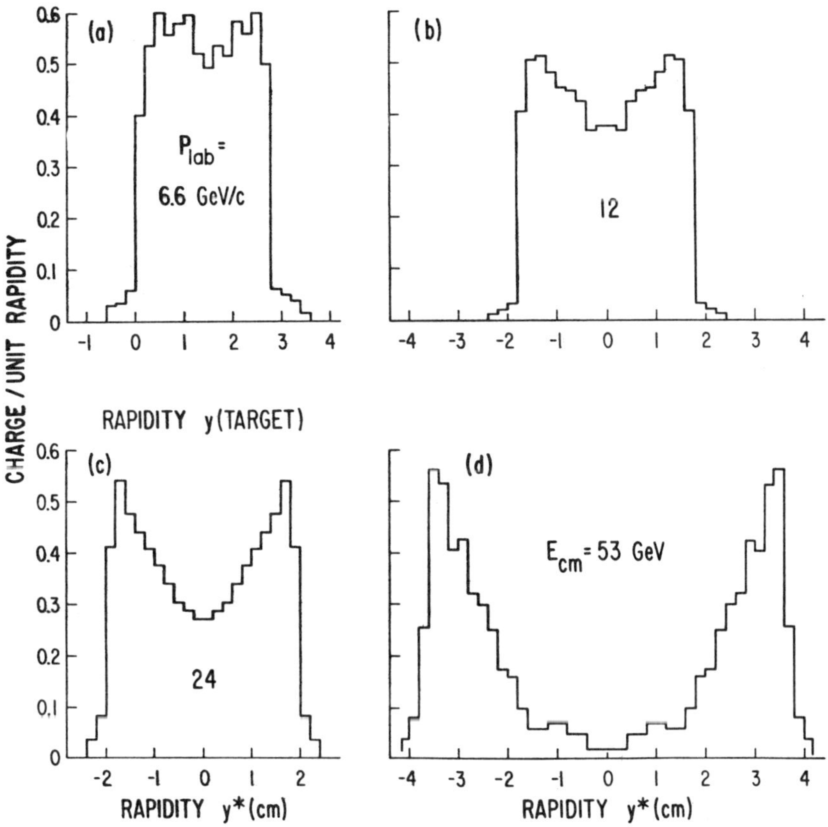

FIG. 30. Differential distribution in rapidity of electric charge for p–p interactions at $P_{\rm lab} = 6{\cdot}6$, 12, 24 GeV/c and $W = 54$ GeV ($P_{\rm lab} \simeq 1500$ GeV/c) (from Sivers, 1973).

inside is essentially one-dimensional. Figure 31 is a schematic diagram of the envisioned situation. The normalized n-particle distributions defined by (E.5) are thought of in much the same way as the corresponding densities in a real fluid. In particular, the essential working hypothesis is that the *correlation length* in rapidity over which a given particle can be influenced by another *is finite*. We see immediately that a number of important results follow directly from this hypothesis. The idea of a finite correlation length and its consequences is contained in the work of Amati, Stanghellini, and Fubini (1962), on the multiperipheral model and in the papers of Wilson (1963, 1970). Wilson attributes the fluid analogy to Feynman.

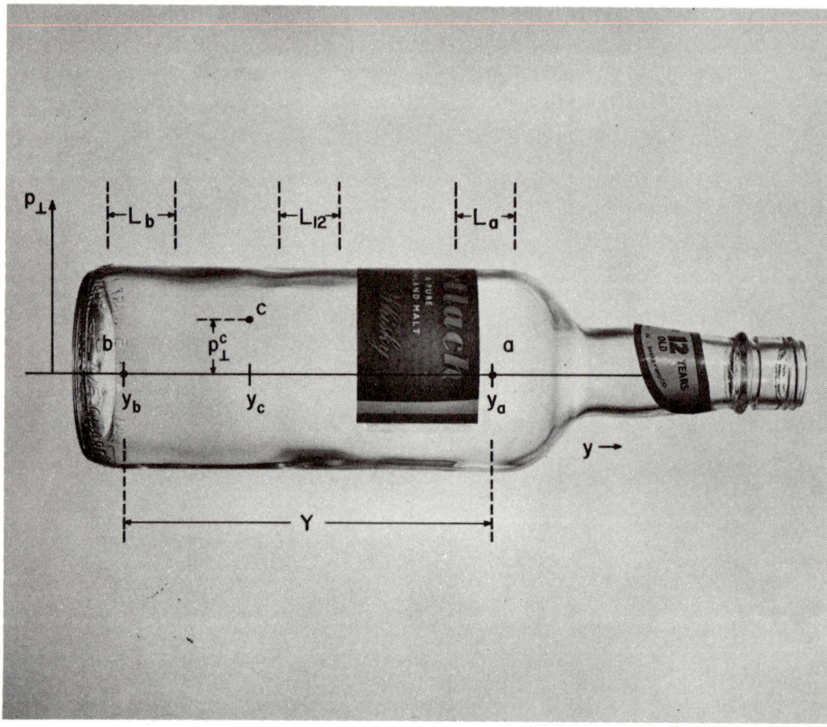

FIG. 31. Feynman–Wilson gas picture. The "gas" of produced particles is confined in phase space to the general region of the "bottle". The length of the bottle is determined by the kinematics, while its radius is governed by the dynamic limitation of $p_\perp$. The finite correlation lengths at each end and in the centre are indicated by L_a, L_b, and L_{12}.

(a) *Limiting fragmentation and scaling at finite x*

Consider the single particle inclusive density,

$$\rho^c_{ab} = \rho^c_{ab}(y_a - y_b, y_c - y_a, \vec{p}_\perp). \tag{70}$$

In general it depends on the indicated three variables, the first being equivalent to the c.m.s. energy W and the second and third to the momentum $\vec{p}$ of particle c in a frame related to the incident particles. Suppose now that particle c is produced "in the vicinity" of particle a, that is, the rapidity difference $(y_c - y_a)$ is finite as $s(Y) \to \infty$. The assumption of a *finite* range of correlation implies that at large enough energies particle c cannot "know" *what type* or *where* on the rapidity axis particle b is. In this circumstance the density (70) must become independent of $Y = y_a - y_b$ and also of particle b:

$$\lim_{\substack{Y \to \infty \\ y_c - y_a \text{ fixed}}} \rho^c_{ab} = f^c_a(y_c - y_a, \vec{p}_\perp) = g^c_a(x, \vec{p}_\perp). \tag{71}$$

The first form in (71) can be recognized as a statement of limiting fragmentation of particle a (Benecke *et al.*, 1969), while the second, equivalent [because of eqn (D.27)], form is a statement of Feynman scaling (Feynman, 1969a, b) in the region $x > 0$.

If particle c is produced "in the vicinity" of particle b then the limit corresponding to (71) has $a \to b$ and $x < 0$.

(b) *Central plateau*

Suppose that particle c is produced in the central region of Fig. 31, that is, many correlation lengths L away from either end. Then it will be unaware of the identity or position in rapidity of either a or b. The density (70) then becomes independent of Y and y and is only a function of $\vec{p}_\perp$:†

$$\lim_{\substack{Y \to \infty \\ y_c \quad y_b \gg L_b \\ y_a - y_c \gg L_a}} \rho^c_{ab} = h^c(\vec{p}_\perp). \tag{72}$$

The invariant cross section in this so-called central region is flat in rapidity and depends on the incident particles a and b only through the factor σ_{ab}.

For p–p interactions at ISR energies the apparent development of a central plateau, as well as evidence for limiting fragmentation, is shown in Fig. 32. Data from a number of experiments at different c.m.s. energies, but all at $p_\perp = 0{\cdot}4$ GeV/c, are plotted as a function of $y_{\text{lab}} = y_c - y_a$. Limiting fragmentation to an accuracy of 10–20% over the ISR range can be seen from the data at $y_{\text{lab}} < 3$ for all particles except possibly antiprotons. Comparison of the dashed lines with the ISR points indicates departures from scaling at $W = 6{\cdot}8$ GeV. The central plateau for $\pi^\pm$, $K^\pm$, and p seems established although a plot of the data versus y^* instead of y_{lab} would indicate some tendency for the plateau to rise with increasing energy by 10–20%. The logarithmic scales in Fig. 32 and the scatter of the data points makes this

† One might think that there could be a possible dependence on y, independent of y_a and y_b, but the Lorentz invariance of ρ^c_{ab} requires that it be a function of $y - y_a$ and $y - y_b$.

difficult to see. The conclusion from Fig. 32 is that (71) and (72) are verified, at least at the 10–15% level.†

(c) *Growth of multiplicity and correlation parameters with energy*

The existence of a central plateau over all but finite regions at either end of the total rapidity interval Y implies that as $s \to \infty$ the multiplicity must grow logarithmically:

$$\langle n_c \rangle = \int \rho^c_{ab} \, d\phi_c \simeq \text{const.} + \ln(s/m_a m_b) \int \rho^c_{ab}(y^* = 0, \vec{p}_\perp) \, d^2 p_\perp .$$

Thus the coefficient a_1 in (68) is given by

$$a_1 = \sum_{c \text{ charged}} \int \rho^c_{ab}(y^* = 0, \vec{p}_\perp) \, d^2 p_\perp \tag{73}$$

Direct computation of the right-hand side of (73) from the charged particle distributions at $\theta_{\text{c.m.s.}} = 90°$ at the ISR gives numbers in reasonable agreement with a_1, as inferred from the multiplicity as a function of energy.

Stodolsky (1973) has remarked on the connection between the coefficient a_1 and the limitation on $p_\perp$. In essence he employs the conservation of energy sum rule (E.19). For simplicity consider that only one kind of particle is produced, that its $p_\perp$ distribution is more or less independent of y^* and that its distribution in y^* can be approximated by a rectangle on the range, $-\frac{1}{2}Y + \Delta < y^* < \frac{1}{2}Y - \Delta$. Then (E.19), evaluated in the c.m.s. for $\mu = 0$, reads

$$\begin{aligned} W &= \int \omega \cosh y^* \rho(Y, y^*, \vec{p}_\perp) \, dy \, d^2 p_\perp \\ &\simeq 2\langle \omega \rangle a_1 \sinh\left(\frac{Y}{2} - \Delta\right) \\ &\simeq \langle \omega \rangle a_1 \exp\left(\frac{Y}{2} - \Delta\right) = \frac{\langle \omega \rangle a_1}{m_N} W e^{-\Delta} \end{aligned} \tag{74}$$

where $\langle \omega \rangle$ is the average value of the transverse mass (D.5) and we have assumed nucleon–nucleon collisions. From (74) we have

$$\langle \omega \rangle \simeq \frac{m_N e^{\Delta}}{a_1} . \tag{75}$$

† In his lectures at Middleton Hall Prof. Sens reported more recent results with greater accuracy (~3–5%) that seem to show the various inclusive cross sections following universal curves in $y_{\text{lab}} = y_c - y_a$, with a continued steady rise as $(y_{\text{lab}})_{\text{max}} = Y/2$ increases. Such behaviour is inconsistent with the short-range correlation picture, provided we are in the asymptotic domain. With $L \sim 2$ and $Y/2 \sim 4$ we might expect deviations of the order of $\exp(-Y/2L) \simeq e^{-2} \simeq 0{\cdot}1^+$. Thus asymptopia may not be available quite yet at the ISR.

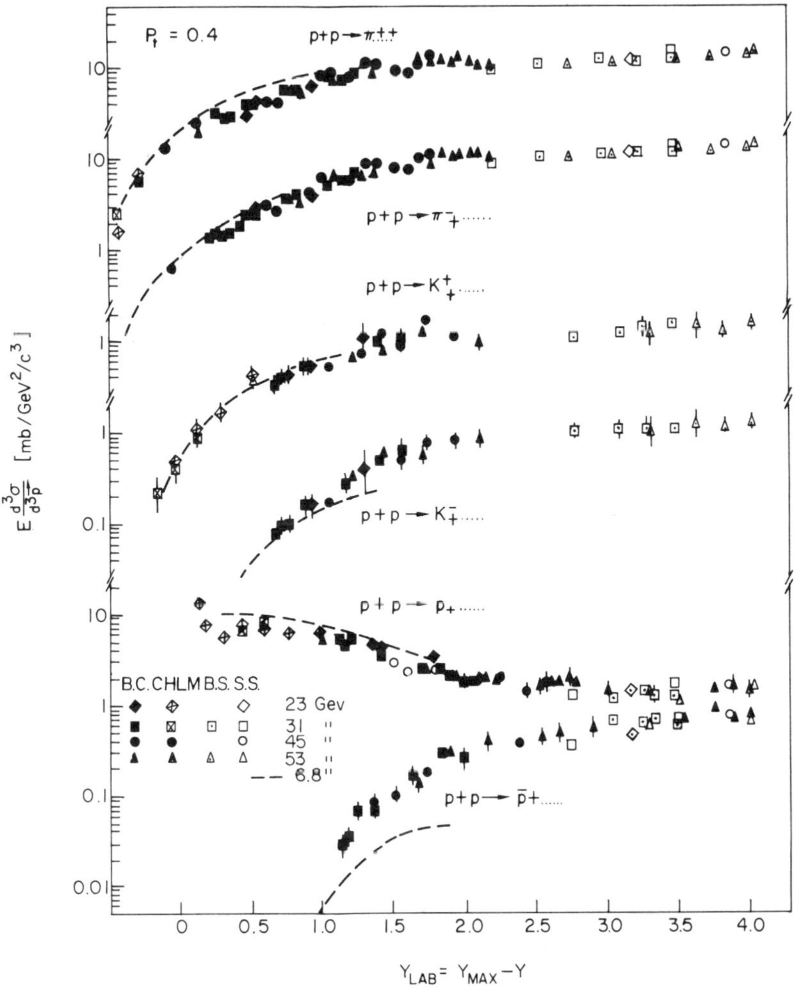

FIG. 32. Complication of single-particle inclusive invariant cross-sections in mb/(GeV)^2 at $p_\perp = 0{\cdot}4$ GeV/c for $p + p \to c$ + anything with $c = \pi^+, \pi^-, K^+, K^-, p, \bar{p}$ at various ISR energies (and 24 GeV/c in the lab) as functions of $y_{\text{lab}} = y_c - y_a$. Note the displaced ordinate scales. The centre of the rapidity scale $(y_a + y_b)/2$ is at $y_{\text{lab}} = 2{\cdot}0^-$, $3{\cdot}2$, $3{\cdot}5$, $3{\cdot}9^-$, and $4{\cdot}0^+$ for $W = 6{\cdot}8$, 23, 31, 45, and 53 GeV, respectively. (Figure from Bussiere, 1973.)

Empirically $a_1 \simeq 2^-$ for charged particles. If most particles are pions and all three charged states are produced equally, we can expect the total multiplicity to have $a_1 \simeq 3$. With $|\Delta| \ll 1$, we find $\langle \omega \rangle \simeq 0{\cdot}3$ GeV, in good agreement with the observed value of $\langle p_\perp \rangle$ for pions. It can be argued that (75) is just an expression of conservation of energy and therefore a definition of Δ in terms of a_1 and $\langle \omega \rangle$. Nevertheless, it is an explicit demonstration of the connection between the multiplicity and the extension of the inclusive cross section in transverse momentum and in rapidity.

The asymptotic energy dependence of the correlation parameters f_k (E.13) in the finite-correlation-length picture follows in the same way as for the multiplicity. Consider the two-particle correlation function $C(1, 2)$. In the Feynman–Wilson gas of Fig. 31, $C(1, 2)$ falls rapidly to zero for $|y_1 - y_2| \gg L_{12}$. Thus in the integral (E.16) defining f_2, integration over $(y_1 - y_2)$ with y_2 fixed will give a finite Y-independent value. The subsequent integration over y_2 will effectively multiply by a factor Y and $f_2 \propto Y = \ln(s/m_a m_b)$. This same behaviour occurs for higher correlation parameters, with all but the last integration yielding a Y-independent result asymptotically and the final integration introducing a factor of Y. Thus the hypothesis of finite range correlations leads to the asymptotic energy dependence,

$$f_k \propto \ln s \tag{76}$$

for all k. Present energies may not be sufficiently large to test this sort of asymptotic statement, especially for the higher correlation parameters. In Section 5 we discuss a mixed description of production that differs in its predictions from (76).

4. RELATION OF THE FEYNMAN–WILSON GAS TO A REGGE DESCRIPTION, THE APPROACH TO SCALING

The framework of Section 3 is an asymptotic one expected to be valid as $s \to \infty$, or better, as $Y \gg L$. At finite energies (5 GeV/c – 500 GeV/c lab momentum) we expect to see departures from the predictions of (71) and (72). The energy dependences of the nonscaling contributions can be anticipated by invoking a Regge description in analogy with the Regge theory of $2 \to 2$ processes. It is customary to speak of the Mueller–Regge description because of Mueller's important paper relating one-particle inclusive cross sections to the discontinuity in M^2 of the forward $3 \to 3$ amplitude (Mueller, 1970). This subject is dealt with in detail by other lecturers and in the references cited at the beginning of this Chapter. I restrict my explicit discussion of Muellerism to Fig. 33 where the standard set of diagrams are displayed. For the theorists who wish to know about the firmness of the theoretical foundations there are papers by Stapp (1971), Tan (1971), and Polkinghorne (1972). Our brief treatment below is based on heuristic arguments of power-law behaviour in the various subenergies with little reference to the details of Regge theory.

(a) *Fragmentation region, triple Regge region*

Equation (71) is the statement of scaling in the fragmentation region of a.

In Regge language the energy dependence of the cross section is expected to be $s^{\alpha(0)-1}$ and the absence of any energy variation is attributed to exchange of the Pomeranchuk trajectory with $\alpha(0) = \alpha_P(0) = 1$. At finite energies other Regge singularities with smaller intercepts will contribute. Thus (71) is generalized to

$$\lim_{\substack{y_c - y_a \text{ fixed} \\ Y \gg 1}} \rho_{ab}^c = \sum_k \left(\frac{s}{s_0}\right)^{\alpha_k(0)-1} g_k(x, \vec{p}_\perp). \tag{77}$$

Note that in (77) we can if we wish replace the variable x with s/M^2 according to (D.29). On the basis of $2 \to 2$ phenomenology where the dominant nondiffractive Regge exchanges have $\alpha(0) \simeq \frac{1}{2}$ we expect that in the fragmentation region a good description of the s dependence will be given by

$$\rho_{ab}^c \simeq g_0(x, \vec{p}_\perp) + \left(\frac{s_0}{s}\right)^{\frac{1}{2}} g_1(x, \vec{p}_\perp). \tag{78}$$

For the reaction $pp \to p +$ anything (78) has been tested over the range from 40 to 400 GeV incident energy at NAL (Sannes *et al.*, 1973, and private communication). In the range $0{\cdot}75 < x < 0{\cdot}95$ and $0{\cdot}2 < |t| < 0{\cdot}5$ eqn (78) is quite consistent with the data; furthermore, the functions g_0 and g_1 seem to have closely the same shape in x and in $p_\perp^2$.

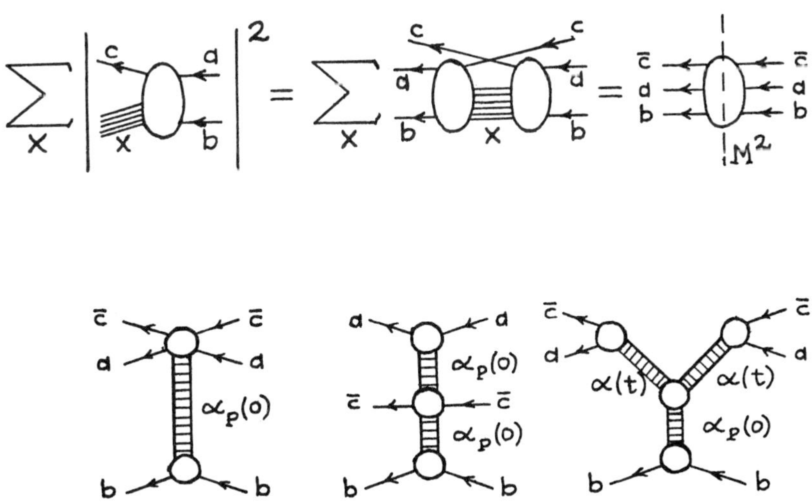

FIG. 33. Schematic diagrams for Regge analysis of inclusive processes. Top line: Conversion of the inclusive cross section for $a + b \to c +$ anything, via unitarity and analytic continuation, into a discontinuity (in M^2, the mass squared of X) of the forward 6-particle amplitude for $ab\bar{c} \to ab\bar{c}$. Bottom line: Various assumed Regge limits, single Regge (limiting fragmentation region), double Regge (central or pionization region), triple Regge limit.

The triple-Regge region indicated by the lower right-hand diagram of Fig. 33 is a subset of the fragmentation region corresponding to s/M^2 large, as well as s and M^2. Since $M^2 \simeq s(1-x)$, the region of large s/M^2 has $x \simeq 1^-$. As $s \to \infty$ this means that there is a large rapidity gap between c and the nearest particle in "anything". Regge behaviour as $s^{\alpha_i(t)}$, where $t = (p_a - p_c)^2$, is expected. The cross section will thus vary as $s^{2\alpha_i(t)-1}$ for fixed M^2, with the s^{-1} factor coming from division by the incident flux. But if M^2 is also large we expect a Regge description in terms of M^2, i.e. $(M^2)^{\alpha_k(0)}$. In order to be a special case of (77) there must be additional M^2 dependence in the form of a factor $(M^2)^{-2\alpha_i(t)}$, leading to the expression

$$\rho_{\text{triple Regge}} \simeq \frac{1}{s} \sum_{i,j,k} \beta_{ijk}(t) \left(\frac{s}{M^2}\right)^{\alpha_i(t)+\alpha_j(t)} (M^2)^{\alpha_k(0)} \tag{79}$$

In (79) we have generalized somewhat by replacing $2\alpha_i(t)$ by $\alpha_i(t) + \alpha_j(t)$ to allow for interference terms and summing over i, j, k. The scaling contributions in (79) come from $\alpha_k(0) = 1$ independently of $\alpha_i(t)$ and $\alpha_j(t)$. The dependence on s/M^2 or x is, of course, dependent on $\alpha_i(t)$ and $\alpha_j(t)$ as well as $\alpha_k(0)$.

The triple-Regge region is of great interest both theoretically and experimentally, as is discussed in detail by other lecturers.

(b) *Central region*

For particle c in the central region of Fig. 31 eqn (72) is the scaling statement. At finite energies we expect secondary contributions depending on the two rapidity differences $y_c - y_b$ and $y_a - y_c$, or equivalently on the subenergies s_{bc} and s_{ac} [see eqn (D.17)]. The general behaviour in s will be $(s_{bc})^{\alpha(0)-1} \propto \exp\{[\alpha(0) - 1](y_c - y_b)\}$ and similarly for s_{ca}. Keeping only the leading secondary contributions, as in (78), we find

$$\lim_{\substack{y_a - y_c \gg 1 \\ y_c - y_b \gg 1}} \rho_{ab}^c \simeq h_0(\vec{p}_\perp) + e^{-(y_c - y_b)/2} h_1^{(1)}(\vec{p}_\perp) + e^{-(y_a - y_c)/2} h_1^{(2)}(\vec{p}_\perp) \tag{80}$$

where the secondary functions $h_1^{(i)}(\vec{p}_\perp)$ depend on the Regge couplings of $\bar{b}b \to$ Reggeon $\to c$ and $\bar{a}a \to$ Reggeon $\to c$. For a symmetric situation such as $pp \to c +$ anything, (80) can be written conveniently in terms of c.m.s. rapidity as

$$\rho_{\text{central}}(s, y^*, \vec{p}_\perp) \simeq h_0(\vec{p}_\perp) + \left(\frac{s_0}{s}\right)^{\frac{1}{4}} h_1(p_\perp) \cosh y^*/2. \tag{81}$$

There are two observations here. The first is that the s dependence of the approach to scaling is $s^{-\frac{1}{4}}$, rather than $s^{-\frac{1}{2}}$. This follows immediately from the fact that the relevant rapidity difference is $Y/2$, not Y. The second observation is that the sign of $h_1(\vec{p}_\perp)$ governs the curvature in rapidity away from $y = 0$. If the approach to the scaling limit in the central region is from above (below) then the curvature of the distribution in y^* is concave upwards (downwards). Experiment for pions or all charged particles shows a rise with increasing energy at $y^* = 0$ and concavity downwards away from $y^* = 0$ (see Fig. 32). This behaviour is quite consistent with (81). The rise shown by a combination of data at $P_{\text{lab}} < 30$ GeV and early ISR data was compatible with a $s^{-\frac{1}{4}}$ variation (Ferbel, 1972). More recently, however, the simple behaviour of (81) has been cast in doubt by more extensive data from the ISR. These results at $y^* \sim 0$ in combination with lower energy data are more consistent with $s^{-\frac{1}{2}}$ than $s^{-\frac{1}{4}}$ (Jarlskog, 1973).

The question of the approach to scaling is obviously an interesting one. The above simple remarks are just the beginnings. Duality and exchange degeneracy arguments can be brought to bear in order to estimate the expected sizes of the secondary contributions. This subject is dealt with in some detail by Roberts in his lectures. There is also the "threshold" aspect exhibited in Fig. 24 wherein appreciable production of heavier particles begins at higher s values and takes longer to approach an approximately scaling limit. The effect at $y^* \simeq 0$ is even more pronounced than in the integrated results shown in Fig. 24. While smallest for pions, this threshold effect is undoubtedly present at a level such that plots of data at $y^* \simeq 0$ from $P_{\text{lab}} < 30$ GeV to ISR energies against $s^{-\frac{1}{4}}$ or $s^{-\frac{1}{2}}$ are of dubious value. Some theoretical support for the importance of a threshold effect comes from application of two-component duality to inclusive distributions with the consequence that the nonleading ($s^{-\frac{1}{4}}$) contributions in (81) must be positive.

(c) *Factorization*

The finite-range correlation picture for the n-particle distributions has implicit in it the idea of factorization. Thus, in (71) the *normalized* density depends on a and c, but not on b, and in (72) is independent of both a and b. This means that the inclusive *cross section* in these regions depends on b (or on a and b) only through the multiplicative factor σ_{ab}. In Regge language this is a natural consequence of the assumption of Pomeranchuk-singularity dominance of the total cross sections and the appearance of the $P - \bar{b}b$ and $P - \bar{a}a$ vertices in the $3 \to 3$ amplitude, as shown in Fig. 33.

These factorization properties have been tested extensively for the fragmentation region, mostly at energies below 30 GeV incident in the laboratory (Chen *et al.*, 1971; Lander, 1971; Berger, Oh, and Smith, 1972; Miettinen, 1972; Lam, 1972; Fry *et al.*, 1972). The overall conclusion is that factorization holds for the scaling contribution in the fragmentation region to 10–15%.

There is also some evidence for factorization of secondary contributions (Miettinen, 1972) and for two-particle distributions (Lam, 1972).

Factorization for the central region is tested only roughly by the analysis of the approach to scaling of Ferbel (1972). More detailed checks will be possible with Serpukhov and NAL data on reactions initiated by pions, K-mesons and antiprotons, as well as protons. There is a high probability that factorization will fail to hold at the level of a few per cent because of the presence of Regge cuts or, in other terms, from the existence of a diffractive contribution associated with the incident particles, as well as a finite-correlation length contribution. See the next section and, in much more detail, Harari's lectures. See also Wilson (1970).

5. BRIEF REMARKS ON A TWO-COMPONENT DESCRIPTION

There have been various specific models devised to describe multiparticle reactions. These are described and references to the literature are given in Frazer *et al.* (1972) and Horn (1972). The short-range correlation picture with its explicit realization in the multiperipheral model of Amati, Stanghellini, and Fubini (1962) and many subsequent versions is perhaps one extreme. At the opposite extreme is the diffractive model in its recent realizations of Adair (1972), Hwa (Hwa, 1971; Hwa and Lam, 1971), and Jacob and Slansky (Jacob and Slansky, 1972; Jacob, Slansky, and Wu, 1972; Berger, Jacob, and Slansky, 1972) in which all the production is assumed to occur by excitation of the incident particles into fireballs or nova that subsequently decay. A more reasonable view is that nature elects to make use of both dynamical mechanisms (Wilson, 1970). There is now support for the presence of a "diffractive" as well as a "multiperipheral" component in a number of experiments. Professor Sens discusses one of these—the correlation of a proton near $x = 1^-$ with the angular distribution of charged particles. Another is the behaviour of the prong cross sections at different energies (Figs 20–22) or equivalently $\langle n \rangle$ and f_2. We will comment briefly on this aspect. It is dealt with in more depth by Harari.

The two-component or two-mechanism model for prong cross section in its simplest form assumes an incoherent superposition of a "diffractive" cross section and a "multiperipheral" cross section for each n value. It is further assumed that the diffractive part contributes most importantly to the low multiplicities and is independent of energy in each topology, while the multiperipheral part is perhaps Poisson-like in its distribution over topologies with a mean multiplicity that grows with energy, but whose total contribution to the inelastic cross section is an energy-independent fraction. Different versions of this two-component model have been discussed by many authors (Białas, Fiałkowski and Załewski, 1972; Fiałkowski, 1972;

Fiałkowski and Miettinen, 1973; Frazer *et al.*, 1973; Harari and Rabinovici, 1973; Quigg and Jackson, 1972; Van Hove, 1973). The variants differ in detail, but agree on the essentials. There is a relatively small, but important, diffractive component that amounts to $20 \pm 5\%$ of the inelastic cross section, with the remainder as multiperipheral. The consequences for the prong cross sections are sketched in Fig. 34 at the bottom. At high energies

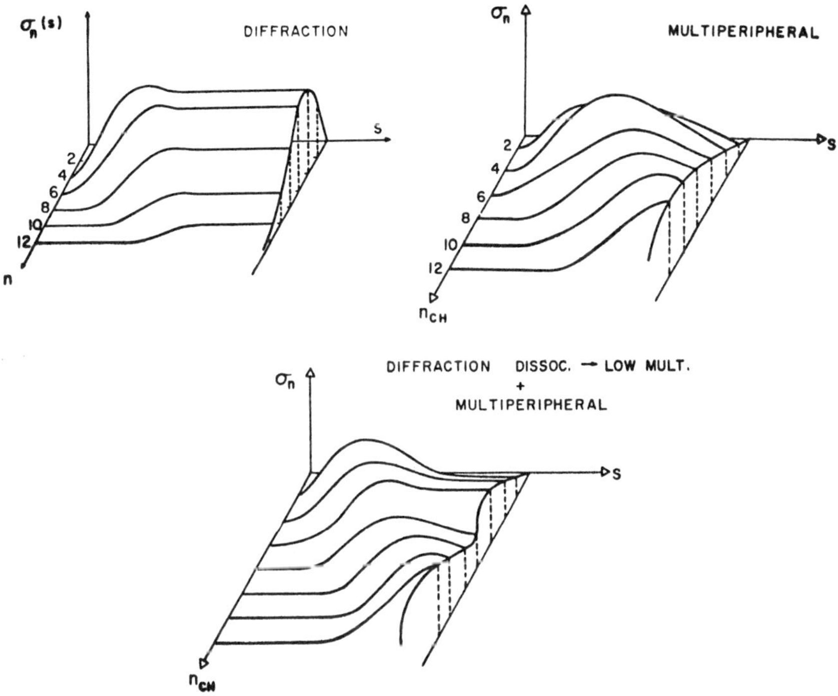

FIG. 34. Prong cross sections as functions of n and s. The expected behaviour for diffractive, multiperipheral and two-component models are shown at upper left, upper right, and lower centre, respectively (from Morrison, 1973).

there should be seen a clear distinction between the constant diffractive component at low prong number and the multiperipheral component that moves out in n roughly proportional to $\ln s$. Estimates indicate that at the highest ISR energy there might be signs of a shoulder at small n.

The behaviour of correlation parameters in a two-component model is of interest. First we note that in the extreme diffractive model the logarithmically growing multiplicity is generated by assuming

$$\sigma_n \propto \frac{1}{n^2} \quad \text{for} \quad n < N \tag{82}$$

where $N \propto s^{\frac{1}{4}}$ (or some more modest power of s). The correlation parameter f_2 involves the sum of $n(n-1)\sigma_n$ and so behaves asymptotically as $f_2 \propto N \propto s^{\frac{1}{4}}$. This is in marked contrast to the result (76) of the multiperipheral (finite-range correlation) model. The purely diffractive model, while in reasonable accord with the data on f_2 in pp collisions at Serpukhov energies and below is in gross disagreement with the NAL bubble chamber data at 200 and 300 GeV.

In the two-component model the diffractive part is confined to low multiplicities and is s-independent. Thus its contributions to $\langle n \rangle$ and f_2 are constant in energy. All the energy dependence is by hypothesis in the other component. Let α_d and $\alpha_m = 1 - \alpha_d$ be the fractions of the two components in the inelastic cross section and let $\langle n \rangle_d$, $\langle n \rangle_m$, f_{2d}, f_{2m} be the mean multiplicities and correlation parameters for each component separately (normalized to the diffractive and multiperipheral parts of the inelastic cross section, respectively). In the two-component model without interference it is easy to show that the multiplicity and the correlation parameter are given by

$$\langle n \rangle = \alpha_d \langle n \rangle_d + \alpha_m \langle n \rangle_m$$

$$f_2 = \alpha_d f_{2d} + \alpha_m f_{2m} + \alpha_d \alpha_m (\langle n \rangle_m - \langle n \rangle_d)^2. \tag{83}$$

Now suppose that $\langle n \rangle_d$, f_{2d} are independent of energy, while $\langle n \rangle_m$, f_{2m} grow asymptotically as $\ln s$. Then we find from (83) the asymptotic behaviour,

$$\langle n \rangle \propto \ln s$$

$$f_2 \propto (\ln s)^2. \tag{84}$$

[For higher correlation parameters one finds $f_k \propto (\ln s)^k$.] The cross term in f_2 (83)—a long-range correlation effect—produces $f_2 \propto (\ln s)^2$, even though the ingredients vary as $\ln s$ at most. The two humps developing in Fig. 34 for increasing s are, of course, just a different manifestation and source of this phenomenon. The many versions of the two-component model have little difficulty fitting the observed correlation parameters f_2 up to the highest NAL energies.

As a final remark on evidence in support of two operative mechanisms in multiparticle production we mention briefly the rapidity correlations observed in pp interactions at ISR energies between charged particles and photons (Dibon *et al.*, 1973). The correlation functions $R(y_1, y_2)$ defined by eqn (E.18) for these data are displayed in Fig. 35. The distributions at the different ISR energies are qualitatively and even quantitatively similar. The open circles show the correlation as a function of y^*_{photon} for $y^*_{\text{ch}} = 0$. There appears to be a fairly important short-range correlation centered about $y^*_{\text{photon}} \simeq y^*_{\text{ch}}$ and having a range of the order of $L \sim$ 2-3. The solid points are

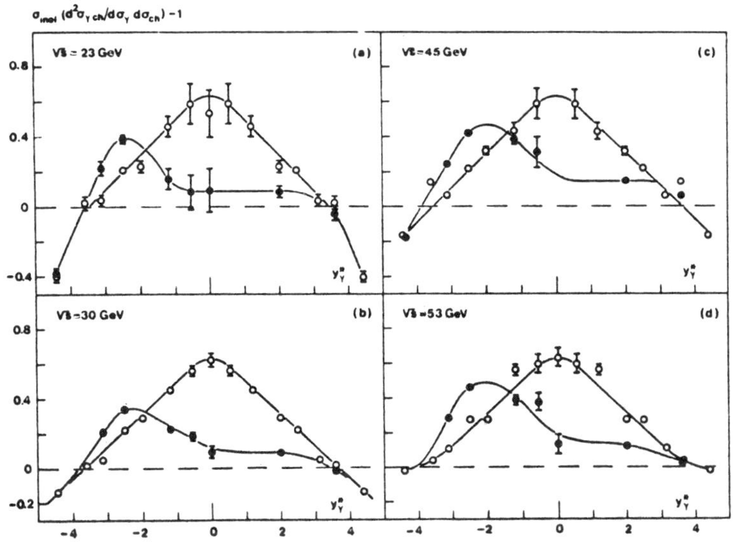

FIG. 35. Rapidity correlation functions $R(y_1, y_2)$ defined by (E.18) for photon-charged particle correlations in p–p collisions at ISR energies ($W = 23$, 30, 45, and 53 GeV) as functions of y^*_{photon}. The open circles correspond to $y^*_{\text{ch}} = 0$ and the solid points to $y^*_{\text{ch}} = -2{\cdot}5$ (from Dibon *et al*, 1973).

for $y^*_{\text{ch}} = -2{\cdot}5$. They exhibit a smaller short-range effect centred at $y^*_{\text{photon}} \simeq y^*_{\text{ch}}$ and also a long-range correlation whose existence is clear, but whose actual magnitude is probably not easy to establish with certainty because of errors in normalization of the various cross sections. In spite of such uncertainties the presence of a long-range correlation component in the data is certain. Integration of such a contribution leads to $f_2 \propto (\ln s)^2$, as we have already seen.

ACKNOWLEDGMENT

It is a pleasure to acknowledge my indebtedness to the co-chairmen of the 1973 Scottish Universities' Summer School in Physics, Professor I. S. Hughes of Glasgow and Professor N. Kemmer of Edinburgh, for the opportunity to participate again in the school. They and their staff made the 1973 school at least as memorable and pleasurable as the first one in 1960. For conversations on physics I owe thanks to R. N. Cahn, M. B. Einhorn, G. L. Kane, H. I. Miettinen, R. G. Roberts, and J. C. Sens. And Dick Roberts made a valuable suggestion.

APPENDIX A

S-MATRIX FORMULAS, CROSS SECTIONS, TWO-BODY KINEMATICS

In what follows the usual units for relativistic particle physics and quantum theory are used: $\hbar = c = 1$; masses, momenta, and energies are expressed in GeV; cross sections are calculated in units of $M^{-2} = (\text{GeV})^{-2}$ and are converted to millibarns by multiplication by the magic factor 0·389. Free particle states are normalized to one particle per unit volume. The phase space for a single particle is thus $d^3p/(2\pi)^3$. The convention on 4-vectors is indicated by $p_\mu = (E, p_x, p_y, p_z)$ and $a \cdot b = a_0 b_0 - \mathbf{a} \cdot \mathbf{b}$.

1. INVARIANT OR FEYNMAN AMPLITUDE $\mathcal{M}$

The invariant amplitude $\mathcal{M}$ is related to the S-matrix through the relation,

$$S_{\beta\alpha} = \delta_{\beta\alpha} - i(2\pi)^4 \delta^{(4)}(p_\beta - p_\alpha)\mathcal{M}_{\beta\alpha}/\sqrt{[\Pi_i 2E_i]} \tag{A.1}$$

where α and β are the labels for the initial and final states and the product of factors $(2E_i)$ is over all the particles in both initial and final states. Implicit in the S-matrix element are the conservation of energy and momentum delta functions. The invariant amplitude has these factors extracted. Its various arguments are thus to be evaluated taking the conservation laws into account. With the presence of the square root of the product of factors $(2E_i)$ and the single particle normalization stated above, the invariant amplitude is Lorentz invariant, as its name implies.

2. DECAY PROCESSES $\alpha \to (1, 2, \ldots, n)$

The transition probability is

$$dw_{\beta\alpha} = \frac{(2\pi)^4}{2E_\alpha} |\mathcal{M}_{\beta\alpha}|^2 \prod_{i=1}^{n} \frac{d^3p_i}{(2\pi)^3(2E_i)} \cdot \delta^{(4)}(p_1 + p_2 + \ldots + p_n - p_\alpha). \tag{A.2}$$

For a two-particle final state, in the rest frame of α,

$$dw_{\beta\alpha} = \frac{1}{32\pi^2} |\mathcal{M}_{\beta\alpha}|^2 \frac{p_{\text{c.m.s.}}\, d\Omega_{\text{c.m.s.}}}{m_\alpha^2} \tag{A.3}$$

where

$$\left(\frac{p_{\text{c.m.s.}}}{m_\alpha}\right)^2 = \frac{1}{4}\left[1 - \left(\frac{m_1 + m_2}{m_\alpha}\right)^2\right] \cdot \left[1 - \left(\frac{m_1 - m_2}{m_\alpha}\right)^2\right].$$

If some observables, e.g. spins, are not detected, averages of the initial state and sums over the final state are understood.

3. TWO-PARTICLE COLLISION CROSS SECTION $\alpha = (1, 2)$; $\beta = (3, 4, \ldots, n + 2)$

The projectile is labelled #1 and the target #2 in what follows. The differential cross section is

$$d\sigma_{\beta\alpha} = \frac{(2\pi)^4 |\mathcal{M}_{\beta\alpha}|^2}{4\sqrt{[(p_1 \cdot p_2)^2 - m_1^2 m_2^2]}} \cdot \prod_{i=3}^{n+2} \frac{d^3 p_i}{(2\pi)^3 (2E_i)} \cdot \delta^{(4)}\left(\sum_i p_i - p_\alpha\right). \tag{A.4}$$

In this form the cross section is manifestly Lorentz invariant. The factor $\sqrt{[(p_1 \cdot p_2)^2 - m_1^2 m_2^2]}$ is called the *flux factor*†. It is obviously invariant and has the useful values,

$$\sqrt{[(p_1 \cdot p_2)^2 - m_1^2 m_2^2]} = \begin{cases} m_2 p_1 \text{ (laboratory or target rest frame)} \\ W p_{\text{c.m.s}} \text{ (centre of mass frame)} \end{cases} \tag{A.5}$$

where W is the total energy in the c.m.s.

For a two-body final state $(m_1 + m_2 \to m_3 + m_4)$, the c.m.s. differential cross section is

$$\frac{d\sigma}{d\Omega_{\text{c.m.s.}}} = \frac{1}{64\pi^2 s} \frac{p'_{\text{c.m.s.}}}{p_{\text{c.m.s.}}} |\mathcal{M}_{\beta\alpha}|^2 \tag{A.6}$$

where $s = m_1^2 + m_2^2 + 2m_2(E_1)_{\text{lab}}$ is the square of the total energy W in the c.m.s. The ratio of the final to initial c.m.s. momenta is

$$\frac{p'_{\text{c.m.s.}}}{p_{\text{c.m.s.}}} = \sqrt{\left\{\frac{[s - (m_3 + m_4)^2][s - (m_3 - m_4)^2]}{[s - (m_1 + m_2)^2][s - (m_1 - m_2)^2]}\right\}}.$$

The differential cross section expressed per unit interval in invariant momentum transfer $t = (p_1 - p_3)^2 = (p_2 - p_4)^2$ is

$$\frac{d\sigma}{dt} = \frac{\pi}{p'_{\text{c.m.s.}} p_{\text{c.m.s.}}} \cdot \frac{d\sigma}{d\Omega_{\text{c.m.s.}}} = \frac{1}{64\pi s p^2_{\text{c.m.s.}}} |\mathcal{M}_{\beta\alpha}|^2. \tag{A.7}$$

Note that $sp^2_{\text{c.m.s.}}$ is just the square of the invariant flux factor (A.5).

For elastic scattering the standard c.m.s. scattering amplitude $f_{\text{c.m.s.}}$ is related to the invariant amplitude by

$$f_{\text{c.m.s.}} = \frac{-1}{8\pi W} \mathcal{M}. \tag{A.8}$$

For inelastic two-body processes conventions vary, but usually an additional factor of $\sqrt{(p'_{\text{c.m.s.}}/p_{\text{c.m.s.}})}$ appears on the right-hand side of (A.8). Then the differential cross section (A.6) is given by the absolute square of $f_{\text{c.m.s.}}$.

† With the advent of colliding beam machines, in which the target is in motion and not necessarily in head-on collision with the projectile, there has arisen some question about the appropriateness of this flux factor. See Newton (1966), p. 220–1, for the generally accepted view and Martin and Spearman (1970), p. 153–4, for a different, non-invariant convention as well as the standard result.

4. OPTICAL THEOREM

The optical theorem that follows from conservation of probability (unitarity of the S matrix) is

$$-\tfrac{1}{2}\,\mathrm{Im}\,\mathcal{M}_{\alpha\alpha} = \sqrt{[(p_1 \cdot p_2)^2 - m_1^2 m_2^2]}\,\sigma_{\text{total}} \tag{A.9}$$

where σ_{total} is the total cross section for the channel α. This can be written in a more familiar form,

$$\sigma_{\text{total}} = \frac{4\pi}{p_1}\,\mathrm{Im}\,f(0^\circ) \tag{A.10}$$

where $f(0^\circ)$ is the forward scattering amplitude (whose square gives the differential elastic scattering cross section per unit solid angle) in any frame moving parallel to the incident particle's direction and p_1 is the incident particle's momentum in that frame.

5. TWO-BODY KINEMATICS

Some kinematic variables have already been defined. We gather here a summary of useful two-body quantities. The general notation is indicated in Fig. A.1 for the process,

$$m_1 + m_2 \rightarrow m_3 + m_4. \tag{A.11}$$

The invariant variables s, t, u are defined by

$$\begin{aligned} s &= (p_1 + p_2)^2 = (p_3 + p_4)^2 \\ t &= (p_1 - p_3)^2 = (p_2 - p_4)^2 \\ u &= (p_1 - p_4)^2 = (p_2 - p_3)^2 \end{aligned} \tag{A.12}$$

with the constraint equation,

$$s + t + u = m_1^2 + m_2^2 + m_3^2 + m_4^2. \tag{A.13}$$

In terms of *laboratory* quantities (the laboratory is the frame where particle 2 is at rest) the invariants are

$$\begin{aligned} s &= m_1^2 + m_2^2 + 2m_2(E_1)_{\text{lab}} \\ t &= m_2^2 + m_4^2 - 2m_2(E_4)_{\text{lab}} \\ u &= m_2^2 + m_3^2 - 2m_2(E_3)_{\text{lab}}. \end{aligned} \tag{A.14}$$

For elastic scattering or reactions in which $m_4 = m_2$ the momentum transfer t simplifies to $t = -2m_2(T_4)_{\text{lab}}$, where T_4 is the kinetic energy of the recoiling particle #4.

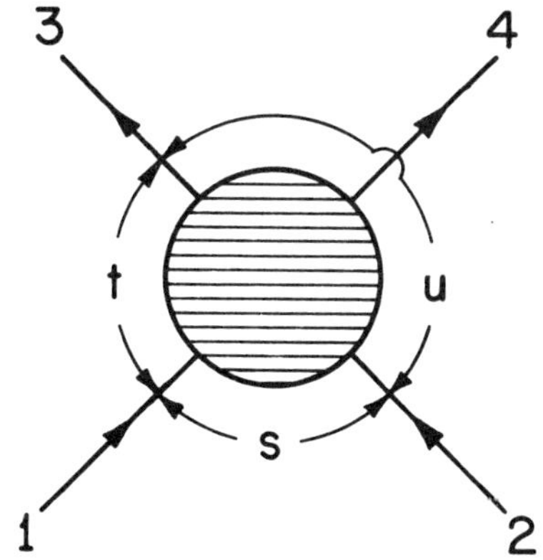

FIG. A.1.

The invariants can also be expressed in terms of the *centre of mass* (c.m.s.) variables. Let the c.m.s. energies and momenta of the particles be E_1, E_2, E_3, E_4 and $p_1 = p_2 = p$, $p_3 = p_4 = p'$. Then $s \equiv W^2 = (E_1 + E_2)^2 = (E_3 + E_4)^2$ is the square of the total c.m.s. energy and

$$\left.\begin{aligned} E_1 &= \frac{W^2 + m_1^2 - m_2^2}{2W}, \qquad E_2 = \frac{W^2 + m_2^2 - m_1^2}{2W} \\ E_3 &= \frac{W^2 + m_3^2 - m_4^2}{2W}, \qquad E_4 = \frac{W^2 + m_4^2 - m_3^2}{2W} \\ p^2 &= \frac{1}{4s}[s^2 - 2(m_1^2 + m_2^2)s + (m_1^2 - m_2^2)^2] \\ p'^2 &= \frac{1}{4s}[s^2 - 2(m_3^2 + m_4^2)s + (m_3^2 - m_4^2)^2] \end{aligned}\right\} \quad \text{(A.15)}$$

and

The two momentum transfer variables are

$$\begin{aligned} t &= t_{\min} - 2pp'(1 - \cos\theta_{\text{c.m.s.}}) \\ u &= u_{\min} - 2pp'(1 + \cos\theta_{\text{c.m.s.}}) \end{aligned} \quad \text{(A.16)}$$

where $\theta_{\text{c.m.s.}}$ is the angle between $\vec{p}_1$ and $\vec{p}_3$ in the c.m.s. and

$$\begin{aligned} t_{\min} &= (E_1 - E_3)^2 - (p - p')^2 \\ u_{\min} &= (E_1 - E_4)^2 - (p - p')^2. \end{aligned} \quad \text{(A.17)}$$

At high energies ($W \gg m_i$) some useful approximations are

$$p \simeq \frac{W}{2}\left(1 - \frac{m_1^2 + m_2^2}{W^2} - \frac{2m_1^2m_2^2}{W^4} + \ldots\right)$$

$$p' \simeq \frac{W}{2}\left(1 - \frac{m_3^2 + m_4^2}{W^2} - \frac{2m_3^2m_4^2}{W^4} + \ldots\right) \tag{A.18}$$

$$t_{\min} \simeq -\frac{1}{s}\left[(m_1^2 - m_3^2)(m_2^2 - m_4^2) + \frac{1}{s}(m_1^2m_2^2 - m_3^2m_4^2) \times (m_1^2 + m_2^2 - m_3^2 - m_4^2) + \ldots\right]$$

and $u_{\min}$ has $m_3 \leftrightarrow m_4$. Note that if $m_1 = m_3$ *or* $m_2 = m_4$ $t_{\min} \propto s^{-2}$ and so approaches zero very rapidly at high energies. For elastic scattering $t_{\min} \equiv 0$, of course, and $u_{\min} = (m_1^2 - m_2^2)^2/s$ without approximation.

The three-dimensional scalar product of the c.m.s. momenta $\vec{p}_1$ and $\vec{p}_3$ can be expressed in terms of the invariants s, t, u:

$$4p_sp_s' \cos\theta_s = t - u + \frac{(m_1^2 - m_2^2)(m_3^2 - m_4^2)}{s} \tag{A.19}$$

where we have changed the notation slightly in order to discuss channels other than the s-channel of Fig. A.1. For the t- and u-channel processes, where t and u are, respectively, the squares of the total c.m.s. energies in the two channels and the other invariants are momentum transfers, the corresponding expressions are

$$4p_tp_t' \cos\theta_t = s - u + \frac{(m_1^2 - m_3^2)(m_2^2 - m_4^2)}{t} \tag{A.20}$$

$$4p_up_u' \cos\theta_u = t - s + \frac{(m_1^2 - m_4^2)(m_3^2 - m_2^2)}{u}. \tag{A.21}$$

The angle θ_t is the angle between $\bar{2}$ and 1 in the t-channel c.m.s. of the process $1 + \bar{3} \to \bar{2} + 4$; θ_u is the angle between $\bar{1}$ and $\bar{3}$ in the process $\bar{3} + 2 \to \bar{1} + 4$. The various momenta are the c.m.s. values in each channel, obtained from (A.15) by substitution.

Appendix B

Partial Waves, Helicity Amplitudes, Impact Parameter Representation

The material in this appendix is well known. It is brought together here mostly to fix notation, not to give derivations.

1. PARTIAL WAVE DECOMPOSITIONS

Because of the convenience of dealing with a Lorentz invariant amplitude we define the scattering (or two-body reaction) amplitude $F_{\beta\alpha}(s, t)$ by

$$F_{\beta\alpha}(s, t) = \frac{-1}{8\pi} \mathcal{M}_{\beta\alpha}(s, t) \tag{B.1}$$

where $\mathcal{M}_{\beta\alpha}$ is the invariant amplitude in (A.1). From (A.7) the differential cross section is

$$\frac{d\sigma_{\beta\alpha}}{dt} = \frac{\pi}{sp^2} |F_{\beta\alpha}(s, t)|^2. \tag{B.2}$$

For particles without spin $F_{\beta\alpha}(s, t)$ has the Rayleigh–Faxen–Holtsmark partial wave expansion,

$$F_{\beta\alpha}(s, t) = \left(\frac{s}{pp'}\right)^{\frac{1}{2}} \sum_{l=0}^{\infty} (2l + 1) a_l^{(\beta\alpha)}(s) P_l(\cos\theta) \tag{B.3}$$

where $\cos\theta$ is related to t via (A.16) and $a_l^{(\beta\alpha)}(s)$ is called the partial wave amplitude.

The optical theorem (A.9) can be translated into a statement about the partial wave amplitudes:

$$\text{Im}\, a_l^{(\alpha\alpha)}(s) = \sum_{\beta} |a_l^{(\beta\alpha)}(s)|^2 + G_l^{(\alpha)}(s) \tag{B.4}$$

where the sum over β is over all the kinematically allowed two-body channels and the function $G_l^{(\alpha)}(s)$ represents the contribution from the open channels with more than two particles. Sometimes (B.4) is written

$$\text{Im}\, a_l^{(\alpha\alpha)}(s) = |a_l^{(\alpha\alpha)}(s)|^2 + \tfrac{1}{4} H_l^{(\alpha)}(s) \tag{B.5}$$

where $H_l(s)$ contains all the contributions to the unitarity sum except for the elastic scattering part. $H_l(s)$ is the partial wave projection of the so-called overlap function (Van Hove, 1964). If the elastic scattering partial wave amplitude is written as

$$a_l^{(\alpha\alpha)}(s) = \frac{1}{2i} (e^{2i\delta_l(s)} - 1), \tag{B.6}$$

then (B.5) implies that

$$|e^{2i\delta_l(s)}|^2 = 1 - H_l^{(\alpha)}(s). \tag{B.7}$$

If only elastic scattering can occur, $H_l = 0$ and the *phase shift* δ_α is real. If other channels are open then δ_α is complex, with positive imaginary part. The partial wave amplitude (B.6) thus lies on the boundary of or inside the unitarity circle, a circle on an Argand diagram of radius 1/2 centred at the point $i/2$.

2. HELICITY AMPLITUDES

If the particles possess spin a convenient set of amplitudes are the helicity amplitudes of Jacob and Wick (1959). The initial and final states are specified by the usual kinematic variables of Appendix (A.5) and also by the helicities λ_j of the particles. The generalization of (B.2) for unpolarized beams is

$$\frac{d\sigma}{dt} = \frac{\pi}{sp^2} \cdot \frac{1}{(2s_1 + 1)(2s_2 + 1)} \sum_{\{\lambda\}} |\langle \lambda_3 \lambda_4 | F(s, t) | \lambda_1 \lambda_2 \rangle|^2. \tag{B.8}$$

The generalization of the partial wave expansion (B.3) is

$$\langle \lambda_3 \lambda_4 | F(s, t) | \lambda_1 \lambda_2 \rangle = \left(\frac{s}{pp'}\right)^{\frac{1}{2}} \sum_j (2j + 1) \langle \lambda_3 \lambda_4 | a_j(s) | \lambda_1 \lambda_2 \rangle \, d^j_{\lambda\mu}(\theta) \tag{B.9}$$

where $d^j_{\lambda\mu}(\theta)$ is a Wigner rotation function (see the Appendix of Jacob and Wick, 1959) and $\lambda = \lambda_1 - \lambda_2$, $\mu = \lambda_3 - \lambda_4$, while j is the angular momentum.

3. HELICITY AMPLITUDES FOR $0^- + \frac{1}{2}^+ \to 0^- + \frac{1}{2}^+$ PROCESSES

An important special case of helicity amplitudes is for the reaction, $0^- + \frac{1}{2}^+ \to 0^- + \frac{1}{2}^+$ or $M + B \to M' + B'$, where M stands for meson and B for baryon. To reduce the number of subscripts we introduce a change in notation from Appendix (A.5). We put

$$\left.\begin{aligned} m_1 &= \mu, & m_2 &= m, & m_3 &= \mu', & m_4 &= m' \\ E_1 &= \omega, & E_2 &= E, & E_3 &= \omega', & E_4 &= E' \\ \vec{p}_1 &= \vec{q}, & \vec{p}_2 &= \vec{p}, & \vec{p}_3 &= \vec{q}', & \vec{p}_4 &= \vec{p}' \end{aligned}\right\} \tag{B.10}$$

In the c.m.s. we have $\vec{p} = -\vec{q}$ and $\vec{p}' = -\vec{q}'$.

The invariant spinor amplitude $F_{\beta\alpha}(s, t)$ is

$$F_{\alpha\beta}(s, t) = \bar{u}_\beta(\vec{p}')[A(s, t) + \tfrac{1}{2}(\not{q} + \not{q}')B(s, t)]u_\alpha(\vec{p}) \tag{B.11}$$

where $\bar{u}_\alpha u_\alpha = 2m$, $\bar{u}_\beta u_\beta = 2m'$ and $A(s, t)$, $B(s, t)$ are two scalar invariant amplitudes. The convection on gamma matrices is that of Bjorken and Drell (1964). A straightforward Pauli reduction leads to the form,

$$F_{\beta\alpha}(s, t) = \langle \chi'_{\lambda'} | f_1(s, t) + f_2(s, t) \vec{\sigma} \cdot \hat{p}' \vec{\sigma} \cdot \hat{p} | \chi_\lambda \rangle \tag{B.12}$$

where

$$\left.\begin{aligned} f_1 &= \sqrt{[(E + m)(E' + m)]}\left[A + \left(W - \frac{m + m'}{2}\right)B\right] \\ f_2 &= \sqrt{[(E - m)(E' - m')]}\left[-A + \left(W + \frac{m + m'}{2}\right)B\right] \end{aligned}\right\} \quad \text{(B.13)}$$

The helicity amplitudes $F_{\lambda'\lambda}$ are explicitly

$$\begin{aligned} F_{++} &= F_{--} = (f_1 + f_2)\cos\tfrac{1}{2}\theta \\ F_{+-} &= -F_{-+} = (f_1 - f_2)\sin\tfrac{1}{2}\theta. \end{aligned} \quad \text{(B.14)}$$

(In these expressions as in all our helicity amplitudes we have specialized the azimuthal angle to $\phi = 0$.)

The differential cross section and polarization are

$$\left.\begin{aligned} \frac{d\sigma}{dt} &= \frac{\pi}{sp^2}(|F_{++}|^2 + |F_{+-}|^2) \\ P &= \frac{2\,\mathrm{Im}\,(F_{++}F^*_{+-})}{|F_{++}|^2 + |F_{+-}|^2} \end{aligned}\right\} \quad \text{(B.15)}$$

Each of the helicity amplitudes $F_{\lambda'\lambda}$, called s-channel helicity amplitudes to distinguish them from t-channel helicity amplitudes obtained by taking helicity projections in the process $M\overline{M}' \to \overline{B}B'$, has a partial wave expansion of the form (B.9).

4. IMPACT PARAMETER REPRESENTATION

At high enough energies, where very many partial waves are important, the partial wave series (B.3) and (B.9) can be converted into more convenient and intuitive integrals over impact parameter. For large j and not too large angles the Wigner rotation functions are approximately,

$$d^j_{\lambda\mu}(\theta) \simeq J_n((2j + 1)\sin\tfrac{1}{2}\theta) \quad \text{(B.16)}$$

where $J_n(z)$ is the nth order Bessel function and $n = \mu - \lambda = (\lambda_3 - \lambda_4 - \lambda_1 + \lambda_2)$ is the net helicity flip. With the approximation of replacing the sum in (B.9) by an integral over $(j + \frac{1}{2})$ and the use of (B.16) we obtain,

$$\langle\lambda_3\lambda_4|F(s, t)|\lambda_1\lambda_2\rangle \simeq \left(\frac{s}{pp'}\right)^{\frac{1}{2}} \int 2(j + \tfrac{1}{2})\,d(j + \tfrac{1}{2})$$
$$\times \langle\lambda_3\lambda_4|a(s, j + \tfrac{1}{2})|\lambda_1\lambda_2\rangle\, J_n(2(j + \tfrac{1}{2})\sin\tfrac{1}{2}\theta).$$

At high energies where t_{min} is negligible, (A.16) shows that

$$-t \equiv q^2 \simeq 4pp' \sin^2 \tfrac{1}{2}\theta. \tag{B.17}$$

We therefore define the impact parameter b by

$$\sqrt{(pp')}b = j + \tfrac{1}{2} \tag{B.18}$$

and write the argument of the Bessel function as

$$2(j + \tfrac{1}{2}) \sin \tfrac{1}{2}\theta = qb.$$

The integral over $(j + \frac{1}{2})$ then becomes the *impact parameter representation*,

$$\langle \lambda_3\lambda_4 | F(s, t) | \lambda_1\lambda_2 \rangle \simeq 2(spp')^{\frac{1}{2}} \int_0^\infty b\, \mathrm{d}b\, J_n(qb) \langle \lambda_3\lambda_4 | a(s, b) | \lambda_1\lambda_2 \rangle. \tag{B.19}$$

For particles with no spin the representation corresponding to (B.3) is

$$F_{\beta\alpha}(s, t) \simeq 2(spp')^{\frac{1}{2}} \int_0^\infty b\, \mathrm{d}b\, J_0(qb)\, a_{\beta\alpha}(s, b). \tag{B.20}$$

The "partial wave" projection formulas complementary to (B.19) and (B.20) are

$$\langle \lambda_3\lambda_4 | a(s, b) | \lambda_1\lambda_2 \rangle = \frac{1}{2(spp')^{\frac{1}{2}}} \int_0^\infty q\, \mathrm{d}q\, J_n(qb)\, \langle \lambda_3\lambda_4 | F(s, t = -q^2) | \lambda_1\lambda_2 \rangle \tag{B.21}$$

and

$$a_{\beta\alpha}(s, b) = \frac{1}{2(spp')^{\frac{1}{2}}} \int_0^\infty q\, \mathrm{d}q\, J_0(qb)\, F_{\beta\alpha}(s, t = -q^2). \tag{B.22}$$

In these integrals over momentum transfer it is assumed that the amplitude falls off rapidly enough in $q = \sqrt{-t}$ that the integrals converge. From (A.18) it can be seen that at high energies where (B.19) and (B.20) are likely to be useful the coefficient of the integrals is

$$2(spp')^{\frac{1}{2}} = s\left(1 - \frac{m_1^2 + m_2^2 + m_3^2 + m_4^2}{2s} + \ldots\right) \simeq s. \tag{B.23}$$

It is sometimes useful to replace (B.20) and (B.22) by an equivalent two-dimensional Fourier transform representation,

$$F_{\beta\alpha}(s, t) = \frac{s}{2\pi} \int \mathrm{d}^2 b \exp(i\vec{q}\cdot\vec{b})\, a_{\beta\alpha}(s, b)$$

and its inverse, (B.24)

$$a_{\beta\alpha}(s, b) = \frac{1}{2\pi s} \int \mathrm{d}^2 q \exp(-i\vec{q}\cdot\vec{b})\, F_{\beta\alpha}(s, t = -q^2)$$

Hence $\vec{q}$ and $\vec{b}$ are two-dimensional transverse vectors. We have used (B.23) to simplify slightly.

For elastic scattering of spinless particles the representation (B.20) leads [via integration of (B.2) over all q^2 and by means of the optical theorem (A.9)] to the following expressions for the elastic and total cross sections:

$$\sigma_{\text{el}} = 4\pi \int_0^\infty \mathrm{d}(b^2)|a(s, b)|^2$$
$$\sigma_t = 4\pi \int_0^\infty \mathrm{d}(b^2)\,\mathrm{Im}\,a(s, b). \tag{B.25}$$

If $a(s, b)$ is written in a form equivalent to (B.6):

$$a(s, b) = \frac{1}{2i}(\mathrm{e}^{2i\delta(s\ b)} - 1) \tag{B.26}$$

and the complex phase shift $\delta(s, b)$ is given by $\delta = \alpha + i\beta$, then the total, elastic and inelastic cross sections become

$$\left.\begin{aligned} \sigma_t &= 2\pi \int_0^\infty \mathrm{d}(b^2)\,[1 - e^{-2\beta}\cos 2\alpha] \\ \sigma_{\text{el}} &= 4\pi \int_0^\infty \mathrm{d}(b^2)\,[\mathrm{e}^{-2\beta}\sin^2\alpha + \tfrac{1}{4}(1 - \mathrm{e}^{-2\beta})^2] \\ \sigma_{\text{inel}} &= \pi \int_0^\infty \mathrm{d}(b^2)\,[1 - \mathrm{e}^{-4\beta}] \end{aligned}\right\} \tag{B.27}$$

Appendix C

Eikonal Approximation

For short-wavelengths scattering can be described in terms of a semi-classical trajectory (localized wave packets) and impact parameters. For the Schrödinger equation this leads to the eikonal approximation for the phase shift as a function of energy and impact parameter. The standard derivation of the eikonal approximation is given in many places (e.g. Gottfried, 1966, p. 113ff; Glauber, 1959, p. 315ff). The eikonal approximation to the wave function leads at high energies to

$$\psi_k(\vec{x}) \simeq \psi_k^{(0)}(\vec{x}) \exp\left[\frac{-i}{\hbar v}\int_{-\infty}^{z} V(\vec{b}, z')\,\mathrm{d}z'\right] \tag{C.1}$$

where $\vec{b}$ is the impact parameter of the incident particle and $V(\vec{x})$ is the scattering potential. The exponent represents the phase accumulated up to the point $(\vec{b}, z)$ by the action of the potential. Straightforward calculation of the scattering amplitude leads to the approximate expression,

$$f(\vec{k}', \vec{k}) \simeq \frac{k}{2\pi i}\int d^2b \exp\left[-i(\vec{k}' - \vec{k})_\perp \cdot \vec{b}\right]\left[\exp\left[2i\Delta(\vec{b})\right] - 1\right] \tag{C.2}$$

where

$$\Delta(\vec{b}) = -\frac{1}{2\hbar v}\int_{-\infty}^{\infty} V(\vec{b}, z)\, dz \tag{C.3}$$

is the eikonal approximation to the phase shift. For a spherically symmetric potential integration over azimuth in (C.2) leads to an expression equivalent to (B.20) with (C.3) as an approximation to the phase in (B.26).

The lowest order approximation to (C.2) in powers of the potential is obtained by expanding the exponential in the integrand. This gives

$$f^{(1)}(\vec{k}', \vec{k}) = -\frac{1}{4\pi}\int d^3x \exp\left[-i(\vec{k}' - \vec{k})_\perp \cdot \vec{x}\right]\frac{2m}{\hbar^2} V(\vec{x}) \tag{C.4}$$

which, for small angle scattering where $\vec{k}' - \vec{k}$ is perpendicular to the incident direction, is just the first Born approximation.

An alternative derivation of (C.2) and (C.3) with closer connection to the relativistic problem is based on the Born series of which (C.4) is the first term. With the definition $U(\vec{x}) = -2mV(\vec{x})/\hbar^2$, and its Fourier transform,

$$U(\vec{q}_1, \vec{q}_2) = \int d^3x \exp\left[-i\vec{q}_1 \cdot x\right] U(\vec{x}) \exp\left[i\vec{q}_2 \cdot \vec{x}\right], \tag{C.5}$$

the $(n + 1)$th term in the Born series,

$$f(\vec{k}', \vec{k}) = \sum_{n=1}^{\infty} f^{(n)}(\vec{k}', \vec{k}), \tag{C.6}$$

can be written formally

$$4\pi f^{(n+1)}(\vec{k}', \vec{k}) = \langle \vec{k}' | UGUG \ldots GU | \vec{k} \rangle$$

where there are $n + 1$ factors of U and n Green functions G. More explicitly, this reads

$$4\pi f^{(n+1)} = \left[\frac{1}{(2\pi)^3}\right]^n \int d^3q_1 \int d^3q_2 \ldots \int d^3q_n\, U(\vec{k}', \vec{q}_1)$$
$$\times \frac{1}{q_1^2 - k^2 - i\varepsilon} \cdot U(\vec{q}_1, \vec{q}_2) \cdot \frac{1}{q_2^2 - k^2 - i\varepsilon} \cdots \frac{1}{q_n^2 - k^2 - i\varepsilon} \cdot U(\vec{q}_n, \vec{k}). \tag{C.7}$$

Each potential factor in (C.7) is replaced by its representation (C.5). Then the eikonal approximation is made by approximating the intermediate momentum factors in all the Green function propagators as follows. Choose the z-axis for all integrations as the incident direction $\vec{k}$ [or the average direction $(\vec{k}+\vec{k}')/2$] and define parallel and perpendicular components of every vector: $\vec{x}=(\vec{x}_{\perp}, z)$, $\vec{q}_n=(\vec{q}_{n\perp}, q_{n\parallel})$. A typical propagator is then approximated by neglecting $q_{\perp}^2$:

$$(q^2-k^2-i\varepsilon)^{-1} \equiv (q_{\perp}^2+q_{\parallel}^2-k^2-i\varepsilon)^{-1} \simeq (q_{\parallel}^2-k^2-i\varepsilon).$$

The amplitude (C.7) then becomes

$$4\pi f^{(n+1)} \simeq \int d^3x \exp(-i\vec{k}'\cdot\vec{x})\, U(\vec{x}_{\perp}, z) \prod_{j=1}^{n} (2\pi)^{-3} \int d^2x_{j\perp} \int dz_j$$
$$\times\, U(\vec{x}_{j\perp}, z_j) \int dq_{j\parallel} \frac{1}{q_{j\parallel}^2-k^2-i\varepsilon} \cdot \int d^2q_{j\perp}$$
$$\times \exp[i\vec{q}_{j\perp}\cdot(\vec{x}_{j-1}-\vec{x}_j)_{\perp} + iq_{j\parallel}(z_{j-1}-z_j)] \exp(i\vec{k}\cdot\vec{x}_n).$$

The integrals over the transverse components of each intermediate momentum $d^2q_{j\perp}$ can now be performed to yield the product of two-dimensional delta functions,

$$\prod_{j=1}^{n} (2\pi)^2\, \delta^{(2)}(\vec{x}_{(j-1)\perp}-\vec{x}_{j\perp}).$$

This is equivalent to the semi-classical straight line path implicit in (C.1). The integrations over the $(n+1)$ different $d^2x_{j\perp}$ are thus reduced to only one:

$$4\pi f^{(n+1)} = \int d^2x_{\perp} \int dz \exp(-i\vec{k}'\cdot\vec{x})\, U(\vec{x}_{\perp}, z) \cdot \prod_{j=1}^{n} (2\pi)^{-1} \int dz_j\, U(\vec{x}_{\perp}, z_j)$$
$$\times \int dq_{j\parallel} \frac{\exp[iq_{j\parallel}(z_{j-1}-z_j)]}{q_{j\parallel}^2-k^2-i\varepsilon} \exp(i\vec{k}\cdot\vec{x}_n). \tag{C.8}$$

The integral over $dq_{j\parallel}$ can be done by contour integration to yield

$$\int dq \frac{\exp[iq(z_{j-1}-z_j)]}{q^2-k^2-i\varepsilon} = \frac{\pi i}{k} \exp(ik|z_{j-1}-z_j|). \tag{C.9}$$

Inspection of the remaining integrals over dz_j in (C.8) with (C.9) inserted shows that each integrand oscillates rapidly at high energies and thus will give a negligible result unless in each successive integral $(z_{j-1}-z_j) \geqslant 0$. With the approximation to small angle scattering, (C.8) can thus be written

$$4\pi f^{(n+1)} \simeq \left(\frac{i}{2k}\right)^n \int d^2x_\perp \exp(i\vec{q}\cdot\vec{x}_\perp) \prod_{j=0}^{n} \int dz_j\, U(\vec{x}_\perp, z_j)$$

$$\times\, \theta(z_0 - z_1)\,\theta(z_1 - z_2)\ldots\theta(z_{n-1} - z_n) \qquad \text{(C.10)}$$

where $\vec{q} = \vec{k} - \vec{k}'$ is the vectorial momentum transfer. The restriction on the ranges of integration has a familiar counterpart in the limitations on the time integrations in the expansion of the S matrix in quantum field theory. Because of the symmetry of the integrand of the $(n + 1)$-dimensional integral the ranges of all the z integrations can be extended to the interval $(-\infty, \infty)$ provided we divide by $(n + 1)!$ Thus (C.10) becomes

$$f^{(n+1)} = \frac{k}{2\pi i}\cdot\frac{1}{(n+1)!}\int d^2x_\perp \exp(i\vec{q}\cdot\vec{x}_\perp)\left[\frac{i}{2k}\int_{-\infty}^{\infty} dz\, U(\vec{x}_\perp, z)\right]^{n+1}. \qquad \text{(C.11)}$$

With the definition of $U(\vec{x})$ and the Born series (C.6) it is directly evident that (C.11) is the $(n + 1)$st term in the expansion of the eikonal formula (C.2).

The derivation of the eikonal approximation in relativistic field theories seems possible in some theories and not in others (see, for example, Abarbanel and Itzykson, 1969; Lévy and Sucher, 1969; Tiktopoulos and Treiman, 1971; Fried, 1971; Cheng and Wu, 1972; Swift, 1972). In spite of the uncertainty of its fundamental basis the eikonal method is an extremely plausible and simple way to impose the requirements of unitarity in the direct channel. The standard recipe for "eikonalization" relies on the connection (C.4) of the lowest order exchange amplitude (the first Born approximation) with the two-dimensional Fourier transform of the phase shift $\Delta(b)$. Explicitly, if the lowest order relativistic amplitude is $F_{\text{Born}}(s, t)$, then the relativistic eikonal phase is [in the notation of (B.24) and (B.26)]

$$\delta_{\text{eikonal}}(s, b) = \frac{1}{2\pi s}\int d^2q \exp(-i\vec{q}\cdot\vec{b})\, F_{\text{Born}}(s, t = -q^2) \qquad \text{(C.12)}$$

and the relativistic eikonal amplitude is

$$F_{\text{eikonal}}(s, t) = \frac{s}{4\pi i}\int d^2b \exp(i\vec{q}\cdot\vec{b})(\exp[2i\delta_{\text{eikonal}}(s, b)] - 1). \qquad \text{(C.13)}$$

The eikonal phase (C.12) is in general complex and describes elastic scattering in the presence of competing processes. In nonrelativistic problems this situation is normally described by a complex optical model potential.

For reference we quote some simple examples of $F_{\text{Born}}(s, t)$ and its two-dimensional Fourier transform, $s\,\delta_{\text{eikonal}}(s, b)$:

$F_{\text{Born}}(s, t = -q^2)$	$s\,\delta_{\text{eikonal}}(s, b)$	
$A(s)\,(q^2 + \mu^2)^{-1}$	$A(s)\,K_0(\mu b)$	(C.14)
$A(s)\,(q^2 + \mu^2)^{-2}$	$A(s)\mu^{-2} \cdot \mu b\,K_1(\mu b)$	(C.15)
$A(s)\exp[-B(s)q^2/2]$	$A(s)(B(s))^{-1}\exp[-b^2/2B(s)]$	(C.16)
$\dfrac{[A(s)]^2}{q(q^2 + 4\mu^2)^{\frac{1}{2}}}\ln\left(\dfrac{\sqrt{(q^2 + 4\mu^2)} + q}{\sqrt{(q^2 + 4\mu^2)} - q}\right)$	$[A(s)\,K_0(\mu b)]^2.$	(C.17)

APPENDIX D
KINEMATICS OF INCLUSIVE PROCESSES

The process $a + b \to c +$ anything is indicated diagrammatically in Fig. D.1. With the masses of a, b, and c given and unpolarized beams, there are three kinematic variables, plus a trivial one (e.g. the azimuthal angle of the momentum of c in the c.m.s.) needed to describe the cross section. These can be simply the total c.m.s. energy W and the components of the momentum of c

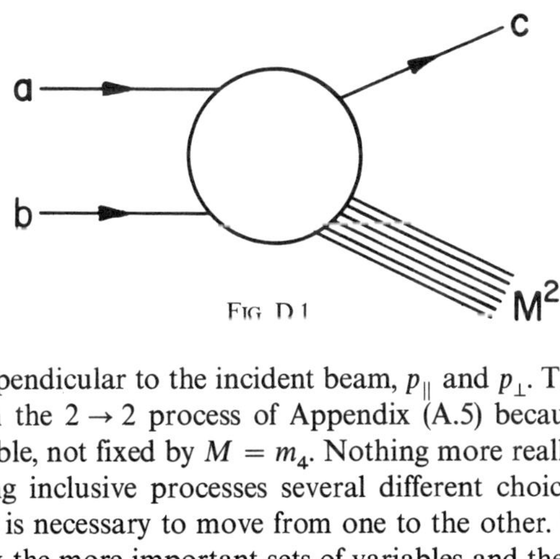

FIG. D.1

parallel and perpendicular to the incident beam, $p_{\parallel}$ and $p_{\perp}$. This is one more variable than in the $2 \to 2$ process of Appendix (A.5) because the missing mass M is variable, not fixed by $M = m_4$. Nothing more really need be said, but in describing inclusive processes several different choices of variables are made and it is necessary to move from one to the other. We summarize in this Appendix the more important sets of variables and their connections. An exhaustive compilation of inclusive kinematics can be found in Kasman (1972).

1. THE VARIABLES (s, t, u, M^2)

The obvious and direct extension of the two-body kinematics is the use of any three of s, t, u, M^2 as variables. The definitions of s, t, u are given in (A.12), which in the present notation are

$$s = (p_a + p_b)^2, \qquad t = (p_a - p_c)^2, \qquad u = (p_c - p_b)^2 \tag{D.1}$$

and

$$M^2 = (p_a + p_b - p_c)^2. \tag{D.2}$$

The constraint equation (A.13) reads

$$s + t + u - M^2 = m_a^2 + m_b^2 + m_c^2. \tag{D.3}$$

2. THE VARIABLES $(s, p_\perp, x)$

In discussing scaling Feynman (1969a, b) introduced a reduced longitudinal momentum variable called x. In the c.m.s. frame let the momentum $\vec{p}$ of particle c have components parallel and perpendicular to the incident direction (that of particle a) denoted by $p_\parallel^*$ and $\vec{p}_\perp$. Then Feynman's x is defined as either

$$x = \frac{p_\parallel^*}{p_{\max}^*} \qquad \text{or} \qquad x = \frac{2p_\parallel^*}{\sqrt{s}} \tag{D.4}$$

where $\rho_{\max}^*$ is the maximum momentum permitted to c in the c.m.s. The two definitions are equivalent at high energies. For simplicity and definiteness we use the second form throughout.

The variables s, $p_\perp$, x (plus the azimuth of $\vec{p}_\perp$) are an equivalent set for inclusive processes. Note one peculiarity of the x variable. If $p_\parallel$ is finite then as $s \to \infty$, $x \to 0$ independent of the particular value of $p_\parallel$. This means that for $s \to \infty$ a finite part of phase space is mapped into $x = 0$. This can cause some conceptual problems, as we will discuss below when we consider cross section formulas.

3. THE VARIABLES $(s, p_\perp, y)$

In the c.m.s. or any other frame K moving uniformly parallel to the incident direction (z-axis) particle c has momentum $\vec{p}$ with components $p_\parallel$ and $\vec{p}_\perp$. There is another Lorentz frame K' moving with a relative velocity β parallel to the z-axis in which particle c has only transverse components of momentum, i.e. $\vec{p}' = \vec{p}_\perp$. In that frame, the energy of particle c is $E' = \omega_c$, where

$$\omega_c = \sqrt{(p_\perp^2 + m_c^2)} \tag{D.5}$$

is sometimes called the transverse or the longitudinal mass and is denoted by m_T, μ, κ by other authors. The energy and momentum of particle c in the frame K can evidently be expressed in terms of $p_\perp$ and β according to

$$\left.\begin{aligned} \vec{p}_\perp &= \vec{p}_\perp \\ p_\parallel &= \omega \sinh y \\ E &= \omega \cosh y \end{aligned}\right\} \tag{D.6}$$

where ω is given by (D.5) and the longitudinal boost or *rapidity* y is related to β by

$$y = \tanh^{-1} \beta. \tag{D.7}$$

Thus $p_{\parallel}$ can be replaced by y, and $(s, p_{\perp}, y)$ can be used as the three kinematic variables. For reference we record two more expressions for y:

$$y = \begin{Bmatrix} \frac{1}{2} \ln \left(\dfrac{E + p_{\parallel}}{E - p_{\parallel}} \right) \\ \ln \left(\dfrac{E + p_{\parallel}}{\omega} \right) \end{Bmatrix} \tag{D.8}$$

The above expressions define the rapidity y in the frame K. What about different frames, e.g. c.m.s. and laboratory? The laws of Lorentz transformations are such that the rapidity y_1 in frame K_1 differs by a constant from the rapidity y_2 in frame K_2, the constant being the longitudinal boost that takes one from K_1 to K_2 according to (D.7). This translation by a constant amount in going from one Lorentz frame to another (along the beam direction) is one of the attractive features of rapidity.

4. RAPIDITY AND RELATED ANGULAR VARIABLES

If $p_{\perp}^2 \gg m^2$ the expression (D.8) for y can be approximated by

$$y \simeq \ln \left(\cot \frac{\theta}{2} \right) - \frac{m^2}{2p_{\perp}^2} \cos \theta \tag{D.9}$$

where $\tan \theta = p_{\perp}/p_{\parallel}$. Thus in the c.m.s. the rapidity is approximately equal to the cosmic ray angular variable,

$$\eta = \ln \left(\cot \frac{\theta_{\text{c.m.s.}}}{2} \right). \tag{D.10}$$

Another cosmic ray angular variable is

$$\eta' = - \ln (\tan \theta_{\text{lab}}) \tag{D.11}$$

where θ_{lab} is the angle of emission of particle c in the laboratory. If c is relativistic in the lab and the c.m.s. motion is also relativistic in the lab, then the two angular variables are related by

$$\eta' = \eta + \ln (\gamma_{\text{c.m.s.}}) \tag{D.12}$$

where $\gamma_{\text{c.m.s.}} = (1 - \beta_{\text{c.m.s.}}^2)^{-\frac{1}{2}} \simeq W/2m_b$.

5. INVARIANT SUB-ENERGIES IN TERMS OF RAPIDITY DIFFERENCES, s AND Y

The invariant sub-energy of particles 1 and 2 is

$$s_{12} = (p_1 + p_2)^2. \tag{D.13}$$

Using (D.6) to represent the momenta and energies this can be written

$$s_{12} = m_1^2 + m_2^2 + 2\omega_1\omega_2 \cosh(y_1 - y_2) - 2\vec{p}_{1\perp} \cdot \vec{p}_{2\perp}. \tag{D.14}$$

Similarly, the invariant momentum transfer,

$$t_{12} = (p_1 - p_2)^2, \tag{D.15}$$

can be written

$$t_{12} = m_1^2 + m_2^2 - 2\omega_1\omega_2 \cosh(y_1 - y_2) + 2\vec{p}_{1\perp} \cdot \vec{p}_{2\perp}. \tag{D.16}$$

If the rapidity difference $y_1 - y_2$ is large, then

$$s_{12} \simeq -t_{12} \simeq \omega_1\omega_2 \exp|y_1 - y_2|. \tag{D.17}$$

If we specialize to the incident particles a and b then we have

$$s = m_a^2 + m_b^2 + 2m_a m_b \cosh Y \tag{D.18}$$

where $Y = y_a - y_b$ is the laboratory rapidity of particle a. At high energies, Y is given approximately by

$$Y \simeq \ln(s/m_a m_b) + O\left(\frac{1}{s}\right). \tag{D.19}$$

In this limit the rapidity of the c.m.s. is

$$Y_{\text{c.m.s.}} \simeq \tfrac{1}{2}(Y - \Delta) \tag{D.20}$$

and the rapidities of particles a and b in the c.m.s. are

$$y_a^* \simeq \tfrac{1}{2}(Y + \Delta), \quad y_b^* \simeq -\tfrac{1}{2}(Y - \Delta) \tag{D.21}$$

where

$$\Delta = \ln(m_b/m_a).$$

This is indicated in Fig. D.2 for $\Delta > 0$. The maximum rapidity interval available to particle c is approximately Y, but depending on its mass it may be slightly larger or smaller than this. If c is lighter than both a and b then the maximum and minimum rapidities for c are indicated on Fig. D.2 with

$$\Delta' = \ln(m_a/m_c)$$

$$\Delta'' = \ln(m_b/m_c) = \Delta + \Delta'.$$

This corresponds to an absolute maximum c.m.s. rapidity for particle c of

$$|y_c^*|_{\max} = \ln(W/m_c). \tag{D.22}$$

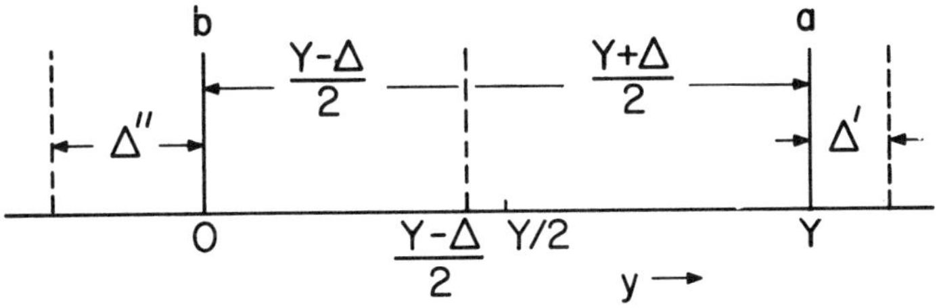

FIG. D.2.

The value (D.22) is attained at $p_\perp = 0$. If particle c has a non-vanishing value of $p_\perp$ the maximum allowed rapidity is smaller. The range of rapidity is restricted to

$$-y^*_{\max}(p_\perp) \leqslant y^*_c \leqslant y^*_{\max}(p_\perp)$$

where at high energies and for $p_\perp \ll W/2$ the extreme is given by

$$y^*_{\max}(p_\perp) \simeq \ln(W/\omega_c).$$

Since most particles have relatively small $p_\perp$ this value is typically smaller than (D.22) by less than unity. At large $p_\perp$, of course, it causes an appreciable shortening of the range of rapidity.

6. PHASE SPACE AND INVARIANT CROSS SECTION

The single-particle inclusive cross section is given by (A.4), integrated over all final state momenta except particle c and summed over all final states β that contain c and are kinematically allowed. The cross section thus appears schematically as

$$\mathrm{d}\sigma^c_{ab} = \frac{\text{(Lorentz invariant)}}{\text{(Flux factor)}} \cdot \frac{\mathrm{d}^3 p_c}{E_c} \tag{D.23}$$

where the flux factor is given by (A.5). It is useful to define the *invariant differential cross section* for $a + b \to c +$ anything as

$$E_c \frac{\mathrm{d}^3 \sigma^c_{ab}}{\mathrm{d}^3 p_c} = F^c_{ab}. \tag{D.24}$$

It is easy to show that the invariant phase space d^3p/E can be written in terms of $\vec{p}_\perp$ and y as

$$\frac{\mathrm{d}^3 p}{E} = \mathrm{d}^2 p_\perp \frac{\mathrm{d}p_\parallel}{E} = \mathrm{d}^2 p_\perp \, \mathrm{d}y. \tag{D.25}$$

Thus the invariant differential cross section is sometimes written as

$$F^c_{ab} = \frac{d^3\sigma}{d^2p_\perp\, dy}. \tag{D.26}$$

7. RELATIONS BETWEEN VARIABLES

We have the three major sets of variables $(s, p_\perp, y)$, $(s, p_\perp, x)$, and (s, t, M^2). We give here the relations among these variables in the high energy limit where terms of order $1/s$ relative to the leading contributions are neglected.

(a) *Rapidity y and Feynman's x*

For finite $|x|$ (not of order $1/\sqrt{s}$), fixed moderate $p_\perp$, and large s the relation between x and y is

$$x = \begin{cases} \dfrac{\omega}{m_a} \exp(y_c - y_a), & x > 0 \\[2ex] -\dfrac{\omega}{m_b} \exp(y_b - y_c), & x < 0 \end{cases} \tag{D.27}$$

where ω is given by (D.5) and $x > 0$ means that particle c is in the same hemisphere as particle a in the c.m.s. For fixed $p_\perp$ the differentials are related by $dy = dx/x$, as follows from (D.25) when E^* is approximated by $p^*_\parallel$.

(b) *Relations between (t, M^2) and $(p_\perp, x)$*

For finite $|x|$, moderate $p_\perp$, and large s the invariant momentum transfer t can be written in terms of $p_\perp$ and x as

$$t = m_a^2(1 - x) + m_c^2\left(1 - \frac{1}{x}\right) - \frac{p_\perp^2}{x} + O(1/s) \tag{D.28}$$

provided $x > 0$. For $x < 0$, the connection is

$$t = -s|x| + m_a^2 + m_c^2\left(1 - \frac{1}{|x|}\right) + m_b^2|x| - \frac{p_\perp^2}{|x|}. \tag{D.28$'$}$$

Exactly at $x = 0$, i.e. for $p^*_\parallel = 0$, the relation is

$$t = m_a^2 + m_c^2 - \omega_c\sqrt{s}. \tag{D.28$''$}$$

Similarly, the square of the missing mass is

$$M^2 = s(1 - |x|) + m_c^2\left(1 - \frac{2}{|x|}\right) - \frac{2p_\perp^2}{|x|} + O(1/s) \tag{D.29}$$

provided $|x|$ is finite. For $p_{\parallel}^{*} = 0$, we have instead

$$M^2 = s - 2(\sqrt{s})\omega_c + m_c^2. \tag{D.29'}$$

It is worthwhile to note that for a given x the minimum t value (obtained by putting $p_{\perp} = 0$ in (D.28)) is

$$t_{\min} = m_a^2(1 - x) + m_c^2\left(1 - \frac{1}{x}\right). \tag{D.30}$$

If we use $x = 1 - M^2/s$ from (D.29), $t_{\min}$ can be written alternatively as

$$t_{\min} \simeq \begin{cases} (m_a^2 - m_c^2)M^2/s, & m_a \neq m_c \\ -m^2(M^2/s)^2, & m_a = m_c = m. \end{cases} \tag{D.31}$$

These last expressions can also be obtained from (A.18) with appropriate approximations. They do not hold too near $x = 1$, i.e. where M^2 is not large compared to the other masses.

(c) *Relations between* (t, M^2) *and* $(p_{\perp}, y)$

From (D.16) it follows that the invariant momentum transfer is given in terms of rapidity by

$$t = m_a^2 + m_c^2 - 2m_a\omega_c \cosh(y_c - y_a). \tag{D.32}$$

For large $|y_c - y_a|$ this becomes

$$t \simeq m_a^2 + m_c^2 - m_a\omega_c \exp|y_c - y_a|. \tag{D.32'}$$

The square of the missing mass can be found in the high energy limit from (D.29) and (D.27) to be approximately

$$M^2 \simeq \begin{cases} s\left[1 - \dfrac{\omega}{m_a}\exp(y_c - y_a)\right], & y_c^* > 0 \\ s\left[1 - \dfrac{\omega}{m_b}\exp(y_b - y_c)\right], & y_c^* < 0. \end{cases} \tag{D.33}$$

(d) *Phase space connections and invariant cross sections*

In the high energy limit the differentials of the three sets of variables are related as follows:

$$\mathrm{d}y\,\mathrm{d}p_{\perp}^2 \simeq \frac{\mathrm{d}x}{x}\mathrm{d}p_{\perp}^2 \simeq \mathrm{d}x\,\mathrm{d}t \simeq \frac{1}{s}\mathrm{d}M^2\,\mathrm{d}t \tag{D.34}$$

This means that the invariant cross section can be written in the alternative forms,

$$F_{ab}^{c} = \frac{d^3\sigma_{ab}^{c}}{dy\, d^2p_{\perp}} \simeq x\frac{d^3\sigma_{ab}^{c}}{dx\, d^2p_{\perp}} \simeq \frac{1}{\pi}\frac{d\sigma_{ab}^{c}}{dx\, dt} \simeq \frac{s}{\pi}\frac{d\sigma_{ab}^{c}}{dM^2\, dt}. \tag{D.35}$$

The factors of π in the last two forms are present to give cross sections per unit azimuthal angle.

The approximate phase space in terms of x and $p_{\perp}^2$ in (D.34) is singular at $x = 0$. The exact expression is $dx\, dp_{\perp}^2/x_0$ where

$$x_0 = \left(x^2 + \frac{4\omega^2}{s}\right)^{\frac{1}{2}}$$

is the scaled energy. For $|x| \gg 2\omega/\sqrt{s}$, $x_0 \simeq |x|$ and the expressions in (D.34) and (D.35) are accurate. As $|x| \to 0$, however, the distinction between $|x|$ and x_0 must be made. Note that in integrating F_{ab}^{c} over any finite interval of some variable one must multiply by the relevant phase space differential from (D.34). Thus the number of particles seen in the interval $\Delta(p_{\perp}^2)$ and Δx, integrated over azimuth, is

$$\Delta N = (\text{flux}) \times (\text{time}) \times F_{ab}^{c} \times \pi \times \Delta(p_{\perp}^2)\frac{\Delta x}{x_0}. \tag{D.36}$$

For very large s, this number is almost singular at $x = 0$ if $F \neq 0$ there.

Appendix E
Distributions, Correlations, Moments, Sum Rules in Inclusive Processes

We have defined the kinematics of inclusive processes in Appendix D. Here we put down definitions of the normalized distributions, correlation functions, moments, and sum rules.

1. Normalized distributions

The n-particle inclusive invariant cross section for the process,

$$a + b \to (1 + 2 + \ldots + n) + \text{anything} \tag{E.1}$$

is defined in conformity to the single particle cross section (D.24) as

$$E_1 E_2 \ldots E_n \frac{d^{3n}\sigma_{ab}(1, 2, \ldots, n)}{d^3p_1\, d^3p_2 \ldots d^3p_n}. \tag{E.2}$$

In order to compress the notation somewhat it is convenient to adopt the notation

$$d\phi_i = \frac{d^3p_i}{E_i}, \tag{E.3}$$

for the Lorentz invariant phase space. Then the invariant cross section (E.2) is

$$\frac{d^n\sigma_{ab}(1, 2, \ldots, n)}{d\phi_1\, d\phi_2 \ldots d\phi_n}. \tag{E.4}$$

It is also convenient to introduce the *normalized distributions*,

$$\rho_{ab}(1, 2, \ldots, n) = \frac{1}{\sigma_{ab}} \frac{d^n\sigma_{ab}(1, 2, \ldots, n)}{d\phi_1\, d\phi_2 \cdots d\phi_n}. \tag{E.5}$$

In (E.5) sometimes σ_{ab} is the inelastic, rather than total, cross section if the elastic scattering contribution is omitted from the appropriate inclusive cross sections. In what follows we will generally omit the subscripts *ab*. The incident particles will be understood to be given.

2. MULTIPLICITIES AND HIGHER MOMENTS

The average number of particles of type i is given by the integral over all phase space of the normalized single-particle inclusive distribution for particles of type i:

$$\langle n_i \rangle = \int \rho(i)\, d\phi_i. \tag{E.6}$$

Higher moments are defined similarly by integrals over all phase space of 2-particle and higher distributions. For example, the second moments are given by

$$\langle n_i n_j - \delta_{ij} n_i \rangle = \int \rho(i, j)\, d\phi_i\, d\phi_j. \tag{E.7}$$

The presence of $\langle n(n-1)\rangle$ for $i = j$ stems from the definition of the inclusive cross section—if there are n particles of type i in a given event, the first one can be picked out in n different ways and the second in $(n - 1)$. The third moment is given by

$$\langle n_i n_j n_k - \delta_{ij} n_i n_k - \delta_{ik} n_i n_j - \delta_{jk} n_i n_j + 2\delta_{ij}\delta_{jk} n_i \rangle = \int \rho(i, j, k)\, d\phi_i\, d\phi_j\, d\phi_k \tag{E.8}$$

and so on.

The analogue in inclusive distributions of the prong cross sections discussed in Section III.2(a) are the n-charged particle distributions summed over all type of charged particles. We denote these normalized distributions by $\rho_{ch}(1, 2, \ldots, n)$ but it should be remembered that 1 stands for *any* charged particle in the phase space element $d\phi_1$, and similarly for the other particles. The various charged particle moments can be defined either through the prong cross sections:

$$\langle n^k \rangle = \sum_n n^k \sigma_n / \sigma, \tag{E.9}$$

or in terms of integrals over the various normalized distributions

$$\left.\begin{aligned} \langle n_{\mathrm{ch}} \rangle &= \int \rho_{\mathrm{ch}}(1)\, d\phi_1 \\ \langle n_{\mathrm{ch}}(n_{\mathrm{ch}} - 1) \rangle &= \int \rho_{\mathrm{ch}}(1, 2)\, d\phi_1\, d\phi_2 \end{aligned}\right\} \tag{E.10}$$

and so on.

3. CORRELATION COEFFICIENTS, CORRELATION FUNCTIONS

The idea of totally uncorrelated production of particles would lead to the prediction that the n-particle distribution is given by

$$\rho(1, 2, \ldots, n) = \rho(1)\, \rho(2) \ldots \rho(n). \tag{E.11}$$

This is not actually attainable on kinematic grounds alone (see Section 4 below), but is a useful norm from which to measure correlations. The analogue for the prong cross sections is a Poisson distribution,

$$\frac{\sigma_n}{\sigma} = \frac{\langle n \rangle^n}{n!} e^{-\langle n \rangle}. \tag{E.12}$$

From (E.11) or (E.12) and the definitions of the moments in Section 2 above it can be shown quite simply that the integral correlation coefficients f_k introduced by Mueller (1971),

$$\begin{aligned} f_1 &= \langle n \rangle \\ f_2 &= \langle n(n-1) \rangle - \langle n \rangle^2 \\ f_3 &= \langle n(n-1)(n-2) \rangle - 3\langle n \rangle \langle n(n-1) \rangle + 2\langle n \rangle^3 \\ &\vdots \end{aligned} \tag{E.13}$$

vanish for $k > 1$ for uncorrelated charged particle production. The correlation coefficients f_k are thus useful as empirical quantities measuring the character of the n-particle distributions. Equation (E.13) defines f_k for the charged particles, but there are obvious generalizations for other situations, e.g. charged and neutral, π^+, K^+, and p, etc.

Sometimes moments for negative prongs instead of all charged prongs are presented. These moments are trivially related to the moments for all charged prongs because of charge conservation. Let Q be the total charge in units of the proton's charge in the initial state. Then for an event with n charged prongs, the number n_- of negative prongs is $n_- = (n - Q)/2$. This leads to the relations,

$$\langle n_- \rangle = \tfrac{1}{2}\langle n \rangle - \frac{Q}{2}$$
$$f_2^- = \tfrac{1}{4} f_2 - \frac{\langle n \rangle}{4} + \frac{Q}{2} \tag{E.14}$$

and corresponding linear combinations for higher moments.

The integral correlation coefficients f_k have their differential counterparts constructed from the normalized n-particle inclusive distributions $\rho(1, 2, \ldots, n)$. The most commonly used one is the two-particle correlation function C(1, 2):

$$C(1, 2) = \rho(1, 2) - \rho(1)\rho(2). \tag{E.15}$$

Evidently, the integral of $C(1, 2)$ over phase space gives f_2:

$$f_2 = \int C(1, 2)\, \mathrm{d}\phi_1\, \mathrm{d}\phi_2. \tag{E.16}$$

Because of limited statistics, often correlations are given in only one variable, say rapidity. The two-particle rapidity correlation function is defined by

$$\bar{C}(y_1, y_2) = \int C(1, 2)\, \mathrm{d}^2 p_\perp^{(1)}\, \mathrm{d}^2 p_\perp^{(2)}. \tag{E.17}$$

From the experimental point of view it is more convenient to define a correlation function $R(y_1, y_2)$ as the ratio of $\bar{C}$ to the integral of $\rho(1)\rho(2)$ over $\mathrm{d}^2 p_\perp^{(1)}\, \mathrm{d}^2 p_\perp^{(2)}$. Thus the function $R(y_1, y_2)$ is

$$R(y_1, y_2) = \frac{\int \rho(1, 2)\, \mathrm{d}^2 p_\perp^{(1)}\, \mathrm{d}^2 p_\perp^{(2)}}{\int \rho(1)\, \mathrm{d}^2 p_\perp^{(1)} \int \rho(2)\, \mathrm{d} p_\perp^{(2)}} - 1. \tag{E.18}$$

The advantage of $R(y_1, y_2)$ is that it measures the fractional correlation and so treats favoured and unfavoured portions of phase space equally, whereas $C(1, 2)$ given by (E.15) or $\bar{C}(y_1, y_2)$ given by (E.17) can be small because $\rho(1, 2)$ and $\rho(1)\rho(2)$ are small, even though $\rho(1, 2)/\rho(1)\rho(2) \neq 1$.

4. ENERGY-MOMENTUM AND OTHER SUM RULES

The strict requirement of conservation of 4-momentum between the initial and final state in every collision event leads to a family of "sum rules" involving inclusive cross sections or normalized densities (Chou and Yang, 1970; DeTar, Freedman, and Veneziano, 1971; Predazzi and Veneziano, 1971). The simplest of these involves the single-particle densities. The 4-momentum of the initial state can be written as

$$(p_a + p_b)^\mu = \sum_{(i)} \int p_i^\mu \rho(i)\, \mathrm{d}\phi_i \tag{E.19}$$

where the sum is over all types of particles and $\mu = 0, 1, 2, 3$. The right-hand side of (E.19) is just the summing up of all the momenta in the final state. If we multiply both sides by σ_{ab} we see that (E.19) is a relation between the 0-particle inclusive cross section and an integral over the 1-particle inclusive cross section. There is a fairly obvious generalization relating the n-particle inclusive cross section to the $(n + 1)$-particle ones. Consider the process $a + b \rightarrow (1 + 2 + \ldots + n)$ + anything with its cross section $\mathrm{d}^n\sigma(1, 2, \ldots, n)/\mathrm{d}\phi_1\, \mathrm{d}\phi_1\, \mathrm{d}\phi_2 \ldots \mathrm{d}\phi_n$. For fixed momenta $p_1, p_2, \ldots, p_n$, the momentum of "anything", namely $P = p_a + p_b - (p_1 + p_2 + \ldots + p_n)$, can be thought of as being built up in the manner of (E.19) by a sum of integrals over the various cross sections for $(n + 1)$ particles in the reaction, $a + b \rightarrow (1 + 2 + 3 + \ldots + n + (n + 1))$ + anything. Thus we have the general sum rule,

$$(p_a + p_b - p_1 - p_2 - \cdots - p_n)^\mu \rho_{ab}(1, 2, \ldots, n)$$
$$= \sum_{(n+1)} \int (p_{n+1})^\mu \rho_{ab}(1, 2, \ldots, n, n + 1)\, \mathrm{d}\phi_{n+1} \tag{E.20}$$

where the sum is over the different particle types chosen as the $(n + 1)$ st particle. Evidently, (E.19) is (E.20) for $n = 0$.

The most useful of these sum rules are the $n = 0$ and $n = 1$ forms. The $n = 0$ one is (E.19). The $n = 1$ expression is

$$(p_a + p_b - p_1)^\mu \rho_{ab}(1) = \sum_{(2)} \int (p_2)^\mu \rho_{ab}(1, 2)\, \mathrm{d}\phi_2.$$

This can be written in terms of the correlation function (E.15) with the aid of (E.19):

$$-p_1^\mu \rho_{ab}(1) = \sum_{(2)} \int p_2^\mu C_{ab}(1, 2)\, \mathrm{d}\phi_2. \tag{E.21}$$

The existence of a nonvanishing value on the left-hand side of (E.21) shows that $C_{ab}(1, 2)$ cannot be identically zero. Particle production cannot be completely uncorrelated emission. Some correlations are imposed merely by energetics.

Similar sum rules can be written for any conserved additive quantity.

Denoting such a quantity by Q (Q = electric charge, z-component of isospin, hypercharge, baryon number, etc.), the sum rule reads

$$Q_a + Q_b = \sum_{(i)} \int Q_i \rho(i)\, \mathrm{d}\phi_i. \tag{E.22}$$

The manner in which different regions of phase space contribute to the sum on the right hand side is discussed for electric charge in Section III.2(f).

Bibliography and References

1. CONFERENCES AND SUMMER SCHOOL PROCEEDINGS (CHRONOLOGICAL ORDER)

Proceedings of the 4th Scottish Universities' Summer School, **Edinburgh,** July 28–August 17, 1963, *Strong Interactions and High Energy Physics*, ed. R. G. Moorhouse (Oliver and Boyd, Edinburgh and London, 1964).

Proceedings of the 7th Scottish Universities' Summer School, **Edinburgh**, August 1–20, 1966, *Particle Interactions at High Energies,* eds. T. W. Preist and L. L. J. Vick (Oliver and Boyd, Edinburgh and London, 1967).

Proceedings of the Third International Conference on High-Energy Collisions, **Stony Brook,** September 5–6, 1969, *High Energy Collisions*, ed. C. N. Yang *et al.* (Gordon and Breach, N.Y., 1969).

Proceedings of the XV International Conference in High Energy Physics, **Kiev**, August 26–September 4, 1970, ed. V. Shelest *et al.* ("Naukova Dumka", Kiev, 1972).

Proceedings of the Colloquium on Multiparticle Dynamics, **Helsinki**, May 25–28, 1971, ed. E. Byckling, K. Kajantie, H. Satz, and J. Tuominiemi (Research Institute for Theoretical Physics, Helsinki, Finland).

Proceedings of the Rochester Meeting of the Division of Particles and Fields of the American Physical Society, **Rochester**, August 30–September 12, 1971, *Particles and Fields—1971*, ed. A. C. Melissinos and P. F. Slattery (A.I.P. Conference Proceedings No. 2, Particles and Fields Subseries No. 1, American Institute of Physics, N.Y., 1971).

Proceedings of the Fourth International Conference on High-Energy Collisions (Stony Brook Series), **Oxford**, April 5–7, 1972, ed. J. R. Smith (Rutherford High Energy Laboratory, Chilton, Didcot, Berks., U.K., 1972).

Proceedings of the III International Colloquium on Multiparticle Reactions, **Zakopane**, June 20–24, 1972 (Institute of Nuclear Research, Warsaw, Report "P" NR 1421/VI/PH).

Proceedings of the XVIth International Conference on High-Energy Physics, **Chicago-Batavia**, September 6–13, 1972, eds. J. D. Jackson, A. Roberts, R. Donaldson (National Accelerator Laboratory, Batavia, Illinois, 1973).

Proceedings of the International Conference on New Results from Experiments on High Energy Particle Collisions, **Vanderbilt**, March 26–28, 1973, ed. Robert S. Panvini (A.I.P. Conference Proceedings No. 12, Particles and Fields Subseries No. 4, American Institute of Physics, N.Y., 1973).

2. REFERENCES

Abarbanel, H. D. I. and Itzykson, C. (1969). *Phys. Rev. Letts.* **23**, 53.
Adair, R. K. (1972). *Phys. Rev.* **D5**, 1105.
Albrow, M. G. *et al.* (1973). *Nucl. Phys.* **B51**, 388.
Allaby, J. V. (1970). Kiev, p. 11–40.
Amaldi, U. (1973). ISR results on Proton–proton elastic scattering and total cross-sections, *In* "Highlights in Particle Physics," ed. A. Zichichi (Editrice Compositori, Bologna, 1973).
Amaldi, U. *et al.* (1973a). *Phys. Letts.* **43B**, 231.
Amaldi, U. *et al.* (1973b). *Phys. Letts.* **44B**, 112.
Amati, D., Stanghellini, A. and Fubini, S. (1962). *Nuovo Cimento* **26**, 896.
Amendolia, S. R. *et al.* (1973). *Phys. Letts.* **44B**, 119.
Ammosov, V. V. *et al.* (1972). *Phys. Letts.* **42B**, 519.
Ammosov, V. V. *et al.* (1973). *Nucl. Phys.* **B58**, 77.
Antinucci, M. *et. al.* (1973) *Lett. Nuovo Cimento* **6**, 121.
Ball, J. S. and Zachariasen, F. (1972). *Phys. Letts.* **40B**, 411
Barbiellini, G. et al. (1972). *Phys. Letts.* **39B**, 663.
Barger, V., Phillips, R. J. N. and Geer, K. (1972). *Nucl. Phys.* **B47**, 29.
Barger, V., Phillips, R. J. N. and Geer, K. (1973). *Nucl. Phys.* **B54**, 463.
Barletta, W. et al. (1973). *Phys. Rev.* **D7**, 3233.
Barloutaud, R., 1973, Oxford, Vol. 1, p. 215.
Benecke, J. *et al.* (1969). *Phys. Rev.* **188**, 2159.
Berger, E. L., Jacob, M. and Slansky, R. (1972). *Phys. Rev.* **D6**, 2580.
Berger, E. L., Oh, B. Y. and Smith, G. A. (1972). *Phys. Rev. Letts.* **28**, 322.
Bertin, A. *et al.* (1972). *Phys. Letts.* **42B**, 493.
Białas, A., Fiałkowski, K. and Załewski, K. (1972). *Nucl. Phys.* **B48**, 237.
Bjorken, J. D. and Drell, S. D. (1964). "Relativistic Quantum Mechanics" Appendix A, (McGraw-Hill, New York).
Blatt, J. M. and Weisskopf, V. F. (1952). "Theoretical Nuclear Physics" (John Wiley & Sons, New York).
Bosetti, P. *et al.* (1973). *Nucl. Phys.* **B54**, 141.
Bourrely, C. and Fischer, J. TH. 1652-CERN, 12 April/73.
Bracci, E. *et al.* (1972a). "Compilation of Cross Sections $I - \pi^-$ and π^+ Induced Reactions". CERN/HERA 72–1, May 1972.
Bracci, E. *et al.* (1972b). "Compilation of Cross Sections $II - K^-$ and K^+ Induced Reactions". CERN/HERA 72–2, October 1972.
Braccini, P. (1973). Oral presentation at the Vanderbilt Conference.
Bronzan, J. (1973). *In*, "Symposium on the Pomeron", Argonne National Laboratory, March 1, 1973, ANL/HEP 7327, p. 33.
Bussiere, A. (1973). Vanderbilt, p. 235.
Chabaud, V. *et al.* (1972). *Phys. Letts.* **38B**, 441, 445, 449.
Chapman, J. W. *et al.* (1972). *Phys. Rev. Letts.* **29**, 1686.
Charlton, G. *et al.* (1972). *Phys. Rev. Letts.* **29**, 515.
Chen, M.-S. *et al.* (1971). *Phys. Rev. Letts.* **26**, 1583.
Cheng, H. and Wu, T. T. (1970). *Phys. Rev. Letts.* **24**, 1456.
Cheng, H. and Wu, T. T. (1972). *Phys. Rev.* **D5**, 3170.
Cheng, H., Walker, J. K. and Wu, T. T. (1973a). *Phys. Letts.* **B44**, 97.
Cheng, H., Walker, J. K. and Wu, T. T. (1973b). *Phys. Letts.* **B44**, 283.
Chiu, C. B. (1972). *In* "Annual Review of Nuclear Sciences", Vol. 22 ed. E. Segré, J. R. Grover and H. P. Noyes, (Annual Reviews, Palo Alto, California).

Chou, T. T. and Yang, C. N. (1970). *Phys. Rev. Letts.* **25**, 1072.
Collins, P. B. D. and Squires, E. J. (1968). "Regge Poles in Particle Scattering", Springer Tracts in Modern Physics, Vol. 45 (ed. G. Hohler). (Springer-Verlag, Berlin).
Cornille, H. and Martin, A. (1972). *Phys. Letts.* **40B**, 671.
Dao, F. T. *et al.* (1972). *Phys. Rev. Letts.* **29**, 1627.
Denisov, S. P. *et al.* (1971). *Phys. Letts.* **36B**, 415, 528.
de Groot, E. H. and Miettinen, H. I. (1973). R.H.E.L. preprint RL-73-003 (April, 1973).
De Tar, C. E., Freedman, D. Z. and Veneziano, G. (1971). *Phys. Rev.* **D4**, 906.
Dibon, H. *et al.* (1973). *Phys. Letts.* **44B**, 313.
Diddens, A. N. (1972). Oxford, Vol. 1, p. 127.
Eden, R. J. (1967). "High Energy Collisions of Elementary Particles". (Cambridge University Press).
Eden, R. J. (1971). *Rev. Mod. Phys.* **43**, 15.
Feinberg, E. L. (1972). *Physics Reports* (*Phys. Letters C*) **5**, 237.
Ferbel, T. (1972). *Phys. Rev. Letts.* **29**, 448.
Feynman, R. P. (1969a). Stony Brook, p. 237.
Feynman, R. P. (1969b). *Phys. Rev. Letts.* **23**, 1415.
Feynman, R. P. (1972). "Photon-Hadron Interactions". (W. A. Benjamin, Reading, Mass.).
Fiałkowski, K. (1972). *Phys. Letts.* **41B**, 379.
Fiałkowski, K. and Miettinen, H. I. (1973). *Phys. Letts.* **43B**, 61.
Foley, K. J. *et al.* (1969). *Phys. Rev.* **181**, 1775.
Fox, G. C. and Quigg, C. (1973). *In* "Annual Review of Nuclear Science". Ed. E. Segré, J. R. Grover and H. P. Noyes. (Annual Reviews, Palo Alto, California), Vol 23, p. 219.
Frazer, W. R. *et al.* (1972). *Rev. Mod. Phys.* **44**, 284.
Frazer, W. R. *et al.* (1973). *Phys. Rev.* **D7**, 2647.
Fried, H. M. (1971). *Phys. Rev.* **D3**, 2010.
Froissart, M. (1961). *Phys. Rev.* **123**, 1053.
Fry, J. R. *et al.* (1973). *Nucl. Phys.* **B58**, 420.
Fubini, S. (1963). Edinburgh, p. 259.
Giacomelli, G. (1972). Chicago-Batavia, Vol. 3, p. 219.
Glauber, R. J. (1959). *In* "Lectures in Theoretical Physics". Vol. 1. Ed. W. E. Brittin. (Interscience, New York.)
Gottfried, K. (1966). "Quantum Mechanics". Vol. 1. (W. A. Benjamin, Reading, Mass.)
Grunberg, G. and Truong, T. N. (1973). *Phys. Rev. Letts.* **31**, 63.
Harari, H. and Rabinovici, E. (1973). *Phys. Letts.* **43B**, 49.
Heckman, P. and Henzi R. (1972). *Phys. Letts.* **41B**, 189.
Horn, D. (1972). *Physics Reports* (*Phys. Letters C*) **4**, 1.
Horn, D. and Zachariasen, F. (1973). "Hadron Physics at Very High Energies". (W. A. Benjamin, Reading, Mass.).
Huson, F. R. (1973). Vanderbilt, p. 60.
Hwa, R. C. (1971). *Phys. Rev. Letts.* **26**, 1143.
Hwa, R. C. and Lam, C. S. (1971). *Phys. Rev. Letts.* **27**, 1098.
Jackson, J. D. (1970). *Rev. Mod. Phys.* **42**, 12.
Jackson, J. D. (1973). Vanderbilt, p. 404.
Jacob, M. (1972). Chicago-Batavia, Vol. 3, p. 373.
Jacob, M. and Slansky, R. (1972). *Phys. Rev.* **D5**, 1847.

Jacob, M., Slansky, R. and Wu, C.-C. (1972). *Phys. Rev.* **D6**, 2444.
Jacob, M. and Wick, G. C. (1959). *Annals of Physics* (*N.Y.*) 7, 404.
Jarlskog, G. (1973). Vanderbilt, p. 169.
Jin, Y. S. and Martin A. (1964). *Phys. Rev.* **135B**, 1375.
Kasman, J. (1972). *In* (M. E. Law *et al.* Eds.). "A Compilation of Data on Inclusive Reactions", Lawrence Berkeley Laboratory Report LBL-80, August, 1972, Part III.
Khuri, N. N. and Kinoshita, T. (1965). *Phys. Rev.* **137B**, 720; **140B**, 706.
Kroll, P. (1973). *Lett. Nuovo Cimento* **7**, 745.
Lam, W. S. (1972). *Phys. Letts.* **40B**, 466.
Lander, R. L. (1971). Rochester, p. 71.
Levy, M. and Sucher, J. (1969). *Phys. Rev.* **186**, 1656.
Lovelace, C. (1972). Chicago-Batavia, Vol. 3, p. 73.
MacDowell, S. W. and Martin, A. (1964). *Phys. Rev.* **135B**, 960.
Martin, A. (1963). Edinburgh, p. 105.
Martin, A. (1966). *Nuovo Cimento* **42A**, 930.
Martin, A. D. and Spearman, T. D. (1970). "Elementary Particle Theory" (North-Holland, Amsterdam).
Michael, C. (1972). Chicago-Batavia, Vol. 3, p. 165.
Miettinen, H. I. (1972). *Phys. Letts.* **38B**, 431.
Morrison, D. R. O. (1970). Kiev, "Review of Quasi-Two-Body Reactions". Sec. 4 and Fig. 4.2 (issued as a separate report ITP-71-66E by the Institute of Theoretical Physics, Academy of Sciences of the Ukrainian SSR, Kiev, 1971).
Morrison, D. R. O. (1972). Oxford, Vol. 1, p. 253.
Morrison, D. R. O. (1973). *Proc. Roy. Soc.* **A335**, 461.
Mueller, A. H. (1970). *Phys. Rev.* **D2**, 2963.
Mueller, A. H. (1971). *Phys. Rev.* **D4**, 150.
Mueller, A. H. (1972). Chicago-Batavia, Vol. 1, p. 347.
Newton, R. G. (1966). "Scattering Theory of Waves and Particles". (McGraw-Hill, New York).
Phillips, R. J. N. and Ringland, G. A. (1972). *In* "High Energy Physics", Vol. 5. Ed. E. H. S. Burhop. (Academic Press, New York and London).
Polkinghorne, J. C. (1972). *Nuovo Cimento* **7A**, 555.
Predazzi, E. and Veneziano, G. (1971). *Lett. Nuovo Cimento* **2**, 749.
Quigg, C. (1971). Rochester, p. 40.
Quigg, C. and Jackson, J. D. (1972). National Accelerator Laboratory report NAL-THY-93 (October, 1972).
Roy, S. M. (1972). *Physics Reports* (*Phys. Letters C*) **5**, 125.
Sannes, F. *et al.* (1973). *Phys. Rev. Letts.* **30**, 766.
Sivers, D. (1973). *Phys. Rev.* **D8**, 2272.
Stapp, H. P. (1971). *Phys. Rev.* **D3**, 3177.
Stodolsky, L. (1973). *Phys. Rev. Letts.* **31**, 139.
Strolin, P. (1973). Oral presentation at the Vanderbilt Conference.
Swift, A. R. (1972). *Phys. Rev.* **D5**, 1400.
Tan, C.-I. (1971). *Phys. Rev.* **D4**, 2412.
Tiktopoulos, G. and Treiman, S. B. (1971). *Phys. Rev.* **D3**, 1037.
Titchmarsh, E. C. (1950). "Theory of Functions" (corrected 2nd ed.). (Oxford University Press).
Van Hove, L. (1964). *Rev. Mod. Phys.* **36**, 655.
Van Hove, L. (1971). *Physics Reports* (*Phys. Letters C*) **1**, 347.

Van Hove, L. (1973). *Phys. Letts.* **43B**, 65.
Wang, C. P. (1969). *Phys. Rev.* **180**, 1463.
Wilson, K. G. (1963). *Acta Phys. Austriaca* **17**, 37.
Wilson, K. G. (1970). Cornell University Report CLNS-131 (November, 1970); also these Proceedings, pp. 701–725.
Wroblewski, A. (1970). Kiev, p. 42.
Yodh, G. B., Pal, Y. and Trefil J. S. (1972). *Phys. Rev. Letts.* **28**, 1005.
Zachariasen, F. (1971). *Physics Reports* (*Phys. Letters C*) **2**, 1.

2. TOPICS IN PARTICLE PHYSICS WITH COLLIDING PROTON BEAMS

J. C. SENS

Foundation for Fundamental Research on Matter, The Netherlands and CERN, Geneva, Switzerland

I. Basic Characteristics of the CERN ISR†

1. ENERGY, INTENSITY OF THE STORED PROTONS

In high energy accelerators with stationary targets the total energy in the centre of mass system is given by

$$E^* \approx \sqrt{(2mE_A)} \tag{I.1}$$

(E_A = accelerator energy). In head-on colliding beam collisions this is

$$E^* = 2E_S \tag{I.2}$$

(E_S = energy of each beam). Colliding beams with energies E_S are thus equivalent to an accelerator with an energy

$$E_A \equiv E_{\text{eq}} = \frac{2E_S^2}{m}. \tag{I.3}$$

If the colliding beams cross at angle α (I.2) and (I.3) become:

$$E^* = 2E\cos\frac{\alpha}{2} \qquad E_{\text{eq}} = m\left[2\frac{E^2}{m^2}\cos^2\frac{\alpha}{2} - 1\right] \tag{I.4}$$

Currently the CERN–ISR runs at one of 5 sets of energies (GeV):

E_1/E_2	$E^* = \sqrt{s}$	E_{eq}
11·7/11·7	23	290
15·3/15·3	31	500
22·5/22·5	45	1100
26·5/26·5	53	1500
31·5/31·5	63	2100

In all other respects storage rings and equivalent accelerators are manifestly different: in storage rings the centre of mass is at rest in the laboratory, hence in particular heavy particles have a chance of being produced with low laboratory-velocity; particle detection can be done with standard, ~25 GeV type techniques, the luminosity is several orders of magnitude lower, only collisions involving p, e^+, e^- are feasible in practice, there are particular problems of detection at very small angles. Figure 1 shows the layout of the storage rings and the accelerator from which they are fed.

The number of stored protons can be estimated by noting that, apart from a loss in the transfer, the phase space density in the accelerator which supplies the particles is equal to the phase space density of the stored beams:

$$\frac{N_S}{(\Delta p\Delta\phi)_S} = \eta\frac{N_A}{(\Delta p\Delta\phi)_A} \tag{I.5}$$

where η = loss factor in the transfer $\approx$ 0·5.

† This is a slightly expanded version of a lecture given at the Erice Summer School 1972.

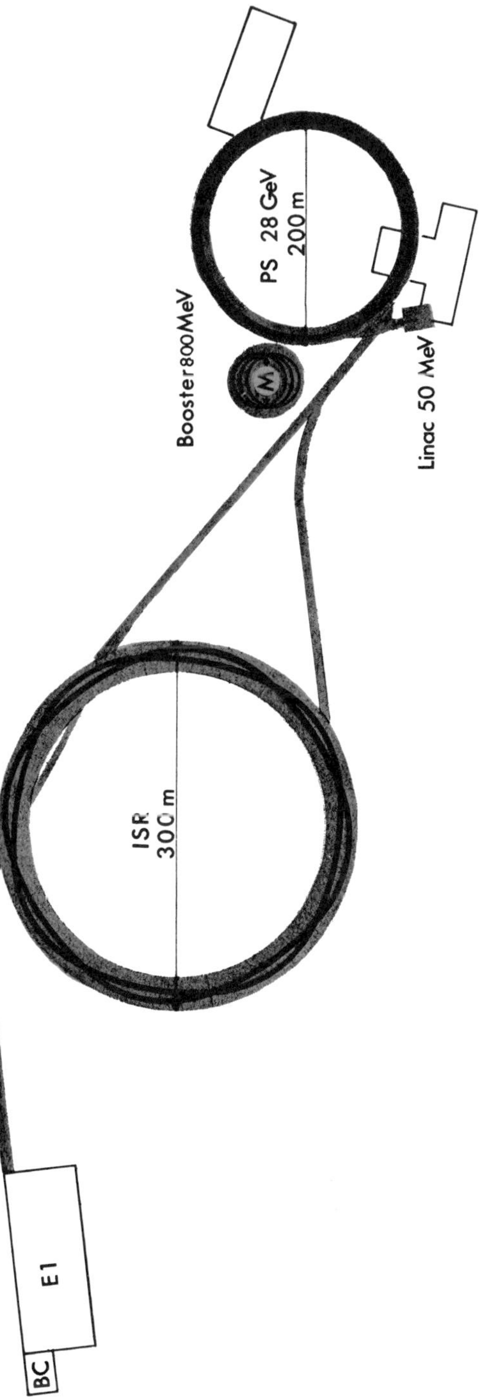

FIG. 1. Layout of the CERN PS and CERN ISR, E1 = Experimental Hall for PS experiments. BC = Hall in which a large bubble chamber (BEBC) is placed.

$\Delta\phi_{S/A}$ = phase width of stored/accelerated particles w.r.t. RF frequency.

$\Delta p_{S/A}$ = momentum spread of stored/accelerated particles.

The CERN PS has

$$\Delta p_A \approx 7 \text{ MeV}/c \qquad N_A = 5 \times 10^{10} \text{ protons/bunch}$$

$$\Delta\phi_A = \tfrac{1}{8} \times 2\pi \qquad (1 \text{ pulse} = 20 \text{ bunches})$$

while at the ISR:

$$\Delta\phi_S = 2\pi \times \frac{\text{period of revolution}}{\text{bunch spacing}} = 2\pi \times \frac{3000 \text{ nsec}}{1000 \text{ nsec}}$$

Hence

$$N_S = 0{\cdot}9 \times 10^{12}\, \Delta p_S$$

With a stored momentum bite of 2% at 25 GeV, we would have $N_S \approx 4{\cdot}5 \times 10^{14}$ protons. This takes 450 PS pulses, or ≈900 sec transfer time; the current is

$$I = \frac{Nec}{2\pi R} \approx 20 \text{ amp} \tag{I.6}$$

These are design figures. In practice one has reached ~18 amp, physics is done with ⩽14 amp beams (July 1973).

2. NUMBER OF INTERACTIONS

The rate of collisions leading to a particular final state is given by:

$$N_c = 2\sigma c \cos^2 \frac{\alpha}{2} \int n_1(x_1 y_1 z) n_2(x_1 y_1 z)\, \mathrm{d}x\, \mathrm{d}y\, \mathrm{d}z \tag{I.7}$$

σ = cross-section for collisions leading to the final state in question $n_{1,2}$ = No. of protons/cm^3 in the beams; c = velocity of the protons; α = crossing angle.

In order to prove this formula[(1)] we define $v_{rel} \equiv$ velocity of a proton in beam 1 w.r.t. restframe of beam 2. Then in the restframe of $2 (c = 1)$:

$$\mathrm{d}N = \sigma\beta_{rel} n_1 n_2\, \mathrm{d}t\, \mathrm{d}V \quad \text{(stat. target)} \tag{I.8}$$

In order to obtain the equivalent of (I.8) valid for any frame we put

$$\mathrm{d}N = A n_1 n_2\, \mathrm{d}t\, \mathrm{d}V \tag{I.9}$$

$\mathrm{d}N$ and $\mathrm{d}t\,\mathrm{d}V$ are Lorentz invariant, hence An_1n_2 is also Lorentz invariant. In the restframe of $2, A = \sigma\beta_{rel}$. Now the densities transform like energies, since $n_i = n_{0i}\gamma = n_{0i}E_i/m$, where n_{0i} is the density in the restframe of particle i. Hence AE_1E_2 and also $AE_1E_2/\mathbf{p}_1\mathbf{p}_2$ are invariants ($\mathbf{p}$ = 4-vector). In the restframe of 2

$$\frac{AE_1E_2}{\mathbf{p}_1\mathbf{p}_2} \rightarrow A = \sigma\beta_{\text{rel}} \tag{I.10}$$

Hence in an arbitrary frame

$$A = \frac{\mathbf{p}_1\mathbf{p}_2\sigma\beta_{\text{rel}}}{E_1E_2} \tag{I.11}$$

and, with (I.9)

$$\mathrm{d}N = \frac{\mathbf{p}_1\mathbf{p}_2\sigma\beta_{\text{rel}}}{E_1E_2} n_1 n_2 \,\mathrm{d}t\,\mathrm{d}V \tag{I.12}$$

Now

$$\mathbf{p}_1\mathbf{p}_2 = m_1 m_2 \gamma_{\text{rel}} = m_1 m_2 (1 - \beta_{\text{rel}}^2)^{-\frac{1}{2}} \tag{I.13}$$

and also

$$\mathbf{p}_1\mathbf{p}_2 = E_1E_2 - \vec{p}_1\vec{p}_2 = E_1E_2(1 - \vec{\beta}_1\vec{\beta}_2) = m_1 m_2 \gamma_1\gamma_2(1 - \vec{\beta}_1\vec{\beta}_2) \tag{I.14}$$

From (I.13), (I.14)

$$\left(\frac{\mathbf{p}_1\mathbf{p}_2}{E_1E_2}\right)^2 \beta_{\text{rel}}^2 = (1 - \vec{\beta}_1\vec{\beta}_2)^2 - (1 - \beta_1^2)(1 - \beta_2^2)$$
$$= (\vec{\beta}_1 - \vec{\beta}_2)^2 - (\vec{\beta}_1 \times \vec{\beta}_2)^2 \tag{I.15}$$

For the ISR, with equal energies in the two beams and $\beta \rightarrow 1$ we have

$$\mathrm{d}N = 2c\sigma\cos^2\frac{\alpha}{2} n_1 n_2 \,\mathrm{d}t\,\mathrm{d}V \tag{I.16}$$

leading to (I.7). For unequal energies, e.g. 11·8 GeV versus 31·4 GeV the correction to (I.16) is less than 0·2 %.

If the protons in each ring are uniformly distributed over the circumference $2\pi R$, the beam height h and the beam width w we have

$$N_c = 2\sigma c\cos^2\frac{\alpha}{2}\left(\frac{N_S}{2\pi Rhw}\right)^2 \frac{w^2 h}{\sin\alpha} \tag{I.17}$$

The last factor is the volume of the "diamond" formed by the two crossing, rectangularly shaped beams.

$$N_c = \frac{\sigma c}{h\,\mathrm{tg}(\alpha/2)}\left(\frac{N_S}{2\pi R}\right)^2 = \frac{\sigma I^2}{e^2 ch\,\mathrm{tg}(\alpha/2)} \equiv L\sigma \tag{I.18}$$

This defines the luminosity L of the machine; if the current I in the two rings are not the same, and if the particle densities are not uniform we have, instead of (I.8):

$$L = \frac{I_1 I_2}{e^2 c \operatorname{tg}(\alpha/2)\, h_{\text{eff}}} \tag{I.19}$$

with

$$\frac{1}{h_{\text{eff}}} = \frac{\int \rho_1(z)\rho_2(z)\,dz}{\int \rho_1(z)\,dz \int \rho_2(z)\,dz} \tag{I.20}$$

$\rho_{1,2}(z)$ = the vertical density distribution in beams 1, 2. At present the effective height $h_{\text{eff}} \approx 5$ mm, $L = 2 \times 10^{28}/\text{cm}^2/\text{sec}/\text{amp}$ in each of the eight intersections of the ISR.

3. APERTURE OF THE ISR

For each particle with momentum p there is an orbit, the equilibrium orbit, which closes on itself after each revolution. The particles oscillate around this equilibrium orbit, both in the horizontal and vertical plane, with amplitudes given by:[(2)]

$$y(s) = a\sqrt{[\beta(s)]} \cos\left[\int \frac{ds}{\beta(s)}\right] + \delta = a\sqrt{[\beta(s)]} \cos(Q\phi(s) + \delta) \tag{I.21}$$

$\beta(s)$ = betatron amplitude function at position s along the azimuth.

$\phi(s)$ = betatron phase function, increases by 2π per revolution.

Q = number of betatron oscillations per revolution.

a, δ = constants.

$\beta(s)$ is fixed by the focussing/defocussing properties of the magnets, which have steep gradients, by magnet imperfections and by the presence of straight sections.

If we now admit particles with different momenta into the ring, then the equilibrium orbit of a particle with momentum $p + \Delta p$ will be horizontally displaced with respect to the one with momentum p by an amount:

$$\Delta x(s) = \alpha_p(s)\frac{\Delta p}{p} \tag{I.22}$$

where $\alpha_p(s)$ is the momentum compaction function at azimuthal position s.

Equations (I.21) and (I.22) determine the aperture required for the stored beams. The vertical aperture is specified by the vertical betatron amplitude $\beta_V(s)$, the horizontal aperture by both $\beta_H(s)$ and the momentum compaction function, $\alpha_p(s)$. Note that in storage rings where Δp can be very large (typically a few hundred MeV at ISR) the horizontal aperture is mainly determined by the momentum compaction. In accelerators where Δp is small (at the CERN PS $\Delta p < 10$ MeV/c) the horizontal betatron amplitude is the dominating factor.

By applying suitable quadrupole magnetic fields to the stored beams, one can make the function $\alpha_p(s) = 0$ locally (Terwilliger scheme). This has recently been applied successfully to several intersections, where the volume of interacting particles (the "diamond") is then reduced from typically $40 \times 5 \times 0{\cdot}5$ cm^3 to just a few cm^3 without loss in luminosity.

4. INJECTION AND STACKING

During the typically 20 minute long period of injection, every 2 seconds a pulse consisting of 20 bunches with $\sim 10^{11}$ accelerated protons per bunch is sent through one of the transfer tunnels to the ISR. The bunch length is ~ 15 nsec, spacing ~ 100 nsec, total time for 20 bunches 2·3 μsec. Near the ISR rings the particles pass through two septum magnets (2·2 m long, 10 kGauss each), whose function is to bend them sideways (and slightly upwards as well) without affecting the particles that are already circulating in the rings. Figure 2 shows the trajectory after the second septum magnet. The trajectory is then made to cross the ISR "injection orbit". At the crossing point a 1·2 m long, $\sim\frac{3}{4}$ kGauss, inflector magnet turns the trajectory by

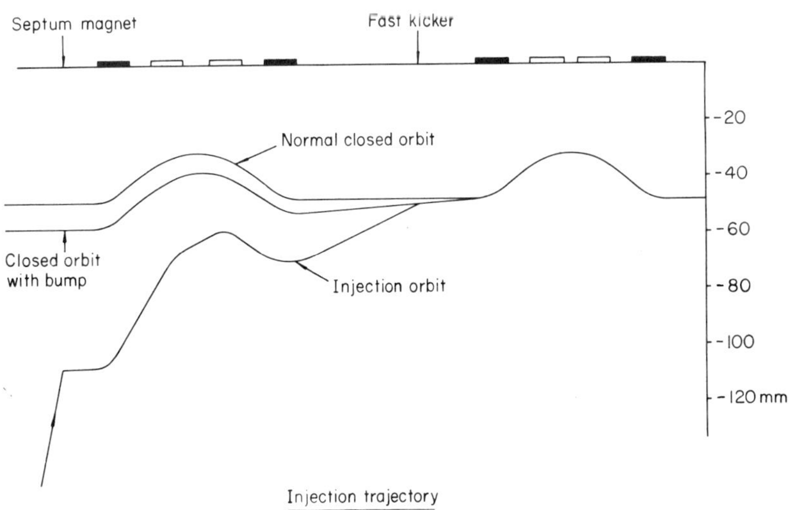

FIG. 2. Trajectory at injection into the ISR.

$\sim$2·5 mrad onto the injection orbit. Since the ISR diameter is 1·5 times the PS diameter, the pulse train fills only $\frac{2}{3}$ of the ISR circumference, hence after the last bunch of the pulse there is $\sim$1 μsec to turn the inflector magnet off before the front end of the train reaches the magnet again after one revolution in the ring.

As soon as the pulse is parked on the injection orbit the R.F. is switched on at 9·54 MHz, corresponding to the bunch spacing on the injection orbit. A slow change in frequency then increases the energy and moves the orbit towards a correspondingly larger radius (see Fig. 3). During this stacking process the R.F. voltage is run down from $\sim$20 kV to a few hundred volts. When the final orbit has been reached the R.F. is switched off. Two kinds of stacking are possible. In the "repetitive stacking" scheme the newly injected pulse is always deposited at the same place near the outside of the vacuum chamber and the previously stacked particles are automatically (as a result of the requirements of phase space density conservation) pushed inwards. In this scheme both the R.F. voltage and the rate of change of frequency remain constant no matter how many pulses have already been injected. In the "non-repetitive" stacking scheme the pulses are put side by side, without disturbing previously stacked particles. In this scheme the R.F. amplitude is not constant and the time dependence of the R.F. frequency is more complicated than in the repetitive scheme. About 4·3 kHz frequency variation will shift the beam over $\sim$6 cm in $\sim$2 seconds. As soon as the R.F. is switched off the beam begins to debunch (as a result of momentum spread and magnet imperfections) and will gradually fill the entire circumference. The system is then ready for the next pulse from the accelerator. The stacked DC beam is to a good approximation not affected by the R.F. voltage which transports the next pulse across the chamber.

5. WORKING LINES

It is well known that in accelerators coupling between horizontal and vertical betatron oscillations leads to beam blow up. This happens if the condition

$$nQ_H + mQ_V = k, \quad n, m, k \text{ integers} \tag{I.23}$$

is satisfied by the Q-values of the machine. In storage rings where the same condition applies, there is the added complication that, as Q varies with radius, typically $\sim$2% over a $\sim$5 cm wide stack, there is a range of values ΔQ_H, ΔQ_V which must all be avoided if a stable stack is to be maintained. The Q-values can be adjusted with 24 pole face windings (extra coils wound around the main bending + focussing magnets of the ring) while in addition the radial dependence of Q is adjustable by means of 8 sextupole magnets, especially installed for this purpose. Figure 4[(3)] shows the Q_H, Q_V diagram

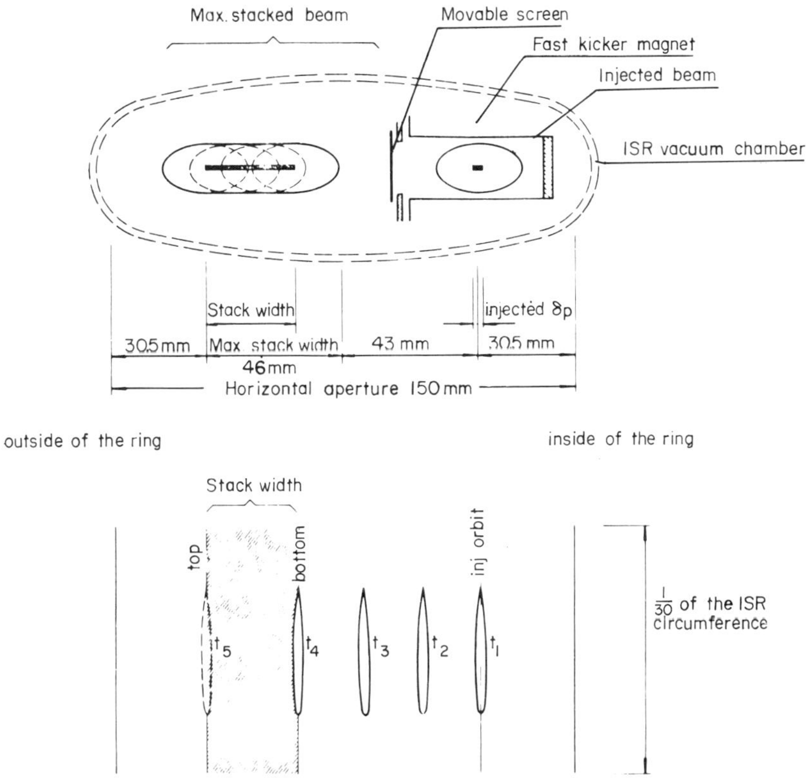

FIG. 3. Inflector magnet and stack.

of the ISR. The straight lines satisfy (I.23). The curved lines ("working lines") indicate actual stacks ("1" = inner edge, CL = centre line, "2" = outer edge). The working line "bare machine" is seen to lay across a third order resonance ($3Q_V = 26$) and was found to give unstable beams. The line "ANNA" was obtained by shifting the "bare machine" line by means of the pole face windings. The line CLEO was obtained by applying sextupole corrections with the effect of rotating the stack in the Q_H, Q_V plot. The line FATA crosses only 5th order resonances. Currents up to 6 amp were reached with the line 15 FA.

The Q_H, Q_V diagram can be probed for resonance lines by injecting just one pulse into one ring and moving it outwards by changing the main magnetic field. An example[4] is shown in Fig. 5. A beam of ~14 mA is injected at −32 mm with respect to the centre of the vacuum chamber. When moving the beam outwards losses occur at given radii. The corresponding Q-values satisfy the following resonance conditions:

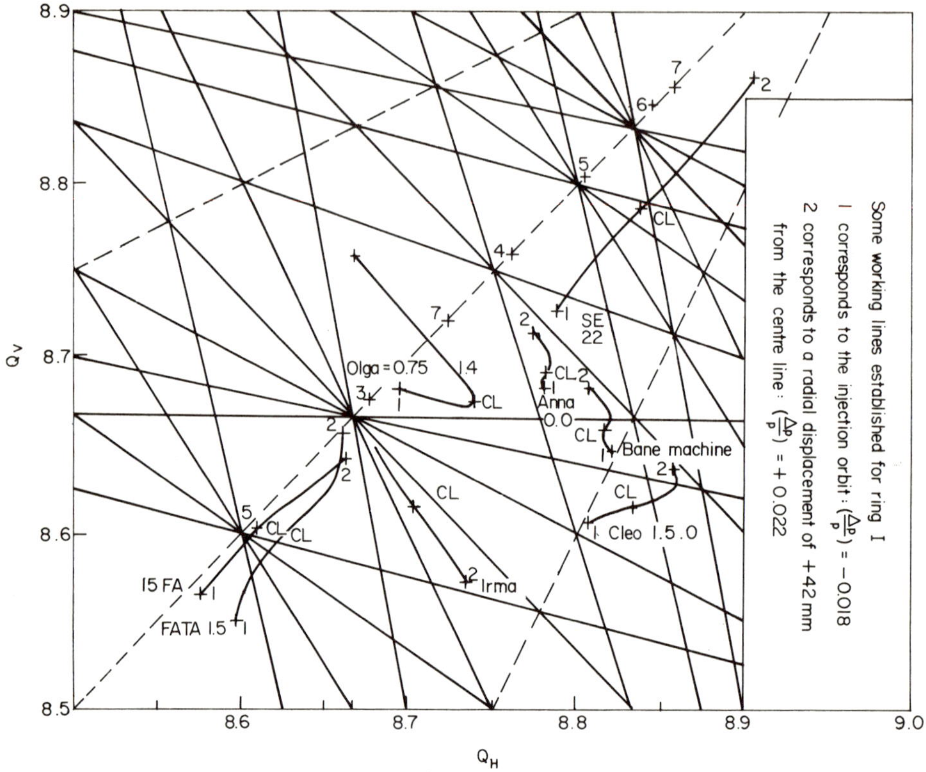

FIG. 4. Examples of working lines in the Q_H, Q_V plot.

Position (mm)	*Residual loss* (%)	*Resonance line*
$-23 \rightarrow -21$	11	$3Q_H + Q_V = 35$
$-15 \rightarrow -11$	14	$3Q_V = 26$
$-5 \rightarrow -3$	23	$2Q_H + 2Q_V = 35$
$+13 \rightarrow +16$	7	$3Q_V + Q_H = 35$ and others
$+32 \rightarrow +34$	7	$4Q_V = 35$

When returning the beam to the inner side further losses occur at some (not all) of the resonance radii.

Over 50 different working lines have been investigated to date.

6. BEAM/GAS INTERACTIONS

Even if an ideal working line could be set up there would still be losses due to the interactions of the circulating protons with the residual gas. These losses

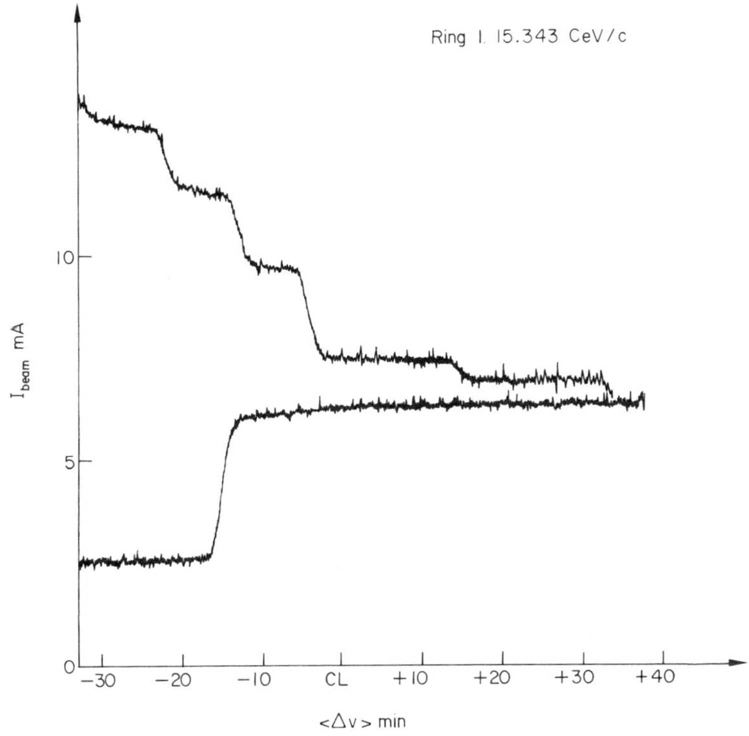

FIG. 5. Probing the Q_H, Q_V diagram for resonances by moving one bunch across the vacuum chamber. The bunch is displaced by changing the main magnetic field of the ISR. CL = centre of vacuum chamber. The vacuum chamber is of elliptical shape (see Fig. 3), its semi long axis is 80 mm.

fall into 3 categories. Nuclear scattering is the least serious. For a nuclear interaction cross-section on the residual gas of 400 mb and a pressure of 10^{-10} Torr the loss is 0·3% per 12 hours. Multiple Coulomb scattering results in a gradual growth of the stack. It has been computed[5] that the height h of the stack varies with time t as

$$h^2(t) = h_0^2 + 6{\cdot}2 \times 10^{-5}\, t$$

(h in mm, t in sec) for a pressure of 10^{-10} Torr. It will thus take ~300 hours to double the height of a ~ 5 mm high stack.

Experiments indicate losses which are very much higher than expected on the basis of nuclear interactions and Coulomb scattering alone. It appears that electrons and negative ions which have been liberated by collisions are attracted to the centre of gravity of the protons and set up coherent electron–proton oscillations. The oscillations are governed by equations of the type

$$\ddot{z}_e = Q_e^2\Omega^2 (z_e - \bar{z}_p)$$

$$\ddot{z}_p + Q_p^2\Omega^2 z_p = Q_p^2\Omega^2 (z_p - \bar{z}_e)$$

i.e. the driving force is provided by the distance of the electron from the mean proton position and v.v. $Q_{e,p}$ are the "bounce" frequencies of one particle in the potential of the other, Ω is the betratron oscillation frequency. Both amplitudes grow, but the electron amplitude grows faster, resulting in a "shaking out" of the electrons from the beam, after which the proton amplitude reduces until sufficient electrons have been attracted towards the beam again. A beat-pattern results. Upon reaching the walls of the vacuum chamber more electrons are liberated and the background increases further. Good vacuum, clean surfaces and clearing electrodes (plates with several kV, to suck away the electrons) are the remedy. Figure 6 shows the striking effects of the clearing electrodes. Considerable theoretical effort is going into understanding these instabilities, see e.g. Ref. 6.

7. INSTRUMENTATION FOR BEAM DIAGNOSTICS

Around the ISR there is a wealth of instruments designed to contribute in one way or another to monitor the behaviour of the beams and vacuum. In

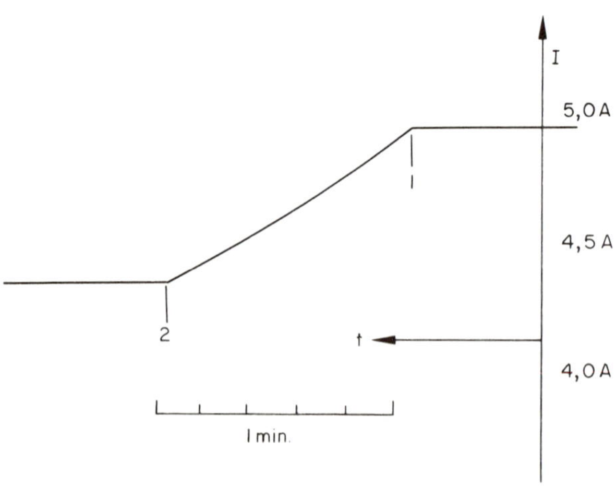

Effect on beam intensity decay of turning clearing fields off (1) and on again (2) in one octant

FIG. 6. Arresting beam decay by means of clearing electrodes.

order to give you a glance at what they are like, four examples are shown in Figs 7, 8, 9, 10.

Figure 7 shows a luminescent screen monitor. It is located in the transfer lines, which connect the proton synchrotron with the ISR. Beam passing through one of the fluorescent screens is made visible on a T.V. screen.

Figure 8 shows secondary emission profile monitor (S.E.M.), also placed in the transfer lines. It consists of a number of aluminium strips. Each strip is connected to an amplifier in which the charges liberated by the beam protons are collected. There are guard electrodes to prevent secondaries escaping from the aluminium from travelling away. There are 2 sets of strips, separated by an isolating electrode.

FIG. 7. Luminescent screen monitor.

Figure 9 shows a Beam probe. This is mounted in the ISR rings themselves. It consists of Ti rods. Charge amplifiers detect at which positions the beam is intercepted.

Figure 10 shows a Beam scraper. Thin foils are moved into the beam and used to scrape away the tails of the stack. This last device has proven very

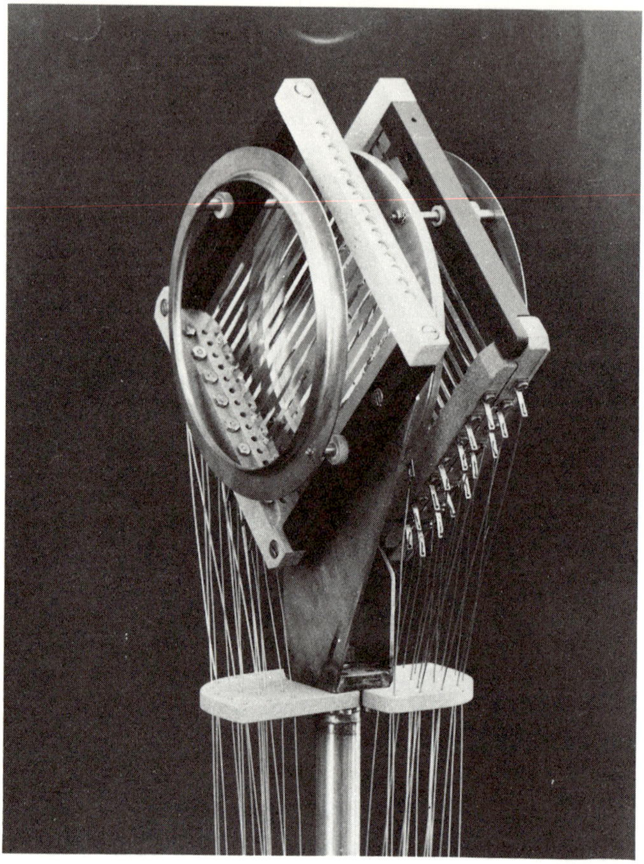

FIG. 8. Secondary emission profile monitor (S.E.M.), used to measure beam profiles in the transfer lines between PS and ISR.

useful in reducing background from low momentum tails. These tails occasionally reach out to the wall of the vacuum chambers and produce high background in the experiments.

II. ISR PERFORMANCE AND DEVELOPMENT

1. THE LUMINOSITY AND ITS MEASUREMENT

The rate of collisions is given by eqn I.7

$$N_c = 2\sigma c \cos^2 \frac{\alpha}{2} \int n_1(x, y, z) n_2(x, y, z) \, dx \, dy \, dz \tag{II.1}$$

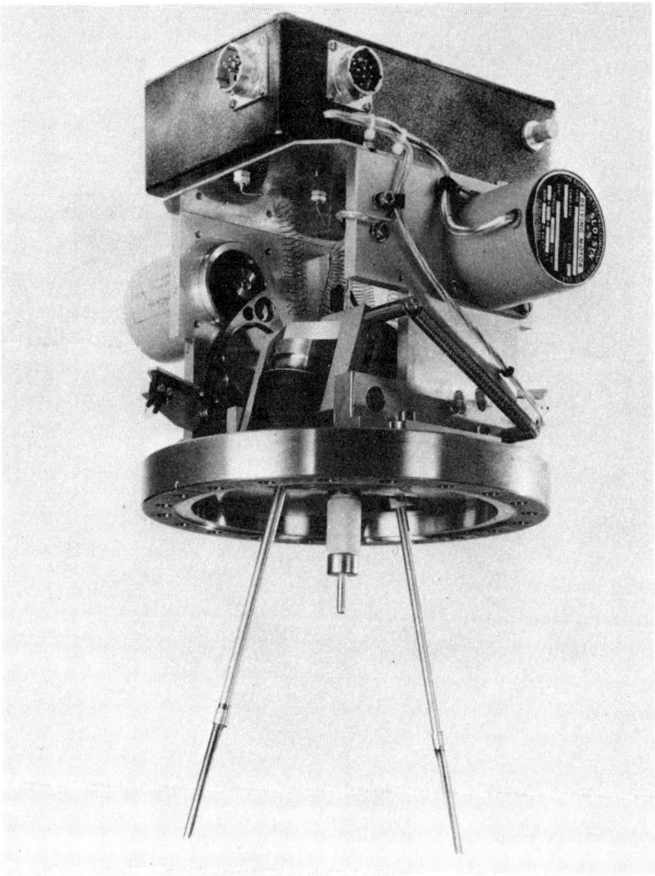

FIG. 9. Beam probe in the ISR.

We introduce one system of coordinates per beam, with z vertical, x along the beam, y perpendicular to the beam. We assume that in the region where collisions occur the density distributions are independent of x. The densities of the beams in their respective frames are then

$$n_1'(y_1, z), \qquad n_2'(y_2, z) \tag{II.2}$$

Transforming to the non-orthogonal y_1, y_2, z frame we have

$$\int n_1 n_2 \, dV = \int n_1'(y_1, z) n_2'(y_2, z) \, dx_1 \, dy_1 \, dz \quad \text{(beam 1 frame)} \tag{II.3}$$

$$= \frac{1}{\sin \alpha} \int n_1'(y_1, z) n_2'(y_2, z) \, dy_1 \, dy_2 \, dz \quad (y_1, y_2, z \text{ frame}) \tag{II.4}$$

Now define

$$S_i(z) = \int n_i'(y_i, z)\,\mathrm{d}y_i\,(i = 1, 2) \tag{II.5}$$

FIG. 10. Beam scraper in the ISR.

Equation (II.1) then becomes

$$N_c = 2\sigma c \frac{\cos^2 \alpha/2}{\sin \alpha} \int S_1(z) S_2(z)\,\mathrm{d}z$$

$$= \frac{c\sigma}{\mathrm{tg}\,\alpha/2} \int S_1(z) S_2(z)\,\mathrm{d}z \equiv L\sigma \tag{II.6}$$

provided the beams are uncoupled, i.e. the distributions in y_1 and y_2 are independent of each other. Note that it is not necessary that within a given beam the distributions in y and z are uncoupled, as they indeed are not in most stacks in practice.

It has been pointed out by v. d. Meer[7] that the overlap integral

$$\int S_1(z)S_2(z)\,\mathrm{d}z \tag{II.7}$$

can be measured by displacing the beams vertically by a distance h and measuring the collision products with a set of counters nearby. In order to keep the background from beam/gas collisions to a minimum the counters are generally put in pairs near the downstream part of the two rings, at several meters from the intersection: a coincidence between two particles emerging back to back from the intersection can kinematically not have been produced by a beam/gas collision. With distance h the counting rate is

$$R(h) = \eta \frac{c\sigma}{\operatorname{tg}\alpha/2} \int S_1(z)S_2(z-h)\,\mathrm{d}z \tag{II.8}$$

where η takes accounts of geometry and efficiency factors. The rate integrated over all steps h is:

$$\begin{aligned}\int_{-\infty}^{+\infty} R(h)\,\mathrm{d}h &= \eta \frac{c\sigma}{\operatorname{tg}\alpha/2} \int\int S_1(z)S_2(z-h)\,\mathrm{d}z\,\mathrm{d}h \\ &= \eta \frac{c\sigma}{\operatorname{tg}\alpha/2} \int S_1(z)\,\mathrm{d}z \int S_2(z-h)\,\mathrm{d}h \\ &= \eta \frac{c\sigma}{\operatorname{tg}\alpha/2} \int S_1(z)\,\mathrm{d}z \int S_2(z)\,\mathrm{d}z \end{aligned} \tag{II.9}$$

which is proportional to (II.7). Hence by measuring $R(0)$ and $\int_{-\infty}^{+\infty} R(h)\,\mathrm{d}h$ one obtains the *normalized* overlap integral:

$$\frac{R(0)}{\int_{-\infty}^{+\infty} R(h)\,\mathrm{d}h} = \frac{\int S_1(z)S_2(z)\,\mathrm{d}z}{\int S_1(z)\,\mathrm{d}z \int S_2(z)\,\mathrm{d}z} \equiv \frac{1}{h_{\mathrm{eff}}} \tag{II.10}$$

in a way which is independent of counter geometry and efficiencies. Equation

II.10 defines the effective height of the system of two intersecting beams. With

$$\int S_i(z)\,\mathrm{d}z = I_i/ec \tag{II.11}$$

where I_i is the current, we have, with (II.6) and (II.10)

$$N_c = \frac{I_1 I_2 \sigma}{e^2 c\,\mathrm{tg}\,(\alpha/2) h_{\mathrm{eff}}} \equiv L\sigma. \tag{II.12}$$

In practice one measures only the quantity $\int_{-\infty}^{+\infty} R(h)\,\mathrm{d}h$ and the currents. The *monitor constant* $K \equiv \eta\sigma$ then follows from (II.9) and (II.11):

$$\int_{-\infty}^{+\infty} R(h)\,\mathrm{d}h = K \frac{I_2 I_2}{e^2 c\,\mathrm{tg}\,\alpha/2} \tag{II.13}$$

Once the constant K is known, then, *at any later time*:

$$R_M = \text{Rate/sec in monitor} = KL \tag{II.14}$$

and hence the counting rate in an experiment, e.g. $pp \rightarrow pp$ is related to the cross-section to be measured, e.g. σ_{EL}, by

$$N_{pp\rightarrow pp} = \frac{R_M}{K}\sigma_{pp\rightarrow pp} \tag{II.15}$$

The conditions for validity are:

The monitor geometry must remain unchanged between the LUM measurement and the data taking period of the experiment.

For each set of ISR-energies there is a different K since for counters at fixed angle the minimum visible (outside pipe) t is different for different beam energies.

Note that the shape, height, intensity of the stack need not remain the same between the LUM measurement and the data taking.

Some further comments:

In practice the displacements are made by small steering magnets mounted on either side of the intersections. Some non-local effects, ~0·3 mm displacements, have been observed in "other" intersections.

No intensity effects have been observed in the range 2/2 to 5/5 A.

For given currents the luminosity can be increased by vertically "shaving" the beam during stacking. The best result obtained this way is

$$L = 4{\cdot}4 \times 10^{30}/\mathrm{cm}^2/\mathrm{sec} = \text{Design } L + 10\%$$

11/12A $\quad h = 3{\cdot}2$ mm $\quad \Delta p/p = 3{\cdot}8\%$

shaving down to 30%. Stacks for 4 hours.

The steering method of v. d. Meer can be checked by measuring the profile of each beam with spark chambers and integrating to obtain the overlap 11.7[8]. The results are

	11·8/11·8	15/15	22/22	26/26
$\dfrac{h_{\mathrm{eff}}(\mathrm{vdM}) - h_{\mathrm{eff}}(\mathrm{profile})}{h_{\mathrm{eff}}(\mathrm{vdM})} =$	2%	−3·5%	−3·3%	+1·5%
	−3%			+4·0%

The current belief is that under "normal" conditions L can be measured to ~5% accuracy, while with special precautions a ~2% level can be reached.

2. LUMINOSITY, LIMITATIONS AND IMPROVEMENTS

Under present (1973) conditions the half life of the luminosity is of order 30 hours. The main limitations are due to (we follow mainly Ref. 9 in this paragraph):

(A) Beam-induced deterioration of the vacuum. This effect has been described above in Section I.6. Negative ions and electrons are attracted towards the beam, set up oscillations there and are shaken out. Positive ions are repelled from the beam and also strike the vacuum chamber walls. The electrostatic potential of a beam of 10 A is 1·5 kV. The pressure rise, as a result of gas molecules liberated from the walls, is proportional to the number of protons and the number of gas molecules: $\Delta P \sim P \cdot I$ (P = pressure, I = current). Obviously there is then a critical current above which the pressure continues to rise. Note that

$$\text{nucleon density in the beam} = \frac{N}{2\pi R \pi a b}$$

$$\text{nucleon density in gas} \frac{6 \times 10^{23}\, P \cdot A}{22{\cdot}2 \times 10^{3} \times 760} = 10^{18} P$$

(A = 28 for air, P = pressure (Torr))

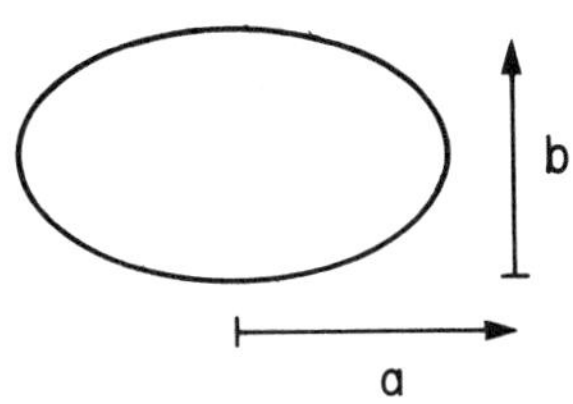

For typical beam sizes the two densities are equal for $P = 2 \times 10^{-10}$ Torr and 5 amp. There are thus as many stationary as moving protons per cm^3 in the "beam area" of the machine.

The current remedies for this problem are:

(1) Bake out of the vacuum chamber at 300°C and 24 hours every time a section is opened up. There have been 60 such bake outs in 1972.
(2) Surface treatment by ions from a glow discharge from a wire in the chamber. This is done in the laboratory prior to mounting in the ISR.
(3) More pumps. 500 additional Titanium sublimation pumps have been installed in 1972.
(4) Beam-induced pumping. Once the walls are clear and the vacuum is good, the beam itself may act as a pump: the beam sends ions off to the walls, but the number of ions desorbed/incident ion becomes less than one. This effect has been observed in some parts of the ISR.

(B) Transverse instability due to the resistivity and inductance of the walls. This is a subject by itself and will not be discussed in detail. Loosely speaking transverse instabilities are caused by coherent oscillations induced by some perturbation. The resulting local density variation is

$$\rho(z, t) = \rho_0 \exp(ikz - wt). \qquad \text{(II.16)}$$

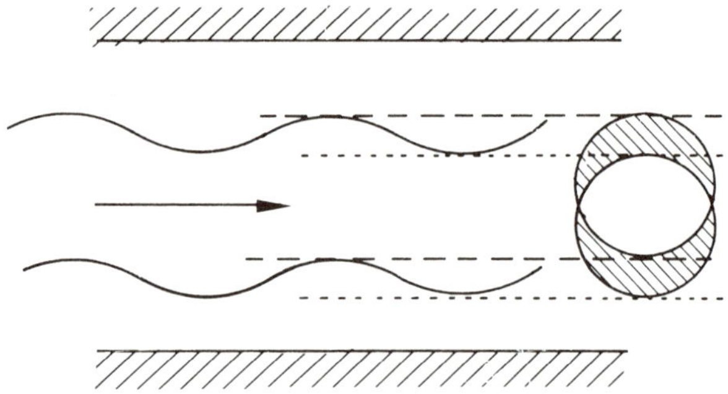

With a vacuum chamber made of metal of finite resistivity and inductance, the equations of motions resulting from this density variation will contain image charges as well. They effectively generate out of phase forces. This in turn implies an imaginary part in the frequency spectrum and an amplitude which grows with time.

The remedy lies in a judicious choice of working lines (see Section I.5). A feed back system has been designed which is intended to automatically counter run-away background build-up due to these effects. This device is being tested.

The current belief is that with the presently available tools or minor additions a factor of ~2 can be gained in luminosity.

The luminosity versus time curve is shown in Fig. 11.

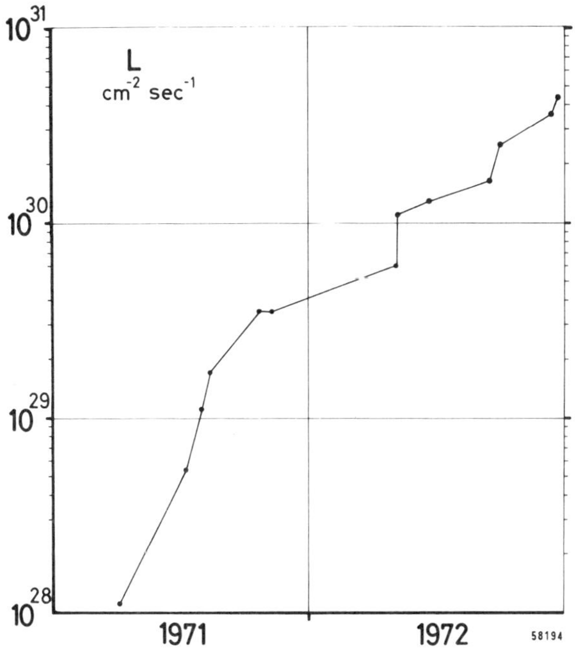

FIG. 11. The "Dow-Jones" curve of the ISR. Luminosity versus time.

3. LOW β SECTION

Recall the expressions for the vertical and horizontal betatron amplitude (Section I.3)

$$y(s) = a\sqrt{[\beta(s)]}\cos(Q\phi(s) + \delta)$$

Now since the luminosity

$$L \sim \frac{1}{h_{\text{eff}}} \sim \frac{1}{\sqrt{\beta_v}} \tag{II.17}$$

it is of interest to consider magnet schemes in which β_v can be made as small as possible in the intersections.

The envelope of the betatron oscillations around 1/8 of the ISR circumference is sketched in Fig. 12. The oscillating character of the curve is the result of focussing forces of the magnets, of the straight sections, etc. The envelope indicates the extent to which the beam is vertically compressed and expanded by the magnet lattice. (There is a corresponding curve for horizontal oscillations).

One scheme[10] to reduce β_v in an intersection is shown in Fig. 13. The normal magnets are indicated by F (focussing) and D (defocussing). By placing 6 more magnets ($Q_1 - Q_6$) in each ring the envelope of Fig. 12, dotted line, is obtained. A gain of a factor five is obtained in luminosity. The price is a narrower intersection (bad for experiments at small angle, no problem for large angle experiments) and the fact that 5 of each 6 magnets must be superconducting to reach this gain.

Design studies are also underway of an intersection in which both the betatron amplitude is reduced and the crossing angle α is made equal to zero.

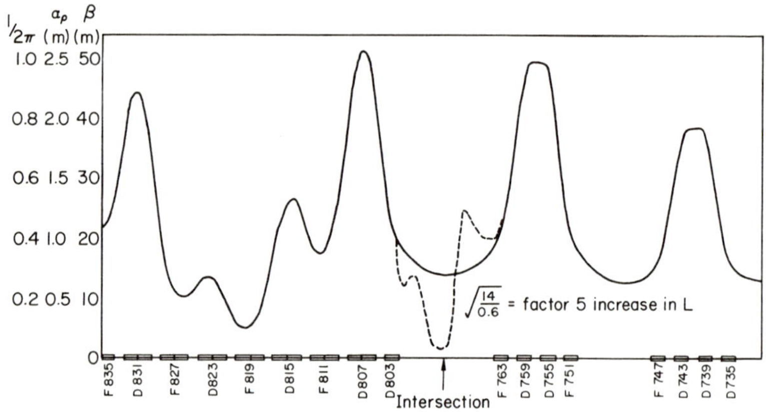

FIG. 12. Vertical betatron oscillation function versus azimuth, without and with low β section. The pattern repeats 8 times around the ISR circumference.

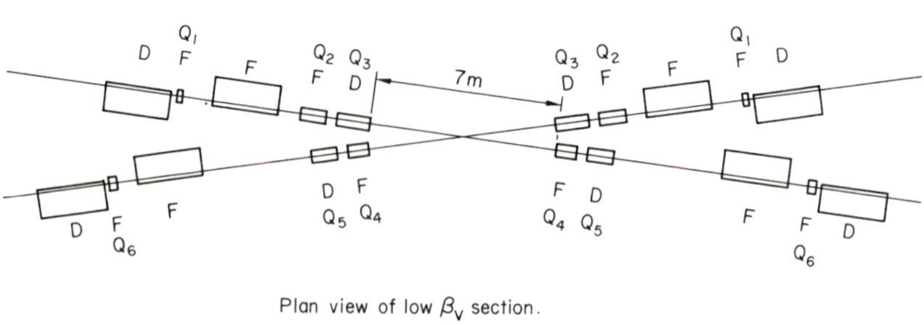

FIG. 13. Possible arrangement of magnets in a low β section. The extra magnets are $Q_1 \rightarrow Q_6$.

Since the luminosity is given by

$$L_\alpha = \frac{I_1 I_2}{e^2 c h_{\text{eff}} \operatorname{tg} \alpha/2} \underset{\alpha \to 0}{\longrightarrow} L_{\alpha=0} = \frac{I_1 I_2}{e^2 c h_{\text{eff}} w/l} \tag{II.18}$$

(w, l = width, length of intersection) one begins to gain as soon as $w/l < \operatorname{tg} \alpha/2 = 0{\cdot}13$. The luminosity then no longer depends on effective height only, but on width and length as well.

4. STOCHASTIC DAMPING OF BETATRON OSCILLATIONS

An entirely different way of reducing the vertical betatron amplitude and thereby gaining in luminosity is based on the detection of the statistical fluctuations in the average position of the beam. Although the method (suggested by S. v. d. Meer) turns out not to be practical with present equipment, it represents a very interesting application of statistical mechanics to particle beams. We follow Ref. 11 step by step.

Liouville's theorem implies that the density in phase space, i.e. betatron oscillations in the case at hand, cannot be influenced by magnetic fields that deflect the particles. This statement is true for an infinite number of particles and also for a finite number of particles if their positions in phase space were not known. It is manifestly not true if each particle could be located separately, since then a correction could be applied to its trajectory, resulting in a modified phase space density. Now if a beam makes coherent betatron oscillations (with all particles at a given section going up and down at the same time) it behaves to some extent as if it were one particle, and hence a correction can be applied by means of a pick-up + deflection system. But even if there are no such "explicitly" coherent betatron oscillations, for a finite number of particles some coherence would make itself felt through the statistical fluctuations in the average beam position. This is like the drunken sailor holding himself to a lamp post: due to random motion his mean *square* distance from the lamp post increases proportional to the number of steps, or time (up to some limit, e.g. neighbouring walls). Note that if his motion were completely coherent it would be the mean distance that increases proportionally with time. Hence we have some measure of the amount of coherence in his irregular motion: it is the amount of head-way he makes compared to how much he would have made had he been sober. Now after his mean square distance has increased for some time his arm begins to feel the strain and pulls him back to a point near the lamp post. If this pull has not sobered him up the random motions will continue and the process repeats itself. As a result of the repeated pulling of his arm the time average of the mean square distance is thus reduced.

Consider a damping system D which will react provided no less than n

particles traverse it. The number n is more or less determined by the rise time of the electronics in D. The particles have oscillation amplitudes $A_1 \ldots A_n$ randomly taken from a distribution function $F(A)$ with $\int F(A)\,dA = 1$. The mean square amplitude is

$$\overline{A^2} = \frac{\Sigma A_i^2}{n} \tag{II.19}$$

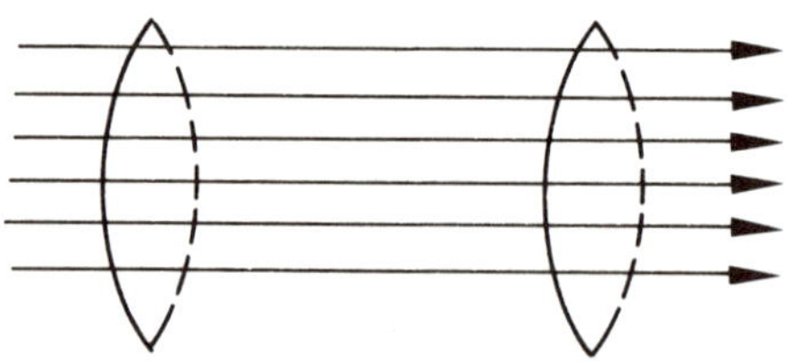

Damping system D

Each particle occupies a point in phase space characterized by

$$x_i = A_i \cos \psi_i, \qquad x_i' = A_i \sin \psi_i \tag{II.20}$$

The centre of gravity of the particles before entering D then has the coordinates

$$\bar{x} = \frac{1}{n} \sum A_i \cos \psi_i, \qquad \bar{x}' = \frac{1}{n} \sum A_i \sin \psi_i \tag{II.21}$$

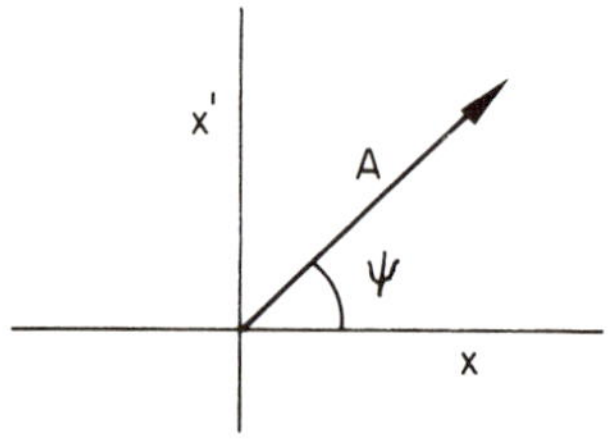

Upon entering D all particles get the same kick, such that the centre of gravity is returned to (0, 0). The new position of particle i is then

$$x_{Ni} = x_i - \bar{x}, \qquad x_{Ni}' = x_i' - \bar{x}'; \tag{II.22}$$

the square of its amplitude

$$A_{Ni}^2 = x_{Ni}^2 + x_{Ni}'^2, \tag{II.23}$$

and the mean square amplitude

$$\overline{A_N^2} = \frac{1}{n}\sum A_{Ni}^2 \tag{II.24}$$

$$= \frac{1}{n}\left[\sum A_i^2 - \frac{1}{n}\left\{(\sum (A_i \cos \psi_i))^2 + (\sum (A_i \sin \psi_i))^2\right\}\right]$$

The action of D thus reduces the mean square amplitude by a fraction

$$= \frac{\sqrt{\overline{A^2}} - \sqrt{\overline{A_N^2}}}{\sqrt{\overline{A^2}}} \approx \frac{\overline{A^2} - \overline{A_N^2}}{2\overline{A^2}} = \frac{\{\sum (A_i \cos \psi_i)\}^2 + \{\sum (A_i \sin \psi_i)\}^2}{2n \sum A^2} \tag{II.25}$$

for $\mathscr{E} \ll 1$. $\mathscr{E}$ is the reduction of a *given* root mean square amplitude, resulting from a *given* distribution of amplitudes A_i. The average value of this reduction, $\overline{\mathscr{E}}$, is found by weighing the amplitudes by their distribution functions and integrating.

$$\overline{\mathscr{E}} = \iint_0^\infty \cdots \iint_0^{2\pi} \cdots \frac{\{\Sigma (A_i \cos \psi_i)\}^2 + \{\Sigma (A_i \sin \psi_i)\}^2}{2n \Sigma A_i^2} \cdot F(A_1) F(A_2) \ldots$$

$$F(A_n) \frac{1}{(2\pi)^n} \mathrm{d}\psi_1 \ldots \mathrm{d}\psi_n \, \mathrm{d}A_1 \ldots \mathrm{d}A_n \tag{II.26}$$

Now since

$$\iint_0^{2\pi} \cdots [\{\sum (A_i \cos \psi_i)\}^2 + \{\sum (A_i \sin \psi_i)\}^2] \, \mathrm{d}\psi_1 \ldots \mathrm{d}\psi_n = (2\pi)^n \sum A^2$$

we have

$$\overline{\mathscr{E}} = \frac{1}{2n} \iint \ldots F(A_1) F(A_2) \ldots F(A_n) \, \mathrm{d}A_1 \, \mathrm{d}A_2 \ldots \mathrm{d}A_n = \frac{1}{2n} \tag{II.27}$$

Hence on average the reduction is independent of the distribution of amplitudes. The average reduction is larger, the smaller the number of particles to which the system can react, i.e. the closer the single particle limit can be approached in the damping system D. After $2n$ passes through the system the r.m.s. amplitude would be reduced by a factor e, provided the sample randomizes again after each passage (i.e. provided the sailor does not sober up as a result of the pulls on his arm). The latter condition is easily met in the ISR since a package of n particles, with a momentum spread of 2%, traversing D changes its population before traversing it again, as a result of the 25 cm longer circumference for the maximum compared with the minimum momentum particles.

Why can this method, although feasible in concept, not be realized in practice? The answer is that the number of particles "seen" within the rise time of any present day electronics is necessarily too large. Suppose the rise time of D is τ and the time for one revolution is T_R (= 3 μsec). Then

$$n = N\frac{\tau}{T_R} \tag{II.28}$$

where N is the total number of particles in the ring. From (II.27) and (II.28) we then have for the damping time constant T_d

$$T_d = 2nT_R = 2n\frac{N\tau}{n} = 2N\tau$$

for a rise time of 0·2 nsec and $N = 4 \times 10^{14}$, $T_d \approx 40$ hours, which is too long for damping practical stacks. Obviously, a short-armed sailor will on average stray less far from the lamp post than his long-armed friend in the same condition.

5. STORAGE OF ANTIPROTONS?

The 400 GeV CERN super proton synchrotron (SPS), presently under construction, will be equipped with a station for extracting external beams which is located at only ~200 meter from the storage rings (see Fig. 14). This raises hopes that one might be able to use the external SPS proton beam to produce a relatively large number of antiprotons (compared to what could be produced at the 25 GeV PS) from an external target and guide them into one of the two storage rings, the other one being filled with protons from the 25 GeV PS as usual. The interest of having high energy (say 11–25 GeV) antiprotons colliding with protons lies in the possibility of measuring the total cross-section $\bar{p}p$ in a range of c.m. energies where the pp cross-section has been shown recently to rise by ~10%. A measurement of $\sigma(\bar{p}p)$ in the same energy range would put strong constraints on current "asymptotic freedom". Figure 15 illustrates the present status of the data on $\sigma(pp)$ and $\sigma(\bar{p}p)$.

The scheme of filling one ring with antiprotons from the SPS has been considered by Hübner, Ref. 12. We shall repeat here the calculation of the expected rate of collisions, with the exception that experimentally measured[13] rates at ISR energies are used instead of models to estimate the expected $\bar{p}$ yields.

Notation:

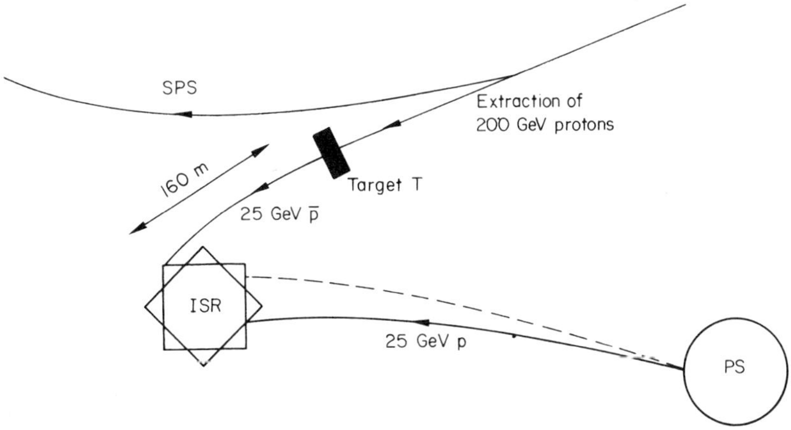

λ_p = line density (No. of particles/cm path length) of the SPS external proton beam
λ_p = line density of the $\bar{p}$ beam produced in target T
n = number of nucleons in T/cm^3
x = thickness of T
$\dfrac{d^2\sigma}{dp\, d\Omega}$ = diff. cross-section for the production of 25 GeV $\bar{p}$ by protons; it is assumed that for reasons of shielding, bending power, etc., the highest *extracted* proton energy is 200 GeV.

The line density of antiprotons produced is then:

$$\lambda_{\bar{p}} = \lambda_p n x \frac{d^2\sigma}{dp\, d\Omega} \Delta p\, \Delta\Omega \qquad \text{(II.29)}$$

After transport from T to ISR we have

$$\begin{aligned} N_{\bar{p}} = 2\pi R \lambda_{\bar{p}} &= \lambda_p 2\pi R n x \frac{d^2\sigma}{dp\, d\Omega} \Delta p\, \Delta\Omega\, \eta_t \cdot \eta_T \cdot \eta_s \\ &= \underset{\text{SPS}}{\underset{\uparrow}{\lambda_p}} \underset{\substack{\text{Target +}\\ \text{transfer}}}{\underset{\uparrow}{\left(\eta_T \frac{d^2N}{dp\, d\Omega} \Delta\Omega\right)}} \underset{\substack{\text{Injection + RF}\\ \text{ISR}}}{\underset{\uparrow}{\left(2\pi R \frac{\Delta p}{p} p\, \eta_t \cdot \eta_s\right)}} \end{aligned} \qquad \text{(II.30)}$$

Here
R = the radius of the ISR; $2\pi R = 10^5$ cm
η_T = the efficiency of the target in producing 25 GeV $\bar{p}$ from 200 GeV p. η_T is determined by the shape of the target, Coulomb scattering, etc. Typically, for a $\phi = 1$ mm wire 20–30 cm long $\eta_T \approx e^{-1}$.

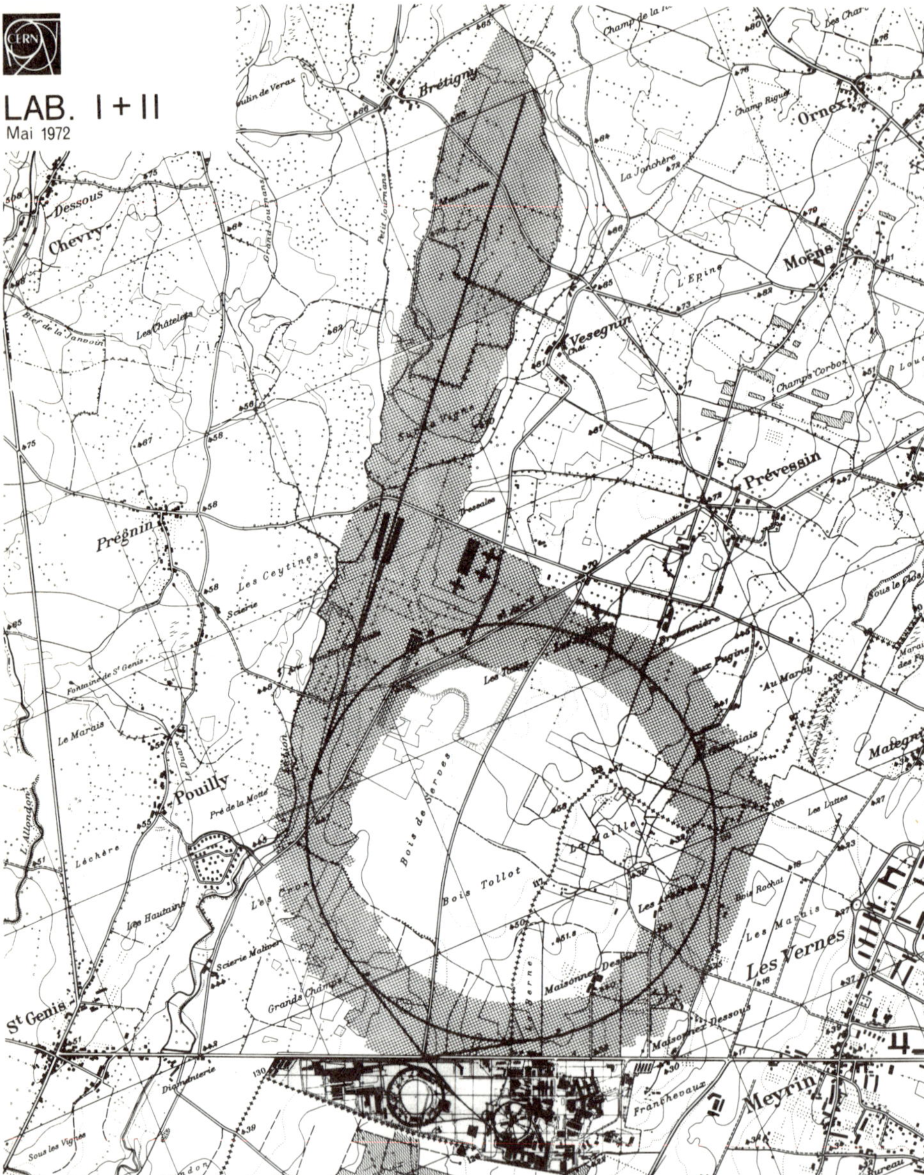

FIG. 14. Layout of the CERN SPS, ISR and PS. The extracted beam emerging from the SPS towards the left passes the ISR at $\lesssim$200 meter.

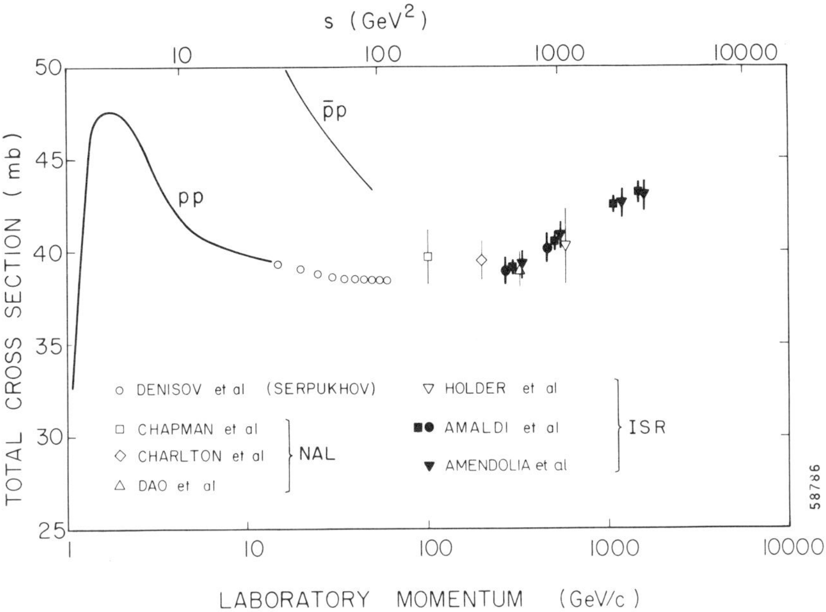

Fig. 15. Present status of total cross-section data on pp and p̄p.

η_t = efficiency for injection into the ISR, determined by alignment, aberrations in the focussing lenses, the ISR inflector magnet, etc. Experience with ISR beam transfer suggests $\eta_t \approx 0{\cdot}35$.

η_s = stacking efficiency: the efficiency for accelerating the antiprotons after arrival in the ISR in order to make a stack (see Section I.4). Estimate $\eta_s \approx 0{\cdot}5$.

$\dfrac{\Delta p}{p} = 2\%$ the maximum momentum bite the ISR can hold.

$\Delta\Omega$ = solid angle seen by the $\bar{p}$ transport system from the target. For a "matched" target/beam transport system this number is typically $\approx \pi \times 10^{-4}$ sr.

In order to calculate the line density λ_p the bunch structure of both the PS, which serves as an injector to the SPS, and the SPS must be taken into account. The PS has 20 bunches/pulse. The repetition rate is ~1 pulse/2 sec and each bunch is ~30 nsec long. With 10^{13} proton/pulse in the PS we have $10^{13}/20 = 5 \times 10^{11}$ protons/bunch. The RF of the SPS is arranged such that 1 SPS bunch is 5 nsec at the beginning of the acceleration cycle and has shrunk to 1 nsec after acceleration. Hence

$$1 \text{ SPS bunch} = 5 \times 10^{11} \times \frac{5 \text{ nsec}}{30 \text{ nsec}} \times \frac{1}{2} = 4 \times 10^{10} \text{ proton/SPS bunch}$$

allowing for an efficiency of $\frac{1}{2}$ in the transfer PS → SPS. After acceleration, when 1 SPS bunch = 1 nsec, we have

$$\frac{4 \times 10^{10}}{10^{-9}} = 4 \times 10^{19} \text{ protons/sec}$$

and

$$\lambda_p = \frac{4 \times 10^{19}}{3 \times 10^{10}} = 1{\cdot}3 \times 10^9 \text{ protons/cm} \qquad \text{(II.19)}$$

in the beam line between SPS and the target T.

In order to estimate the yield of antiprotons produced by 200 GeV SPS protons, we use information recently obtained in single particle production experiments at the ISR. At $s = 2820\ \text{GeV}^2$, $p_T = 0{\cdot}17$, $x = 2p_L^*/\sqrt{s} = 0{\cdot}12$, the invariant diff. cross-section has been measured to be

$$\frac{E}{p^2}\frac{d^2\sigma}{dp\, d\Omega} = 0{\cdot}6\ \text{mb/GeV}^2$$

To translate this into the required yield we assume scaling down to $s =$ 378 GeV^2, equivalent to 200 GeV p_{lab}. This is a rough approximation since there is no *exact* scaling for $\bar{p}$ in the ISR energy range. Furthermore we take the p_T dependence of ISR $\bar{p}$ data to extrapolate from $p_T = 0{\cdot}17$ to $p_T = 0\ \text{GeV}/c$:

$$\frac{E}{p^2}\frac{d^2\sigma}{dp\, d\Omega} = A \exp(-Bp_T^2) \text{ with } B = 2{\cdot}8$$

$$\rightarrow \left.\frac{E}{p^2}\frac{d^2\sigma}{dp\, d\Omega}\right|_{p_T=0} = 0{\cdot}65\ \text{mb/GeV}^2$$

Noting that:

$$p_{\text{lab}} = 200\ \text{GeV} \rightarrow \gamma_{\text{cm}} = 10{\cdot}3\ \beta_{\text{cm}} = 0{\cdot}99$$

$$x = 0{\cdot}12 \rightarrow p_{\bar{p}}^* = 1{\cdot}2\ \text{GeV}/c \rightarrow p_{\bar{p}}^{\text{lab}} = 28\ \text{GeV}/c$$

close enough to the 25 GeV/c considered above, we then evaluate the invariant cross-section at $s = 378\ \text{GeV}^2$ and $p_{\bar{p}}^{\text{lab}} = 28\ \text{GeV}/c$, $p_T = 0$ in the lab system:

$$\left(\frac{d^2\sigma}{dp\, d\Omega}\right)_{\text{lab}} = 0{\cdot}65 \times \frac{p^2_{\bar{p}\,\text{lab}}}{E_{\bar{p}\,\text{lab}}} = 18\ \text{mb/GeV}/c/\text{sr}.$$

Hence the yield of antiprotons is

$$\frac{d^2N}{dp\, d\Omega} = \frac{1}{\sigma_{\text{tot}}}\frac{d^2\sigma}{dp\, d\Omega} = 0{\cdot}45\ \bar{p}/\text{GeV}/c/\text{sr/interacting proton.} \qquad \text{(II.20)}$$

Substitution into eqn (II.30) then gives for the number of antiprotons storable in the ISR:

$$N_{\bar{p}} = 7 \times 10^8 = 35\mu\text{A} \tag{II.21}$$

resulting in a luminosity

$$L = \frac{I_1 I_2}{e^2 c \operatorname{tg}(\alpha/2)\, h_{\text{eff}}} = 0{\cdot}5 \times 10^{25}/\text{cm}^2/\text{sec} \tag{II.22}$$

assuming a 20 A proton beam in the other ring. This is six orders of magnitude lower than the maximum luminosity for pp collision. The number of $\bar{p}p$ collisions is

$$N_{\text{coll}\,\bar{p}p} = 0{\cdot}5 \times 10^{25} \times 40 \times 10^{-27} = 0{\cdot}2 \text{ events/sec.} \tag{II.23}$$

It is estimated in Ref. 12 that the time needed to store the antiprotons into the ISR is of the order of three hours.

It must be stressed that since this is a lecture, not a design study, the above calculation is meant to illustrate *how one can go about* getting the right answer, not to present a verdict as to whether storing antiprotons is feasible or not. The only definite conclusion is that the rate is very much on the edge of what one can do experiments with, even if the combined efficiency factors could be increased by an order of magnitude, as is perhaps possible in rigorous optimization of the parameters (primary energy, target dimensions, aperture of the ISR inflector, etc.). On the other hand the rates obtained from estimates such as the one above do not *a priori* exclude the possibility of $\bar{p}p$ colliding beam experiments and hence make it of great interest to pursue this tantalizing problem in more detail.

III. One Particle Inclusive Distributions

This topic has been the subject of numerous reviews. The intention here is to write down the Mueller–Regge formulation of single particle production, to bring out some features of data taking and analysis, and to show recent results.

1. THREE PARTICLE OPTICAL THEOREM

The[14] two particle optical theorem states that, in order to compute the total cross-section for $a + b \rightarrow \ldots$, it is not necessary to evaluate

$$\sum (\text{all ampl. } a + b \rightarrow \ldots)$$

but, instead, it is sufficient to know the quantity

$$\text{Im (elastic ampl. } a + b \rightarrow a + b \text{ at } t = 0)$$

This is summarized in statements such as

$$\sigma_T = \frac{1}{s}\,\mathrm{Im}\,\langle i|\,T\,|i\rangle = \frac{1}{s}\,\mathrm{Im}\,F(s, t = 0) = \frac{1}{s}\,\mathrm{DISC}_s\,T_{ab}(s) \qquad \text{(III.1)}$$

in obvious notations, or pictorially as:

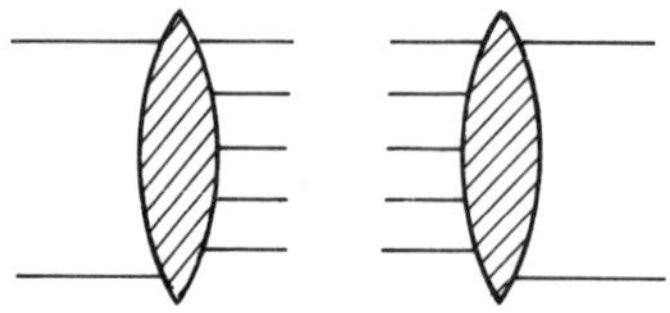

The three particle optical theorem states that, in order to compute the differential cross-section for the inclusive production of particle c for $a + b \rightarrow c + \ldots$ it is not necessary to evaluate

$$\Sigma\,(\text{all ampl. } a + b \rightarrow c + \ldots) = \Sigma\,(\text{all ampl. } a + b + \bar{c}(p, \theta) \rightarrow \ldots)$$

but, instead it is sufficient to know the quantity

$$\mathrm{Im}\,(\text{elastic ampl. } a + b + \bar{c} \rightarrow a + b + \bar{c} \text{ at } t = 0)$$

i.e.

$$E_c \frac{\mathrm{d}^3\sigma}{\mathrm{d}p_c^3} = \frac{1}{s}\,\mathrm{Im}\,\langle ab\bar{c}|\,T\,|ab\bar{c}\rangle = \frac{1}{s}\,\mathrm{Im}\,F(s_{ac}, s_{bc}, M^2)$$

$$\equiv \frac{1}{s}\,\mathrm{DISC}_{M^2}\,T_{ab\bar{c}}(s_{ac}, s_{bc}, M^2) \qquad \text{(III.2)}$$

pictorially:

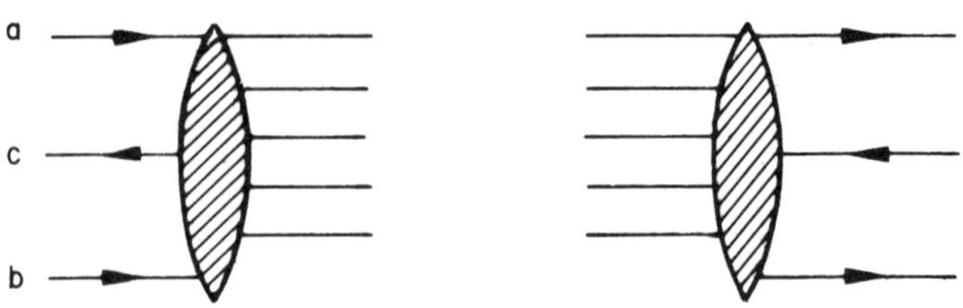

Writing $F(s, t)$ as a sum of Reggeon-exchange amplitudes we have

$$F(s,t) = \sum_i \gamma(t) \frac{1 \pm e^{-i\pi\alpha_i(t)}}{\sin \pi\alpha_i(t)} s^{\alpha_i(t)}$$

Residue Signature Energy
(Im $= \pm 1$) dependence

This expansion is valid at high energy, hence the rule: whenever the energy (subenergy) $\to \infty$, replace blob (sub-blob) by Reggeons.

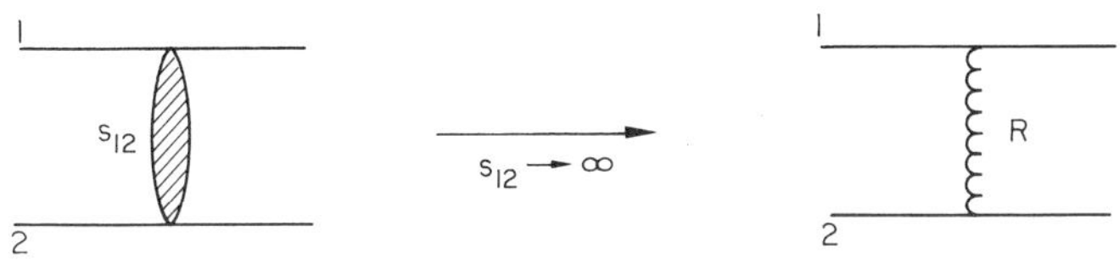

This then gives rise to two regions: the central region and the fragmentation region. In the *central region* (CR) both subenergies $s_{ac} \to \infty$ and $s_{bc} \to \infty$ as $s \to \infty$ and hence

$$F(s_{ac}, s_{bc}, M^2) \to \sum \gamma_{aR}\gamma_{bR'}\gamma_{RR'C}(s_{ac})^{\alpha_R}(s_{bc})^{\alpha_{R'}} \qquad \text{(III.3)}$$

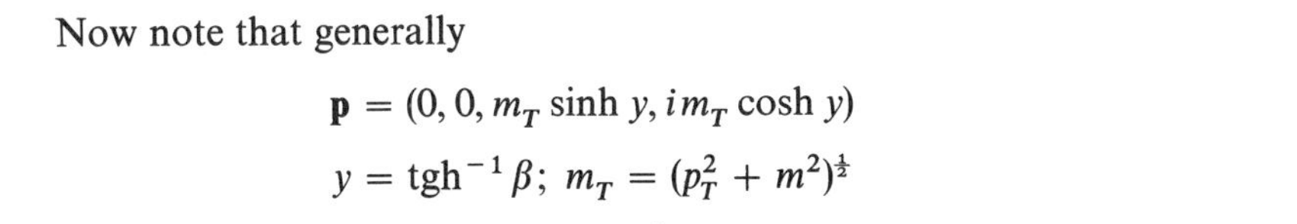

Now note that generally

$$\mathbf{p} = (0, 0, m_T \sinh y, i m_T \cosh y)$$

$$y = \mathrm{tgh}^{-1} \beta; \quad m_T = (p_T^2 + m^2)^{\frac{1}{2}}$$

$$\vec{p} = m_T \sinh y \approx \pm\tfrac{1}{2} m_T \exp(\pm y)$$

$$s_{ac} = (\mathbf{p}_a - \mathbf{p}_c)^2 \approx 4p_a p_c \approx m_a m_c \exp(y_a - y_c)$$

$$s_{bc} = (\mathbf{p}_b - \mathbf{p}_c)^2 \approx 4p_b p_c \approx m_b m_c \exp(y_c - y_b)$$

$$s = (\mathbf{p}_a + \mathbf{p}_b)^2 \approx 4p_a p_b \approx m_a m_b \exp(y_a - y_b)$$

Hence $s = s_{ac}s_{bc}/m_c^2$

Hence the first term in (III.3), with $R = R'$, gives for (III.2)

$$E_c \frac{d^3\sigma}{dp_c^3} = \frac{1}{s} \sum \gamma_{aR}\gamma_{bR}\gamma_{RRc}(s_{ac}^{\alpha_R})(s_{bc}^{\alpha_R}) = G_{RRc}(p_{T_c}^2)s^{\alpha_R - 1} \tag{III.4}$$

Hence for Pomeron exchange only, there would be neither s nor y-dependence in the invariant cross-section in the central region: the Mueller–Regge expansion results in Feynman scaling. The next term would be R = Reggeon, R' = Pomeron

$$E_c \frac{d^3\sigma}{dp_c^3} = \frac{1}{s}\left[\gamma_{aR}\gamma_{bP}s_{ac}^{\alpha_R}s_{bc}^{\alpha_P} + \gamma_{bR}\gamma_{aP}s_{ac}^{\alpha_P}s_{bc}^{\alpha_R}\right]G_{PRc}(p_{T_c}^2)$$

For $s_{ac} \approx s_{bc}$, $\alpha_P \approx 1$ and $a = b$ this is

$$E_c \frac{d^2\sigma}{dp_c^3} = \gamma_{\alpha R}\gamma_{\alpha P}\left(\frac{1}{s^{\frac{1}{2}}}\right)^{\alpha_P - \alpha_R} H_{PRc}(p_{Tc}^2, y_c) \tag{III.5}$$

Since $\alpha_R \approx \frac{1}{2}$ the approach to scaling goes approximately as $s^{\frac{1}{4}}$.

In the fragmentation region (FR), where particle c is considered to be a break up product of, say, particle b we have

$$s_{ac} \to \infty \qquad s_{bc} \not\to \infty \quad \text{as} \quad s \to \infty$$

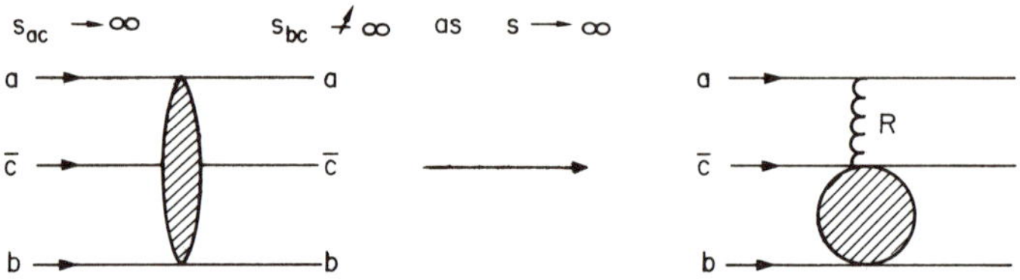

and hence only one amplitude is now Reggeized, the other exchange is still represented by a blob.

$$E_c \frac{d^3\sigma}{dp_c^3} = \frac{1}{s} \sum \gamma_{\alpha R}\, s_{ac}^{\alpha_R}\, \phi_{bcR}(y_b - y_c, p_{T_c}^2) \tag{III.6}$$

Regge Blob function
ampl.

Now $s = m_a m_b \exp(y_b - y_a)$

$$s_{ac} = m_a m_c \exp(y_c - y_a + y_b - y_b)$$

$$= \frac{s}{m_b} m_c \exp(y_c - y_b) \to s \exp(y_c - y_b) \tag{III.7}$$

and thus

$$E_c \frac{d^3\sigma}{dp_c^3} = \sum s^{\alpha_R - 1} G(y_b - y_c, p_{T_c}^2) \tag{III.8}$$

In particular if R = Pomeron,

$$E_c \frac{d^3\sigma}{dp_c^3} = G(y_b - y_c, p_{T_c}^2)$$

Hence the invariant cross-section in FR is a function of $y_b - y_c$ and, to leading order, independent of s: the Mueller–Regge expansion results in limiting fragmentation (Yang) and Feynman scaling in FR.

The next term is R = Reggeon, and from (III.8) we then have that the approach to scaling in FR goes as $s^{-\frac{1}{2}}$.

2. EXPERIMENTS AND RESULTS ON INCLUSIVE SINGLE PARTICLE PRODUCTION

There is no such thing as "typical apparatus" around the ISR; different particles (π^0, K^0 versus charged particles) and different angular or rapidity ranges require different techniques. One piece of equipment, with which some of the problems that arise in the data analysis can usefully be illustrated is a two-magnet spectrometer, placed on a rotating table, with which angles between 36° and 90° can be scanned.[15] It is shown in Fig. 16. There are scintillation hodoscopes $H_1H_2H_3$ forming the trigger and 6 spark chambers with magnetostrictive readout. The chambers are used to get straight line segments from the spark locations. A *track* is considered "good" if the appropriate segments meet in both magnets. An *event* is considered "good" if the track matches the hodoscope pattern. In this selection a number of real events get lost, since for example, some of the spark chambers did not fire when they should have. How is it that one does not loose normalization in this selection process?

In order to normalize data, i.e. turn numbers of tracks, characterized by a momentum $p \pm dp$ and angle of production $\theta \pm d\theta$ into mb/sr/GeV/c one requires a "monitor", a set of counters which intercepts a fixed fraction of the collisions products, which remains unchanged in geometry, momentum cut-off, etc., throughout the experiment, and whose counting rate has a known relation to the luminosity:

$$R_M = \text{mon. rate/sec} = KL \tag{III.9}$$

This relation is obtained in a luminosity measurement as discussed in Section II.

If during the time the spectrometer is taking data, we also count the events in the monitor ΣR_M we obtain, with K, the required relation between events and cross-section. Now after each trigger (e.g. an $H_1H_2H_3$ coincidence) there is a transfer of the chamber + counter information to magnetic tape. One of the scalers whose content is transferred is the monitor scaler. It usually contains some 10 to 100 counts per event. After the transfer the scalers are reset to zero and ready to start counting again until the next transfer. Hence the pattern is

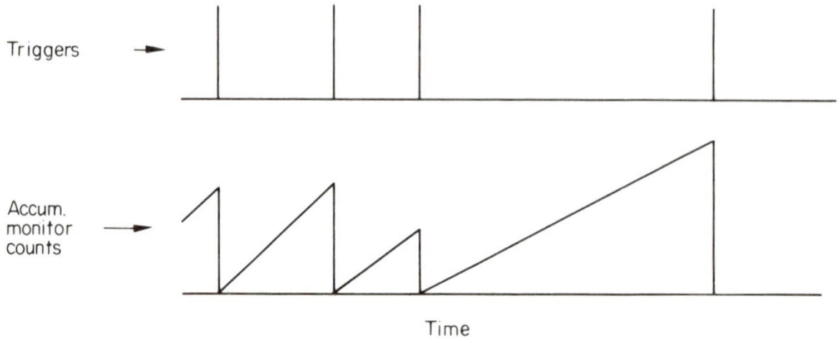

Now there are three kinds of "events"

(a) Triggers + reconstructable tracks. These events are binned by momentum and angle; the content of a bin is related to the cross-section by

$$\frac{d^2\sigma}{dp\, d\Omega}\Delta p\Delta \cos\theta = \frac{F(p, \theta)K}{\Sigma R_M}\frac{dN(p, \theta)}{\Delta p\Delta \cos\theta}\Delta p\Delta \cos\theta$$

where $F(p, \theta)$ in the acceptance (see below)

(b) Triggers + non-reconstructable tracks, e.g. a good event, but a failing spark chamber. In this case one rejects the event as well as the monitor counts accumulated since the previous event.

(c) False triggers, e.g. a reconstructed track does not project through one of the trigger counters. In that case the trigger does not correspond to an event, one rejects the event but retains the monitor count accumulated since the previous trigger, i.e. the transfer was accidental and should not have taken place.

A fraction of the events are not due to beam–beam but to beam–gas events. One way of correcting for this background is to run with one beam only and make a subtraction. If the background is low it is sufficient to examine the "diamond": the overlap region of the two beams. This distribution will have tails extending beyond the region of overlap, due to beam/gas events. Extrapolation of the tails into the diamond proper gives the correction. The beam/gas ratio can be as low as 4%.

Systematic errors can arise e.g. from misalignment of the spark chambers. This is most easily checked by reconstructing tracks with the fields off and

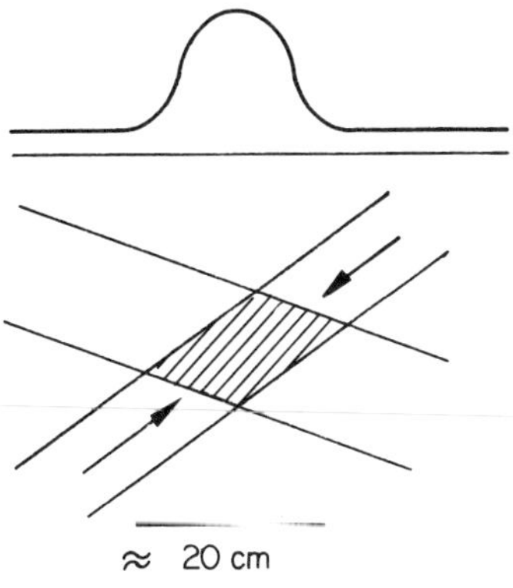

Diamond

requiring the tracks to be straight lines. A further check is provided by the momentum which is measured in each magnet separately. Consistency requires that

$$\delta p = \frac{p_1 - p_2}{p_1 + p_2}$$

has a standard deviation (s.d.) which agrees with calculations based on

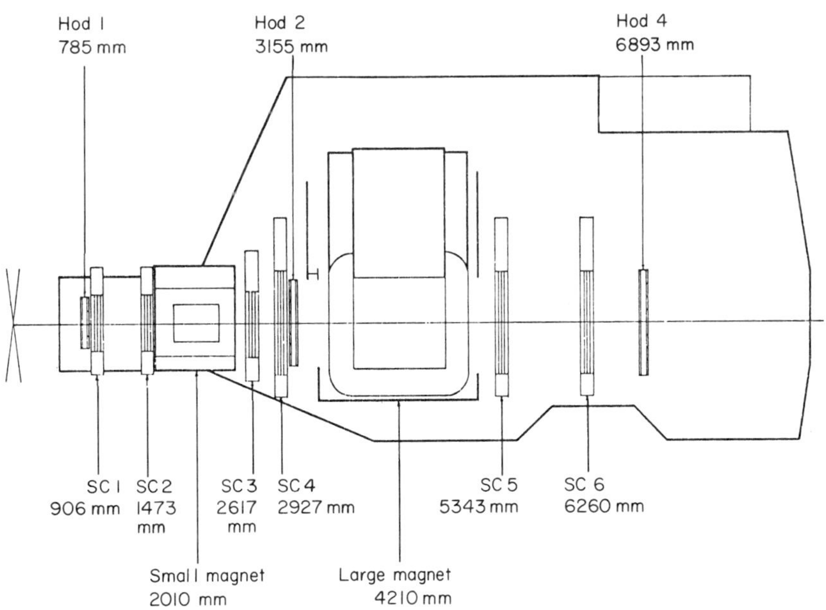

FIG. 16. Layout of a spectrometer for measuring single particle distributions at 90° w.r.t. the colliding beams.

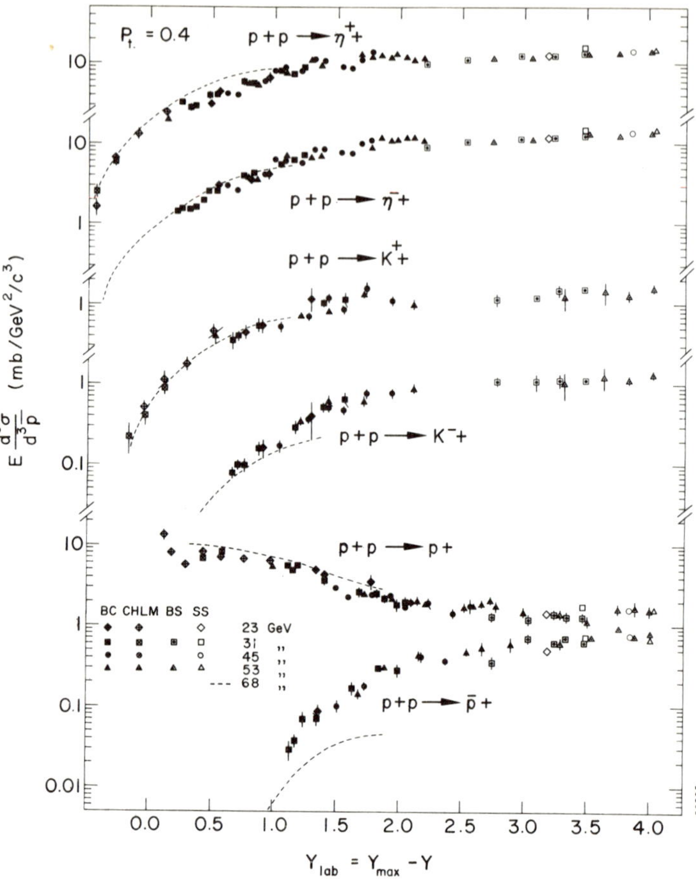

FIG. 17. Summary of single particle data at $p_T = 0{\cdot}4$ GeV. From Ref. 16.

measuring errors, multiple scattering etc. The momentum dependence of this standard deviation is sketched in the diagram. Finally the acceptance function $F(p, \theta)$ is calculated by generating a number of events distributed over the diamond and tracking them through the apparatus, taking account of multiple scattering, absorption, decay, etc.

Results of single particle inclusive experiments are shown in Figs 17–32.

(a) To an accuracy of approximately $\pm 15\%$ there is a flat central plateau

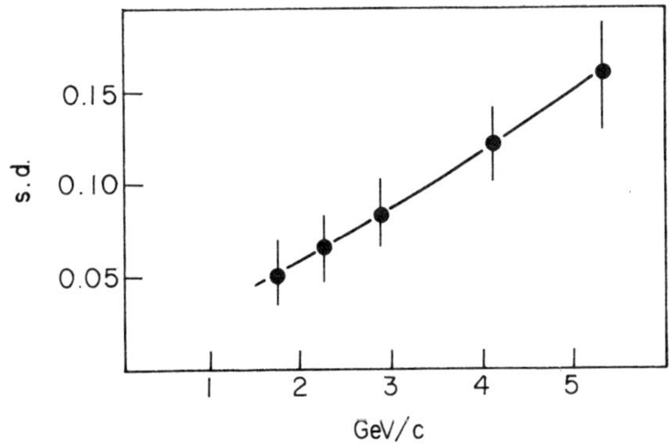

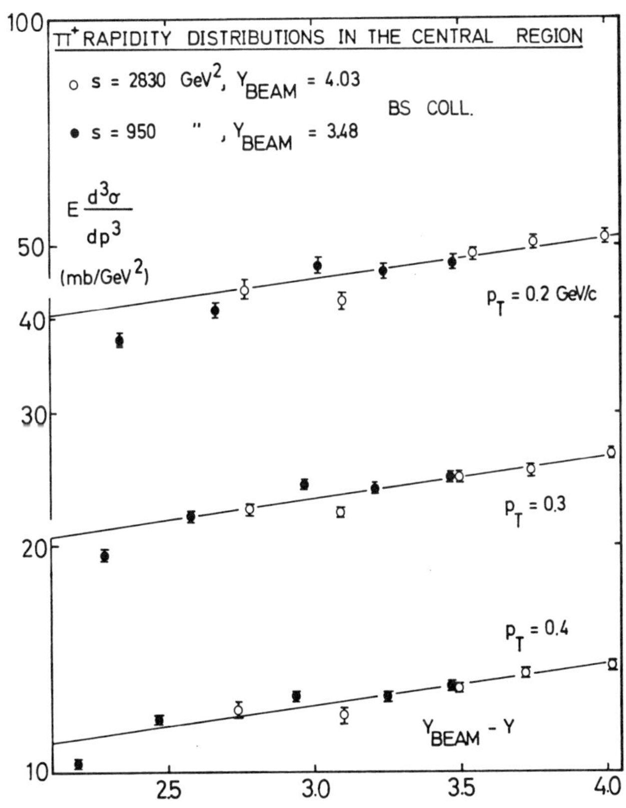

FIG. 18. Invariant diff. cross-sections for the production of π^+ versus $y_{\text{beam}} - y$ at $s = 2830$ GeV^2 and at $s = 950$ GeV^2. Data at $p_T = 0{\cdot}2, 0{\cdot}3, 0{\cdot}4$ GeV/c.

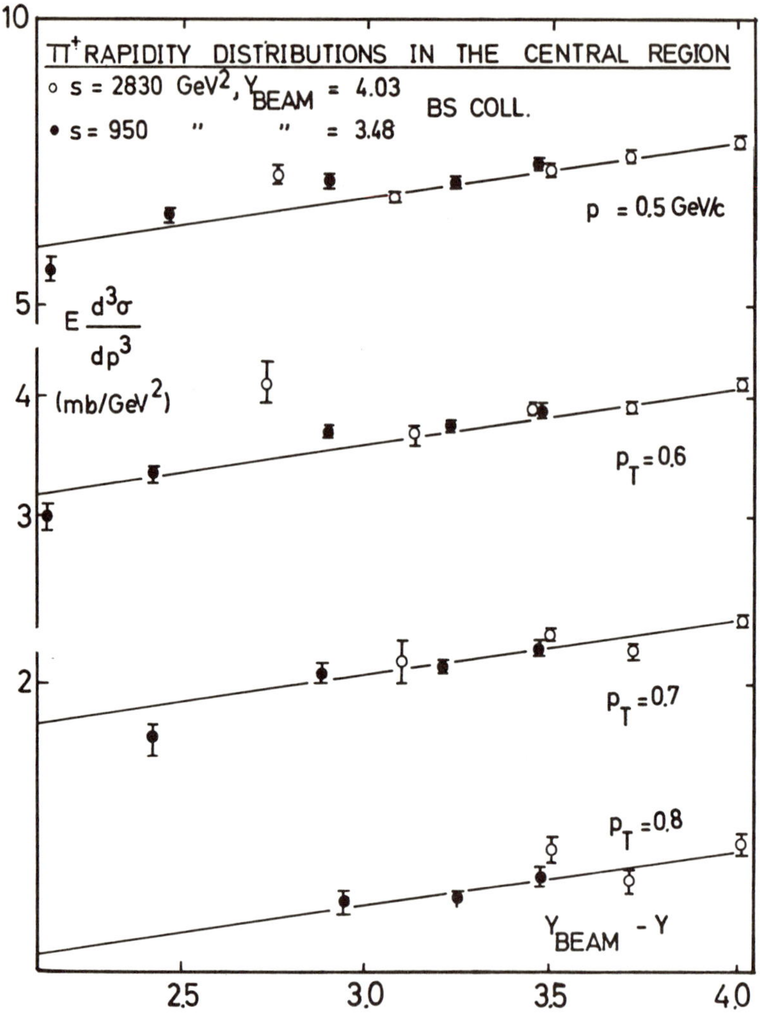

FIG. 19. Same as Fig. 3 for p_T = 0·5, 0·6, 0·7, 0·8 GeV/c.

in y, in accordance with Mueller–Regge Pomeron exchange. See Fig. 17, Ref. 16.

(b) Recent data[17] in which the accuracy is increased and on which more systematic checks (see above) have been made, have revealed a slight dependence on rapidity at fixed s, and a dependence on s at fixed y. Neither PP exchange, nor RR exchange alone would result in a

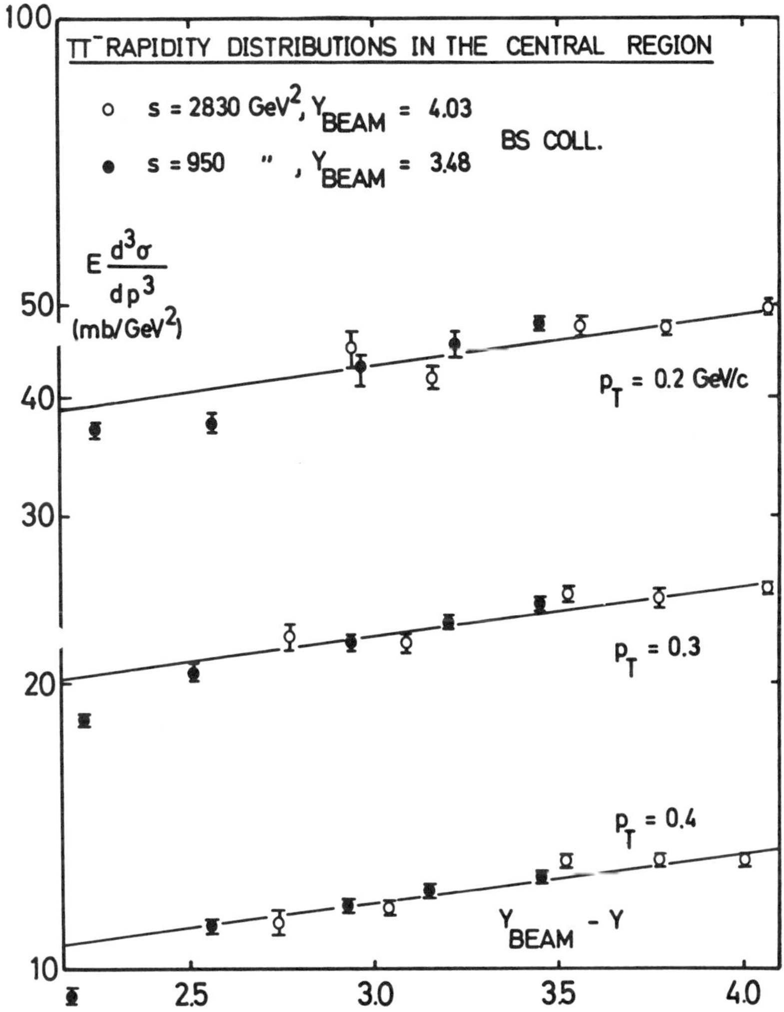

FIG. 20. Same as Fig. 3, for π^-.

y-dependence, see eqn (III.4). The data suggest, but by no means prove, a dependence on y and s such that

$$E_c \frac{d^3\sigma}{dp^3} = f(y, p_T, s) = f(y_B - y, p_T) \sim A(p_T) \exp(-D(y_B - y))$$

with $D \approx 0{\cdot}15 \pm 0{\cdot}03$. ($y_B$ = rapidity of the beam). Such a dependence is characteristic for *fragments* produced with Pomeron exchange, see eqn (III.8), and thus suggests that fragments penetrate all the way down to $y = 0$ on the rapidity plot. See Figs 18, 19, 20, 21.

(c) As to the approach to scaling, the data are consistent with various forms for the s-dependence. The form $s^{-\frac{1}{4}}$ is one possibility, but there are others, e.g. $s^{-\frac{1}{2}}$, see Fig. 22.

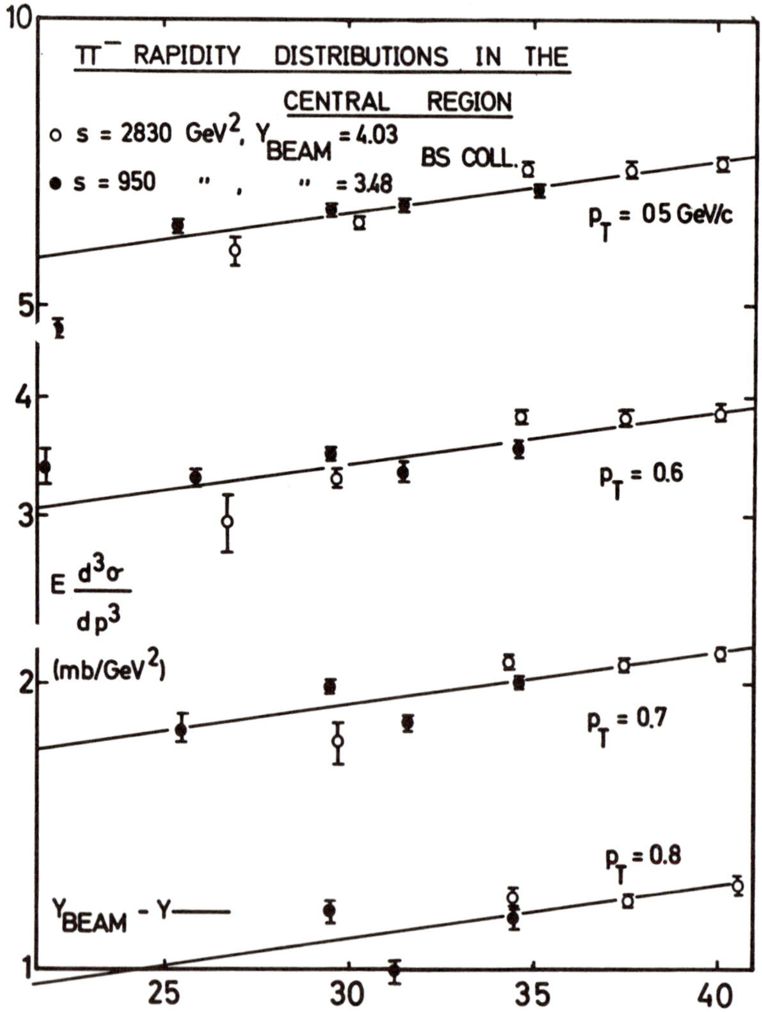

FIG. 21. Same as Fig. 4, for π^-.

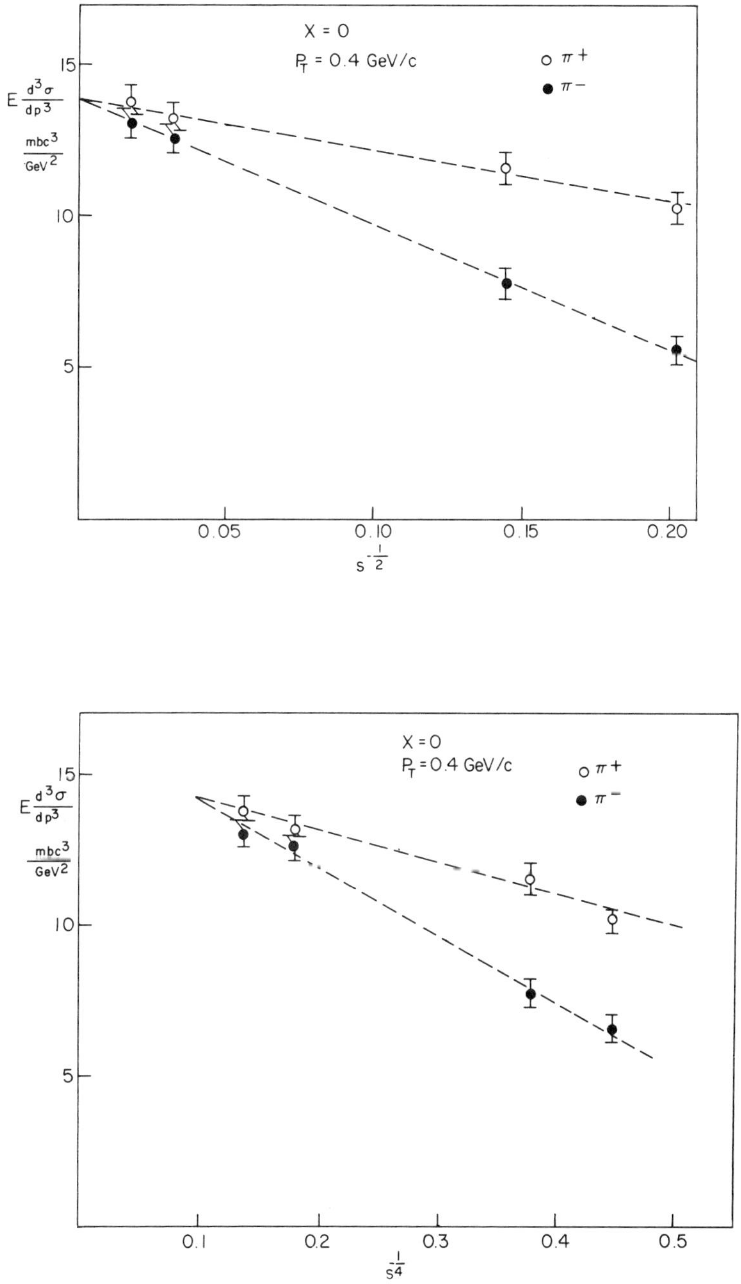

FIG. 22. Approach to scaling in the central region versus $s^{-\frac{1}{2}}$ (above) and $s^{-\frac{1}{4}}$ (below).

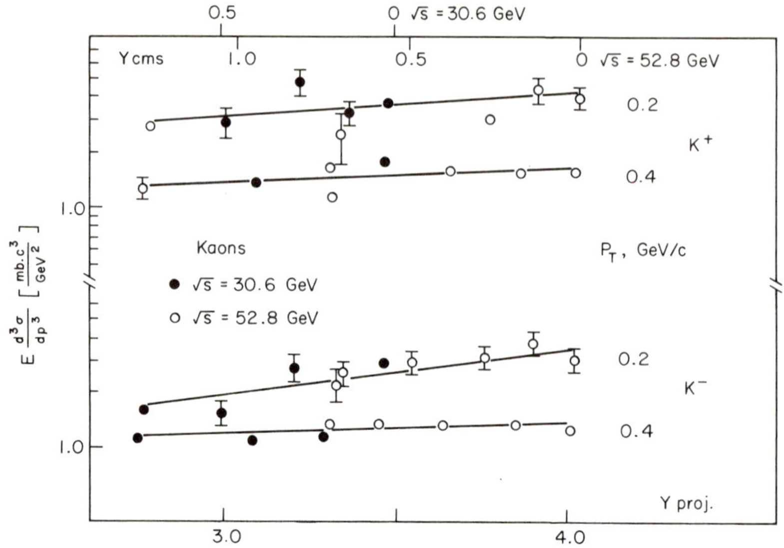

FIG. 23. Central region, K^+ and K^-.

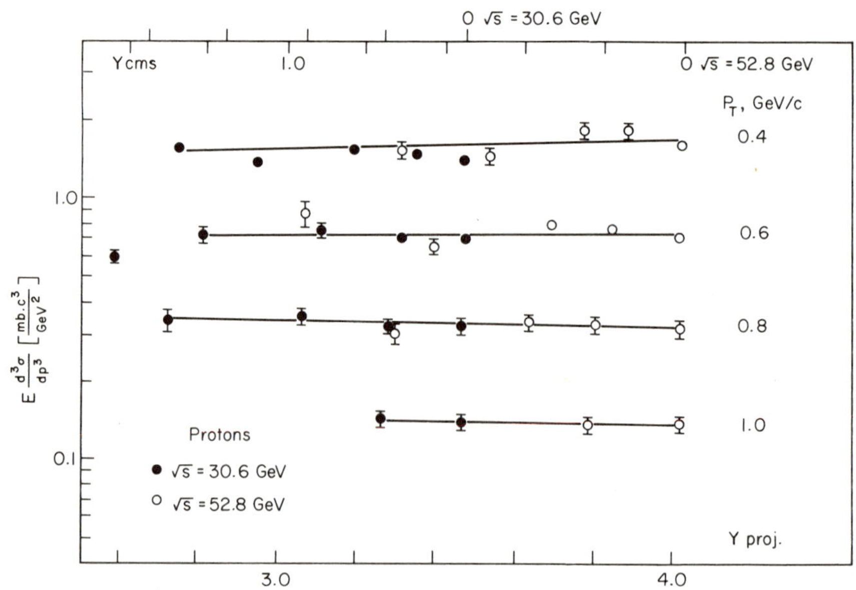

FIG. 24. Central region, protons.

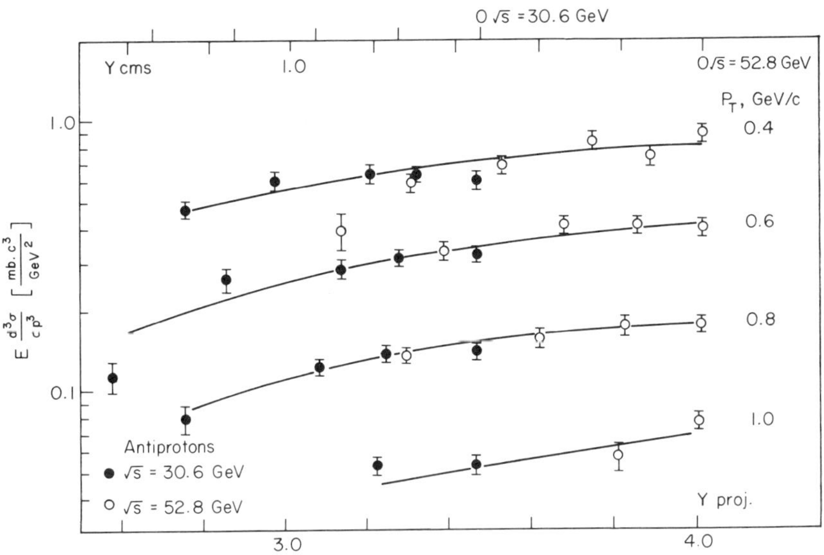

FIG. 25. Central region, antiprotons.

(d) Data on K^+ and p are consistent with no y-dependence, K^- is inconclusive and $\bar{p}$ data are definitely sloping. See Figs 23, 24, 25.

(e) In the fragmentation region, π^- data at fixed $y_B - y = 0{\cdot}36$ versus p_T show good evidence for scaling in the projectile frame, from $s = 47\ \text{GeV}^2$ upwards and out to $p_T \approx 1\ \text{GeV}/c$, provided the low s data are represented in the variable $p_L/p_{L\,\max}(0)$ or, better still, $p_L/p_{L\,\max}(p_T)$, i.e. the one that takes the phase space boundary into account[18]. See Fig. 26.

(f) In the fragmentation region, K^- and $\bar{p}$ at $y_B - y = 0{\cdot}36$ scale in the ISR energy range, but differ appreciably from data at low s, see Figs 27 and 28, Ref. 19.

(g) The approach to scaling in the fragmentation region is consistent with $s^{-\frac{1}{4}}$. This is shown in Fig. 29 (Ref. 19), where the K^- and $\bar{p}$ yields are expressed at ratios to the scaling (see Fig. 26) π^- yields. Other forms of s-dependence are likewise possible, however.

(h) For π^+ we have a behaviour similar to that of π^-. Figure 30 shows π^+ data at fixed $x = 0{\cdot}38$ versus p_T and Fig. 31 data at fixed $p_T = 0{\cdot}8$ versus x.[20] There is no discernable energy dependence as long as $p_T < 1\ \text{GeV}/c$.

(i) For protons, the approach to scaling at fixed (x, p_T) is illustrated in Fig. 32, showing NAL[20] data versus s, compared with a curve $A + B/\sqrt{s}$. More about protons later.

In summary one might say that the absence of energy dependence in inclusive single particle spectra continues to hold to an impressive degree, although the so called central region is perhaps less central (at ISR energies) than was thought originally.

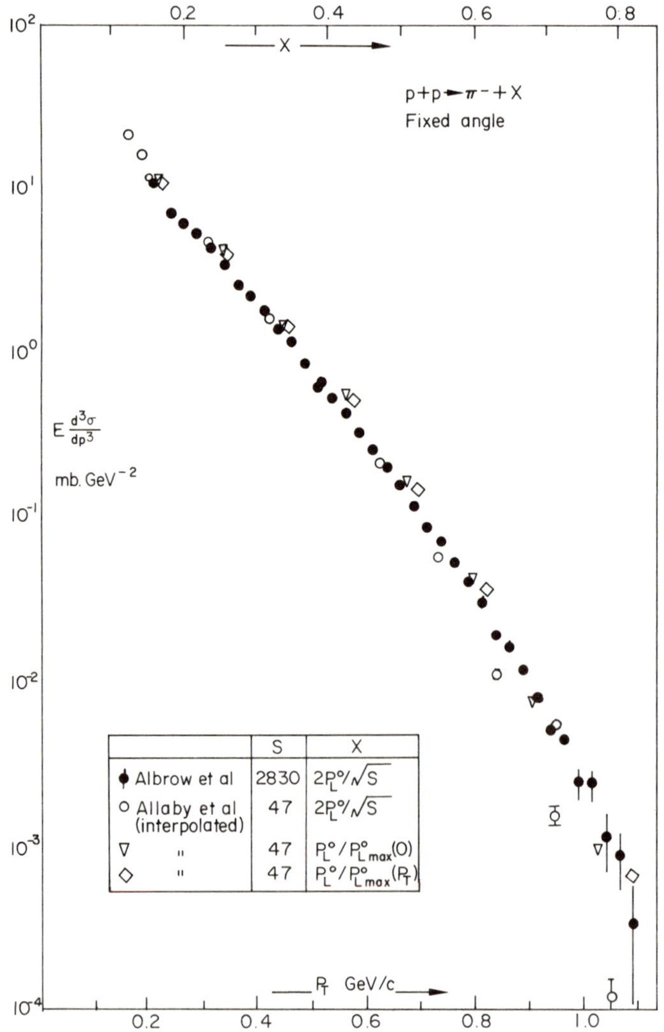

FIG. 26. Fragmentation region, data at $y_{\text{beam}} - y \approx 0{\cdot}36$, at $s = 47\ \text{GeV}^2$ and $s = 2830\ \text{GeV}^2$.

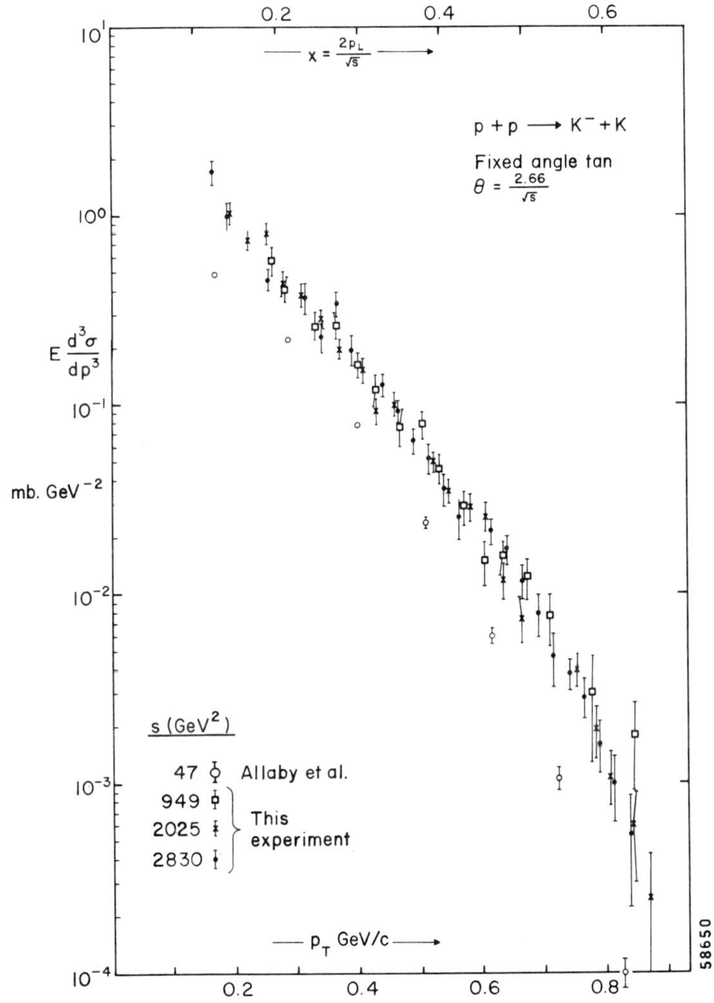

FIG. 27. Fragmentation region, data at $y_{beam} - y \approx 0{\cdot}36$ at s for inclusive K^- production.

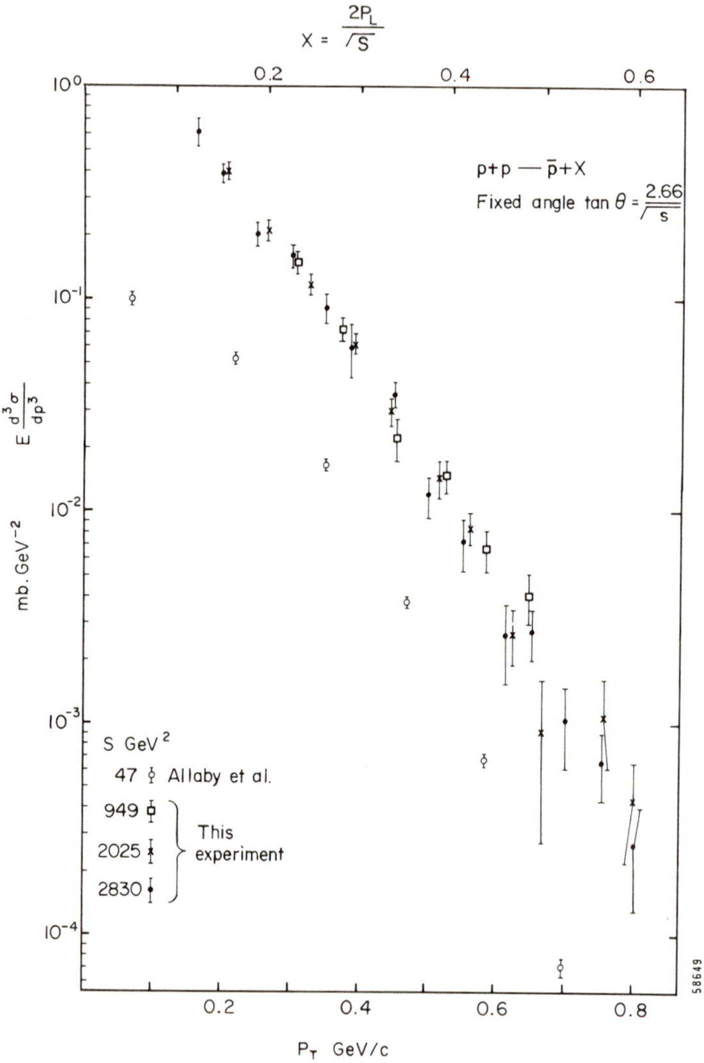

FIG. 28. Same as Fig. 12, for $\bar{p}$.

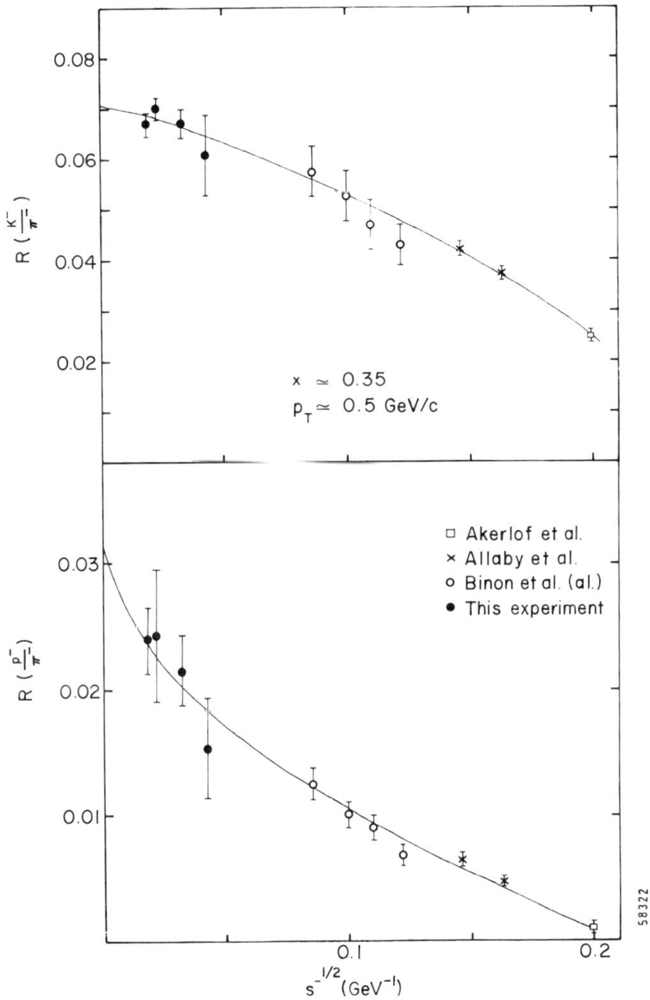

FIG. 29. s-dependence of K^-/π^- and $\bar{p}/\pi^-$ versus $s^{-\frac{1}{2}}$ at $p_T \approx 0{\cdot}5$ GeV/c showing the approach to scaling for K^- and $\bar{p}$ fragments.

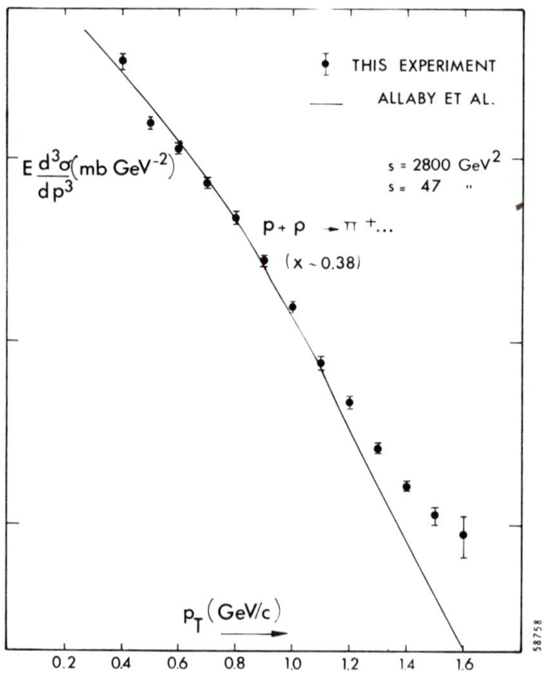

FIG. 30. p_T-dependence at $s = 47$ and $2800\ \mathrm{GeV}^2$ for π^+ fragments, showing how fragments scale as long as $p_T \lesssim 1\ \mathrm{GeV}/c$ (Ref. 6).

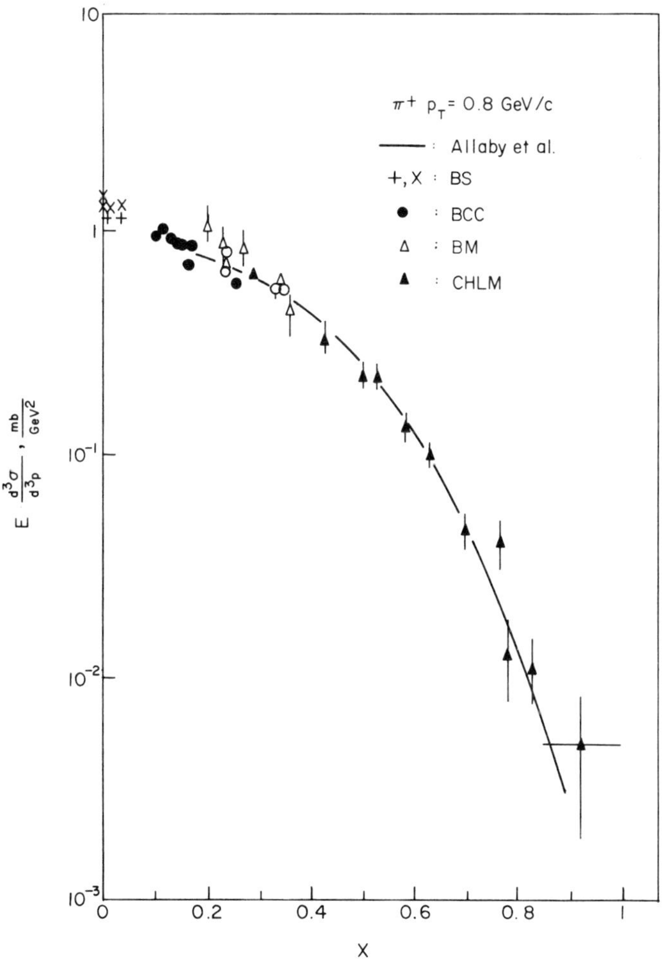

FIG. 31. x-dependence at $p_T = 0{\cdot}8$ GeV/c and various $550 < s < 2800$ GeV^2, showing how fragments scale at all measured values of x.

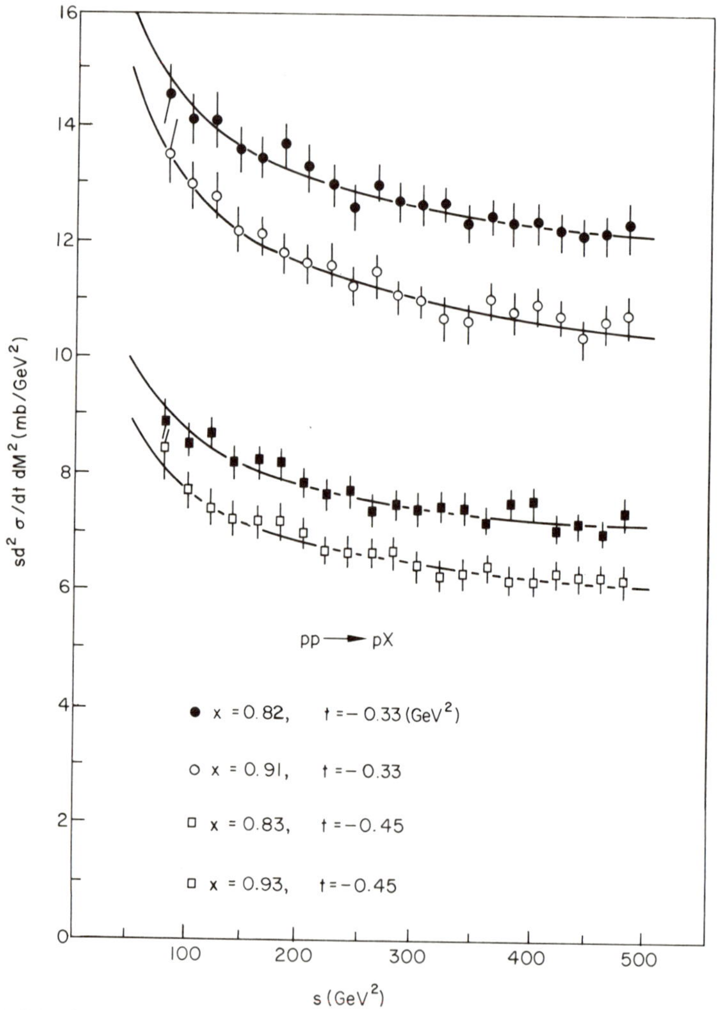

FIG. 32. NAL-data (Ref. 21) on protons versus s at fixed x and t showing how scaling is approached "from above" for proton fragments.

IV. Summary of Low Energy Missing Mass Data

1. HISTORY

Missing mass experiments in proton–proton collisions on stationary targets have been performed since 1960. In summarizing the results we follow here Ref. 20. A (hopefully complete) list of references follows below.

Brief Summary of Low Energy ($s < 50\,\text{GeV}^2$) *Missing Mass Experiments*

Chadwick	BNL	(1960), PRL **4**, 611	$\Delta(1236)$, $N(1520)$
Cocconi	CERN	(1961), PRL **7**, 450	$N(1688)$, $N(1470)$?
Chadwick	BNL	(1962), PR **128**, 1823	
Diddens	CERN	(1962), PRL **9**, 111	
Cocconi	CERN	(1964), PL **8**, 134	
Bellettini	CERN	(1965), PL **18**, 167	Confirms $N(1470) 21 < s < 50$ small $t < 0{\cdot}01$
Anderson	BNL	(1966), PRL **16**, 855	$N(2190)$ $13 < s < 58$ $t < 1$
Foley	BNL	(1967), PRL **19**, 397	$t < 0{\cdot}2$
Ankenbrandt	Berkeley	(1968), PR **170**, 1223	$8 < s < 15$
Allaby	CERN	(1968), PL **28B**, 229	$s = 38$ $t < 6$
Blair	NIMROD	(1969), N.C. **63A**, 529	$7{\cdot}3 < s < 17$ small t
Edelstein	BNL	(1972), PR **D5**, 1073	$13 < s < 58$
Allaby	CERN	(1972), *Nucl. Phys. B*	$s = 47$ $t < 6$

....

Albrow	ISR	(1973), *Nucl. Phys.* **B51**, 388	$s = 2030$
Albrow	ISR	(1973), *Nucl. Phys.* **B54**, 6	$s = 960$
Bromberg	NAL	(1973), Preprint	$s = 190$
Dao	NAL	(1973), PRL **30**, 34	$s = 560$
Sannes	NAL	(1973), PRL **30**, 766	$80 < s < 480$
Barish	NAL	(1973), APS April 1973	$s = 380$
Childress	NAL	(1973), PRL (submitted)	$s = 380$

Note the jump in s, made with the start-up of NAL and ISR (dotted line).

2. TYPICAL APPARATUS

Missing mass experiments require a precision measurement of momentum and angle of the produced proton in

$$p + p \rightarrow p + X$$

The apparatus of Allaby *et al.* with which the latest low s data have been obtained is described in Ref. 22. We shall list the salient features of this equipment below. The layout is shown in Fig. 33.

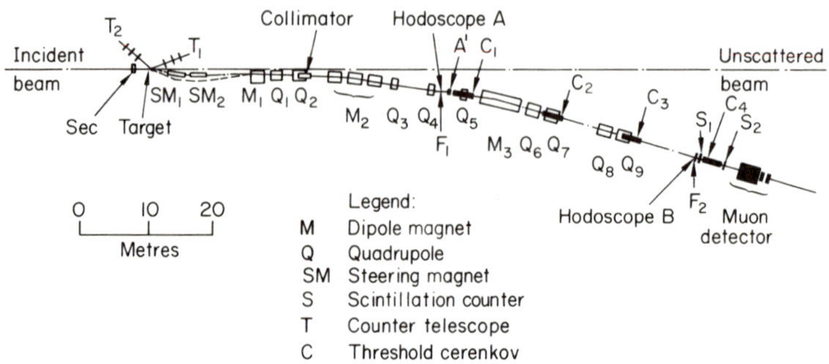

FIG. 33. Spectrometer for the measurement of $pp \rightarrow pX$ on a stationary target at energies <25 GeV.

Slow extracted beam; 350 msec spill; $10^{12} p$/burst.
Liq. H_2 target; spotsize $\sim$ 4 mm $\varnothing$, $\pm 0{\cdot}5$ mrad, 10–20 cm.
Spectrometer with axis fixed at $\theta = 37$ mrad.
Calibration of inc. flux by Na^{24} activity induced in Al-foil by beam.
Q_1, Q_2 image target to F1.
M_2 bends 120 mrad. At F1 dispersion = 1·6 cm for $\Delta p/p = 1{\cdot}0\%$.
Q_6, Q_7 make beam parallel for use of C.M_3 bends 120 mrad.
Q_3, Q_4, Q_5 project central plane of M_2 onto central plane of M_3.
F_2 is final focus, made by Q_8, Q_9.
Acceptance hor. $\pm 1{\cdot}5$ mrad, vert. $\pm 4{\cdot}5$ mrad. $\Delta p/p = \pm 1{\cdot}5\%$.
Displ. in target $\pm 1{\cdot}5$ cm.
Acc. defined by collimator of $\varnothing$ = 8 cm; $\Delta\Omega = 1{\cdot}3 \times 10^{-5}$ sr.
$SM_1 SM_2$, M_1 for $\theta \neq 37$ mrad.
Angular range 12 → 152 mrad. consequence $p(\theta - 37) = \text{const.}$
Magnet calibration to $\Delta p/p = \pm 0{\cdot}1\%$.
Overall length 86 m.

3. FEATURES OF THE RESULTS

Figure 34 shows the data obtained with the apparatus of Fig. 33: the diff. cross-section $d^2\sigma/dt\,dM^2$ versus M^2 at fixed lab angle (Ref. 22). Note that the horizontal scale starts above the location of the elastic peak. Since the angle for each curve is fixed, $|t|$, the 4-momentum transfer, changes over the

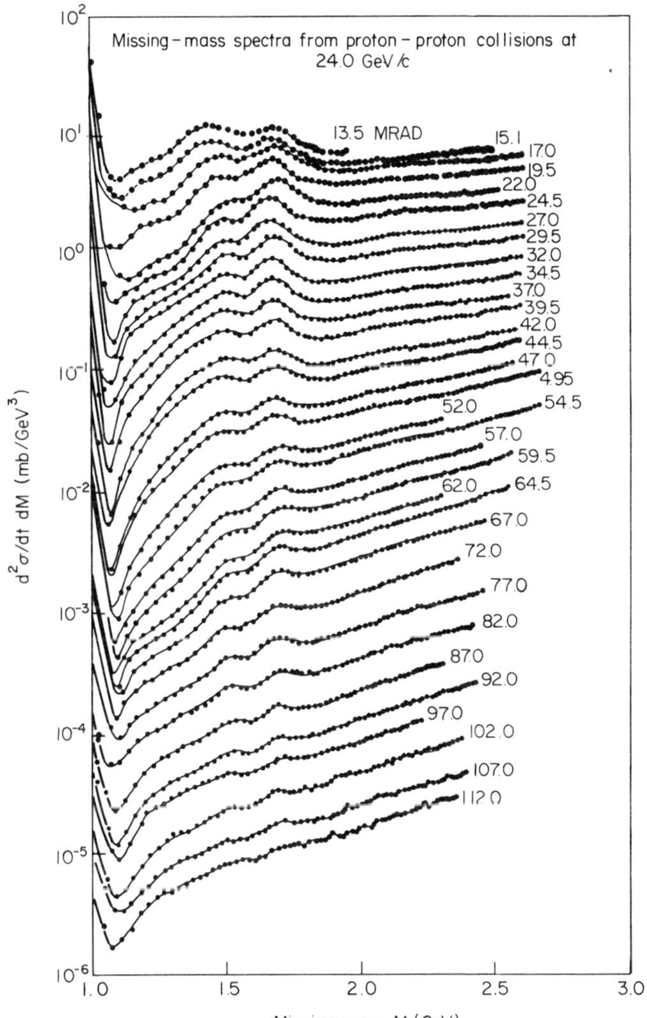

FIG. 34. Missing mass spectra taken at 24 GeV/c incident momentum. The elastic peaks are off the scale to the left.

mass range covered by each curve. A valley is seen below the elastic peak, followed by a gradual rise with several bumps superimposed.

Figure 35 shows the structure in more detail. There is evidence for several bumps. Fitting the data with a polynomial describing the continuum (dash–dotted line in Fig. 35) and Breit–Wigner formulae to describe the resonances (dotted line in Fig. 35) the following results were obtained:

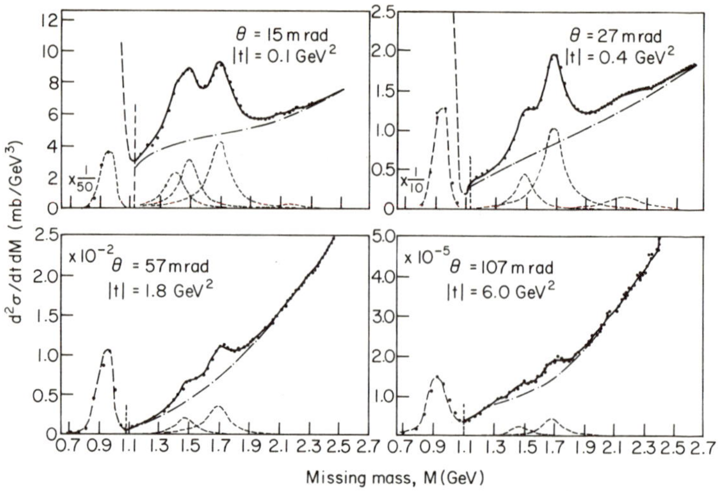

FIG. 35. $pp \rightarrow pX$ at 24 GeV/c, details. The lines are fits to the background + resonances.

	Slope (GeV^{-2})	dσ/dt(t = 0) mb/GeV^2	σ (mb)
Δ(1236)	6·4 ± 0·8	0·18 ± 0·05	0·057 ± 0·017
N(1400)	24·2 ± 2·5	11·6 ± 4·2	0·96 ± 0·35
N(1520)	4·84 ± 0·24	0·91 ± 0·19	0·37 ± 0·07
N(1688)	5·12 ± 0·08	2·35 ± 0·36	0·92 ± 0·14
N(2190)	3·01 ± 0·17	0·16 ± 0·05	0·108 ± 0·033

Some features of these data are:

At $|t| \approx 0{\cdot}1$, the N(1400), N(1520), N(1688) are excited with comparable cross-section. The N(2190) and Δ(1236) are excited with $\frac{1}{10}$ the strength.

The isolation of the Δ(1236) is complicated by the fact that it is located in the valley between the elastic peak and the rising inelastic continuum. Its cross-section decreases strongly with energy, $\sim s^{-2{\cdot}5}$. Its dependence on t and s is shown in Fig. 36.

The N(1400) has a steep dependence on t and possibly an (elastic-like) diffraction minimum + secondary maximum, see Fig. 37.

The N(1520) and N(1688) have a t-dependence similar to that of elastic scattering, at $t > 2\,\text{GeV}^2$. No dip is seen in the N(1500) and N(1688) t-distributions, neither at $t \approx 1{\cdot}4\,\text{GeV}^2$ nor anywhere else. See Fig. 38.

The N(1520) and N(1688) are weakly dependent on s at small t, strongly at large t. See Figs 39, 40.

For completeness the t-distr. of the N(2190) is shown in Fig. 41.

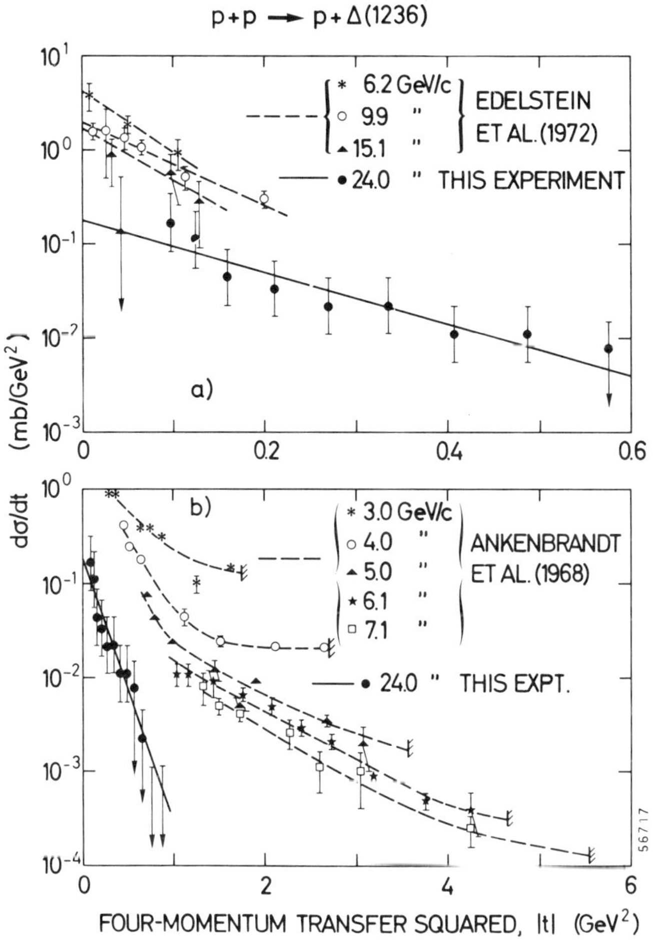

FIG. 36. Differential cross-section for the production of Δ(1236) at 24 GeV/*c*. Lower energy data are also shown.

The integrated cross-sections for *N*(1450), *N*(1520), *N*(1688), *N*(2190) are independent of *s*.

Comparing the production of *N*(1688) by pions and protons one finds factorization:

$$\frac{\pi p \to \pi N(1688)}{\pi p \to \pi p} = \frac{pp \to pN(1688)}{pp \to pp}$$

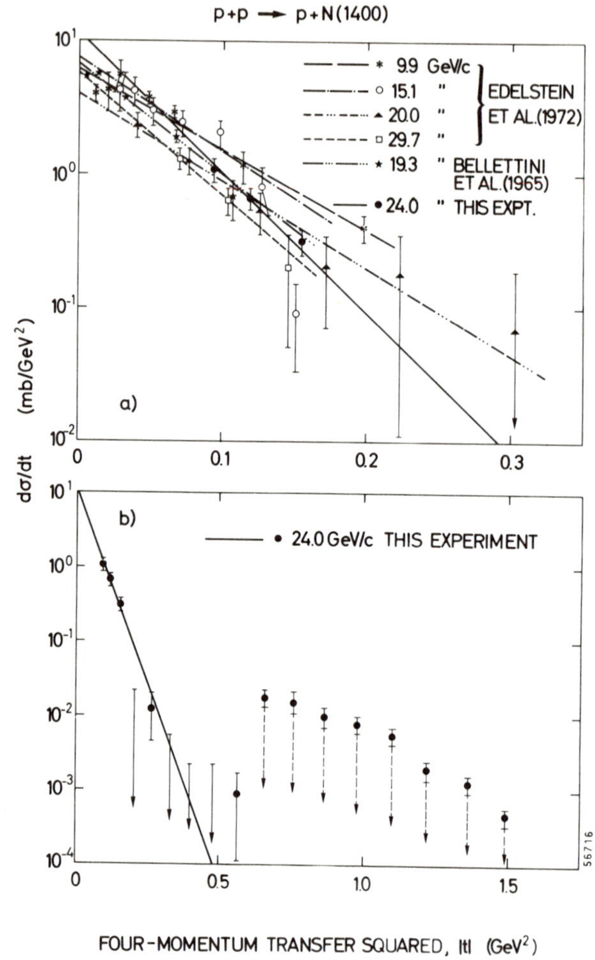

FIG. 37. Diff. cross-section for the production of $N(1400)$. The broken lines represent the uncertainty, additional to the statistical errors, due to the fitting procedure.

The main *conclusions* to be drawn from low s missing mass data are thus:

(1) Dominance of *minimal exchange*, pointing to diffraction dissociation.

$\sigma(I = \frac{1}{2})$ is constant with increasing s (P-exchange)
$\sigma(I = \frac{3}{2})$ falls with increasing s ($\pi \rightarrow$ exchange)

(2) The data support the Morrison–Gribov rule stating that for diffractively

produced states the exchange of spin and parity are limited by the constraint $\Delta P = (-1)^{\Delta J}$:

$N(1520)$	$N(1688)$	$N(2190)$
$J^P = \frac{3}{2}$	$\frac{5}{2}^+$	$\frac{7}{2}^-$

(3) There seem to be important "multiple scattering" effects at large t.
(4) The $N(1688)$ cross-section in πp and pp is consistent with a factorizable Pomeron.

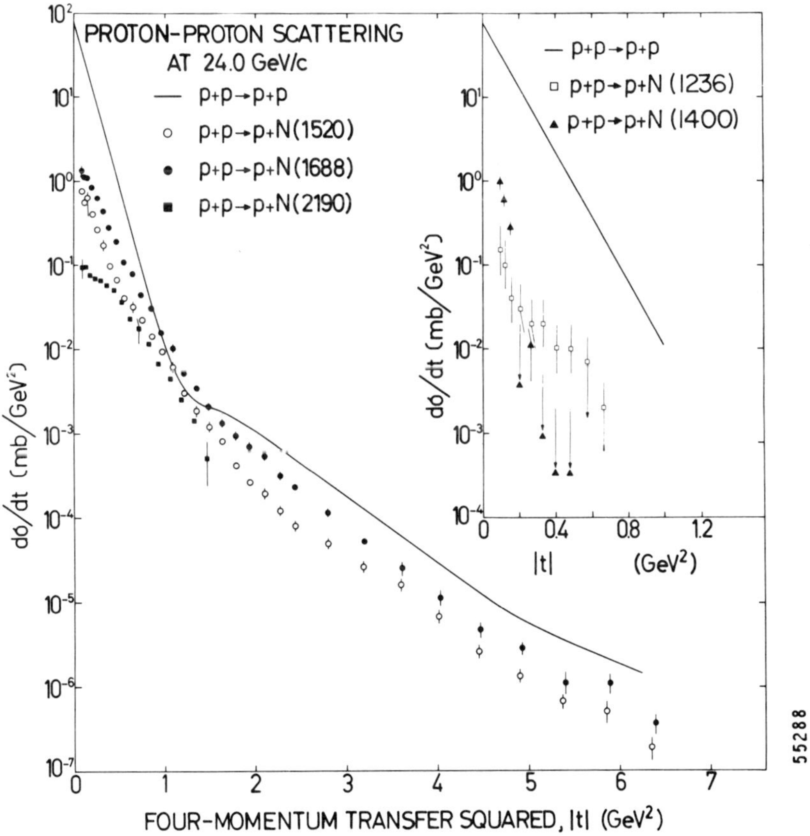

FIG. 38. Differential cross-section for the production of $N(1520)$ and $N(1688)$ at 24 GeV/c.

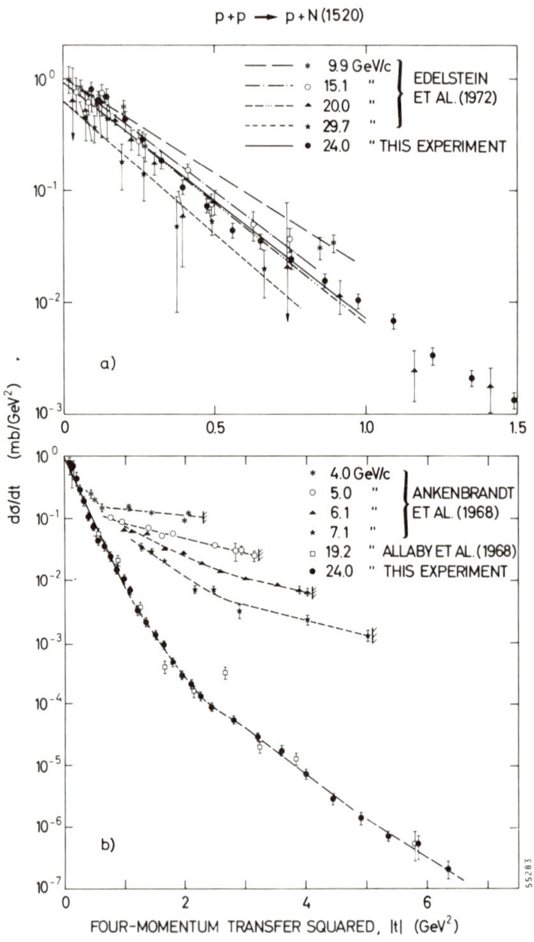

FIG. 39. $N(1520)$ production at various s.

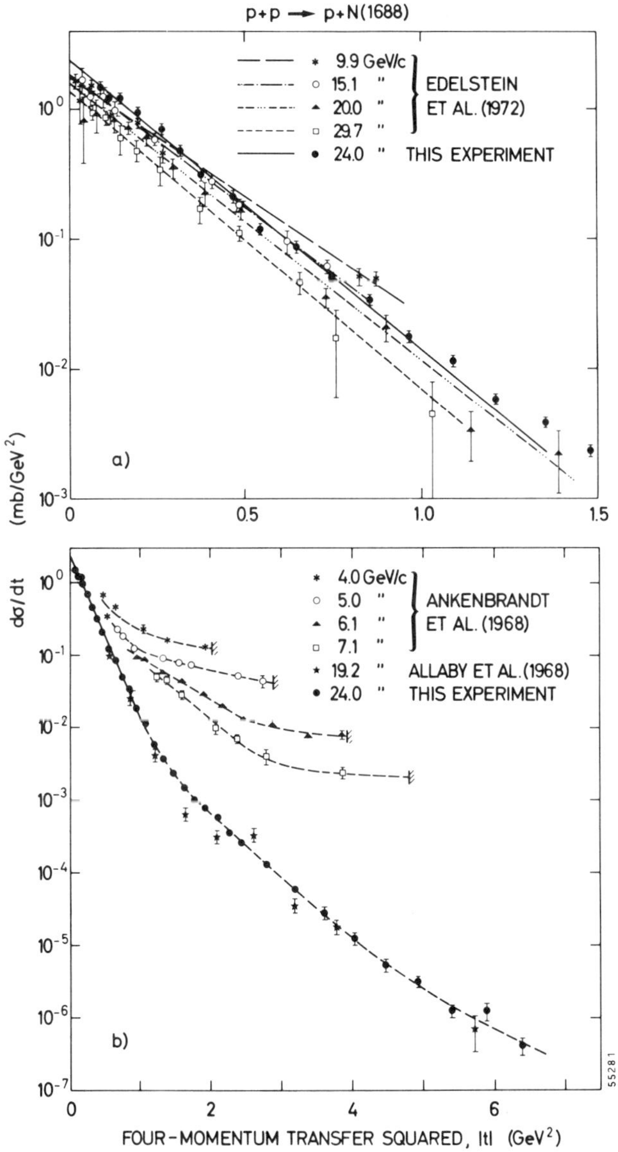

FIG. 40. $N(1688)$ production at various s.

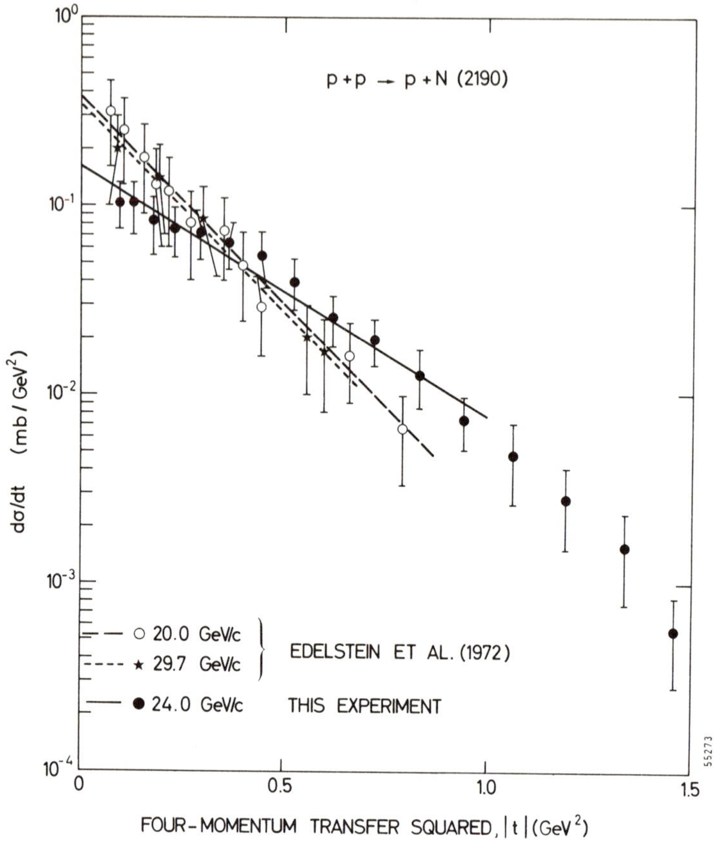

FIG. 41. $N(2190)$ production.

V. HIGH ENERGY ($s > 100$ GeV^2) MISSING MASS EXPERIMENTS

This section, and the next, deals with missing mass experiments at NAL and ISR. First, some kinematical relations relevant to MM experiments with stationary targets and with colliding beams will be listed. This is followed by a discussion of three experiments and of some results. More complete summaries of results can be found in various recent review articles.[23]

1. KINEMATICS

$$\text{Consider} \quad p + p \to p + X \qquad (m = \text{proton mass})$$

$$1 + 2 \to 3 + 4 \tag{V.1}$$

where 3 is the particle observed and 4 stands for all the other particles combined. Then ($\mathbf{p} = p, iE/c$)

$$-s = (\mathbf{p}_1 + \mathbf{p}_2)^2 \qquad -t = (\mathbf{p}_1 - \mathbf{p}_3)^2$$

$$-M^2 = (\mathbf{p}_1 + \mathbf{p}_2 - \mathbf{p}_3)^2 = (p_1 + p_2)^2 - m^2 - 2\mathbf{p}_3(\mathbf{p}_1 + \mathbf{p}_2) \qquad \text{(V.2)}$$

In the c.m.: $\vec{p}_1 + \vec{p}_2 = 0$.

Hence $M^2 = s + m^2 - 2\sqrt{s}E_3^*$ (E_3^* = energy of 3 in c.m. system)

and

$$M^2 = s + m^2 - 2\sqrt{s}\sqrt{\left(\frac{x^2 s}{4} + p_T^2 + m^2\right)} \approx s(1 - x) \qquad \text{(V.3)}$$

where $x \equiv 2p_{3L}^*/\sqrt{s}$. The maximum value of x is attained when $M^2 = m^2$:

$$x_{\max} = \left[1 - \frac{4(p_T^2 + m^2)}{s}\right]^{\frac{1}{2}} \qquad \text{(V.4)}$$

Note that $x_{\max}$ depends on s. For example, for $p_{\text{lab}} = 100$ GeV/c and $p_T =$ 1 GeV/c, $x_{\max} = 0{\cdot}98$.

The dependence of M^2 on the momenta involved (p_1, p_2, p_3) is obtained from:

$$M^2 = -(\mathbf{p}_1 + \mathbf{p}_2 - \mathbf{p}_3)^2 =$$
$$3m^2 - 2\vec{p}_1\vec{p}_2 + 2E_1E_2 + 2\vec{p}_1\vec{p}_3 - 2E_1E_3 + 2\vec{p}_2\vec{p}_3 - 2E_2E_3 \qquad \text{(V.5)}$$

Putting $\delta p_1/p_1 \approx \delta E_1/E_1 = \mathscr{E}$ one gets by differentiating

$$\delta M^2 = \mathscr{E}(-2\vec{p}_1\vec{p}_2 + 2E_1E_2 + 2\vec{p}_1\vec{p}_3 - 2E_1E_3)$$
$$= \mathscr{E}(M^2 - 3m^2 - 2\vec{p}_2\vec{p}_3 + 2E_2E_3) \qquad \text{(V.6)}$$

At this point we note that in missing mass experiments the momentum of 3 is close to its maximum value and the angle is small. We take 3 to emerge near to the direction of 1;

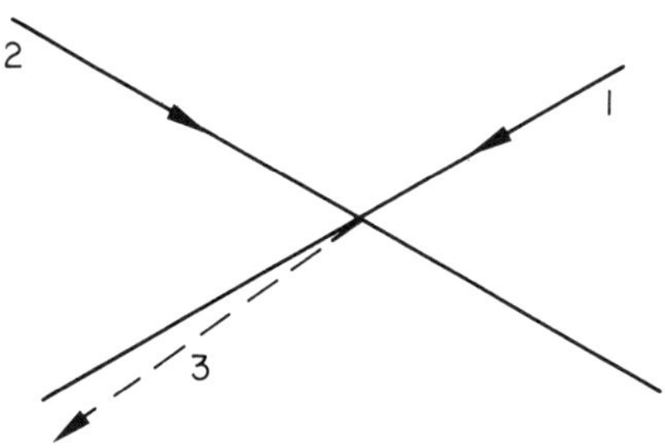

Then (V.6) becomes:

$$\delta M^2 \approx \mathscr{E}(M^2 - 3m^2 + 4E_2E_3)$$

With

$$s = (\mathbf{p}_1 + \mathbf{p}_2)^2 \approx 4E_1E_2 \approx 4E_2E_3$$

we then have

$$\delta M^2 \approx \mathscr{E}(M^2 - 3m^2 + s) \approx \mathscr{E}s$$

or

$$\delta M^2_{p_1} = s\,\frac{\Delta p_1}{p_1} \tag{V.7}$$

For example, for $s = 200\ \text{GeV}^2$ and $\Delta p/p = 1\%$, $\delta M^2 = 2\ \text{GeV}^2$.

The dependence in M^2 on the uncertainty in p_2 can be calculated by putting $\delta p_L/p_L \approx \delta E_2/E_2 = \mathscr{E}$ and differentiating (V.5) with respect to p_2:

$$\delta M^2 \approx \mathscr{E}(M^2 - 3m^2)$$

and thus

$$\delta M^2_{p_2} \approx M^2\,\frac{\Delta p_2}{p_2} \tag{V.8}$$

Likewise, for the dependence of the resolution in M^2 on the accuracy with which the produced proton is measured we have

$$\delta M^2_? = \mathscr{E}(M^2 - 3m^2 - s)$$

$$\delta M^2_{p_3} \approx -\,s\,\frac{\Delta p_3}{p_3} \tag{V.9}$$

Note that we work with invariants, hence for missing mass experiments on stationary target we have

$$\delta M^2_{p_1} = s\,\frac{\Delta p_1}{p_1}, \qquad \delta M^2_{p_2} = 0, \qquad \delta M^2_{p_3} = -\,s\,\frac{\Delta p_3}{p_3} \tag{V.10}$$

Hence in either case, colliding beams or stat. target, there is a strong dependence on the knowledge of the momentum of one of the incident protons, resp. the incident protons, and of the produced proton; there is little or no sensitivity to the momentum of the "target" proton.

Next, we consider the energy loss. For stationary targets we have from (V.5)

$$p_2 = 0 \qquad M^2 = (E_1 + m - E_3)^2 - (\vec{p}_1 - \vec{p}_3)^2$$
$$= t + m^2 + 2m(E_1 - E_3)$$

Hence for small t

$$\Delta E = E_1 - E_3 = \frac{M^2 - m^2}{2m} \quad \text{Stat. target} \tag{V.11}$$

e.g. $E_1 = 300\,\text{GeV}$, $M = 1{\cdot}5\,\text{GeV}$, $\Delta E = 730\,\text{MeV}$ or 0·25%.

For colliding beams we get

$$M^2 = ((E_1 - E_3) + E_2)^2 - ((\vec{p}_1 - \vec{p}_3) + \vec{p}_2)^2$$
$$\approx t + m^2 + 4E_2(E_1 - E_3)$$

for small crossing angle and (as always) high momenta. Thus

$$\Delta E = E_1 - E_3 = \frac{M^2 - m^2}{2m}\frac{m}{2E_1} \quad \text{Coll. beams} \tag{V.12}$$

e.g. 11/8/11·8 GeV, $M = 1{\cdot}5\,\text{GeV}$, $\Delta E = 31\,\text{MeV}$ or 0·25%.

Hence in colliding beam experiments we have a very much smaller energy loss for given s, than in stationary target experiments, although the percentage loss is the same.

Let us look a little closer at the differences between colliding beam and stationary target experiments on diffraction dissociation. There is single diffraction dissociation in which either the beam or the target proton breaks up or double diffraction dissociation. So far we have compared single diffraction dissociation of either beam in colliding beam experiments with *target* diffraction dissociation in the case of stationary targets.

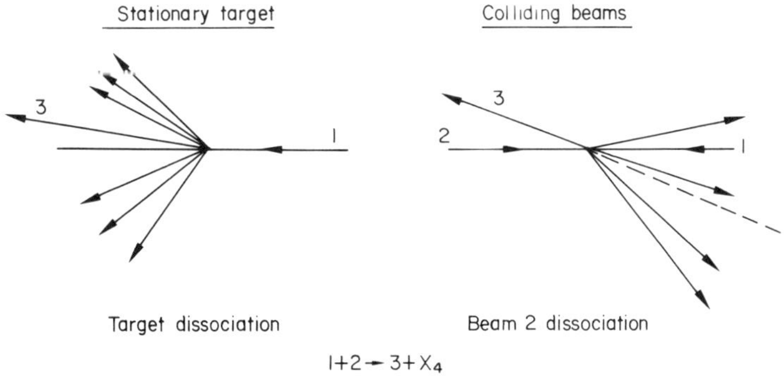

Obviously in the latter case there is the other, non symmetric, possibility of beam dissociation, with a single recoil proton coming off the target. In that case

$$-t = (\mathbf{p}_1 - \mathbf{p}_3)^2 = (\mathbf{p}_2 - \mathbf{p}_4)^2 = 2mT_4 \tag{V.13}$$

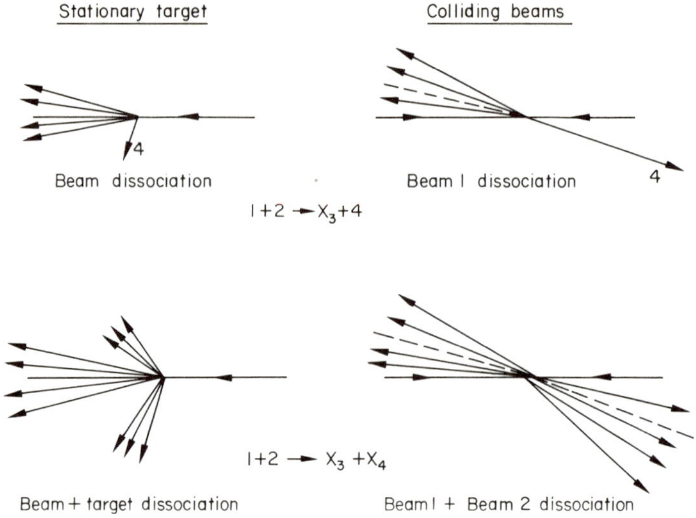

where T_4 is the kinetic energy of the recoil proton, and

$$M^2 = m^2 + 2p_1p_4 \cos\theta_{14} - 2(m + E_1)T_4 \tag{V.14}$$

In recoil detection the sensitivity of M^2 to the momentum of the incident beam is

$$\delta M^2_{p_1} \approx M^2 \frac{\Delta p_1}{p_1}\bigg|_{\text{Recoil of stat. target}} \tag{V.15}$$

Comparing this with (V.10) one sees a reduction in sensitivity by a factor s/M^2, i.e. typically 20. The sensitivity to angle and kinetic energy of the recoil proton is sketched in Fig. 42. For example, at $M^2 \approx 20\,\text{GeV}^2$ and $t \approx 0{\cdot}15\,\text{GeV}/c^2$, $a \pm 5\%$ measurement of the kinetic energy results in $\delta M^2 \approx \pm 0{\cdot}6\,\text{GeV}^2$. The sensitivity to angle is $\delta M^2 \approx \pm 1\,\text{GeV}^2$ per $\pm 0{\cdot}25°$. Comparing this with colliding beam experiments at the same s:$a \pm 0{\cdot}5\%$ measurement of the momentum of the produced proton results in (see (V.9)) $\delta M^2 \approx \pm 3\,\text{GeV}^2$, i.e. a factor of 5 worse although the accuracy of the energy measurement is a factor 10 higher. *Hence missing mass experiments with colliding beams have intrinsically (i.e. independent of the quality of the detectors) a worse mass resolution than experiments on stationary targets in which angle and momentum of the recoil proton are measured.*

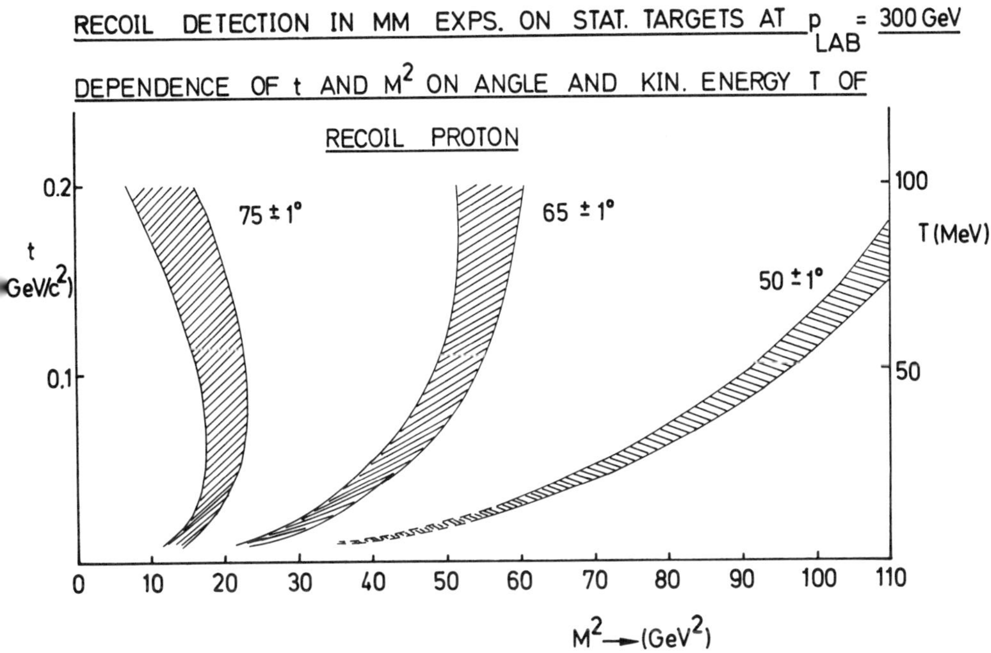

FIG. 42. Recoil detection in missing mass experiments on stationary targets. Dependence of t and M^2 on angle and kinetic T of the recoil proton. Data for $p_{lab} = 300$ GeV.

Finally there is the question of minimum t. We have

$$-t = (\mathbf{p}_1 - \mathbf{p}_3)^2 = 2m^2 - 2E_1E_3 + 2p_1p_3 - 4p_1p_3 \sin^2 \theta/2$$

$$= 2m^2 - 2\frac{(E_1E_3 - p_1p_3)(E_1E_3 + p_1p_3)}{E_1E_3 + p_1p_3} - 4p_1p_3 \sin^2 \theta/2$$

$$= m^2(2 - x - 1/x) - 4p_1p_3 \sin^2 \theta/2 \approx p_1p_3\,\theta^2 \qquad \text{(V.16)}$$

Note that (V.16) does not depend on p_2. In colliding beam machines the minimum angle at which positively charged particles can be detected is not zero, and is in fact rather large, due to the dimensions of the primary beams and the vacuum chamber in which they move. Typically $\theta_{lab} = 35$ mrad. Example: $(p_3 \approx p_1)\, p_1 = 15{\cdot}5$ GeV/c, $p_2 = 15{\cdot}5$ GeV/c, $s = 960$ GeV^2, $t_{min} = p_1p_3\theta^2 = 15^2 \times (35 \times 10^{-3})^2 = 0{\cdot}28$. But since t_{min} does not depend on p_2 we can reach a smaller t_{min} by running with unequal energies: $p_1 = 11$GeV/c, $p_2 = 22$ GeV/c, $s \approx 960$ GeV^2, $t_{min} = 11^2 \times (35 \times 10^{-3})^2 = 0{\cdot}15$ GeV/c^2.

For stationary targets and observing the fast proton (target dissociation), assuming $\theta_{min} = 1{\cdot}5$ mrad and $p_1 = 300$ GeV/c we have $t_{min} = 0{\cdot}20$ GeV/c^2. On the other hand, by measuring recoils one has (see (V.13)) $t = 2\,mT$. Using solid state devices or proportional counters one can measure down to (for example) 100 keV. Thus $t = 0{\cdot}0002\ \text{GeV}^2$. This is another advantage of recoil detection on stationary targets.

2. MISSING MASS EXPERIMENTS AT NAL AND ISR

Two counter experiments have recently been performed at NAL in which angle and momentum of the target recoil in

$$p + p \rightarrow X + p$$

is detected, from which the diff. cross-section for excitation of mass X is then deduced.

The experiment of Sannes *et al*[24] has been done by letting the circulating beam in NAL run through a hydrogen jet which is turned on twice at precisely defined times during the acceleration cycle, once at a fixed instant near the beginning of each cycle, the other at some chosen later time; the incident energy is varied by varying the latter switch-on time, while the relative normalization is obtained from the ratio of the yields during the two intervals. The acceleration ramp lasts $\sim 2{\cdot}5$ sec, the jet is pulsed for 2×250 msec. The proton density in the jet is $2 \times 10^{-7}\ \text{g/cm}^3$. The beam profile is $\sim 2 \times 3$ mm. The target × beam intensity in each burst is monitored by detecting elastically scattered events, at 86° and with $T = 11{\cdot}6$ MeV, in a suitably placed solid state detector. Since the differential elastic cross-section is independent of s (apart from a 2·2% correction for shrinkage) this detector provides the absolute normalization (up to a geometrical factor) through:

$$\left(\frac{dN}{dt}\right)_{\text{finite } t} \rightarrow \left(\frac{dN}{dt}\right)_{t=0} = L\left(\frac{d\sigma}{dt}\right)_{t=0} = L\frac{\sigma_T^2}{16\pi}$$

Using Serpukhov data (arrow to the first step); Using Serpukhov + ISR data (±2·5%) (arrow to the last step)

The target recoil is detected by range, see Fig. 43. A coincidence $C_1C_2C_3C_4C_5\overline{C}_6$ signals a proton stopping in absorber No. 2, with a momentum $560 < p_4 < 660$ MeV/c; a coincidence $C_1C_2C_3C_4C_5C_6\overline{C}_7$ signals a stop in absorber No. 3, with $660 < p_4 < 780$ MeV/c.

The experiment is run in one of two ways:

(a) keep p_4, θ_4 fixed and vary p_1 in the range $40 < p_1 < 260$ GeV/c. Since

$s = 2mE_1$, $-t = 2mT_4$, $M^2 = s(1 - (E_4 - p_4 \cos\theta_{14})/m)$

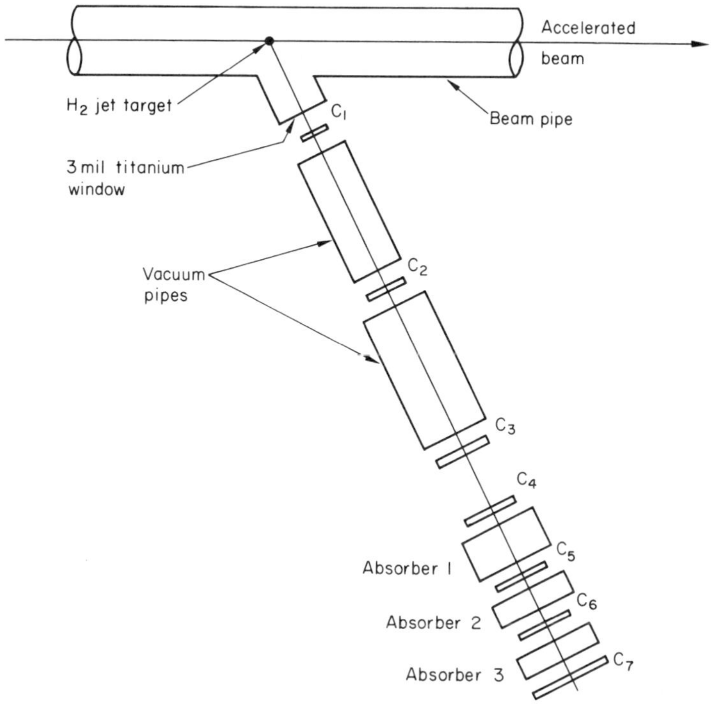

FIG. 43. Missing mass experiment of Sannes *et al.*, at NAL. The missing mass is determined by angle and range of the recoil proton.

t and M^2/s are determined by p_4 and θ_4. For example:

p_4	θ_4	t	$M^2/s = 1 - x$	Proton stops in:
0·6	55·6°	−0·33	0·18	absorber 2
0·7	55·6	−0·45	0·17	absorber 3

(b) keep s and p_4 (i.e. t) fixed and vary θ_4 by rotating the whole set up. In this way a range $0{\cdot}06 < M^2/s < 0{\cdot}21$ is covered at fixed t.

The final normalization is made by extrapolating down to $s = 56$ and 47 GeV^2 (BNL and CERN data).

The data show an inclusive differential invariant cross-section whose s-dependence is in good agreement with the form $A + B/\sqrt{s}$ at several pairs (x, t), see Fig. 32 in Section III. Furthermore the invariant cross-section at fixed t and s shows a minimum at $x \approx 0{\cdot}9$ followed by a rise up to the highest measured $x \approx 0{\cdot}93$. The shape does not depend much on s.

A second counter experiment,[25] also detecting recoil protons has been done using an array of solid state detectors. 18 pairs of detectors are placed

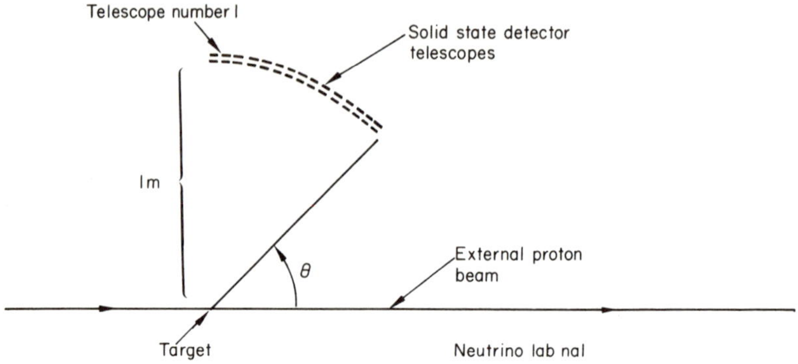

FIG. 44. Missing mass experiment of Childress *et al.*, at NAL. The missing mass is determined by angle and kinetic energy of the recoil proton.

on a circle of ~1 m radius around a CH_2 or C target in an extracted proton beam at NAL (see Fig. 44). They are uniformly spaced between 48° and 89° with respect to the beam, each counter pair subtending ~0·4°. Each pair consists of two detectors, the first $\frac{1}{2}$ mm thick, the second 5 mm thick, both 1 cm^2 in area. The first measures dE/dx, the second the total energy, if the kinetic energy T is less than 30 MeV, or otherwise the energy loss.

In the latter case T can be deduced from dE_1/dx and dE_2/dx up to $T =$ 100 MeV. The relative calibration of the different pairs is monitored by means of elastic scattering: at each angle elastically scattered protons have a definite kinetic energy (see eqn (V.14) for $M = m$) and t, e.g. at $\theta = 75°$ $T = 143$ MeV, $t = 0{\cdot}28$ GeV^2 (see Fig. 42). The diff. cross-section versus M^2 for fixed t is obtained by taking the rates at fixed T, i.e. t, in the various pairs, hence the importance of relative normalization.

The data on protons are obtained by subtracting the data on C from those on CH_2. The normalization necessary for this procedure is obtained from deuteron and triton production which is the same in both targets. In the region of elastic events the C-background is small, but in the inelastic region the pp-spectrum is obtained by subtracting 90% of the data. Dead time corrections are relatively large since solid state detectors are inherently slow and the experiment is mounted very near to an intense primary beam. The absolute normalization is by scaling the integrated elastic peaks to bubble chamber results on $d\sigma/dt$ at the same energy. This is expected to be accurate to $\lesssim 30\%$. The missing mass resolution, obtained from the shape of the elastic peaks is $\delta M^2 = 1$ GeV^2 (FWHM) at $|t| \approx 0{\cdot}03$, 2 GeV^2 at $t = 0{\cdot}11$ and 4 GeV^2 at $t = 0{\cdot}17$. This is thus the *measured* resolution and includes the effects of finite resolution in θ and T.

At ISR missing mass experiments are being done with the set up shown in

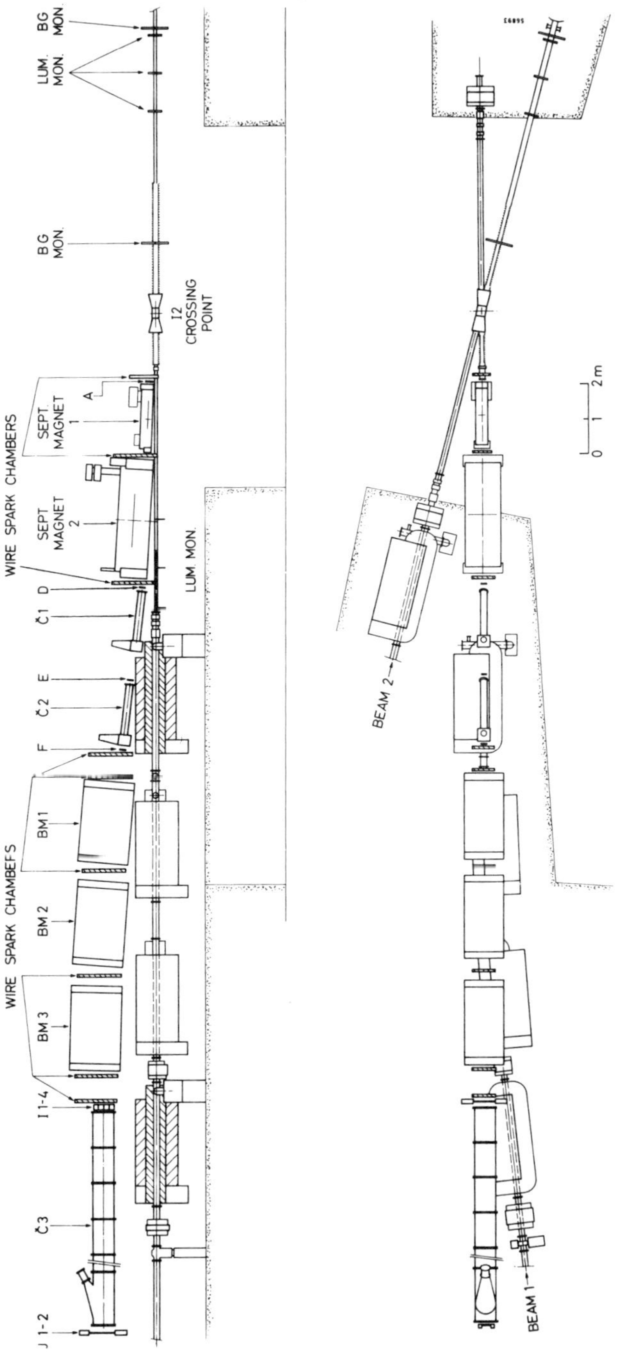

FIG. 45. Missing mass experiment of Albrow *et al.*, at ISR. The missing mass is determined by angle and momentum of the proton in the spectrometer.

Fig. 45 (Refs 26, 27). The spectrometer consists of 5 magnets, 3 Cerenkov counters and a number of spark chambers and scintillators placed between the magnets. The axis of the spectrometer is at 85 mrad in the vertical plane through the intersecting beams, but other angles can be reached by moving the first two magnets and Cerenkov counters up and down. In this way particles produced at angles other than 85 mrad are guided towards the fixed part of the spectrometer. This feature is similar to that of the spectrometer mentioned in Section IV (see Fig. 33). In order to reach the smallest possible angle the first magnets are placed very close to the primary beams. Current-carrying septa cancel the strong stray field which would otherwise perturb the beams. The minimum angle is then determined by the available "free" space along the downstream beams and the thickness of the septa. In practice at ISR $35 < \theta < 180$ mrad. At a given setting the spectrometer accepts $\Delta\Omega = 10^{-4}$ sr, $\Delta p/p \approx \pm 15\%$. The momentum resolution is $\delta p/p \approx 0{\cdot}8\%$ (FWHM), related to δM^2 as in eqn (V.9). Particles are identified by the Cerenkov counters and by time of flight.

Aspects of the data analysis have been discussed in Section III. The normalization is obtained by means of monitor counters after calibration with the v. d. Meer beam steering method (Section II).

A point particular to colliding beam experiments is the very broad range of momenta in the primary beams. This is because in filling the rings the successive pulses are put side by side by means of slight acceleration with an RF cavity (see Section I). The result is a ~5 cm wide ribbon (~3 mm high) in which the momentum varies from inside to outside by ~2%. Locating the point of origin in the diamond then gives the primary momenta of the event in question. The data can then all be scaled to, say, the central values of the primary momenta, by use of the relation:

$$\Delta x = \alpha\, \Delta p/p$$

where x is the coordinate "across" the beam and α is the known momentum compaction function. In this way the lack of knowledge of the primary momentum is reduced from, say, 500 MeV/c at 26 GeV/26 GeV to ~60–100 MeV/c and, with eqn (V.7), $\delta M^2 \approx 50\ \text{GeV}^2 \rightarrow 7\text{–}10\ \text{GeV}^2$. The remaining spread is due to the smearing out of the momentum/position relation by the betatron amplitudes. The spread can be further reduced by stacking discontinuously: in that case the only remaining uncertainty in momentum is that of the original bunch as it came from the accelerator which feeds the storage rings. This spread is of the order of 5–10 MeV/c. Hence by running with discontinuous stacks the effect of the spread in primary momentum on δM^2 is reduced to a level below that contributed by the error in the momentum measurement of the secondary proton. Figure 46 shows data obtained with a stack consisting of three separate bunches. The top histogram shows

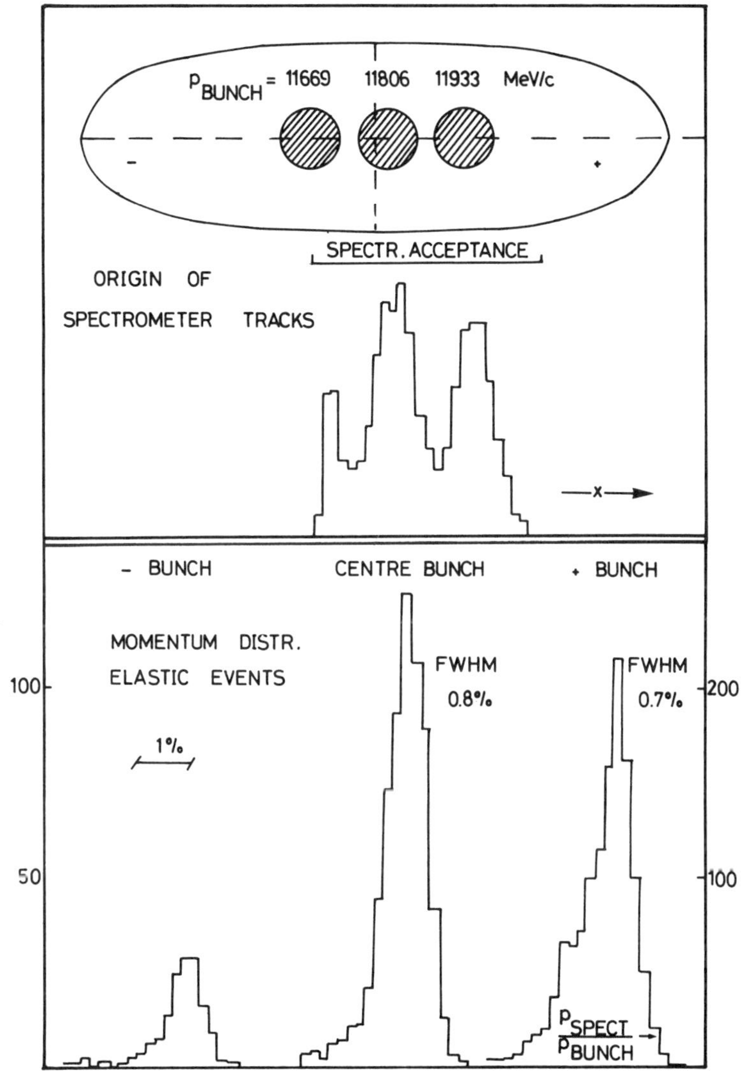

FIG. 46. Data on (quasi)-elastic scattering obtained with a discontinuous stack at ISR. Top histogram: distribution of origins of tracks in the spectrometer of Ref. 5. Bottom histograms: momentum distribution of (quasi)-elastic events originating from each of the three bunches. The left bunch is at the edge of the spectrometer acceptance and hence contains less events than the others.

the distribution of points of interaction across the vacuum chamber, as reconstructed from spark chamber tracks in the spectrometer of Ref. 26. The histograms below show the distribution in momentum of events in the

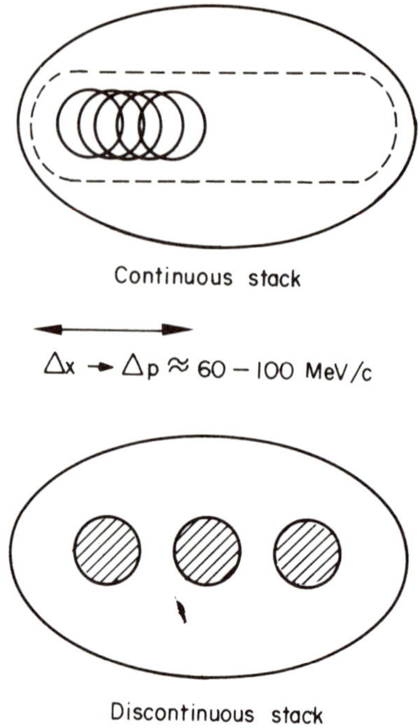

spectrometer, originating from each of the three bunches. A loose collinearity requirement on a particle going off in opposite direction has been applied in order to enrich the fraction of elastic events in the samples. The FWHM thus obtained is then a measure for the resolution of the spectrometer itself.

3. A FEW RESULTS

The data on $pp \to pX$ show the following features:

(1) At fixed medium x and p_T or t the data approach a scaling limit from above in the range $100 < s < 360\ \mathrm{GeV}^2$ and do not vary appreciably with energy for $s > 550\ \mathrm{GeV}^2$, see Fig. 47 and Section III, Fig. 32.

(2) At high x a peak develops indicative of single diffraction dissociation in which the observed proton remains intact while the other one breaks up into a mass $M^2 \approx s(1 - x)$.

(3) The shape and magnitude of the forward peak varies little with energy. Fig. 48.

These features will be analysed in the next lecture.

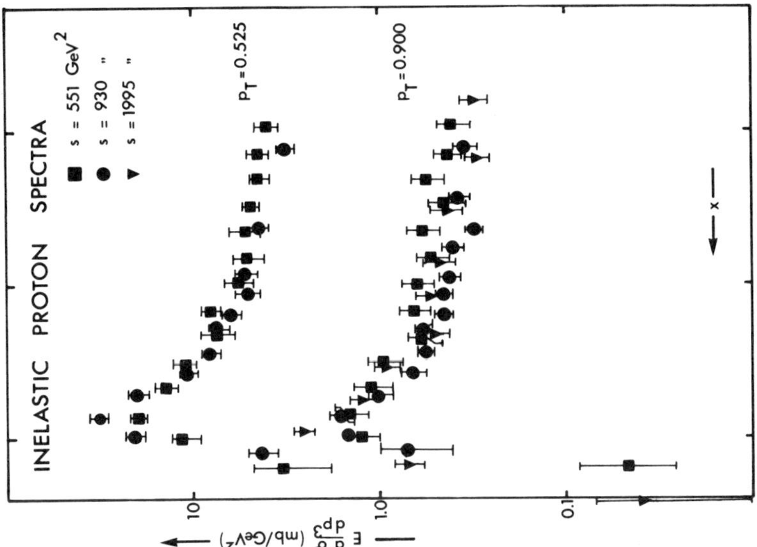

FIG. 48. Inelastic proton spectra at various ISR energies and fixed p_T, at high x.

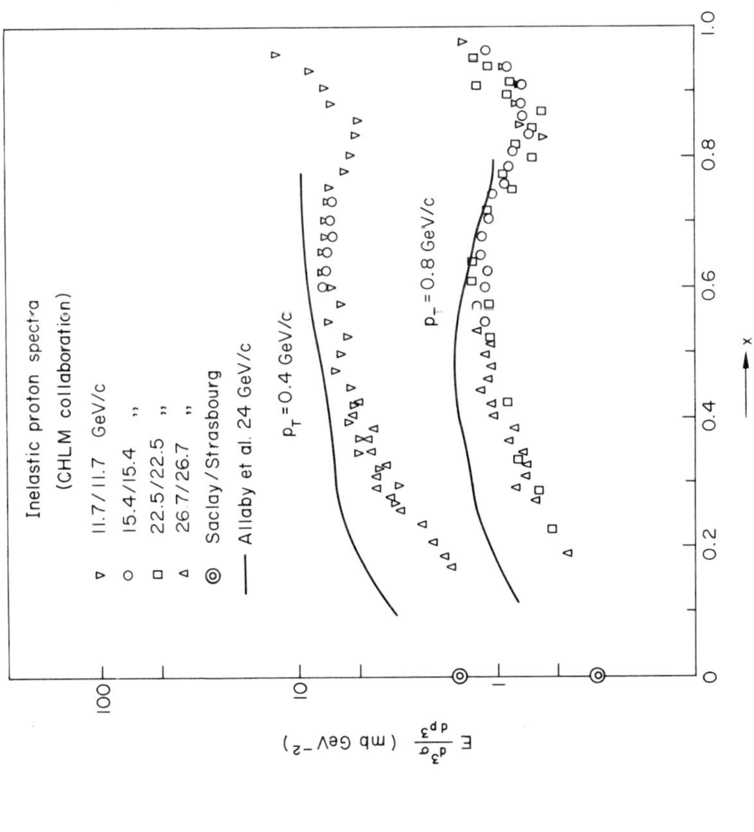

FIG. 47. Proton spectra at various ISR energies and fixed p_T, versus x.

VI. High Energy Missing Data and the Triple-Regge Limit

In this section the ISR high energy missing mass data will be considered in the light of Reggeon-exchange.

1. The triple-Regge limit

Let us recall the Mueller–Regge picture of Section III. Using the 3 particle optical theorem and Reggeizing the exchange if the appropriate (sub) energy $\rightarrow \infty$ as $s \rightarrow \infty$ we have obtained the central and fragmentation regions. We now go to the extreme end of the fragmentation region, i.e. there where one of the incident protons does not break up in the interaction with the other proton but retains its identity, just changing slightly its angle and momentum. This is the triple–Regge limit. Pictorially:

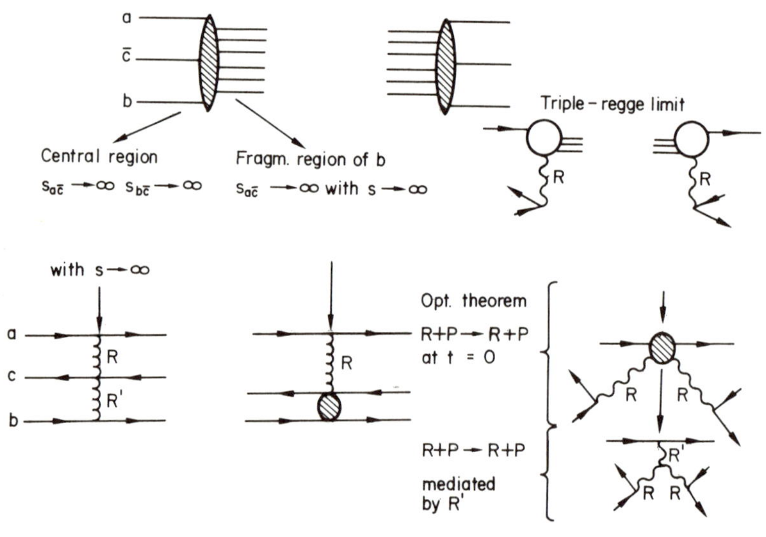

Putting the contributions of each vertex together into

$$E_c \frac{\mathrm{d}^3\sigma}{\mathrm{d}p_c^3} = \frac{1}{s}\,\mathrm{DISC}\,T_{ab\bar{c}}(s_{a\bar{c}}, s_{b\bar{c}}, M^2)$$

one obtains

$$E_c \frac{d^3\sigma}{dp_c^3} = \frac{m_0^2}{16\pi^2 s} \sum_{iij} G_{iij}(t) \left(\frac{s}{M^2}\right)^{2\alpha_i(t)} \left(\frac{M^2}{m_0^2}\right)^{\alpha_i(0)} \tag{VI.1}$$

To first order there are 4 terms: PPP, RRP, PPR and RRR. Their contributions are sketched in Fig. 49. The invariant cross-section at $t = 0$ depends on x as listed below. Also indicated are the dependence on s of the integrated cross-section at $t = 0$.

	$\sigma_{inv}(t = 0)$	$\int_0^{X_{max}} d\sigma$
PPP	$(1 - x)^{-1}$	$\sim \ln s/2m^2$
RRP	const.	$\sim 1 - 2m^2/s \to$ const. $s \to \infty$
PPR	$\frac{1}{\sqrt{s}}(1 - x)^{-\frac{3}{2}}$	$\sim$ const.
RRR	$\frac{1}{\sqrt{s}}(1 - x)^{-\frac{1}{2}}$	$\sim \frac{c}{s} + \frac{c'}{\sqrt{s}} \to 0$ $s \to \infty$

Figure 49 and the table above show the trend of the various terms; there are two-scaling, two non-scaling terms. Only PPP has both a peak at large x and, after integration, rises with increasing s, as a result of the s-dependence of the limit of integration $X_{max} \approx 1 - 2m^2/s$.

2. PARAMETRIZATION OF THE ISR MISSING MASS DATA

The observations on inclusive proton spectra, both at NAL and ISR, and in particular the occurrence of a pronounced peak in the inelastic data have met with great interest. Most simply stated the reason is that in observing the lone, unexcited proton in one c.m. hemisphere one measures the probability of raising the proton in the other hemisphere to an energy M, at a momentum transfer $\sqrt{t}$, directly; this in contrast to, e.g. the observation of inclusively produced pions whose relation to the basic exchange of forces is "second hand", i.e. requires either an additional decay mechanism

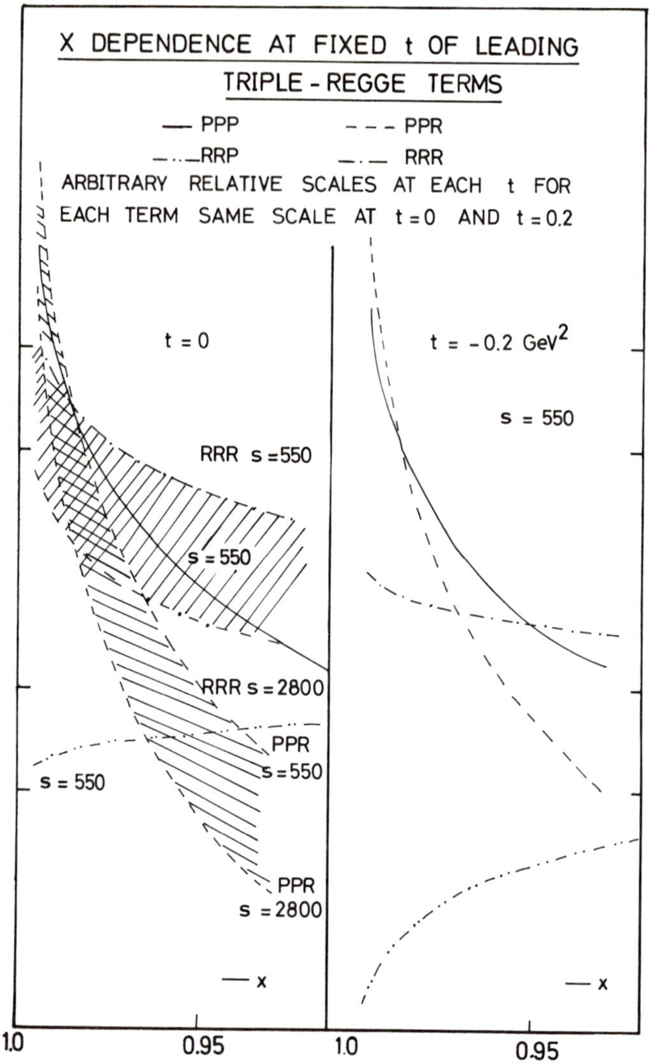

FIG. 49. Leading terms in the triple-Regge expansion.

or information on the correlations with other particles produced simultaneously. Diffraction dissociation, of which the asymmetrically shaped inelastic peak is a manifestation, is responsible for approximately half of the average multiplicity of the secondaries. For this half it provides a qualitative understanding of the population distribution on the rapidity plot. The M^2, t and s-dependence of its cross-section has direct bearing on the nature of the Pomeron.

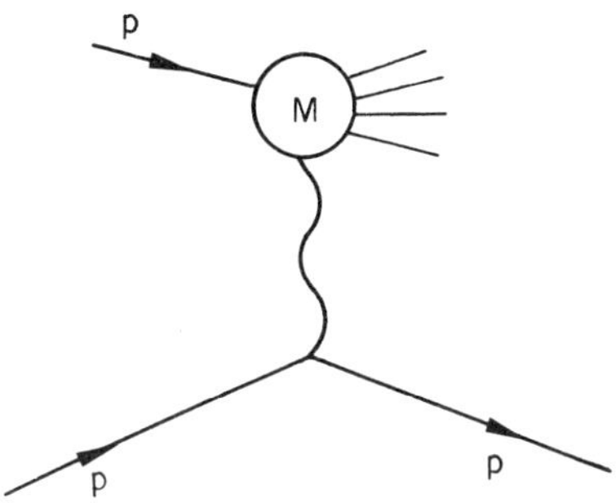

Triple-Regge fits to ISR missing mass data have been made by several authors[28]. The following is a summary of a recent, unpublished, analysis, and of some older results, by Albrow *et al.*[29]. The fits were made on the following data:

$s = 1995\ \text{GeV}^2$	$0{\cdot}5 < x < 0{\cdot}82^5$	$0{\cdot}7 < p_T < 1{\cdot}2\ \text{GeV}/c$
$s = 551\ \text{GeV}^2$	$5 < M^2 < 30\ \text{GeV}^2$	$0{\cdot}15 < t < 1{\cdot}25\ \text{GeV}/c$
$s = 930\ \text{GeV}^2$	$7 < M^2 < 50\ \text{GeV}^2$	$0{\cdot}45 < t < 1{\cdot}65\ \text{GeV}/c$

The data at $s = 1995\ \text{GeV}^2$ are shown in Fig. 50, a sample of the data at $s - 930\ \text{GeV}^2$ is shown in Fig. 51.

In fitting the medium x data ($x < 0{\cdot}82^5$) at $s = 1995\ \text{GeV}^2$ it has been assumed that the cross-section there is dominated by just one term in (VI.1), the scaling RRP term. In view of the energy independence of the data (see Section V, Fig. 47) non scaling terms such as RRR or interference terms may be expected not to contribute more than, say, 20% to the cross-section. For the RRP term, with $R = f =$ effective meson trajectory, we have

$$E\frac{\mathrm{d}^3\sigma}{\mathrm{d}p^3} = \frac{m_0^2}{16\pi^2 s} G_{ffP}(t)\left(\frac{s}{M^2}\right)^{2\alpha_f(t)}\left(\frac{M^2}{m_0^2}\right)^{\alpha_P(0)} \qquad \text{(VI.2)}$$

With $m_0^2 = 1$ and $\alpha_P(0) = 1$

$$E\frac{\mathrm{d}^3\sigma}{\mathrm{d}p^3} = \frac{G_{ffP}(t)}{16\pi^2}\left(\frac{M^2}{s}\right)^{1-2\alpha_f(t)} \qquad \text{(VI.3)}$$

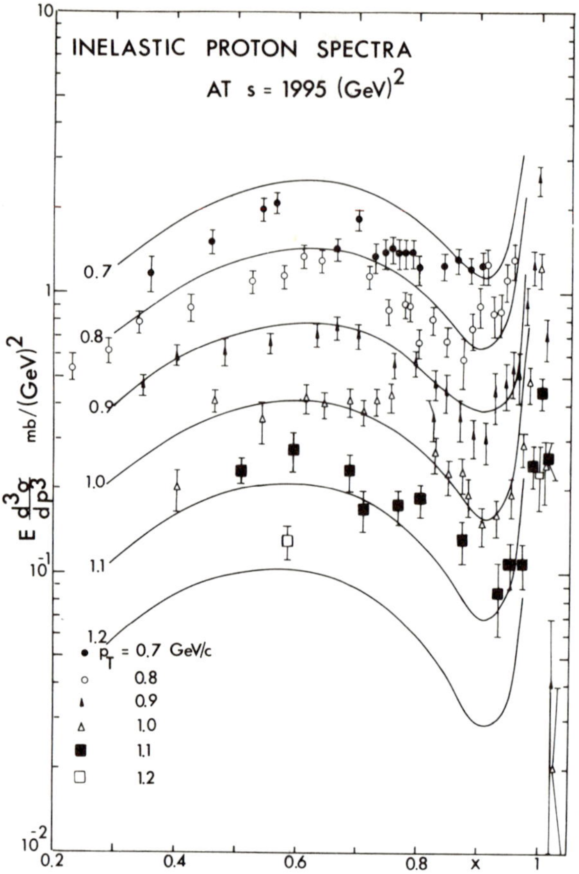

FIG. 50. Data on $pp \to pX$ at $s = 1995\ \text{GeV}^2$. The points at $0{\cdot}5 < x < 0{\cdot}82^5$ were used to obtain the RRP triple-Regge cross-section.

A best fit to the data at $s = 1995\ \text{GeV}^2$ then results in

$$\alpha_f = 0{\cdot}45 + 0{\cdot}75\,t \qquad \text{(VI.4)}$$

for an effective meson trajectory. The trajectory is shown in Fig. 52, the dotted line there indicates the sensitivity to a change of slope. Note that this result is independent of the absolute normalization. Substituting (VI.4) into (VI.2) one obtains for $G_{ffP}(t)$

$$G_{ffP}(t) \approx 1100\ \text{mb/GeV}^2 \qquad \text{(VI.5)}$$

independent of t. The latter result is dependent on normalization.

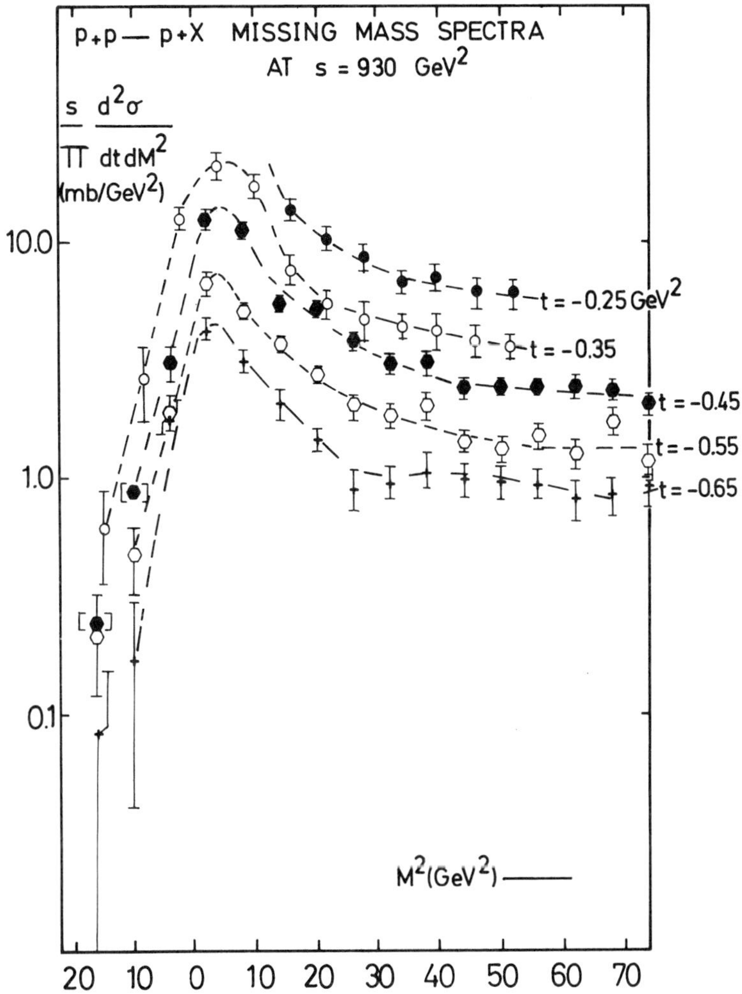

FIG. 51. Data on $pp \to pX$ at $s = 930\,\text{GeV}^2$. The points at $7 < M^2 < 50$ at $0{\cdot}45 < t < 1{\cdot}65$ were used to obtain the PPP triple-Regge cross-section. Elastically scattered protons have been removed.

The analysis of the forward peaks at $s = 551$ and $930\,\text{GeV}^2$ is less straight forward:

(a) A substantial fraction of the raw data in the peak is due to elastic scattering. They are separated from the inelastics by requiring a second particle to emerge collinearly in the opposite hemisphere. The effect of applying a collinearity cut is shown in Fig. 53. The

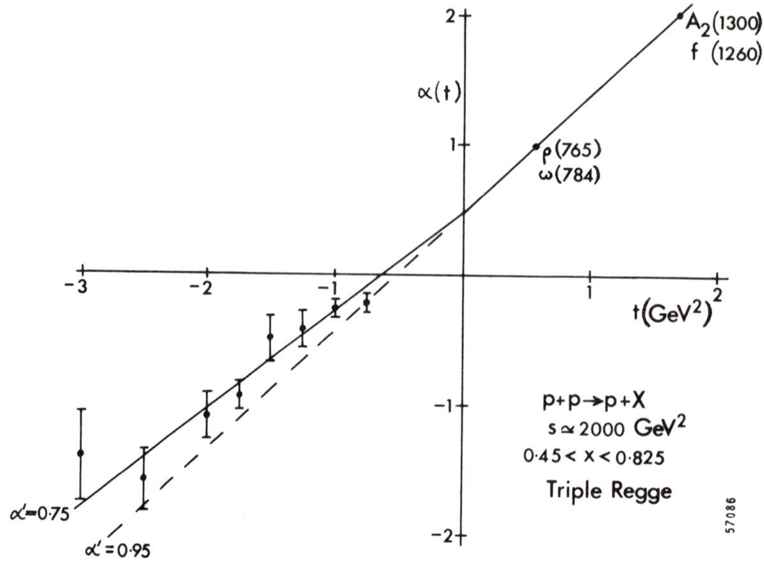

FIG. 52. Effective meson trajectory obtained from the data of Fig. 2.

elastic peak thus obtained is used to check the momentum scale. The t-dependence of the elastic data is checked to be in agreement with published results.

(b) The momentum spread in the primary beams and the measuring errors on the secondary proton momentum (see Section V.1) modify the shape of the peak. The elastic scattering data in (a) show a peak whose width is a measure for the overall resolution of the equipment. The result, expressed as δx, δM^2 and standard deviation of the momentum curve for elastic events:

s_1	δM^2	δx (FWHM)	St. dev. in p
551 GeV^2	11 GeV^2	2%	0·85%
930 GeV^2	9·6 GeV^2	1%	0·42%

These data are used to fold resolution functions of the specified widths (corresponding to Gaussian with standard deviations of 4·7 and 3·9 GeV^2 resp.) into the triple-Regge formula (VI.2).

(c) At high x the spectrum is composed partly of diffractively scattered protons, partly of high momentum fragmentation protons. The latter contribution can be subtracted by assuming its x-dependence to be specified by the scaling RRP triple-Regge term, as suggested by the data at s = 1995 GeV^2 above. The effect of subtracting a "background" of the form of eqn (VI.2) with trajectory (VI.4) is shown in Fig. 54.

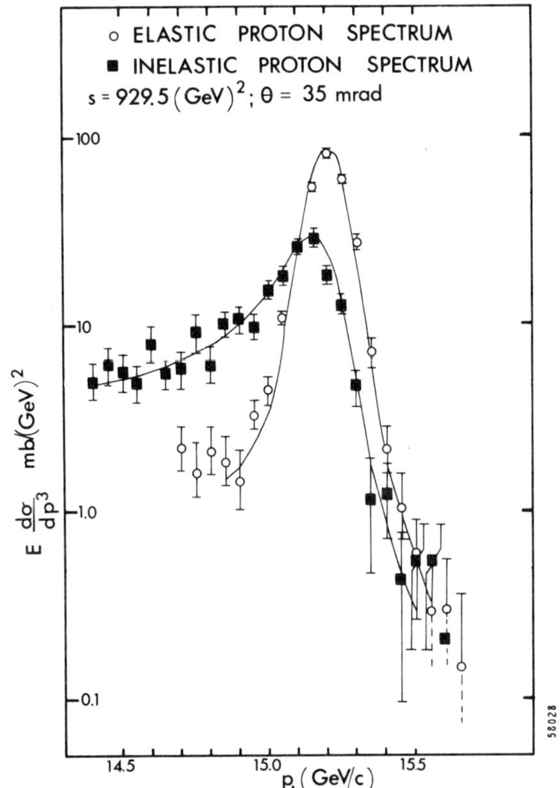

FIG. 53. Separation of elastic from inelastic events by means of applying a cut on the angle between the proton in the spectrometer and one other particle going into the opposite hemisphere.

With these problems out of the way we now look at the data and note that (1) there is little if any dependence on s; (2) at the lowest t measured the slope of the M^2 distribution is $B(t = 0{\cdot}15) = 0{\cdot}98 \pm 0{\cdot}06$. This suggests dominance of the PPP term, since at $t = 0$:

$$\text{PPP} \qquad \sigma_{\text{inv}} \sim \frac{1}{1-x} \sim \left(\frac{M^2}{s}\right)^{-1}$$

$$\text{PPR} \qquad \sigma_{\text{inv}} \sim \frac{1}{\sqrt{s}} \sim \frac{1}{(1-x)^{\frac{3}{2}}} \sim \frac{1}{\sqrt{s}}\left(\frac{M^2}{s}\right)^{-1\cdot5}$$

$$\text{RRR} \qquad \sigma_{\text{inv}} \sim \frac{1}{\sqrt{s}} \sim \frac{1}{(1-x)^{\frac{1}{2}}} \sim \frac{1}{\sqrt{s}}\left(\frac{M^2}{s}\right)^{-0\cdot5}$$

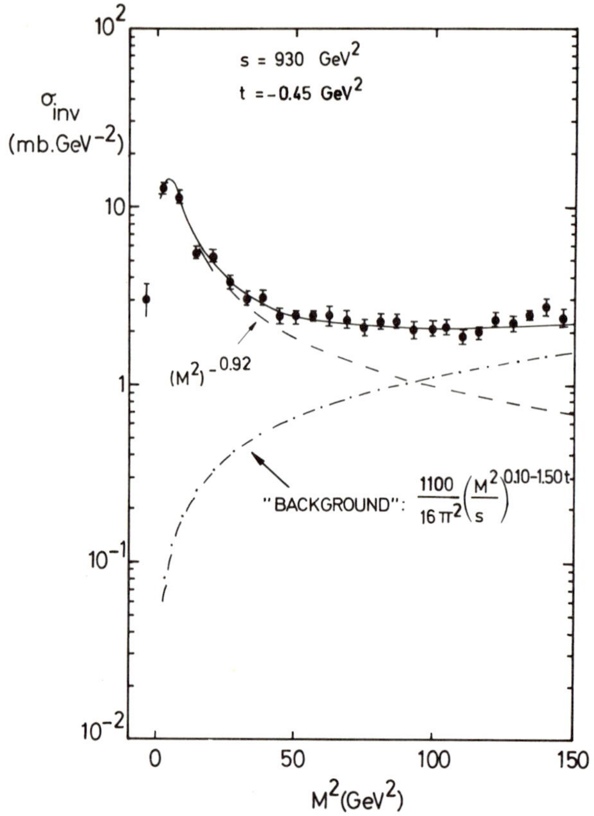

FIG. 54. Decomposition of the forward peak in $pp \to pX$ into contribution arising from leading protons and from fragmentation protons ("background").

while interference terms such as PRP would have $B(0) \sim 0{\cdot}5$ (although this term would scale). If we then assume PPP to be the only important term we have

$$E\frac{d^3\sigma}{dp^3} = \frac{G_{PPP}(t)}{16\pi^2}\left(\frac{M^2}{s}\right)^{1-2\alpha_P(t)} \qquad \text{(VI.6)}$$

Fitting this to the data, and taking into account points a, b, c above, gives the result:

$$\alpha_P(t) = 1{\cdot}0 + 0{\cdot}2t \qquad \text{(VI.7)}$$

as the effective Pomeron trajectory, see Fig. 55. Note again that this result is independent of the normalization. The data at both energies, $s = 551$ and 930 GeV^2, are consistent with the same trajectory.

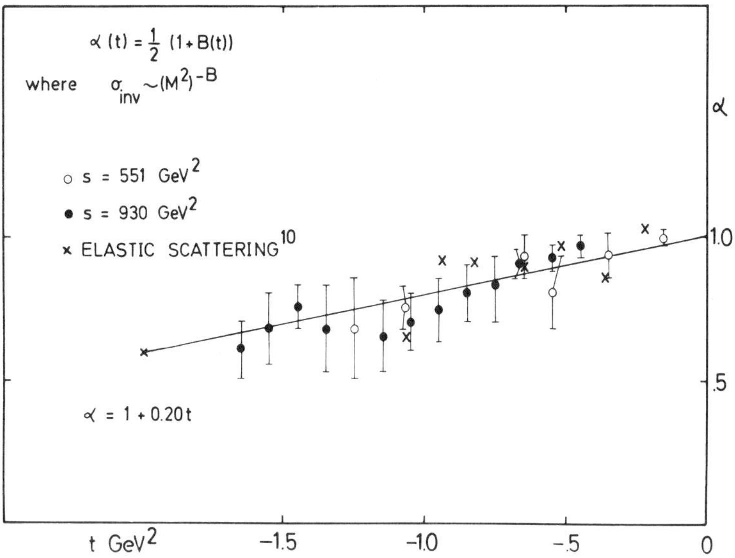

FIG. 55. Effective Pomeron trajectory obtained by fitting the PPP triple-Regge term to the $pp \rightarrow pX$ at $s = 551$ and $930\ \mathrm{GeV}^2$ after correction for elastic events, effects of finite resolution, and contributions by fragmentation protons, Ref. 3.

Encouraged by the results we now look for confirmation or denial in other processes. If elastic scattering were dominated by Pomeron exchange we would parametrize the differential cross-section with the expression

$$\frac{d\sigma}{dt} = f(t)\, s^{2\alpha_P(t) - 2} \tag{VI.8}$$

Using recent data at the same energies, $s = 551$ and $s = 930\ \mathrm{GeV}^2$, we then find the crosses in Fig. 55, also suggesting that the Pomeron trajectory has a small, but non-zero slope.

It is of interest to check the whole triple-Regge idea in reactions where quantum number exchange restricts the possible exchanges. One case is $pp \rightarrow \pi^+ X$. The only possible terms are now RRP and RRR with R = nucleon trajectory. Since RRR → const. and RRP → $1/\sqrt{s}$ (see table above) we expect the scaling limit to be approached as $A + B/\sqrt{s}$, in accord with data (see Section III, Fig. 32).[3] A fit[30] to data (see Fig. 56) at fixed $p_T = 0{\cdot}8\ \mathrm{GeV}/c$ versus x at $s = 2000\ \mathrm{GeV}^2$ results in

$$\alpha_N = -0{\cdot}35 + 1{\cdot}0\, t, \tag{VI.9}$$

a not unreasonable value for the nucleon trajectory. Note that experimentally such data are difficult to obtain due to the very steep drop in

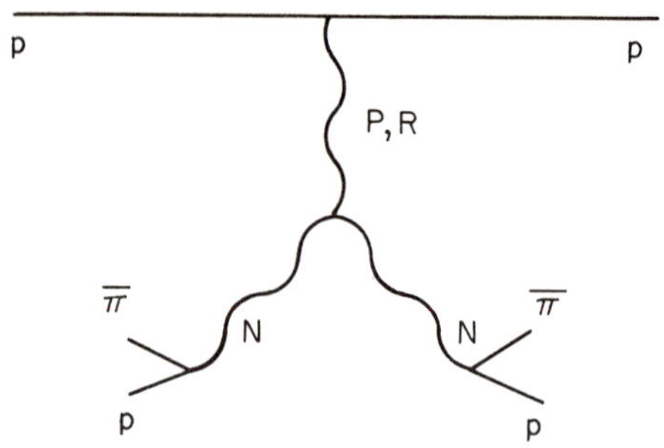

cross-section with increasing x in the absence of PPP and PPR contributions.

Recently,[31] data on $pp \to nX$ at $s = 530$, 930, 2025 and 2800 GeV^2 in limited regions of x and p_T have been analysed with a single RRP term dominating the spectrum at all x. As expected the data show no forward peak. The data, shown in Fig. 57, are not complete enough to map out a full trajectory but they are consistent with an effective meson trajectory given by

$$\alpha_R = 0 + 1{\cdot}5\,t \qquad \text{(VI.10)}$$

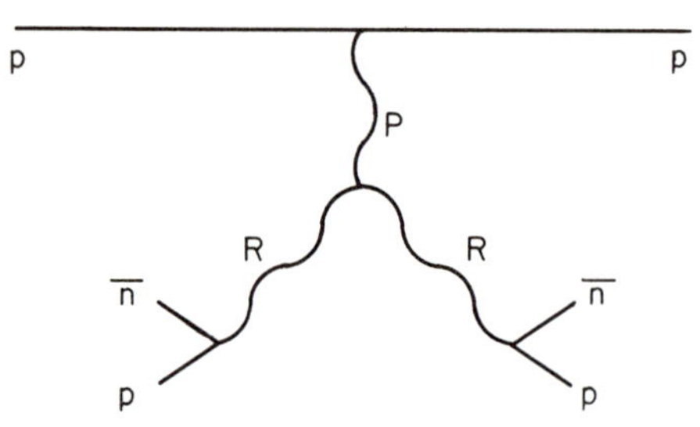

This is suggestive of the π-dominance at the $\overline{n}p$ vertex. Note that this is in contrast, but not in contradiction, to the effective meson trajectory (VI.4) which is dominated by ρ exchange.

In conclusion one might say that in spite of the many assumptions and the very incomplete data, the classical Regge-exchange picture of strong interactions, carried as close to its asymptotic limits as present apparatus permits, provides a consistent description of particle production data and leads to "reasonable" effective Pomeron, nucleon and meson trajectories.

3. PROCESSES CONTRIBUTING TO THE RISE IN σ_{tot}

The data discussed in Section VI.2 on $pp \to pX$ at $s = 551$ and 930 GeV^2 permit one further conclusion. Either by substituting the trajectory (VI.7)

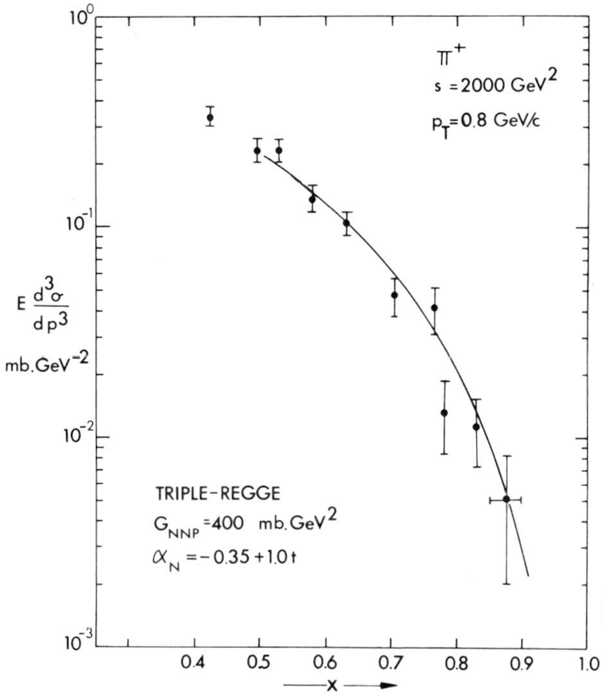

FIG. 56. Inclusive π^+ production versus x at $p_T = 0{\cdot}8$, compared with a fit to the $R(=N)\ R(=N)$ P triple-Regge term. (M. G. Albrow *et al.*, unpublished).

into (VI.6) or by using the "experimental" numbers for the slopes of the M^2 distributions at fixed t on a log–log plot one obtains the quantity $G_{\mathrm{PPP}}(t)$ as a function of t. The result is shown in Fig. 58 ("trajectory" data above, "measured slope" data below). The result at the two energies is the same, both in slope and in magnitude. (This latter conclusion depends on the normalization).* The t-dependence is non-exponential and can be parametrized in several ways, e.g. as

$$G_{\mathrm{PPP}}(t) = A \exp (Bt + Ct^2) \qquad \text{(VI.11)}$$

with

$$A = 150\,\mathrm{mb/GeV^2} \qquad B = 4{\cdot}3\,\mathrm{GeV^{-2}} \qquad C = 1{\cdot}09\,\mathrm{GeV^{-4}}$$

or as

$$G_{\mathrm{PPP}}(t) = A, \exp B_1 t + A_2 \exp B_2 t \qquad \text{(VI.12)}$$

* Note added in proof: Recent new data at $s = 551\,\mathrm{GeV^2}$ has thrown doubt on the correctness of the normalization of the $s = 551\,\mathrm{GeV^2}$ data considered here. Statements concerning the magnitude of $G_{\mathrm{PPP}}(t)$ should thus be treated with caution for the time being.

with

$$A_1 = 140 \quad A_2 = 8{\cdot}8\,\text{mb/GeV}^2 \quad B_1 = 4.3 \quad B_2 = 0.76\,\text{GeV}^{-2}$$

(The latter fit is drawn as the solid line in Fig. 58).

The quantity $G_{PPP}(t)$ is of fundamental interest. It links various processes and prescribes their behaviour at energies above the presently available

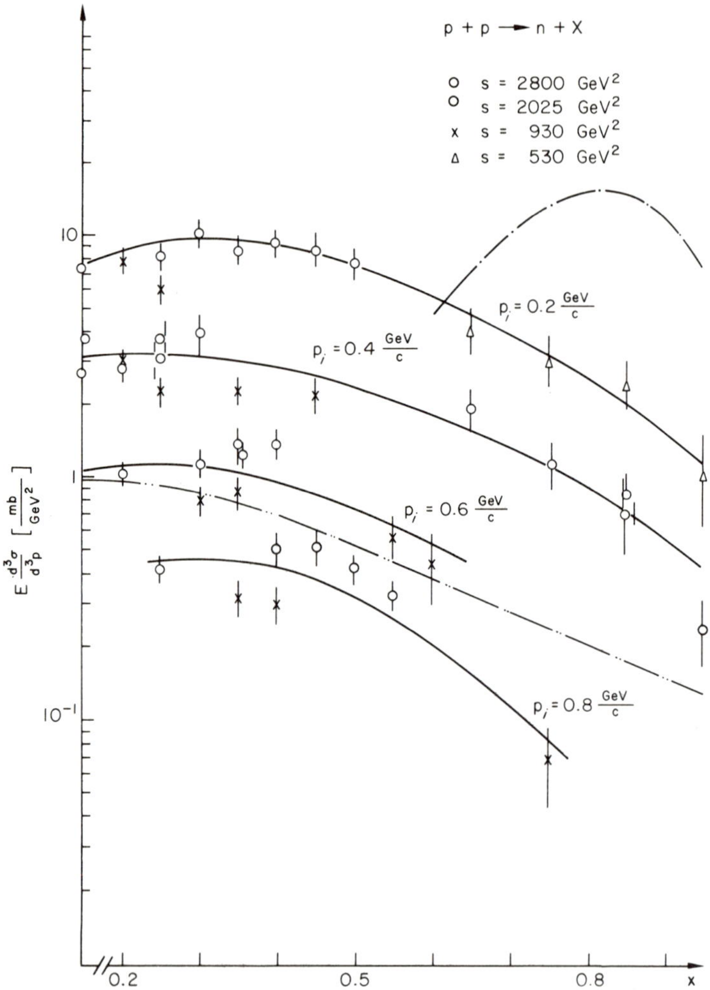

FIG. 57. Data of Ref. 4 on $pp \to nX$ at various s and p_T. These data have been used to extract an effective meson trajectory, by fitting to the RRP triple-Regge term. The result suggests π-exchange dominance in $pp \to nX$.

ones. Inspection of the triple-Regge diagram shows that $G_{\text{PPP}}(t)$ is the product of 4 factors:

$$G_{\text{PPP}}(t) = |\beta_{ppP}(t)|^2 \beta_{ppP}(0) g_{\text{PPP}}(t, t, 0) \tag{VI.13}$$

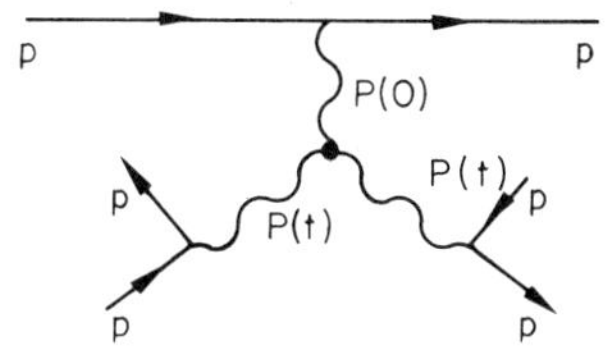

Now in elastic scattering

$$\frac{d\sigma}{dt}(t) = \frac{1}{16\pi} |\beta_{ppP}(t)|^2 |\beta_{ppP}(t)|^2 s^{\alpha_P(t) - 2} \tag{VI.14}$$

while the total cross-section is given by

$$\sigma_T = \beta_{ppP}(0) \beta_{ppP}(0) s^{\alpha_P(0) - 1} \tag{VI.15}$$

and thus

$$g_{\text{PPP}}(t, t, 0) = G_{\text{PPP}}(t) \left[16\pi\sigma_T \frac{d\sigma}{dt}(t) s^{2 - 2\alpha_P(t)} \right]^{-\frac{1}{2}} \tag{VI.16}$$

Hence the triple-Pomeron coupling constant g_{PPP} can be determined from a combination of data on elastic scattering, the total cross-section and diffractive inelastic scattering. In putting in various recent data (details omitted here) it turns out that the t-dependence of $G_{\text{PPP}}(t)$ is largely cancelled by the t-dependence of the bracket in (VI.16), yielding a triple Pomeron coupling constant $g_{\text{PPP}}(t) \approx 0{\cdot}4\,\text{GeV}^{-1}$, not or very weakly dependent on t in the range $0{\cdot}15 < t < 0{\cdot}8\,\text{GeV}^2$ (above $t \approx 0{\cdot}8$ the dip at $t = 1{\cdot}4\,\text{GeV}^2$ affects the results).

This result can be put into a different form by rewriting the triple Pomeron expression for the cross-section, with (VI.2) and (VI.13):

$$E\frac{d^3\sigma}{dp^3} = \frac{1}{16\pi^2} |\beta_{ppP}(t)|^2 \left(\frac{M^2}{s}\right)^{\alpha_P(0) - 2\alpha_P(t)} \sigma_{pP}(M^2, t) \tag{VI.17}$$

with

$$\sigma_{pP}(M^2, t) = \beta_{ppP}(0) g_{\text{PPP}}(t) (M^2)^{\alpha_P(0) - 1} \tag{V.18}$$

(only the PPP-term has been retained for clarity; other terms could be included).

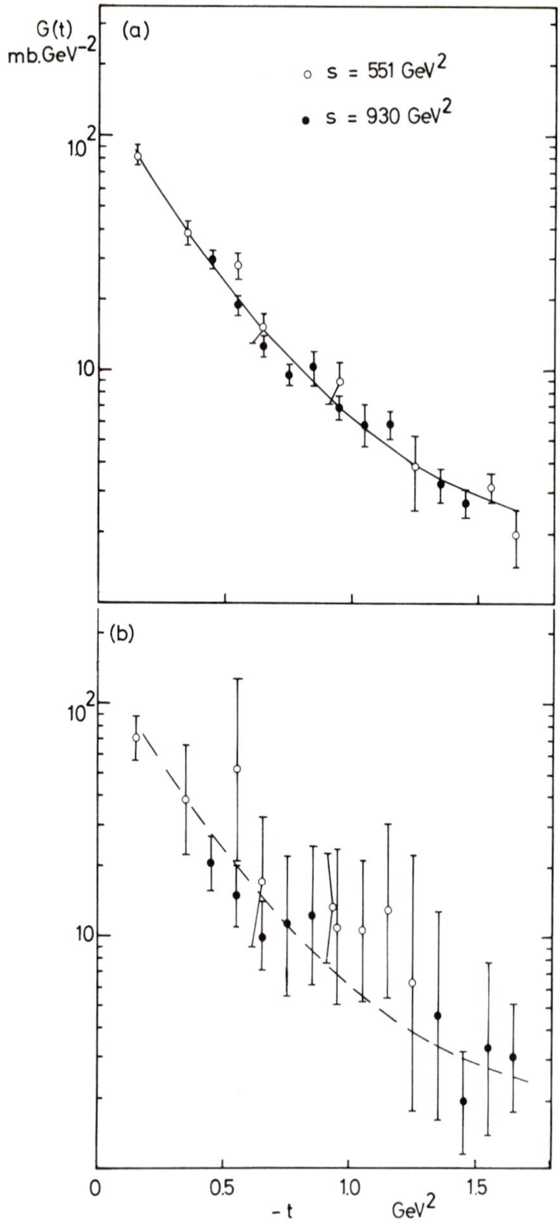

FIG. 58. The "overall' PPP coupling constant as a function of t, extracted from $pp \rightarrow pX$ data at $s = 551$ and 930 GeV^2, see text. See eqn (VI.13) for a break-down of $G_{PPP}(t)$ into its components.

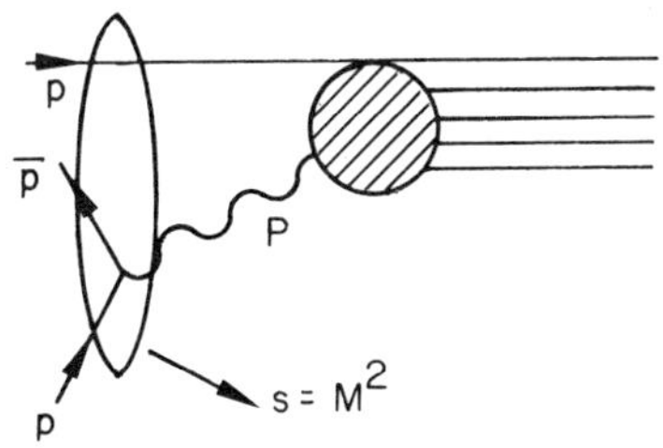

$\sigma_{pP}(M^2, t)$ is the Pomeron-proton total cross-section at total energy $s = M^2$ and with mass$^2 = t$.

Since dependence on t in (VI.18) enters only through $g_{PPP}(t)$ the result above (constant g_{PPP}) seems to say that there is no strong dependence on the "Pomeron mass". As to the dependence on M^2: if PPP is a good description of the data, then $\sigma(M^2, t)$ should be independent of M^2. Explicit evaluation of (VI.17) and (VI.14) using the elastic and inelastic (with peaks corrected for fragmentation protons as in Section VI.2) data at $s = 551$ GeV^2 and at

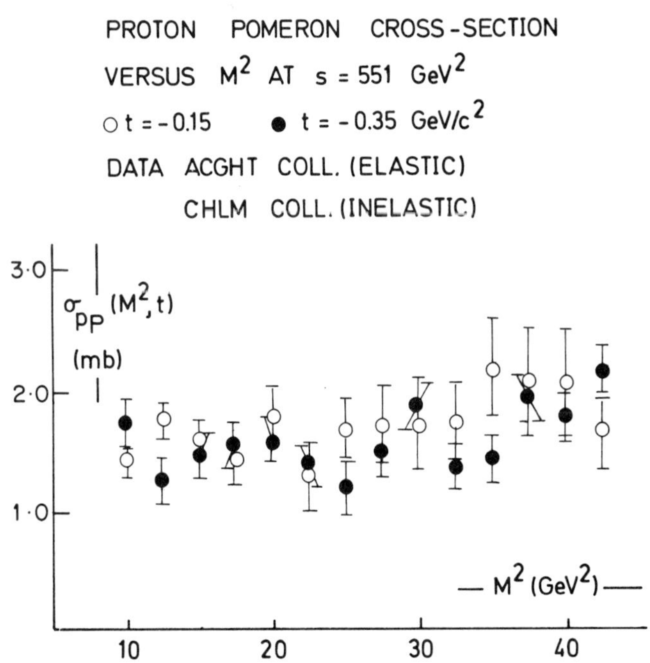

FIG. 59. Total effective Pomeron-proton cross-section as a function of the total energy M^2 and the "Pomeron mass" t, deduced from data on elastic ($pp \to pX$) scattering, see eqns (VI.14) and (VI.17) in the text.

$t = -0{\cdot}15$ and $-0{\cdot}35\ \mathrm{GeV}^2$ gives the points shown in Fig. 59. The result is independent of energy (M^2), as is pp scattering in the same energy (s) range.

This view of the interaction in terms of the total proton–Pomeron cross-section has another interesting consequence. In the limit where this picture is valid (high s, high M^2, small t) the average multiplicity of events produced in proton–Pomeron interactions should depend in the same way on the total energy $s = M^2$ as the overall average multiplicity in proton–proton interaction depends on s. In particular the coefficient A in

$$\langle n\rangle_D = A \ln M^2 \qquad \langle n\rangle = A \ln s$$

should be the same. There is evidence in NAL bubble chamber experiments that this is in fact the case.

One remaining question is the value of $g_{\mathrm{PPP}}(t)$ at $t = 0$ and how this value is approached. To get a feeling for how this might go we extrapolate the PPP cross-section obtained from the ISR data to the t-range covered by the recoil detection experiment at the same s at NAL (see Section V). The RRP term has been added in order to account for the contribution of the fragmentation protons at large M^2. The result is shown in Fig. 60. It is seen that in the mass range in which the PPP fit has been made ($5 < M^2 < 30\ \mathrm{GeV}^2$) the agreement between the NAL data and the PPP + RRP curve is reasonably good. One might interpret this as saying that down to $|t| = 0{\cdot}056\ \mathrm{GeV}^2$ there is no sign of any turn over in $G_{\mathrm{PPP}}(t)$. If this trend were to continue down to $t = 0$ we would have that

$$g_{\mathrm{PPP}}(0) \approx 0{\cdot}4\ \mathrm{GeV}^{-1}$$

This result is tentative and needs confirmation; ISR cannot easily contribute to its solution, due to the relatively short straight sections, making momentum measurements at small t very difficult to do.

The increase in σ_{tot} with s observed recently naturally brings up the question as to which one(s) of the processes contributing to σ_{tot} causes this dependence on energy. The rise in σ_{tot} is $4{\cdot}1 \pm 0{\cdot}7$ mb between $s = 551$ and $s = 2788\ \mathrm{GeV}^2$. Subtracting the (rising) elastic cross-section we are left with a rise of $3{\cdot}3 \pm 0{\cdot}7$ mb to be accounted for by inelastic processes.

Consider first diffraction dissociation. We have seen that the inelastic peak in $pp \to pX$ roughly scales and that at small t the dependence on mass, $\mathrm{d}\sigma/\mathrm{d}M^2$, suggests something close to $1/M^2$. This suggests PPP dominance and hence we have proceeded to parametrize the data in this way *after subtraction of the contribution of fragmentation protons and in a mass range where (a) the contribution of fragments is reasonably small,* $M^2 < 30\ \mathrm{GeV}^2$, *and (b) where* PPP *dominates over any* PPR *terms, direct resonances or whatever,* $M^2 > 5\ \mathrm{GeV}^2$. The result thus represents a parametrization of the PPP content of the diffraction peak and hence by integrating it we shall

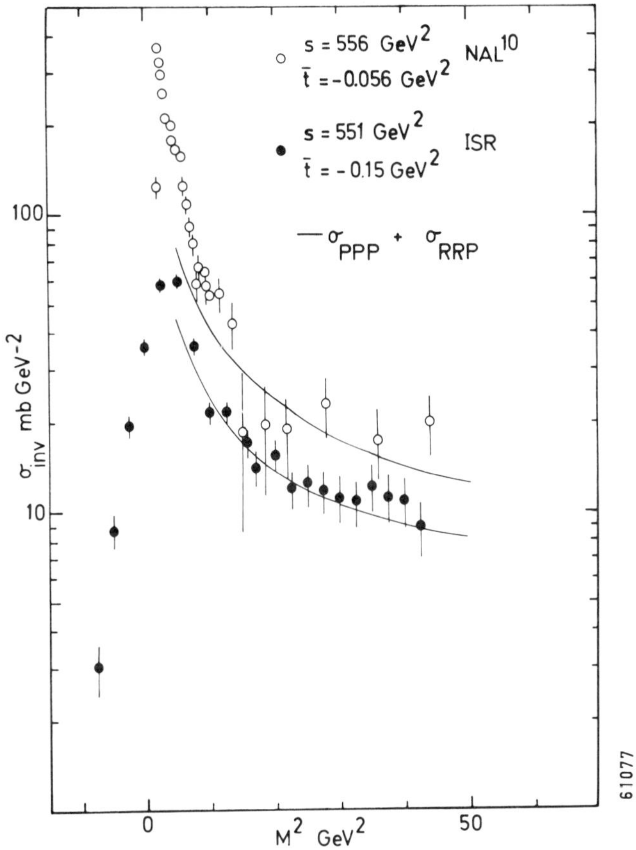

FIG. 60. The PPP cross-section obtained from ISR data (Ref. 3) at $0{\cdot}15 < t < 1{\cdot}25\ \text{GeV}^2$, extrapolated down to $t \approx 0{\cdot}056\ \text{GeV}^2$ and compared to NAL data (Ref. 32) at some total energy. The PPP line has been obtained in the range $5 < M^2 < 30\ \text{GeV}^2$. At smaller masses other contributions enter, as suggested by the large peak at low t.

get the triple-Pomeron part of the cross-section for diffraction dissociation. This is

$$E\frac{d^3\sigma}{dp^3} = \frac{s}{\pi}\frac{d^2\sigma}{dt\,dM^2} = \frac{G_{PPP}(t)}{16\pi}\left(\frac{M^2}{s}\right)^{1-2\alpha_P(t)}$$

$$\sigma = \frac{1}{16\pi}\int\int G_{PPP}(t) z^{-1-2\alpha' t}\,dt\,dz$$

with $z = M^2/s$. The limits of integration are ($m = m_p$)

$$z_{\min} = \frac{m^2}{s} \qquad z_{\max} = \frac{(\sqrt{s} - m)^2}{s}$$

$$-t_{\min} = \frac{m^2}{x}(1 - x)^2 \approx m^2\left(\frac{M^2}{s}\right)^2 \qquad -t_{\max} = \infty$$

There are some delicate points about this integral, concerning the relevant range in mass and the influence of very small t. We shall not belabour this further here. To get an estimate we assume $t_{\min} \approx 0$ and $G_{PPP}(t)$ as given by eqn (VI.12). Then

$$\sigma_i = \frac{A_i}{32\pi\alpha_P} \ln \frac{B_i - 2\alpha' \ln (M^2_{\min}/s)}{B_i - 2\alpha' \ln (M^2_{\max}/s)}$$

and this results in

$$\sigma(2788\ \text{GeV}^2) - \sigma(551\ \text{GeV}^2) \approx 1{\cdot}5\ \text{mb}.$$

Restricting the integration to e.g. $2 < M^2 < 50\ \text{GeV}^2$ results in 1·2 mb. Note that although we have obtained our PPP parameters in a restricted range of M^2 and t, we now extend the range in making the integral. Our tentative conclusion is that the PPP contribution to the diffraction cross-section is somewhere in the range 1 to 2 mb, and hence does not account for the entire rise in σ_{inel}.

Other contributions to the rise in σ_{tot} could originate from:

(1) $pp \to pX$ with $M_x^2 \lesssim 3\ \text{GeV}/c$ at small t. In this range terms other than PPP may contribute (see Fig. 60).
(2) π^+ and π^- production at $X \approx 0$ and small p_T. At $p_T \gtrsim 0{\cdot}1$ a small rise is observed.
(3) At very large p_T there is a rise with increasing s in the invariant diff. cross-section for π-production.
(4) $\bar{p}p$ production at $x \approx 0$.

Data are not yet good enough to make quantitative estimates of the various contributions.

Acknowledgement

I am indebted to Mrs. M. Keller for taking excellent care of the manuscript.

References

1. L. D. Landau and E. M. Lifshitz. "The Classical Theory of Fields (12, p. 37, revised second edition); applied to ISR e.g. in J. F. v. d. Veen, M.Sc. Thesis, University of Utrecht (1973, unpublished).

2. E. D. Courant, H. S. Snyder, *Ann. Phys.* **3**, 1 (1958).
3. This figure and its description is taken from J. P. Gourber and K. N. Henrichsen. In "Proceedings of the 8th International Conference on High Energy Accelerators". CERN 1971, p. 518.
4. ISR running-in reports 1971.
5. B. de Raad, "Beam-gas Interactions with ISR". CERN 68/ISRU/103.
6. E. Keil, CERN 72-14 (Lectures Academic Training Programme).
7. S. v. d. Meer, CERN Internal Report, ISR-PO/68-31 (1968).
8. Amendolia *et al.*, contribution to the Int. Conf. on Instrumentation for High Energy Physics, Frascati, 8–12 May 1970.
9. W. Schnell, operating results from ISR, ISR–RF/73-7.
10. See e.g. K. Johnsen, "Summary of Present Thinking on High Luminosity Insertions in the ISR." CERN/ISR-DI/73-18.
11. S. v. d. Meer, "Stochastic Damping of Betatron Oscillations in the ISR." CERN/ISR-PO/72-31 (1972).
12. K. Hübner, "The Storage of Antiprotons in the ISR". CERN/ISR-TH/73-19.
13. M. G. Albrow *et al. Nucl. Phys.* **B56**, 333 (1973).
14. See e.g. K. Gottfried. "Lectures Academic Training Program," CERN/TH 1615, for a much more detailed discussion.
15. British-Scandinavian collaboration, B. Alper *et al.*, preprint.
16. M. Antinucci *et al. Lett. Nuovo. Cimento* **6**, 121, 1973.
17. B. Alper *et al.* Submitted to *Phys. Letts.*
18. A. Diddens, Lectures IVth Seminar on Theoretical Physics, GIFT, Barcelona, April 1973.
19. M. G. Albrow *et al.*, *Nucl. Phys.* **B56**, 333, 1973.
20. E. Lillethun, Lectures at the XIII Cracow School of Theoretical Physics, Zacopane, June 1973.
21. F. Sannes *et al. Phys. Rev. Letts.* **30**, 766, 1973.
22. Allaby *et al. Nucl. Phys.* **B52**, 316 (1973).
23. L. Foa, invited paper Aix-en-Provence Conference 1973; M. Jacob, Lectures Louvain 1973; E. Berger, Lectures 1973. J. C. Sens, invited paper, Conf. Recent advances in Particle Physics, N.Y., 1973.
24. F. Sannes *et al. Phys. Rev. Letts.* **30**, 766, 1973.
25. S. Childress *et al.* NAL preprint.
26. M. G. Albrow *et al. Nucl. Phys.* **B51**, 388 (1972); *Nucl. Phys.* **B54**, 6 (1973).
27. M. G. Albrow *et al.* To be published *in Nucl. Phys.*
28. A. Capella, SLAC-PUB-1198 (1973); A. B. Kaidalov *et al.* Submitted to *Phys. Lett.*; J. Gabarro and C. Pajares, LPTHE 73/17; R. G. Roberts and D. P. Roy, RL-73-065; M. H. Albrow *et al. Nucl. Phys.* **B51**, 388 (1972); and others.
29. M. G. Albrow *et al. Nucl. Phys.* **B51**, 388 (1972).
30. M. G. Albrow *et al.* Unpublished.
31. J. Engler *et al.* Contribution to the Aix-en-Provence Conference.
32. Childress *et al.* NAL preprint.

3. PHENOMENOLOGY OF INCLUSIVE REACTIONS

R. G. ROBERTS

Rutherford High Energy Laboratory, Chilton, U.K.

I. INTRODUCTION

The study of strong interactions during the last two years has been dominated by one main topic, namely multiparticle production and the attempts to learn the dynamics which govern the cross-sections, multiplicities, correlations, etc., of such processes. The source of all this interest has been the large flow of detailed information which has been coming from the new high energy accelerators; the ISR at CERN and the 300 GeV accelerator at NAL. Although the ISR is able to reach much higher centre of mass energies than NAL, the latter is able to provide more varied information and the roles of the two machines are very complementary in nature. At the same time other machines, Serpukhov and the lower energy accelerators, have provided important data which allow the energy dependence of many effects to be studied over the widest possible range. We shall see, in fact, that some high-energy models for particle production provide plausible explanations of data on quasi-two-body scattering at energies as low as 8 GeV/c.

One of the most striking features of the ISR data is the huge number of particles which are produced in the collisions. The *average* number of particles in the final state is around seventeen and it is not unusual for 30 particles or so to be produced. Of course it is virtually impossible to experimentally detect such a state and absolutely impossible to identify all the particles and detect their momenta. The compromise usually reached is to identify and measure one or may be two particles of a particular type, e.g. π^+ and to ignore the remainder of the junk associated with that particle.

The reaction is therefore of the type

$$a + b \to c + X$$

or

$$a + b \to c + d + X$$

(where $X \equiv$ sum over all possible states)

and such reactions are known as inclusive reactions to distinguish them from exclusive reactions where all final state particles are measured. Apart from the experimental simplicity (!) such reactions can be handled much more easily, simply because there are only a few variables to consider. For the case $a + b \to c + X$ there are only three independent variables including the incoming energy and so even a theorist has some hope of understanding the data. The most inclusive reaction of all is one where no final state particles are measured, i.e. $a + b \to X$ and is just a total cross-section depending on only the incoming energy.

Our understanding of the behaviour of total cross-sections comes almost entirely from using the optical theorem to translate the study to that of elastic scattering amplitudes. The systematic behaviour of these, in turn, can be largely explained by Regge theory and duality. In this way, for example, we understand why certain total cross-sections show far less variation with energy than others.

The content of these lectures is mainly a generalization of this approach to inclusive reactions and it is quite remarkable how much progress has been made from this direction in explaining features of both the high and lower energy data. This is not to say that this is the *only* profitable approach and I shall perhaps touch on others later on. However the theme of these talks is the application of the ideas of Regge theory and duality to inclusive reactions.

Our interest in inclusive reactions was sparked off by a prediction due to Feynman[(1)] and Yang and collaborators[(2)] that the cross-sections would exhibit scaling at high energies. By this we mean that the inclusive cross-section would be a function of the ratio of two variables, one of which is the energy, rather than the energy directly. This is analogous to Bjorken's prediction that cross-sections for $e + p \to e + X$ would show scaling in the deep inelastic region but let me immediately remark that the two scaling phenomena are quite distinct and separate.

The arguments of Feynman and Yang were based on physical intuition, e.g. by analogy with e.m. radiation from a classical current, in Feynman's case. I shall not describe their basis for the scaling prediction—instead we shall see how scaling follows naturally in the Regge picture once we accept that *total* cross-sections approach some asymptotic limit.

The breakthrough which allowed us to carry over Regge-duality ideas to inclusive cross-section was achieved by Mueller[(3)] who generalized the usual optical theorem so that inclusive cross-sections could be related to certain forward 3-body scattering amplitudes. These in turn could be described in terms of Regge exchange, etc.

Before describing this generalized optical theorem and the consequences which follow from it, we shall first have to go through some kinematics and list some definitions to help us later on.

II. Kinematics and Definitions

Here we shall consider only the *single*-particle inclusive reaction $a + b \to c + X$ (let us also call "a" beam, "b" target).

In the target frame

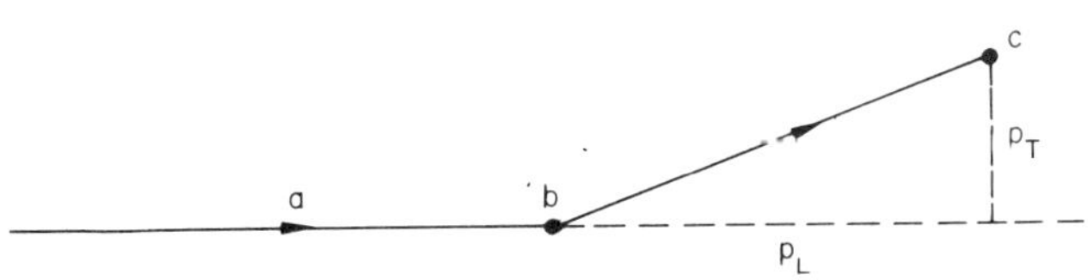

The longitudinal and transverse momenta of c, p_L and p_T are defined above. Also similarly in the beam frame or centre of mass frame, in each case only p_L changes, p_T is invariant to boosts along the beam direction. p_L in c.o.m. is usually denoted p_L^*.

One of the most striking features of all multiparticle processes is the strong tendency for p_T to remain small even as the energy increases. No matter how much we raise the incoming energy the average value of p_T stays around 0·3 GeV with an exponential distribution in p_T^2. Although we have mentioned the "high" multiplicities observed at ISR, nevertheless the actual growth in multiplicity is *slow*, for example consistent with ln s. This, taken together with the smallness of p_T, implies that the bulk of the incoming energy is converted into longitudinal motion of the "few" particles in the final state.

Writing the momentum vector of c

$$p_c = (E, p_L, p_T)$$

we write

$$E^2 = \kappa^2 + p_L^2$$

where $\kappa = (m_c^2 + p_T^2)^{\frac{1}{2}}$ is called the "longitudinal mass" of c.

Ignoring what happens in the transverse plane, the situation is one-dimensional with κ playing the role of the mass of c at a fixed p_T.

Despite the fact that there are only 3 independent variables for $a + b \to c + X$ there are many choices, each of which is aptly suited to discuss a particular approach or particular region of phase space.

A frequently used variable is x, defined as

$$x = \frac{p_L^*}{p_L^{*\max}} \xrightarrow[s \to \infty]{} \frac{2}{\sqrt{s}} p_L^*. \tag{1}$$

Thus x varies from -1 to $+1$.

Another longitudinal variable is "rapidity":

$$y = \sinh^{-1}\left(\frac{p_L}{\kappa}\right)$$

$$= \ln\left(\frac{E + p_L}{\kappa}\right)$$

$$= \frac{1}{2}\ln\left(\frac{E + p_L}{E - p_L}\right). \tag{2}$$

Naturally y depends in which frame p_L is measured but one attraction of this variable is its *additive* nature under boosts along the longitudinal axis.

Suppose we work in the target frame, Y = rapidity of beam

$$p_a = (m_a \cosh Y, 0, 0, m_a \sinh Y)$$

$$p_b = (m_b, 0, 0, 0)$$

$$s = (p_a + p_b)^2 \xrightarrow[s \to \infty]{} m_a m_b\, e^Y, \text{ i.e. } Y \sim \log\frac{s}{m_a m_b}$$

$$p_c = (\kappa \cosh y, \underline{p}_T, \kappa \sinh y)$$

and so

$$\ln\frac{\kappa}{m_b} < y < Y + \ln\frac{m_a}{\kappa}. \tag{3}$$

Hence the y plot grows logarithmically in s. To get a comparative feeling for the x and y plots we can put $p_T = 0$ and look at $pp \to p$ at ISR.

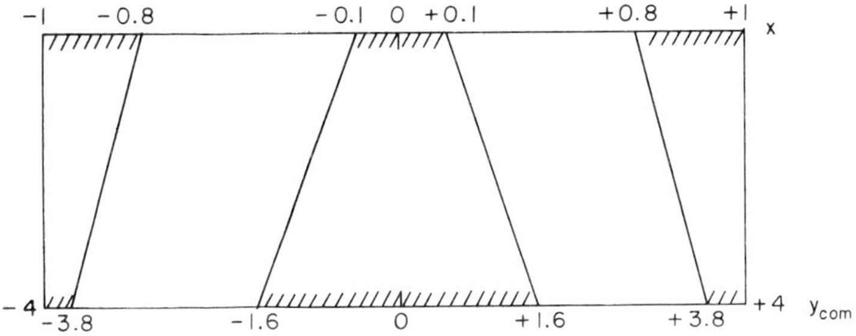

Thus the "central" shaded region covers only 10% of the x plot but 40% of the rapidity plot. The "edge" shaded regions each cover 10% of the x plot but only 2·5% of the y plot. Clearly it pays to use the right longitudinal variable depending upon which region of phase space one is discussing The particular model one has in mind is naturally better expressed in one variable rather than another. The central region corresponds to the region where the produced particle c is slow in the c.o.m. frame and so rapidity is the more natural variable to use. Outside of this centre region, the particles tend to be slow in either the target or projectile frame and may be thought of as "fragments" of the target or projectile. One then talks of the "fragmentation" regions of target and projectile and x is the most natural variable, (or p_L^{target}, p_L^{proj}). These definitions will be tightened up as we proceed.

For most of our discussion we shall be using invariant quantities suitable for Regge theory.

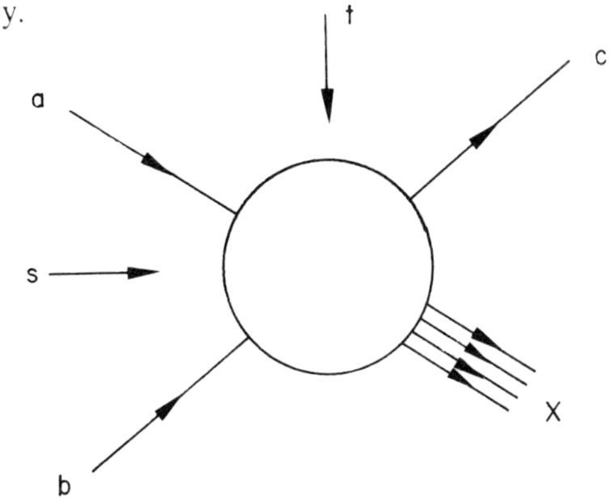

$$s = (p_a + p_b)^2 \quad t = (p_a - p_c)^2 \qquad u = (p_b - p_c)^2.$$

Also the missing pass squared $M^2 = (p_a + p_b - p_c)^2$ and $s + t + u = m_a^2 + m_b^2 + m_c^2 + M^2$. In the projectile fragmentation region, s, t and M^2 would be suitable variables since t is finite there while s, u and M^2 would be the variables to discuss the target fragmentation region. In the central region t and u *both* become large.

In the target system

$$u = m_b^2 + m_c^2 - 2m_b E$$

$$M^2 = (m_a^2 + m_b^2 + m_c^2) + 2[E^a_{\text{lab}} m_b - (m_b + E^a_{\text{lab}})E + p^a_{\text{lab}} p_L] \tag{4}$$

In the c.o.m. frame

$$u = m_b^2 + m_c^2 - 2E_b E^* - 2pp_L^*$$

$$M^2 = s - 2(\sqrt{s})E^* + m_c^2. \tag{5}$$

From the last line we see that if $p_L^* \gg p_T$ then

$$\frac{M^2}{s} \simeq 1 - \frac{2}{\sqrt{s}} p_L = 1 - x \tag{6}$$

For a fixed s, therefore, the extremities of the x plot are populated by low missing mass, i.e. by low multiplicity final states. Also we see that in this region close to $|x| \simeq 1$, $|u| = p_T^2/x$.

Having defined Lorentz invariant variables we now define Lorentz invariant cross-sections.

Define

$$f(s, p_c) = E_c \frac{d^3\sigma}{d^3 p_c} \tag{7}$$

The factor E_c comes because the LIPS is d^3p_c/E_c. Similarly the two-particle inclusive cross-section is

$$f(s; p_c, p_d) = E_c E_d \frac{d^6\sigma}{d^3p_c\, d^3p_d}. \tag{8}$$

Written in terms of our variables above the single particle invariant cross-section is:

$$f(s, p_c) = \frac{E_c}{\pi} \frac{d^2\sigma}{dp_L\, dp_T^2} = \frac{1}{\pi} \frac{d^2\sigma}{dy\, dp_T^2}$$

$$= \frac{\lambda^{\frac{1}{2}}(s, m_a^2, m_b^2)}{\pi} \frac{d^2\sigma}{dt\, dM^2}$$

i.e.

$$f \to \frac{1}{\pi} \frac{\mathrm{d}^2\sigma}{\mathrm{d}t\,\mathrm{d}(M^2/s)} \quad \text{as} \quad s \to \infty \tag{9}$$

or

$$f \to \frac{s}{\pi} \frac{\mathrm{d}^2\sigma}{\mathrm{d}t\,\mathrm{d}u}. \tag{10}$$

If we integrate $f(s, p_c)$ w.r.t. the invariant phase space, what do we get?

Take a concrete example. Suppose we were studying $pp \to \pi^+ + X$ in a bubble chamber experiment. We obtain an event with two π^+ on it. We measure *each* π^+ and count *both* in the inclusive cross-section. This implies that when we do our integration, the cross-section for $pp \to 2\pi^+$ + everything else must be weighted by a factor 2. So we get the sum

$$1 \times \sigma(pp \to one\,\pi^+ + \dots) + 2 \times \sigma(pp \to two\,\pi^+ + \dots) + 3 \times \sigma(pp \to 3\pi^+ + \dots) + \dots$$

Now the average multiplicity $\langle n_{\pi^+} \rangle$ *is defined as*

$$\langle n_{\pi^+} \rangle = \frac{\Sigma n \times \sigma(pp \to n\pi^+ + \dots)}{\Sigma\, \sigma(pp \to n\pi^+ + \dots)}$$

and so the answer to our question is

$$\int f(s, p_c) \frac{\mathrm{d}^3 p_c}{E_c} = \langle n_c \rangle\, \sigma^{\mathrm{ab}}_{\mathrm{tot}} \tag{11}$$

where $\langle n_c \rangle$ is the average multiplicity for c-type particles. Similarly for the two-particle inclusive

$$\int f(s; p_c, p_d) = \langle n_c(n_c - 1) \rangle \sigma^{\mathrm{ab}}_{\mathrm{tot}} \tag{12}$$

In addition we can get useful relations by the constraint of conservation of energy, momentum, charge, etc. For example, if we weight *each* event by the energy of the detected particle, E_c, before integrating, the answer is independent of the average multiplicities, if we sum over all types of particles:

$$\sum_c \int f(p_c)\, \mathrm{d}^3 p_c = (\sqrt{s}) \sigma^{\mathrm{ab}}_{\mathrm{tot}}. \tag{13}$$

III. Mueller's Optical Theorem

This theorem[3] is the basis on which the Regge approach to inclusives rests. It has exactly the same role as the ordinary optical theorem has for total cross-sections. Indeed it is just a generalization of the optical theorem.

Let us briefly recall the optical theorem. The total cross-section $ab \to X$ has the amplitude

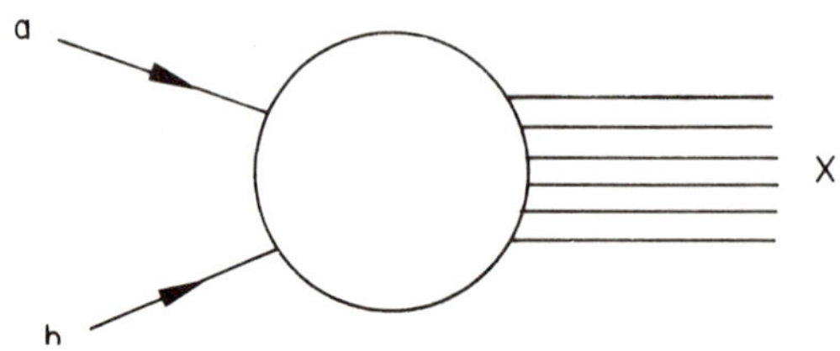

which we multiply by its complex conjugate and sum over all possible intermediate states

$\sigma_{tot}^{ab}(s) \sim \sum$

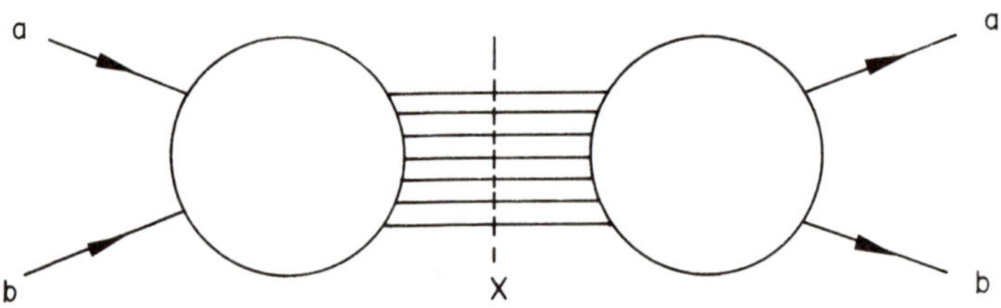

which gives the discontinuity of the elastic amplitude $ab \to ab$ in the forward direction:

$\sigma_{tot}^{ab}(s) \sim \mathrm{Im}$

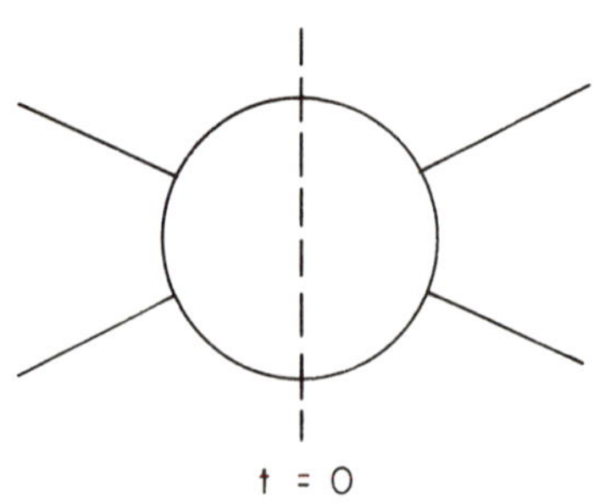

The great advantage of the theorem is that a seemingly complicated situation (i.e. sum of exclusive processes all with varying energy dependences) is converted into a manageable one, i.e. the energy dependence of just a single physical amplitude.

Let us make a list of how one exploits the optical theorem to learn about total cross-sections—these are the features we generalize to learn about inclusives.

(1) Regge pole exchange in t channel

$\sigma^{ab}_{tot}(s) \sim$ 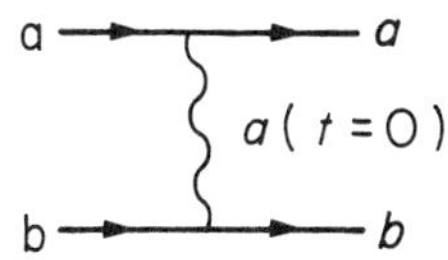 $\sim s^{\alpha(0)-1}$

Leading Regge poles: Pomeron $\alpha(0) = 1$

mesons: $\rho - f - \omega - A_2$ $\alpha(0) = \frac{1}{2}$

hence we expect, in general,

$$\sigma^{ab}_{tot}(s) \simeq A + \frac{B}{\sqrt{s}} \tag{14}$$

Note (*a*) If, for some reason, meson exchange is absent ($B = 0$) then $\sigma_{tot.} \sim$ constant for *all* energies.

Note (*b*) When meson-exchange *is* allowed, the approach to an asymptotic limit is like $s^{-\frac{1}{2}}$.

(2) Duality (Harari–Freund version)[4]

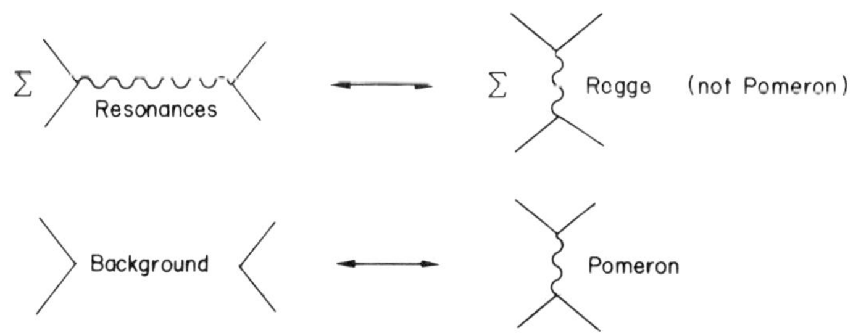

Thus, if the quantum numbers of the s-channel are exotic then we have only Pomeron exchange, i.e. $B = 0$ above and so we expect σ^{pp}_{tot}, $\sigma^{K^+p}_{tot}$ to be independent of energy from very low energies ($\sim 3\,\text{GeV}^2$). Other total cross-sections like K^-p, $\pi^{\pm}p$, etc., we therefore expect to fall to asymptotic limit like $1/\sqrt{s}$. ($B > 0$ since it is $= \Big|$ 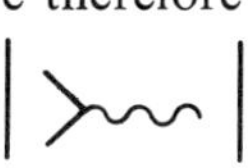 $\Big|^2$).

(3) Factorization

i.e. 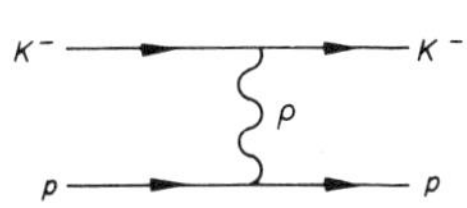 $= \gamma^{\rho}_{K\bar{K}}(t)\,\gamma^{\rho}_{p\bar{p}}(t)\;s^{\alpha_\rho(t)}$

In this way certain total cross-sections can be related. Taking the case of Pomeron exchange we get

$$\sigma_{\text{tot}}^{\pi\pi}(\infty)\sigma_{\text{tot}}^{NN}(\infty) = [\sigma_{\text{tot}}^{\pi N}(\infty)]^2 \tag{15}$$

If the meson Regge-exchanges also factorize we can take combinations of total cross-sections to isolate particular exchange, e.g.

$$\sigma_{\text{tot}}^{\bar{p}p} + \sigma_{\text{tot}}^{\bar{p}n} - \sigma_{\text{tot}}^{pp} - \sigma_{\text{tot}}^{pn}$$

gives pure ω exchange.

Now let us see how all these are generalized for the case of inclusive reactions. As stated before, Mueller[3] generalized the optical theorem. In this case the amplitude for the inclusive reaction is

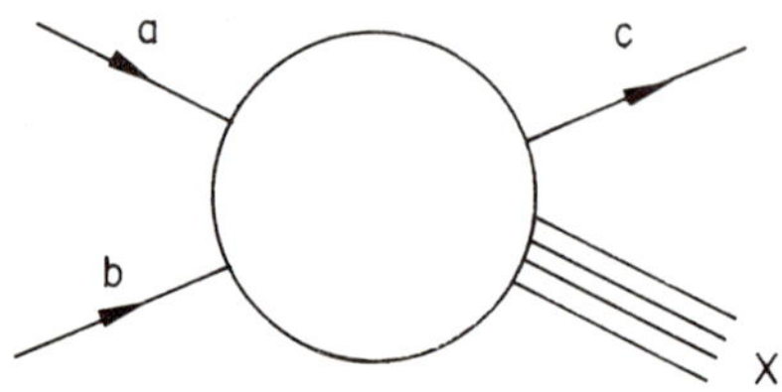

which we then "square" and sum over all possible intermediate states

$$f(s, p_c) \sim \sum_X$$

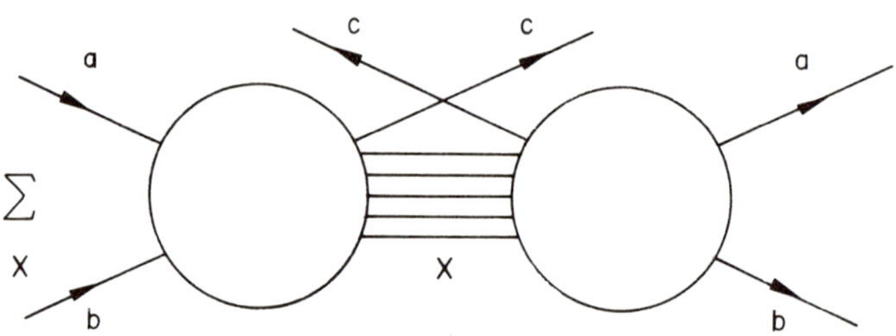

which is, in turn, proportional to the discontinuity (in M^2) of the forward amplitude $ab\bar{c} \to ab\bar{c}$

i.e. $f(s, p\ ) \sim$

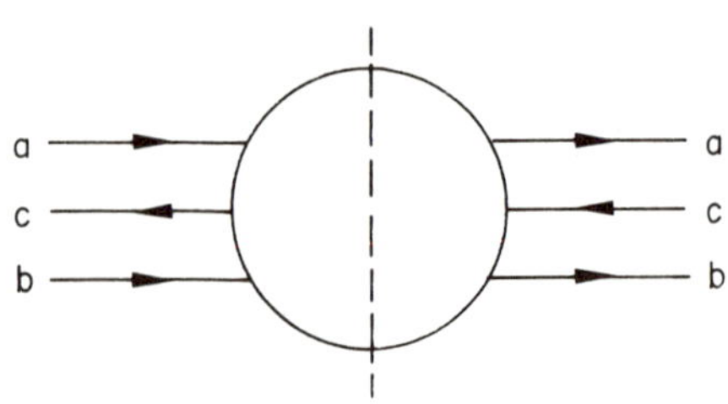

In contrast to the situation in the optical theorem, our forward amplitude is unphysical since the sub-energies $s_{a\bar{c}}$, $s_{b\bar{c}}$ are negative (t and u). The situation is, in fact, rather complicated since the relevant discontinuity is

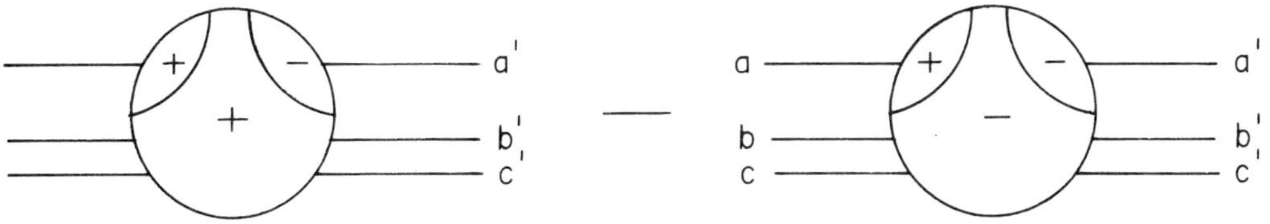

i.e.

$$f(s, p_c) \sim T(s_{ab} + i\varepsilon, M^2 + i\varepsilon, s_{a'b'} - i\varepsilon) - T(s_{ab} + i\varepsilon, M^2 - i\varepsilon, s_{a'b'} - i\varepsilon). \tag{16}$$

The proof is complicated but field theory experts[5] give us assurances that the result is very plausible, to say the least. Next let us look at the 3 body amplitude in different regions of phase space.

1. FRAGMENTATION REGION (Single-Regge Reaction)

For example, consider $b \xrightarrow{a} c$ (this means b fragmenting to c under impact of a). As $s \to \infty$ this means p_L^{lab} is kept fixed and small, so is p_T. This in turn means u fixed and small and M^2/s fixed and small as $M^2 \to \infty$. For the $ab\bar{c} \to ab\bar{c}$ amplitude, M^2 is the total 3-body energy, u is the (mass)^2 of the $b\bar{c}$ system and M^2/s is a quantity involved in the orientation of b relative to the $a\bar{c}$ system. The appropriate graph is therefore

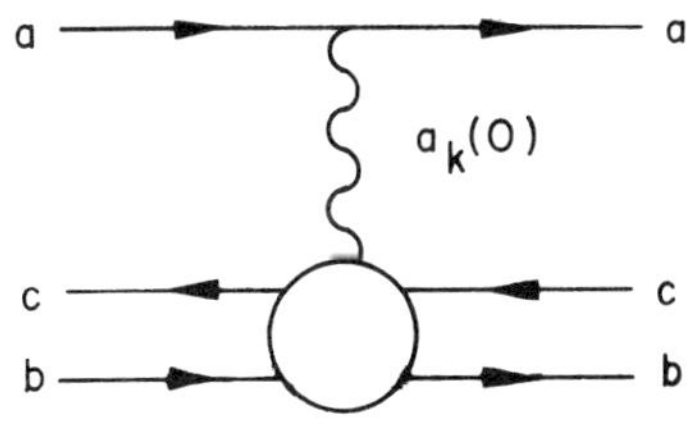

where some Regge poles $\alpha_i(t = 0)$ are exchanged. We expect the energy behaviour then to be $(M^2)^{\alpha_k(0)-1}$ but since M^2/s is finite we can write

$$f(p_c) = \sum_k F_k\left(u, \frac{M^2}{s}\right) s^{\alpha_k(0)-1}. \tag{17}$$

So taking Pomeron and leading meson exchange we get

$$f(p_c) = F_{\mathbf{P}}\left(u, \frac{M^2}{s}\right) + F_M\left(u, \frac{M^2}{s}\right)\frac{1}{\sqrt{s}} \tag{18}$$

Just as in *total* cross-sections, the Pomeron exchange term gives the limiting cross-section, while meson exchange tells us how that limit is approached, (again like $1/\sqrt{s}$). The first term is called the "scaling" contribution, the second, the "non-scaling" piece.

A particular sub-region of the fragmentation region is reached by letting $s/M^2 \to \infty$ while still keeping $M^2 \to \infty$.

This allows a Regge expansion within the "blob" in the above graph and the appropriate graph is a triple-Regge graph. We shall discuss this region in some detail later on.

If we proceed right to the edge of the fragmentation region by relaxing the M^2 large condition, so s/M^2 is very large, we have the *normal* Regge region corresponding to the amplitude

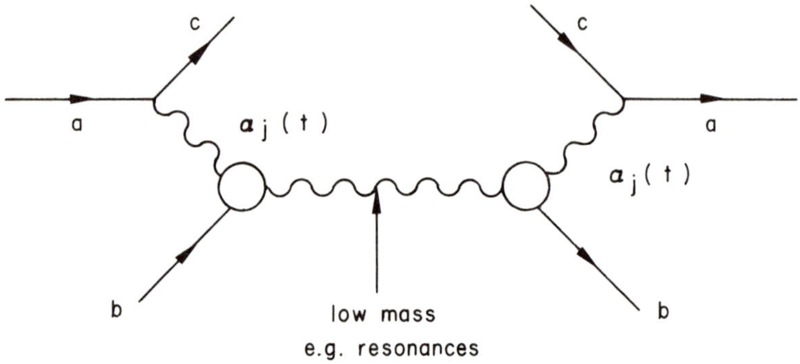

Again we shall discuss this later.

2. CENTRAL REGION (Double-Regge Region)

Here particle c is slow in neither the projectile or target frame; i.e. t and u are both large. p_T^2 is still small and in this approximation we can write

$$t \sim -(\sqrt{s})(E^* - p_L^*)$$
$$u \sim -(\sqrt{s})(E^* + p_L^*) \tag{19}$$

i.e. $\kappa^2 = ut/s$ is finite. The c.o.m. rapidity is then

$$y^* = \tfrac{1}{2} \ln u/t.$$

For u, t large, the relevant Regge graph for $ab\bar{c} \to ab\bar{c}$ is clearly

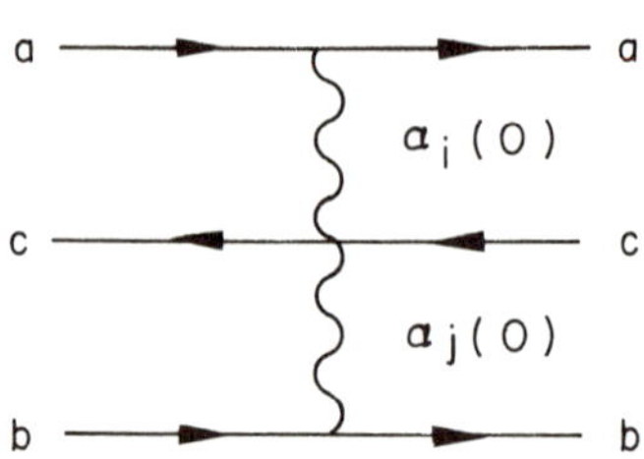

The Regge expression for this double-Regge graph is

$$f(s, p_c) = \sum_{ij} \beta_{ij}(\kappa)|t|^{\alpha_i(0)-1}|u|^{\alpha_j(0)-1}. \tag{20}$$

Note that the β's can depend only on κ. If the Pomeron is exchanged in both legs, then $f(p_c)$ is independent of energy—this is "scaling in the central region".

$$f(s, p_c) \to \beta_{\mathbf{PP}}(\kappa) \tag{21}$$

Thus the cross-section not only is independent of energy but independent of y^*. Thus we expect a y^* distribution like

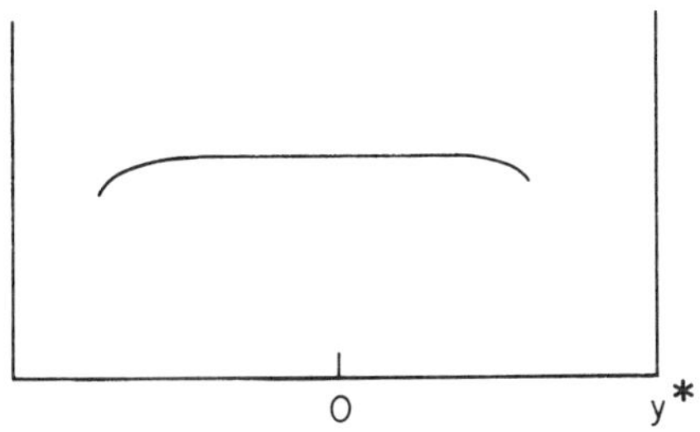

i.e. the inclusive cross-section is expected to show a plateau in the central region.

What about the approach to scaling? This we expect to get from the next most important graphs:

non-scaling

$$\begin{aligned} f(s; p_c) &= \beta_{\mathbf{P}M}(\kappa)\{|t|^{-\frac{1}{2}} + |u|^{-\frac{1}{2}}\} \\ &= \beta_{\mathbf{P}M}(\kappa)\frac{u^{\frac{1}{2}} + t^{\frac{1}{2}}}{(ut)^{\frac{1}{2}}} = \beta_{\mathbf{P}M}(\kappa)\frac{(u^{\frac{1}{2}} + t^{\frac{1}{2}})}{(ut)^{\frac{1}{4}}} \times \frac{1}{(ut)^{\frac{1}{4}}} \\ &= \beta_{\mathbf{P}M}(\kappa)\cosh(\tfrac{1}{2}y^*)(\kappa^2 s)^{-\frac{1}{4}} \\ &= \bar{\beta}_{\mathbf{P}M}(\kappa)s^{-\frac{1}{4}}\cosh(\tfrac{1}{2}y^*) \end{aligned} \tag{22}$$

so the approach to scaling would be expected to be like $s^{-\frac{1}{4}}$ at fixed values of y^*.

From our experience in two-body Regge phenomenology we might expect the above expansion to be valid only for u, t both $>5\,\mathrm{GeV}^2$ which means that it is only at ISR energies and above that we reach the possibility of exploring the central region.

The different kinematic regions are well illustrated in a $t-u$ plot at fixed s.[11] Figure 1 shows 10000 events for $K^-p \to \bar{K}^0 + X$ at 14 GeV/c. Notice

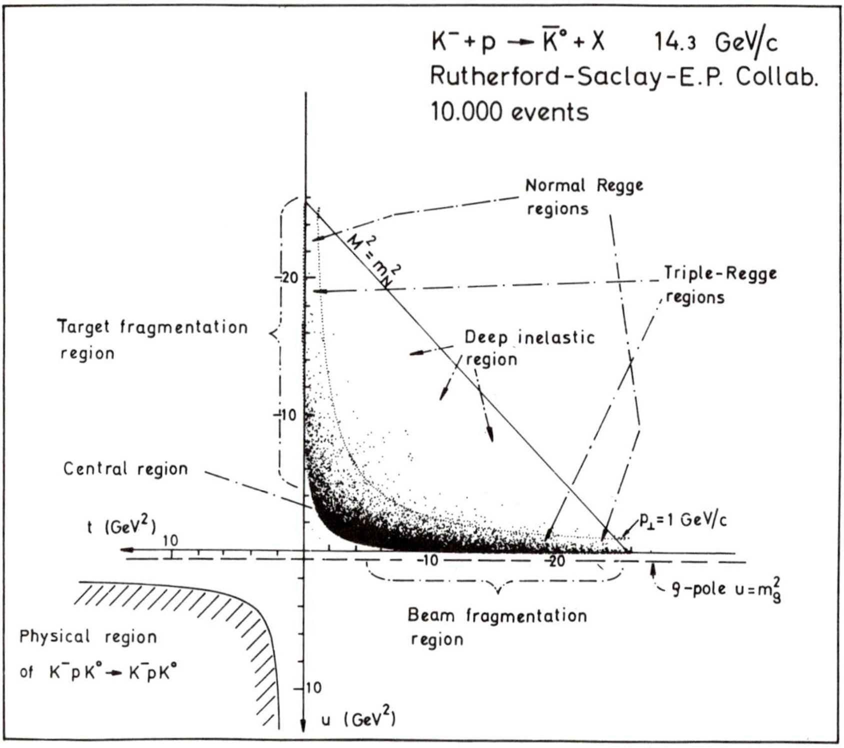

FIG. 1. Distribution of events in the t–u-plane for the reaction $K^- + p \to \bar{K}^0 + X$ at 14·3 GeV/c. For details see the text.

the strong concentration of events near the kinematic limit which is the effect of suppression of large p_T. The beam fragmentation region is heavily populated, in contrast to the target fragmentation region. This is because $K^- \to \bar{K}^0$ proceeds via $\rho - A_2$ exchange (note the proximity of the ρ-pole in Fig. 1) while $p \to \bar{K}^0$ can proceed only via exchange of exotic quantum number and is comparatively suppressed.

IV. Duality

We saw earlier the connection between the energy dependence of total cross-sections and the quantum numbers of the initial state. When *ab* was exotic we expected $\sigma_{\text{tot}}^{\text{ab}}$ to be flat from quite low energies and this is well known for K^+p and *pp* (forgetting the rise seen at highest available energies!).

If we directly carry over the same reasoning to inclusive cross-sections we would expect, from Mueller's theorem, constancy from low energies (*early* scaling) if our incoming state $ab\bar{c}$ had exotic quantum numbers. This in fact was the early scaling criterion first proposed by Chan, Hsue, Quigg and Wang.[6] For some time a controversy raged over different criteria for early scaling and even today the dust has not finally settled. From the strictly theoretical viewpoint there is still a choice of such rules. In the end it has to be data itself which determines which criterion is correct. To discuss this and other developments later on we must recall first two-component duality for total cross-sections and then see how this generalizes to a seven (or nine) component duality for inclusive processes.

(a) 2-*component duality*

Consider the total cross-section $ab \to X$.

The amplitude can be separated into two parts:

(i) one which can be represented by planar dual graphs. The sum of which we write schematically as

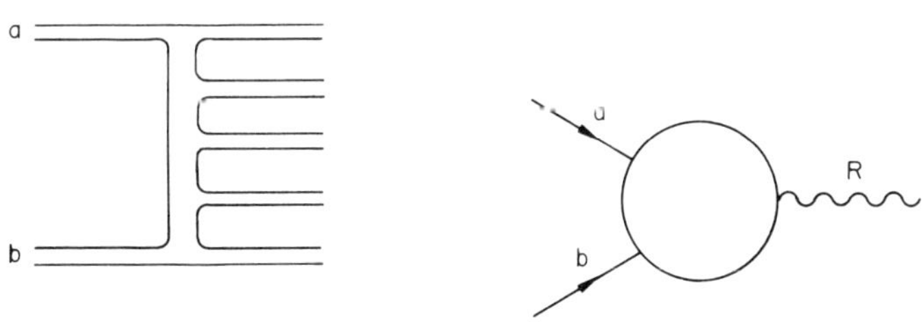

which we square to get

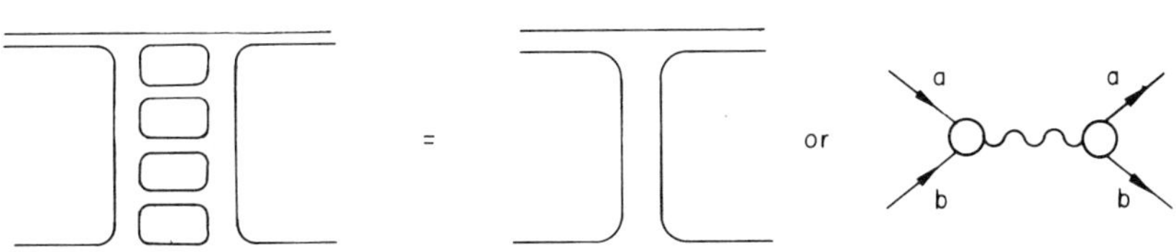

i.e. this is $\sigma_{\text{tot}}^{\text{res}}$.

The convention is to distort this centre graph to

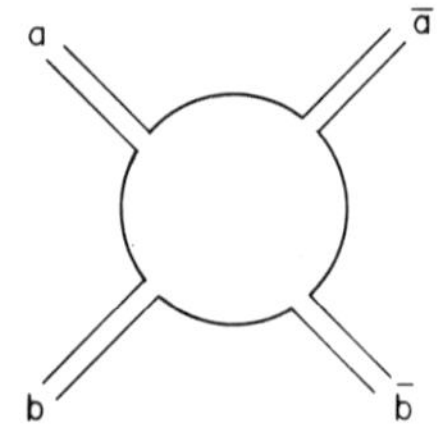

and abbreviate this to

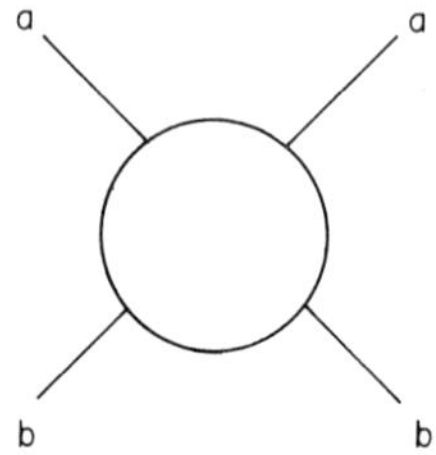

i.e. if we cut it in either the s or t channels we have quark-antiquark, i.e. resonances dual to Regge exchange.

(ii) the second part corresponds to non-planar graphs like

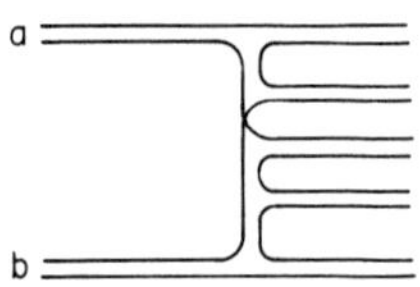

Such graphs can always be "untwisted" to give

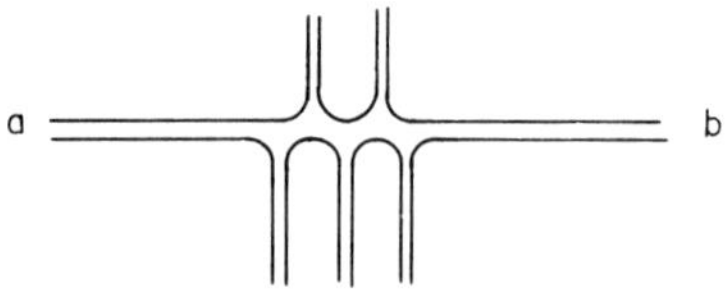

which we schematically write as

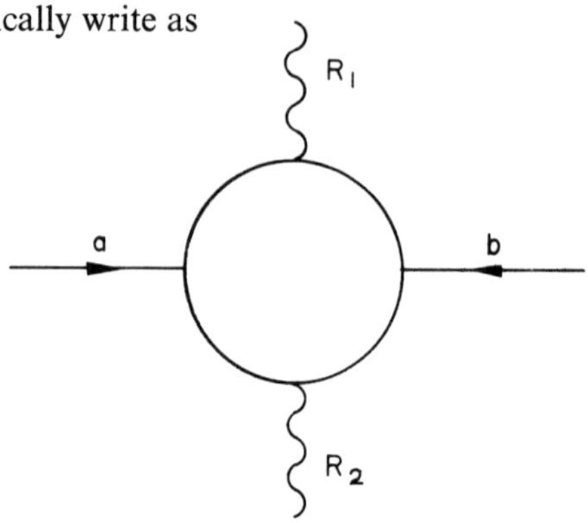

Squaring this gives

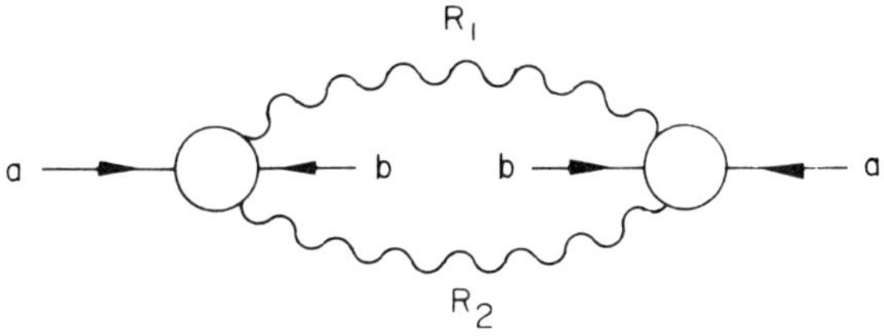

for which the duality graph is

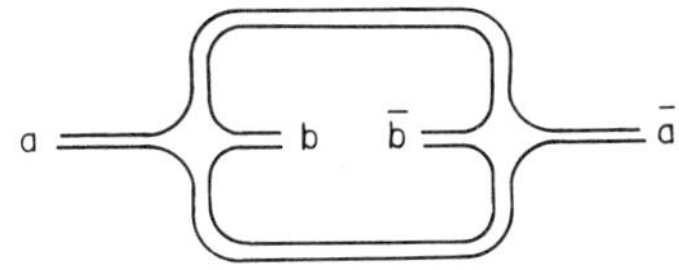

which can be abbreviated to

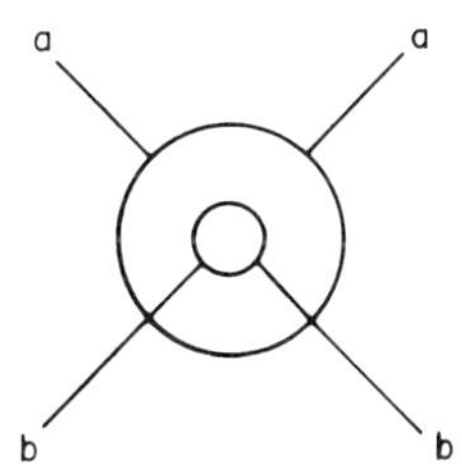

We see that two quark lines which *begin* with b also *end* at $\bar{b}$ (similarly for a and $\bar{a}$). This implies the exchange of vacuum quantum numbers (VQN) in the t-channel. At the same time we see that cutting the graph in the s-channel gives an exotic state. We then associate this second component as Pomeron exchange dual to non-resonant "background". The first component alone gives variation of the total cross-section with energy—therefore to guarantee "early" energy independence we simply insist on ab having exotic quantum numbers. Although Regge-poles could be exchanged in the second component (e.g. f-exchange) we choose to work in the small width (Γ) approximation. This implies the first component behaves as Γ^2 compared to the Γ^4 behaviour of the second and so the effect of the Regge exchange from the latter is relatively small. In the same spirit, we shall assume that even the Pomeron part is small compared to the first component when the energy is very low.

(b) *Seven* (*or nine*) *component duality*

For $ab \rightarrow cX$ we first consider the resonant part of X. This gives us three graphs:

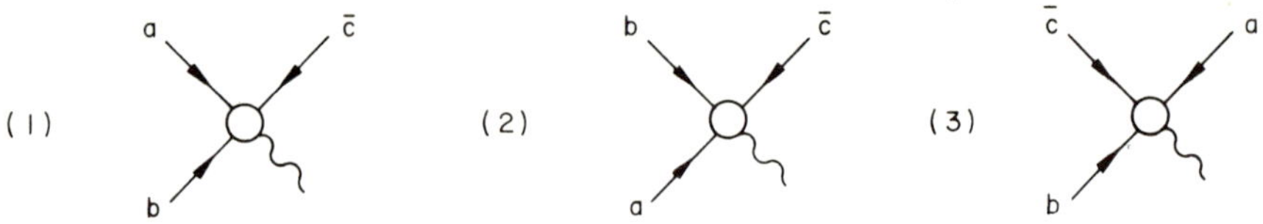

and four when X is exotic:

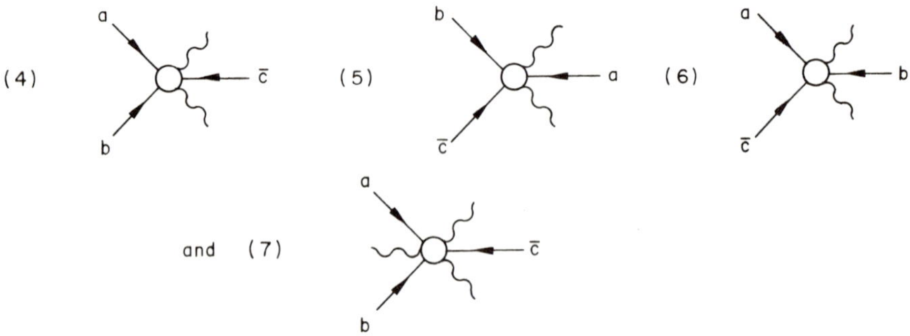

In graph 1, the *non*-exotic channels are $ab\bar{c}$, ab, $a\bar{c}$
In graph 2, the *non*-exotic channels are $ab\bar{c}$, ab, $b\bar{c}$
In graph 3, the *non*-exotic channels are $ab\bar{c}$, $a\bar{c}$, $b\bar{c}$
In graph 4, the *non*-exotic channels are ab
In graph 5, the *non*-exotic channels are $b\bar{c}$
In graph 6, the *non*-exotic channels are $a\bar{c}$
In graph 7, no channel is *non*-exotic.
Now we square each of these graphs and obtain

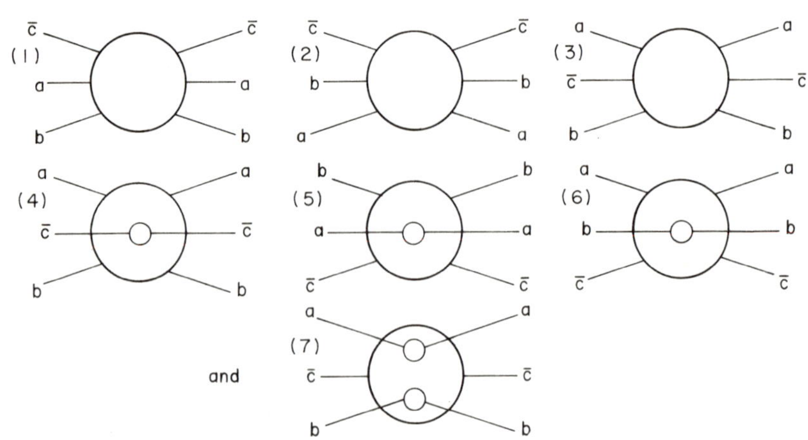

We interpret these graphs as before, so that particles lying on *different* quark loops are connected by Pomeron exchange and those on the *same* quark loop by Regge exchange.

The next thing to realize is that all seven components do not contribute with the same importance to a given region of phase space. For example, consider the fragmentation $b \xrightarrow{a} c$ where $u = s_{b\bar{c}}$ is small. From the first set of graphs the dominant contributions will come from graphs where the b and $\bar{c}$ legs are adjacent, i.e. graphs (2), (3) and (5). Energy dependence comes only from (2) and (3) since we see that from the second set of graphs, (5) corresponds to Pomeron exchange in the $a\bar{a}$ channel.

Therefore if we insist only on the constraint $ab\bar{c}$ *exotic*, then (2) and (3) are removed and we are left with just one energy independent graph. Hence we achieve early scaling in the fragmentation region $b \xrightarrow{a} c$.

However, had we chosen a reaction where the $b\bar{c}$ channel was *exotic* then graph (5) as well as (2) and (3) would be absent and the fragmentation cross-section would be expected to vanish. In fact we often find cross-sections for which $b\bar{c}$ is exotic, like $p \to K^-$ to be an order of magnitude smaller than those for which $b\bar{c}$ is not exotic, e.g. $p \to K^+$. *In the fragmentation region region* $b \xrightarrow{a} c$, *we therefore expect early scaling if* $ab\bar{c}$ *is exotic, provided* $b\bar{c}$ *is not exotic.* Notice that we have essentially two-component duality in this case for the pseudo-two-body scattering amplitude

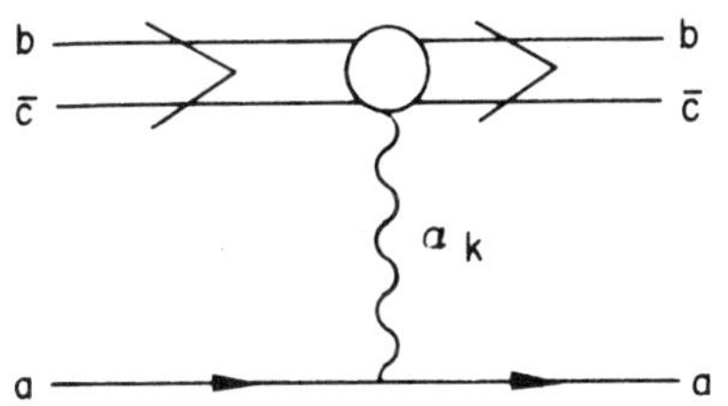

i.e. $(b\bar{c}) + a \to (b\bar{c}) + a$

The value of $s_{b\bar{c}}$ is really not very far from the physical pole corresponding to the first resonance in $\alpha_{s_{b\bar{c}}}$ and so we might not expect much difference in the Regge exchanges α_k as we go slightly off-shell.

In the central region we have both $t = s_{a\bar{c}}$ and $u = s_{b\bar{c}}$ large. Component (7) gives the purely scaling term since we see from the second set of graphs that it corresponds to

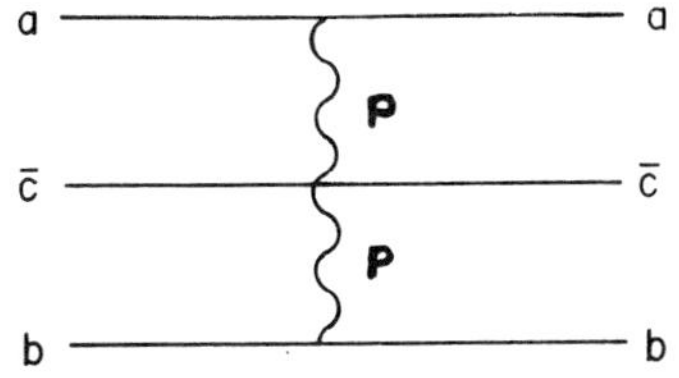

while the leading non-scaling terms are (5) and (6) corresponding to

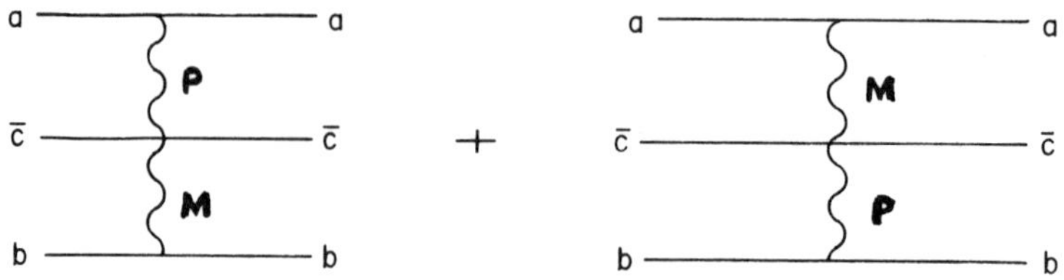

Looking back at the first set of graphs we see that to kill these energy dependent contributions we require $a\bar{c}$ and $b\bar{c}$ exotic. *Thus to get early scaling in the central region the $a\bar{c}$ and $b\bar{c}$ channels must be exotic.*

The $ab\bar{c}$ exotic rule was not the only criterion for early scaling originally proposed. For the fragmentation $b \xrightarrow{a} c$ these rules were concerned with eliminating the energy dependent graphs (4) and (6) which we did not consider for kinematic reasons; e.g. *ab* and $a\bar{c}$ exotic in addition to $ab\bar{c}$.

Notice that in this seven component scheme, as in the two-component scheme, all components are formed from the *squares* of production amplitudes and, consequently, are all *positive*. So like total cross-sections, we expect inclusive cross-sections to approach their asymptotic limits from *above*. We shall see, later, that some central region cross-sections *increase* to their scaling limits. In our approach such *negative* contributions could arise from

(a) Interference terms which we have neglected.
(b) Higher-order corrections to the seven components, corresponding to absorptive corrections, e.g. Pomeron–Pomeron cut.
(c) Non-leading exchanges, e.g. daughters—which are related to *s*-channel threshold effects.[18]

So far we have considered only non-diffractive inclusive reactions, i.e. $a \xrightarrow{b} c$ with non-Pomeron exchange in the $a\bar{c}$ channel. Suppose[8] *a* and *c* are the same particle so Pomeron exchange is expected to dominate. In addition to our seven components we would then have

and similarly for Pomeron exchange in the $b\bar{c}$ channel we have

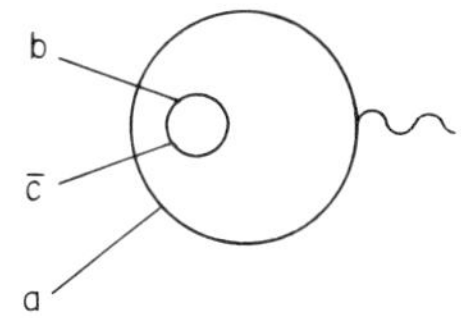

which on squaring gives

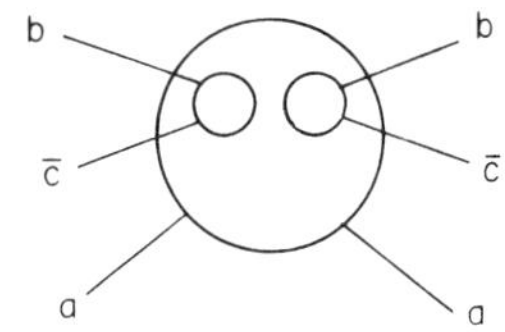

If we cut these graphs down the middle, we find a quark–antiquark pair, i.e. resonances in the M^2 channel. At the same time we see, in the $a \xrightarrow{b} c$ case, that the Pomeron is exchanged in the $b\bar{b}$ channel. What this suggests therefore is an abnormal type of duality in which diffractive resonances are dual to the Pomeron, instead of usual Regge exchange. We shall return to this intriguing situation later on.

V. Phenomenology of Fragmentation Region

A very powerful aid for phenomenology is factorization, by which the inclusive cross-sections for several reactions can be connected. Assuming that both the Pomeron and the leading meson exchanges factorize, we can write for the fragmentation $b \xrightarrow{a} c$

$$f(p_c) = \gamma_a^{\mathbf{P}} \Gamma_{b\bar{c}}^{\mathbf{P}} \left(u, \frac{M^2}{s} \right) + \sum_{i = \rho, \omega, f, A_2} \gamma_a^i \Gamma_{b\bar{c}}^i \left(u, \frac{M^2}{s} \right) s^{-\frac{1}{2}} \tag{23}$$

i.e.

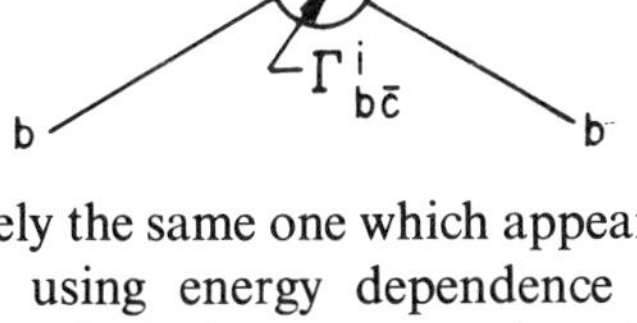

The top vertex is precisely the same one which appears in *total* cross-sections, and can be extracted using energy dependence and factorization. For example, if we compare inclusive cross-sections for reactions with only Pomeron exchange and divide by σ_{tot} we should get equality for all such cases.

A very nice exercise has been done by Miettinen[9] who looked at six examples of the fragmentation $p \xrightarrow{a} \pi^-$, where $a = p, \pi^+, K^+, \pi^-, K^-, \gamma$. In Fig. 2 are plotted the fragmentation cross-sections divided by $\sigma_{\underline{tot}}^{ab}(\infty)$, evaluated at $p_L^{lab} = 0$, against $1/\sqrt{s}$. Three are examples of $ab\bar{c}$ exotic ($b\bar{c}$ not

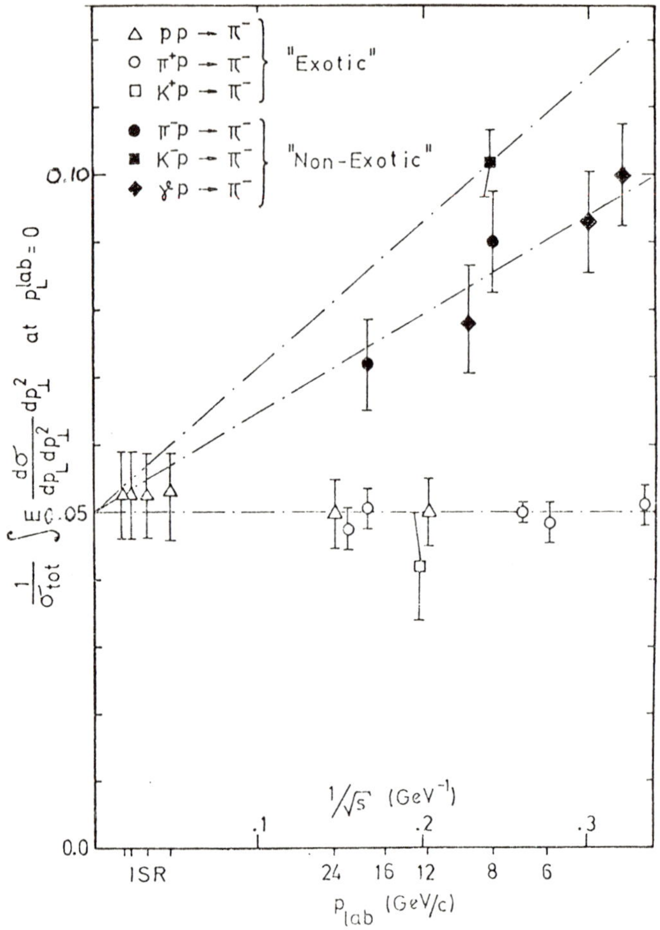

FIG. 2. Comparison of the exotic cross-sections $p \xrightarrow{a} \pi^-$, $a = p, \pi^+, K^+$ with the non-exotic ones $p \xrightarrow{a} \pi^-$, $a = \pi^-, K^-, \gamma$. Taken from Ref. 9.

exotic) and one can see how beautifully constant these cross-sections are with a variation in energy between 6 and 1 500 GeV/c. This is impressive support for not only the early scaling criterion but also for Pomeron factorization since all three cross-sections appear to lie on the *same* constant line. The early scaling phenomenon is even more impressive than for total cross-section since it appears that for non-exotic reactions the relative proportion of the non-scaling part is much larger. (Comparing $p \xrightarrow{\pi^-} \pi^-$ and $p \xrightarrow{\pi^+} \pi^-$ in Fig. 2 shows the non-scaling to scaling contribution ratio to be around three times the corresponding ratio for total cross-sections.) The figure also demonstrates that the non-scaling contribution to non-exotic reactions is positive and fully consistent with an $s^{-\frac{1}{2}}$ decrease in accordance with our previous discussion.

Compared with total cross-sections, fragmentation cross-sections have a greater freedom in two respects. Firstly, freedom of quantum numbers—in the case of σ_{tot} only two possibilities are exotic (pp and K^+p) while here we have many examples, for example $\pi^+p \to \pi^-X$, $K^+p \to \pi^+X$, $\pi^+p \to \Lambda X$, $\pi^+p \to K^0X$, $K^+p \to \pi^-X$, $pp \to \pi^+X$, $K^+p \to \Lambda X$, $pp \to \Lambda X$, $pp \to \pi^-X$, In each case, the fragmentation cross-sections have been checked for energy independence and the only slight exceptions are $pp \to K^0X$, $pp \to K^+X$, $pp \to \Lambda X$ where a small *rise* is seen. Secondly, fragmentation cross-sections in addition to being functions of energy are functions of two other variables and statements about scaling, etc., should hold at *each* value of x and p_T^2.

Figure 3 is a fine example of scaling, comparing $pp \to \pi^+$ at 24 GeV/c with the ISR data.[10]

Let us turn our attention to the meson exchange terms above. As in the case of total cross-sections, the meson exchanges should cancel each other out when the direct channel is exotic—this is exchange degeneracy. As we have just seen the evidence in favour of exchange degeneracy is very good and if we now combine this with the assumption of factorization of these secondary trajectories, we can derive relations between various non-exotic and exotic fragmentation cross-sections.[11]

For example consider $p \xrightarrow{a} \pi^-$ for various choices of a.

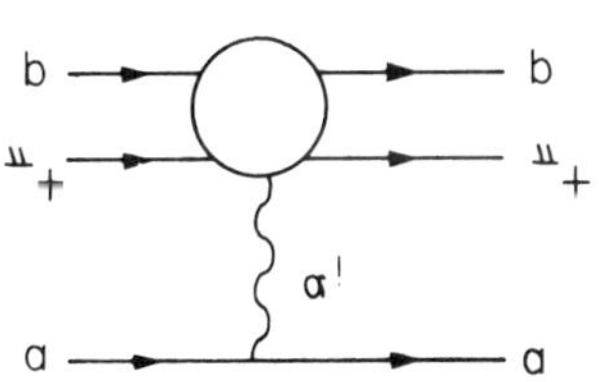

Furthermore let us consider only the meson-exchange ($\rho - \omega - f - A_2$) part of the cross-section $s^{\frac{1}{2}}f^M(p_c)$. Let $a = K^+, K^0, \pi^+, K^-, \pi^-$ and γ.

From the situation in two-body scattering, exchange degeneracy gives

$$\gamma_\pi^f = \gamma_\pi^\rho = \gamma_\pi$$

$$\gamma_K^f = \gamma_K^\rho = \gamma_K^{A_2} = \gamma_K^\omega = \gamma_K$$

(i) $a = K^+$ $\quad s^{\frac{1}{2}}f^M = \gamma_K(\Gamma^f - \Gamma^\rho - \Gamma^\omega + \Gamma^{A_2})$ $\qquad$ (definitions)

(ii) $a = K^0$ $\quad s^{\frac{1}{2}}f^M = \gamma_K(-\Gamma^f - \Gamma^\rho + \Gamma^\omega + \Gamma^{A_2})$ (isospin on vertices from (i)) (isoscalars change sign)

(iii) $a = \pi^+$ $\quad s^{\frac{1}{2}}f^M = \gamma_\pi(\Gamma^f - \Gamma^\rho)$ (SU(3) from (i))

(iv) $a = \pi^-$ $\quad s^{\frac{1}{2}}f^M = \gamma_\pi(\Gamma^f + \Gamma^\rho)$ (odd signatures change sign from (iii))

(v) $a = K^-$ $\quad s^{\frac{1}{2}}f^M = \gamma_K(\Gamma^f + \Gamma^\rho + \Gamma^\omega + \Gamma^{A_2})$ (odd signatures change from (i))

(vi) $a = \gamma$ $\quad s^{\frac{1}{2}}f^M = \gamma_\gamma^f\Gamma^f + \gamma_\gamma^{A_2}\Gamma^{A_2}.$

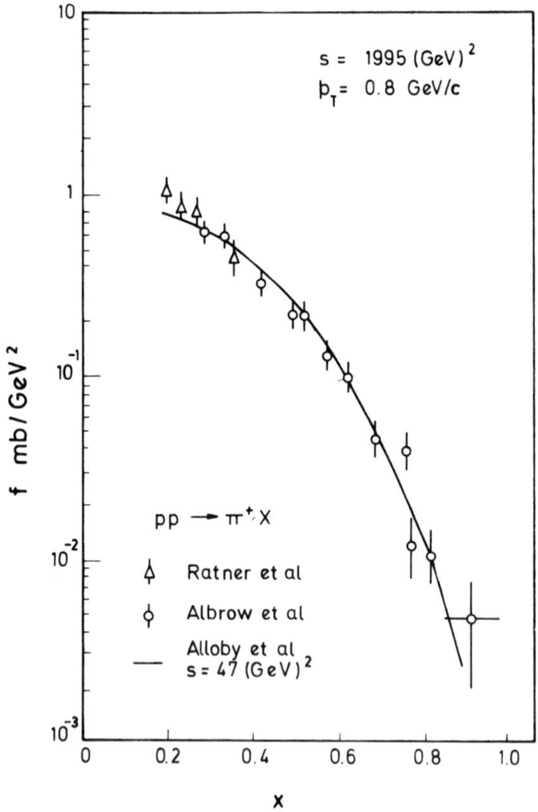

FIG. 3. Comparison of the accelerator and ISR cross-sections for the exotic reaction $pp \to \pi^+X$, as a function of x. Taken from Ref. 46.

Cases (i), (ii) and (iii) all have $ab\bar{c}$ exotic and so in each case $f^M = 0$. This gives $\Gamma^f = \Gamma^\rho = \Gamma^\omega = \Gamma^{A_2} = \Gamma^M$ and so for

$$a = \pi^- \quad s^{\frac{1}{2}} f^M = 2\gamma_\pi \Gamma^M$$

$$a = K^- \quad s^{\frac{1}{2}} f^M = 4\gamma_K \Gamma^M$$

and so we can write

$$f(p \xrightarrow{\pi^-} \pi^-) = f(p \xrightarrow{\pi^+} \pi^-) + \frac{\gamma_\pi}{2\gamma_K}[f(p \xrightarrow{K^+} \pi^-) - f(p \xrightarrow{K^-} \pi^-)]. \tag{24}$$

This result, first obtained by Miettinen,[11] is well satisfied experimentally giving strong support to the factorization and exchange degeneracy properties of secondary trajectories. The γ_π and γ_K are got from total cross-sections and agree well with the SU(3) result $\gamma_\pi = 2\gamma_K$. Hence the above equation is simply the result

$$f^{\text{Non-scaling}}_{p \xrightarrow{\pi} \pi^-} = f^{\text{Non-scaling}}_{p \xrightarrow{K} \pi^-} \tag{25}$$

which is the analogue of the total cross-section result

$$(\sigma^{\pi^- p}_{\text{tot}})_{\text{Non-Pomeron}} = (\sigma^{K^- p}_{\text{tot}})_{\text{Non-Pomeron}}. \tag{26}$$

Similarly one can derive relations for $p \xrightarrow{a} \pi^+$ cross-sections and $p \xrightarrow{a} \Lambda$, $p \xrightarrow{a} K^0$.

From (vi) we see that for $a = \gamma$ the non-scaling contribution is

$$(\gamma^f_\gamma + \gamma^{A_2}_\gamma)\Gamma.$$

This implies that the photoproduction cross-section is given in full by

$$f(p \xrightarrow{\gamma} \pi^-) = \frac{\gamma^{\mathbf{P}}_\gamma}{\gamma^{\mathbf{P}}_K} f(p \xrightarrow{K^+} \pi^-) + \frac{(\gamma^f_\gamma + \gamma^{A_2}_\gamma)}{2\gamma^f_K}$$

$$\times [f(p \xrightarrow{K^-} \pi^-) - f(p \xrightarrow{K^+} \pi^-)] \tag{27}$$

where all the required couplings can be got from total cross-sections on γp and K^+p, K^-p. This result together with other similar results was derived by Chan, Miettinen and Lam,[12] who found it well obeyed by photoproduction data at 9 GeV/c. Figure 4 shows a test of the relation carried out by Fry *et al.*[13]

In all this we have treated the fragmentation blob, Γ, just like the other vertex γ. We have not discussed its shape, i.e. dependence on u and M^2/s—this we shall leave to a later section on triple-Regge region. Also we have not discussed the situation when the fragmentation blob has vacuum quantum numbers—this again is left to the section on diffractive reactions.

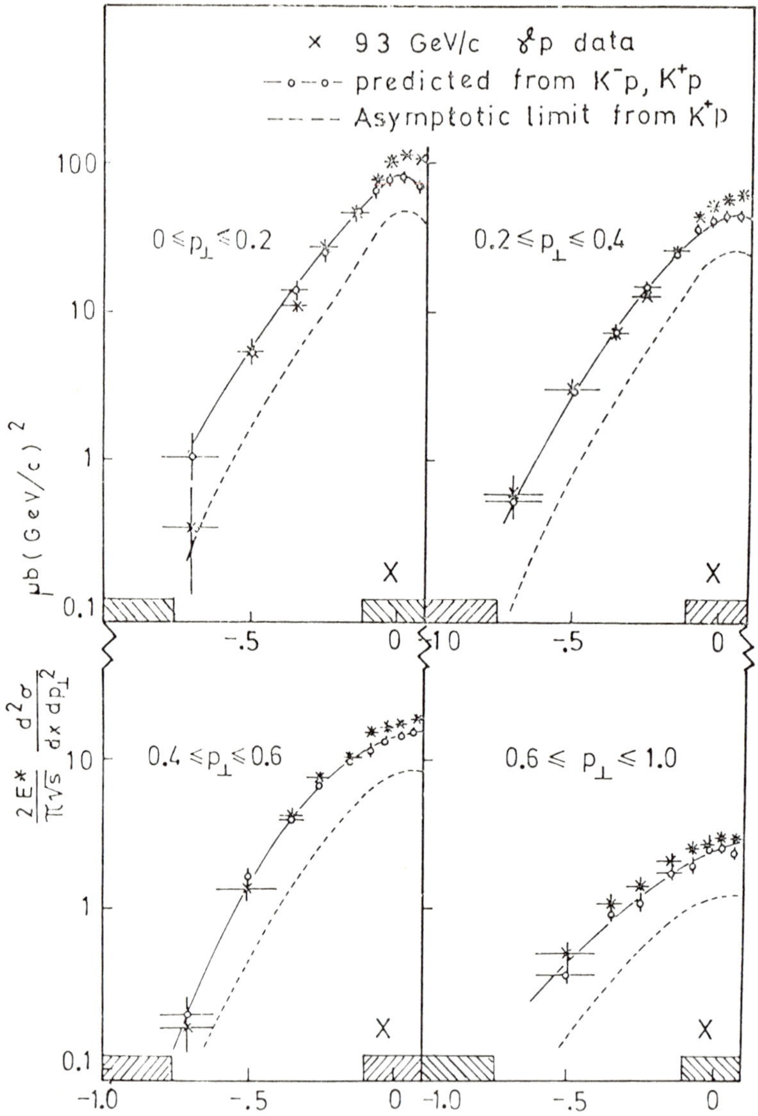

FIG. 4. Comparison of the cross-sections for $\gamma p \to \pi^- X$, at 9·3 GeV/c, predicted from K^-p and K^+p reactions with data. The smooth curve shows the trend of the predicted points. The region of validity for the comparison is between the hatched areas. Taken from Ref. 13.

While discussing factorization it is worth mentioning that more stringent tests can be obtained by studying polarization in inclusive reactions and there has been some effort in this direction.[14] If we measure the polarization of a in the fragmentation $b \xrightarrow{a} c$ then the fragmentation blob $\Gamma^i_{b\bar{c}}(u, M^2/s)$ should not care what value the polarization takes. Conversely if either b or c is polarized, the "polarization blob" should be independent of particle a. One drawback is the paucity of data on polarized inclusive reactions but hopefully this will improve.

VI. Central Region Phenomenology

As discussed in an earlier section the only possibilities of studying the central region is at ISR or the upper energies at NAL. The appropriate Regge graph we saw is the double Regge graph and we also saw that in the scaling limit (i.e. double Pomeron exchange) the cross-section is expected to develop a "plateau" since it becomes independent of the rapidity y also.

One of the striking successes of the ISR was the confirmation of such a plateau though we hastily add that this is a triumph not for the Regge approach in particular since it is almost a natural companion to the logarithmic growth of multiplicity. In Fig. 5 we see the data on $pp \to \pi^+, \pi^-, K^+, K^-$ in the fragmentation and central regions for $p_T = 0{\cdot}4$. An essentially flat distribution can be seen for y between 2 and 4 which means on the full plot a plateau from -2 to $+2$. It is not perfectly flat, there being a rise of about 10%, which has encouraged some people[16] to fit with a power law dependence. Nevertheless, we feel this to be clear evidence for a scaling plateau. One can see how this shape is built up as the energy increases from 12, 24 GeV to ISR by looking at Fig. 6.

This figure also demonstrates an unmistakable fact—namely that the cross-sections in the central region *rise* as the energy increases to their eventual asymptotic limit. This is in contrast to the fragmentation and total cross-sections we have discussed up to now. Looking back at Fig. 5 we can see that the magnitude of the rise depends strongly on which particle we are detecting. For pions the rise is not too big. At $x = 0$, small p_T, $pp \to \pi^+ X$ rises by about 50% from 24 GeV to highest ISR; $pp \to \pi^- X$ rises by about 70% in the same range. However we see that the rise for K^- and $\bar{p}$ is large. In the latter case the $x = 0$ cross-section rises essentially by an order of magnitude. How then can we understand such a behaviour? Firstly we note that we expected early scaling in the central region when $b\bar{c}$ and $a\bar{c}$ was exotic. The two cases where this is satisfied, $pp \to \bar{p}$ and $pp \to K^-$ are showing most variation with energy and secondly, the non-scaling piece is negative where again we expected a *positive* piece. This means that there is some huge negative non-scaling piece in component 7.

How then can we understand such behaviour? If we forget the Mueller approach for a moment and think of the multiperipheral process which produces a $\bar{p}p$ pair in the central region then it is not hard to see why the

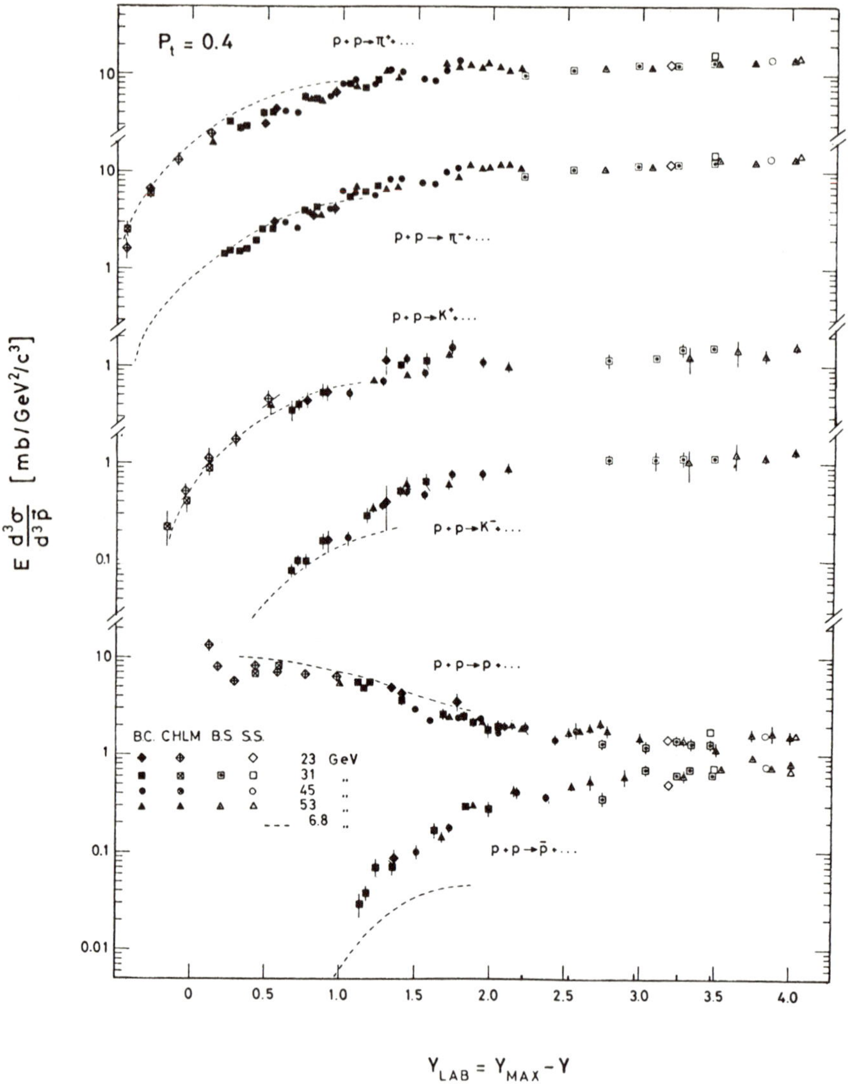

FIG. 5. Compilation of data on the rapidity distributions, in the beam proton rest frame, of pions, kaons, protons, and antiprotons for a fixed transverse momentum p_T of 0·4 GeV/c. The dashed curve is the data at 24 GeV, Ref. 45. Taken from Ref. 15.

cross-section should rise. Of course there is plenty of energy to produce $\bar{p}p$ but when we constrain the momentum transfers to be small this naturally suppresses the production of massive particles in the centre. This suppression is most severe at lower energies and at small x. So this is simply a threshold

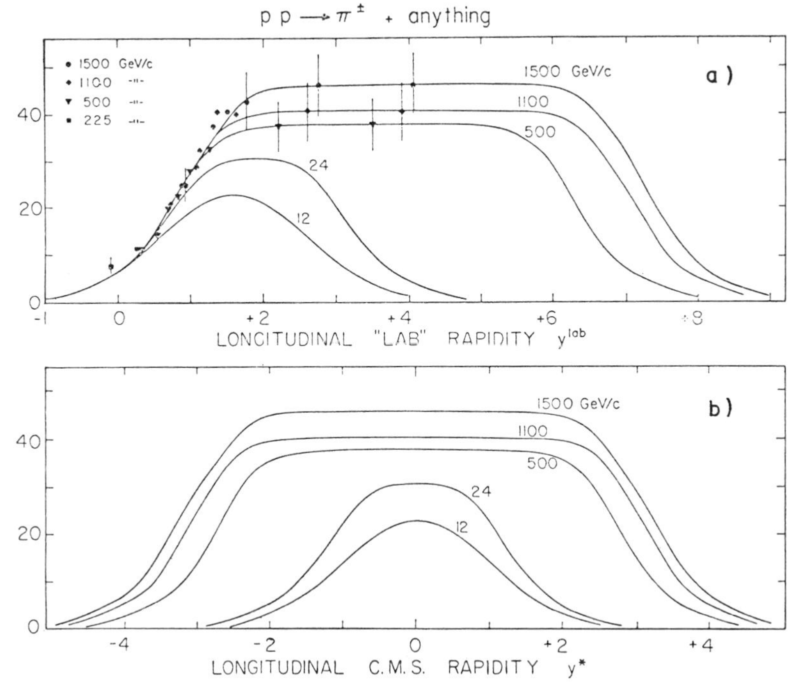

FIG. 6. Rapidity plots for $pp \to \pi^{\pm}X$. (a) in the lab. system, (b) in the c.o.m. system. Taken from Ref. 17.

effect and the rise in cross-section has nothing to do with the non-scaling behaviour predicted from Regge.

Cancschi[(18)] has considered the graph

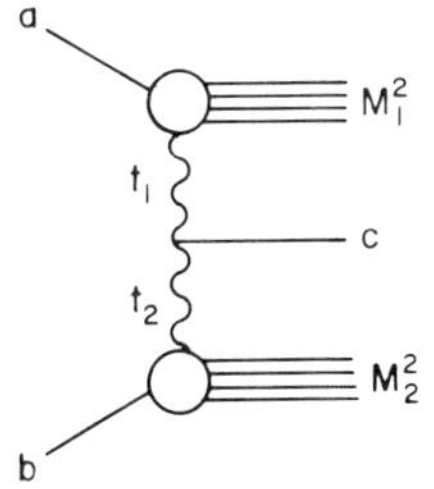

which on squaring gives

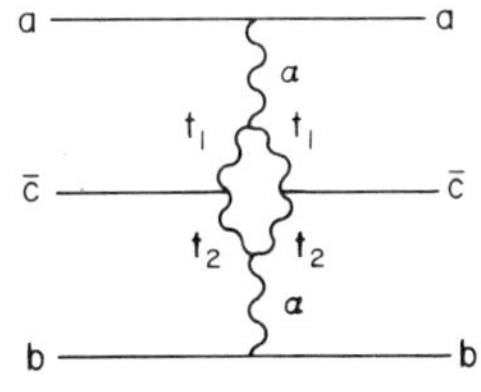

where we assume the M^2 dependence to be given by exchange of a Regge pole α. This gives a behaviour $(M_1^2)^\alpha$ but when we do the loop integration in t_1 and t_2 we get a corresponding expansion in $t = s_{a\bar{c}}$, i.e.

$$(M_1^2)^\alpha \to t^\alpha - c_1 t^{\alpha-1} + \dots \quad (28)$$

This is equivalent to the exchange of α and its daughters in the $a\bar{a}$ channel. Note the negative sign in front of $t^{\alpha-1}$. When α is the Pomeron, we get non-scaling behaviour

$$\underset{\text{non-scaling}}{f(x=0)} \simeq -c(|t|^{-1} + |u|^{-1}) \sim -2c's^{-\frac{1}{2}} \cosh y^* \tag{29}$$

Calculations of the increase in the $\bar{p}$ cross-section in the central region based on this threshold effect appear to be consistent with the data.[19]

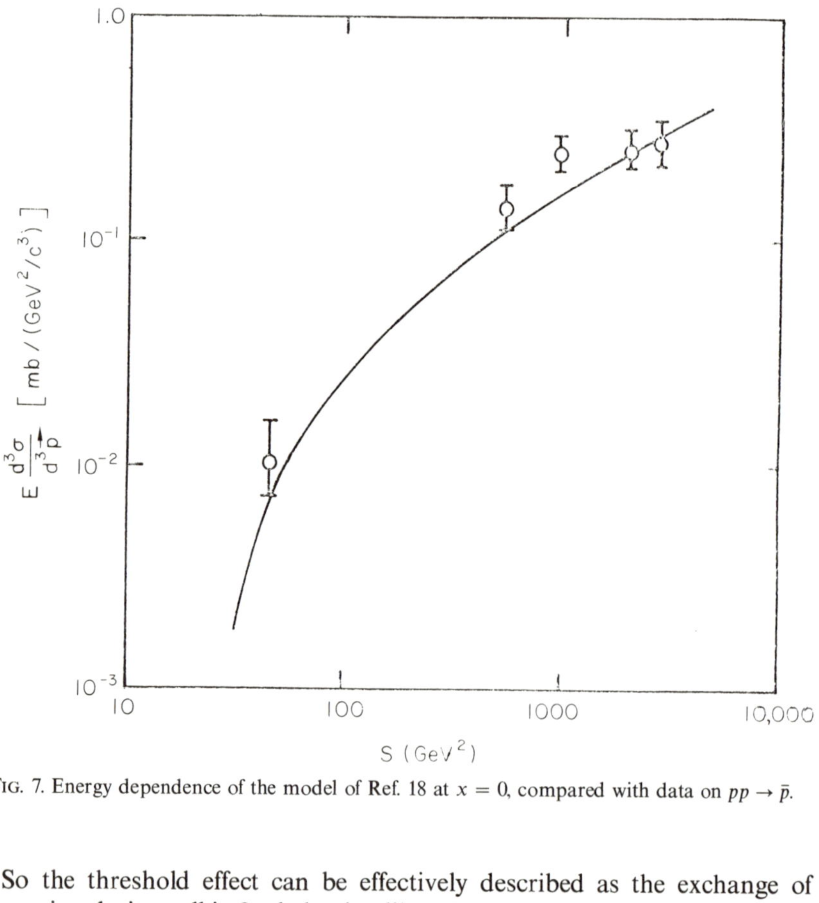

FIG. 7. Energy dependence of the model of Ref. 18 at $x = 0$, compared with data on $pp \to \bar{p}$.

So the threshold effect can be effectively described as the exchange of some singularity, call it Q—behaving like a Pomeron daughter, with *negative* residue.

So if we want to do phenomenology in the central region we have first to admit the presence of this threshold behaviour and if we wish still to pursue the Mueller approach then this effect can be parameterized by the exchange of some effective singularity with low intercept, taking the above result of Caneschi.[18] This was the line of approach adopted by Chan, Miettinen, Roy and Hoyer,[20] who labelled this effective singularity Q.

Another school of thought[21] however has insisted that the rise of central cross-sections is consistent with the simple Mueller–Regge picture with exchange only of Pomeron and meson exchanges provided that the sign of non-scaling piece is taken *negative*. This attitude can be firmly knocked on the head by inspecting the relevant data.

Figure 8 shows the π^+, π^- data at ISR and at 12, 24 GeV/c, plotted against $s^{-\frac{1}{4}}$ and $s^{-\frac{1}{2}}$ in turn. In the first case, an extrapolation to infinite energies shows a clear violation of the Pomeranchuk theorem. The $s^{-\frac{1}{2}}$ fit looks rather good and is consistent with equality of π^+ and π^- asymptotically.

Let us follow the approach of Chan *et al.*,[20] who parametrize the threshold behaviour in terms of an exchange (call it Q) in the double Regge graph. Since this exchange must be present in component (7) it must have the quantum numbers of the vacuum (i.e. Q is something like a daughter of the Pomeron, i.e.

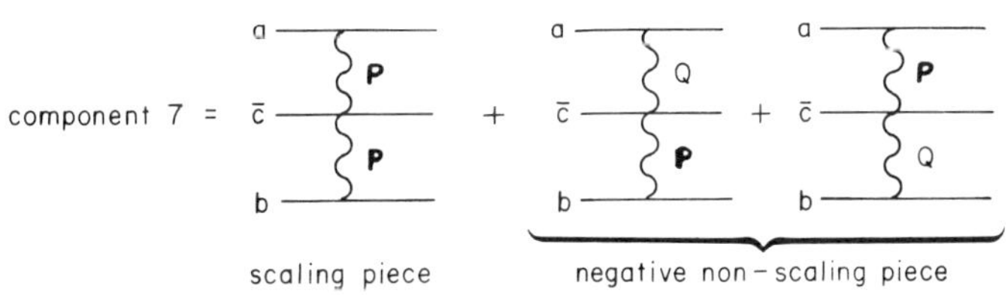

Not knowing anything about Q we can arrange to avoid component (7) by looking at the difference of particle and anti-particle cross-sections. If we can isolate the components (5) + (6), i.e.

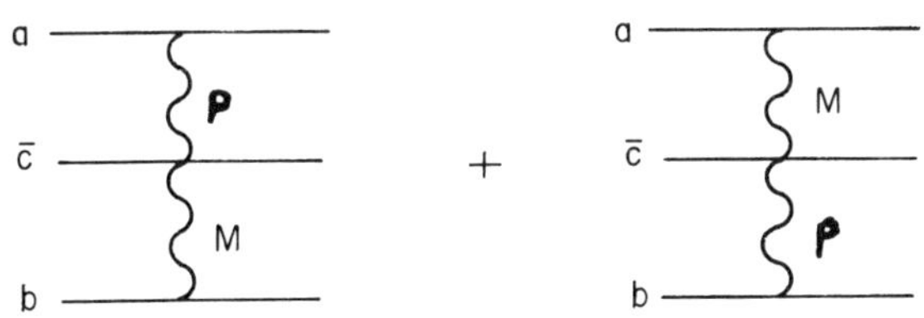

then we expect to retrieve the $s^{-\frac{1}{4}}$ behaviour, i.e.

$$\begin{aligned} \Delta_c &= f(ab \to cX) - f(ab \to \bar{c}X) \\ &= \beta_{\mathbf{PM}}(\kappa)\, s^{-\frac{1}{4}} \cosh\left(\tfrac{1}{2}y^*\right). \end{aligned} \tag{30}$$

It is now interesting to test directly whether such behaviour is consistent

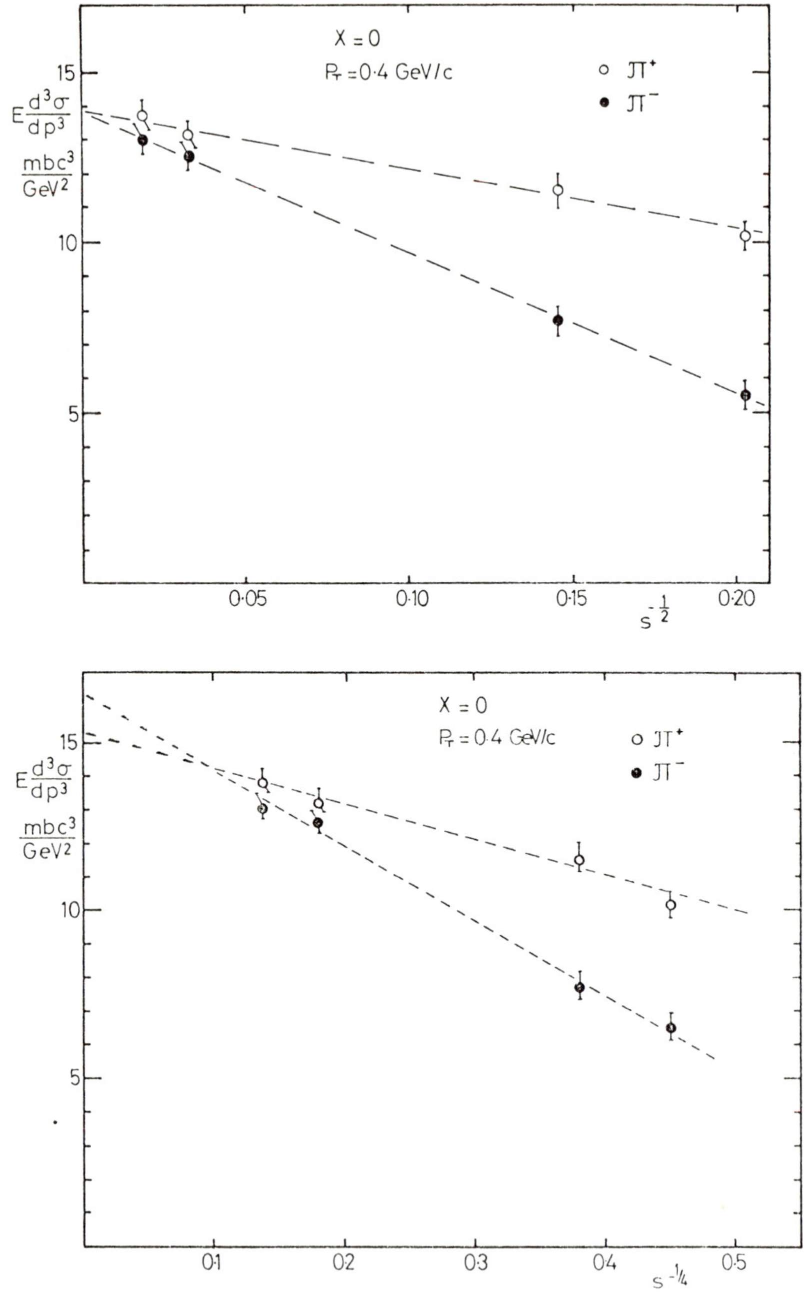

FIG. 8. The invariant production cross-section for pions at $x = 0$ and $p_T = 0{\cdot}4$ GeV/c as a function of (a) $s^{-\frac{1}{2}}$ and (b) $s^{-\frac{1}{4}}$. Taken from Ref. 15.

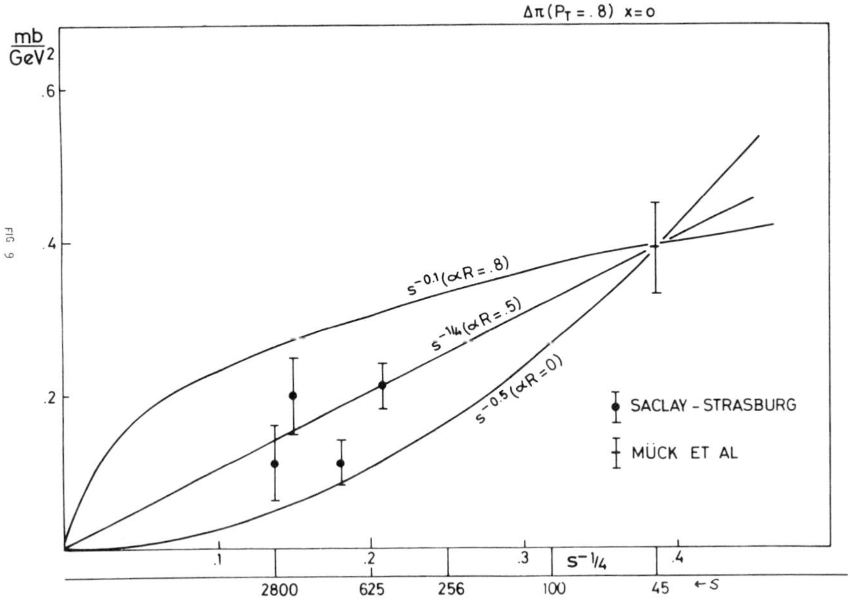

FIG. 9. The cross-section difference Δ_p $(=f(pp \to \pi^+ X) - f(pp \to \pi^- X))$ on an $s^{-\frac{1}{4}}$ plot. The straight line prediction corresponds to eqn (30). The power behaviours $s^{-0\cdot1}$ and $s^{-0\cdot5}$, corresponding to an $\alpha_R(0) = 0\cdot8$ and 0, are also shown for comparison. Taken from Refs 20, 22.

with data. Figures 9, 10, 11 are the results of an analysis by Miettinen and Roy.[22, 23] The data on Δ_π and Δ_p at a fixed value of $y^* = 0$ are shown first and favour the $s^{\frac{1}{4}}$ dependence. The errors are large and one cannot rule out an exponent between $-0\cdot1$ and $-0\cdot5$. Figure 11 shows a plot of $s^{\frac{1}{4}}\Delta_p$ which favours a linear dependence on cosh $(\frac{1}{2}y^*)$.

With the assumption of such an object one can more readily understand other features of central cross-sections. For example, experimentally the following inequality is true

$$f(\pi^+ p \to \pi^+ X) > f(\pi^+ p \to \pi^- X)$$

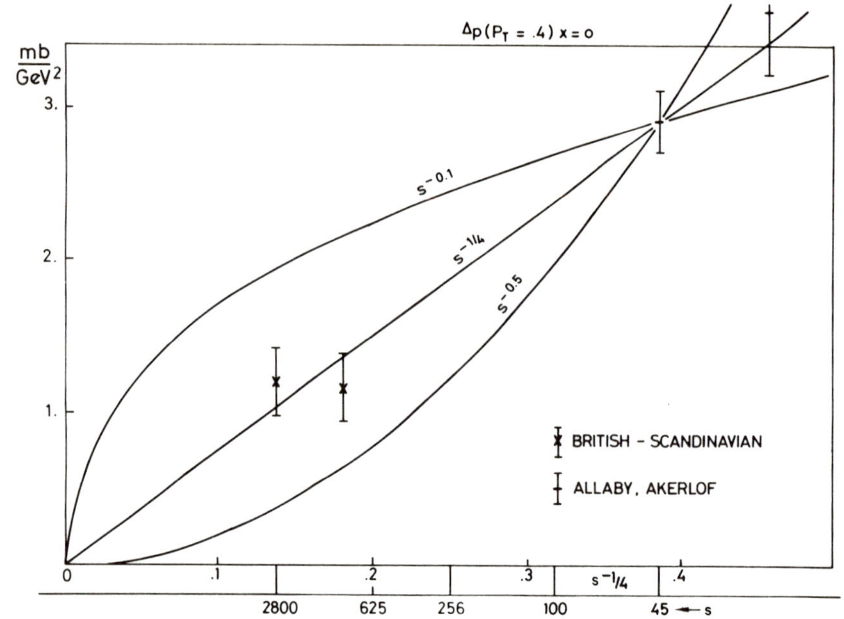

FIG. 10. Same as Fig. 9 for the cross-section Δp. Taken from Refs 20, 22.

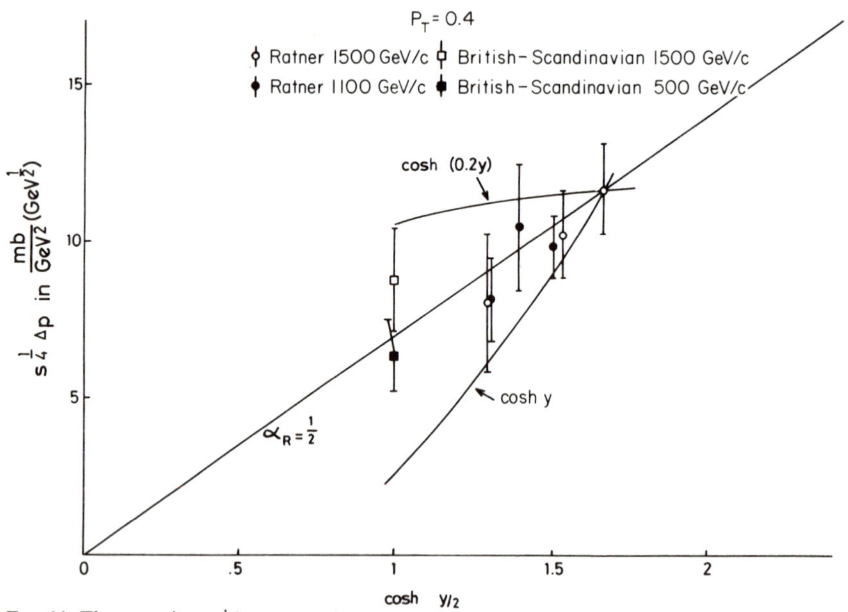

FIG. 11. The quantity $s^{-\frac{1}{4}}\Delta_p$ on a cosh $Y/2$ plot. The straight line corresponds to eqn (27). The behaviours cosh (0·2Y) and cosh (Y), corresponding to $\alpha_R(0) = 0{\cdot}8$ and 0, are also shown for comparison.

Since $a\bar{c}$ is exotic in (b), meson exchange is absent in the upper part.

Therefore (a) − (b) = $\beta_{MP}(\kappa)t^{-\frac{1}{2}}$.

Furthermore the central vertex is the same as for $pp \to \pi^{\pm}$. Without Q, the *rise* in $pp \to \pi^{\pm}$ has to be accounted for by $\beta_{MP}(\kappa) < 0$. Consequently (a) − (b) < 0. With the assumption that $\beta_{PQ}(\kappa)$ is responsible for the $pp \to \pi^{\pm}$ rise we can revert to $\beta_{MP} > 0$ and so get the desired result.

One can proceed further and develop the ideas of duality and factorization for the Regge exchanges in both the upper and lower parts of the graph. In this way, many cross-section *differences* for $\sigma(ab \to c) - \sigma(ab \to \bar{c})$ can be related, e.g.

$$\begin{array}{ll} \Delta(pp \to \pi^+) & 2 \\ \Delta(\pi^+ p \to \pi^+) & 3 \\ \Delta(\pi^- p \to \pi^+) & \sim -2 \\ \Delta(K^+ p \to \pi^+) & \frac{7}{4} \\ \Delta(K^- p \to \pi^+) & -\frac{3}{4} \end{array}$$

Experimentally there seems to be some evidence in favour of the first three numbers from 24 GeV/c data.

VII. Triple-Regge Behaviour and Application to Data

When discussing the fragmentation region (or single Regge region) we parametrized the cross-section, in eqn (18), as

$$f = \frac{d^2\sigma}{dt\, d(M^2/s)} = \sum_k F_k\left(t, \frac{s}{M^2}\right) s^{\alpha_k(0)-1} \tag{31}$$

where F_k denotes the "blob"

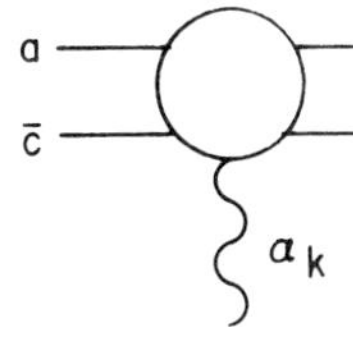

As far as phenomenology was concerned F_k was treated just as a normal factorizable residue function, no comment being made about its dependence on either t or s/M^2. In this case we had considered M^2 to be large; if we go to the particular region of fragmentation phase space where in addition s/M^2 is large, then we can learn what the dependence on s/M^2 should be in this region.

If we begin by assuming s/M^2 large and $s \gg t$ and keeping M^2 moderate then we have essentially the near forward pseudo-two body process $a + b \to c + X$ for which the appropriate graph is

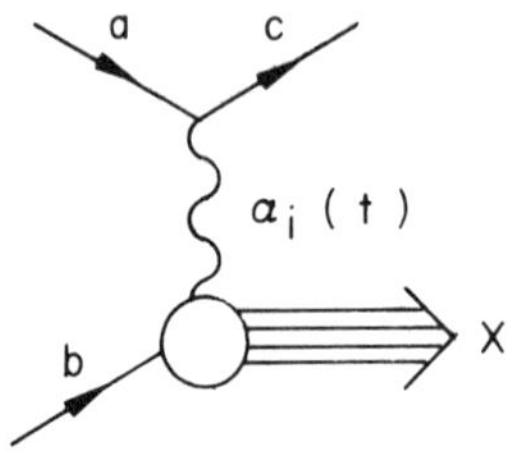

corresponding to exchange of the leading Regge trajectories in the t-channel.

which on squaring and summing over all states in X gives

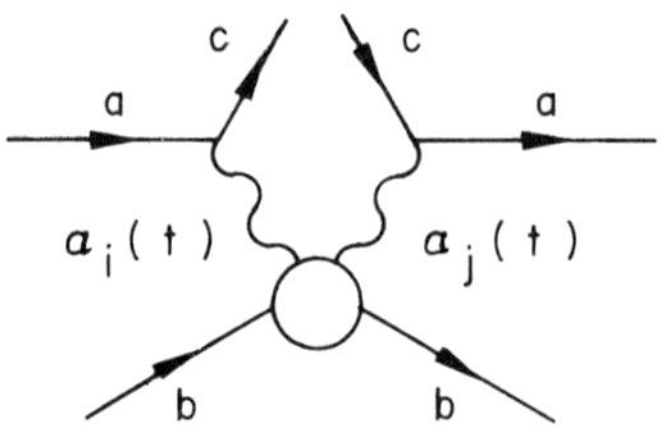

summed over i and j

$$f = \frac{d^2\sigma}{dt\, d(M^2/s)} = \frac{1}{s} \sum_{ij} \beta^i_{a\bar{c}} \xi_i(t) \beta^j_{a\bar{c}} \xi^*_j s^{\alpha_i(t)+\alpha_j(t)} \operatorname{Im}[A_{ib\to jb}(M^2, t)]. \tag{32}$$

The quantity $A_{ib\to jb}(M^2, t)$ is usually referred to as the forward Reggeon-particle scattering amplitude—strictly speaking, it is the analytic continuation of the maximum helicity flip amplitude in the centre-of-mass of the crossed channel $b\bar{b} \to \bar{\alpha}_i \alpha_j$

$\beta^i_{a\bar{c}}$ is the coupling of

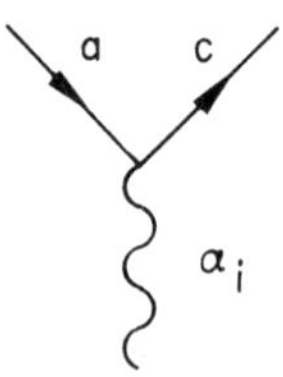

and $\xi_i(t)$ is the signature factor

$$(\tau_k + e^{-i\pi\alpha(t)}) \times \frac{1}{\sqrt{\pi}} \Gamma(1 - \alpha_i(t))$$

(this normalization we choose now, for use later on).

When we allow M^2 to become large (but still keeping s/M^2 large) the asymptotic behaviour of $A_{ib\to jb}$ is controlled by exchange of Regge poles $\alpha_k(0)$

which can couple to $b\bar{b}$ and $\alpha_i\bar{\alpha}_j$

$$\operatorname{Im} A_{ib\to jb}(M^2, t) = \beta^k_{b\bar{b}} g^k_{ij}(t)\,(M^2)^{\alpha_k(0)-\alpha_i(t)-\alpha_j(t)} \tag{33}$$

(The $-[\alpha_i + \alpha_j]$ term in the exponent is connected to the fact that $A_{ib\to jb}$ corresponds to maximal helicity flip of the Reggeon legs).

Combining (32) and (33)

$$\frac{d^2\sigma}{dt\,d(M^2/s)} = \sum_{ijk} G_{ijk}(t)\left(\frac{M^2}{s}\right)^{\alpha_k(0)-\alpha_i(t)-\alpha_j(t)} s^{\alpha_k(0)-1} \tag{34}$$

where $G_{ijk}(t) = \beta^i_{a\bar{c}}\beta^j_{a\bar{c}}\xi_i(t)\xi^*_j(t)g^k_{ij}(t)\beta^k_{b\bar{b}}$.

Expression (34) is called the triple-Regge formula since it describes the graph

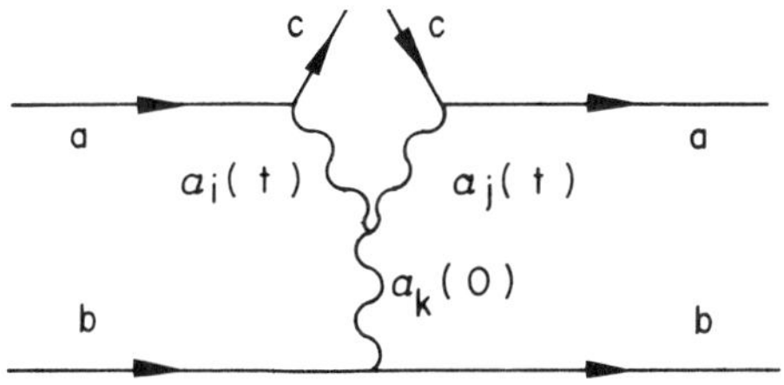

Comparing with (31) we see that the fragmentation residue has been expanded as

$$F_k\left(t, \frac{M^2}{s}\right) = \sum_{ij} G_{ijk}(t)\left(\frac{M^2}{s}\right)^{\alpha_k(0)-\alpha_i(t)-\alpha_j(t)} \tag{35}$$

and we can now see the necessity of $-[\alpha_i + \alpha_j]$ term in (33) in order to ensure consistency with the fact that F_k has to be a function of the ratio M^2/s.

Now we turn to a discussion of how the triple-Regge expression (34) can be applied to experimental data. As we saw this formula is strictly valid only if both quantities M^2 *and* s/M^2 are simultaneously large (i.e. >5). If we adhere to this constraint then we see that at energies below NAL, the region of applicability is very small. We shall see later on that there is a justification for relaxing the M^2 condition based on duality.

In this section we discuss only those cases where $\alpha_i(t)$ $\alpha_j(t)$ are not the Pomeron trajectory. There have been many papers applying triple-Regge (TR) formulae to data and we shall mention only a few of them here. The basic approach has been the following: First look at a reaction where the quantum numbers of a and c are such that the t-channel is presumably dominated by one or two (exchange–degenerate) trajectories. Secondly, since such an analysis looks usually only at data at *one* energy, assume a value for $\alpha_k(0)$—this is usually taken as 1 (large M^2 dominated by Pomeron). Finally, study the M^2 dependence and extract a value of $\alpha(t)$ and see if this agrees with the trajectory one expected from the quantum numbers.

For example, if one looks at $K^- \xrightarrow{p} \bar{K}^0$ one expects ρ and A_2 exchange in the t-channel. Chliapnikov *et al.*[24] and Paler *et al.*[25] have analysed their data and find reasonable trajectories. Figure 12 shows the latter's

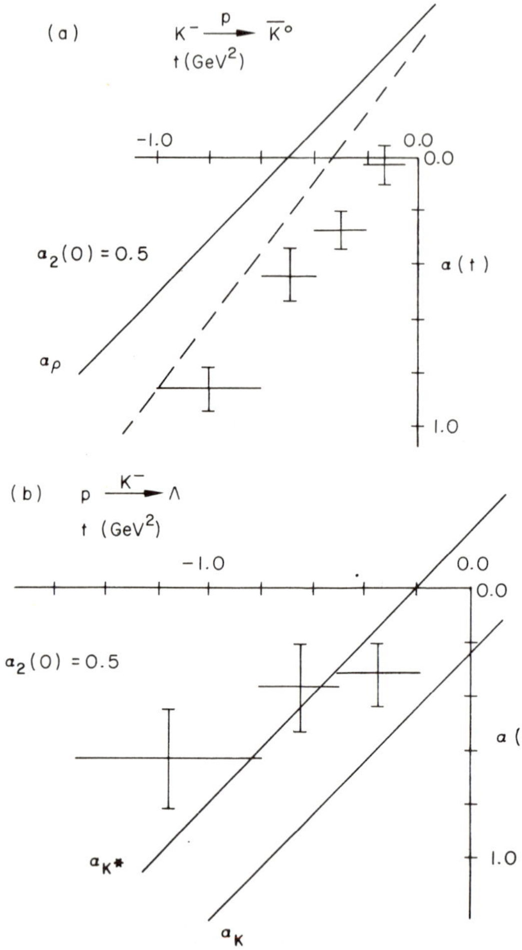

FIG. 12. Effective ρ and K^* trajectories extracted from M^2 dependence of $K^- \xrightarrow{p} \bar{K}^0$ and $p \xrightarrow{K} \Lambda$ at 14 GeV/c, taken from Ref. 25.

results together with their resulting trajectory for $p \xrightarrow{K^-} \Lambda$ which should resemble the K^* trajectory. For a situation like this where $ab\bar{c}$ is not exotic, one expects $\alpha_k(0)$ to be between $\frac{1}{2}$ and 1. They took $\alpha_k(0) = \frac{1}{2}$ but remember that a choice of 1 moves up the trajectory only by 0·25. In this analysis the range of M^2 was $4 \to 13\,\text{GeV}^2$ and $3 \to 7\,\text{GeV}^2$ and since the experiment was at 14 GeV the corresponding range of s/M^2 was $2{\cdot}2 \to 7$ and $4 \to 9{\cdot}5$.

Another nice example is the reaction $\gamma p \to \pi^- X$ which has been measured at energies up to 9·3 GeV.[(26)] The analysis of the photon fragmentation region gives a trajectory which one would expect from a pion with slope $\alpha' \simeq 1$.

Now let us turn to baryon exchange and let us divide the data into two classes separated by different kinematics.

(a) heavy particle → light particle

(b) light particle → heavy particle.

Measurements in the (a) group are, for example $p \xrightarrow{p} \pi^{\pm}$ or $p \xrightarrow{\pi} \pi^-$ and in (b), $\pi^- \xrightarrow{p} p$ or $K^- \xrightarrow{p} \Lambda$.

Triple-Regge analyses of group (a) have been done by many people, for example Refs (26), (27), (28), (29). While for the second case there are Refs (25) and (30). The essential difference in the results obtained for the two groups is summarized in Fig. 13, which is the result obtained by Paler *et al.* for $K^- \xrightarrow{p} \Lambda$ compared with the corresponding results on $p \to \pi^+$. In both cases we expect the t-channel trajectory to be N_α with intercept $\approx -\frac{1}{2}$. Consistently the analysis of $p \to \pi^+$ has found trajectories with much too low an intercept, while group (b) has found consistency with the expected $\alpha_N(t)$.

Chan Hong-Mo has suggested a possible reason for this behaviour. Because of the different kinematic situations for (a) and (b):

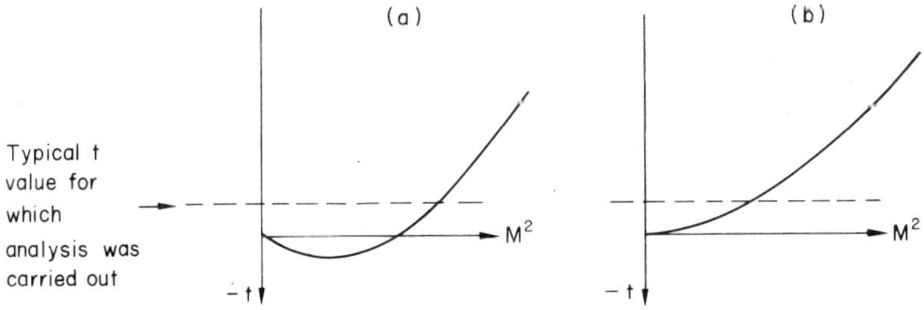

In the first case we are not close to t_{min}. The triple-Regge formula receives its contributions from graphs of the type:

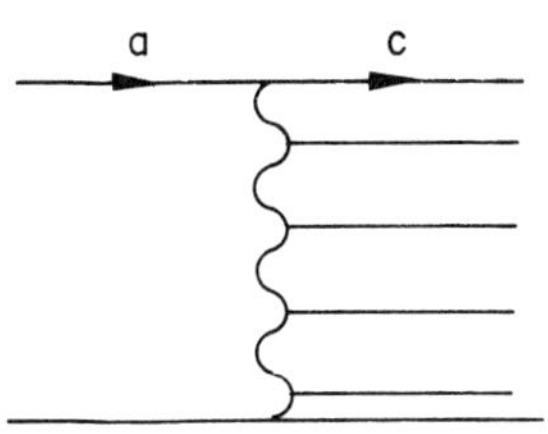

which dominate near to t_{min} as in the case in (b)

However for case (a) we might therefore be contaminated by graphs of the form

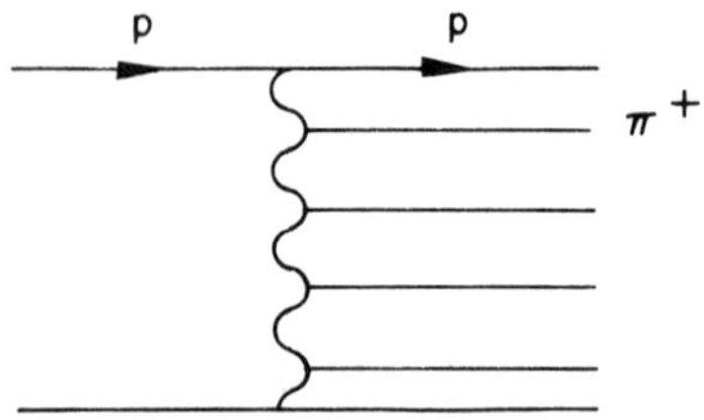

in which the π^+ is not the top particle in the ladder, and so such a graph is *not* a genuine contributor to the triple-Regge behaviour of $p \xrightarrow{p} \pi^+$.

Another nice example is $p \xrightarrow{\pi^+} \Delta^{++}$ which has been measured at CERN at 8 and 16 GeV.[31] For t not very small we expect $\rho - A_2$ exchange in the t-channel. Since the experiment is done at two widely spaced energies, $\alpha_k(0)$ can be determined rather than assumed ($\alpha(0) = 1$ can be justified only in the case X exotic). Again one sees an effective $\alpha(t)$ which is reasonably close to the expected $\rho - A_2$ trajectory with intercept $\frac{1}{2}$.

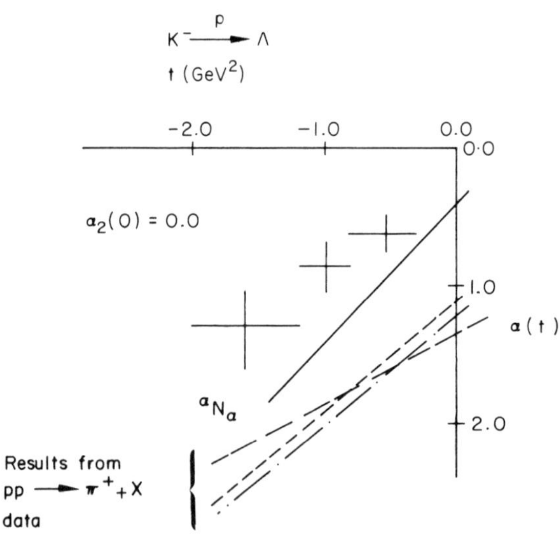

FIG. 13. Effective N trajectory extracted from M^2 dependence of $K^- \xrightarrow{p} \Lambda$ at 14 GeV/c, taken from Ref. 25.

Triple-Regge analyses have been carried out for processes involving Pomeron exchange but this topic is best dealt with after introducing the Finite Mass Sum Rules.

VIII. Finite Mass Sum Rules

In two body scattering, the development of the ideas of duality began with the writing down of the Finite Energy Sum Rules. In the last section we have been discussing the form of the amplitude for Reggeon-particle scattering and so it is natural to ask whether similar sum rules can be stated in this case where the "dispersing variable" is the missing mass M^2 rather than the incoming energy s; hence the name finite *mass* sum rule (FMSR).[32, 33] Of course we need to consider only one special case where the momentum transfer is zero, since we are considering *forward* Reggeon-particle scattering. The procedure we follow is that of Kwiecinski and Einhorn *et al.*[32] Again we begin by assuming s/M^2 large and $s \gg t$ so that the inclusive cross-section is given by eqn (32) represented by the graph immediately before it.

The next step is to write down the large M^2 behaviour of $A_{ib\to jb}(M^2, t)$—this is just eqn (33).

To write a dispersion relation for $A_{ib\to jb}(M^2, t)$ we need to know its analytic properties. The assumption has to be made that these are the same as for forward two-body scattering. This is supported by analyses of the six-point function in dual resonance models.

As in the two-body case, we need to consider both the right- and left-hand cuts of $A_{ib\to jb}(M^2, t)$. To do this, it is first convenient to define an anti-symmetric variable ν (analogous to $s - u$ in the two-body case) given by

$$\nu = p_b \cdot (p_a - p_c) = \tfrac{1}{2}(M^2 - t - m_b^2).$$

I.e. $\nu \to -\nu$ as

$$(\alpha_i + b \to \alpha_j + b) \to (\bar{\alpha}_j + b \to \bar{\alpha}_i + b)$$

or equivalently as

$$(a + b \to c + X) \to (c + b \to a + X)$$

$\downarrow$	$\downarrow$
The absorption part of this gives the right-hand cut in $A_{ib\to jb}(\nu, t)$	The absorption part of this gives the left-hand cut in $A_{ib\to jb}(\nu, t)$

The large ν behaviour is of course given by the triple-Regge behaviour and we get, taking contributions from left- and right-hand cuts

$$A_{ib\to jb}(\nu, t) = \sum_k \beta^k_{b\bar{b}} g^k_{ij}(t)\, \nu^{\alpha_k(0) - \alpha_i(t) - \alpha_j(t)}$$
$$\times \{\tau_i \tau_j \tau_k + \exp[-i\pi(\alpha_k(0) - \alpha_i(t) - \alpha_j(t))]\} \qquad (36)$$

which is not surprising except, perhaps, for the triple-Regge signature factor $\tau = \tau_i \tau_j \tau_k$. This is simply the signature τ_k of the Regge trajectory α_k when $i = j$ but when i, j have opposite signature $\tau = -\tau_k$.

To write down the FMSR one then simply has to take a combination of amplitudes which is guaranteed to be antisymmetric in ν:

$$\int_0^N \mathrm{d}\nu\, \nu^n [\mathrm{Im}\, A(\nu, t) + (-1)^{n+1}\, \mathrm{Im}\, A(-\nu, t)]$$

$$= \sum_k \beta^k_{b\bar{b}} g^k_{ij}(t)\, [1 + (-1)^{n+1}\, \tau_i \tau_j \tau_k] \frac{N^{\alpha_k(0) - \alpha_i(t) - \alpha_j(t) + n + 1}}{\alpha_k(0) - \alpha_i(t) - \alpha_j(t) + n + 1} \tag{37}$$

i.e.

$$\int_0^N \mathrm{d}\nu\, \nu^n \left[s \frac{\mathrm{d}^2\sigma}{\mathrm{d}t\, \mathrm{d}M^2} (ab \to cX) + (-1)^{n+1}\, s \frac{\mathrm{d}^2\sigma}{\mathrm{d}t\, \mathrm{d}M^2} (cb \to aX) \right]$$
$$= \sum_{ijk} G_{ijk}(t)\, [1 + (-1)^{n+1}\, \tau_i \tau_j \tau_k]$$
$$\times\, s^{\alpha_i(t) + \alpha_j(t) - 1} \frac{N^{\alpha_k(0) - \alpha_i(t) - \alpha_j(t) + n + 1}}{\alpha_k(0) - \alpha_i(t) - \alpha_j(t) + n + 1}. \tag{38}$$

If we take a combination which is *symmetric* in ν, i.e. symmetric under "$s \leftrightarrow u$" for ab scattering, then the contribution from residues of the nonsense wrong signature *fixed* poles has to be included:

$$\int_0^N \mathrm{d}\nu\, \nu^n \left[s \frac{\mathrm{d}^2\sigma}{\mathrm{d}t\, \mathrm{d}M^2} (ab \to cX) + (-1)^n s \frac{\mathrm{d}^2\sigma}{\mathrm{d}t\, \mathrm{d}M^2} (cb \to aX) \right]$$
$$= \sum_{ij} s^{\alpha_i(t) + \alpha_j(t) - 1} \{ H^{(n)}_{ij}(t) + G_{ijk}(t) [1 + (-1)^n \tau_i \tau_j \tau_k]$$
$$\times \frac{N^{\alpha_k(0) - \alpha_i(t) - \alpha_j(t) + n + 1}}{\alpha_k(0) - \alpha_i(t) - \alpha_j(t) + n + 1} \Big\}. \tag{39}$$

This is just the same for two-body scattering. The fixed poles arise from the presence of a non-zero third double-spectral function which is symmetric under "$s \leftrightarrow u$". In the language of Veneziano 4 pt. functions, the terms $V(s, t)\, V(u, t)$ give the usual Regge pole behaviour, while $V(s, u)$ gives rise to the fixed pole behaviour. However if we can arrange matters such that either s or u for the ab scattering is exotic, then $V(s, u)$ vanishes and hence the fixed pole contribution vanishes for the resonant part of the amplitude (Finkelstein Selection Rule).

The "odd" sum rule (38) is the analogue of the "good" FESR, while (39) is the analogue of the Schwarz sum rule.

IX. Applications of FMSR

To try to estimate the triple-Regge couplings by applying the triple-Regge expression (34) directly to data is difficult since its validity really requires both s/M^2 and M^2 large. The low M^2 data can now be used to feed into the l.h.s. of the FMSR and with a suitable cut-off N, the triple-Regge couplings obtained. N has to be chosen above the resonances—say $M^2 \simeq 4\ \text{GeV}^2$ but the error obtained on the estimate of the couplings will naturally be $O(M^2_{\text{cut off}}/s)$ which is not small for $p_{\text{lab}} < 25\ \text{GeV}/c$.

Next, what about duality for Reggeon-particle scattering? The FMSR provide a method for testing whether the Harari–Freund two-component duality holds for αb scattering. According to the seven component duality scheme presented in the earlier lectures we saw that for the case of a normal Reggeon for $\alpha(t)$, the usual 2 component picture is expected to hold, i.e. resonances in the missing mass dual to Regge in the $b\bar{b}$ channel, non-resonating background dual to the Pomeron. In fact we shall proceed assuming this to be the case—the assumption being justified *a posteriori*. On the other hand, we also saw previously that when $\alpha(t)$ is a Pomeron, the duality situation is controversial. Einhorn *et al.*[8] have suggested that for Pomeron-particle scattering the resonances are dual to the *Pomeron*. Therefore we make no duality assumption for diffraction dissociation but rather try to resolve the question by applying FMSR to the data.

Related to this, is the controversy over the smallness or otherwise of the triple-Pomeron couplings. There are strong arguments for believing that when $t = 0$, $g_{ppp}(t) = 0$. Generally these arguments follow from unitarity but one can regard the zero as arising from a nonsense wrong signature point of the amplitude $A_{ih\to jh}(M^2, t)$. This amplitude corresponds to maximum helicity flip in the $b\bar{b}$ channel, one Reggeon with helicity $\alpha_i(t)$, the other with $-\alpha_j(t)$. We therefore expect the usual nonsense zeroes at $\alpha_i(t) + \alpha_j(t) - 1$, $\alpha_i(t) + \alpha_j(t) - 2$, etc. If we are looking at the even signature exchange $\alpha_i(t) + \alpha_j(t) - 1$ is a N.W.S. point and we expect a zero of the form $(\alpha_k(0) - \alpha_i(t) - \alpha_j(t) + 1)$. Thus $g_{ppp}(0) = 0$ if $\alpha_p(0) = 1$. But we also expect fixed poles at exactly these nonsense points which in general cancel the zeroes. Abarbanel and Green[34] have pointed out that the discontinuity at the branch point of the cut generated by the exchange of $\alpha_i(t)$ and $\alpha_j(t)$ is proportional to the residue of the fixed pole. The unitarity argument of Bronzan and Jones requires the discontinuity to vanish at the branch point, hence the fixed pole residue vanishes leaving the triple-Pomeron zero at $t = 0$ uncancelled.

It has been conjectured by many people that the triple-Pomeron vertex would therefore be small even for $t \neq 0$. This can be precisely tested by evaluating the FMSR for diffractive processes.

Next consider the even moment sum rule (39). In ordinary two body scattering the fixed pole contributions have no physical interpretation and only tend to be a nuisance. However the fixed pole contributions to Reggeon-particle scattering have a direct physical significance.

Writing

$$H_{ij}^{(n)}(t) = \beta_{a\bar{c}}^{i}\xi_{i}(t)\beta_{a\bar{c}}^{j}\xi_{j}R_{ij}^{(n)}(t)$$

then $R_{ij}^{(0)}(t)$ is precisely the vertex for the cut generated by the exchange of α_i and α_j, according to the Gribov calculus.[35] The cut is generated by convoluting $R_{ij}^{(0)}(t)$ with itself:

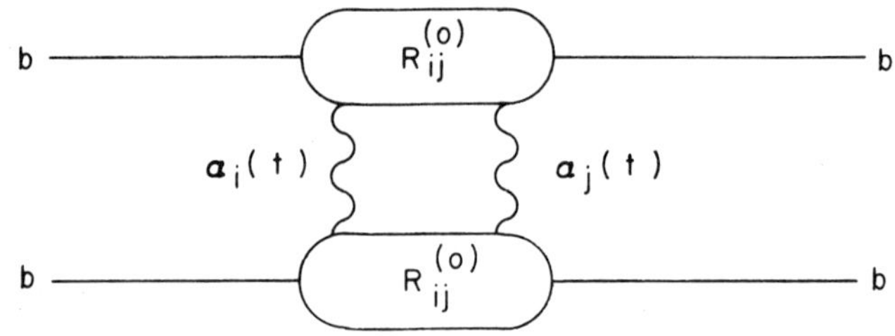

Notice that since the momentum transfers for α_i and α_j are the same, we can calculate the cut contribution only in the forward direction. At least in principle, we can use data on inclusive reactions and via the FMSR calculate the two-Reggeon contribution to a total cross-section.

Essentially the only other prescription for calculating Regge–Regge cuts is the eikonal procedure. In this model we have a sum over low mass resonances (including the incoming particle itself) instead of the fixed pole contribution, i.e.

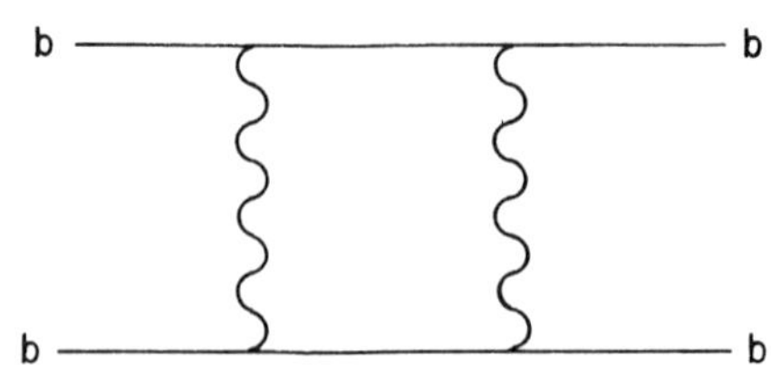

Actually this cannot really give rise to a cut since, if we dualize the intermediate states, we get

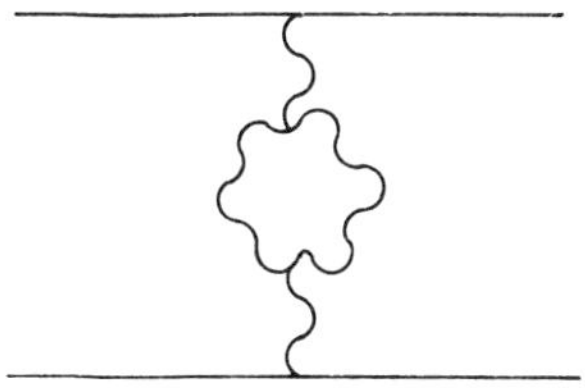

which can be regarded as a correction to simple Regge pole exchange. We mentioned earlier that the fixed poles could be connected with the $V(s, u)$ term—which is represented by the non-planar graph:

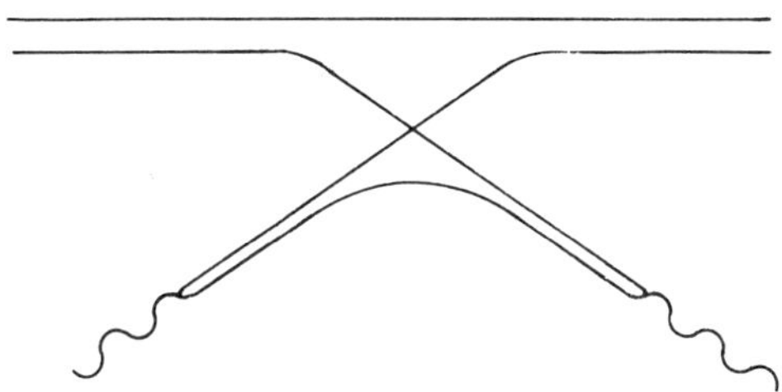

It is the non-planar nature of the "vertex" which is essential in generating j-plane cuts, e.g. the famous Mandelstam cut graph.

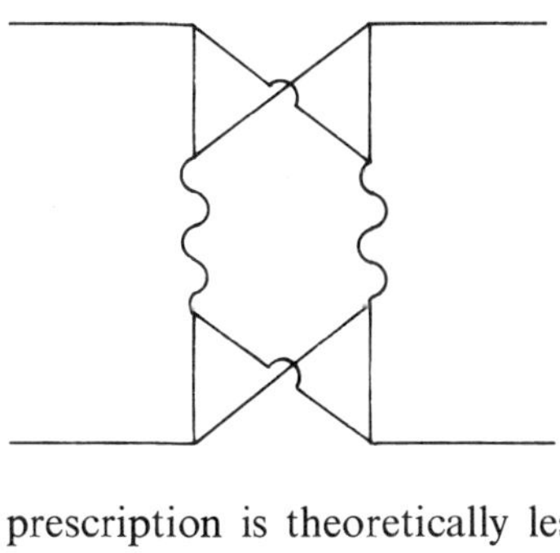

Even if the eikonal prescription is theoretically less justified, almost all previous cut estimates are based on it. What is therefore interesting to compare is the estimate based on the Gribov calculus with that based on the eikonal model. This amounts to comparing, in eqn (39) with $n = 0$, the size of $H_{ij}^{(0)}$ and the low mass integral on the l.h.s., which can be done using the $n = 1$ sum rule (38) in conjunction.†

A special form of a FESR is one in which the leading Regge terms, for one

† From eqn (9) we see that the fixed pole residue is given by (low mass integral)—(Regge terms (positive)). Thus the magnitude of the cut could in general be smaller than the eikonal prescription because of the minus sign attached to the Regge terms, i.e. the Gribov–Migdal lower bound is not generally valid.

reason or another, are absent. This is called a superconvergent rule. Similarly we can form analogous superconvergent FMSR.[36, 37] The low mass integral, which is essentially a sum of resonance production reactions, is then equal to zero, thus providing a relation between such quasi-two-body processes.

To get a genuine superconvergence relation one should arrange things to ensure exotic quantum numbers in the $b\bar{b}$ channel and therefore, to leading order, the Regge terms in the FMSR vanish.[37]

Consider $\rho\pi$ elastic scattering; by taking the combination

$$A_{\rho^-\pi^+} + A_{\rho^+\pi^+} - 2A_{\rho^0\pi^+} \tag{40}$$

we guarantee pure $I = 2$ in the crossed ($\pi\pi \to \rho\rho$) channel.

For example, these amplitudes could be extracted from inclusive data by looking at the *natural* parity exchange contributions of

$$p \xrightarrow{\pi^+} \Delta^{++}, \quad p \xrightarrow{\pi^+} \Delta^0, \quad p \xrightarrow{\pi^+} \Delta^+. \tag{41}$$

Using isospin to relate the $\rho p\Delta$ vertices, and taking $n = 1$ for the odd sum rule (8) we get

$$\int_0^N \nu \frac{\mathrm{d}M^2}{2} \left\{ \frac{1}{3} \frac{\mathrm{d}^2\sigma}{\mathrm{d}t\,\mathrm{d}M^2} (p \xrightarrow{\pi^+} \Delta^{++}) + \frac{\mathrm{d}^2\sigma}{\mathrm{d}t\,\mathrm{d}M^2} (p \xrightarrow{\pi^+} \Delta^0) - \frac{\mathrm{d}^2\sigma}{\mathrm{d}t\,\mathrm{d}M^2} (p \xrightarrow{\pi^+} \Delta^+) \right\} = 0 \tag{42}$$

Next we take the resonance contributions only to the missing mass, the assumed resonances in the $\rho\pi$ channel being π, ω, A_2. Using isospin we can relate all to the reactions $\pi^+ p \to (\text{resonance})^0 \Delta^{++}$ and obtain finally:

$$-t \frac{\mathrm{d}\sigma_\rho}{\mathrm{d}t} (\pi^+ p \to \pi^0 \Delta^{++}) - (m_\omega^2 - t - m_\pi^2) \frac{\mathrm{d}\sigma_\rho}{\mathrm{d}t} (\pi^+ p \to \omega \Delta^{++}) + (m_{A_2}^2 - t - m_\pi^2) \frac{\mathrm{d}\sigma_\rho}{\mathrm{d}t} (\pi^+ p \to A_2^0 \Delta^{++}) = 0 \tag{43}$$

(the ρ subscript is to remind us that we are taking only the natural parity exchange contribution.)

Similarly one gets

$$-t \frac{\mathrm{d}\sigma_\rho}{\mathrm{d}t} (\pi^- p \to \pi^0 n) - (m_\omega^2 - t - m_\pi^2) \frac{\mathrm{d}\sigma_\rho}{\mathrm{d}t} (\pi^- p \to \omega^0 n) + (m_{A_2}^2 - t - m_\pi^2) \frac{\mathrm{d}\sigma_\rho}{\mathrm{d}t} (\pi^- p \to A_2^0 n) = 0. \tag{44}$$

Before attempting to find whether data support this relation, it is worth pointing out the novelty of connecting reactions involving particles which belong to completely different SU (3) classifications. The "elastic" reaction is strongly damped at small t, hence we have approximately

$$\frac{d\sigma_\rho}{dt}(\pi^- p \to A_2^0 n) \simeq \frac{M_\omega^2}{M_{A_2}^2}\frac{d\sigma_\rho}{dt}(\pi^- p \to \omega n)$$
$$\simeq \frac{1}{3}\frac{d\sigma_\rho}{dt}(\pi^- p \to \omega n). \tag{45}$$

From density-matrix analysis of ω production we know that natural parity exchange accounts for about 60% of the reaction. Equation (49) *predicts* the natural parity contribution to A_2^0 production and when we compare this with the measured differential cross-section we find the prediction accounts for about only 30%. Thus we note an example of a systematics which we shall understand later on; high mass resonances are produced via an increasing proportion of *un*natural parity exchange.

Since density matrix elements for A_2^0 have not been measured we have to do a little juggling before testing eqn (44). We have $\rho_{11} + \rho_{1-1}$ for A_2^- production and using factorization we can write

$$\frac{d\sigma_\rho}{dt}(\pi^- p \to A_2^0 n) = \frac{d\sigma_{\rho,f}}{dt}(\pi^- p \to A_2^- p) \times \frac{(d\sigma_\rho/dt)(\pi^- p \to \pi^0 n)}{(d\sigma_{\text{no Pomeron}}/dt)(\pi^- p \to \pi^- p)} \tag{46}$$

This last quantity involves using a model for πN scattering (we used Barger and Phillips 5 pole fit). At last we can test eqn (44), the results being shown in Fig. 14 for $p_{\text{lab}} \simeq 7$ GeV.

X. Semi-Local Duality for Reggeon-Particle Scattering

We are, one by one, generalizing features of duality for two-body scattering to the case of Reggeon-particle scattering. The next concept we carry over is that of semi-local duality. If we write FESR for the resonance part of the two-body scattering amplitude

$$\int_0^N \operatorname{Im} f_{\text{res}}(\nu, t)\nu^n \, d\nu = \sum \beta(t)\frac{N^{\alpha+n+1}}{\alpha + n + 1} \tag{47}$$

then we assume that resonance fluctuations in $\operatorname{Im} f_{\text{res}}(\nu, t)$ average out near relatively small energy intervals, i.e.

$$\int_{N_1}^{N_2} \operatorname{Im} f_{\text{res}}(\nu, t)\nu^n \, d\nu = [\text{r.h.s. of (47)}]_{N=N_1}^{N=N_2} \tag{48}$$

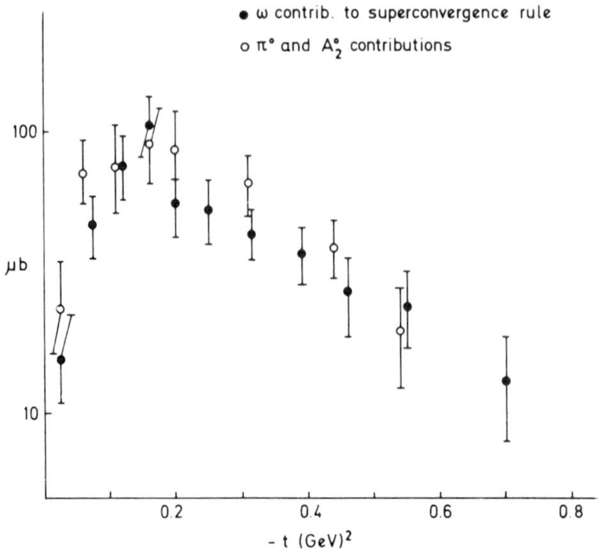

FIG. 14. Experimental values of contributions to the superconvergence relation (44), evaluated from the data taken from Ref. 37.

or

$$\langle \text{Im}\, f_{\text{res}}\, \nu, t\rangle = \text{Im}\, f_{\text{Regge}}(\nu, t). \tag{49}$$

This can be seen in Fig. 15 where Im $A'^{(-)}(\nu, 0)$ for πN is plotted against ν. It can be seen that the contributions from various resonances fluctuate strongly but, *on the average*, the ρ-Regge pole term gives a good description.

But perhaps the most striking demonstration of semi-local duality was when Dolen–Horn–Schmid[38] showed that by adding the contributions of just two or three resonances to $\int \text{Im}\, \nu B^{(-)}(\nu, t)\, dt$ one could easily produce the famous W.S. zero at $\alpha(t) = 0$ at $t \simeq -0{\cdot}5$ characteristic of ρ exchange. The Born contribution (elastic scattering) is negative and rapidly gets more so as $-t$ increases. The contribution from the resonances, dominantly Δ, is positive but also decreases with $-t$. Thus as t moves through $-0{\cdot}4 \to -0{\cdot}6$ the contributions from just one or two resonances cancel out and produce the zero due to ρ exchange in the t-channel.

Next we shall see how these ideas about semi-local duality can be carried over to inclusive reactions—following the development of Hoyer[37] *et al.*

In the case of inclusive reactions we are looking only at the *forward* scattering of Reggeon-particle. If semi-local duality holds in this case, then we can expect the leading Regge exchange in the $b\bar{b}$ channel to interpolate

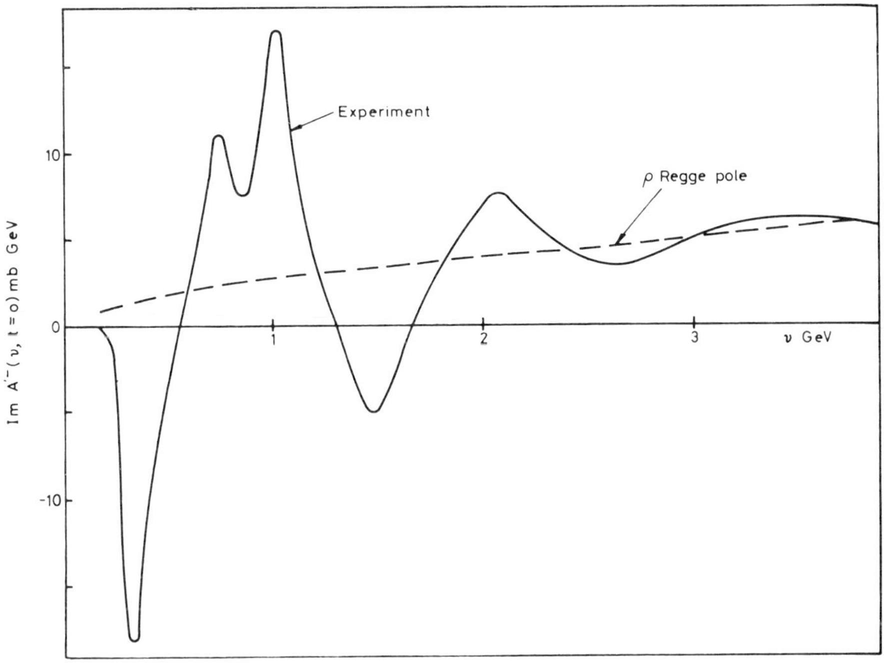

FIG. 15. An illustration of duality, for the πN amplitude A'^{-}. Experimentally, Im A'^{-} fluctuates at low energy but is given in the mean by the smooth ρ Regge pole term. From Ref. 38.

the resonances in the missing mass in an average sense. I.e. if no fixed poles are present

$$\left\langle \frac{d\sigma}{dt}(a + b \to c + \text{Resonances}) \right\rangle \sim (\nu_{res})^{\alpha_k(0) - 2\alpha_i(t)} \tag{50}$$

From our experience in the case of two-body scattering the averaging is to be done over a range of about 1 GeV^2 in M^2.

For the quasi-two body reaction:

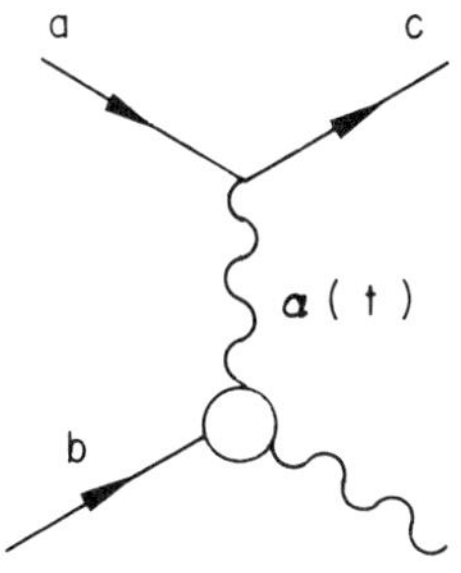

eqn (50) tells how the cross-section, at fixed s and t, varies with the mass of the produced resonance. We essentially have a four-point function where in addition to knowing the dependence on s and t, we know the dependence on the mass of one of the external legs.

In the situations we shall now consider we shall assume the validity of the two-component duality, i.e. resonances in missing mass are dual to the leading Regge exchange (meson, with $\alpha_k = \frac{1}{2}$) and background dual to Pomeron ($\alpha_k = 1$). So we can write

$$\int_0^N \mathrm{d}\nu\, \nu^n \frac{\mathrm{d}\sigma_\alpha}{\mathrm{d}M^2\,\mathrm{d}t}(a + b \to c + \text{res}) \sim g_{\alpha\alpha}^M(t) s^{2\alpha(t)-2} N^{\frac{1}{2}-2\alpha(t)+n+1} \tag{51}$$

In particular if there are no fixed poles present

$$\left\langle \frac{\mathrm{d}\sigma_\alpha}{\mathrm{d}t}(a + b \to c + \text{res.}) \right\rangle \sim (\nu_{\text{res}})^{\frac{1}{2}-2\alpha(t)}. \tag{52}$$

This use of Harari–Freund duality greatly reduces the number of triple-Regge terms since we need never consider Pomeron exchange in the $b\bar{b}$ channel. Also since we are isolating one trajectory $\alpha(t)$, the number of terms is reduced. In fact we need not consider interference terms since we always look at exchange degenerate pairs like π–B or ω–f.

Looking at resonance production cross-sections offers several advantages. Very often there are no data on the general inclusive reaction, e.g. $\pi^- p \to n +$ anything but there are excellent data on the quasi-two-body reactions, e.g. $\pi^- p \to \rho n$ or fn or gn. Again for the FMSR to be valid, s must be much larger than M^2 and by looking at M^2 in the resonance region and keeping s large this is automatically guaranteed.

Incidentally we are assuming everywhere that we really have genuine Regge pole–particle scattering. The FMSR are of course not valid if cuts were also important in the t-channel. To avoid such problems we try to choose examples where such cuts are thought to be rather small. For many cases there is evidence that poles are dominating the $a\bar{c}$ channel from measurements of the decay density-matrix elements.

(i) *π-exchange*

To test the above formula we should isolate a particular exchange trajectory $\alpha(t)$.

For the reactions

$$\begin{aligned} &\pi^- p \to (\rho^0, f, g^0) n \qquad &\text{(a)} \\ &K^- p \to (K^{*0}, K^{**0}) n \qquad &\text{(b)} \end{aligned} \tag{53}$$

The small t-region is known to be dominated by π-exchange (energy dependence, big ρ_{00}).

For (53a) we are studying $\pi^+\pi^-$ scattering and for (53b) π^+K^-. In either case the u-channel for the Reggeon-particle scattering is exotic; hence no fixed poles enter since we are looking only at s-channel resonances. So we may consider $n = 0$ as well as $n = 1$ in

$$\left\langle \nu_{\text{res}}^n \frac{d\sigma_\pi}{dt} (\pi^-(K^-)p \to X_{\text{res}}^0 n) \right\rangle = f(t) s^{2\alpha_\pi(t) - 2} (\nu_{\text{res}})^{\frac{1}{2} - 2\alpha_\pi(t) + n} \tag{54}$$

The experimental data for the l.h.s. for both reactions and for $n = 0, 1$ are compared with the ν behaviour given by r.h.s. of (54) and are shown in Figs 16 and 17, where we can see that semi-local duality appears to be working reasonably well.

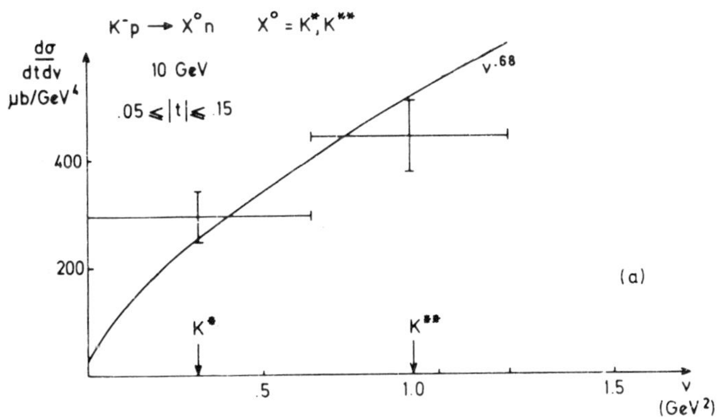

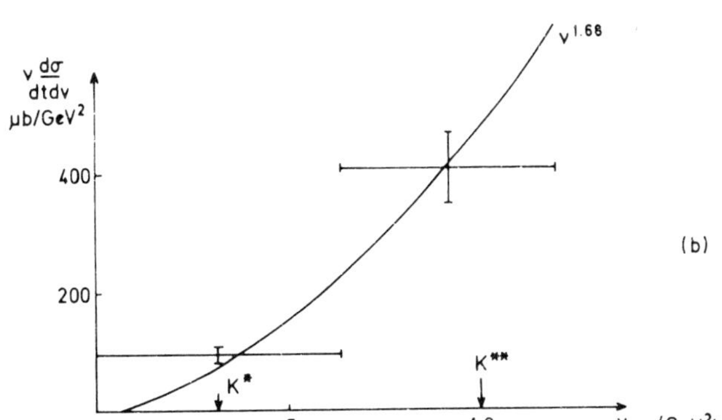

FIG. 16. Experimental $\nu^n(d\sigma_\pi/dt)(K^-p \to K^*, K^{**})n)$ at 10 GeV/c for $0{\cdot}05 \leqslant |t| \leqslant 0{\cdot}15$ compared with the π-exchange Regge contribution. (a) $n = 0$, (b) $n = 1$. Taken from Ref. 37.

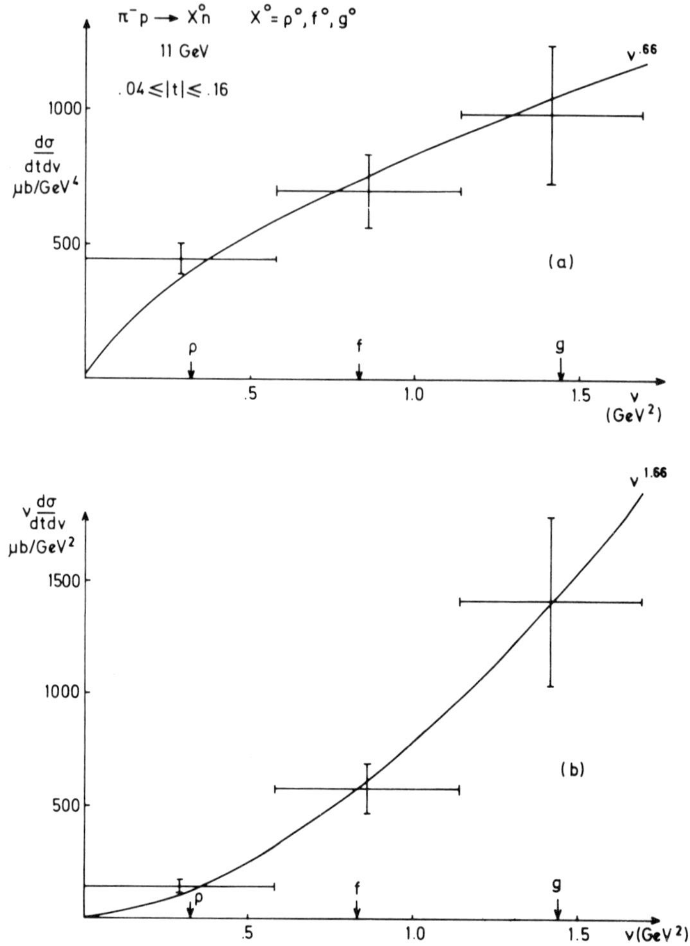

FIG. 17. Experimental $v^n(d\sigma_\pi/dt)(\pi^- \to (\rho^0, f^0, g^0)n)$ at 11 GeV/c for $0{\cdot}04 \leqslant |t| \leqslant 0{\cdot}16$ compared with the π-exchange Regge contribution. (a) $n = 0$, (b) $n = 1$. Taken from Ref. 37.

(ii) *Natural parity exchange*

We may consider f–ω exchange in the t-channel for the reactions

$$\begin{aligned} &\pi^- p \to (\pi^-, \rho^-, A_1^-, A_2^-, g^-)p \qquad &(a) \\ &K^- p \to (K^-, K^{*-}, Q^-, K^{**-})p \qquad &(b) \end{aligned} \tag{55}$$

To isolate the isospin zero exchange, the correct combinations with the charge–exchange reactions were formed. Estimates for the non-diffractive

Q^- cross-section used the observed cross-over in Q, $\bar{Q}$ production. In contrast to the previous case, neither the s or u channel for the Reggeon-particle scattering is exotic, so only $n = 1$ can be compared.

The only essential difference between (i) and (ii) is the value of the intercept of the particular trajectory—this makes the crucial difference when we come to consider:

(iii) *Ratio of unnatural/natural parity exchange*

In (i) we had π (unnatural parity) exchange, while in (ii) we had $f - \omega$ (natural parity) exchange.

Consequently from eqn (51), we get the ratio

$$\frac{\langle \nu \, \mathrm{d}\sigma_{\pi,B}/\mathrm{d}t[\pi^-(K^-) \to X^0_{\text{res}} n]\rangle}{\langle \nu \, \mathrm{d}\sigma_{f,\omega}/\mathrm{d}t[\pi^-(K^-) \to X^-_{\text{res}} p]\rangle} \sim (\nu^2_{\text{res}})^{2\alpha_{f,\omega}(t) - 2\alpha_{\pi,B}(t)} \sim M^2 \tag{56}$$

In other words, the cross-section, at fixed s and t, should show the unnatural parity contribution rapidly increasing w.r.t. the natural parity contribution as the mass of the produced resonance increases. This is the explanation behind the strong decreasing proportion of ρ-exchange in going from $\pi^- p \to \omega n$ to $\pi^- p \to A_2^0 n$, that we saw in the section on superconvergent relations. In fact we can see how it drops from almost 70% to $\sim$40% since semi-local duality predicts

$$\frac{\sigma_{\text{unnat}}(\pi^- p \to A_2^0 n)}{\sigma_{\text{nat}}(\pi^- p \to A_2^0 n)} \simeq \left(\frac{m^2_{A_2}}{M^2_{\omega}}\right) \frac{\sigma_{\text{unnat}}(\pi^- p \to \omega n)}{\sigma_{\text{nat}}(\pi^- p \to \omega n)} \simeq 3 \times \frac{1/3}{2/3}$$
$$= 1\tfrac{1}{2} \tag{57}$$

Hence only about $\frac{2}{5}$ of A_2^0 production should be via ρ-exchange. We can directly test the prediction (57) by plotting the l.h.s. as a function of ν, both for π^- and K^- induced reactions. In Fig. 18 we show the situation for 8 GeV/c in $\pi^- p$ reaction, the first bin is the ratio

$$\frac{\mathrm{d}\sigma_\pi}{\mathrm{d}t}(\pi^- p \to \rho^0 n) \Bigg/ \left[\frac{\mathrm{d}\sigma_f}{\mathrm{d}t}(\pi^- p \to \pi^- p) + \frac{\mathrm{d}\sigma_\omega}{\mathrm{d}t}(\pi^- p \to \rho^- p)\right];$$

the second bin is

$$\frac{\mathrm{d}\sigma_\pi}{\mathrm{d}t}(\pi^- p \to f n) \Bigg/ \frac{\mathrm{d}\sigma_f}{\mathrm{d}t}(\pi^- p \to A_2^- p);$$

the third bin

$$\frac{\mathrm{d}\sigma_\pi}{\mathrm{d}t}(\pi^- p \to g^0 n) \Bigg/ \frac{\mathrm{d}\sigma_\omega}{\mathrm{d}t}(\pi^- p \to g^- p).$$

(iv) *Antishrinkage*

We can express the t-dependence in equation (30) as

$$\left\langle \frac{d\sigma}{dt}(a + b \to c + \text{res}) \right\rangle \sim \exp\{[B - 2\alpha' \log \nu_{\text{res}}]t\} \tag{58}$$

i.e. we obtain a logarithmic antishrinkage as the resonance mass increases. This means that higher mass resonances are expected to be less peripherally produced:

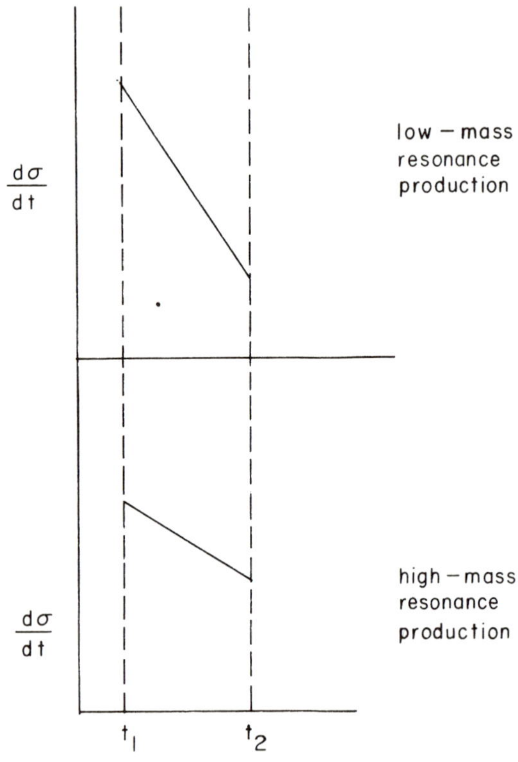

This means that if we form the ratio

$$\frac{(d\sigma/dt)(a + b \to c + \text{res})|_{t=t_2}}{(d\sigma/dt)(a + b \to c + \text{res})|_{t=t_1}} = R(\nu_{\text{res}})$$

then $R(\nu_{\text{res}})$ should be increasing like

$$(\nu_{\text{res}})^{-2\alpha'(t_1 - t_2)}$$

There is some evidence in favour of this antishrinkage.

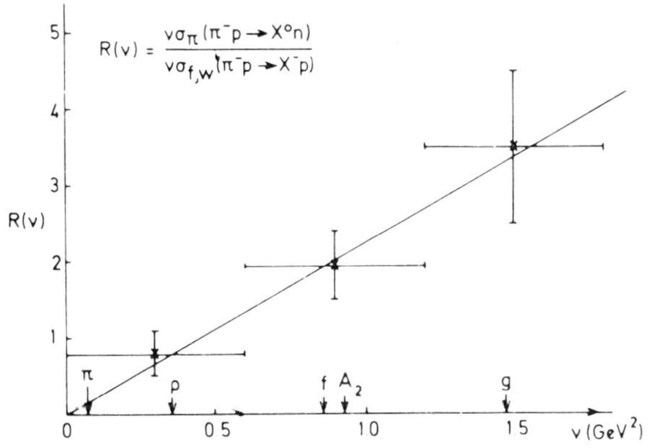

FIG. 18. Ratio of $\pi/f - \omega$ exchange to the reaction $\pi^- N \to XN$ at 8 GeV/c for $0{\cdot}05 \leqslant |t| \leqslant 0{\cdot}25$. Taken from Ref. 37.

In Fig. 19 the ratio $R(\nu_{\text{res}})$ is plotted for $t_1 = -0{\cdot}2$, $t_2 = -0{\cdot}6$ for the 10 GeV K^-p data, the two bins being K^{*-} + non-diffractive K^- and K^{**-} + non-diffractive Q^-. Similar results are obtained for the π^-p reactions but now f and ω exchanges must be separately treated.

Note that we are not discussing diffractive processes here. There is certainly antishrinkage seen in such reactions whose behaviour is much stronger than logarithmic and whose origin is far from obvious.

(v) *Background*

We can likewise write an expression analogous to (52) but for the background component in missing mass

i.e. $$\left\langle \frac{d\sigma_\alpha}{dt}\left(a + b \to c + \begin{matrix}\text{non-resonating}\\ \text{background}\end{matrix}\right)\right\rangle \sim (M^2)^{1-2\alpha(t)}. \tag{59}$$

In principle this equation could be used to separate the resonance and background components in the large M^2 regions. It seems that eqn (59) gives a behaviour consistent with the data although any hard conclusions are difficult to make.

(vi) *Baryon-exchange*[39]

So far we have looked only at meson exchange in the t-channel; we now turn to the case where $\alpha(t)$ is an anti-baryon and we can consider the duality properties of baryon–antibaryon annihilation. There is a well known inconsistency in the case of meson resonances formed by such annihilation for if we require the s-channel to correspond to a quark-antiquark pair then the t-channel *has* to be exotic

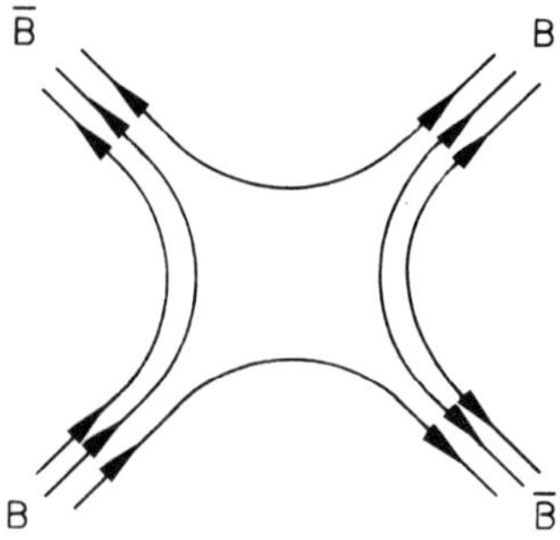

However all the established meson-resonances lie below $N\bar{N}$ threshold and so the prediction cannot be tested by direct annihilation experiment. The trick is to let the anti-baryon be a Reggeon—whose (mass)^2 is then negative, then mesons of any mass can be produced in the direct channel.

For example, consider $\pi^- p \to p_{\text{fast-forward}} + X^-$ which is assumed to proceed via Δ exchange in the u-channel, i.e.

$$s\frac{d^2\sigma}{dt\,dM^2} \propto$$

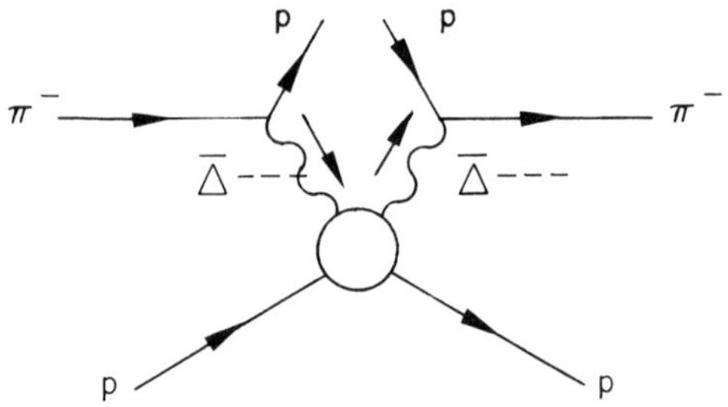

i.e. $\propto$ Reggeized antibaryon baryon scattering in the forward direction.

Separating the missing mass spectrum into resonance and non-resonating background, we expect from the semi-local duality considerations of the previous section:

$$\left\langle\frac{d\sigma}{du}(\pi^- p \to p + \text{(meson resonance)})\right\rangle \sim (M_{\text{res}}^2)^{\alpha_k(0) - 2\alpha_\Delta(u)}$$
$$\left\langle\frac{d\sigma}{du}(\pi^- p \to p + \text{background})\right\rangle \sim (M^2)^{1 - 2\alpha_\Delta(u)} \tag{60}$$

where $\alpha_k(0)$ is the intercept of the trajectory to which the meson-resonances are dual.

The above reaction has been studied at 16 GeV/c where the resonances observed were π^-, ρ^-, A_1^-, A_2^-, g^- and in the region $u \sim -0{\cdot}2$ where we expect $\alpha_\Delta \simeq 0$. The experimentalists carried out a resonance-background separation so one can check both parts of eqn (60).

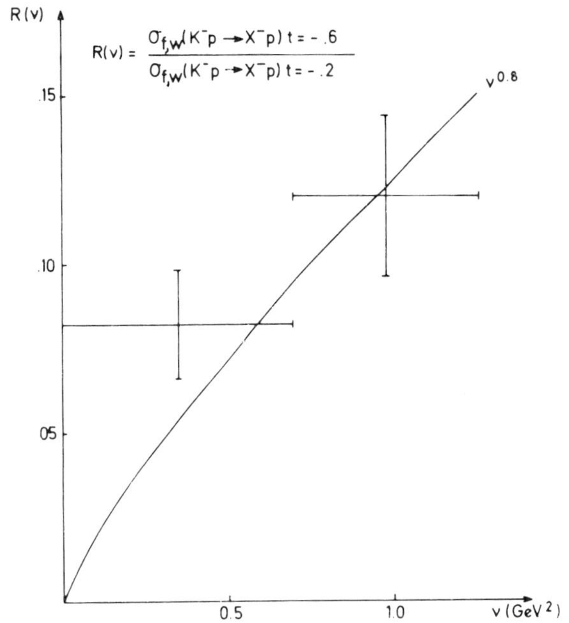

FIG. 19. Ratio of $f - \omega$ exchange contributions to the reaction $K^-p \rightarrow X^-p$ at 10 GeV/c at two values of t, $-0{\cdot}2$ and $-0{\cdot}6$ compared with the $f - \omega$ exchange Regge term. Taken from Ref. 37.

Figure 20 shows the M^2 distribution for each part.

The background in fact looks linear in M^2, which is consistent with (60) when $\alpha_{\Delta} = 0$. Actually we would expect, on the duality argument basis, some background contribution from meson exchange corresponding to exotic resonance in the missing mass channel. This could well be present since our phenomenology will not be very sensitive to such a contribution.

The resonance cross-sections are divided into three bins corresponding to (π^-, ρ^-), (A_1^-, A_2^-), g. The striking feature is the *fall* of the resonance cross-section with M^2, remembering that a normal Regge intercept of $\frac{1}{2}$ would give a *rise* with M^2. In fact a fit with $(M^2)^{\alpha_k(0)}$ gives $\alpha_k(0) = -0{\cdot}5 \pm 0{\cdot}3$. So the analysis seems to give support to the duality diagram suggestion of meson-resonances in baryon–antibaryon scattering being dual to exotic exchange.

These systematics are given further support by looking at $K^-p \rightarrow \Lambda + X^0$, where

$$X^0_{\text{res}} = \pi^0, \eta^0, \rho^0, \omega^0, f, A_2^0 \tag{a}$$

and

$$\pi^- p \to \Lambda + X^0 \tag{61}$$

where

$$X^0_{\text{res}} = K^0, K^{*0}, K^{**0} \tag{b}$$

which are proportional to $\bar{p}p$ and $\bar{Y}^*p$ forward scattering. A similar analysis to that above on data around 4·5 GeV gives $\alpha_k(0) = -1{\cdot}4$ and $-0{\cdot}8$.

We can also look at baryon-exchange but for *baryon* resonance production, e.g.

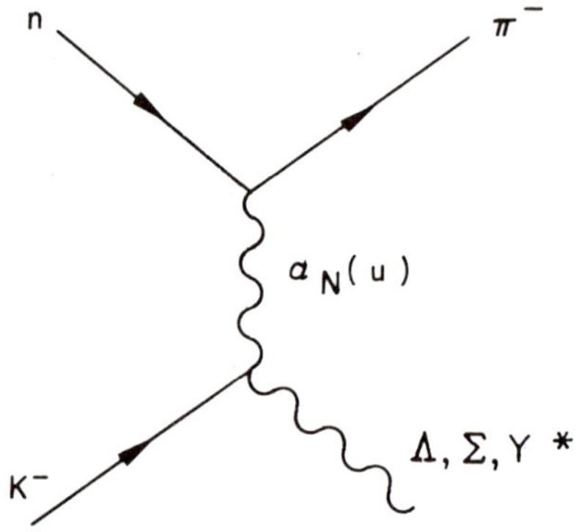

In this case we have Reggeized baryon–meson scattering and so except "normal" duality, i.e. $\alpha_k(0) = \frac{1}{2}$.

On the other hand, the same reaction proceeds in the forward direction via K^* exchange:

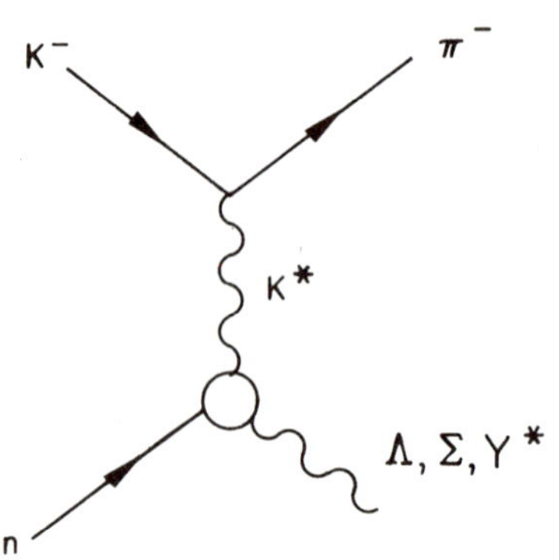

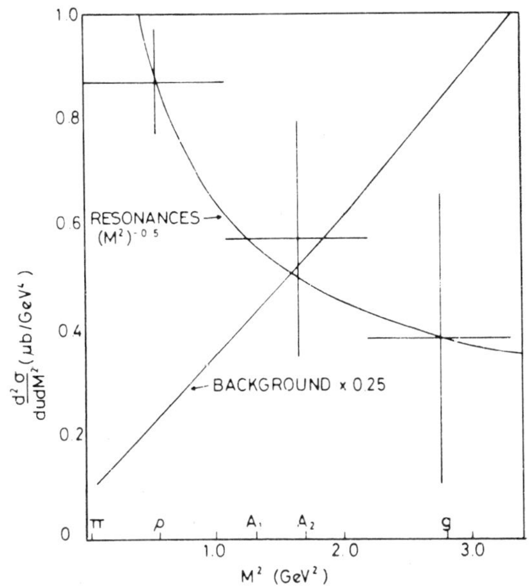

FIG. 20. Resonance and background cross-sections for $\pi^- p \to pX^-$ at 16 GeV/c. The resonance curve corresponds to $\alpha_R(0) = -0{\cdot}5$ while the background requires $\alpha_R(0) = 1$, $u = -0{\cdot}2$. Taken from Ref. 39.

with again $\alpha_k(0) = \frac{1}{2}$. Therefore if we compare the reaction in the forward and backward directions we expect

$$\frac{\langle d\sigma/dt \text{ forward}\rangle}{\langle d\sigma/dt \text{ backward}\rangle} \sim (M^2)^{2\alpha_{K^*}(t) - 2\alpha_N(u)} \sim M^2. \tag{62}$$

The data, even if it is rather sparse at 3 → 5 GeV does tend to confirm this behaviour. Such a prediction is, of course, in the same spirit as that for the ratio unnatural/natural parity exchange we encountered previously.

One of the nice features of studying Reggeon-particle is that one is not always tied down to a nucleon "target". For example, when studying π-exchange we were considering meson–meson forward scattering. If we wish to make further tests of duality diagrams we could study reactions (60a) and (60b) but instead in the *forward* direction. In one case, (a), we would have $K^* - \bar{K}$ scattering and in the other, (b) $K^* - \pi$ scattering. The meson resonances in the first case would be dual to the $w + f'$ trajectory ($\alpha_k(0) \sim 0$) and in the second case to the $\rho - f$ trajectory. Data are not too good on these reactions, the energies being rather low. However there does seem to be an indication for an $(M^2)^{\frac{1}{2}}$ difference in the two cases.

(vii) *Semi-local factorization*

In the entire fragmentation region $a \xrightarrow{b} c$ we have exchange of Regge poles $\alpha_k(0)$ in the $b\bar{b}$ channel, i.e.

$$s\frac{d^2\sigma}{dt\,dM^2} =$$

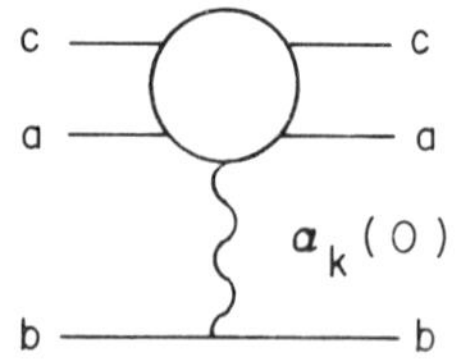

As we saw very early on there is very good evidence to support the idea that both exchange of Pomeron *and* secondary trajectories *factorize.* If this is true not only when M^2 is large but also down to the resonance region of M^2 then we obtain strong relations correcting various resonance production processes. This is another consequence of duality being valid in the semi-local sense.

Still assuming the Harari–Freund view of duality we need consider just meson exchange in the $b\bar{b}$ channel and then the relations become particularly simple.

For example, let us consider the two charge–exchange reactions

$$\begin{aligned} \pi^- p &\to X^0 n \qquad &(a)\\ \gamma p &\to X^+ n \qquad &(b) \end{aligned} \tag{63}$$

The resonances in the first case are

$$\begin{aligned} X^0 &= (\pi^0, \eta), (\rho^0, \omega), (f, A_0^2), g^0 \qquad &(a)\\ X^+ &= \pi, \quad \rho^+, \quad A_2^+, \quad g^+ \qquad &(b) \end{aligned}$$

Using this idea of semi-local factorization we then obtain

$$\frac{\nu\,(d\sigma/dt)(\pi^- p \to \rho^0 n)}{\nu\,(d\sigma/dt)(\gamma p \to \rho^+ n)} = \frac{\nu\,(d\sigma/dt)(\pi^- p \to f n)}{\nu\,(d\sigma/dt)(\gamma p \to A_2^+ n)} = \frac{2\gamma_{f\pi\pi}}{\gamma_{f\gamma\gamma}} \tag{64}$$

Data for the photoproduction reactions are very poor and only limited tests can be made of (64). Within the rather large errors the data do agree with the claims made. Of course, one can also connect π and K induced resonance production reactions in an exactly similar way, but the novelty is missing in this case since such reactions are connected anyway by SU(3). In the above relations only a higher symmetry model, e.g. quark model could connect the photon and pion induced processes.

In fact one can in principle relate the production of *meson* resonances with that of *baryon* resonances—we shall meet such an example in the next section.

XI. DIFFRACTION

Lastly we consider the case of Pomeron exchange in the t-channel. In the inclusive reaction $a \xrightarrow{b} c$ we look at examples where particles a and c have the same quantum numbers:

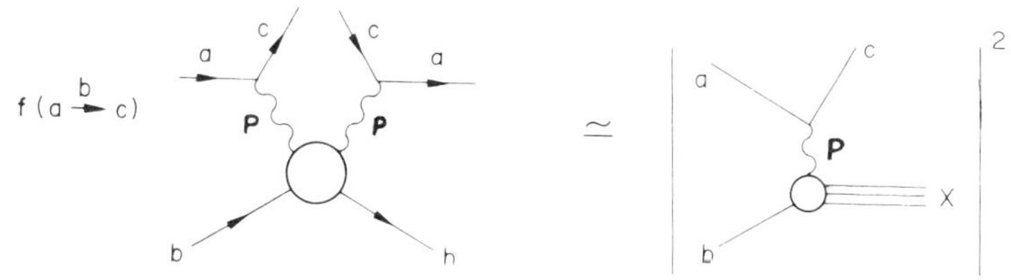

So in this section we shall be discussing, Pomeron-particle total cross-section or Pomeron-particle elastic scattering.

The whole subject of diffractive scattering is perhaps the most intriguing problem in strong interactions. In particular I feel that ISR and NAL have been producing extremely interesting data during the last year from which we have made considerable progress in understanding the nature of the Pomeron.

We shall discuss how the language developed in previous sections can be used to ask and deal with questions about the Pomeron. Especially we shall be discussing triple-Regge phenomenology and so we could sub-title this section "Triple-Pomeron and all that".

Some time ago it was realized that there was a problem associated with the triple-Pomeron coupling. If the Pomeron is really a factorizable pole with intercept $\alpha_p(0) = 1$ then, by using the sum rule (13) for example, one was led to a total cross-section rising like $\ln \ln s$ in contradiction with the original assumption. A possible way to restore the consistency of the Pomeron is to assume that the triple-Pomeron coupling $G_{PPP}(t)$ vanishes when $t = 0$.[40]

As far as the phenomenology of triple Regge was concerned, all the early attempts to fit data assumed that not only did it vanish at $t = 0$, but could be neglected even when $t \neq 0$ but small. The important triple-Regge term was *assumed* to be PPM. In such analyses there was another major assumption—that the usual two-component duality was still valid in this case, i.e.

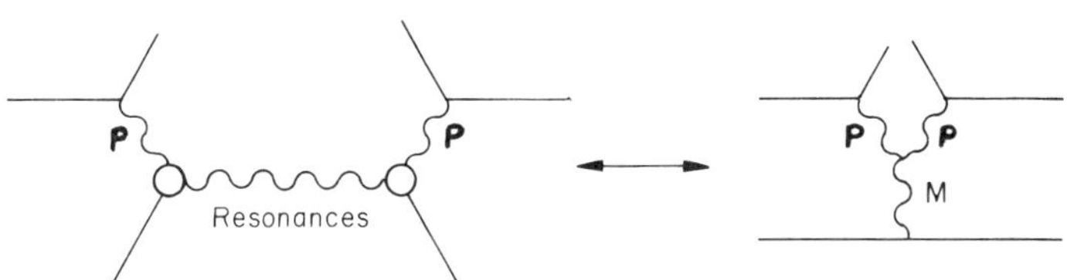

and

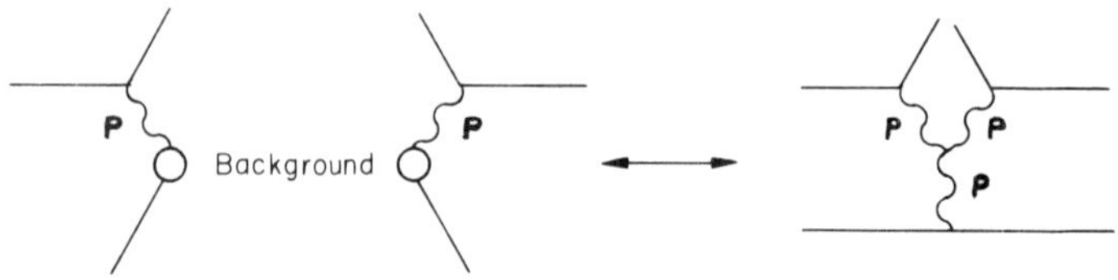

Putting these two assumptions together implied that for t close to zero, the missing mass for reactions like $pp \rightarrow pX$ was in fact dominated by resonances at high energies. Indeed there even appeared to be some evidence in favour of such systematics.(41)

Before the NAL and ISR data on the proton spectrum appeared, there were many phenomenological fits to data on $pp \rightarrow pX$ and $\pi^- p \rightarrow pX$ in which satisfactory agreement could be obtained with just a PPM term and the non-diffractive MMP term.(42) Such fits were carried out at essentially one energy and not a great deal could be learnt and gradually it was realised that information on PPP and PPM couplings required a study of the energy dependence.(43)

As we saw in our discussion of 7 or 9 component duality, there are reasons for believing that perhaps the usual two-component duality above does not hold. Therefore let us attempt to look at data without assuming this and also without any prejudice about the vanishing or otherwise of the triple-Pomeron coupling. An analysis of this nature was carried out by Chan *et al.*(44) whose motivation was to investigate the duality properties of diffractive resonances.

The essential point of this analysis is the emphasis on studying the energy dependence of the proton spectrum. For the reaction $\pi^- p \rightarrow pX$ there is a wide range of energy, from 5 GeV/c to 40 GeV/c. However this high quality data is measured only in one t-interval around $t = -0{\cdot}25$.

Let us assume the possibility of four terms, PPP, PPM, MMP and MMM. We do not include interference terms such as PMP (this assumption is hard to justify). However the analysis would be impossibly difficult with so many unknown couplings and such interference terms lie outside the scope of our nine component duality.

The full triple-Regge expansion is then

$$\frac{d^2\sigma}{dt\, d(M^2/s)} = F_{\mathbf{P}}(t, M^2/s) + F_M(t, M^2/s)\,\frac{1}{\sqrt{s}} \tag{65}$$

where

$$\begin{aligned} F_{\mathbf{P}}(t, M^2/s) &= G_{\text{PPP}}(t)\,(M^2/s)^{1-2\alpha_P(t)} + G_{\text{MMP}}(t)\,(M^2/s)^{1-2\alpha_M(t)} \\ F_M(t, M^2/s) &= G_{\text{PPM}}(t)\,(M^2/s)^{\frac{1}{2}-2\alpha_p(t)} + G_{\text{MMM}}(t)\,(M^2/s)^{\frac{1}{2}-2\alpha_M(t)} \end{aligned} \tag{66}$$

How do the shapes of each look like? At $t = -0{\cdot}25$ we have, taking a slope of $\frac{1}{2}$ for the Pomeron;

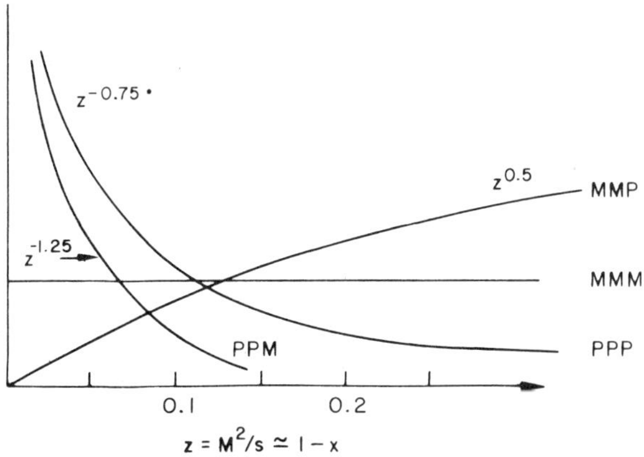

The important thing to realize from this is that the very low M^2/s region is dominated by Pomeron exchange in the t-channel while the relatively larger region is dominated by meson exchange.

For the strength of couplings corresponding to this figure we could say that for $M^2/s < \sim 0{\cdot}1$, diffraction scattering is dominant while for $M^2/s > \sim 0{\cdot}1$, non-diffractive scattering takes over. This separation corresponds to an x-value of $\sim 0{\cdot}9$.

However to separate the PPP and PPM terms on M^2/s dependence would be virtually impossible since they are so similar. Clearly the only way is by energy dependence since one term scales while the other falls away like $1/\sqrt{s}$. So we can schematically write a table:

	Scaling contribution	Non-scaling contribution
small M^2/s i.e. $\lvert x \rvert$ close to 1	PPP	PPM
larger M^2/s (still $< \sim 0{\cdot}3$)	MMP	MMM

On this basis we could just go ahead and use the expression (65) to separate the left- and right-hand columns at each value of M^2/s and t and then fit each part to the M^2/s dependence to get all four contributions. However to get as good energy separation as possible means that we must use relatively *low* energy data, i.e. $p_{\text{lab}} < 25$ GeV. This, in turn, means that the low M^2/s region where diffraction is important is the low M^2 region, i.e. resonance region, which is outside the scope of the triple Regge formula. This means we are forced to use the Finite Mass Sum Rules to evaluate the four contributions.

The method is quite straightforward. Taking our odd sumrule with $n = 1$ and $a = c$ we get:

$$I_1(t, N, s) = \int_0^N \mathrm{d}(v/s)\,(v/s)\,\frac{\mathrm{d}^2\sigma}{\mathrm{d}t\,\mathrm{d}(v/s)}$$
$$= I_1^{\mathbf{P}}(t, N) + I_1^{M}(t, N)\,(1/\sqrt{s}) \qquad (67)$$

with

$$I_1^{\mathbf{P}}(t, N) = G_{\mathrm{PPP}}(t)\frac{N^{3-2\alpha_p(t)}}{3 - 2\alpha_p(t)} + G_{\mathrm{MMP}}(t)\frac{N^{3-2\alpha_M(t)}}{3 - 2\alpha_M(t)} \qquad (68)$$

and

$$I_1^{M}(t, N) = G_{\mathrm{PPM}}(t)\frac{N^{2\cdot 5-2\alpha_p(t)}}{2\cdot 5 - 2\alpha_p(t)} + G_{\mathrm{MMM}}(t)\frac{N^{2\cdot 5-2\alpha_M(t)}}{2\cdot 5 - 2\alpha_M(t)} \qquad (69)$$

and this should be valid for a cut-off N in the range $N \geqslant 4/s$ and $N \leqslant \frac{1}{4}$.

The procedure consists of plotting the missing mass spectrum at each energy as a function of v/s (this means a given resonance, such as the A_1, moves to the left as the energy increases). Then at each value of v/s (t is fixed) we separate the integral I_1 into the scaling and non-scaling piece, according to (67) and then each part is fitted to the M^2 dependence of (68) and (69). Finally with the values of the $G_{iij}(t)$ obtained, we substitute into the original triple-Regge formula (65) and (66) and compare with the data. Such a result is shown in Fig. 21. In this particular fit the Pomeron and meson trajectories were taken as

$$\alpha_p(t) = 1 + \tfrac{1}{2}t$$
$$\alpha_M(t) = 0\cdot 2 + t$$

The low intercept of the meson trajectory is not unexpected. The MMP and MMM terms are the dominant contributions as v/s gets bigger, i.e. as the Regge-expansion variable s/M^2 gets smaller and so lower-lying singularities are expected to contribute.

The impressive feature of Fig. 21 is the success with which the resonance region is described by the triple-Regge expression. This is another demonstration of semi-local duality in operation. To see this more clearly we can plot the first moment of the cross-section and include the elastic-scattering contribution—this is done in Fig. 22.

The next thing to ask is how important are each of the contributions from the four terms. This show in Fig. 23 at 16 and 40 GeV/*c*. Essentially the result is that the scaling contributions PPP and MMP are far more important. The smallness of the non-scaling piece PPM has strong consequences as far as duality is concerned. If normal two-component duality is valid for diffractive resonances we should expect this part of the spectrum to be described, on average, by the PPM term. If, in fact, we naively separate the

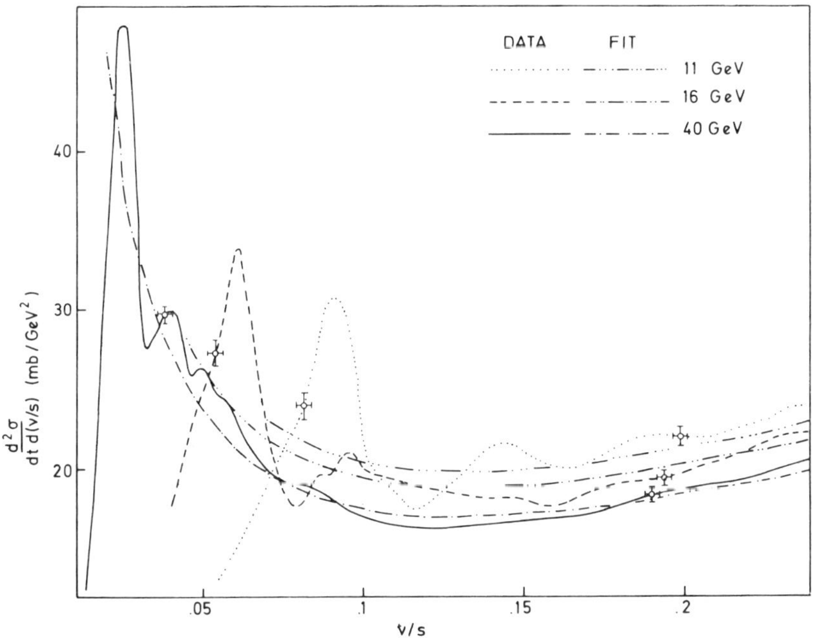

FIG. 21. Cross-sections for $\pi^- p \to pX$ at 11, 16, 40 GeV/*c* compared with results of the Regge fit described in the text. Typical errors are shown for the experimental curves. Taken from Ref. 44.

resonances from the background and compare the former with the non-scaling terms, as shown in Fig. 24, we find a strong mis-match. Up to $M^2 = 2{\cdot}7\ \text{GeV}^2$, the resonance contribution is considerably greater than the PPM + MMM contribution. This implies that the diffractive resonances are, at least partly, dual to the Pomeron.

This is some support for the 8th and 9th components suggested by Einhorn *et al.*[8] which we discussed earlier. We saw in the reaction $a \xrightarrow{b} c$ that when the Pomeron is exchanged in the *t*-channel, we have the graph:

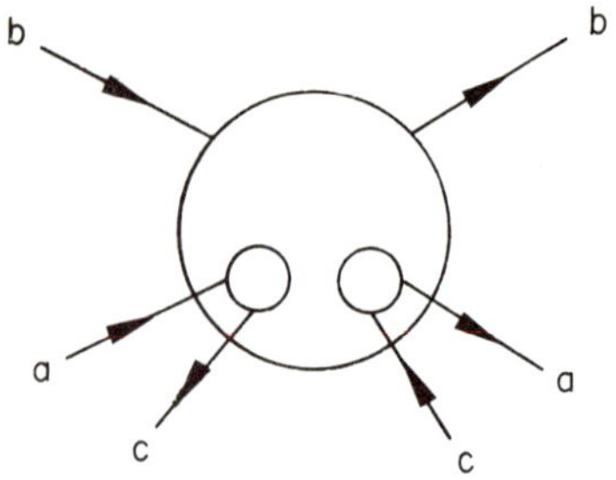

which implies resonances in the M^2-channel and Pomeron in the $b\bar{b}$ channel. This represents diffractive resonances being described by the PPP term. In this scheme there is no room for the PPM term, at least to leading order. Although we found the PPM contribution to be indeed small it should be added that the "abnormal" duality proposed by Einhorn *et al.* is still somewhat controversial. We should also add some cautious remarks to the above analysis:

(1) The resonance-background separation is open to some question. These diffractive resonances do not sit on a *simple* background and there may be large errors in the estimate of resonance cross-section.
(2) The bumps in the missing mass A_1, A_3 are perhaps not really resonances. Our conclusion may be turned around to say that this is further evidence against an interpretation of genuine resonances.
(3) The above analysis was carried out for $0{\cdot}17 \leqslant |t| \leqslant 0{\cdot}34$.

The A_1 has an anomalously high slope in $d\sigma/dt$ which meant that our conclusions about duality properties of resonances were using only a fraction of the resonant cross-section. In fact, our triple-Regge expressions cannot hope to describe, in average, such violent t-behaviour.

Nevertheless the exercise nicely demonstrates the dual properties of the missing mass spectra. We can furthermore ask what semi-local factorization has to say in this case of diffractive scattering. If we consider only small M^2/s with assumed Pomeron exchange only, we can relate the missing mass spectra of $\pi^- p \to pX$ and $pp \to pX$:

$$f(p \xrightarrow{p} p) = \frac{\gamma_{\mathrm{P}p}}{\gamma_{\mathrm{P}\pi}} f_p(p \xrightarrow{\pi^-} p) + \frac{\gamma_{fp}}{\gamma_{f\pi}} f_M(p \xrightarrow{\pi^-} p) \frac{1}{\sqrt{s}} \tag{70}$$

This is shown in Fig. 25.

If we just simply compare

$$\frac{f(p \xrightarrow{p} p)}{\sigma_{\mathrm{tot}}(pp)} \quad \text{and} \quad \frac{f(p \xrightarrow{\pi} p)}{\sigma_{\mathrm{tot}}(\pi p)}$$

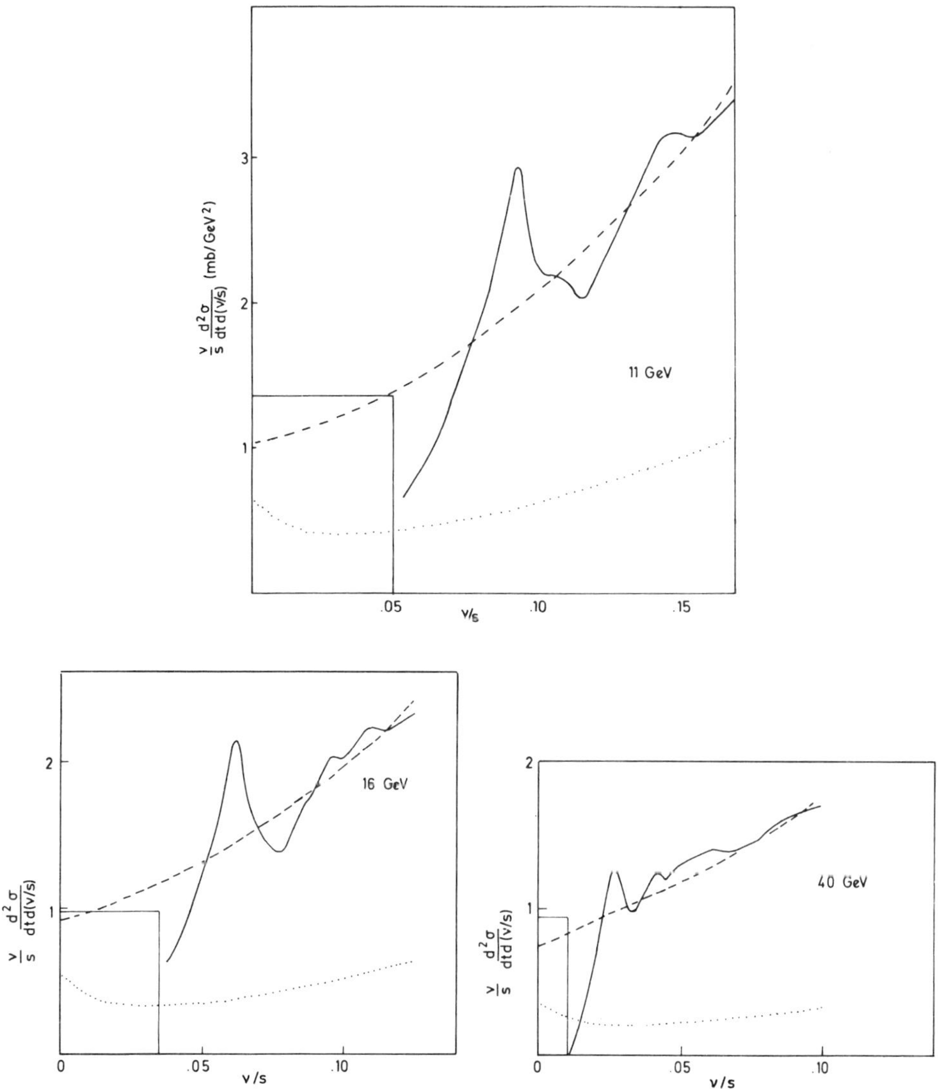

FIG. 22. First moments of cross-sections at 11, 16 and 40 GeV/c; the histogram denotes π-contribution. Dashed curves represent results of Regge fit; dotted curves represent contribution from meson-exchange term only. Taken from Ref. 44.

over the M^2 spectrum, then we are approximately repeating the last figure. Note that we are comparing a *baryon* resonance spectrum with a *meson* spectrum and it is quite remarkable how good the agreement is in quite a local sense, see Fig. 26.

Now let us turn to the data on $pp \to pX$. What we saw in Figs 25 and 26 was the famous data by Allaby *et al.*[45] at 24 GeV. One of the most exciting pieces of data to come from the ISR is the measurement of the same reaction at ISR by the CERN–Holland–Lancaster–Manchester collaboration.[46] In this experiment the proton spectrum was measured over the whole fragmentation region at various energies corresponding to $s = 551$, 930, 1995 GeV^2. Figures 27 and 28 demonstrate clearly the most interesting features which are observed.

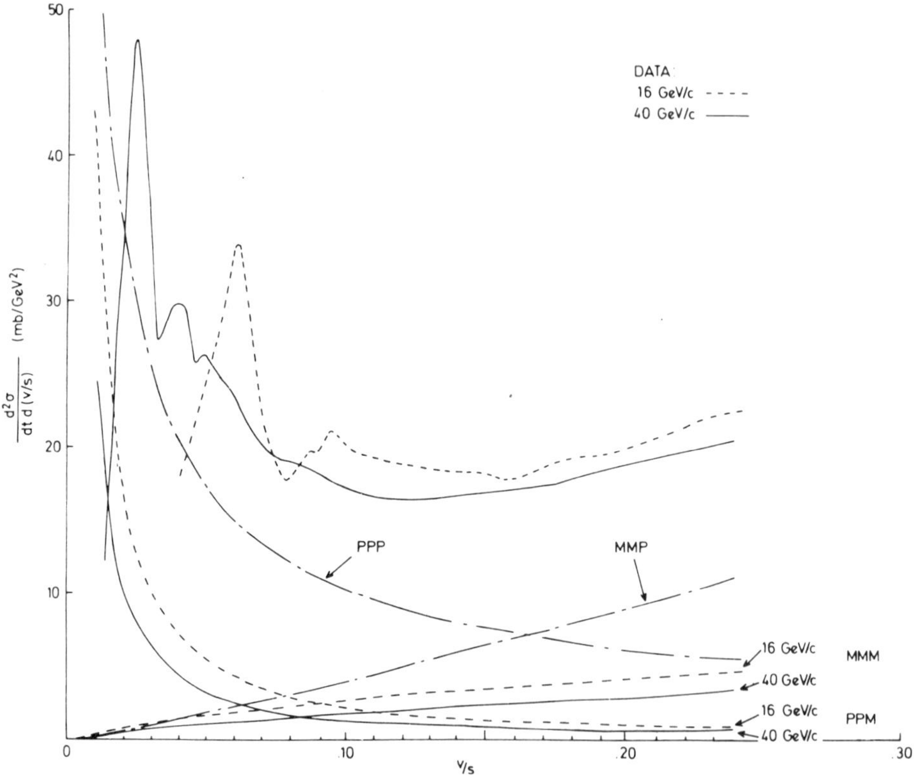

FIG. 23. Cross-sections at 16 and 40 GeV/c. The contributions from the four terms in the Regge fit are shown. The PPP and MMP terms are energy independent; the PPM and MMM terms behave as $s^{-\frac{1}{2}}$. Taken from Ref. 44.

Firstly there is unmistakable evidence for diffractive inelastic scattering. The huge peak seen for $|x|$ close to 1 (or M^2/s close to zero) we have learnt to be the characteristic feature of Pomeron exchange in the t-channel.

Next we learn that the energy dependence of this peak is rather weak.

Figure 27 shows that there is only a factor 2 roughly between $s = 50$ and $s = 1000$. This, once again, means dominance of Pomeron in the $b\bar{b}$ channel. (If we had meson exchange dominant we would expect a factor 5.) Also Fig. 28 shows that, within the ISR range, the diffractive peak scales.

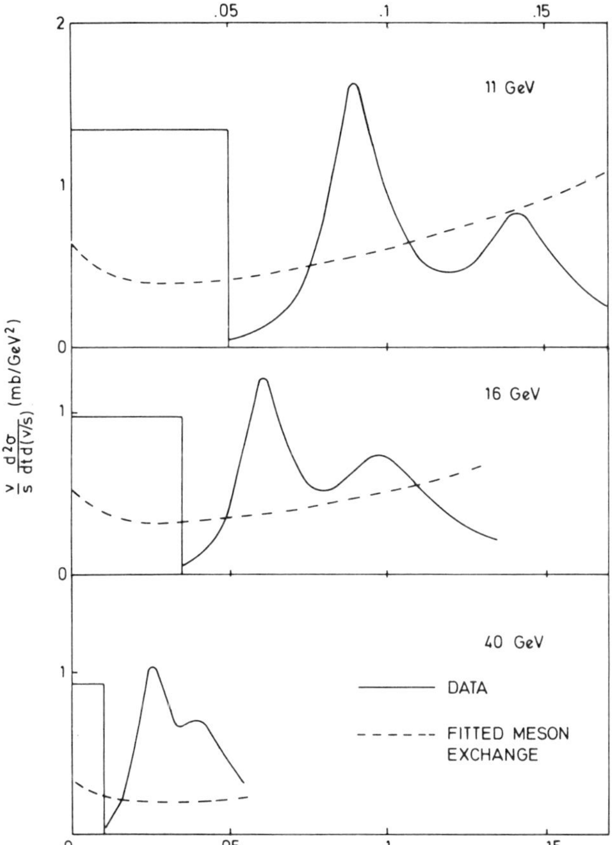

FIG. 24. Resonance contributions of the cross-sections at 11, 16 and 40 GeV/c obtained by the resonance-background separation described in the text. Histogram is the pion contribution. Dashed curve is the meson-exchange contribution of the Regge fit. Taken from Ref. 44.

Furthermore the "dip" at $x \sim 0{\cdot}85$ means we have diffraction out to $M^2/s \sim 0{\cdot}15$ or $M \sim 12$ GeV! This means that the "mass cut-off" of the diffractive cluster is not independent of the incoming energy, but rather $M^2_{\text{cut-off}}$ seems to be proportional to s, allowing diffractive production of very heavy clusters. We are therefore studying Pomeron-proton σ_{tot} for "P_{lab}" up to 75 GeV!

We can repeat our previous exercise now with $pp \to pX$ and do a triple Regge analysis over a wide energy range. If we choose $s = 47$ and 930 we find excellent fits to the data for several t-values, the results being shown in Figs 29 and 30.[47] The curves correspond to a slope of the Pomeron of $\frac{1}{4}$,

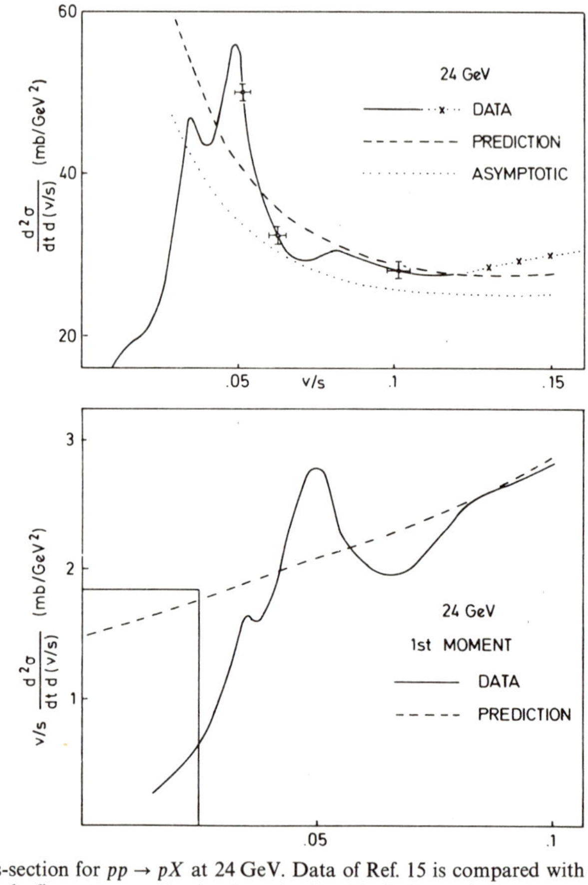

FIG. 25. Cross-section for $pp \to pX$ at 24 GeV. Data of Ref. 15 is compared with the prediction obtained from the fit to $\pi^- p \to pX$ using factorization. Typical experimental errors are indicated. Taken from Ref. 44.

in fact this is the value one gets from fitting to the M^2 dependence in the region (see Fig. 31). For instance if the slope was $\frac{1}{2}$ then the PPP term gives constant at $t = -1{\cdot}0$, and as Fig. 30 shows, the diffraction, still *peaks* at this t value.

Now we can begin to look at the t-dependence of the triple-Regge couplings, in particular it is very interesting to discover if the PPP coupling shows any evidence of turning over in the small t range, indicating that $G_{\mathrm{PPP}}(t) = 0$ when $t = 0$. At ISR the minimum t-value which can be attained experimentally depends on s. For $s = 2000$ the value is $-0{\cdot}6$, at $s = 930$ it is $-0{\cdot}25$ and at $s = 560$ they can get down to $t = -0{\cdot}16$.

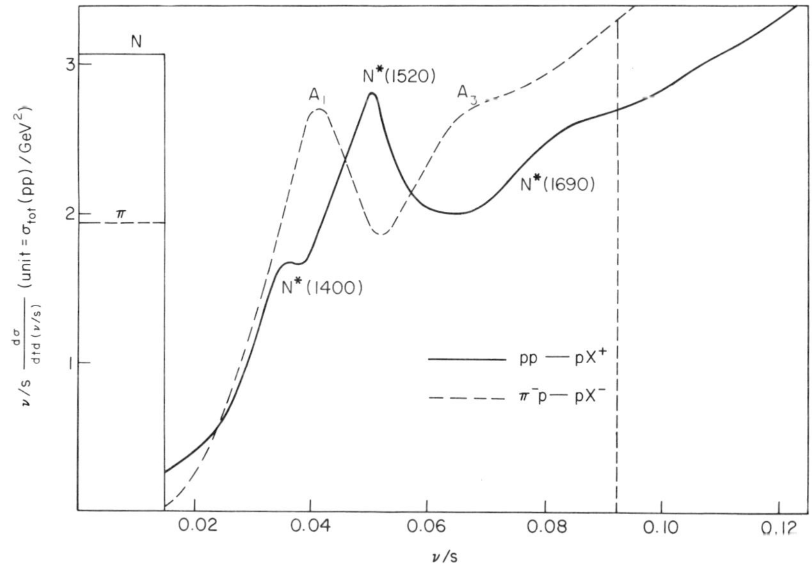

FIG. 26. First moments of cross-sections for $pp \to pX$ and $\pi^- p \to pX$ at $24 \to 25$ GeV/c and $t = -0{\cdot}25$, divided by relevant total cross-sections. Dashed vertical line indicates region of equal areas formed by both curves.

Having evaluated the PPP coupling at several values of t, we fit an exponential giving the approximate result

$$G_{\mathrm{PPP}}(t) \approx 5\,\mathrm{e}^{5t}$$

This t dependence is essentially that of the two Reggeon vertices so that the Pomeron-particle total cross-section is independent of the "mass" of the Pomeron and has a value

$$\sigma_{\mathrm{tot}}(\mathbf{P}p) \approx 7{\cdot}5 \text{ mb.}$$

We are now gathering a lot of information on $pp \to pX$ since NAL has been supplementing the ISR and "low" energy data by filling the "intermediate" region. The first data of this type came from the gas jet experiment[48] and subsequently from the various bubble chamber collaborations at 100,[49] 200[50] and 300[51] GeV/c and the polythene + carbon experiment at 200 GeV/c.[52]

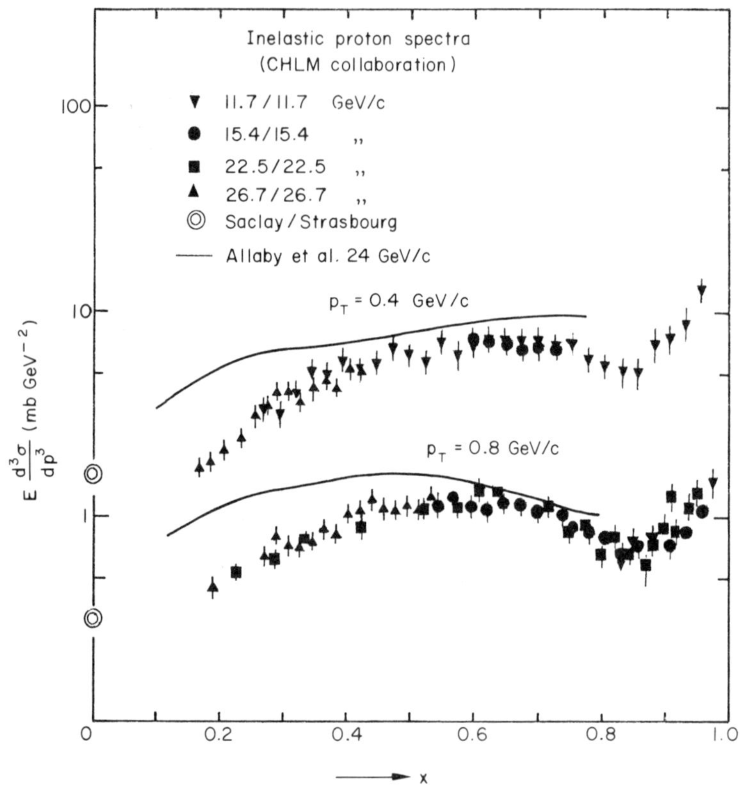

FIG. 27. Inelastic proton spectra at $p_T = 0{\cdot}4$, $0{\cdot}8$ GeV. Taken from Ref. 46.

The virtue of the gas-jet experiment[48] was that it really covered a huge energy range, 40 to 250 GeV/c. In this way the data at a fixed t and x value is plotted against $s^{-\frac{1}{2}}$ and it can be seen how *linear* the points are. On the same graph we have plotted all the present data on $pp \to pX$ at these values of t and M^2/s together with the $\pi^- p \to pX$ data[53] we used in our previous analysis.

The purpose of this figure is to illustrate the difficulty of trying to carry out a precise Regge analysis of the proton spectrum. In the curves shown in

Figs 29 and 30, only the data at $s = 47$ and $930\ \mathrm{GeV}^2$ were used. We then see that such an analysis would give $f(p_c) \sim 7\ \mathrm{mb}\,.\,\mathrm{GeV}^{-2}$ at these values of t and M^2/s. However, using other data we could get as much as $13\ \mathrm{mb}\,.\,\mathrm{GeV}^{-2}$. The slope of the graph gives essentially the PPM term and the intercept the PPP term, and clearly the resulting values of these contributions depends very much on the selection of data. A weak point in the NAL gas jet experiment is the uncertainty in the overall normalization and

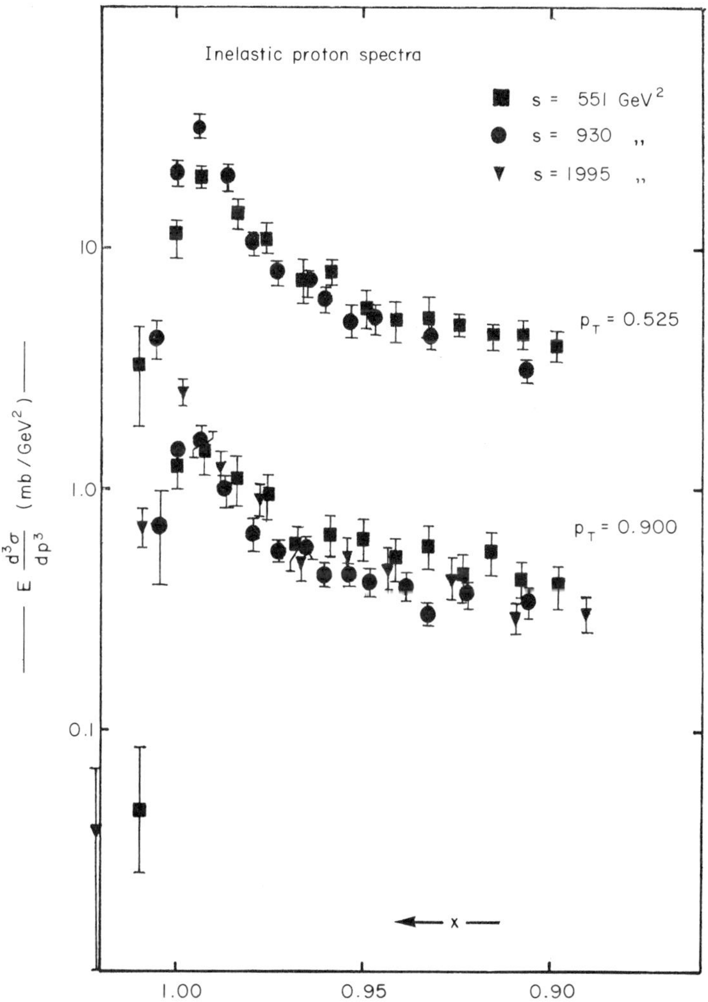

FIG. 28. Proton spectra for $x > 0{\cdot}9$ at $s = 551, 930, 1995\ \mathrm{GeV}^2$. Taken from Ref. 46.

so these data can be bodily shifted up or down. There is also some doubt about the normalization of the BNL 28 GeV data. Since the slope of gas-jet data is the same as that of πp data, my own prejudice is that the inconsistencies are eliminated if the ISR data is renormalized by around 30–40 %. Whether this is the cause or not—all I wish to emphasize is that a very precise Regge analysis cannot be performed with the data in the present state.

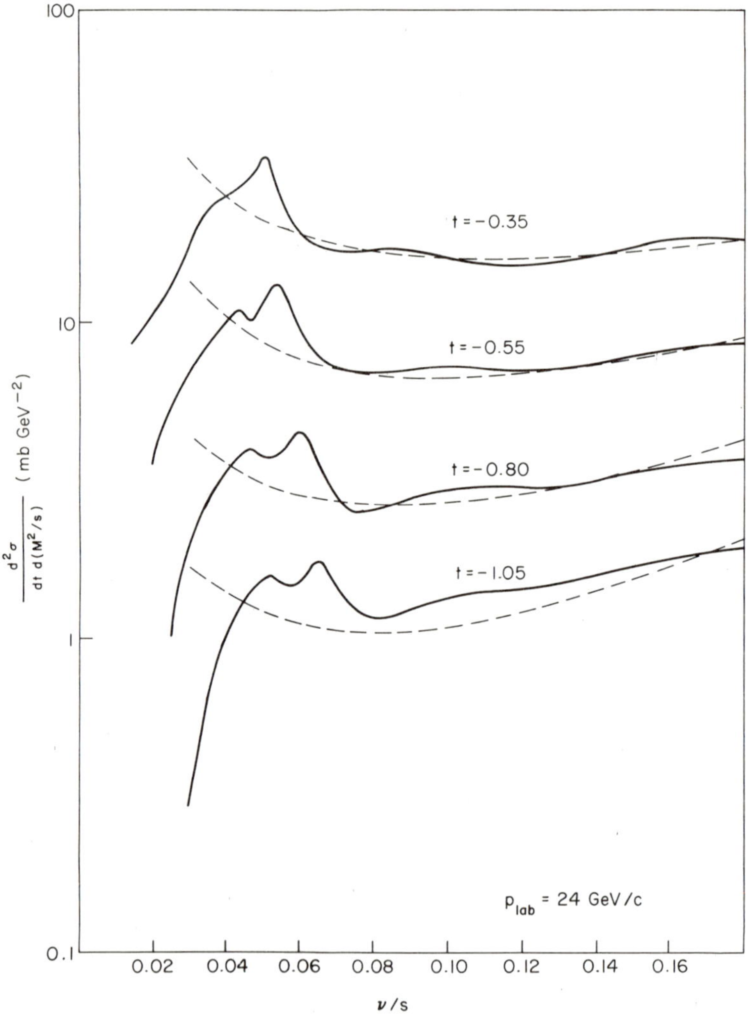

FIG. 29. Cross-sections for $pp \to pX$ at 24 GeV/c (Ref. 45). Dashed lines correspond to triple Regge fit of Ref. 47.

Let us return to the question of t-dependence. Figure 33 shows the p_T^2 plot at 102 GeV for $M^2/s < 0{\cdot}15$—this is essentially a t distribution in this range of M^2/s. There is no evidence at all for a turnover at small t (thus suggesting the failure of the PPP coupling to vanish at $t = 0$). The only suggestion of such a zero is from the Franzini[52] preliminary data but the errors are very large and the turnover seen near $t = -0{\cdot}08$ is contradictory with the bubble chamber data.

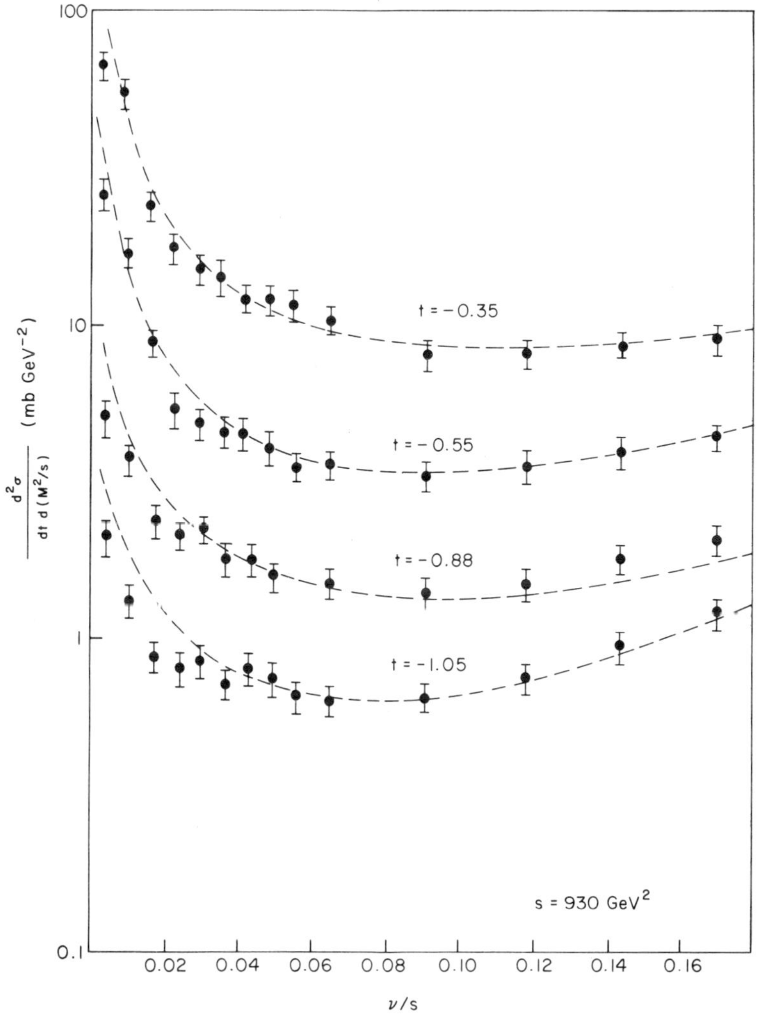

FIG. 30. Cross-sections for $pp \to pX$ at $s = 930\ \text{GeV}^2$ (Ref. 46). Dashed lines correspond to triple-Regge fit of Ref. 47.

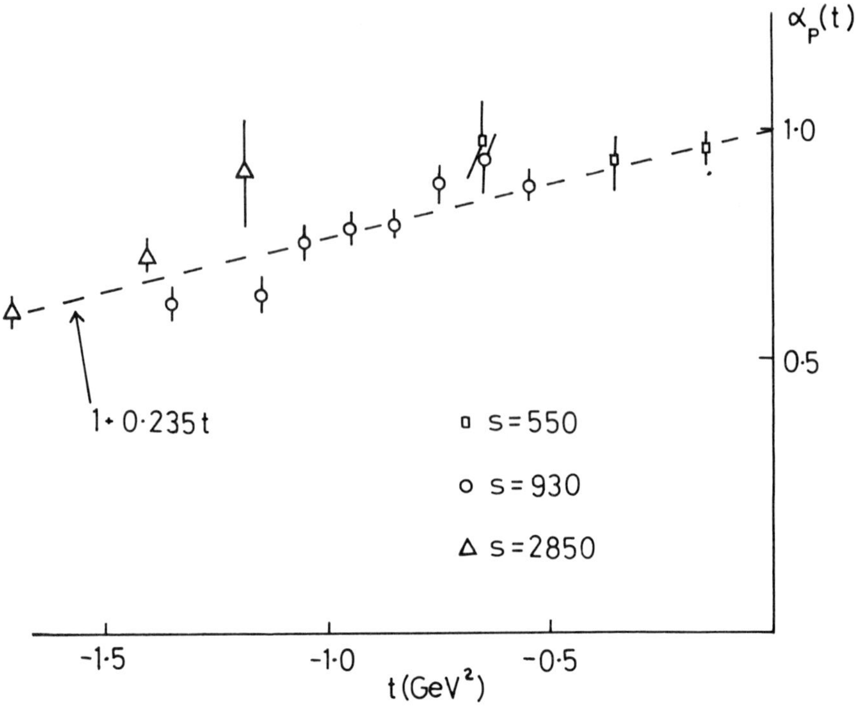

FIG. 31. Effective Pomeron trajectory obtained by fitting the M^2 dependence of the ISR data on $pp \to pX$ (Ref. 46) in the diffractive region.

Now that we have an estimate of the t-dependence we can proceed to integrate the diffractive peak over the whole t-range and thereby learn the size of the inelastic diffractive cross-section described by the graph:

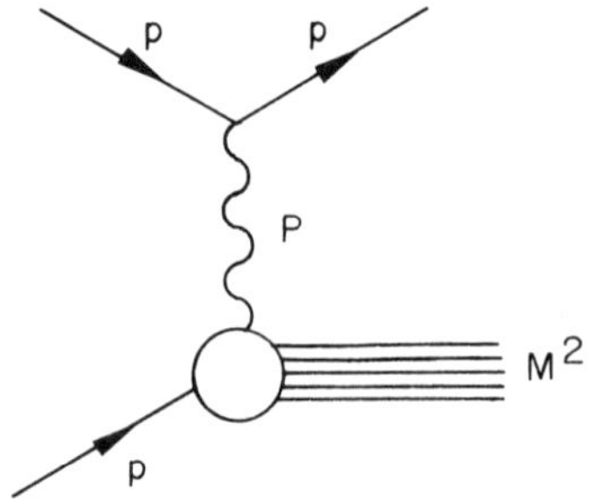

p_{lab} (GeV/c)		x-selection	σ (mb)
100	NAL	$\lvert x\rvert \geqslant 0{\cdot}87$	$6{\cdot}8 \pm 1{\cdot}0$
300	NAL	$\geqslant 0{\cdot}96$	$5{\cdot}5 \pm 1{\cdot}1$
300	NAL	$\geqslant 0{\cdot}91$	$7{\cdot}3 \pm 1{\cdot}1$
300	ISR	$\gtrsim 0{\cdot}90$	$5{\cdot}4 \pm 1{\cdot}0$
500	ISR	$\gtrsim 0{\cdot}90$	$5{\cdot}0 \pm 1{\cdot}0$

To this should be added contributions from

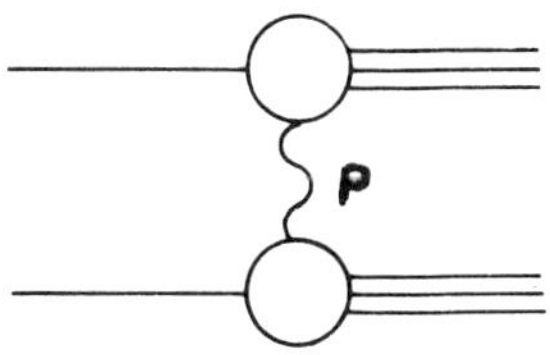

which, if factorization holds, gives another 15% roughly. Altogether one arrives at a figure of about 7 mb for the inelastic diffraction, compared with 7 mb for the elastic and 25 mb for the non-diffractive cross-section. That the

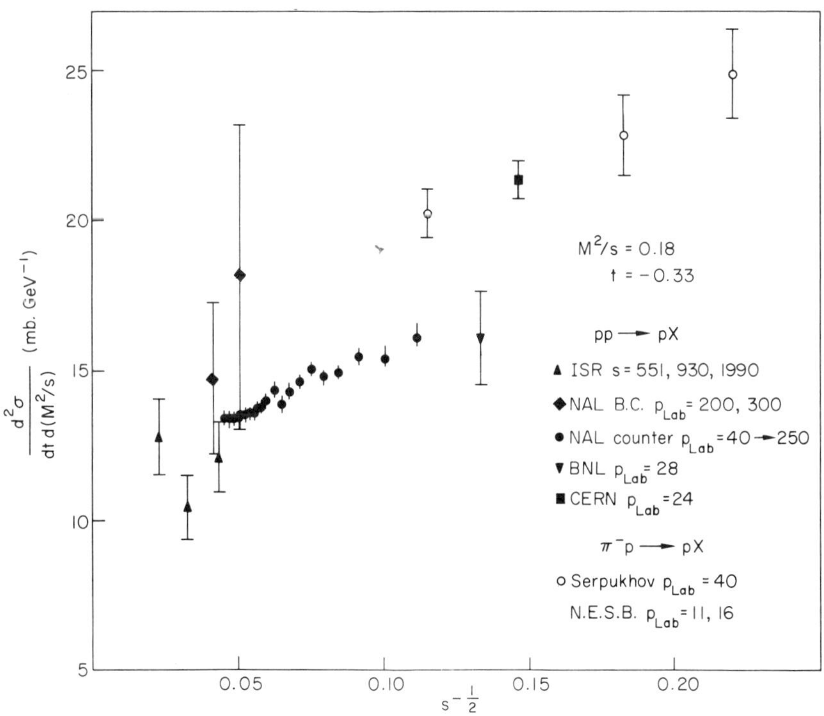

FIG. 32. Data on $pp \to pX$ and $\pi^- p \to pX$ at $t = -0{\cdot}33$ and $M^2/s = 0{\cdot}18$ plotted against $1/\sqrt{s}$.

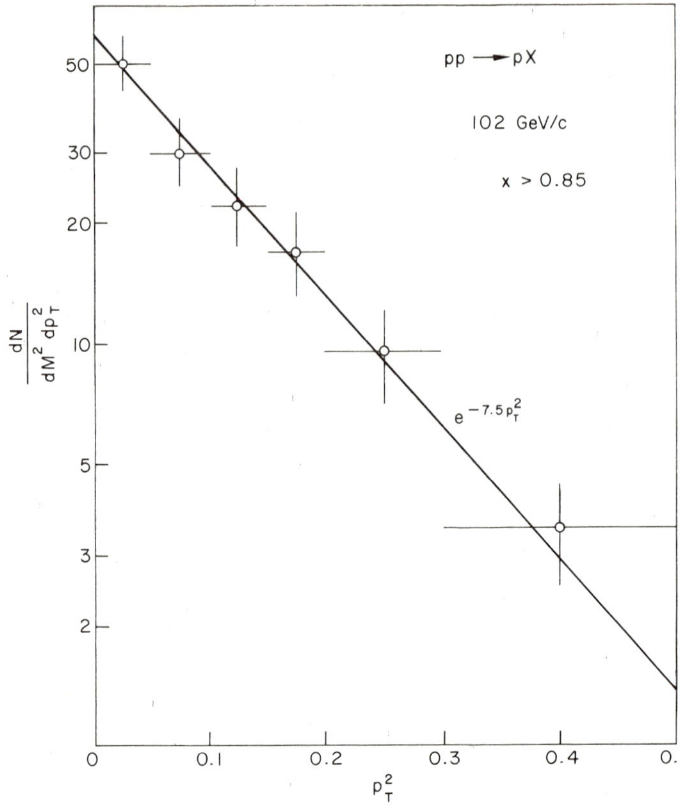

FIG. 33. p_T^2 distribution of $pp \to pX$ at 102 GeV/c (Ref. 49).

inelastic cross-section can be split up into two parts, diffractive and non-diffractive, each with separate mechanism is the basis of the successful two-component model.[(54)] This model has, for one thing, accounted for the broadening of the multiplicity distributions as the energy increases from Serpukhov to NAL energies. The idea is that such a broadening resulted from adding a multiplicity distribution from the diffractive part which populated the low multiplicity region and remained constant with energy, with a Poisson-like distribution whose centre moved up like ln s. This led to a dispersion D of the overall distribution going up like ln s instead of like $(\ln s)^{\frac{1}{2}}$ for a pure Poisson. Eventually, it was predicted, the multiplicity distribution would separate into the two components leaving a dip to develop at intermediate multiplicities—the basic assumption being that the diffractive part would not change with energy and be restricted to very low multiplicities.

From our discussion in this section we have more or less arrived at a good description of the diffractive component of the two-component model which is itself given by the sum of the PPP and PPM contributions

i.e.

$$\frac{d^2\sigma^{\text{diff}}}{dt\, dM^2} = [G_{\text{PPP}}(t)\,(M^2)^{1-2\alpha_p(t)} + G_{\text{PPM}}(t)\,(M^2)^{\frac{1}{2}-2\alpha_p(t)}]\, s^{2\alpha_p(t)-2}. \quad (71)$$

Let us now explore some of the consequences of this triple-Regge model for diffraction scattering. One thing is to integrate each term in the above expression to see what energy dependence we get for diffraction.

Let us consider two possibilities for the PPP coupling:

assume $$G_{\text{PPM}}(t) = A\, e^{\beta t}$$

$$G_{\text{PPP}}(t) = C\, e^{Dt}$$

or $$G_{\text{PPP}}(t) = C'(-t)\, e^{D't} \quad (72)$$

If we neglect t_{min} for the moment, then integrating from $-\infty$ to 0 in t gives

$$\frac{d\sigma_{\text{PPM}}}{dM^2} = \frac{AM^{-3}}{B + 2\alpha'_p \ln s - 2\alpha'_p \ln M^2} \to \frac{A}{B} M^{-3} \quad \text{if } \alpha'_p = 0$$

$$\frac{d\sigma^{\text{coupled}}_{\text{PPP}}}{dM^2} = \frac{CM^{-2}}{D + 2\alpha'_p \ln s - 2\alpha'_p \ln M^2} \to \frac{C}{D} M^{-2} \quad \text{if } \alpha'_p = 0 \quad (73)$$

$$\frac{d\sigma^{\text{decoupled}}_{\text{PPP}}}{dM^2} = \frac{C'M^{-2}}{[D' + 2\alpha'_p \ln s - 2\alpha'_p \ln M^2]^2} \to \frac{C'}{D'^2} M^{-2} \text{ if } \alpha'_p = 0$$

M^2 is now integrated from some M_0^2 to εs where ε is some cut-off $\ll 1$. This gives

$$\sigma_{\text{PPM}} \simeq \frac{2A\, M_0^{-1}}{B + 2\alpha'_p \ln s - 2\alpha'_p \ln M_0^2} \to \frac{2A}{B\, M_0} \quad \text{if } \alpha'_p = 0$$

$$\sigma^{\text{coupled}}_{\text{PPP}} = \frac{C}{2\alpha'_p} \ln\left[\frac{D + 2\alpha'_p \cdot \ln(s/M^2)}{D - 2\alpha'_p \cdot \ln\varepsilon}\right] \to \frac{C}{D} \ln\frac{\varepsilon s}{M_0^2} \quad \text{if } \alpha' = 0$$

$$\sigma^{\text{decoupled}}_{\text{PPP}} = \frac{C'}{2\alpha'_p}\left[\frac{1}{D' - 2\alpha'_p \ln\varepsilon} - \frac{1}{D' + 2\alpha'_p \ln(s/M_0^2)}\right] \to \frac{C'}{D'^2} \ln\frac{\varepsilon s}{M_0^2}$$

$$\text{if } \alpha'_p = 0. \quad (74)$$

(Note that the actual values must be multiplied by 2 to include the "backward" peak also).

Se we see firstly that σ_{PPM} is essentially *constant*, apart from a shrinkage factor. This is naïvely the sort of behaviour one expects for a diffractive cross-section. The results for the PPP are not surprising either. The ln ln s growth in $\sigma_{PPP}^{\text{coupled}}$ was the result obtained earlier which led to the inconsistency of a factorizable Pomeron pole. The decoupled solution corresponds to a [const − (const/ln s)] behaviour, which represents a pole + "absorptive" cut systematics. In this case the cross-section asymptotically approaches a constant rather than increases indefinitely as for the coupled PPP coupling.

Recently the measurements of pp total cross-sections at ISR indicate that there is a rise of something like 4 mb between NAL and ISR energies(55) and it was suggested by Chew(56) and several others(57) that this rise may be associated with the dominance of the PPP contribution in inelastic diffraction scattering. To test whether this is in fact possible one can feed in values for A, B, C, D or C', D' taken from triple-Regge fits and in this way one easily gets a rise of 2·5 mb in the same energy interval. Taking into account the possibility of double dissociation in addition, it is not unlikely that the bulk of the rise of σ_{tot}^{PP} can be attributed to diffractive production of heavy mass clusters.

It is therefore interesting to ask what multiplicity distributions are predicted by this model for inelastic diffraction—this has been investigated by Roberts and Roy(58) and Frazer and Snider.(59) We have already seen that PPP is the important term in diffraction—this in turn means that the "cut-off" mass of the excited cluster increases proportionally to s and so we may expect that the multiplicity of inelastic diffraction scattering continues to *increase* with energy. This allows the possibility that there will always be an "overlap" between the diffractive and "multiperipheral" components and that the famous dip in the multiplicity distribution may not occur.

If the Pomeron factorizes we would expect the average multiplicity of the excited cluster to be given by

$$\langle n_-^D \rangle \simeq a \ln M^2 \tag{75}$$

(− means we are referring to negative particle multiplicity) where the coefficient is the same as for the multiperipheral part of PP collisions—this means the cut-off will increase at the same rate as the centre of the Poissonian component.

To see the different features of multiplicity distributions from the PPM and PPP let us, for the moment, put the multiplicity of the cluster *equal* to the average multiplicity and forget about dispersion.

i.e.
$$M^2 = e^{-n/a} \tag{76}$$

Substituting into (73) and using

$$\frac{d\sigma}{dn} = \frac{M^2}{a}\frac{d\sigma}{dM^2}$$

we get

$$\frac{d\sigma_{PPM}}{dn_-} = \frac{(A/a)\,e^{-n_-/2a}}{B + 2\alpha'_p \ln s - 2\alpha'_p(n_-/a)} \to \frac{A}{aB} e^{-n_-/2a} \quad \text{if } \alpha'_p = 0$$

$$\frac{d\sigma_{PPP}^{\text{coupled}}}{dn_-} = \frac{(C/a)}{D + 2\alpha'_p \ln s - 2\alpha'_p(n_-/a)} \to \frac{C}{aD} \quad \text{if } \alpha'_p = 0 \tag{77}$$

$$\frac{d\sigma^{\text{decoupled}}}{dn_-} = \frac{(C'/a)}{[D + 2\alpha'_p \ln s - 2\alpha'_p(n_-/a)]^2} \to \frac{C'}{(aD')^2} \quad \text{if } \alpha'_p = 0.$$

Thus, qualitatively we have

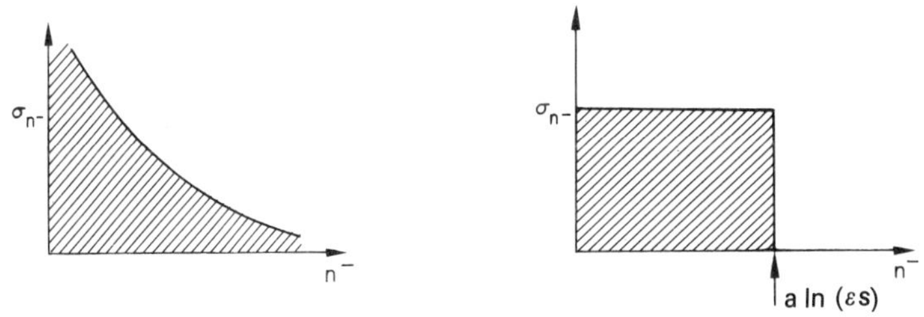

This means that PPM behaves just as a diffractive cross-section would naïvely be expected to—it cuts-off at a fixed multiplicity and does not vary with energy. In fact an earlier version of the model[59] considered only PPM contribution. However the PPP, which we know to be dominant, populates low *and* intermediate multiplicities. Furthermore as s increases, the "tail" moves further out like ln s, and this in turn gives the rise in cross-section.

To do the calculation properly one needs to know the average multiplicity of the excited cluster as a function of M^2. If the Pomeron factorizes we expect $\langle n_-^D \rangle$ at large M^2 to be similar to $\langle n_- \rangle$ for pp scattering at large s. For the low M^2 region (up to $M^2 \simeq 50$ GeV^2) we now have recent data from NAL on the average multiplicity from Pomeron-particle scattering,[60] where a linear rise with M is observed. There is no need to introduce the arbitrary cut-off ε in M^2/s. If we allow for $t_{\min}$ effects then this automatically cuts off the M^2/s integration since $t_{\min}$ is a function essentially of x only. The only other input required is the values of the parameters A, B, etc., which is taken directly from a triple-Regge fit.[47]

In Fig. 34 we see the distribution n_-^D at NAL and ISR energies. Remember

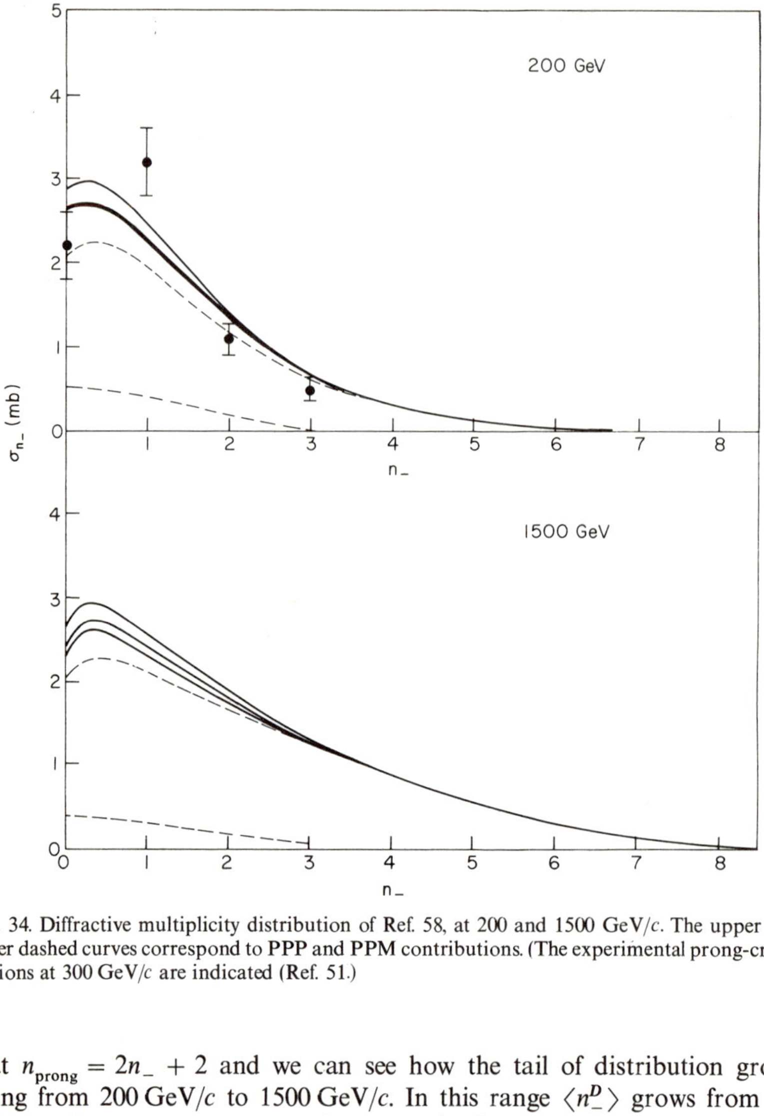

FIG. 34. Diffractive multiplicity distribution of Ref. 58, at 200 and 1500 GeV/c. The upper and lower dashed curves correspond to PPP and PPM contributions. (The experimental prong-cross-sections at 300 GeV/c are indicated (Ref. 51.)

that $n_{\text{prong}} = 2n_- + 2$ and we can see how the tail of distribution grows going from 200 GeV/c to 1500 GeV/c. In this range $\langle n_-^D \rangle$ grows from 1·2 to 2·1 or the average prong number grows by 2. When this is added to non-diffractive component (we take the *same* Poisson distribution for n_- as earlier two-component models) as shown in Fig. 35 it can be seen that this growth of the diffractive multiplicity is sufficient to "fill in the dip" predicted by the previous models. Even at a super high energy of 5000 GeV/c no significant dip has developed.

Our "model" for diffraction scattering is essentially described by one

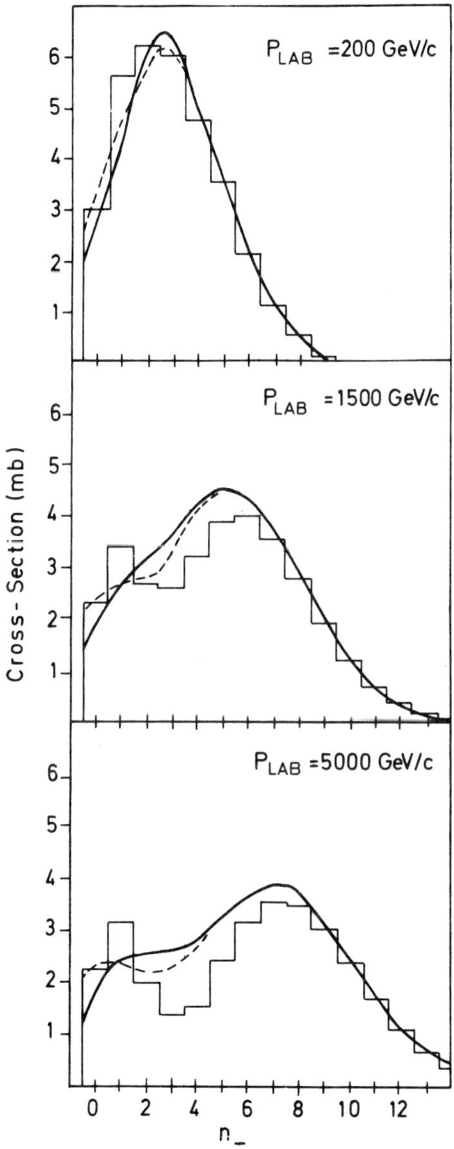

FIG. 35. Full multiplicity distributions of Ref. 58 at $p_{\mathrm{lab}} = 200$, 1500 and 5000 GeV/c. Histogram corresponds to old two-component model predictions.

graph, since we neglect the PPM term to first approximation. That is

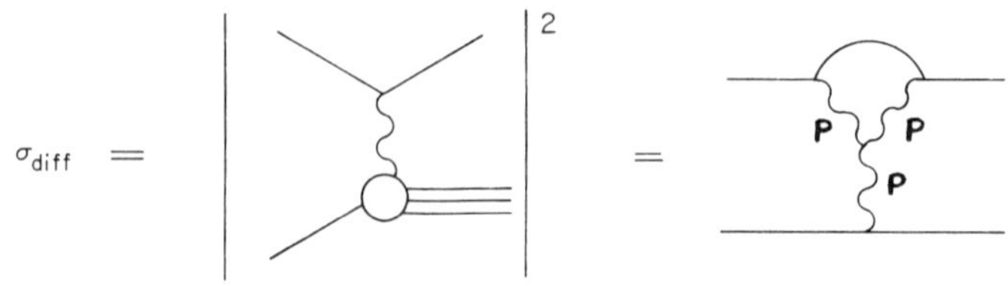

which is *linear* in the triple Pomeron coupling G_{PPP} and the cross-section (in the limit of $\alpha'_p \to 0$) goes like ln s.

There are other graphs one may consider, e.g. double dissociation

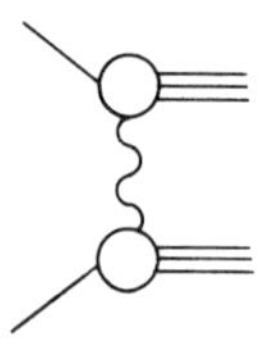

which on squaring is

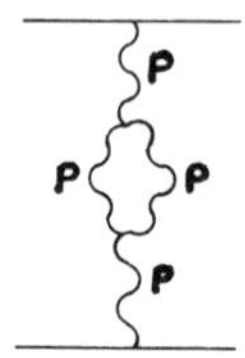

which is quadratic in the triple-Pomeron coupling. It is not hard to show that this contribution behaves as $(\ln s)^2$. Remember, however, that the relative importance of this graph to the single dissociation graph is rather small even at high energies. t_{min} is a very strong damping factor for double dissociation. Anyway, this suggests a perturbative approach to inelastic diffraction scattering which has been suggested by Bishari and Koplik,[61] by Frazer *et al.*[62] and by Amati *et al.*[57] in which the expansion variable is G_{PPP} ln s, essentially. Another graph which is quadratic in G_{PPP} is double Pomeron exchange (for which we have no direct evidence as yet):

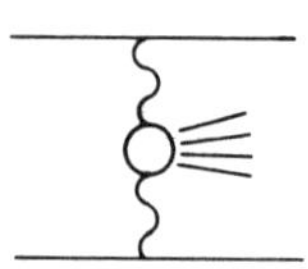

which on squaring is

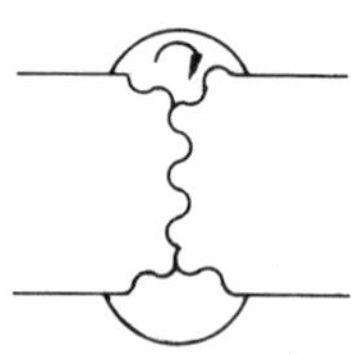

In fact it would be interesting to see, at ISR for example, how important such contributions are.

XII. Correlations

In this section we discuss briefly two-particle inclusive reactions, i.e. $a + b \to c + d + X$. The cross-section

$$f''(s; p_c, p_d) = E_c E_d \frac{d^6\sigma}{d^3p_c\, d^3p_d} \tag{78}$$

is a function of six variables which conveniently can be listed as s, y_c, y_d, p_T^c, p_T^d and ϕ the azimuthal angle between the transverse momenta of c and d. Let us begin with longitudinal correlations.

One way of studying multiparticle production was to pursue the analogy of the final state with a one-dimensional liquid.[(64)] In this analogy the density of the fluid is given by

$$\rho = \frac{1}{\sigma}\frac{d\sigma}{dy_c}$$

and the normalized quantity

$$\frac{1}{\sigma}\frac{d^2\sigma}{dy_c\, dy_d}$$

is the probability of finding one molecule at y_c, while one molecule is fixed at y_d. In such an analogy it is very natural to define a correlation function $C(y_c, y_d)$ given by

$$C(y_c, y_d) = \frac{1}{\sigma}\frac{d^2\sigma}{dy_c\, dy_d} - \frac{1}{\sigma}\frac{d\sigma}{dy_c}\frac{1}{\sigma}\frac{d\sigma}{dy_d}.$$

Different people use different values of σ i.e. *total* cross-section, total *inelastic* cross-section. Since we shall be using the Mueller–Regge approach again, it turns out we should use σ_{tot}.

We have seen already that

$$\frac{1}{\sigma}\frac{d\sigma}{dy} \text{ when integrated yields } \langle n\rangle \text{ and } \frac{1}{\sigma}\frac{d^2\sigma}{dy_c\, dy_d}$$

gives $\langle n(n-1)\rangle$ then we see that the integral of the correlation function is f_2 where f_2 is the famous quantity $\langle n(n-1)\rangle - \langle n\rangle^2$ which is exactly zero for a pure Poisson multiplicity distribution.

The generalization of Mueller's optical theorem would tell us that the quantity f'' is given by the discontinuity of the forward amplitude:

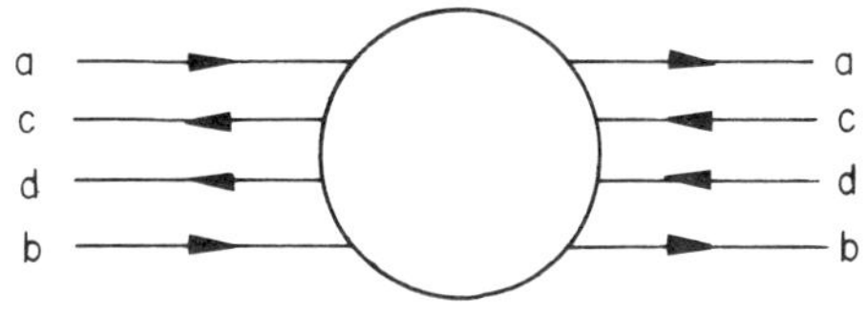

As in the single inclusive case, different Regge limits of this amplitude will be applicable to different regions of phase space for c and d.

If, for example, the "sub-energies" are large, i.e. $t_c = s_{a\bar{c}}$, $u_c = s_{b\bar{c}}$, $t_d = s_{b\bar{d}}$, $u_d = s_{a\bar{d}}$ and $s' = s_{cd}$ all large but $\kappa_c = t_c s'/u_d$ and $\kappa_d = t_d s'/u_c$ fixed, then the appropriate Regge limit is given by

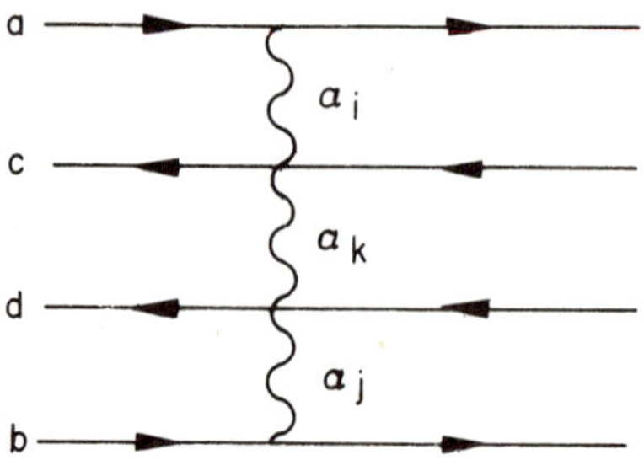

i.e.

$$f'' = \sum_{ijk} \beta_{ijk}(\kappa_c, \kappa_d)\, t_c^{\alpha_i - 1}\, t_d^{\alpha_j - 1}\, s'^{\alpha_k - 1}. \tag{79}$$

For c, d *both* in the central region, the leading behaviour is given by $i = j = \alpha_{\mathbf{P}} = 1$, and so

$$f'' = \beta_{\mathrm{PPP}}(\kappa_c, \kappa_d) + \beta_{\mathrm{PMP}}(\kappa_c, \kappa_d)\, s'^{-\frac{1}{2}}.$$

Since

$$y_c = \tfrac{1}{2} \ln \frac{u_c}{t_c} \qquad \text{and} \qquad y_d = -\tfrac{1}{2} \ln \frac{u_d}{t_d}$$

then

$$s' = (\kappa_c \kappa_d)^{\frac{1}{2}} \left(\frac{u_c u_d}{t_c t_d} \right)^{\frac{1}{2}} = (\kappa_c \kappa_d)^{\frac{1}{2}}\, \mathrm{e}^{(y_c - y_d)} \tag{80}$$

and so

$$f'' = \beta_{\mathrm{PPP}}(\kappa_c, \kappa_d) + \bar{\beta}_{\mathrm{PMP}}(\kappa_c, \kappa_d)\, \mathrm{e}^{-\frac{1}{2}(y_c - y_d)} \tag{81}$$

Thus we see that the two-particle cross-section is a function of the rapidity *difference* $|y_c - y_d|$ rather than as a function of y_c and y_d themselves. This is analogous to the result that $f(p_c)$ is independent of y_c in the central region, i.e. the central plateau.

Assume Pomeron factorization

$$\begin{aligned} \beta_{\mathrm{PPP}}(\kappa_c, \kappa_d) &= \gamma_{\mathrm{P}a}\, \gamma_{\mathrm{P}b}\, \beta^c_{\mathrm{PP}}(\kappa_c)\, \beta^d_{\mathrm{PP}}(\kappa_d) \\ \beta_{\mathrm{PMP}}(\kappa_c, \kappa_d) &= \gamma_{\mathrm{P}a}\, \gamma_{\mathrm{P}b}\, \beta^c_{\mathrm{PM}}(\kappa_d)\, \beta^d_{\mathrm{MP}}(\kappa_d) \end{aligned} \tag{82}$$

$$\therefore\ C(y_c, y_d) = \beta^c_{\mathrm{PP}}(\kappa_c)\, \beta^d_{\mathrm{PP}}(\kappa_d) + \beta^c_{\mathrm{PM}}(\kappa_c)\, \beta^d_{\mathrm{PM}}(\kappa_d)\, \mathrm{e}^{-\frac{1}{2}(y_c - y_d)}$$

$$-\left(\frac{1}{\gamma_{Pa}\gamma_{Pb}}\right)^2 [\gamma_{Pa}\gamma_{Pb}\beta^c_{PP}(\kappa_c)]\,[\gamma_{Pa}\gamma_{Pb}\beta^d_{PP}(\kappa_d)] \tag{83}$$

$$= \beta^c_{PM}(\kappa_c)\,\beta^d_{PM}(\kappa_d)\,e^{-\frac{1}{2}(y_c-y_d)}. \tag{84}$$

Not only does the correlation function depend only on the difference $y_c - y_d$ but there is a prediction that the function vanishes as this difference becomes large.

Again in the liquid analogy, there is the hypothesis of short range correlations in which C tends exponentially to zero:

$$C_2(y_c, y_d) \sim \exp[-(1/\lambda)\,|y_c - y_d|] \tag{85}$$

where λ is called the "correlation length".

So we see that this hypothesis is indeed predicted by the Mueller–Regge approach. Furthermore the correlation length is predicted to be around $\lambda = 2$.

At ISR there is the possibility of testing this hypothesis since the plateau at $s = 3000$ is 4 units long in rapidity and so we can explore both $|y_c - y_d| \gtrless 2$.

So far there have been two such experiments:

(i) Pisa–Stony Brook collaboration[65] measures correlations between charged particles.
(ii) CERN–Hamburg–Vienna[66] collaboration measures correlations between charged and neutrals.

In Fig. 36, $R(y_c, y_d)$ is plotted as a function of $(y_c - y_d)$ where

$$R = \frac{\sigma_{in}\,d^2\sigma/dy_c\,dy_d}{d\sigma/dy_c\,.\,d\sigma/dy_d} - 1.$$

Note that this definition uses $\sigma_{inelastic}$ not σ_{tot}, the latter should strictly be used in our approach. At each energy two values of y_d were chosen, $y_d = 0$ and $y_d = -2{\cdot}5$.

One notices firstly the approximate energy independence which we expected and secondly the rapid decrease when $|y_c - y_d|$ becomes large. In fact the decrease is quite compatible with our expected exponential $\exp(-|y_c - y_d|/2)$, which can be seen from Fig. 37 and Fig. 38.

In the case when c and d form an exotic sub-system, i.e. only Pomeron exchange in the centre, we expect zero correlation. Recent data from NAL[67] on $pp \to \pi^-\pi^- X$ indicate that although the correlation is not zero ($\sim 0{\cdot}2$) it shows no structure as a function $y_c - y_d$, being virtually flat.

If we consider one particle in the fragmentation region (of a say) and the other in the central region, then the appropriate graph for $f''(s, y_c, y_d)$ is

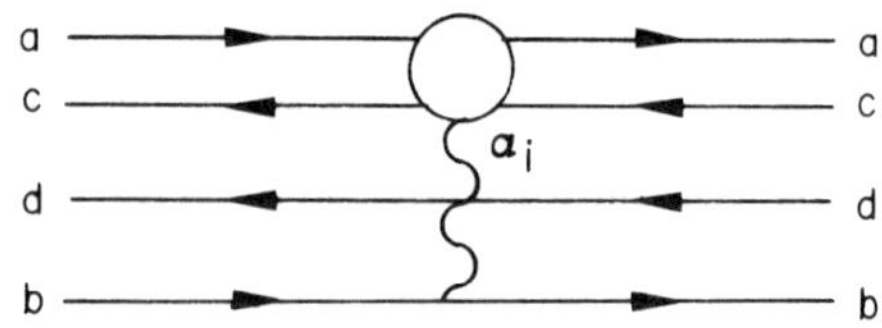

Again, if we assume Pomeron exchange for α_j and Pomeron and meson exchange for α_i we get, assuming factorization

$$f''(s; p_c, p_d) = F_P(t_c, x_c)\, \beta^d_{PP}(\kappa_d)\, \gamma_{Pb} + F_M(t_c, x_c)\, \beta^d_{PM}(\kappa_d)\, \gamma_{Pb}\, s'^{-\frac{1}{2}} \quad (86)$$

In the scaling limit,

$$f^{\text{frag.}}(s, p_c) = F_P(t_c, x_c)\, \gamma_{Pb}$$

and

$$f^{\text{cent.}}(s, p_d) = \gamma_{Pa} \beta^d_{PP}(\kappa_d)\, \gamma_{Pb}$$

so that

$$C(y_c, y_d) = \gamma^{-1}_{Pa}\, F_M(t_c, x_c)\, \beta^d_{PM}(\kappa_d)\, e^{-\frac{1}{2}(y_c - y_d)} \quad (87)$$

so the correlation function depends only on $|y_c - y_d|$ and x_c which is essentially $|y_c - y_a|$. Furthermore as the rapidity gap between c and d becomes large (~ 2), C depends only on $|y_c - y_a|$. This effect can be seen in Fig. 36

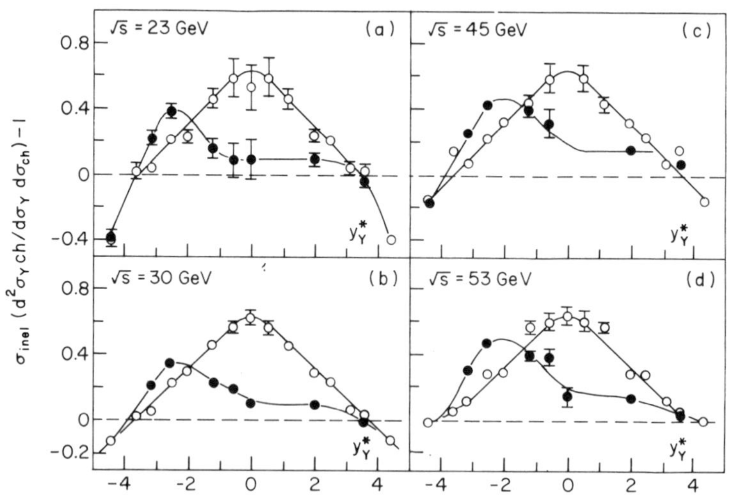

FIG. 36. Two-particle rapidity correlation $F_{\gamma ch}(y^*_\gamma, y^*_{ch}) = \sigma_{incl}(d^2\sigma_{\gamma ch}/d\sigma_\gamma d\sigma_{ch}) - 1$ in the inclusive reaction $pp \to \gamma + ch + \text{anything}$, as a function of c.m. rapidity for $y_{ch} = 0$ (open symbols) and $y_{ch} = -2{\cdot}5$ (full symbols). Taken from Ref. 66.

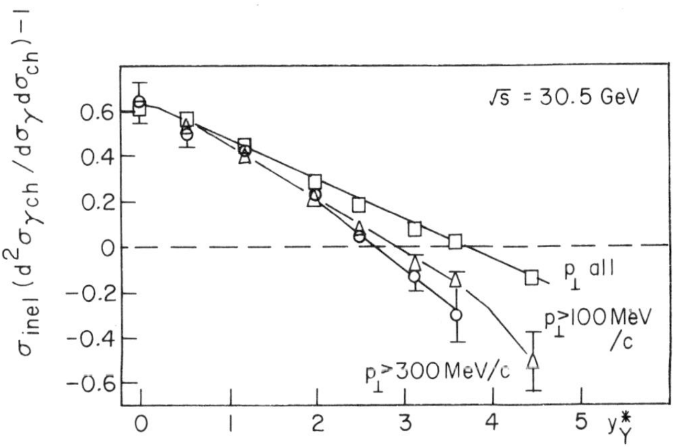

FIG. 37. Two-particle correlation $F_{\gamma ch}$ for three ranges of transverse momentum $p_{\gamma\perp}$. The correlation length is decreasing with increasing $p_{\gamma\perp}$. Taken from Ref. 66.

and in the CHLM[68] data in Fig. 39, where R for protons is plotted in coincidence with a charged particle at $y_d = 0$. In the lab. frame y_d takes on a different value at each energy and Fig. 39 shows that the correlation depends only on $|y_c - y_a|$ and not on y_d or s.

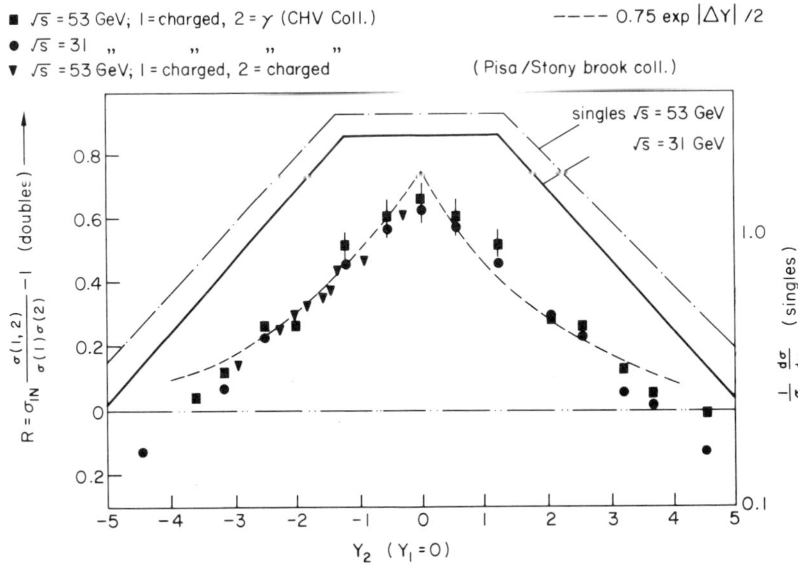

FIG. 38. Correlation function $R(y_c, y_d)$ for fixed $y_c = 0$ versus y_d in the c.o.m. frame. Data from Ref. 65 and Ref. 66. Dashed line corresponds to $\lambda = 2$.

Note that when only Pomeron exchange is allowed in both α_i and α_j, the correlation function vanishes. For example, if we measure the cross-section for $pp \to \pi_{\text{frag.}} + K^-_{\text{cent.}} + X$ then meson exchange is forbidden. These are essentially the same arguments which predicted the scaling of the single particle inclusive cross-section $pp \to K^-_{\text{central}} + X$ and so such a correlation measurement will help to teach us something about the "threshold singularity" we discussed earlier.

Finally, if we consider c in the fragmentation region of a and d in the fragmentation region of b, for which the appropriate graph is

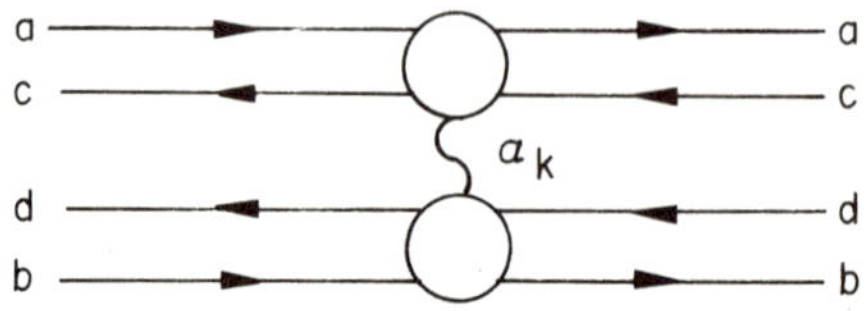

then we get the result that the correlation function depends only on $|y_a - y_c|$ and $|y_b - y_d|$ while if α_k is the Pomeron, then we get *no* correlation.

Thus we see that the Regge–Mueller analysis of two particle inclusive reactions leads to the results of the Short Range Correlation Hypothesis. It is important to realize that the analysis chose $\sigma = \sigma_{\text{tot}}$, while in fact our comparison with data used $\sigma = \sigma_{\text{inel}}$. Having subtracted σ_{elastic}, it is natural to subtract $\sigma_{\text{diffractive}}$. The comparison with data suggests that the short range

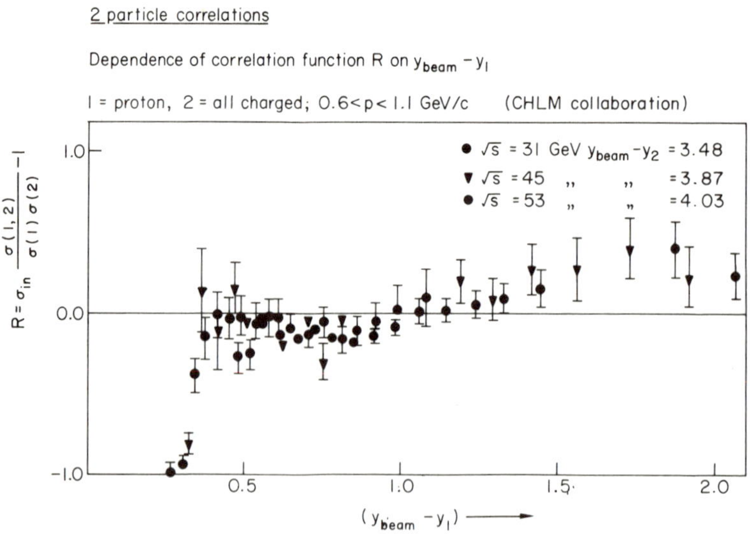

FIG. 39. Correlation function $R(y_c, y_d)$ plotted against $(y_a - y_c)$. Taken from Ref. 68.

correlation hypothesis is valid for the *non*-diffractive part of the cross-section. However the ISR data also indicate the presence of *long*-range correlations.

A scheme in which the multiparticle production is divided into two parts: (i) diffractive giving rise to long-range correlations and (ii) "pionization" in which correlations are all short range ($\lambda \simeq 2$) seems to have the best chance of explaining the data presently available. This is the so-called two-component model, mentioned earlier, which is dealt with fully by Harari.[54]

To see briefly the correlations arising from each component we refer to Fig. 40 where the original Pisa–Stony Brook data is shown together with the

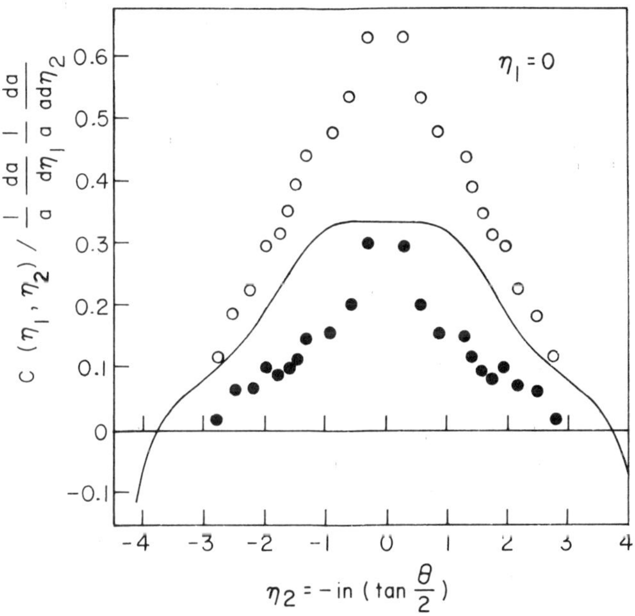

FIG. 40. The two-particle correlation function $C_2(\eta_1, \eta_2)$ for all charged particles at $\sqrt{s} = 31$ GeV. The experimental data are shown by open circles. The solid line gives the predicted contribution of long range correlations and a model for the diffractive component. The difference (dots) between the data points and the solid line should be given by short range correlations of the pionization component. Taken from Ref. 69.

diffractive component in the model of Pirilä and Pokorski[69] and the difference between the two, i.e. the predicted short-range pionization component.

Azimuthal correlations are also of great interest. In the Mueller approach such correlations arise only from the exchange of $M \neq 0$ Lorentz poles (M = Toller quantum number) which could be cuts or from daughters of

$M = 0$ poles. Clearly the measurement of azimuthal correlations at very high energies will give much information. It is interesting to know whether transverse momentum is conserved *locally* or not, i.e. if $p_T = 0$ within rapidity gaps $|y_c - y_d| \ll 2$. There is already some indication[66] that long range azimuthal correlations exist, at least for particles with large p_T values.

For a full discussion of correlations in multiparticle reactions, the reader is recommended to the excellent lectures by Le Bellac.[70]

XIII. CONCLUSIONS

We have seen that the application of Regge ideas to inclusive reactions via Mueller's optical theorem has proved to be a particularly fruitful approach to the study of multiparticle production. It turns out to be more successful in certain regions of phase space than in others. The fragmentation region is more 'Regge-like" than the central region because at present energies the latter case still has important threshold effects.

Clearly the Mueller–Regge approach is not sufficient to give a complete understanding of multiparticle phenomena. We still need explicit models to explain important features such as multiplicity distributions, correlation structure, etc. Much progress has been made with a crude two-component model in which the cross-section is separated into the diffractive and non-diffractive parts. We have seen furthermore that the triple-Regge formalism is a very useful language in which to discuss diffractive inelastic scattering and be taken as an accurate parametrization of the diffractive component. By making simple assumptions about the decay of the diffractive fireball we can then obtain a very explicit model for diffractive production. In particular we then have a prediction for the long-range correlation and we can then subtract this from the experimental correlation to learn something about the non-diffractive or short-range component—such as the size of produced clusters. This illustrates the point that the Mueller–Regge inclusive approach can be used as a basis for building rather explicit models of multi-particle production.

An attractive bonus of applying the Mueller–Regge formalism is the possibility of investigating Reggeon-particle elastic scattering and of testing the predictions of duality. This meant that we could use experimental data to study meson–meson scattering and also baryon–antibaryon scattering in the meson resonance energy range.

During the next year or so we can expect even more exciting data on high energy multiparticle reactions from NAL and from ISR. If such data turn out to be as stimulating as that already produced by these accelerators then we are in for a busy two years or so.

ACKNOWLEDGEMENTS

Most of the material presented in these lectures I have learnt from my colleagues at the Rutherford Laboratory and I am therefore deeply grateful to D. P. Roy, Hannu Miettinen, Hong-Mo Chan and Paul Hoyer for my education. I especially thank Hannu Miettinen and D. P. Roy for their advice in preparing these talks.

REFERENCES

1. R. P. Feynman. *Phys. Rev. Lett.* **23**, 1414 (1969).
2. J. Benecke, T. T. Chou, C. N. Yang and E. Yen. *Phys. Rev.* **188**, 2159 (1969).
3. A. H. Mueller. *Phys. Rev.* **D2**, 2963 (1970).
4. H. Harari. *Phys. Rev. Lett.* **20**, 1395 (1968); P. G. O. Freund. *Phys. Rev. Lett.* **20**, 235 (1968).
5. H. P. Stapp. *Phys. Rev.* **D3**, 3177 (1970); J. C. Polkinghorne. *Nuovo Cim.* **7A**, 555 (1972).
6. Chan Hong-Mo, C. S. Hsue, C. Quigg and J.-M. Wang. *Phys. Rev. Lett.* **26**, 672 (1971).
7. D. Gordon and G. Veneziano, *Phys. Rev.* **D3**, 2116 (1971); G. Veneziano, *Lett. Nuovo Cim.* **1**, 681 (1971).
   S.-H. H. Tye and G. Veneziano, *Phys. Lett.* **38B**, 30 (1972); CERN preprint TH-1552 (1972).
8. M. B. Einhorn, M. B. Green and M. A. Virasoro. *Phys. Lett.* **37B**, 292 (1971); *Phys. Rev.* **D6**, 1675 (1972), *Phys. Rev.* **D7**, 102 (1973).
9. H. I. Miettinen, "Regge Phenomenology of Inclusive Reactions", Doctorate thesis, University of Helsinki (1973).
10. M. G. Albrow *et al.* (CHLM collaboration at ISR). *Nucl. Phys.* **B51**, 388 (1973).
11. H. I. Miettinen. *Phys. Lett.* **38B**, 431 (1972).
12. Chan Hong-Mo, H. I. Miettinen and W. S. Lam. *Phys. Lett.* **40B**, 112 (1972).
13. J. R. Fry *et al. Nucl. Phys.* (to be published).
14. H. D. I. Abarbanel and D. I. Gross. *Phys. Rev. Lett.* **26**, 732 (1971); *Phys. Rev.* **D5**, 699 (1972); R. J. N. Phillips, G. Ringland, R. Worden. *Phys. Lett.* **40B**, 239 (1972).
15. G. Jarlskog. Review talk Nashville Conference, 1973.
16. P. Carruthers, Minh Duong-van. *Phys. Lett.* **41B**, 597 (1972).
17. D. R. O. Morrison. Review lecture, Royal Society meeting, March, 1973.
18. L. Caneschi. CERN preprint TH 1704.
19. E. J. Squires and D. M. Webber. *Lett. Nuovo Cim.* **7**, 193 (1973).
20. Chan Hong-Mo, H. I. Mietinnen, D. P. Roy and P. Hoyer. *Phys. Lett.* **40B**, 406 (1972).
21. T. Ferbel. *Phys. Rev. Lett.* **29**, 448 (1972).
22. D. P. Roy. Review talk, Rutherford Laboratory annual winter meeting, Rutherford preprint RPP/T/41 (1973).
23. H. I. Miettinen, D. P. Roy. unpublished.
24. P. Chliapnikov *et al. Phys. Lett.* **35B**, 581 (1971).
25. K. Paler *et al. Phys. Lett.* **43B**, 437 (1973).
26. K. C. Moffat *et al. Phys. Rev.* **D5**, 1603 (1972).

27. C. Risk. *Phys. Rev.* **D5**, 1685 (1972).
28. G. Ranft and J. Ranft. *Sov. Jou. Nucl. Phys.* **15**, 314 (1972).
29. M. S. Chen, L. L. Wang and T. F. Wong. *Phys. Rev.* **D7**, 1667 (1972).
30. J. Randa, Univ. of Illinois preprint (1972); M. Bishari, H. J. Yesian, *Phys. Lett.* **38B**, 312 (1972).
31. Aachen *et al.* Collaboration, CERN.
32. J. Kwiecinski. *Lett. Nuovo Cim.* **3**, 619 (1972); M. B. Einhorn, J. Ellis, J. Finkelstein. *Phys. Rev.* **D5**, 2063 (1972).
33. P. Oleson, preprint CERN-TH-1376 (1971); A. I. Sanda, *Phys. Rev.* **D6** (1972), 280.
34. H. D. I. Abarbanel and M. B. Green. *Phys. Lett.* **38B**, 90 (1972).
35. H. D. I. Abarbanel. NAL preprint THY-28 (1972).
36. J. Finkelstein. *Phys. Rev.* **D6**, 931 (1972).
37. P. Hoyer, R. G. Roberts and D. P. Roy. *Nucl. Phys.* **B56**, 173 (1973).
38. R. Dolen, D. Horn and C. Schmid. *Phys. Rev.* **166**, 1768 (1968).
39. P. Hoyer, R. G. Roberts and D. P. Roy. *Phys. Lett.* **44B**, 258 (1973).
40. Full references to decoupling of Pomeron can be found in A. H. Mueller, talk at Batavia Conference, Vol. 1, p. 347, (1972).
41. P. H. Frampton and P. V. Ruuskanen. *Phys. Lett.* **38B**, 78 (1972); J. Dias de Dens and W. S. Lam. *Phys. Lett.* **38B**, 220 (1972); P. D. Ting and H. J. Yesian. *Phys. Lett.* **35B**, 321 (1971).
42. S. D. Ellis and A. I. Sanda. *Phys. Rev.* **D6**, 1347 (1972); *Phys. Lett.* **41B**, 87 (1972).
43. F. E. Paige and L. L. Wang. *Nucl. Phys.* **B46**, 477 (1972).
44. Chan Hong-Mo, H. I. Miettinen and R. G. Roberts. *Nucl. Phys.* **B54**, 411 (1973).
45. J. V. Allaby *et al. Nucl. Phys.* **B52**, 316 (1973).
46. M. G. Albrow *et al. Nucl. Phys.* **B51**, 388; *ibid* **B54**, 6 (1973).
47. H. I. Miettinen and R. G. Roberts. (Unpublished).
48. F. Sannes *et al. Phys. Rev. Lett.* **30**, 766 (1973).
49. Michigan-Rochester collaboration at NAL at 100 GeV/c.
50. ANL-NAL collaboration at NAL at 200 GeV/*c*.
51. NAL-UCLA collaboration at NAL at 300 GeV/*c*.
52. P. Franzini *et al.* presented at Vanderbilt Conference (1973).
53. A. Antipov *et al. Phys. Lett.* **40B**, 147 (1972); D. Bowen *et al. Phys. Rev. Lett.* **30**, 332 (1973).
54. For full details see H. Harari's lectures at this school.
55. U. Amaldi *et al. Phys. Lett.* **B44**, 112 (1973).
56. G. F. Chew. *Phys. Lett.* **44B**, 169 (1973); LBL preprint 1556.
57. A. B. Kaidalov *et al.* Leningrad preprint (1973); D. Amati, L. Caneschi, M. Ciafalloni. *Nucl. Phys.* (to be published); W. R. Frazer and D. R. Snider. *Phys. Lett.* **45B**, 136 (1973); A. Capella, Min-Shih Chen, M. Kugler and R. D. Peccei. SLAC preprint (1973).
58. R. G. Roberts and D. P. Roy. Rutherford Laboratory preprint 73-065 T64 (1973).
59. K. Kajantie and P. V. Ruuskanen. *Phys. Lett.* **45B**, 149 (1973).
60. F. T. Dao *et al.* NAL-Pub-73/22 to be published in *Phys. Lett.*
61. M. Bishari and J. Koplik, *Phys. Lett.* **44B**, 175 (1973).
62. W. R. Frazer, D. R. Snider and Chung-I-Tan, University of California. San Diego preprint UCSD-10P10-127 (1973).
63. D. P. Roy and R. G. Roberts. *Phys. Lett.* **40B**, 555 (1972).
64. K. Wilson. Cornell preprint (1970). See Ch 10 of this book
65. Pisa–Stony Brook collaboration. (To be published).

66. H. Dibon *et al.* *Phys. Lett.* **44B**, 313 (1973).
67. S. Barish *et al.* Nal-Conf. 73/25-EXP (1973).
68. M. G. Albrow *et al.* *Phys. Lett.* **44B** 518 (1973).
69. P. Pirila and S. Pokorski. *Phys. Lett.* **B43**, 502 (1973).
70. M. Le Bellac, Lectures at XIIIth School of Theoretical Physics, Krakow, 1973.

4. MULTIHADRON PRODUCTION AT HIGH ENERGIES: PHENOMENOLOGY AND THEORY

HAIM HARARI

Weizmann Institute of Science, Rehovot, Israel and CERN, Geneva, Switzerland

INTRODUCTION

These lectures are devoted to a phenomenological and theoretical discussion of the physics of multihadron processes at high energies. We concentrate on those mechanisms which are presumably responsible for most of the observed cross-sections at laboratory energies of hundreds of BeV's. We therefore restrict our attention to small transverse momentum, large multiplicity, final states.

We introduce the subject in Section I, by recalling some of the useful lessons of the last 12 years of theoretical and phenomenological studies of two-body reactions. Many of these lessons may enable us to generalize useful ideas to the multihadron case, while avoiding a repetition of the many mistakes which were a necessary part of the progress in our understanding of two-body reactions. We review the two component description of two-body scattering amplitudes and generalize it to the multihadron case, defining a diffractive and a non-diffractive part of the production amplitude.

In Section II we present a general description of multihadron processes from the viewpoint of the two component model. We confront the model with various data, emphasizing some of the crucial experimental tests for the existence of the two mechanisms.

Section III is devoted to a detailed discussion of the diffractive component, in view of the recent discovery of a diffractive inclusive proton cross section, which obeys scaling. We develop a simple model for the diffractive part and discuss many of its predictions. Of particular interest, is the possible relation of the model to the rising total cross-section, recently observed at the ISR.

In Section IV we present a discussion of the statistical properties of the Feynman gas. We define the pressure and surface tension of the gas, deduce them from the present data on multiplicity cross-sections, and compute them in various theoretical models. We then briefly speculate on possible deeper applications of the statistical analogy of the Feynman gas.

Throughout our phenomenological discussion of the first lectures we repeatedly emphasize the fact that the Pomeron term in the total cross-section (or a constant fraction of this term) is "built" through the unitarity relation by the non-diffractive contributions. If this is the case, we should be able, in principle, to compute the properties of the Pomeron from the known properties of the non-Pomeron trajectories (or, in other words, from the known spectrum of hadrons). In particular we should study the question of the Pomeron intercept. This is the central topic of Section V: "Why does the Pomeron have a unit intercept?"

Section VI is devoted to a discussion of the two important consistency questions related to the Pomeron: (i) Its consistency with itself, (ii) its consistency with unitarity and the Froissart bound. We discuss these questions from the viewpoint of absorption models for multihadron production amplitudes, emphasizing the possibility that the (important) theoretical consistency problems may be completely unrelated to the phenomenology of the present (and the next?) generation of experiments.

Our main aim in these notes (particularly in the last two lectures), is to present some of the most interesting problems confronting hadron physics, and to provoke new ideas which are so badly needed. We do not try to present solutions to most of these problems. We would like to be able to do so, but—like everyone else—we cannot!

I. Lessons From Two-Particle Final States and Their Generalization to Multihadron Production

1. THE TWO MECHANISMS OF TWO-BODY REACTIONS

Our present understanding of the phenomenology of hadronic reactions with two-body or quasi two-body final states involves two distinct classes of reaction mechanisms.

The first class includes the formation of resonances in the s-channel and the exchange of particles, resonances, or Regge trajectories in the t-channel. All of these mechanisms, commonly known as "non-diffractive" are related to each other by various principles of hadron dynamics such as analyticity, crossing and duality. In principle, a complete understanding of the hadronic spectrum of states (i.e. knowledge of all masses and coupling constants) should enable us to calculate the scattering induced by such mechanisms.

The second class is more mysterious. These are the so-called "diffractive contributions" or the "Pomeron exchange contributions". Their origin is not related in any simple way to any known properties of the hadronic spectrum. Neither the strength nor the intercept of the Pomeranchuk singularity are related in any simple way to existing hadrons, and within the framework of two-body phenomenology, diffraction (or Pomeron exchange) is essentially a code-name for a phenomenon that can be explained only in terms of oversimplified geometrical and optical analogies but not in terms of hadron dynamics.

The diffractive and non-diffractive mechanisms are then the two components present in a two-body scattering amplitude. Let us remind ourselves of some of their general features, emphasizing those aspects that will prove relevant and useful to our discussion of multihadron processes.

2. THE NON-DIFFRACTIVE COMPONENT: R

The non-diffractive component ("R-component") of an amplitude for the process $a + b \to c + d$ may include contributions from:

(i) *s-channel resonances* (*Fig.* 1*a*)

Their contributions are prominent at low energies (below, say 2 BeV), but they are presumably important even at very high energies, in spite of our inability to isolate single resonances at such energies. s-channel resonances are presumably absent in exotic channels (such as K^+p, pp, $\pi^+\pi^+$, etc.).

(ii) *t-channel exchanges* (*Fig.* 1*b*)

The exchanged object can be a Regge trajectory, corresponding to known physical hadrons. Minor contributions (especially at low energies) are expected from two-particle (or two-Reggeon) cuts representing the exchange of two hadronic states. The clearest experimental characteristic of the Reggeon exchange is the energy dependence of the amplitude—$s^{\alpha R}$, $\alpha_R < 1$. The resulting cross-section behaves like $s^{2\alpha R-2}$, always falling to zero like a power, as $s \to \infty$. The highest known non-Pomeron trajectories are around $\alpha_R \sim \frac{1}{2}$. *Consequently, a non-diffractive cross section will fall with energy at least as fast as* s^{-1}. Whenever the t-channel quantum numbers are exotic ($\pi^- p \to \pi^+\Delta^-$, $K^- p \to \pi^+\Sigma^-$, $\pi^-\pi^+ \to \pi^+\pi^-$, etc.) single Reggeon exchange is forbidden and the amplitude disappears very quickly as the energy increases.

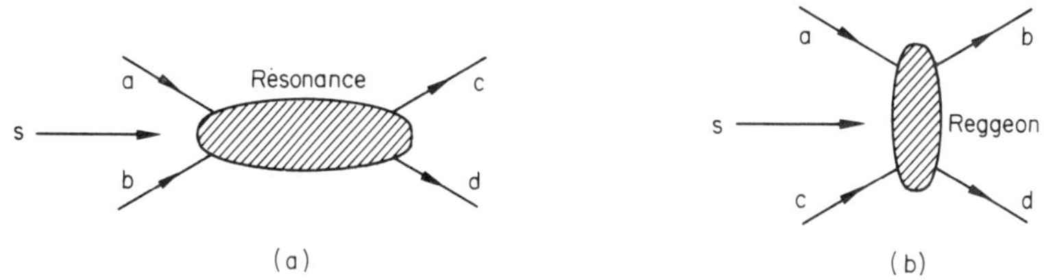

FIG. 1. s-channel resonances and t-channel exchanges.

Duality[1] tells us that the sum of s-channel resonances and the sum of t-channel exchanges each provides us with a complete description of the R-component of the two-body scattering amplitude (Fig. 2). We remember, however, that this complete description is obtained only when we sum over all (infinite number of) resonance in the s-channel or over all (infinite number of) exchanges in the t-channel.

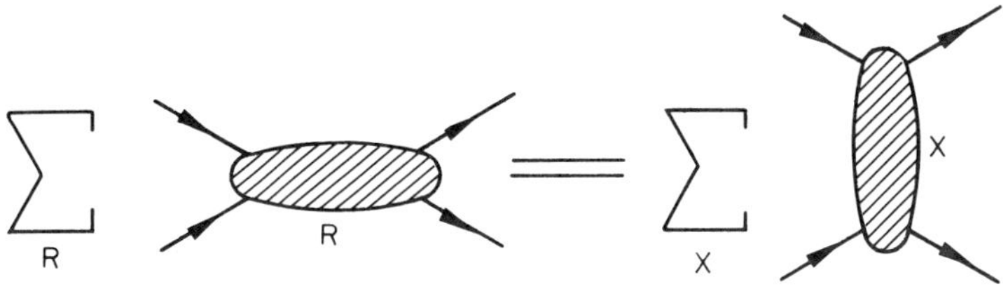

FIG. 2. Duality.

Duality diagrams[2] (Fig. 3) give us a simple quark model manifestation of the possibility of simultaneously having a set of non-exotic s-channel resonances as well as a set of non-exotic t-channel exchanges, describing the R-component of a two-body reaction. The planar diagram (Fig. 3a) shows the duality between s-channel resonances and t-channel exchanges. The imaginary part of the physical scattering amplitude in the s-channel is given by this term. However, for physical scattering in the u-channel, the *imaginary* part of the amplitude is given by the non-planar diagram (Fig. 3b) (which would become planar if we perform the $s \leftrightarrow u$ transformation). The *real* part of both large s and large u scattering will have contributions from both diagrams (a resonance contribution to the imaginary part is much more localized in energy than its contribution to the real part). If we are now interested in scattering at large s, the imaginary part of the amplitude will be given by Fig. 3(a) while the real part has contributions both from 3a and 3b. In simple dual resonance models,[3] Fig. 3(a) represents the term $(-s)^{\alpha}$ while Fig. 3(b) gives the s^{α} term. The full amplitude is then proportional to $s^{\alpha} \pm (-s)^{\alpha} = (1 \pm e^{-i\pi\alpha})s^{\alpha}$ where the $\pm$ sign represents the signature of the exchanged object. Note that the contributions of the two diagrams to the *cross-section* are equal in a dual model.

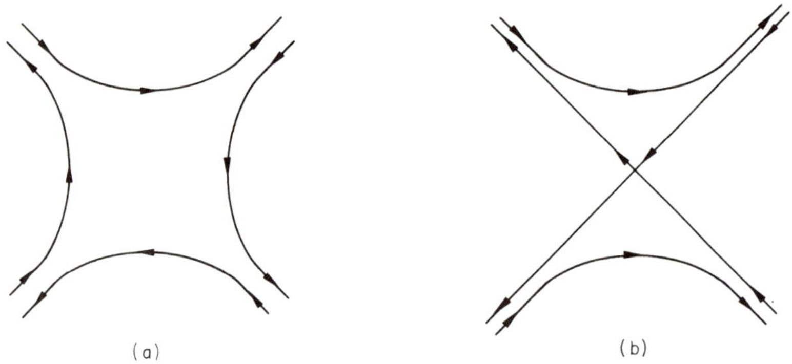

FIG. 3. Duality diagrams.

3. THE DIFFRACTIVE COMPONENT: P

The diffractive, Pomeron exchange term ("P-component") in two-body reactions has the following properties:

(i) It contributes only (or mostly) to the imaginary part of the amplitude.

(ii) It contributes only in cases in which no internal quantum numbers are exchanged in the t-channel (hence it carries no charge, isospin, strangeness or baryon number; it is positive under charge conjugation and approximately invariant under SU (3)).

(iii) Its energy dependence is approximately given by s^{α_P} (where $\alpha_P \sim 1$) in the amplitude, leading to *cross sections which are constant in energy* (modulo terms of log s).

(iv) Representing zero t-channel quantum numbers, the P-component is indifferent to the quantum numbers of the s-channel. Hence it contributes equally to exotic and non-exotic s-channel processes of the same "family" (e.g. $K^-p \to K^-p$ and $K^+p \to K^+p$; $\pi^+\pi^- \to \pi^+\pi^-$ and $\pi^+\pi^+ \to \pi^+\pi^+$). Consequently it cannot be dual to s-channel resonances. The P-component is therefore "non-resonant background" from the s-channel point of view.[4]

(v) The most physical description of the P-component is perhaps its impact parameter representation. Motivated by ideas borrowed from optical diffractive scattering, we believe that the P-component (which is necessarily the dominant term in high energy elastic scattering) is represented by an absorbing disk with a typical radius of the order of 1 Fermi or so (Fig. 4). Whether the disk is "black" or "grey" at the center, whether it is "smoothly fading away" at large b or has a sharp edge, and most importantly—whether its radius is energy independent or expanding logarithmically, we do not know.

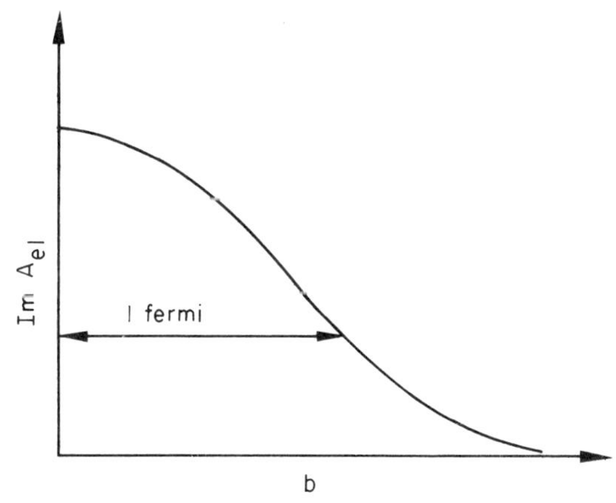

FIG. 4. Impact parameter description of the P-component.

All of these observed phenomenological features of the P-component can be well imitated by postulating[5] the existence of a t-channel singularity at $\alpha = 1$ in the complex j-plane, carrying no internal quantum numbers and having an even signature. Whether this singularity is a simple pole, or a much more complicated object, we do not know.

The duality diagram or quark model description of the P-component is trivially given by Fig. 5, where no quarks are transmitted between the interacting hadrons.

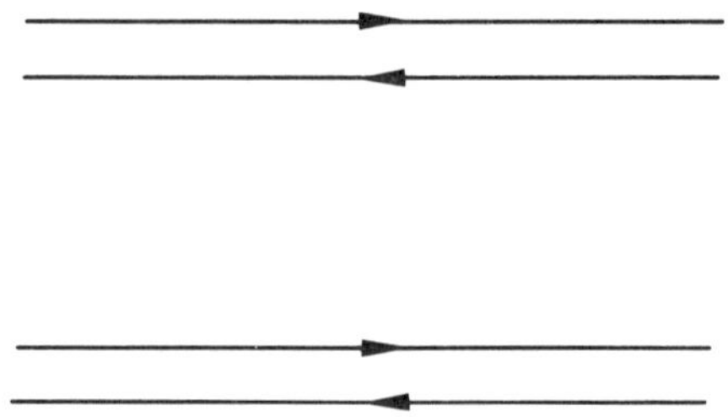

FIG. 5. Duality diagram for Pomeron exchange.

4. ABSORPTIVE CORRECTIONS TO THE NON-DIFFRACTIVE PART

We have several reasons to believe that the R-component of two-body scattering amplitudes must include, in addition to the contributions of simple Reggeon exchanges, important corrections due to absorption. Among the various theoretical and phenomenological arguments which support the introduction of such terms we mention here only a few:

(i) If high energy two-body scattering amplitudes were described only in terms of t-channel poles, we would have many predictions based on the factorization properties of such poles. Many of these predictions fail miserably and necessitate the introduction of new singularities with roughly the same intercepts as the known poles. In the absence of additional observed particles which could correspond to such singularities (if they were simple poles) the most obvious candidate for the new singularities are the cuts generated by the exchange of one Reggeon and a Pomeron.

(ii) Intuitive arguments based on analogy to nuclear and atomic systems lead us to believe that most of the cross-section of any inelastic two-body reaction would come from impact parameters corresponding to the external part ("surface") of the hadron, while high multiplicity inelastic collisions are mostly produced by hitting the centre of the hadron. The exchange of a single t-channel pole produces in most cases amplitudes which are dominated by small ($b \sim 0$) impact parameters. The application of an absorption correction which takes into account the large probability of creating high multiplicity final states at small b rectifies the situation and leaves us with an R-component which is more peripheral.

(iii) There is no reason to assume that in strong interaction physics, the exchange of a given pole is not corrected significantly by initial and final state interaction between the particles involved in the reaction.

(iv) If the R-component can also be described by s-channel resonances it has to reflect the experimental fact that the peripheral resonances (having $l \sim p_{cm}R$ where $R \sim 1$ fermi) are usually the strongest resonances observed. This could happen in a t-channel picture only if there are substantial corrections to the pole terms, which tend to decrease their small-b contributions.

Since absorption corrections reflect the properties of elastic scattering we have here an important piece of the R-component which is influenced by the properties of the P-component. In particular the question of the energy dependence of the effective radius of the P-component in its impact parameter representation is of crucial importance. Absorption corrections always change the b-dependence (and hence the t-dependence) of the amplitude. Their effect on the energy dependence is much more subtle. If the elastic absorption correction to an inelastic amplitude has an absorption radius which is *constant* as a function of energy, the energy dependence of the "absorbed" amplitude will be very *similar* to that of the "unabsorbed" one (up to log s terms). However, if we start by an inelastic amplitude behaving like s^{α}, and completely absorb all its contribution for $b < b_0$ where the absorption radius $b_0 \propto \log s$, we may end up in some cases with a completely *different* energy dependence s^{β} where $\beta < \alpha$. This remark is of special importance for models which tend to violate unitarity. Suppose that a certain model gives us an amplitude of the form:

$$A(s, b) = s^{\alpha} \exp\left(-\frac{b^2}{R^2}\right) \tag{1}$$

where $\alpha > 1$, thus violating the Froissart bound. If the unitarity bound for each impact parameter is $A(s, b) \leqslant 1$, we find that our amplitude violates it whenever:

$$\alpha \log s - \frac{b^2}{R^2} \geqslant 0 \tag{2}$$

or

$$b^2 \leqslant \alpha R^2 \log s. \tag{3}$$

We now see (Fig. 6) that unitarity will be saved if and only if we "absorb" all impact parameters below $b_0 = R(\alpha \log s)^{\frac{1}{2}}$. If we do that, the energy dependence of the absorbed amplitude will not be s^{α} with $\alpha > 1$, but will be bounded by s (or by $\alpha = 1$). This "miracle" happens as a result of an absorption correction with a logarithmically expanding radius! The log s terms in the absorption radius are probably related to log s terms in the energy dependence of the P-component. Those terms may therefore be extremely important!

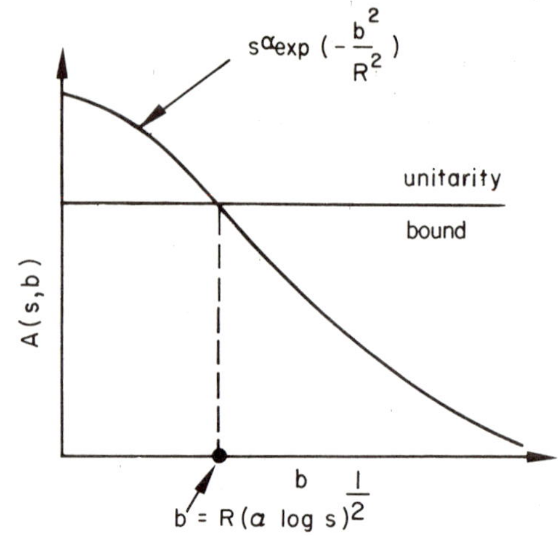

FIG. 6.

5. TWO COMPONENT DUALITY—A BRIEF SUMMARY

Summarizing the points that we have mentioned so far, we have the usual description of hadronic two-body amplitudes in terms of two component duality[(4)] (Table I). The imaginary part of the amplitude is given by the two components, according to the properties outlined in the table. The real part can be obtained from the imaginary part, using dispersion relations.

Two remarks should be made at this point:

(i) The best way of studying the *R* and *P* components is to isolate them from each other. The *P* component is easily isolated in processes which have an exotic *s*-channel. The *R*-component is easily isolated in inelastic reactions involving quantum number exchanges, or by studying the difference between particle and antiparticle elastic processes.

(ii) *The one unambiguous feature which distinguishes most explicitly between the two components is their completely different energy dependence.* This is the simplest and safest experimental determination of the presence or absence of each component in any given situation.

6. THE *R*/*P* RATIO

Having introduced the two components, we would like to study the relations among them. The simplest phenomenological question one can ask is: "what is the relative strength of the two components?"

TABLE I

Im A (a+b → c+d) =	R +	P
s – channel	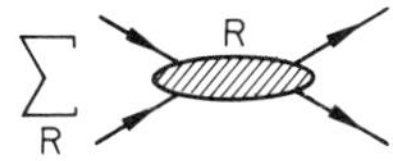	Non resonant background
t – channel	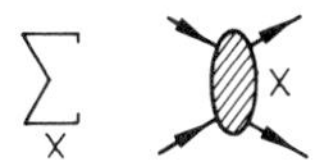 Including Regge Pole exchange and absorptive corrections (Pole – Pomeron cuts)	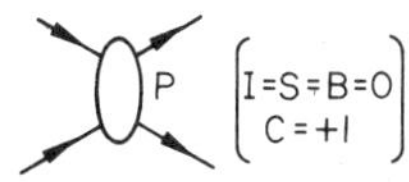
Duality diagrams	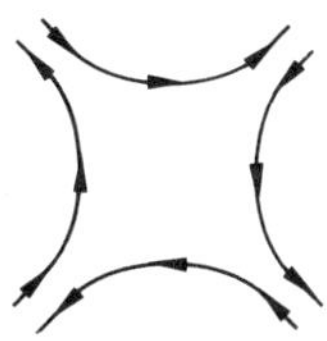	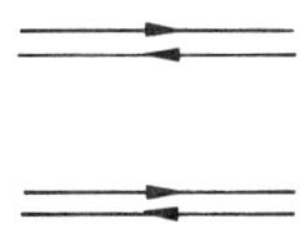
Impact parameter description:	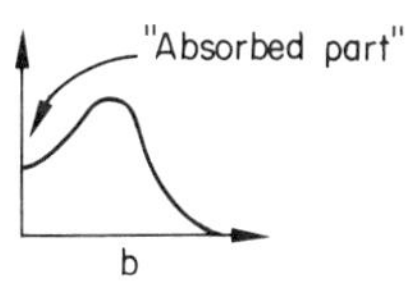	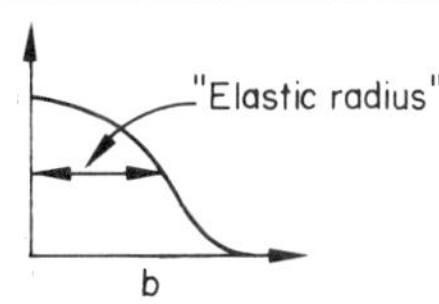
Energy dependance (module log s terms)	s^{u} $\alpha < 1$	s^{α} $\alpha = 1$

The answer is, obviously, energy dependent. If $R \propto s^{\alpha_R}$, $P \propto s^{\alpha_P}$, we have:

$$R/P \propto s^{\alpha_R - \alpha_P} \sim s^{-\frac{1}{2}} \tag{4}$$

Experimentally, it seems that the relation[6]

$$\frac{R}{P} \sim \frac{1}{2\sqrt{v}} \tag{5}$$

where v is the laboratory energy, crudely describes (up to factors of 2 or 3) the relative strength of the two components in most cases. We therefore conclude that whenever both components are allowed to contribute (e.g. elastic scattering with non-exotic s-channel quantum numbers) we have:

$$P \gg R \quad \text{for} \quad \nu \gg 1 \text{ BeV}$$

$$P \sim R \quad \text{for} \quad \nu = 1\text{–}2 \text{ BeV}$$

The situation for $\nu < 1$–2 BeV is complicated. The crucial question is whether the P-component exists at all at very low energies. It is clearly not dominant there. Consequently it is not easy to isolate it or detect it. We have, however, several phenomenological indications, hinting that below 1 BeV or so, the P-component is simply not present. In other words, the P-component which contributes an energy independent cross-section above, say, $\nu = 2$ BeV, is damped by a "threshold" effect below that energy.

One of these indications comes from a detailed analysis[7] of πN scattering at low energies within the framework of two component duality. According to the usual prescription, we define:

$$\text{Im}\, P = \text{Im}\, A_{\text{el}} - \sum_{\text{res}} \text{Im}\, A_{\text{res}} \tag{6}$$

where A_{el} is the full elastic amplitude and A_{res} is the resonance contribution to the amplitude. Using the known πN elastic phase shifts we can then compute Im P. The result, shown in Fig. 7, indicates that below $\nu \sim 1$ BeV the Pomeron term is absent.

A similar hint is offered by the energy dependence of $\sigma_{\text{tot}}(K^+p)$ and $\sigma_{\text{tot}}(pp)$. These processes have exotic s-channel quantum numbers. σ_{tot} should therefore reflect the P-term, beginning at low energies. Experimentally (Figs 8 and 9) both cross sections are approximately constant above $P_{\text{lab}} \sim$ 1·5 BeV. However, below that energy, both show a significantly smaller cross-section with a dip below 1 BeV. The constant P-component is certainly not present below 1·5 BeV!

It is interesting to note that in all three cases the "onset" point of the Pomeron corresponds approximately to the opening of significant inelastic contributions to σ_{tot}.

We therefore conclude that, for energies below 1 or 2 BeV, $R > P$, where P may even be absent for sufficiently low energies.

7. CAN WE RELATE THE R AND P COMPONENTS?

The usual duality treatment of the R- and P-components involves simple linear relations which cannot determine the relative scale of the two components. In particular, even if we assume that the R-component is fully determined by the properties of the hadronic spectrum, we cannot use these properties to compute the P-component within the framework of the two-body amplitudes. Our only hope of predicting the properties of P, starting from those of R, is to use the unitarity condition. Its simplest form, the optical theorem, reads:

$$\operatorname{Im} A_{\mathrm{el}} \propto \sigma_{\mathrm{el}} + \sum_{n} \sigma_n \tag{7}$$

We can now substitute $\operatorname{Im} A_{\mathrm{el}} = R + P$ and $\sigma_{\mathrm{el}} \propto |R + P|^2$. σ_n is the cross section for producing n hadrons. What we have then is a quadratic relation between the R and P terms, which involves, in addition, the n-particle cross sections for all n-values. It is clear from this relation that any hope which we may have for relating R and P necessitates a generalization of these two mechanisms for the case of the multihadron production amplitude. Such a generalization, involving the same basic features of the R and P mechanisms may lead us to a deep connection between the two otherwise unrelated components of hadronic scattering.

At this point we must remark that we have already encountered another connection between the R and P terms. We have stated above that the R-component includes absorptive corrections which are due to the cuts generated by exchanging a Reggeon and a Pomeron. Such corrections are another manifestation of unitarity, they are non-linear and they involve the P-component in determining the full features of the R-component. However, from the theoretical point of view the most interesting challenge is the understanding of the intercept, slope, strength and analytic properties of the Pomeron by starting from the known hadrons. Our best hope for progress in this direction are relations like eqn (7), and in order to use these, we finally have to turn our attention to the multihadron cross sections σ_n and their corresponding amplitudes.

8. GENERALIZATION TO MULTIHADRON PRODUCTION: THE NON-DIFFRACTIVE COMPONENT M

The generalization of the *non-diffractive* component to multihadron production involves any combination of direct channel resonances and t-channel non-Pomeron exchanges. Any one of the schematic diagrams of Fig. 10 represents a possible non-diffractive mechanism.

These include simple multiperipheral diagrams (Fig. 10a), multiperipheral diagrams involving the production of resonances which then decay (Fig. 10b), single Reggeon exchange leading to a final state of two high mass resonances which then gradually decay to lower and lower states (Fig. 10c), a single s-channel resonance cascading down a chain of lighter resonances (Fig. 10d), and many similar possibilities which are not shown here. Duality tells us that any one of these diagrams, when properly summed over all possible (infinite number of) resonances R_i and all possible infinite number of exchanges

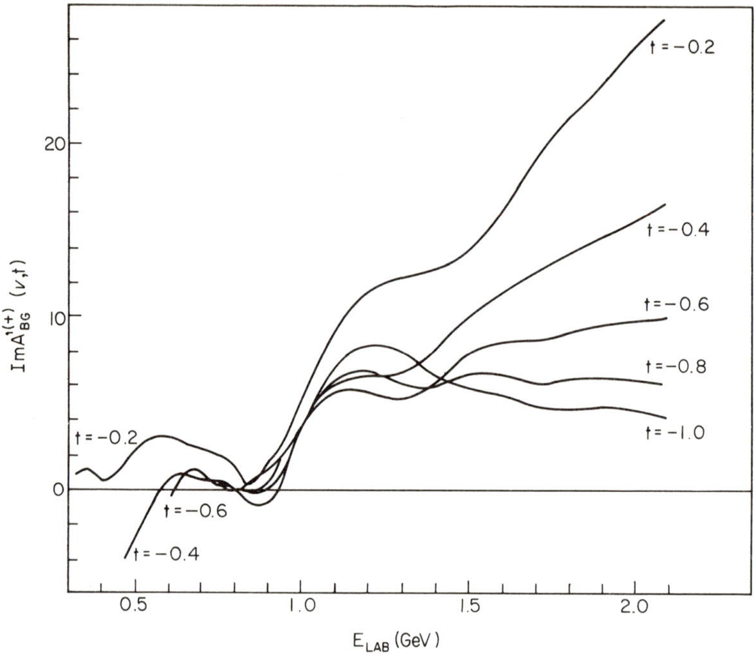

FIG. 7. The Pomeron contribution to πN elastic scattering (assuming two-component duality).

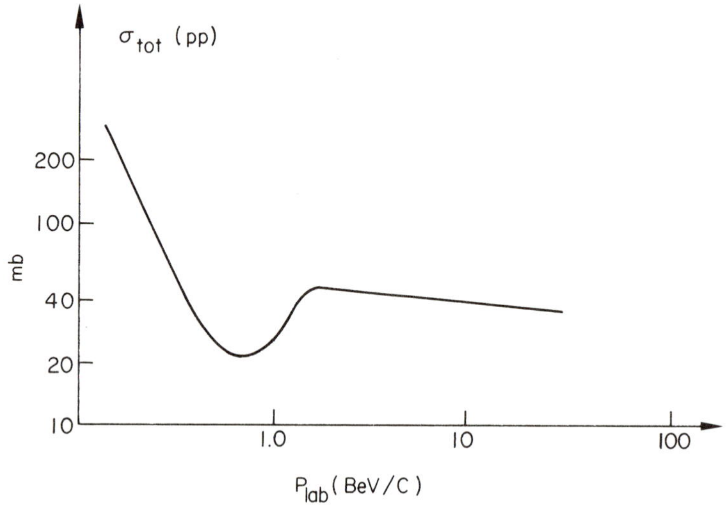

FIG. 8. Total pp cross section.

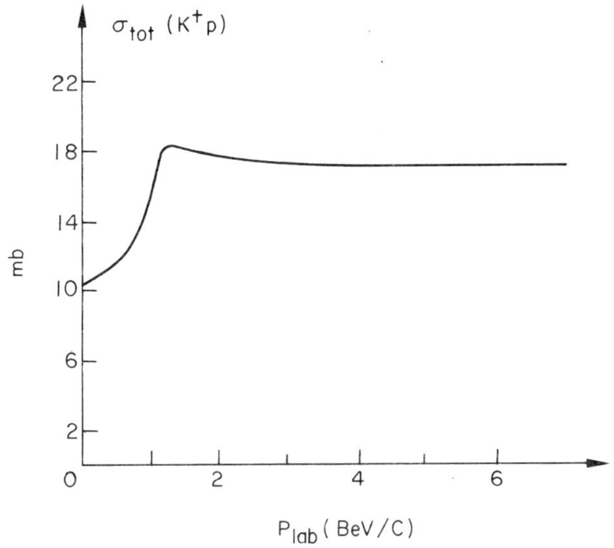

FIG. 9. Total K^+p cross section.

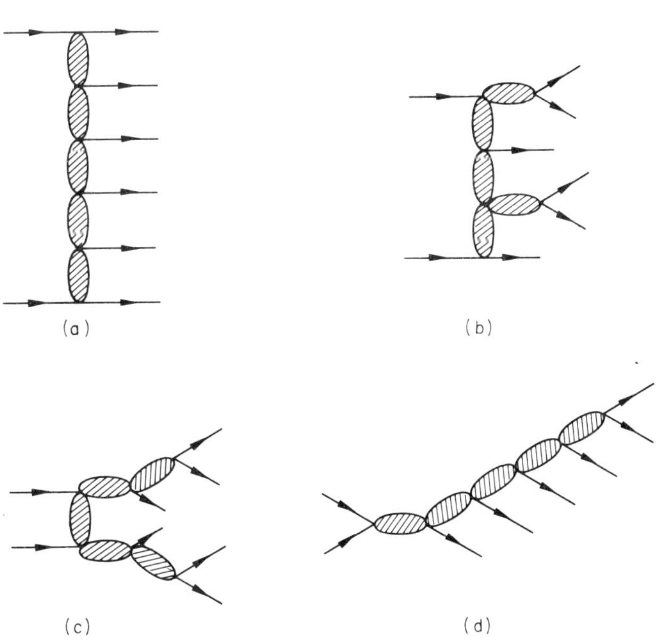

FIG. 10. The non-diffractive *M*-component.

x_i gives a complete description of the non-diffractive term of the $2 \to n$ particle amplitude. In general, however, we may also consider absorption corrections in the initial state or final state, double-Regge exchanges, etc. The crucial requirement for the non-diffractive component is that we should *not* be able to factorize the diagram into the form shown in Fig. 11.

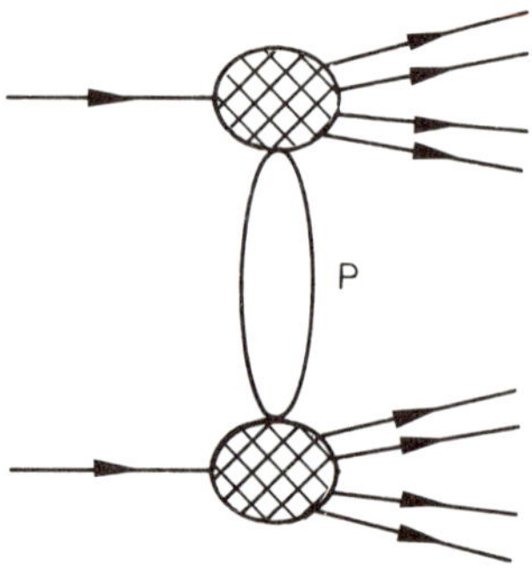

FIG. 11. The diffractive D-component.

We denote the non-diffractive component by M (for multiperipheralism) although we do not necessarily assume a multiperipheral model.

The one experimental feature which enables us to isolate the M-component most easily is the fact that its contribution to any specific cross section σ_n for an n-particle final state behaves at high energies as:

$$\sigma_n \to s^{2\alpha_R - 2} \quad \text{(modulo log } s \text{ terms)} \tag{8}$$

where α_R is the leading non-Pomeron exchange. This energy dependence is easily distinguished from the approximately constant energy dependence of diffractive contributions, and very much like in the two-body case—it is the unmistakable signature of a non-diffractive mechanism.

The duality diagrams for the M-component are shown in Fig. 12. Like in the two-body case the planar diagram (Fig. 12a) is the only one representing a set of non-exotic s-channel resonances, contributing to the imaginary part of the amplitude. However, non planar diagrams (such as Fig. 12b) contribute to the real part and, through it, to the cross-section σ_n.

9. GENERALIZATION TO MULTIHADRON PRODUCTION: THE DIFFRACTIVE COMPONENT D

Generalizing the diffractive component to the multihadron production case is schematically done by considering the diagram of Fig. 11 which represents a wide variety of physical mechanisms. Those include single diffraction excitation of a hadron into a resonance state (Fig. 13a), double diffraction

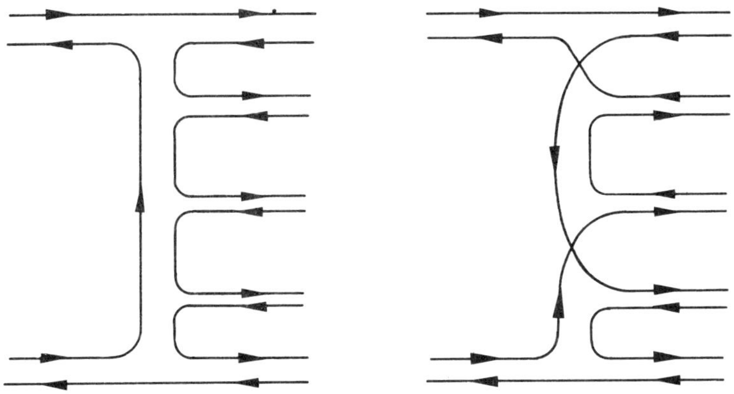

FIG. 12. Duality diagrams for the M-component.

excitation (Fig. 13b), multiperipheral diagrams involving one Pomeron somewhere along the ladder (Fig. 13c) and, possibly, mechanisms involving the exchange of more than one Pomeron (Fig. 13d). Absorptive corrections of various sorts are probably present here as much as in the non-diffractive component.

The diffractive component of multihadron production (hereafter denoted by D) leads us immediately to many difficulties, particularly in those cases where too many Pomerons are being exchanged, but also in other cases. We will return to these difficulties several times in these lectures, and emphasize them in Section VI. At this point, however, we only remark that the words "Pomeron exchange" as well as the simple symbol representing it in Figs 11 and 13 may represent an extremely complicated object which may manifest itself in somewhat different forms at different energies or different processes (especially if several j-plane singularities are involved).

The D-component has, however, one easily identifiable feature—its contribution to any cross section for any given final state is approximately constant in energy for sufficiently large s-values, (unlike the M-component which falls like $s^{2\alpha_R - 2}$). Phenomenologically—this is the clear signature of a diffractive mechanism and of Pomeron exchange.

10. THE UNITARITY RELATION BETWEEN THE TWO COMPONENTS

Having defined the M and D components, we may now formally express the amplitude for a hadronic reaction $a + b \to n$ hadrons as:

$$T_{2 \to n} = M_n + D_n \tag{9}$$

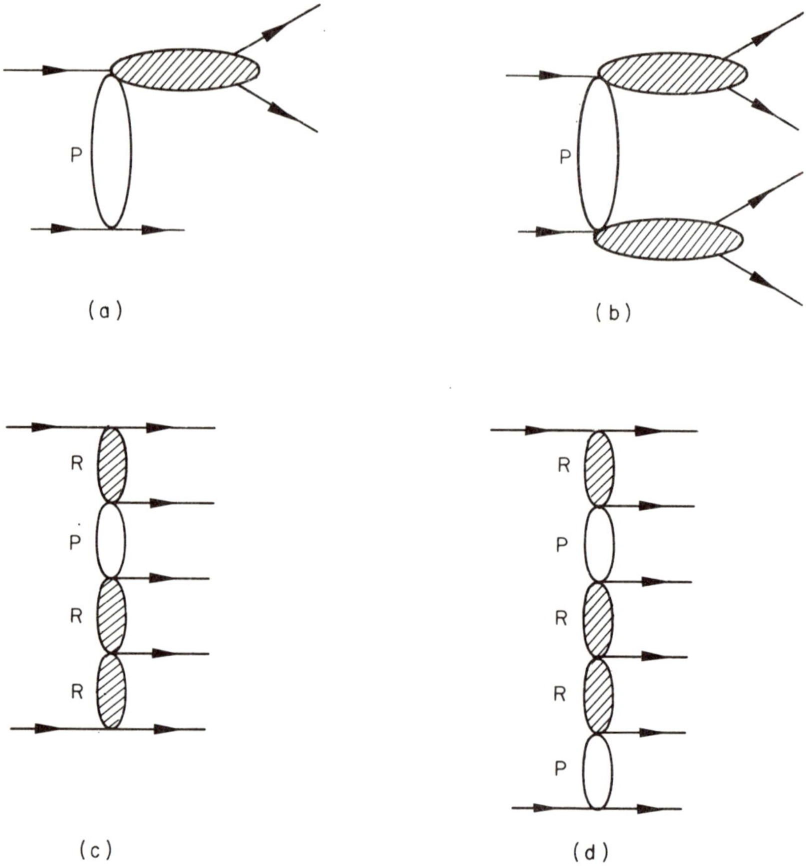

FIG. 13. Diffractive mechanisms.

The n-particle cross section, σ_n, will then take the form

$$\sigma_n = \text{“}M_n^2\text{”} + \text{“}D_n^2\text{”} + \text{“}M_n \cdot D_n\text{”} \tag{10}$$

where the first two terms schematically represent the squared contributions of the two components to the cross section, while the third term represents their interference. Using a similar notation for the elastic amplitude, we write

$$\operatorname{Im} A_{\text{el}} = R + P \tag{11}$$

$$\sigma_{\text{el}} = \text{“}R^2\text{”} + \text{“}P^2\text{”} + \text{“}R \cdot P\text{”} \tag{12}$$

The optical theorem then takes the form

$$R + P = \text{“}R^2\text{”} + \text{“}P^2\text{”} + \text{“}R \cdot P\text{”} + \sum_n \text{“}M_n^2\text{”} + \sum_n \text{“}D_n^2\text{”} + \sum_n \text{“}M_n \cdot D_n\text{”} \tag{13}$$

We roughly know the high energy dependence (modulo factors of log s) of every single term in eqn (13). On the left-hand side $R \to s^{\alpha_R - 1}$, $P \to$ const; on the r.h.s.: "R^2" and every "M_n^2" $\to s^{2\alpha_R - 2}$; "P^2" and every "D_n^2" $\to$ const; "$R \cdot P$" and every "$M_n \cdot D_n$" $\to s^{\alpha_R - 1}$. We do not know, *a priori*, which terms on the r.h.s. contribute how much to which of the two terms on the l.h.s. More importantly, while we know the $s \to \infty$ behaviour of every "M_n^2", we do not know the $s \to \infty$ behaviour of Σ_n "M_n^2". Being a sum of an infinite number of terms, it may easily have a completely different asymptotic energy dependence than any of its single terms.

We may safely assume that the quantity ("P^2" + Σ_n "D_n^2") which is a sum of constant terms (modulo log s corrections) will itself approach a constant (modulo log s) when $s \to \infty$. The reason is simple—this sum is bounded below by every one of its constant terms and is bounded above by the $\log^2 s$ behaviour allowed by the Froissart bound. We may then safely assume that ("P^2" + Σ_n "D_n^2") builds up at least part of the P-component on the l.h.s. of eqn (13). The R-component on the l.h.s. is certainly accounted for by parts of ("R^2" + Σ_n "M_n^2") and possibly by parts of the interference terms ("$P \cdot R$" + Σ_n "$D_n \cdot M_n$"). However, it is perfectly possible that the latter sums may also produce a term which is constant in energy as $s \to \infty$, thereby also contributing a constant fraction of the P-component on the l.h.s. of eqn (13). It is not difficult to construct examples of infinite series of terms, each falling as $s \to \infty$ like an inverse power of s, while the sum approaches a constant. The simplest example of this kind which will serve us again and again throughout these lectures is the Poisson series. We postulate:

$$|M_n|^2 = \frac{(c_1 \log s)^n}{n!} s^{2\alpha_R - 2} \tag{14}$$

and find:

$$\sum_n^{\infty} |M_n|^2 = s^{2\alpha_R - 2 + c_1} \tag{15}$$

If $c_1 = 2 - 2\alpha_R > 0$, $\sum_n^{\infty} |M_n|^2 \xrightarrow[s \to \infty]{}$ const. in spite of the fact that each $|M_n|^2 \to 0$.

If we now temporarily ignore, for simplicity, the $R \cdot P$ and $M \cdot D$ interference terms, we find that our unitarity equation is broken into several pieces. This is shown in Table II.

Figure 14 schematically shows the energy dependence of the various terms (again, ignoring the interference contributions).

Our analysis immediately leads to the following questions:

1. Do we indeed have an energy independent term in the sum Σ_n "M_n^2"?
2. If so, how large is it? In other words—what fraction of the constant P-component in Im A_{el} is built by this non-diffractive term?

TABLE II.

σ_{tot}	$=$	Im A_{el}
$(\text{“}P^2\text{”} + \Sigma_n \text{“}D_n^2\text{”}) \underset{s\to\infty}{\to} \text{const}$	$\Longrightarrow$	$P \underset{s\to\infty}{\to} \text{const}$
$(\text{“}R^2\text{”} + \Sigma_n \text{“}M_n^2\text{”}) \underset{s\to\infty}{\to} \left\{\begin{matrix}\text{const}\\ +\\ s^{\alpha_R - 1}\end{matrix}\right.$	$\Longrightarrow$	$R \underset{s\to\infty}{\to} s^{\alpha_R - 1}$

3. If Σ_n “M_n^{2}” $_{s\to\infty}$ const, we have at our disposal an equation expressing the P-component (or a constant fraction of it) entirely in terms of non-diffractive contributions which, in principle at least, can be determined by the properties of the known hadrons, their trajectories, etc. If that is the case, can we derive the properties of the P-component? Why is $\alpha_P = 1$? What type of singularity controls the P-component?

4. If we then “understand” the Pomeron, we should be able to insert it into every piece of the D-component and see how “P^2” + Σ_n “D^2” produces

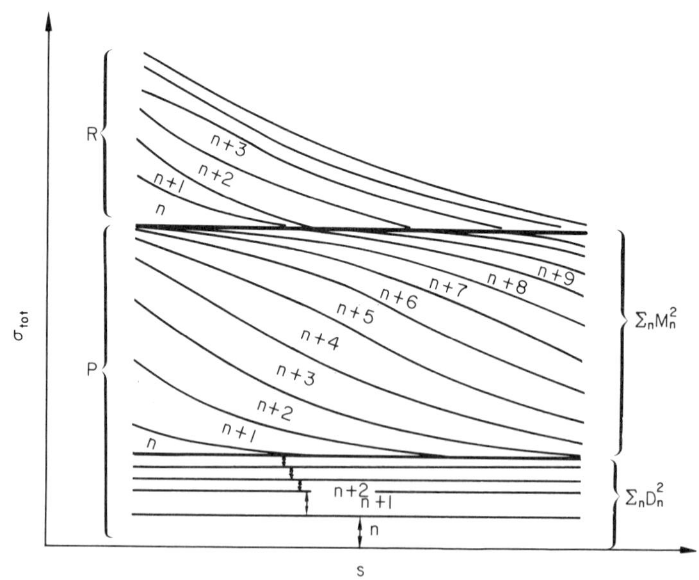

FIG. 14. Schematic description of the energy dependence of the various components of the total cross section.

another piece of the P-component. Here we face questions such as: Is the Pomeron self consistent? Is the Pomeron contribution to every D_n such that Σ_n "D_n^2" obeys the Froissart bound?

The four sets of basic questions listed above form the starting point of the rest of these series of lectures. We discuss questions 1 and 2 in Sections II, III and IV, presenting a phenomenological approach to the two-component picture. We then present our own viewpoint concerning questions 3 and 4 in Sections V and VI respectively. We caution the reader, however, that while we feel that we know the answers to questions 1 and 2, we are certain that we do not yet know the answers to the much deeper questions 3 and 4.

II. A Two Component View of Multihadron Production

1. INTRODUCTION

In this lecture we present a description of high energy multihadron reactions in terms of the diffractive and non-diffractive components introduced in our previous lecture. We discuss many of the detailed predictions of the two component model and compare them with experiment, with particular emphasis on those experimental observations which actually prove the presence of the two mechanisms. Our treatment of the details of the diffractive part, in this lecture, is very crude and superficial, since our main aim is to emphasize the interplay of the two mechanisms. Section III is entirely devoted to a careful detailed discussion of the diffractive part.

2. THE NON-DIFFRACTIVE M-COMPONENT: GENERAL PROPERTIES OF MULTIPLICITY DISTRIBUTIONS

The non-diffractive part was introduced in the previous lecture through the various mechanisms of Fig. 10. The common feature of all of these mechanisms is their asymptotic behaviour as a function of energy. We have:

$$\sigma_n \xrightarrow[s \to \infty]{} s^{2\alpha_R - 2} \quad \text{(modulo } \log s \text{ terms)} \tag{16}$$

where $\alpha_R(\sim\frac{1}{2}?)$ is the intercept of the leading non-Pomeron singularity. We shall assume that the sum of these terms approaches a constant as $s \to \infty$.

$$\sum_n^\infty \sigma_n \xrightarrow[s \to \infty]{} \text{const.} \quad \text{(modulo } \log s \text{ terms)} \tag{17}$$

In principle, the sum of non-diffractive cross sections should also produce the R-component of Im A_{el} (namely the term in σ_{tot} which behaves like $s^{\alpha_R - 1}$). However, most of our comparison with experiment will be done for the case of pp scattering, for which the R-component in σ_{tot} vanishes due to the exotic

s-channel quantum numbers. We will therefore ignore this term and assume the approximate constancy of $\sum_n^\infty \sigma_n$ for pp reactions.

Figure 15 shows the energy dependence of the various σ_n's, assuming the oversimplified Poisson expression:

$$\sigma_n = \frac{(g \log s)^n}{n!} s^{-g} \tag{18}$$

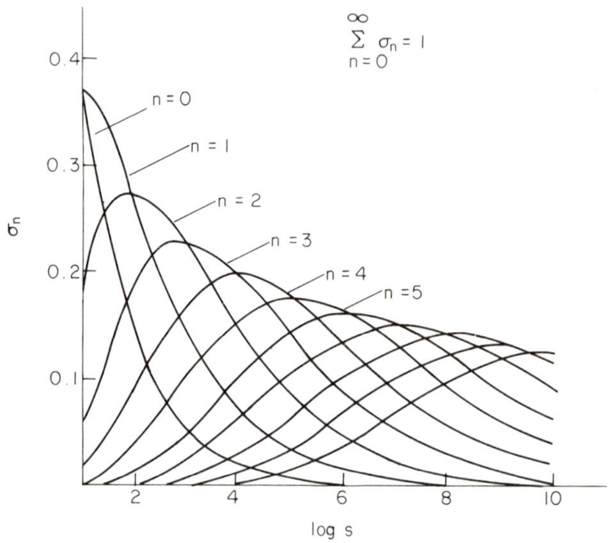

FIG. 15. Energy dependence of multiplicity cross sections σ_n assuming a Poisson distribution. $\sum_n \sigma_n$ is normalized to one, and $g = 1$.

normalized in such a way that $\sum_{n=0}^{\infty} \sigma_n = 1$. We see that for higher values of n, the maxima of σ_n occur at higher energies, but the sum is always constant. The true form of σ_n is probably more complicated. However, regardless of the explicit expression of the σ_n's, several general remarks should be made here with respect to their qualitative features (see Fig. 15).

(a) If the features of the figure are right, i.e. if the sum is really constant, we must be able to calculate the properties of the Pomeron (especially its intercept) using only information on the non-diffractive component.

(b) The subtle balance among many rising and falling cross sections, the sum of which is constant, dictates a relation between the average multiplicity $\langle n \rangle$ and α_R—the rate by which every σ_n falls to zero. It is qualitatively clear that for lower values of α_R, every σ_n will fall faster as $s \to \infty$ and more new cross sections for new multiplicities will have to "open up" at lower energies.

An increase (decrease) in σ_R is therefore always somehow accompanied by a smaller (larger) average multiplicity.

(c) The constancy of $\Sigma_n \sigma_n$ is guaranteed by a careful balance between the increasing high multiplicity cross sections and the decreasing low multiplicity cross sections at any given energy. In order to understand the way in which this subtle balance emerges, we have to study not only the asymptotic $s \to \infty$ behaviour of every σ_n but also its behaviour when it is still rising towards its maximum. In fact, for most simple models, at every energy (no matter how high!), approximately half of the sum $\sum_{n=0}^{\infty} \sigma_n$ is given by those cross sections which are still rising. The asymptotic behaviour of σ_{tot} (or the intercept of the Pomeron) are equally influenced by the asymptotic behaviour and by the "threshold" behaviour of the σ_n's.

The simplest mathematical example for the energy dependence of the non-diffractive σ_n's is given by the Poisson distribution mentioned above (eqns (14), (18)). Let us see how this equation comes about. If particles are emitted independently of each other we would expect, at each energy, a multiplicity distribution of the form:

$$\sigma_n = \frac{\lambda^n}{n!} \cdot f(\lambda, s) \tag{19}$$

where λ is the average multiplicity and $f(\lambda, s)$ is independent of n and determines the overall normalization. If the average multiplicity increases logarithmically (as it does in multiperipheral models and, presumably, in nature) we have:

$$\langle n \rangle = \lambda = g \log s \tag{20}$$

where g represents the squared coupling constant in a multiperipheral model. We then get:

$$\sigma_n = \frac{(g \log s)^n}{n!} f(g, s). \tag{21}$$

The function $f(g, s)$ can now be computed from:

$$\sigma_{\text{tot}} = \sum_n^{\infty} \sigma_n = f(g, s) \sum_n^{\infty} \frac{(g \log s)^n}{n!} = f(g, s) \cdot s^g. \tag{22}$$

Hence:

$$f(g, s) = \sigma_{\text{tot}} \cdot s^{-g} \tag{23}$$

and finally:

$$\sigma_n(s) = \frac{(g \log s)^n}{n!} s^{-g} \cdot \sigma_{\text{tot}}. \tag{24}$$

The average multiplicity is, by definition,

$$\langle n \rangle = \frac{\Sigma n\sigma_n}{\Sigma \sigma_n} = g \log s \tag{25}$$

and the rate of fall of σ_n as $s \to \infty$ is s^{-g}, yielding:

$$-g = 2\alpha_R - 2 \tag{26}$$

We see that the constant g appears twice in eqn (24): Once as the coefficient of $\log s$ in the expression for the average multiplicity (a coefficient which we shall denote by c_1):

$$c_1 \equiv \lim_{s\to\infty} \frac{\langle n \rangle}{\log s} = g \tag{27}$$

and once as the power of energy by which every σ_n falls (eqn (26)). Both of these numbers (c_1 and α_R) are measurable and we can compare their experimental values with the relation[8]

$$c_1 = 2 - 2\alpha_R \tag{28}$$

obtained here. Note that c_1 is the coefficient of $\log s$ for the average multiplicity *of the non-diffractive component*, while α_R determines the energy dependence *of the non-diffractive component* of every σ_n. We must separate the physical cross-section into its two components before we can determine c_1 and α_R. We will return to this point towards the end of this lecture.

3. THE M-COMPONENT: INCLUSIVE CROSS-SECTIONS AND CORRELATIONS

Consider the inclusive pion cross-section

$$\frac{d^3\sigma}{dy\, dp_T^2}(a + b \to \pi + \text{anything}),$$

where y is the rapidity and p_T the transverse momentum. We can define a single particle "density":

$$\rho(y) = \frac{1}{\sigma_{tot}} \int \frac{d^3\sigma}{dy\, dp_T^2}\, dp_T^2 = \frac{1}{\sigma_{tot}} \frac{d\sigma}{dy} \tag{29}$$

The density $\rho(y)$ obeys the sum rule:

$$\langle n \rangle = \int \rho(y)\, dy \tag{30}$$

where $\langle n \rangle$ is the pion multiplicity. Actually we should define densities $\rho_c(y)$ for each inclusive process $a + b \to c +$ anything. The sum of all integrated $\rho_c(y)$ would then give the correct total multiplicity.

In most non-diffractive models the y-dependence of $\rho(y)$ is given by the famous "plateau"[(9)] (Fig. 16). The height of the plateau is then given by $\rho(0) = c_1$.

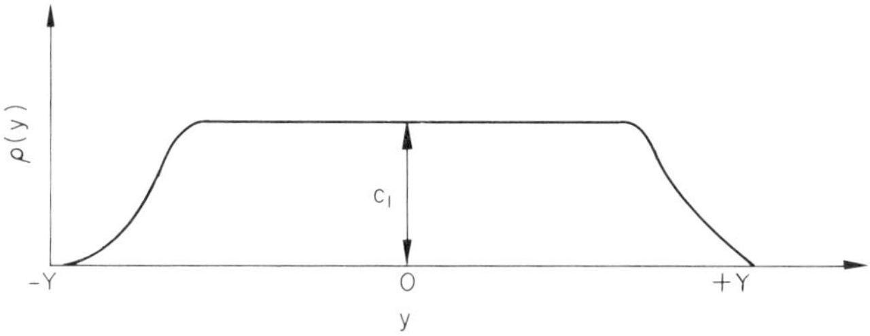

FIG. 16. The hadronic "plateau" in the inclusive cross-section.

The two-particle density $\rho_{cd}(y_c y_d)$ can be similarly defined, starting from the double inclusive cross-section:

$$\rho_{cd}(y_c, y_d) = \frac{1}{\sigma_{\text{tot}}} \frac{d^2\sigma}{dy_c\, dy_d} (a + b \to c + d + \text{anything}) \tag{31}$$

The two-particle correlation function is then given by:

$$C_{cd}(y_c, y_d) = \rho_{cd}(y_c, y_d) - \rho_c(y_c)\, \rho_d(y_d) \tag{32}$$

The integrated correlation for $c = d =$ pion obeys the relation:

$$\int C_{cd}(y_c, y_d)\, dy_c\, dy_d = \langle n(n-1)\rangle - \langle n\rangle^2 \equiv f_2 \tag{33}$$

This is the definition of the "multiplicity moment" f_2. If $C_{cd}(y_c, y_d)$ is bounded by a constant everywhere in the $y_c - y_d$ plane, $f_2 \underset{s\to\infty}{\leqslant} \log^2 s$. This is the case in most reasonable models. However, if we assume that all two-particle correlations are occurring between particles with "neighbouring" rapidities, we find that the function $C_{cd}(y_c, y_d)$ is non-vanishing only when $y_c \sim y_d$. This "short range correlation" assumption is illustrated in Fig. 17 where $C_{cd} \neq 0$ only within a "correlation length" L of the $y_c = y_d$ line. If, in this case, $C_{cd}(y_c, y_d)$ is bounded by a constant, we find:

$$\int C_{cd}(y_c, y_d)\, dy_c\, dy_d \propto LY \tag{34}$$

Hence:

$$f_2 \xrightarrow[s\to\infty]{} c_2 \log s \tag{35}$$

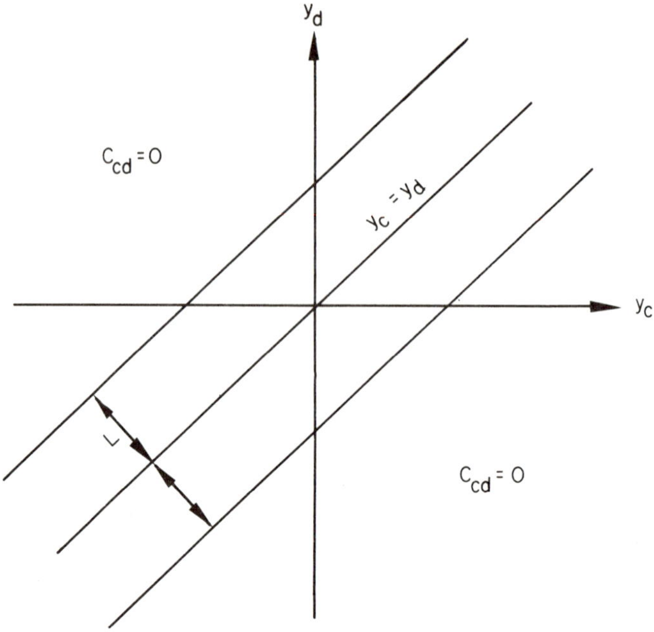

FIG. 17. The integrated correlation f_2 receives contributions from the region near the line $y_c = y_d$ in short-range-correlation models.

This is a common property of all short-range-correlation models. A similar argument applies to any m-particle correlation function $C_{12\ldots m}(y_1, y_2, \ldots, y_m)$. Its integral f_m obeys:

$$f_m \xrightarrow[s\to\infty]{} c_m \log s \text{ (for SRC models).} \tag{36}$$

A trivial exercise shows that for the Poisson distribution (eqn (24)):

$$f_2 = \frac{\Sigma\, n(n-1)\,\sigma_n}{\Sigma\,\sigma_n} - \left(\frac{\Sigma\, n\sigma_n}{\Sigma\,\sigma_n}\right)^2 = 0 \tag{37}$$

This is, of course, expected. The Poisson distribution represents uncorrelated events, and it is self-evident that integrated correlation functions vanish ($f_m = 0$ for any m). The f_m parameters can therefore be viewed as measures of the deviations of a multiplicity distribution from the Poisson expression.

It is important to notice that while most non-diffractive models have short range correlations, it is quite possible, in principle, to have a model with σ_n's falling like powers of s, but with long range correlations. The words "non-diffractive component" and "short range correlation component" are not synonymous, although for most applications, including those of the present lecture, we will not distinguish between them.

4. MULTIPLICITY DISTRIBUTIONS: THE GENERATING FUNCTION

A complete knowledge of all σ_n-values at a given energy enables us to determine the average multiplicity $\langle n \rangle$ (which we will sometimes denote as f_1) as well as all the multiplicity moments f_m defined as the integrated m-particle correlation functions. The connection between the σ_n's and f_m's is most easily given by the generating function:[10]

$$Q(z) = \sum_{n}^{\infty} z^n \sigma_n \tag{38}$$

A complete knowledge of all σ_n's, obviously completely determines the function $Q(z)$. The multiplicity moments f_m are expressed by $Q(z)$:

$$\langle n \rangle = f_1 = \left. \frac{\mathrm{d} \log Q(z)}{\mathrm{d}z} \right|_{z=1} \tag{39}$$

$$f_m = \left. \frac{\mathrm{d}^m \log Q(z)}{\mathrm{d}z^m} \right|_{z=1} \tag{40}$$

If we happen to know the f_m's we can easily recover the function $Q(z)$ through the expansion:

$$\log Q(z) = \sum_{m=1}^{\infty} \frac{f_m}{m!} (z - 1)^m \tag{41}$$

and the σ_n's are then readily obtained:

$$\sigma_n = \left. \frac{1}{n!} \frac{\mathrm{d}^n Q(z)}{\mathrm{d}z^n} \right|_{z=0} \tag{42}$$

The generating function $Q(z)$ provides us with a simple way of deducing the f_m's from the σ_n's and vice versa. It is analogous to the grand partition function in statistical mechanics and we will discuss it in detail in Section IV. Here it suffices to say that any model in which the f_m's are assumed to be known, will yield a prediction for the σ_n's through equations (41), (42).

For a Poisson distribution (normalized to $\sigma_{tot} = 1$):

$$\sigma_n = \frac{(g \log s)^n}{n!} s^{-g} \tag{43}$$

We have

$$Q(z) = \sum_n \frac{(zg \log s)^n}{n!} s^{-g} = s^{g(z-1)} \tag{44}$$

and

$$\log Q(z) = g(z - 1) \log s \tag{45}$$

The proportionality of $\log Q(z)$ and $\log s$, however, is a much more general result, true in all short range correlation models. In such models $f_m \xrightarrow[s\to\infty]{} c_m \log s$. Hence, using eqn (41)

$$\log Q(z) \xrightarrow[s\to\infty]{} \left\{ \sum_{m=1}^{\infty} \frac{c_m}{m!}(z-1)^m \right\} \log s. \tag{46}$$

5. THE DIFFRACTIVE D-COMPONENT: AN EXTREMELY CRUDE DESCRIPTION

The diffractive component of a multihadron production amplitude was defined in section I in terms of Fig. 11. In order to discuss it in detail we have to study the "blobs" by which the Pomeron excites the initial particles into multihadron final states. We will devote Section III entirely to this topic. At this point we only make the extremely crude assumption (to be modified later) that for the diffractive component:

$$\sigma_n \xrightarrow[s\to\infty]{} \text{const (for every } n) \tag{47}$$

$$\sum_n \sigma_n \xrightarrow[s\to\infty]{} \text{const} \tag{48}$$

$$\langle n \rangle \xrightarrow[s\to\infty]{} \text{const} \tag{49}$$

$$f_m \xrightarrow[s\to\infty]{} \text{const } (m = 2, 3, \ldots) \tag{50}$$

Many of these quantities probably involve $\log s$ terms, but for the purpose of our two component analysis in this Section we will ignore this logarithmic behaviour, assuming that over the energy range for which multiplicity data presently exist ($s < 600$ BeV^2) these $\log s$ terms are not crucial.

We emphasize, however, that we believe that the particles produced in any single vertex of the Pomeron exchange diagram (Fig. 11) *do not* show an isotropic distribution in their overall c.m. frame, *do not* have a n^{-2} multiplicity distribution and *do not* have f_m values which increase like a power of s. All of these properties (which we rule out) are necessary concoctions of models[11] in which the *entire* total cross section is accounted for by the diffractive component.

Such models are ruled out by the data and we do not discuss them here. Our diffractive component relates mostly to low multiplicity events, and whenever sufficiently large masses are produced in a given vertex of Fig. 11, the distribution of the final particles is probably not isotropic at all.

6. THE M/D RATIO: WHICH COMPONENT IS LARGER?

Having introduced two components, each contributing a constant total cross-section, we would like to find out which component is actually dominant. A quantitative answer to this question will be given later in this lecture, when we consider the available experimental multiplicity distributions, inclusive cross-sections etc. At this point we would like, however, to present a qualitative hint related to the M/D ratio.

The observed experimental distribution of final pions at very high energies (say $200 < s < 3000\ \text{BeV}^2$) is more or less uniform as a function of the rapidity y, while the average transverse momentum is $\langle p_T \rangle \sim 350$ MeV. The squared subenergy s_{12} of two particles with rapidities y_1, y_2, transverse momenta p_{T1}, p_{T2}, and masses m_1, m_2 is given by:

$$s_{12} = 2m_{T1}m_{T2}\cosh(y_1 - y_2) - 2\vec{p}_{T1}\cdot\vec{p}_{T2} + m_1^2 + m_2^2 \tag{51}$$

where

$$m_{Ti} = \sqrt{(p_{Ti}^2 + m_i^2)} \tag{52}$$

For two "neighbouring" pions in the rapidity plot, the average rapidity difference at high energies is:

$$\langle y_1 - y_2 \rangle = \frac{\log s}{\langle n \rangle} \tag{53}$$

A reasonable guess for the value of this ratio can be obtained by fitting the charged multiplicity in pp collision to a $\log s$ form and assuming that the total number of pions is roughly 3/2 the number of charged pions. This gives

$$\langle y_1 - y_2 \rangle \sim 0{\cdot}4 \tag{54}$$

(corresponding to $\langle n \rangle \xrightarrow[s\to\infty]{} 2{\cdot}5 \log s$).

If we now ignore azimuthal correlation (i.e. assume $\langle \vec{p}_{T1}\cdot\vec{p}_{T2}\rangle = 0$) and use $\langle p_{T1}^2 \rangle = \langle p_{T2}^2 \rangle = 0{\cdot}35^2 = 0{\cdot}12\ \text{BeV}^2$; $m_1 = m_2 = m_\pi$, we find that the average subenergy of a pair of neighbouring pions is:

$$\langle s_{12} \rangle \sim 0{\cdot}35\ \text{BeV}^2 \tag{55}$$

The average invariant mass of such a pair of pions is approximately 600 MeV, independent of the overall energy of the process (since $\langle y_1 - y_2 \rangle$ and $\langle p_T \rangle$ are presumably s-independent). In Section I we have seen that at such low energies the R/P ratio is larger than one. In fact—we have seen that the P-component may be altogether absent for sufficiently low energies. We therefore conclude that the "average" pair of neighbouring pions will always "prefer" to exchange a "non-Pomeron", and only those rare neighbouring pairs of pions with subenergies which are much larger than average may have "reasons" to "prefer" a Pomeron exchange.

While admitting that this argument is crude and highly nonrigorous, we nevertheless conclude that the basic physical process which takes place at any given segment of the rapidity axis (or of the multiperipheral chain) resembles the physics of $s < 1$ BeV^2, and significantly prefers the R-component over the P-component. Consequently, we dare to "guess" that the non-diffractive M-component accounts for a significant portion of the total cross section.

We shall see that the quantitative analysis verifies this suspicion.

The small invariant mass scale for two neighbouring particles is actually responsible for a large number of interesting phenomena:

(a) The π/K ratio is large and it remains large at high energies. The relevant energy scale for this ratio is that of the small subenergies, and, relative to them, the π–K mass difference is significant. If $\langle s_{12} \rangle$ would be of the order of 10 or 100 BeV^2, the π/K ratio would probably be much closer to one.

(b) It is actually ridiculous to discuss simple t-channel exchange models for π–π interactions at $s \sim 0{\cdot}35$ BeV^2. In that sense the multiperipheral model is totally inadequate and it would be much more reasonable to introduce π–π resonances, 3π-resonances, etc. Duality, however, tells us that, as far as the average properties and gross features of the amplitude are concerned, a t-channel exchange may represent the amplitude correctly even at low energies. To the extent that we are using the multiperipheral model, we should use it only in that sense.

(c) In the same way that Reggeon exchange is as important as (or more important than) Pomeron exchange at such low s-values, baryon exchange is as important as meson exchange. In fact, the usual hierarchy of exchanges in two body reactions at a few BeV's may be completely reversed for $s \ll 1$ BeV^2. Any specific multi-peripheral model should take this into account.

7. A GUIDED TOUR IN THE n–log s PLANE

At this point it might be helpful to study the general behaviour of the two components of the multiplicity cross sections σ_n, in the $n - \log s$ plane (Fig. 18). The non-diffractive part falls like $s^{2\alpha_R - 2}$ for fixed n and $s \to \infty$; it falls roughly like $\lambda^n/n!$ for fixed s and $n \to \infty$ (ignoring the kinematic limit which actually prevents n from growing indefinitely for fixed s; that limit is very distant and the multiplicity distribution dies out long before it is ever reached). For any fixed-s line or fixed-n line, the non-diffractive part has a maximum around the line $n = c_1 \log s$. In a correlation-free model, the height of this maximum would decrease like $n^{-\frac{1}{2}}$ as $n \to \infty$ (along that line), while the width of the peak increases like $n^{\frac{1}{2}}$.

The diffractive component tends to constant as $s \to \infty$ for fixed n, and probably falls at least as fast as the non-diffractive part as $n \to \infty$ for fixed s.

Consequently cross-sections in the region $n \ll c_1 \log s$ are dominated by the D-component while for $n \gg c_1 \log s$ the M-component presumably dominates. The $n \sim c_1 \log s$ region is responsible for most of the cross-section, and our analysis of the data will show that, at least at the presently

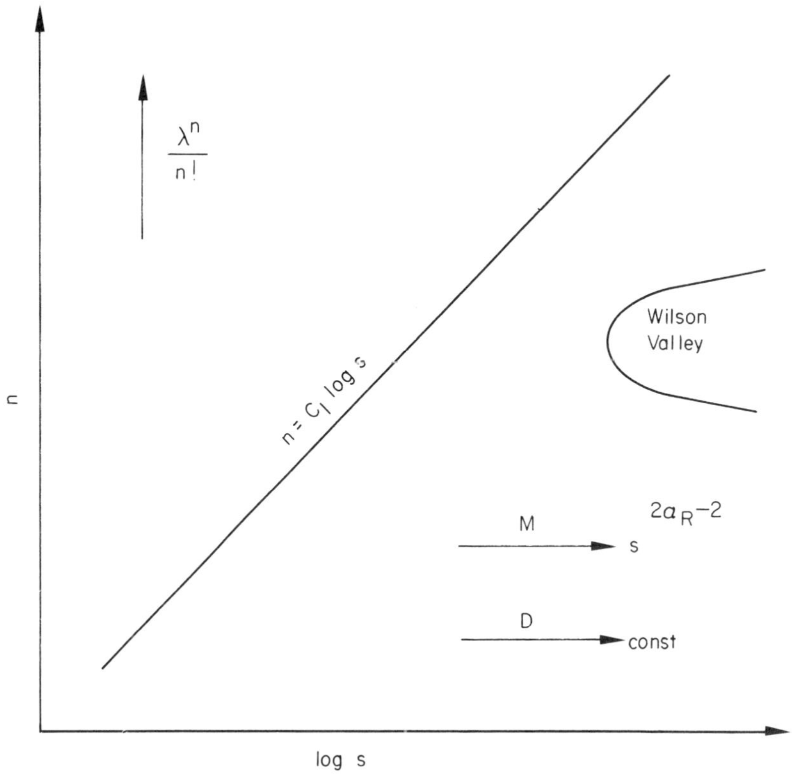

FIG. 18. The n–log s plane.

available energy range, it is dominated by the M-component. Between the $n = c_1 \log s$ "ridge" of the M-component and the $n \ll c_1 \log s$ "diffraction land" a valley may develop. Such a valley was predicted by Wilson.[(12)] If it exists, it would appear as a dip in the multiplicity distribution for fixed energy, for sufficiently large energies.

8. INCLUSIVE PROTON CROSS SECTIONS NEAR $x = 1$ AND THE TWO COMPONENTS

The inclusive cross section for $p + p \rightarrow p +$ anything near $x = 1$ ($x = 2p_l/\sqrt{s}$, $p_l =$ the longitudinal momentum of the detected proton) may help us to identify the contributions of the two components of multihadron production. In a diffractive mechanism such a proton is emitted through the process shown in Fig. 19a. On the other hand, an $x \sim 1$ proton produced non-diffractively would be due to a mechanism in which a non-Pomeron exchange transforms one of the initial protons into the detected proton (Fig. 19b). The x and s dependence of the inclusive proton spectrum near $x \sim 1$, for $s \to \infty$, is given in both cases by the triple-Regge formula:

$$\frac{d\sigma}{dx} \propto s^{\alpha_2 - 1} (1 - x)^{\alpha_2 - 2\alpha_1} \tag{56}$$

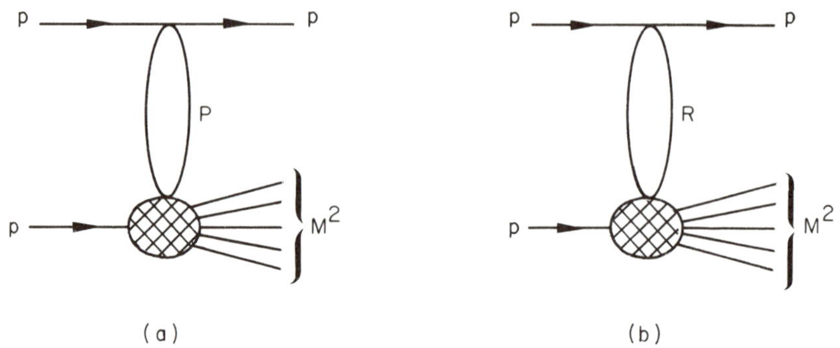

FIG. 19. Diffractive and non-diffractive production of protons near $x = 1$.

Here α_1 is the object exchanged in figure 19 ($\alpha_R \sim \frac{1}{2} + t$ or $\alpha_P \sim 1 + \alpha'_P t$ where α'_P is small) and α_2 is the power behaviour describing the total cross-section for the "scattering" of the α_1-pole on the lower proton of Fig. 19, as a function of $M^2 = s(1 - x)$:

$$\sigma_{\text{tot}}^{\alpha_1 + p} \xrightarrow[M^2 \to \infty]{} (M^2)^{\alpha_2 - 1} \tag{57}$$

For $\alpha_2 = 1$ we obtain scaling:[9]

$$\frac{d\sigma}{dx} \propto (1 - x)^{1 - 2\alpha_1}, \text{ independent of } s \tag{58}$$

The observed scaling cross-section would then behave roughly like $(1 - x)^{-1}$ near $x = 1$ for a diffractive mechanism, and like $(1 - x)^{1 - 2\alpha_R(t)}$ for the non-diffractive component, where $1 - 2\alpha_R(t) \geqslant 0$, for $t \leqslant 0$. Hence, a *peak near $x \sim 1$ reflects the D-component while a dip near $x \sim 1$ or a flat* $d\sigma/dx$ (*for $t \sim 0$*) *reflects the M-component.*

9. WHY DO WE BELIEVE THAT BOTH COMPONENTS ARE PRESENT?

The most convincing argument for the existence[12] of the two components in multihadron production is the simple fact that both components are known to exist in two body reactions, both can be generalized in a simple and straightforward way to the multiparticle case, and there is absolutely no reason why both would not be present in the data.

However, in addition to this unbeatable argument we have several direct indications from the data:

(a) The strongest experimental evidence for the existence and importance of the M-component at present energies comes from the observed energy dependence of the multiplicity cross-sections for pp interactions. These cross-sections[13] are not energy independent, and for $n_{ch} \leqslant 8$ they are already decreasing for the highest measured s-values (Fig. 22). The general pattern of the n and s dependence of these cross-sections is actually similar to the qualitative description of Fig. 15 and it shows the unmistakable characteristics of the M-component.

(b) The strongest experimental evidence for the existence of the D-component are the inclusive proton data[14] near $x = 1$ which show a clear peak near $x = 1$ for $d\sigma/dx$ ($pp \to p +$ anything). Such a peak (Fig. 20) is a clear signature of a diffractive mechanism.

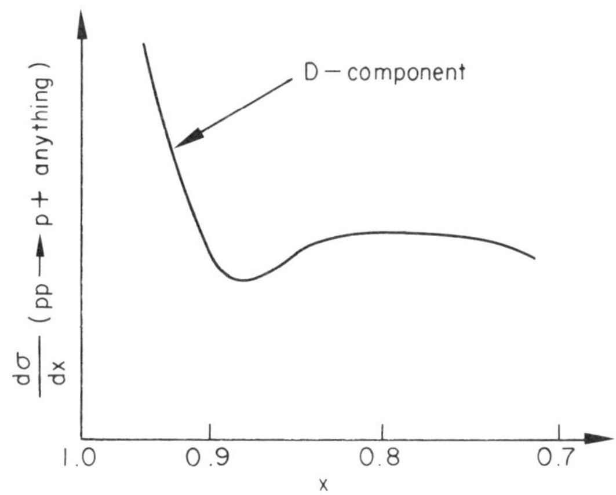

FIG. 20. Schematic description of the inclusive proton cross-section near $x = 1$.

There are many additional direct and indirect pieces of experimental evidence for the existence of the two components, but the points mentioned above are sufficiently convincing. We will return to some of the other points when we discuss our two component fit to the multiplicity distribution.

10. A SIMPLE FRAMEWORK FOR A TWO-COMPONENT FIT TO MULTIPLICITY DISTRIBUTIONS

We assume that every multiplicity cross-section has two components:[12, 15]

$$\sigma_n = m_n + d_n \tag{59}$$

We neglect the M–D interference term. The usual excuse for doing so is the statement that the M-component contributes mostly to large multiplicities and to events which populate the entire rapidity axis, while the D-component contributes mainly to low multiplicity events which are concentrated in the fragmentation regions. Qualitatively this is probably correct, but to the best of our knowledge no careful analysis of the possible effects of such interference terms has been carried out so far. We believe that this point should be studied more carefully in the near future.

We now define:

$$\sum_n \sigma_n = \sigma_{\text{inel}} \tag{60}$$

$$\sum_n m_n = M\sigma_{\text{inel}}; \sum_n d_n = D\sigma_{\text{inel}} \tag{61}$$

Hence:

$$M + D = 1 \tag{62}$$

M (or D) is perhaps the most interesting single parameter in the two component description, since it measures the relative strength of the two components.

The multiplicities of the two components are defined as:

$$\langle n \rangle_m = \frac{\Sigma n m_n}{\Sigma m_n}; \qquad \langle n \rangle_d = \frac{\Sigma n d_n}{\Sigma d_n} \tag{63}$$

Since:

$$\langle n \rangle = \frac{\Sigma n \sigma_n}{\Sigma \sigma_n} \tag{64}$$

we get:

$$\langle n \rangle = M\langle n \rangle_m + D\langle n \rangle_d \tag{65}$$

The full multiplicity is simply given as a weighted average of the two partial multiplicities. The multiplicity moments are defined as:

$$f_{2m} = \langle n(n-1) \rangle_m - \langle n \rangle_m^2; \qquad f_{2d} = \langle n(n-1) \rangle_d - \langle n \rangle_d^2 \tag{66}$$

Hence:

$$f_2 = Mf_{2m} + Df_{2d} + MD(\langle n \rangle_m - \langle n \rangle_d)^2. \tag{67}$$

Note that f_2 is *not* given by a weighted average of the partial f_2's. In fact, if $f_{2m} \underset{s\to\infty}{\to} \log s, f_{2d} \underset{s\to\infty}{\to}$ const, the leading energy dependence of f_2 will be neither $\log s$ nor a constant. Assuming $\langle n \rangle_m \to \log s$, we will get $f_2 \underset{s\to\infty}{\to} \log^2 s$ as a result of the nonlinear term in eqn (67).

11. A FIT OF THE pp MULTIPLICITY DISTRIBUTIONS[16]

We are now ready to fit the existing data. We attempt to fit all available multiplicity cross-sections in the range $s \geqslant 100$ BeV^2. There are five available pp experiments[13] at $p_{\text{lab}} = 50$, 69, 102, 205 and 303 BeV/c. We will denote by σ_n the cross-section for producing n negative particles (or $2n + 2$ prongs). Thus $n = 0, 1, 2, 3, \ldots$. We then make the following assumptions:

(i) The D-component contributes mostly to low multiplicities. We assume that at these energies it contributes only to $n = 0, 1, 2$ (i.e. 2, 4, 6 prongs). We therefore choose d_0, d_1 and d_2 as free parameters and assume:

$$d_m = 0 \text{ (for } m \geqslant 3) \tag{68}$$

(ii) We assume that for the non-diffractive cross-sections m_n, the average multiplicity and the first moment grow as $\log s$ (a necessary requirement for any short range correlation model). We assume:

$$\langle n \rangle_m^- = c_1 \log s/s_1 \tag{69}$$

$$f_{2m}^- = c_2 \log s/s_2 \tag{70}$$

where c_1, c_2, s_1, s_2 are free parameters. We further assume that, for the M-component,

$$f_k^- = 0 \qquad \text{for} \quad k \geqslant 3 \tag{71}$$

This means that we allow the negative multiplicity distribution to deviate from a Poisson distribution by a correlation parameter f_2 but not by more complicated correlations. Having selected an explicit form for all the f_k^- parameters for the M-component we can now compute $Q_M(z)$, using eqn (41). We can then derive the explicit form of all $m_n(s)$, expressed in terms of the parameters c_1, c_2, s_1, s_2.

At this point we have expressed all d_n's and m_n's (hence all σ_n's) in terms of seven parameters—three (d_0, d_1, d_2) for the D-component and four (c_1, c_2, s_1, s_2) for the M-component. Using these parameters we can then fit all the available data for $\sigma_n(s)$.

Our best fit[16] is shown in Figs 21–23. It is obviously a very good fit, but given the number of parameters, the achievement is not so great. What is more important, is the set of parameters chosen by the fit and the many interesting implications and predictions which follow from their values. The parameters given by the fit are:[16]

$d_0 = 2{\cdot}0$ mb $(\pm 0{\cdot}3)$	$c_1 = 1{\cdot}0$ $(\pm 0{\cdot}05)$
$d_1 = 2{\cdot}2$ mb $(\pm 0{\cdot}3)$	$c_2 = 0{\cdot}35$ $(\pm 0{\cdot}15)$
$d_2 = 0{\cdot}9$ mb $(\pm 0{\cdot}2)$	$s_1 = 15$ BeV^2 (± 3)
	$s_2 = 200$ BeV^2 (± 50)

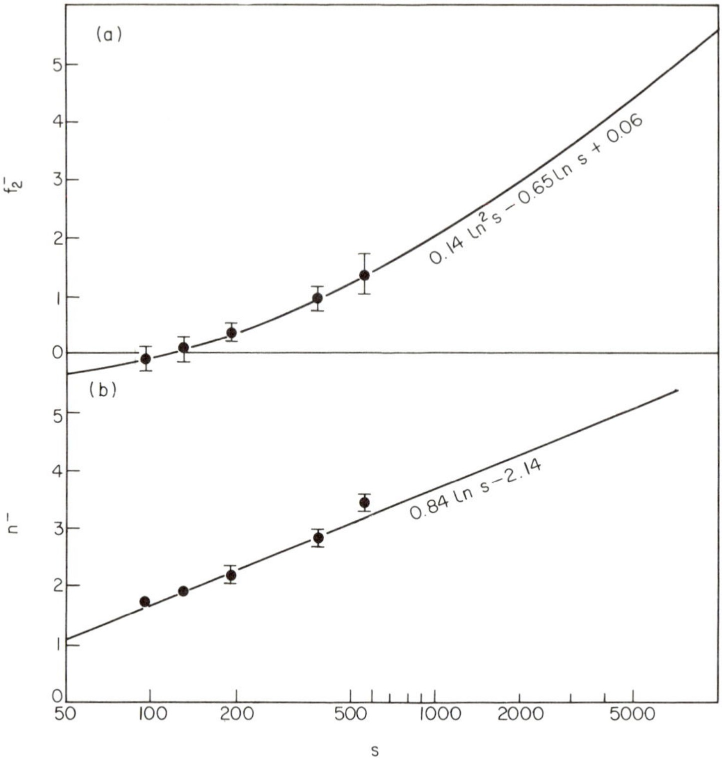

FIG. 21. A two component fit to the average multiplicity and the f_2 parameter.

The error values represent the range within which the parameters can be varied without destroying the successful fit. These parameters lead to the following conclusions:

(i) $M = 0{\cdot}84$, $D = 0{\cdot}16$. We learn that the non-diffractive component is by far the dominant one at present energies. Other fits[17] gave M-values ranging between 0·7 and 0·85. We find that the total contribution of the M-component to the inelastic pp cross-section is approximately 27 mb while the D-component contributes about 5 mb for the energy range $100 < s < 600$ BeV^2.

(ii) $\langle n \rangle^- = 0{\cdot}84 \log s - 2{\cdot}14$. Note that c_1 is *not* the coefficient of $\log s$ in the expression of the full negative multiplicity. We have:

$$\langle n \rangle_m = c_1 \log s/s_1 = 1{\cdot}0 \log s/15 \tag{71}$$

$$\langle n \rangle = M\langle n \rangle_m + D\langle n \rangle_d = Mc_1 \log s + \text{const.} \tag{72}$$

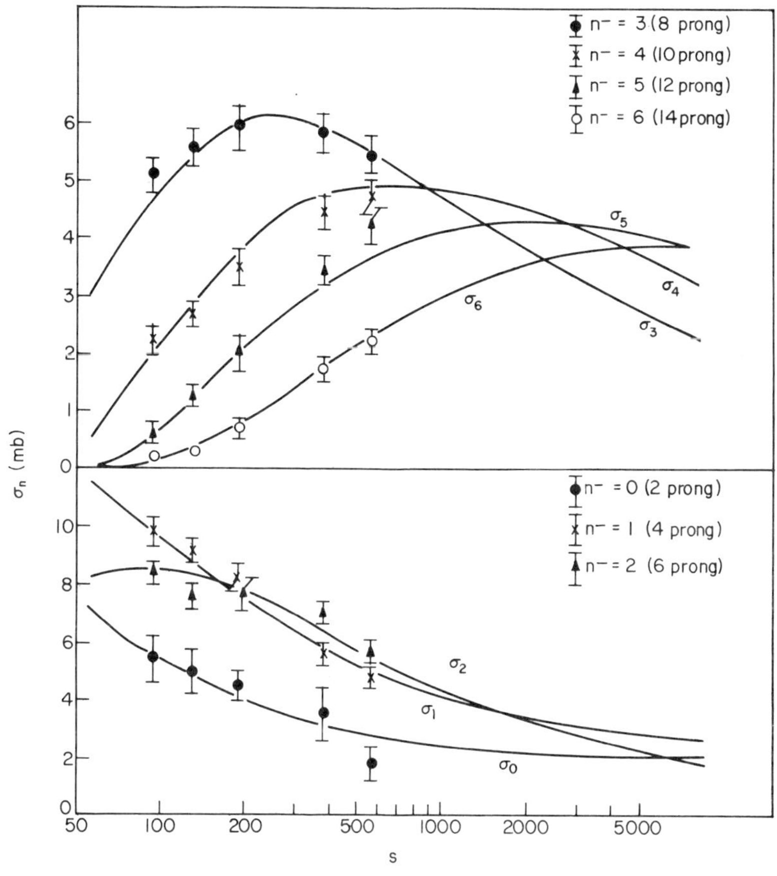

FIG. 22. A two component fit to the energy dependence of the multiplicity cross sections.

(iii) $f_2^- = 0{\cdot}14 \log^2 s - 0{\cdot}65 \log s + 0{\cdot}06$. The coefficient of the $\log^2 s$ term is MDc_1^2. The asymptotic ratio $f_2/\langle n \rangle^2$ is determined only by M (or D):

$$\frac{f_2}{\langle n \rangle^2} \underset{s \to \infty}{\rightarrow} \frac{D}{M} \tag{73}$$

Wroblewsky suggested an empirical low, relating the dispersion parameter

$$\mathscr{D} = \sqrt{(\langle n^2 \rangle - \langle n \rangle^2)} = \sqrt{(f_2 + \langle n \rangle)} \tag{74}$$

to $\langle n \rangle$. He found:

$$\frac{\langle n \rangle}{\mathscr{D}} = 2 \tag{75}$$

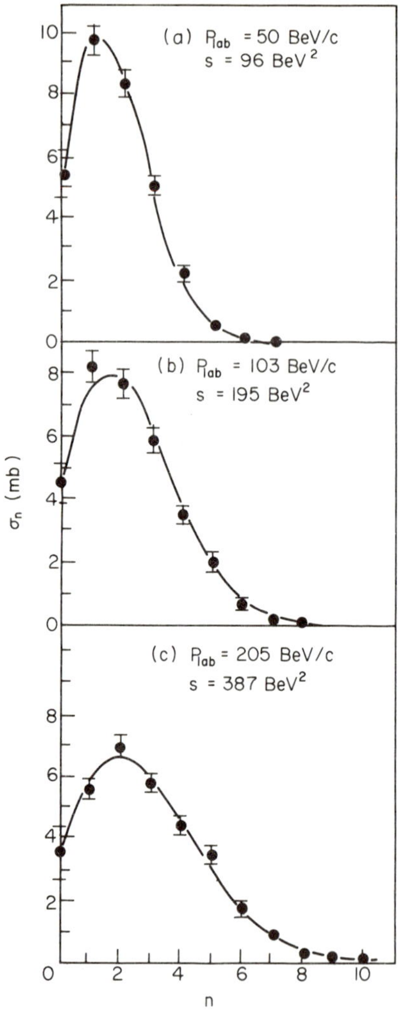

FIG. 23. A two component fit to the n-dependence of the multiplicity cross sections.

over a wide energy range. In our two component approach:

$$\frac{\langle n \rangle}{\mathscr{D}} \underset{s\to\infty}{\to} \frac{\langle n \rangle}{\sqrt{f_2}} \to \sqrt{\left(\frac{M}{D}\right)} = 2{\cdot}2 \tag{76}$$

Figure 24 shows the data for $\langle n \rangle/\mathscr{D}$ as well as our fit. It is clear from the figure that the two component approach provides a natural explanation for the

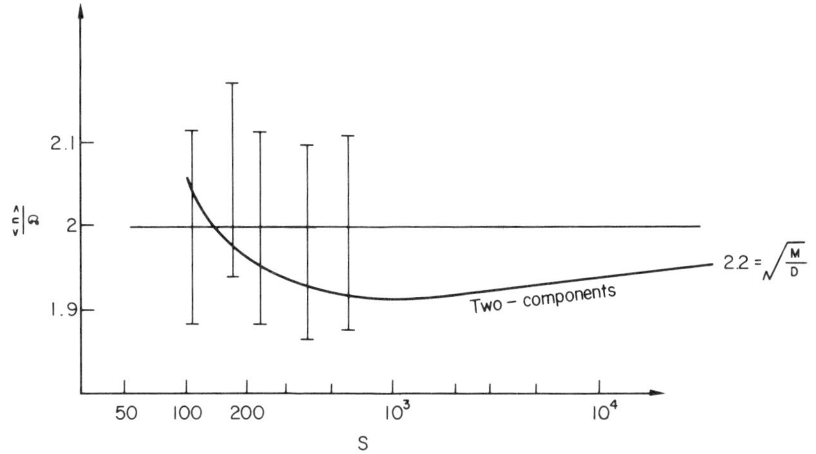

FIG. 24. A two component fit to $\langle n \rangle / \mathscr{D}$.

success of Wroblewsky's law. Note that $\langle n \rangle / \mathscr{D} = 2$ corresponds to $M = 0{\cdot}8$, a perfectly reasonable value.

(iv) The factor s_1 determines the energy scale by which $\langle n \rangle_m$ reaches its asymptotic behaviour. The scale factor s_2 determines the energy scale by which f_{2m} reaches the same behaviour. It is intuitively clear that s_2 should be significantly larger than s_1, and the values obtained by fitting the data confirm this.

(v) A peak near $x = 1$ is observed in the inclusive cross-section for $pp \to p +$ anything (Fig. 20). We have already remarked that such a peak must come from the D-component. It is interesting to note that the total cross-section represented by this peak[14] corresponds to 5–10 mb, consistent with our value of $\Sigma_n d_n$.

(vi) $\langle n \rangle_d = 0{\cdot}78$. A typical diffractive event will have 0·78 negative particles, hence, 3·6 charged particles. If, as the data seem to indicate, most diffractive events have a single proton near $x = 1$ and a group of particles with $x < 0$, the average number of charged particles at $x < 0$ "recoiling against" a leading $x \sim 1$ proton is predicted to be 2·6. The data show[18] that a proton at $x = 0{\cdot}99$ is accompanied on the average by $2{\cdot}8 \pm 0{\cdot}5$ charged particles.

(vii) The asymptotic energy dependence of every m_n is given by:

$$m_n \underset{s \to \infty}{\to} (\log s)^n \, s^{-c_1 + \frac{1}{2}c_2} \tag{74}$$

$$m_n \underset{s \to \infty}{\to} s^{2\alpha_R - 2} \tag{75}$$

we get:

$$2\alpha_R - 2 = -c_1 + \tfrac{1}{2}c_2 \qquad (76)$$

This is the generalization of eqn (28) to the case of $f_2 \neq 0$. Substituting our values of $c_1 = 1{\cdot}0$, $c_2 = 0{\cdot}35$ we get $\alpha_R = 0{\cdot}59$, a reasonable value for the leading non-Pomeron exchange. We must remember, however, that the $s^{2\alpha_R - 2}$ behaviour is expected to hold for a "true" non-diffractive multiplicity cross-section m_n where n is the total number of particles, including *neutrals*. The relation between the power behaviour of the "true" m_n and our m_n (which is a prong cross-section) is model dependent. However, if we assume that the distribution is symmetric between positive, negative and neutral particles, we find:

$$1 - \alpha_R^{\text{true}} = \tfrac{3}{2}(1 - \alpha_R^{\text{charged}}) \qquad (77)$$

Hence, our calculation yields: $\alpha_R^{(\text{true})} = 0{\cdot}4$. Note also that eqn (76) could be interpreted either for n = negative prongs or for n = charged prongs. For negative prongs we had $c_1 = 1{\cdot}0$, $c_2 = 0{\cdot}35$ and $\alpha_R = 0{\cdot}59$. For charged prongs we must have the same $\alpha_R[\sigma(2n + 2 \text{ charged}) = \sigma(n \text{ negative})]$. However the coefficients c_1 and c_2 are different. Applying charge conservation we discover, however, that if $f_3^- = f_4^- = \ldots = f_m^- = 0$, we get $f_3^{\text{ch}} \neq 0$, $f_4^{\text{ch}} \neq 0$. Equation (76) should then read:

$$2\alpha_R - 2 = -c_1^{\text{ch}} + \tfrac{1}{2}c_2^{\text{ch}} - \tfrac{1}{6}c_3^{\text{ch}} + \tfrac{1}{24}c_4^{\text{ch}} \qquad (78)$$

where $c_1^{\text{ch}} = 2{\cdot}0$, $c_2^{\text{ch}} = 3{\cdot}4$, $c_3^{\text{ch}} = 4{\cdot}2$, $c_4^{\text{ch}} = 4{\cdot}2$ can be computed from $c_1^- = 1{\cdot}0$, $c_2^- = 0{\cdot}35$, $c_3^- = c_4^- = 0$ using only charge conservation. Substituting the values of c_i^{ch} in eqn (78) we recover the same value of α_R as in eqn (76).

The lesson is clear—any set of assumptions about correlations for negative particles uniquely determines the charge particle correlations, and the two sets of correlations are always consistent with the same value of α_R. The α_R values obtained from our fit are in reasonable agreement with the expected value $\alpha_R \sim 0{\cdot}5$.

(viii) Extrapolating the fit to higher energies, the predicted multiplicity distribution for the highest ISR energy, is shown in Fig. 25. According to the figure the "Wilson dip" should be observed above $s > 3000$ BeV^2. However, our crude treatment of the diffractive part, ignoring possible $\log s$ terms in $\Sigma_n d_n$, $\langle n \rangle_d$ and f_{2d} is not expected to be appropriate over such a wide energy range. It may be a valid approximation for the region of our fit ($100 < s < 600$ BeV^2) but the $\log s$ terms may alter the situation significantly for $s \sim 3000$ BeV^2. We return to this point in our more sophisticated discussion of the diffractive part in Section III.

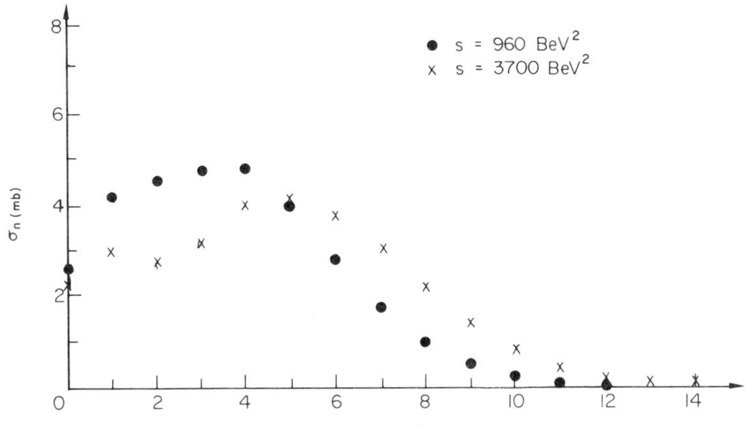

FIG. 25. Predictions of a naive two-component fit for multiplicity distributions at high energies.

12. SUMMARY

The general, oversimplified two-component description presented in this lecture seems to account for most of the gross features of the data in the fitted energy range. There are very clear experimental indications for the existence of the two components and it is certain that neither of them would be sufficient for a qualitative description of the available data. Both are necessary!

The largest fraction of the total inelastic cross-section is contributed by the *M*-component. This verifies the assertion that a large physical cross-section is involved in the unitarity relation between the sum of non-diffractive contributions and the Pomeron term in Im A_{el}. Experimental, phenomenological and theoretical studies of the way in which the m_n cross-sections add up to an energy independent quantity might clarify the role played by the Pomeron in hadron dynamics. We return to this subject in Section V.

Our treatment of the *D*-component was very crude in this Section. A closer inspection of this component uncovers many interesting features. The next Section is devoted to this topic.

III. THE DIFFRACTIVE COMPONENT AND ITS POSSIBLE RELATION TO THE RISING CROSS-SECTION†

1. INTRODUCTION

Our two component analysis of multiplicity distributions in the range $100 < s < 600\ \text{BeV}^2$ led us to the conclusion[16] that the total inelastic

† This section is entirely based on work done in collaboration with E. Rabinovici.

diffractive cross section was approximately 5–6 mb. Other studies[17] of the same data, using slightly different sets of assumptions but a very similar general framework gave values ranging between 5 and 10 mb. At the same time the diffractive inclusive proton cross-section near $x = 1$ (and symmetrically, near $x = -1$) seem to correspond[14] to a cross-section of the same order of magnitude: 5 to 10 mb. The $x \sim 1$ peak in the proton spectrum (Figs 20 and 27) obviously corresponds to events in which the detected proton is produced through the Pomeron exchange mechanism of Fig. 26a rather

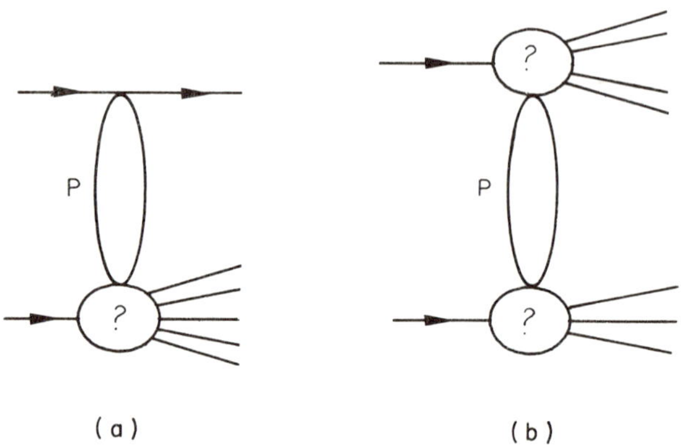

FIG. 26. Single and double diffraction excitation.

than the more general diffractive mechanism of Fig. 26b. We will therefore assume that most of the cross-section of the D-component is generated by the single diffraction excitation mechanism.

We further assume (temporarily) that the Pomeron in Fig. 26a is a simple pole at $\alpha(0) = 1$. This assumption is certainly inappropriate from a theoretical point of view and it leads to self-consistency problems and to difficulties with unitarity. However, we hope that, phenomenologically, it may serve as a good approximation throughout the ISR energy region ($s < 3000$ BeV^2) and we will return to the self-consistency questions in Section VI.

2. THE POMERON–PROTON TOTAL CROSS-SECTION

Having assumed that the D-component is controlled by the mechanism of Fig. 26a, and that the proton is a simple pole, we have to consider the Pomeron–proton scattering process at the lower vertex of the figure. What is the M^2 dependence of σ_{tot}^{Pp}? If we assume that a trajectory α_2 dominates this cross-section,we have

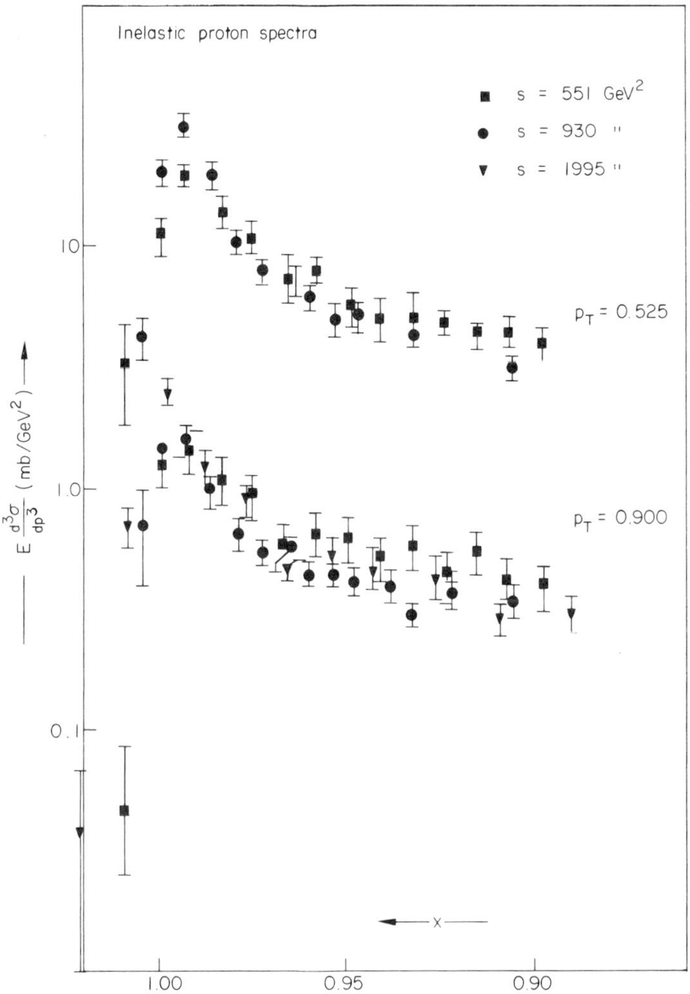

FIG. 27. Inclusive proton cross-sections as a function of x.

$$\sigma_{\text{tot}}^{Pp} \xrightarrow[M^2 \to \infty]{} (M^2)^{\alpha_2 - 1} \tag{79}$$

and the inclusive diffractive cross-section for a single proton near $x = 1$ is given by (here we ignore the slope of the Pomeron):

$$\frac{d\sigma}{dx} = s^{\alpha_2 - 1}(1 - x)^{\alpha_2 - 2} \tag{80}$$

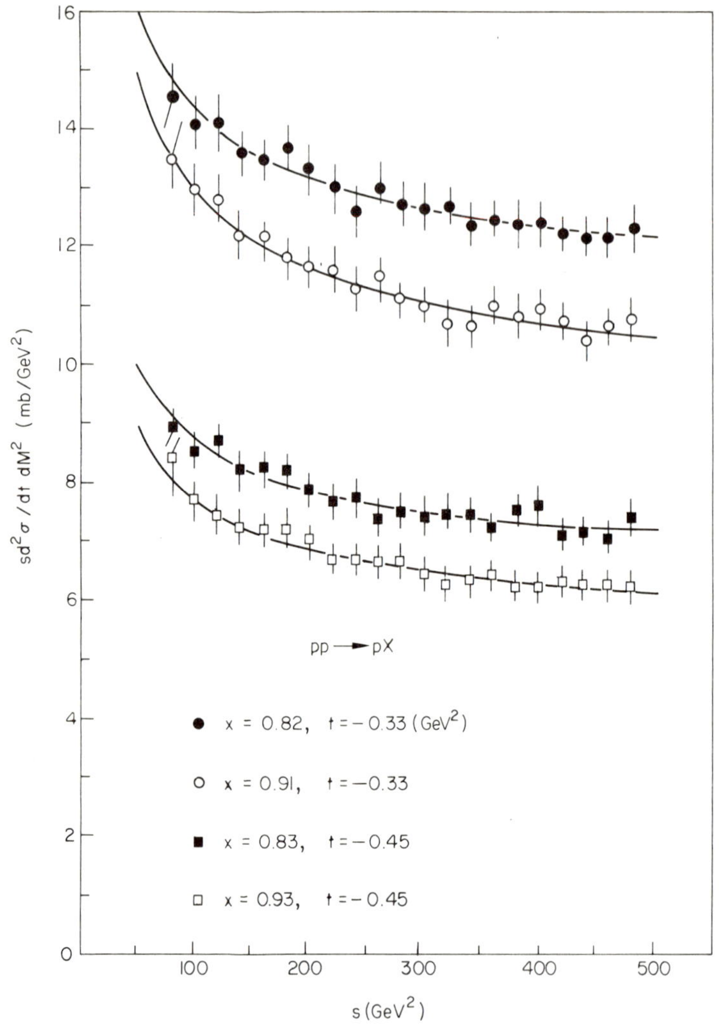

FIG. 28. Inclusive proton cross-sections as a function of energy. The lines represent a fit of the form $c_0 + c_1/\sqrt{s}$.

For $\alpha_2 = 1$ (constant σ_{tot}^{Pp}) we have:

$$\frac{\mathrm{d}\sigma}{\mathrm{d}x} = \frac{1}{1 - x} \tag{81}$$

while for $\alpha_2 = \frac{1}{2}$:

$$\frac{\mathrm{d}\sigma}{\mathrm{d}x} = \frac{1}{\sqrt{s}} \frac{1}{(1 - x)^{\frac{3}{2}}} \tag{82}$$

The x-dependence of the inclusive proton cross-section was measured[14] at several energies at the ISR. A comparison of the results (Fig. 27) at different energies indicates a small, if any, s-dependence of the cross-section at fixed x in the range between $s = 551$ and $s = 930$ BeV^2. An experiment at NAL[19] measured the detailed energy dependence of the inclusive cross-section at fixed $x = 0{\cdot}93$ (Fig. 28). The observed behaviour is well fitted by an expression of the form $c_0 + (c_1/\sqrt{s})$ indicating contributions corresponding to both $\alpha_2 = 1$ and $\alpha_2 = \frac{1}{2}$. However, the $\alpha_2 = \frac{1}{2}$ contribution is smaller than the $\alpha_2 = 1$ term for $s > 100$ BeV^2, and it is practically negligible for $s > 500$ BeV^2.

The $\alpha_2 = \frac{1}{2}$ term corresponds to the so-called PPR term (Fig. 29a). It leads to diffractive cross-sections for which:

$$d_n \underset{s\to\infty}{\to} \text{const} \tag{83}$$

$$\langle n \rangle \underset{s\to\infty}{\to} \text{const} \tag{84}$$

$$\sum_n d_n \underset{s\to\infty}{\to} \text{const} \tag{85}$$

in agreement with our crude description of the D-component in the previous lecture. Since this term represents a term in σ_{tot}^{Pp} which decreases for $M^2 \to \infty$ we will refer to it as "low mass diffraction excitation".

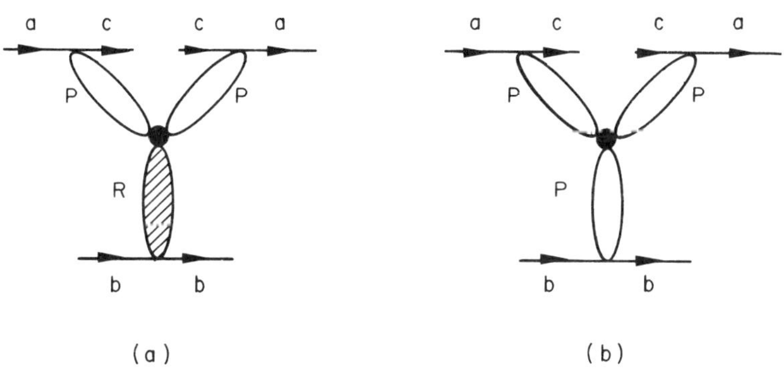

FIG. 29. The *PPR* and *PPP* contributions to inclusive cross-sections.

The $\alpha_2 = 1$ term corresponds to $\sigma_{\text{tot}}^{Pp} \to$ const and we will refer to it as "high mass diffraction excitation" since it allows the excitation of arbitrarily high masses M^2 for sufficiently high energies. This term is the famous *PPP* "triple Pomeron" term (Fig. 29b). Its contribution to the *pp* cross-section obeys:

$$d_n \underset{s\to\infty}{\to} \text{const} \tag{86}$$

$$\sum_n d_n \underset{s\to\infty}{\to} \log s \tag{87}$$

The logarithmic increase of the total diffractive cross-section due to the "high mass" "triple Pomeron" term follows from the $1/(1-x)$ behaviour of $d\sigma/dx$ near $x = 1$ (eqn 81). If the Pomeron pole has a slope in t, we actually get

$$\sum_n d_n \underset{s\to\infty}{\to} \log \log s$$

However, for $\alpha'_p < 0{\cdot}5$ and $s < 3000\ \text{BeV}^2$ the $\log s$ approximation is valid. We will return to this point when we discuss the rising total cross-section.

3. The multiplicity distribution for Pomeron–proton scattering or: a two-component description of Pp scattering

Assuming that $\sigma_{tot}^{Pp} \sim$ const at high M^2, we may now study the breakdown of this constant cross-section into the various multiplicity cross-sections. The problem that we face here is very similar to the one we discussed in Sections I and II for pp scattering. Given a constant σ_{tot} we may imagine two mechanisms which build it:

(i) Every multiplicity cross-section is constant due to a diffractive mechanism. The sum of all constant multiplicity cross-sections gives a constant σ_{tot}.

(ii) Every multiplicity cross-section rises towards a maximum, falls like a power of energy and is due to a non-diffractive mechanism. The sum of all multiplicity cross-sections is constant.

In general, as is the case for pp scattering, both options are used by nature. In the case of Pp scattering we probably have two components: a D-component (Fig. 30a) and an M-component (Fig. 30b). Note that both of these components are part of the overall D-component of p–p scattering! In

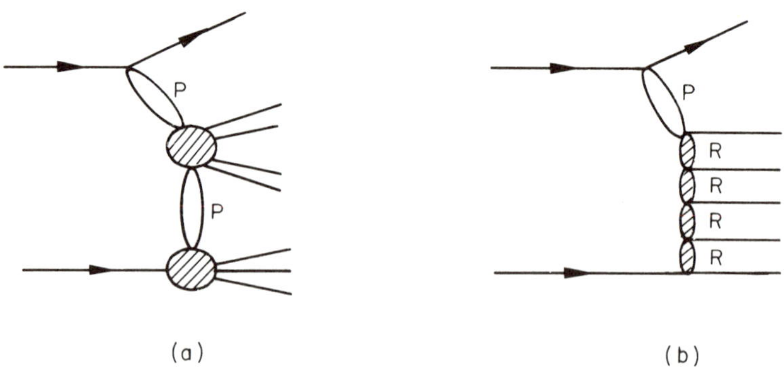

FIG. 30. D and M components for Pomeron-proton scattering.

principle, we may select a fixed Pomeron "mass" (i.e. fixed t-value), and measure $\sigma_n^{Pp}(M^2)$ for all n and M^2 values. We can then perform a two-component fit similar to the one described in Section II.

However, the present data cannot justify such an enterprise. What it does teach us, however, is that the M^2 dependence of the average charged multiplicities corresponding to protons near $x \sim 1$ (which are mostly diffractively produced) is similar to that observed in pp collisions.[20] This experimental fact, as well as our theoretical prejudices, lead us to believe that, as in pp scattering, most of $\sigma_{tot}^{Pp}(M^2)$ is given by its M-component, namely by mechanisms like that of Fig. 30b. In that case:

$$\langle n \rangle^{Pp} \underset{M^2 \to \infty}{\to} \log (M^2) \tag{88}$$

The overall multiplicity for a diffractive event will then be given by averaging the multiplicities over the entire allowed region of M^2. This leads to:

$$\langle n \rangle_d \underset{s \to \infty}{\to} \log s \tag{89}$$

The diffractive multiplicity due to a Pomeron exchange followed by a non-diffractive Pomeron-proton scattering, and corresponding to a constant total Pomeron-proton cross-section, increases logarithmically with energy. In that respect it differs from our crude assumption concerning $\langle n \rangle_d$ in the previous lecture. (Nevertheless our fits are essentially unaffected since the $\log s$ increase of $\langle n \rangle_d$ is probably insignificant for $100 < s < 600$.)

4. AN EXPLICIT MODEL FOR THE D-COMPONENT

We are now in position to assume an explicit simple model for most of the diffractive component of pp scattering. Following the discussion above, we assume:[22]

(1) The D-component is dominated by single excitations (Fig. 26a).
(2) The Pomeron is a simple pole at $\alpha(0) = 1$.
(3) $\sigma_{tot}^{Pp}(M^2) \sim$ const.
(4) $\sigma_{tot}^{Pp}(M^2)$ decreases like a power of M^2, and is due to the mechanism of Fig. 30b. In fact in order to be specific let us neglect correlations between the particles emitted in the Pp "collision" and assume the Poisson distribution:

$$\sigma_n^{Pp}(M^2) = K \frac{(c_1 \log M^2)^n}{n!} (M^2)^{-c_1} \tag{90}$$

where K is the constant value of $\sigma_{tot}^{Pp}(M^2)$. Note that the constant c_1 must be the same as in the multiplicity formula for the M-component of pp scattering, since it reflects the coupling constant g in the multiperipheral ladder, or the height $\rho(0)$ of the central plateau (normalized to σ_{tot}). It is also related to $\alpha_R(c_1 = 2 - 2\alpha_R)$. All of these parameters $(g, \rho(0), \alpha_R)$ are independent of the colliding particles in all short range correlation models. They therefore have the same values for pp and Pp interactions (or, for πp, Kp, γp, etc.).

(5) For convenience we assume, in addition, that the Pomeron pole has a vanishing slope. This is inessential as we will see later in this lecture.

(6) We ignore all questions of self-consistency of the Pomeron, unitarity bounds, absorption corrections, etc. We return to these in Section VI.

Every one of the assumptions is, at best, approximately correct. However, there is a good probability that many of the general features of the D-component are correctly described by such a picture. We therefore proceed to study the experimental consequences of this set of assumptions.

5. PREDICTIONS OF THE MODEL: MULTIPLICITIES

Suppose that the no-correlation assumption (eqn (90)) refers to the cross-section for producing n negative particles. The pp diffractive multiplicity cross-section is then given by:

$$d_n = \int_{s_0}^{s} \frac{\mathrm{d}\sigma_n}{\mathrm{d}M^2}\,\mathrm{d}M^2 = \int_{s_0}^{s} \frac{1}{M^2}\,\sigma_n^{Pp}(M^2)\,\mathrm{d}M^2 \tag{91}$$

where s_0 is the lowest M^2 value for which the triple Pomeron term contributes (kinematically $s_0 = M_N^2$, but we have seen in Section I that P-exchange is actually absent at low energies. A more realistic s_0 would probably be, at least, $s_0 = 3$–$4\ \text{BeV}^2$). The upper integration limit is kinematically the full energy s (ignoring M_N^2/s terms) but, again, the Pomeron can probably contribute only above a minimal subenergy or rapidity gap. This means that the upper limit should really be βs where β is a constant, probably of order 0·5 or so.

Substituting eqn (90) in eqn (91) and performing the integration, we get:

$$d_n = \frac{K}{c_1}\left[\sum_{l=0}^{n} \frac{(c_1 \log s_0)^l}{l!}\, s_0^{-c_1} - \sum_{l=0}^{n} \frac{(c_1 \log s)^l}{l!}\, s^{-c_1}\right]. \tag{92}$$

What is the asymptotic s-dependence of d_n for fixed n? The first sum in eqn (92) is energy independent. The second sum is a sum of a *finite* number of terms decreasing like a power of s. Consequently, for $s \to \infty$, $d_n \to$ const and the approach fo the constant is from below. How is this constant reached for different n-values? For larger n-values, the number of terms in the sum is larger. Consequently, the second sum in eqn (92) will disappear at a higher energy, and the constant asymptotic value of σ_n will be reached at a higher energy.

The total diffractive cross-section is given by:

$$\sum_n d_n = K \log s/s_0 \tag{93}$$

This can be obtained either by summing eqn (92) over all n-values (from 0 to ∞) or (more easily) from:

$$\sum_n d_n = \int_{s_0}^{s} \frac{1}{M^2} \sigma^{Pp}(M^2)\, \mathrm{d}M^2 = K \log s/s_0 \tag{94}$$

The way by which the constant d_n's add up to their logarithmically increasing sum is shown in Fig. 31.

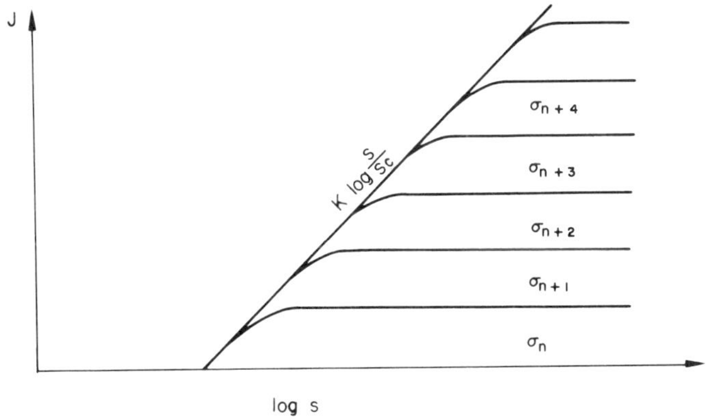

FIG. 31. The logarithmically increasing diffractive cross-section and its breakdown to multiplicity cross-sections.

The average multiplicity can be easily calculated in many ways. We find:

$$\langle n \rangle_d = \frac{\sum nd_n}{\sum d_n} \underset{s\to\infty}{\to} \frac{1}{2} c_1 \log s \tag{95}$$

As we have already anticipated in eqn (89), $\langle n \rangle_d$ increases logarithmically with energy. We now discover, however, that the coefficient of $\log s$ is $\frac{1}{2} c_1$, where c_1 was the coefficient of $\log s$ in the non-diffractive multiplicity $\langle n \rangle_m \to c_1 \log s$. The reason for this (apart from explicitly calculating $\langle n \rangle_d$, say, from eqn (92)) can be easily understood in the following way. The full rapidity axis, in a process described by Fig. 30b, has to be "shared" by the single Pomeron and the many Reggeons. The Pomeron, corresponding to an energy independent cross-section is "indifferent" to the size of the rapidity segment that it "uses". It has no reason to prefer a larger or a smaller rapidity interval. On the average, it will "use" half of the rapidity axis, with a constant probability of using any rapidity length between 0 and the full rapidity $\log s$. The rest of the rapidity axis (half of it, on the average) is then taken by the multiperipheral chain of Fig. 30b, which populates it with particles with the canonical density $\rho(y) = c_1$ of a non-diffractive mechanism. What we have, then, is the same average density of particles along the rapidity axis as in a normal non-diffractive event, except that, on the average, only half of the rapidity axis is populated. The average diffractive multiplicity is, then, half of the corresponding non-diffractive multiplicity. The argument, however, ignores edge effects corresponding to finite (energy independent) segments of the rapidity axis. Hence—only the coefficient of $\log s$ (but not the full $\langle n \rangle_d$) is given by half the non-diffractive value.

6. PREDICTIONS OF THE MODEL: INCLUSIVE PION CROSS-SECTION

The above argument immediately leads to a determination of the inclusive pion cross-section due to our diffractive mechanism. For a fixed M^2 (i.e. a fixed portion of the rapidity axis taken up by the Reggeons) the pion distribution will form a "plateau" over that rapidity range (Fig. 32). Since all such rapidity intervals are equally probable, the overall inclusive pion cross-section will be given by Fig. 33a. However, this inclusive cross-section

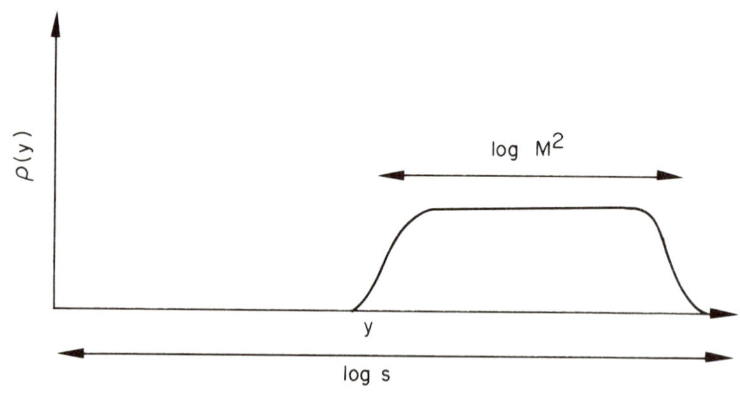

FIG. 32. Schematic representation of inclusive pion spectra coming from single high mass diffraction excitation for a fixed value of M^2 and s.

represents only events in which the "unexcited" proton was on the left while the excited one was on the right. An equal number of events correspond to the opposite situation, and its inclusive pion spectrum is shown in Fig. 33b. The total inclusive cross-section is then given by the sum which shows a central plateau with a height (normalized to $\Sigma_n d_n$) of $\frac{1}{2}c_1$ (Fig. 33c).

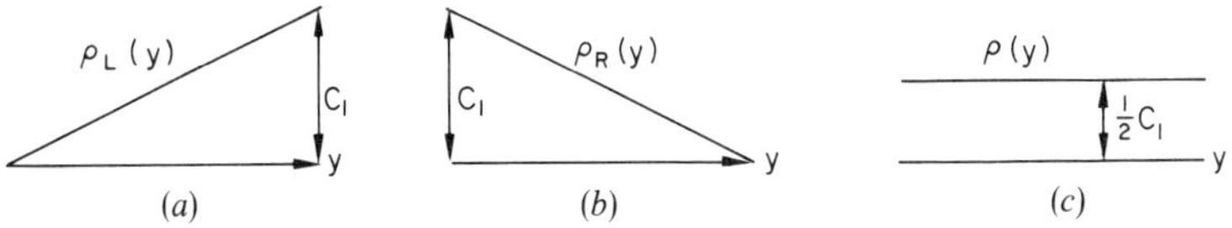

FIG. 33. Schematic representation of inclusive pion spectra coming from single high mass diffraction excitation: (a) For protons near $x = -1$; (b) For protons near $x = 1$; (c) Total inclusive distribution.

The existence of this plateau is an interesting result for a diffractive mechanism which is usually considered to be dominated by the fragmentation region. The plateau here is generated by the fact that diffraction excitation of high mass M^2 is not strongly damped relative to low-mass excitations and hence it spreads over large rapidity intervals.

7. PREDICTIONS OF THE MODEL: CORRELATIONS

We have assumed that the particles produced in a given Pomeron-proton collision were uncorrelated. What are the implications of this for the integrated diffractive correlation f_{2d}?

We can compute f_{2d} directly from eqn (92). We can also compute it from the inclusive two-particle cross-section. If the single pion inclusive cross-section for events with a single proton at the left edge of the rapidity axis is given by Fig. 33a, the two pion inclusive cross-section for similar events is given as a function of y_1, y_2 by Fig. 34. The correlation function C_{12} was defined in Section II as

$$C_{12}(y_1, y_2) = \rho_{12}(y_1, y_2) - \rho_1(y_1)\,\rho_2(y_2) \tag{96}$$

while:

$$f_{2d} = \int C_{12}(y_1, y_2)\,\mathrm{d}y_1\,\mathrm{d}y_2 \tag{97}$$

Figure 34 tells us that the integrated ρ_{12} is given by the volume of a pyramid:

$$\int \rho_{12}(y_1, y_2)\,\mathrm{d}y_1\,\mathrm{d}y_2 = \tfrac{1}{3}c_1^2 \log^2 s \tag{98}$$

Hence:

$$f_{2d} = \tfrac{1}{3}c_1^2 \log^2 s - [\tfrac{1}{2}c_1 \log s]^2 = \frac{c_1^2}{12} \log^2 s \tag{99}$$

The overall correlation corresponding to protons near $x = -1$ and near $x = 1$ is, of course, the same as that for $x = -1$ only, since it is always normalized by the corresponding cross-section.

We therefore find that although the Pomeron-proton "collision" was assumed to involve no correlations, the resulting diffractive cross-section not only has correlations, but they increase like $\log^2 s$, reflecting *long range* effects. The physical reason for that is easy to understand. All pions in any given event are produced at the same side of the rapidity axis, leaving a rapidity gap at the other end (the gap may be very small, but on the average it is half of the full rapidity.). Consequently, once a pion is detected in a given

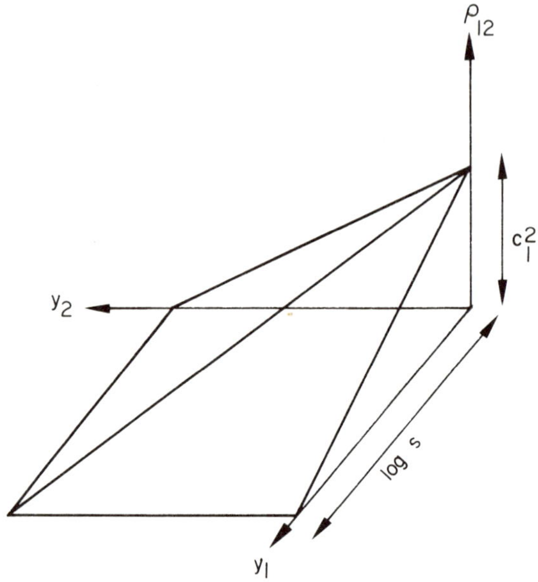

FIG. 34. Inclusive two pion cross-section.

rapidity, we have significant information on the probable location of other pions. Such information is, of course, what we mean by the word "correlation". Some of this information relates to distant points in rapidity, (e.g. if the rapidity axis covers 20 units, $0 \leqslant y \leqslant 20$, and a pion is detected at $y = 15$, we have high probability of finding pions between $19 < y < 20$ and very small probability for pions at $0 < y < 1$, although both rapidity regions are very distant from the detected pion.) Hence we have long range correlations.

8. DIFFRACTIVE MULTIPLICITIES IN THE n-log s PLANE

We have already discussed the energy dependence of the diffractive multiplicity cross-section d_n, for a fixed value of n. We have seen that for

$n \gg c_1 \log s, d_n \sim 0$ while for $n \ll c_1 \log s$, $d_n \to$ const. A closer look at eqn (92) shows that for sufficiently large n the constant $s \to \infty$ limit of d_n is independent of n. Furthermore, the rise of every d_n from zero to its asymptotic constant value is such that for $n = c_1 \log s$, $d_n = \frac{1}{2} \lim_{s \to \infty} d_n(s)$.

What is the n-dependence of $d_n(s)$ at fixed s? Figure 35 describes its qualitative features. d_n starts at zero (or a small value), increases around $n = c_1 \log s_0$ to a constant, and falls to zero around $n = c_1 \log s$. For $\log s_0 \ll \log s$ the

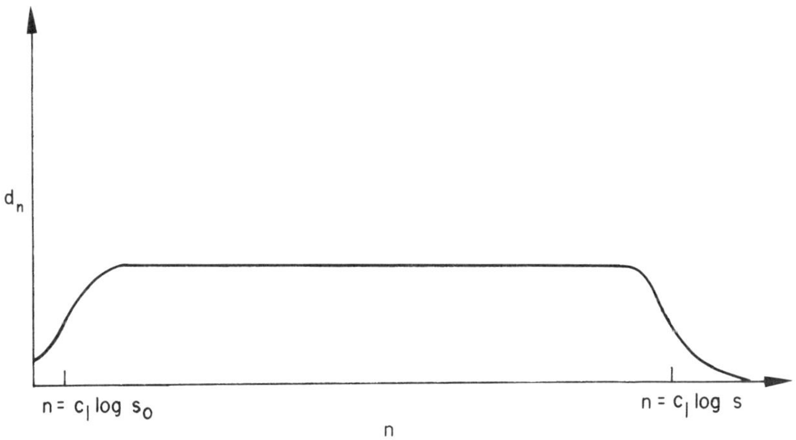

FIG. 35. n-dependence of d_n at fixed s.

multiplicity distribution is predicted to be roughly constant for most of the $n < c_1 \log s$ region and to vanish rapidly beyond it. This, of course, leads us again to the conclusion:

$$\langle n \rangle_d \underset{s \to \infty}{\to} \tfrac{1}{2} c_1 \log s \tag{100}$$

The predicted n and s dependence of $d_n(s)$ is summarized in Fig. 36. A comparison of this figure with Fig. 18 shows that the region for which the "Wilson dip" was predicted for a naive two component model (with $\langle n \rangle_d \underset{s \to \infty}{\to}$ const) corresponds to the constant "plateau" in Fig. 36. In other words—the "boundary" of the diffractive cross-section d_n (the line $n = c_1 \log s$) coincides with the "ridge" of the non-diffractive cross-section m_n. Consequently, it seems that a diffractive model of the type described here may not show a Wilson dip. However, this conclusion is somewhat dependent on the highest M^2-value allowed for a given s, in a single excitation mechanism ($M^2_{\max} = \beta s$). If the highest "allowed" value of $\beta = M^2/s$ is sufficiently small

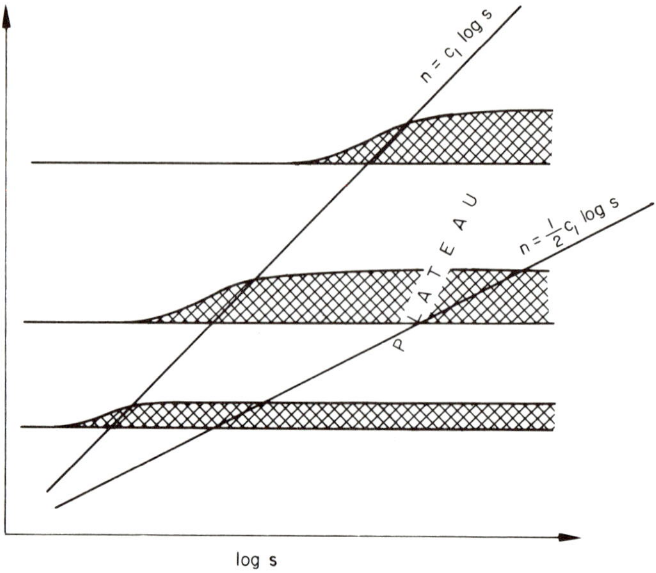

FIG. 36. "High mass" diffraction excitation in the n-log s plane.

(or the smallest rapidity gap corresponding to Pomeron exchange in Fig. 30b is sufficiently large) a "Wilson dip" may still develop. It would be extremely interesting to check this experimentally.

9. THE DIFFRACTIVE INCLUSIVE PROTON CROSS-SECTION AND THE RISING TOTAL CROSS-SECTION

We have already seen that the x-dependence of the diffractive inclusive proton cross-section near $x = 1$ is given by:

$$\frac{d\sigma}{dx} = \frac{1}{1-x}. \tag{101}$$

The full inclusive cross-section as a function of x and p_T can probably be parameterized as:

$$\frac{d^2\sigma}{dx\,dp_T^2} \sim \frac{c\,e^{ap_T^2}}{1-x}. \tag{102}$$

The contributions of such events to the total cross-section is:

$$\sigma(s) = 2\int_{\bar{x}}^{x_0}\int_0^{\infty} \frac{c\,e^{ap_T^2}}{1-x}\,dp_T^2\,dx \tag{103}$$

where $x_0 = 1 - (s_0/s)$ is the highest x-value allowed for the diffractive process at a given energy, s_0 is the same as in eqn (91) and $\bar{x}$ represents the smallest allowed x for a single diffractive excitation, obeying $\bar{x} = 1 - \beta$. The factor 2 comes from adding the regions near $x = 1$ and $x = -1$. The integral in eqn (103) increases like $\log s$ for $s \to \infty$ and causes an increase in the total cross-section. The increase of σ_{tot} between energies s_1 and s_2 is given by:

$$\Delta\sigma(s_1, s_2) = 2\int_{x_1}^{x_2}\int_0^{\infty} \frac{c\, e^{p_T^2}}{1 - x}\, dp_T^2\, dx = \frac{2c}{a}\log\frac{s_2}{s_1} \tag{104}$$

where $x_{1,2} = 1 - (s_0/s_{1,2})$. Figure 37 illustrates the way in which the $x \sim 1$ diffractive events increase the total cross-section. The inclusive proton cross-section is essentially s-independent for fixed x. However, for higher and higher values of s the inclusive cross-sections extends to x-values closer and closer to $x = 1$. The contribution of the "new" range of x produces a "new" cross-section for higher energies, leading to an increase in σ_{tot}.

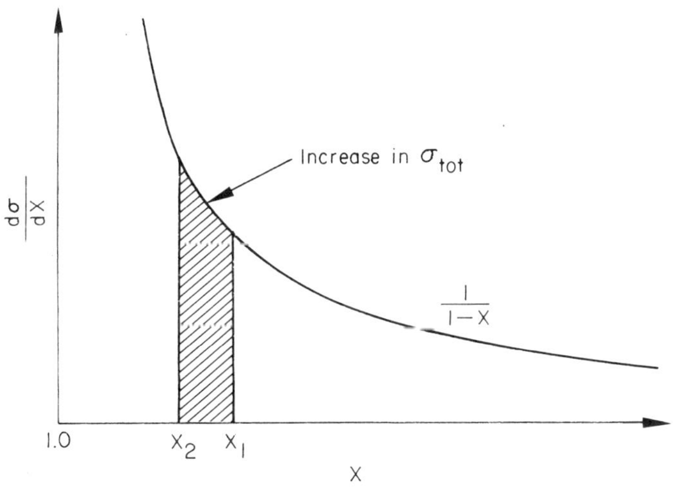

FIG. 37. An extension of the "allowed" x-range near $x = 1$ together with scaling lead to an increasing total cross-section.

The next question is, of course, whether this mechanism can account for the observed increase[21] in σ_{tot} in the range $500 < s < 2800$ BeV^2. In order to answer this question we have to deduce the parameters c and a from the data for $p + p \to p +$ anything. Various attempts to compute these parameters[22] lead to results for σ_{tot} $(s = 2800) - \sigma_{tot}(550)$ which are anywhere between 1 mb and 2·5 mb. The uncertainty stems from several sources:

(i) Errors in the overall normalization of the inclusive data.
(ii) Uncertainty in the determination of the slope parameter a (in $e^{ap_T^2}$).
(iii) Sensitivity to the inclusive cross-section at small t-values for which no data are presently available.
(iv) Possible small effects of the Pomeron slope α'_P which was neglected here (we discuss its effect in section 11).
(v) Possible background from non-diffractive contributions near $x \sim 1$.

Any one of these uncertainties cannot be very large but their combined effect may be considerable.

The observed increase of the total *inelastic* cross-section between $s = 550\ \text{BeV}^2$ and $s = 2800\ \text{BeV}^2$ is approximately 3 mb.[21] We therefore face the tantalizing possibility that almost the entire increase of σ_{inel} is given by the single diffractive excitation events. On the other hand, it is quite possible that these events contribute only a third of the observed increase in σ_{inel}. This question is extremely important. It could (and should) be settled by experiments in the near future.

10. THE D-COMPONENT AND THE RISING CENTRAL PLATEAU

We have already stated (Section III.6 and Fig. 33) that the inclusive pion spectrum generated by the diffractive mechanism forms a central plateau along the rapidity axis. The height of the plateau is energy independent ($\rho(y) = \frac{1}{2}c_1$) when normalized to the total cross-section. Since the total cross-section for our D-component increases like $\log s$, the absolute height of the central plateau should also increase like $\log s$. If we make the two strong assumptions that the entire increase in the inelastic cross-section and the entire increase of the height of the central plateau in a certain energy range are both due to the D-component (and are given by the mechanism of Fig. 30b) we can relate the two effects.

Since our D-component gives

$$\rho(y = 0) = \frac{\int [d^2\sigma/(dy\, dp_T^2)]\, dp_T^2}{\sigma_{\text{tot}}} = \tfrac{1}{2}c_1 \tag{105}$$

we conclude that the increase in $d^2\sigma/(dy\, dp_T^2)$ is given by $\frac{1}{2}c_1$ times the increase in σ_{tot}. Using the observed increase of 3 mb or so for σ_{inel} between $s = 550$ and $s = 2800$, and the observed e^{bp_T} ($b \sim 6$–$7\ \text{BeV}^{-1}$) behaviour of the inclusive pion cross-section in the central plateau, we predict that the invariant inclusive cross-section for $pp \to \pi^- +$ anything increases according to:

$$E\frac{d^3\sigma}{dp^3}\bigg|_{90°, s_2} - E\frac{d^3\sigma}{dp^3}\bigg|_{90°, s_1} \sim 6\, e^{6p_T} \ln\frac{s_2}{s_1} \tag{106}$$

in the ISR energy range. The comparison of this result with the data is difficult in view of the recently observed deviations from a constant central plateau.[23] However, if we ignore this problem, we find that the above prediction accounts only for approximately 30%–50% of the increasing height of the central plateau, with fairly large uncertainties both in the data and in our theoretical considerations.

11. POSSIBLE EFFECT OF $\alpha'_p \neq 0$

All our calculations in this lecture were considerably simplified by our assumption that the Pomeron pole has a vanishing slope. What would be the effect of such a slope?

The inclusive proton cross-section near $x = 1$ will assume the form:

$$\frac{d^2\sigma}{dx\, dp_T^2} = c\, e^{at}(1 - x)^{1-2\alpha_P(t)} \tag{107}$$

where

$$\alpha_p(t) = 1 + \alpha'_p t \tag{108}$$

Hence

$$\frac{d^2\sigma}{dx\, dp_T^2} = \frac{c}{1 - x} \exp\{[a - 2\alpha'_p \log(1 - x)]t\} \tag{109}$$

The contribution of the diffractive peak near $x = 1$ to the increasing total cross-section will then be given by replacing the integrand of eqn (104) by eqn (109). The resulting cross-section is given by:

$$\Delta\sigma_{\text{tot}}(s_1, s_2) = \frac{c}{\alpha'} \log\left[\frac{a + 2\alpha'_p \log s_2/s_0}{a + 2\alpha'_p \log s_1/s_0}\right] \tag{110}$$

The asymptotic energy dependence of this expression is $\log\log s$, in clear contrast to the $\log s$ behaviour of eqn (104). However, the argument in the squared brackets of eqn (110) is of order one for any reasonable set of values of the parameters, and for present and near future energy ranges. Consequently, we may use the approximation $\log(1 + \gamma) \sim \gamma$ for $\gamma \ll 1$. Doing so, we find:

$$\Delta\sigma_{\text{tot}} = \frac{2c}{a + 2\alpha'_p \log \bar{s}/s_0} \log s_2/s_1 \tag{111}$$

where $\bar{s}$ is some average energy between s_1 and s_2. We can now define

$$\bar{a} = a + 2\alpha'_p \log \bar{s}/s_0 \tag{112}$$

corresponding to the average slope (in t) of the inclusive cross-section over the range $s_1 < s < s_2$. Equation (111) then reads:

$$\Delta\sigma_{tot} = \frac{2c}{\bar{a}} \log s_2/s_1 \tag{113}$$

consistent with eqn (104). The corrections to eqn (113) due to α'_P are of order $((\alpha'/\bar{a}^2) \log s_2/s_1$ relative to the leading term (typically 5% over the ISR range).

The lesson of this exercise is quite important—the theoretical log log s behaviour predicted for σ_{tot} in the case $\alpha'_P \neq 0$ is, for all practical purposes at present and near future energies, given to a good approximation by a log s term.

The effects of $\alpha'_P \neq 0$ on the central plateau of the inclusive pion distribution is quite interesting. Instead of a flat central plateau $\rho(y) =$ const. we find:

$$\rho(y) \propto \log y + \log(Y - y) \tag{114}$$

where $Y = \log s$ and y is the laboratory frame rapidity. We therefore find a small deviation from a flat plateau with a "hump" in the centre of the rapidity axis. Numerically, this effect is, again, very small.

The effects of $\alpha'_P \neq 0$ on the multiplicity, correlations, and other quantities are usually insignificant. We find additional log s terms in various expressions but for all practical purposes the $\alpha_P = 0$ expressions discussed in the previous sections are approximately valid.

12. CONSISTENCY PROBLEMS OF THE POMERON—WITH ITSELF AND WITH UNITARITY

Throughout our discussion of the diffractive component we assumed that the Pomeron is a simple pole at $\alpha(0) = 1$. If so, $\sigma_{tot}(s) \underset{s\to\infty}{\rightarrow}$ const. On the other hand, the same assumption yields $\sigma^{Pp}_{tot}(M^2 \underset{M^2\to\infty}{\rightarrow}$ const. leading to an asymptotic log s term (or log log s for $\alpha'_P \neq 0$) for $\sigma^{Pp}_{tot}(s)$. Hence the pole assumption is not self-consistent. Furthermore, if the Pomeron is an $\alpha(0) = 1$ pole, the consideration of mechanisms such as the one in Fig. 38, leads to cross-sections which grow faster than the $\log^2 s$ limit allowed by the Froissart bound.

All of these questions are extremely important in any theoretical considerations of the nature of the Pomeranchuk singularity. We believe that, as $s \to \infty$, the unitarity (Froissart bound) problem is solved by including corrections which may be described as absorption corrections or s-channel unitarity effects. The complete description of the Pomeron will then involve a quite complicated set of singularities in the complex j-plane. However, the direct conflict with unitarity may happen only at extremely high energies. Consequently, at present energies, we may *hope* that the single pole approxi-

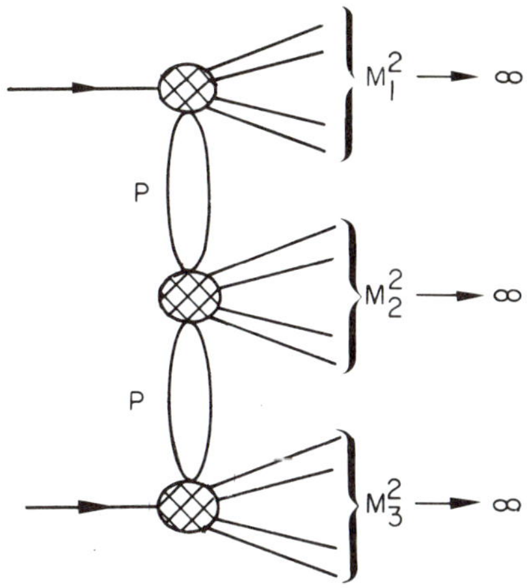

FIG. 38. Assuming that the Pomeron is an $\alpha(0) = 1$ pole, this diagram leads to a violation of the Froissart bound.

mation[22] of the complicated Pomeron may be valid, and its first correction, the $\log s$ (or $\log \log s$) term induced in σ_{tot} by high mass diffraction (or the triple Pomeron term) is the one important correction term which we have to worry about. This was the philosophy behind our simple phenomenological analysis of the present lecture. We will devote Section VI to the theoretical consistency problems mentioned here, but our hope is that these problems have no direct connection to the present phenomenological analysis.

Why do we feel that the conflict with unitarity may come only at very high energies? Figure 39 shows[24] the impact parameter description of the elastic pp amplitude at $s = 550$ and $s = 2800$ BeV^2 (the two energy values, between which σ_{tot} increases by 4 mb). The figure also includes the value corresponding to "full absorption" (or to a vanishing S-matrix). We see that with the present rate of growth of σ_{tot} we are extremely far from any conflict with this limit. The proton is only 75% "black" at ISR energies, and the rate by which it becomes "darker" is relatively slow.

This argument does not prove that absorption corrections are irrelevant at present energies. On the contrary—they are probably relevant even at 5 BeV. However, it is quite possible to have models with "temporary" energy behaviour which may look as if it eventually leads to a conflict with unitarity. We have a lot of time (and energy) before this happens, and by then many additional effects may become important.

13. SUMMARY

Let us now summarize the lessons of our phenomenological discussion of the components of multihadron production processes at high energies.

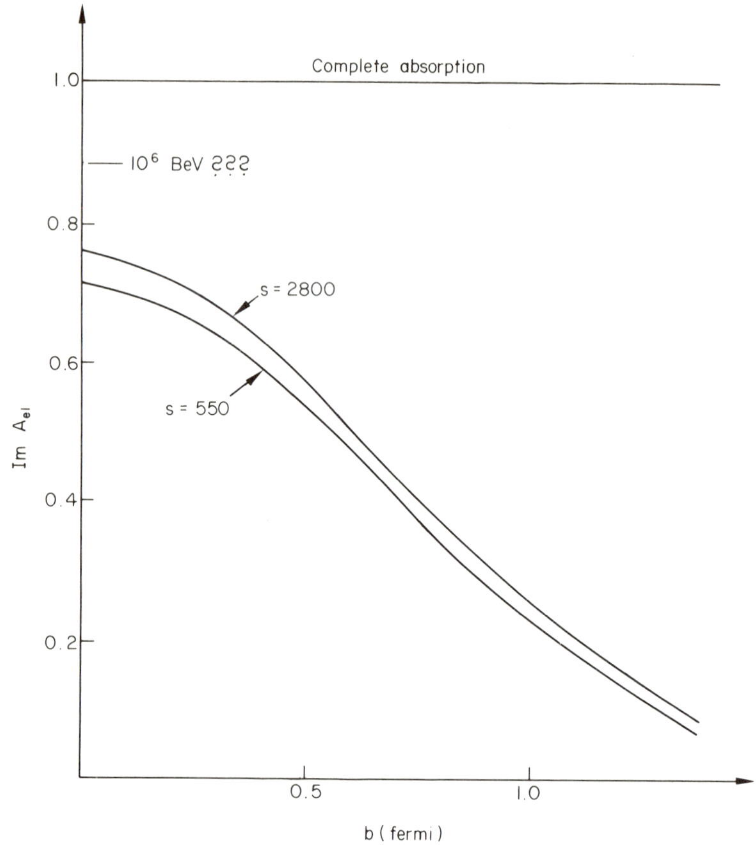

FIG. 39. Impact parameter description of elastic pp scattering. The 10^6 BeV point is calculated assuming a log s rise of Im $A_{el}(b = 0)$.

1. The imaginary part of the forward elastic amplitude or the total cross-section for pp scattering has two components—an R-component (Fig. 40a) and a P-component (Fig. 40b).

2. The R-component (Fig. 40a) behaves at $s \to \infty$ like $s^{\alpha_R - 1}$. It is built (through unitarity) by a sum of non-diffractive terms (Fig. 40c).

3. The P-component (Fig. 40b) behaves at $s \to \infty$ as a constant (modulo $\log s$ terms). It is built (through unitarity) by two components—a non-diffractive M-component (Fig. 40d) and a diffractive D-component (Fig. 40e).

4. The M-component (Fig. 40d) includes multiplicity cross-sections m_n, decreasing like $s^{2\alpha_R - 2}$ as $s \to \infty$. Their sum produces a constant.

5. The D-component (Fig. 40e) includes double diffraction excitation (Fig. 40f) and single diffraction excitation (Fig. 40g).

TABLE III. Summary of the components of multihadron production

Component		σ_{tot}	σ_n	$\langle n \rangle$	Estimated contribution for $100 < s < 3000$ BeV^2
R	(40a, 40c)	$s^{\alpha_R - 1}$	$s^{2\alpha_R - 2}$	log s	<1 mb
P (40b) — M	(40d)	const	$s^{2\alpha_R - 2}$	log s	24–27 mb
P (40b) — D (40e) — Double diffraction excitation	(40f)	Includes many different (log s)^n terms		log s	1–2 mb
P (40b) — D (40e) — Single diffraction excitation (40g) — PPR	(40h, 40j)	const	const	const	1–2 mb
P (40b) — D (40e) — Single diffraction excitation (40g) — PPP (40i) — D	(40k)	Part of Double Diffraction Excitation			
P (40b) — D (40e) — Single diffraction excitation (40g) — PPP (40i) — M	(40l)	log s	const	log s	3–8 mb

6. The double diffraction excitation (Fig. 40f) includes several terms corresponding to high or low mass excitations in each vertex. The leading contribution to σ_{tot} behaves like $\log^2 s$ but is probably of the order of 1–2 mb at $s \sim 3000\ \text{BeV}^2$. We have therefore neglected double excitations in our discussions.

7. The single diffractive excitation (Fig. 40g) may be described in terms of Pomeron–proton elastic scattering at the lower vertex. The elastic Pomeron–proton amplitude has two components—an R-component (Fig. 40h) and a P-component (Fig. 40i).

8. The R-component (Fig. 40h) involves a PPR vertex. It corresponds to $\sigma^{Pp} \underset{M^2\to\infty}{\to} (M^2)^{\alpha_R - 1}$ and is sometimes referred to as "low mass diffraction excitation". Its contribution to σ^{Pp}_{tot} is constant as $s \to \infty$. The R-exchange in Pp elastic scattering is built through unitarity from many non-diffractive terms (Fig. 40j).

9. The P-component (Fig. 40i) involves a PPP vertex. It corresponds to $\sigma^{Pp} \underset{M^2\to\infty}{\to}$ const and is sometimes referred to as "high mass diffraction excitation". Its contribution to σ^{Pp}_{tot} behaves like $\log s$ as $s \to \infty$. The P-exchange in Pp elastic scattering is built through unitarity from two components: a D-component (Fig. 40k) and an M-component (Fig. 40l).

10. The D-component (Fig. 40k) involves diffractive mechanisms in specific Pp production cross-sections. It can be viewed as a part of the double diffraction excitation mechanism (Fig. 40f) and is probably very small for $s < 3000\ \text{BeV}^2$.

11. The M-component (Fig. 40l) of the Pp total cross-sections involves the non-diffractive contributions to Pp scattering. This term probably dominates the diffractive part (Fig. 40e) and it was the subject of our discussion in this Section.

In Table III we present a summary of the various components, including the asymptotic energy dependence of their contributions to the total cross-section and the multiplicity cross-section, and their average multiplicity. We also present a rough estimate of the size of each component at the ISR energy range. Note that the double diffraction excitation term is the term which would include the potentially disastrous contributions which might eventually violate the Froissart bound but are presently extremely small.

Finally, Fig. 41 provides us with a schematic description of the way in which a total hadronic cross-section is built by its many final states.

We conclude that any phenomenological description of pp scattering at the energy range $100 < s < 3000\ \text{BeV}^2$ should include the M-component (discussed in Section II), the high mass single diffraction excitation (the triple Pomeron term) discussed in the present Section, as well as smaller contributions from low mass diffraction single excitation and from double

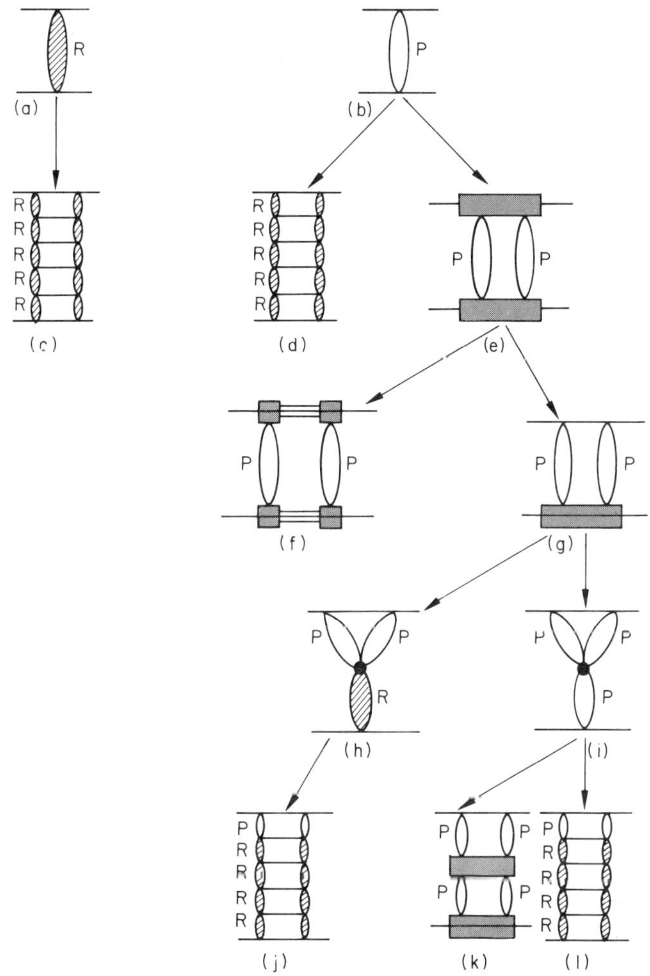

FIG. 40. The various components of the contribution of multihadron production to the forward elastic amplitude.

diffraction excitation. An improved two-component model should allow the last terms as part of the *D*-component. Its general qualitative conclusions will not differ by much from the ones reached in Section II, with the exception of the rise in σ_{tot} and in the central plateau, discussed in this Section. The multihadron data are not yet sufficiently rich to justify a detailed fit involving all the terms mentioned above.

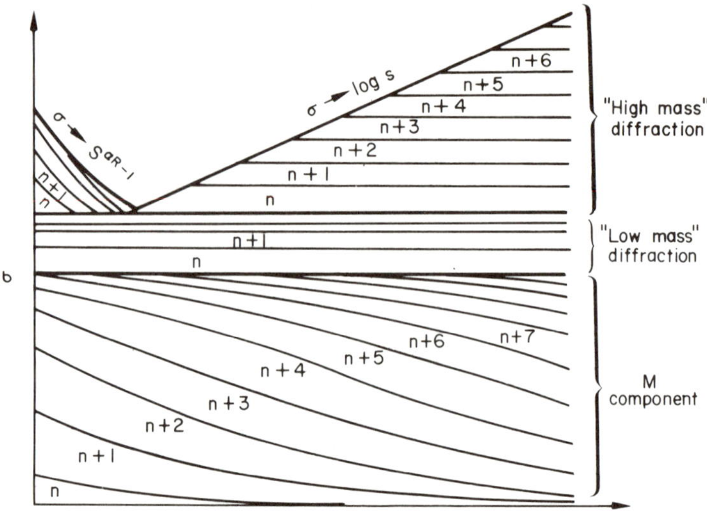

FIG. 41. Schematic description of the energy dependence of the various components of the total cross-section (including "high mass" diffraction excitations).

IV. Statistical Mechanics of the Feynman Gas: A Mathematical Toy or a Useful Analogy?

1. The Feynman Gas Analogy

The Feynman gas analogy[12] treats the three dimensional space defined by the rapidity y and the two components of transverse momentum $\vec{p}_T$ as a "volume of gas" populated by particles produced in hadron collisions. For most purposes it is actually convenient to integrate over the transverse momentum variables and to consider a "one dimensional volume" of gas with length $Y = \log s$. The multiplicity cross-section $\sigma_n(s)$ is analogous to the partition function, the normalized inclusive cross-section

$$\rho(y) = \frac{1}{\sigma_{\text{tot}}} \int \frac{\mathrm{d}^3\sigma}{\mathrm{d}y\,\mathrm{d}p_T^2}\,\mathrm{d}p_T^2 \tag{115}$$

is the gas density, and the generating function

$$Q(z, s) = \sum z^n \sigma_n(s) \tag{116}$$

defined in Section II, corresponds to the grand partition function. The variable z corresponds to the fugacity. For $z = 0$ and $z = 1$ $Q(z)$, has a simple meaning:

$$Q(0) = \sigma_0; \qquad Q(1) = \sigma_{\text{tot}} \tag{117}$$

For $z \ll 1$, $Q(z)$ is strongly dominated by $n \ll \bar{n}$. For $z \gg 1$, $Q(z)$ reflects the properties of σ_n for $n \gg \langle n \rangle$.

In statistical mechanics the logarithm of the grand partition function obeys the following relation:

$$\log Q(z) = p(z)V + S(z) \tag{118}$$

where V is the volume, $p(z)$ is the pressure, and $S(z)$ is proportional to the surface tension. The pressure is essentially defined as:

$$p(z) = \lim_{V \to \infty} \frac{\log Q(z)}{V} \tag{119}$$

Since the volume of the one dimensional Feynman gas is given by $\log s$, we may define the "pressure" of the Feynman gas as:

$$p(z) = \lim_{s \to \infty} \frac{\log Q(z, s)}{\log s} \tag{120}$$

We may also define the "surface tension" of the Feynman gas as the s-independent term in $\log Q(z)$. Hence we have

$$\log Q(z, s) = p(z) \log s + S(z) \tag{121}$$

These expressions for $p(z)$ and $S(z)$ implicitly assume that as $s \to \infty$, $\log Q(z)$ is proportional to $\log s$. This condition (analogous to the requirement of the existence of a thermodynamical limit) is not *a priori* guaranteed in multi-hadron production. However, we have already seen in Section II (eqn (46)) that it is valid in all short range correlation models.

The functions $p(z)$ and $S(z)$ can be computed for any model which makes specific predictions for $\sigma_n(s)$, provided that the thermodynamic limit exists. These functions can also be directly deduced from experimental data on $\sigma_n(s)$. A knowledge of $\sigma_n(s)$ enables us to compute the function $Q(z, s)$, from which $p(z)$ and $S(z)$ are easily calculated. We have to remember, however, that any extraction of $p(z)$ and $S(z)$ from experimental multiplicity cross-sections uses a finite energy range. As a result, the obtained functions may differ from their "true" form which would be derivable only at $s \to \infty$. They may then reflect an "asymptotic" s-dependence over a certain high energy range, without guaranteeing that at still higher energies the same behaviour prevails.

Our ability to compute $p(z)$ and $S(z)$ both from the data and from theoretical models makes these quantities a useful tool for confronting theoretical ideas with the data. It should be stressed, however, that the entire information content of these two functions is included in the $\sigma_n(s)$ cross-sections. Thus, the thermodynamic analysis may be easier, it may be prettier, and it may be capable of emphasizing aspects which are otherwise obscured, but it is not carrying any new fundamental message.

We will return to this question and to possible deeper utilization of the statistical properties of the Feynman gas in the closing section of the present Section.

2. A TRIVIAL EXAMPLE: A POISSON MULTIPLICITY DISTRIBUTION

Our first exercise will be to derive the functions $p(z)$ and $S(z)$ for the case of a non-diffractive M-component of a constant total cross-section. We define the constant σ_{tot} as our unit, and assume the simplest model for $\sigma_n(s)$, namely—the Poisson distribution:

$$\sigma_n(s) = \frac{(g \log s/s_0)^n}{n!}\left(\frac{s}{s_0}\right)^{-g} \tag{122}$$

The function $Q(z)$ is given by:

$$Q(z) = \sum_n \frac{(zg \log s/s_0)^n}{n!}\left(\frac{s}{s_0}\right)^{-g} = \left(\frac{s}{s_0}\right)^{g(z-1)} \tag{123}$$

Hence:

$$\log Q(z) = g(z-1)\log s - g(z-1)\log s_0 \tag{124}$$

The pressure and surface tension are given by:

$$p(z) = g(z-1) \tag{125}$$

$$S(z) = -g(z-1)\log s_0 \tag{126}$$

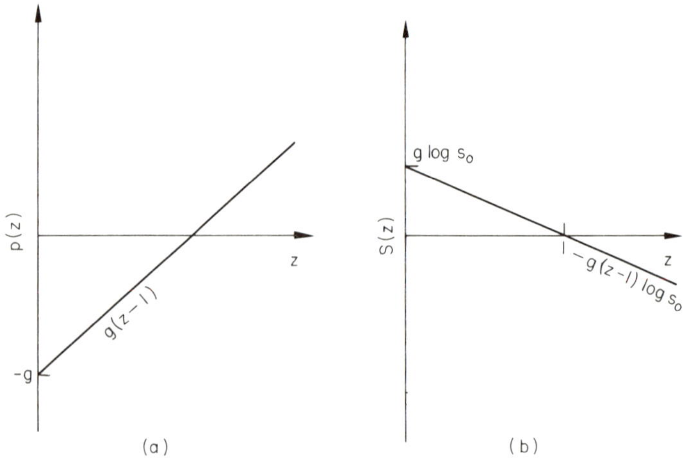

FIG. 42. The pressure and surface tension of a non-diffractive Poisson-type Feynman gas.

and are shown in Fig. 42. The pressure at any given value of z "measures" the power of s, which determines the $s \to \infty$ behaviour of $Q(z)$. Hence, $p(0)$ measures the energy dependence of $Q(0) = \sigma_0$, while $p(1)$ measures the energy dependence of σ_{tot}. For any non-diffractive model (not only for the Poisson distribution) we conclude:

$$p(0) = 2\alpha_R - 2 \quad (127)$$

$$p(1) = \alpha_P - 1 \quad (128)$$

In our simple Poisson example we essentially assumed $2\alpha_R - 2 = -g$; $\alpha_P = 1$. Consequently, we automatically have: $p(0) = -g$, $p(1) = 0$.

3. A MORE TRIVIAL EXAMPLE: LOW MASS DIFFRACTION EXCITATION

A diffractive component of the type discussed in Section II would have

$$d_n \xrightarrow[s\to\infty]{} \text{const} \quad (129)$$

$$\sum_n d_n \xrightarrow[s\to\infty]{} \text{const} \quad (130)$$

In the terminology of Section III this would correspond to "low mass diffraction excitation". Since every d_n is assumed to be energy independent, we have:

$$Q(z, s) = \sum_n z^n d_n \,(\text{independent of } s) \quad (131)$$

Hence,

$$\log Q(z, s) = 0 \,.\, \log s + \log\left[\sum_n z^n d_n\right] \quad (132)$$

and

$$p(z) = 0 \quad (133)$$

$$S(z) = \log\left[\sum_n z^n d_n\right] \quad (134)$$

If we, again, normalize $\Sigma_n d_n = 1$, we have $S(1) = 0$, $S(0) = \log d_0 < 0$. The detailed form of $S(z)$ depends on the n dependence of d_n. However, qualitatively the pressure and surface tension are given by Fig. 43.

A low mass diffractive excitation mechanism will tend to populate only the two fragmentation regions of the rapidity axis. It is therefore not surprising that the physics of such a mechanism is entirely given by the surface tension. It is also not surprising that the pressure vanishes. Since most of the volume is empty, we can hardly have any pressure!

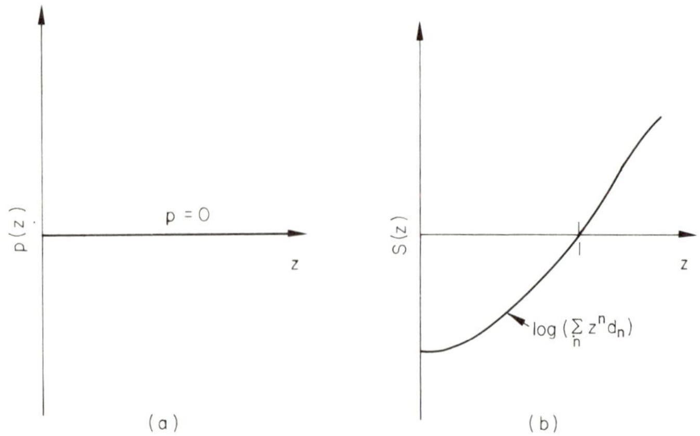

FIG. 43. The pressure and surface tension corresponding to "low mass" diffraction excitation.

4. THE PRESSURE AND SURFACE TENSION FOR THE SIMPLEST TWO-COMPONENT MODEL

Let us now study an extremely oversimplified two-component model in which:

$$\sigma_n = m_n + d_n \tag{135}$$

$$\sum_n \sigma_n = 1; \quad \sum_n m_n = M; \ \sum_n d_n = D \tag{136}$$

$$m_n = M \frac{(g \log s/s_0)^n}{n!} \left(\frac{s}{s_0}\right)^{-g}. \tag{137}$$

$$d_n = \text{const.} \tag{138}$$

Such a model includes a correlation-free M-component (as in Section IV.2) and a "low mass" excitation D-component (as in Section IV.3).

The function $Q(z)$ is given by:

$$Q(z) = \sum_n z^n \sigma_n = M\left(\frac{s}{s_0}\right)^{g(z-1)} + \sum_n z^n d_n. \tag{139}$$

In order to extract the pressure and the surface tension we have to study the $s \to \infty$ behaviour of $Q(z)$. We have to distinguish between two regions:

(i) $z > 1$

In this region the first term of eqn (139) dominates as $s \to \infty$. We have:

$$\log Q(z) \underset{\substack{s\to\infty \\ z>1}}{\to} g(z-1)\log s + \log M - g(z-1)\log s_0 \tag{140}$$

Yielding:

$$\left.\begin{aligned} p(z) &= g(z-1) \qquad &(141)\\ S(z) &= \log M - g(z-1)\log s_0 \qquad &(142)\end{aligned}\right\} z > 1.$$

(ii) $z < 1$

In this region the second term of eqn (139) dominates as $s \to \infty$. We have:

$$\log Q(z) \underset{\substack{s\to\infty\\ z<1}}{\longrightarrow} \log \left[\sum_n z^n d_n\right] \tag{143}$$

$$\left.\begin{aligned} p(z) &= 0 \qquad &(144)\\ S(z) &- \log \left[\sum_n z^n d_n\right] \qquad &(145)\end{aligned}\right\} z < 1.$$

The z-dependence of $p(z)$ and $S(z)$ is shown in Fig. 44. We see that for $z > 1$ the curves behave as if the process were completely non-diffractive while for $z < 1$ the diffractive component alone determines the curves. This is easily understood if we remember that $z \gg 1$ reflects the physics of multiplicities $n \gg \langle n\rangle$ while $z \ll 1$ reflects the mechanism responsible for multiplicities below the average. By taking the $s \to \infty$ limit we sharpen this separation and we obtain a pressure and a surface tension which look as if they underwent a "phase transition" at $z = 1$. Note that $S(z)$ is actually discontinuous at $z = 1$. We have:

$$S(1+\varepsilon) = \log M - g\varepsilon \log s_0 \underset{\varepsilon\to 0}{\rightarrow} \log M \tag{146}$$

$$S(1-\varepsilon) = \log \left[\sum_n (1-\varepsilon)^n d_n\right] \underset{\varepsilon\to 0}{\rightarrow} \log D \tag{147}$$

$$S(1) = 0 \text{ (from eqn (139))} \tag{148}$$

The theoretical curves of Fig. 44 show some interesting features of the simple two-component model. Do the data show these features?

5. EXPERIMENTAL MEASUREMENTS OF THE PRESSURE AND SURFACE TENSION OF THE FEYNMAN GAS

We may now use the Serpukhov and NAL accelerators as barometers, and measure the pressure of the Feynman gas.(25) All we have to do is to take the available measurements of $\sigma_n(s)$ for $p_{\text{lab}} = 50, 69, 102, 205, 303$ BeV/c and compute $Q(z, s)$. We then fit $Q(z, s)$ for every value of z, by an expression of the form: $a \log s + b$. The resulting a and b parameters represent the pressure and surface tension at that particular value of z.

The results of this exercise are shown in Figs 45 and 46. We see that the "experimental" pressure and surface tension do not resemble the predictions of the M-component (Fig. 42) or of the D-component (Fig. 43). However, they

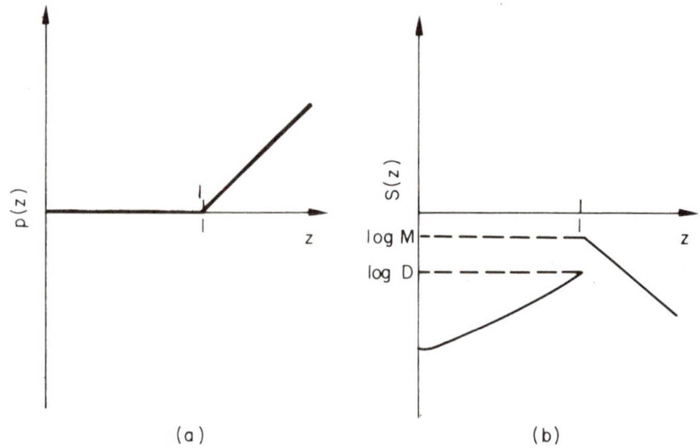

FIG. 44. The pressure and surface tension of the Feynman gas for the simplest two-component model.

are also different from the prediction of the two component model (Fig. 44). The experimental curves show clear signs of deviation from the M-component prediction for $z < 1$, but they actually lie about halfway between the two component prediction and a pure M-component prediction. This clearly indicates that there is *some* truth in the two component model, but why didn't we get the predicted breaks at $z = 1$?

The answer is simple. The "true" $p(z)$ and $S(z)$ curves are obtained for $s \to \infty$. As long as s is finite we are getting only an "effective pressure" and an "effective surface tension" corresponding to the finite energy range in which measurements are made. At that range the expression for $Q(z)$ (eqn (139)), even if it is exactly correct (and it is not!), is not entirely dominated by its first term for $z > 1$ and by its second term for $z < 1$ (as it would be at $s \to \infty$). Hence, the transition between $z > 1$ and $z < 1$ is "softened" by finite energy effects. As we perform our fits at higher and higher energies, these effects should disappear and we should asymptotically reach the curves of Fig. 44.

We may verify this by doing a "theoretical experiment". We may take our phenomenological two component fit of Section II, assume that the data all the way to $s \to \infty$ are given by that fit (a certainly false assumption) and compute $p(z)$ and $S(z)$ for various finite energy ranges. The results are shown in Figs 47 and 48. It is clear from the figures that while the $s \to \infty$ limit may give a sharply changing curve at $z = 1$, fits at lower and lower energy ranges show a smoother transition at $z = 1$. In fact, the curves corresponding to the energy range $100 < s < 600 \text{ BeV}^2$ for which data are available are in good agreement with the "experimental" $p(z)$ and $S(z)$. This is hardly surprising since the theoretical curves of Figs 47 and 48 used parameters fitted

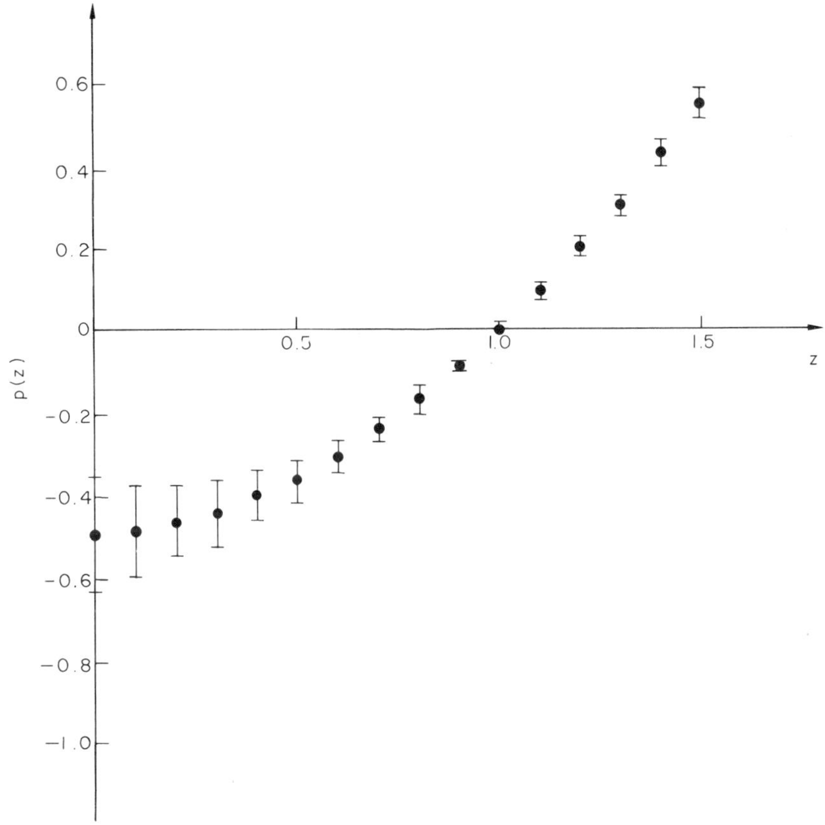

FIG. 45. The "experimental" pressure of the Feynman gas.

to the data which yielded Figs 45 and 46. The agreement here is just another manifestation of the agreement seen in Figs 21–23 in Section II.

6. THE PRESSURE IN A MORE REALISTIC M-COMPONENT

Our discussion of the pressure of the M-component of the Feynman gas was obviously oversimplified by assuming a Poisson-type multiplicity distribution, thus neglecting all correlations.

The most general model for an M-component with short range correlations would allow all moments f_m of the multiplicity distributions to be non-vanishing, but would imply:

$$f_m \underset{s \to \infty}{\to} c_m \log s \tag{149}$$

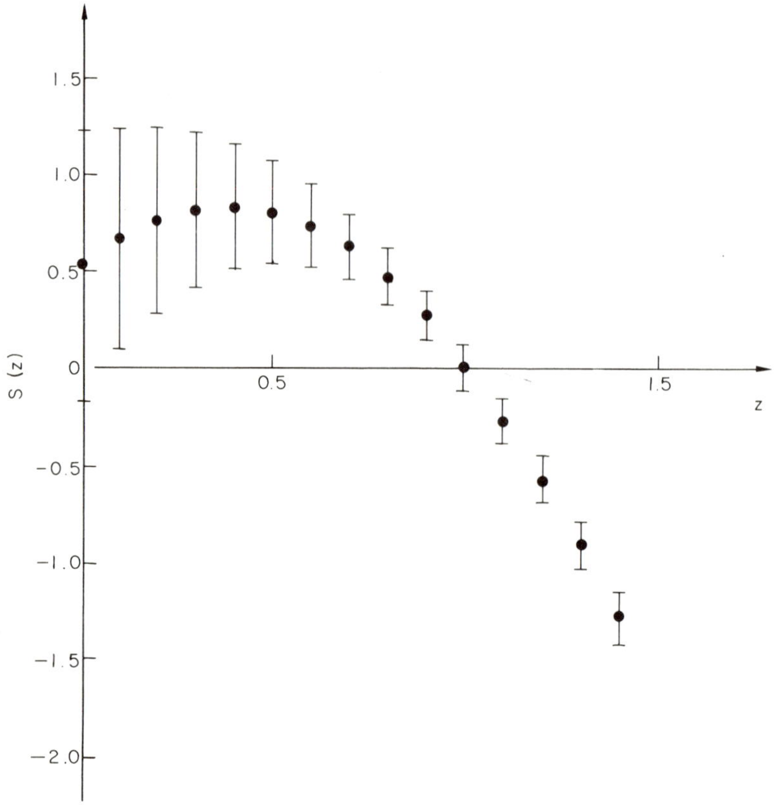

FIG. 46. The "experimental" surface tension of the Feynman gas.

We have seen in Section II that:

$$\log Q(z, s) = \sum_{m=1}^{\infty} \frac{f_m}{m!}(z-1)^m \tag{150}$$

Hence:

$$\log Q(z, s) \underset{s\to\infty}{\rightarrow} \sum_{m=1}^{\infty} \frac{c_m}{m!}(z-1)^m \log s \tag{151}$$

and:

$$p(z) = \sum_{m=1}^{\infty} \frac{c_m}{m!}(z-1)^m \tag{152}$$

The Poisson example is, of course, a special case of this relation with $c_1 = g$; $c_m = 0 (m \geqslant 2)$. Equation (152) is actually valid only when $\sigma_{\text{tot}} = \text{const}$ or

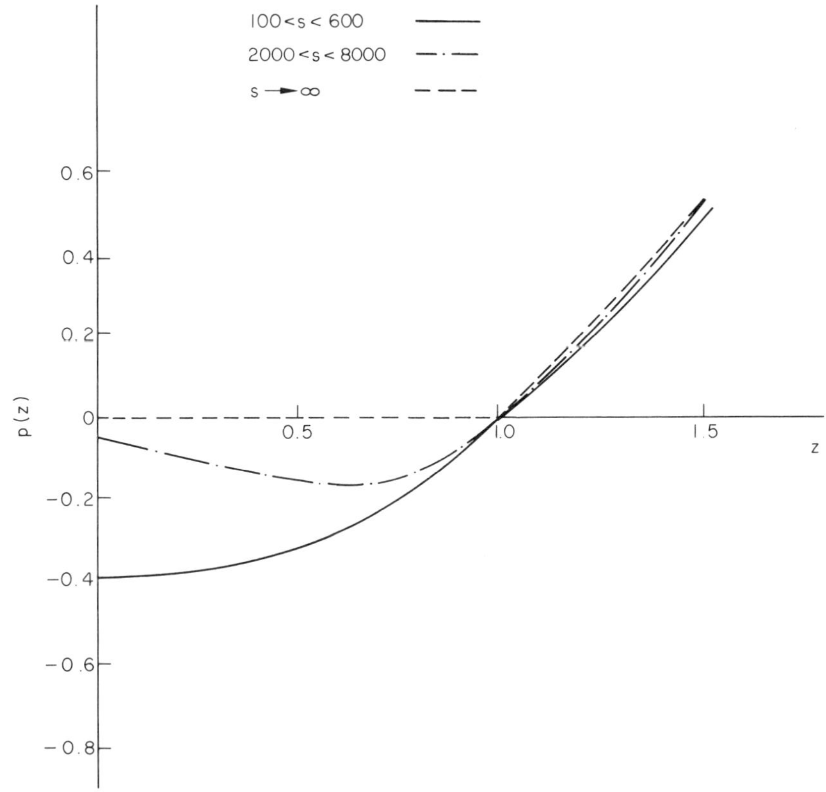

FIG. 47. The pressure of the Feynman gas, as predicted by a naive two-component model, for several energy ranges.

when we normalize $Q(z)$ to the total cross-section (namely—if we measure σ_n in units of σ_{tot}). In a more general case we would have:

$$\sigma_{tot} \underset{s\to\infty}{\to} s^{\alpha_P - 1} \tag{153}$$

Hence:

$$Q(1) \propto s^{\alpha_P - 1}; \qquad \log Q(z=1) \underset{s\to\infty}{\to} (\alpha_P - 1)\log s \tag{154}$$

$$p(z=1) = \alpha_P - 1 \tag{155}$$

Equation (152) would then read:

$$p(z) = p(1) + \sum_{m=1}^{\infty} \frac{c_m}{m!}(z-1)^m \tag{156}$$

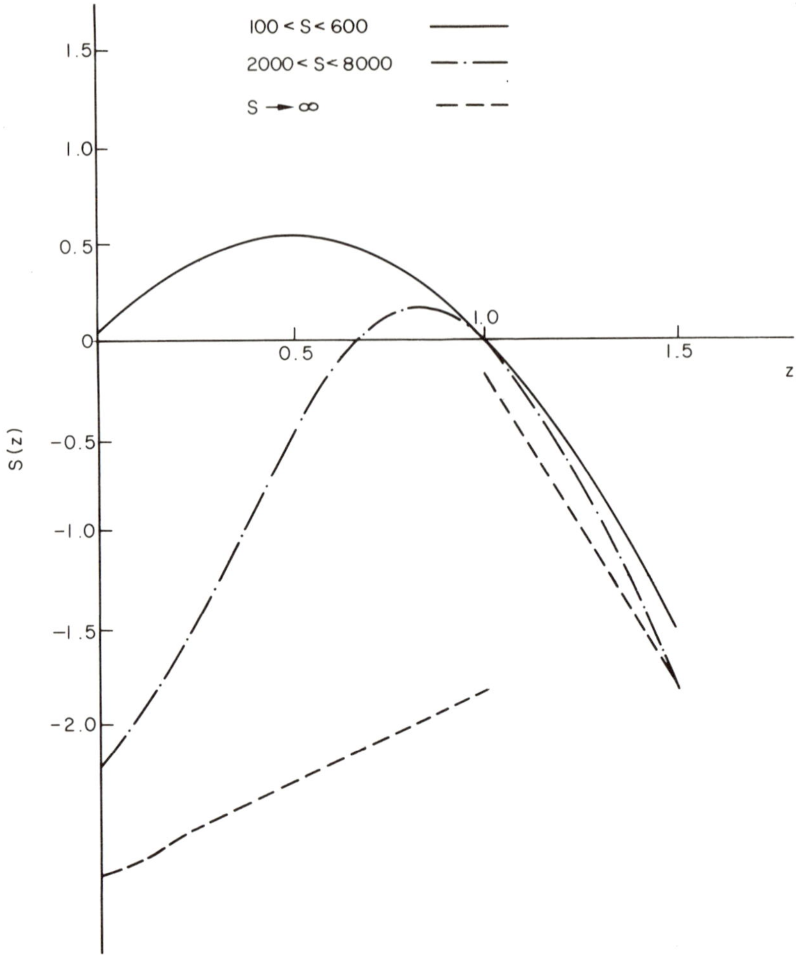

FIG. 48. The surface tension of the Feynman gas, as predicted by a naive two-component model for several energy ranges.

Since, in general, $p(z) = 2\alpha_R - 2$, we have:

$$2\alpha_R - 2 = \alpha_P - 1 + \sum_{m=1}^{\infty} (-1)^m \frac{c_m}{m!} \tag{157}$$

This is the source of eqn (78) which we have used in Section II for relating α_R to the c_i. The general form of the pressure in a short range correlation model is therefore, in general, not linear in z. The various powers of z in the expansion of $p(z)$ reflect the correlation effects. In the case of our two component fit in Section II, for $c_1 = 1{\cdot}0$, $c_2 = 0{\cdot}35$, we have:

$$p(z) = (z - 1) + \tfrac{1}{2} \times 0{\cdot}35\,(z - 1)^2 \tag{158}$$

Figure 49 indicates that the deviation of this expression from a straight line is extremely small.

7. THE PRESSURE IN A MORE REALISTIC *D*-COMPONENT

The pressure shown in Fig. 43 and discussed in Section IV.3 for the *D*-component corresponds to the naïve oversimplified *D*-component which we have discussed in Section II. However, in Section III we introduced a more realistic *D*-component, allowing for high mass diffraction excitation and leading to a logarithmically increasing σ_{tot} and $\langle n \rangle_d$. What would be the pressure of such a *D*-component?

In order to extract the pressure we have to compute the $Q(z, s)$ function corresponding to the multiplicity distribution of eqn (92). We can do it either directly or, more easily by computing $Q(z, M)^2$ for Pomeron–proton scattering (using eqn (90)) and then integrating over M^2. We get:

$$Q(z, s) = \frac{s^{c_1(z-1)} - s_0^{c_1(z-1)}}{z - 1} \tag{159}$$

$$p(z) = \lim_{s \to \infty} \frac{\log Q(z, s)}{\log s} = \begin{cases} c_1(z - 1) & z > 1 \\ 0 & z < 1 \end{cases} \tag{160}$$

We see that the "improved" *D*-component alone gives us a pressure curve with a "break" at $z = 1$. The curve is actually identical to the one obtained from the simplest two component model (Fig. 44) although $Q(z, s)$ is different from that of eqn (139). The physical reason for this strange behaviour can be easily understood. The *D*-component was described in Section III as a mechanism in which, on the average, half of the rapidity axis is "taken" by the Pomeron link and the other half is "taken" by an ordinary multiperipheral chain, producing a Feynman gas with the usual density $\rho(y) = c_1$ but a smaller volume (Fig. 32). Typically, an event with $n \ll \langle n \rangle = \frac{1}{2}c_1 \log s$ will correspond to a large "empty segment" of the rapidity axis and a small "volume of gas" contained in the remaining part of the y-axis. In this case the pressure vanishes, since most of the volume is empty. On the other hand, for $n \gg \bar{n}$ the "empty" portion will be extremely small (and tending to zero for sufficiently large n) while most (or all) of the volume will be taken by a "normal" Feynman gas. Hence, for $z > 1$ (or $n \gg \langle n \rangle$) the "high mass excitation" *D*-component behaves very much like the M-component. The only difference between the two is the object exchanged in the first link of the multi-

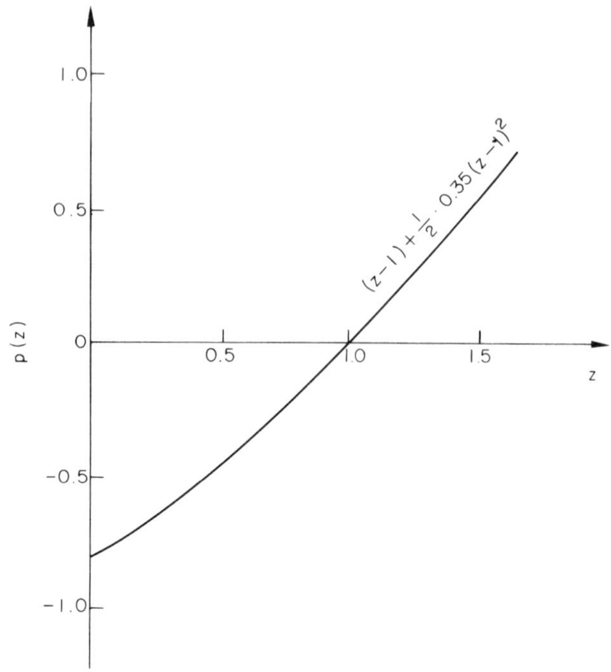

FIG. 49. The pressure $p(z) = (z - 1) + \frac{1}{2} \cdot 0{\cdot}35(z - 1)^2$ corresponding to the M-component of our fit in lecture II.

peripheral chain (compare Fig. 40d and 40l), but when $n \gg \langle n \rangle$ the sub-energy s_1 corresponding to that link is small, and when s_1 is small there is no significant qualitative difference between P-exchange and R-exchange.

8. SUMMARY: THE PRESSURE OF THE FEYNMAN GAS

We have learned that the pressure curve for $z > 1$ reflects the properties of events with multiplicities much above average, while the $z < 1$ region reflects the extremely low multiplicities. We actually expect the $z > 1$ region to represent the M-component (since even if the D-component contributes there, it would have the same gross features as the M-component). The $z < 1$ region should eventually show zero pressure, corresponding to a large gap in the rapidity axis (or a large empty volume) induced by Pomeron exchange. The qualitative behaviour of $p(z)$ is then expected to approach that of Fig. 44a at sufficiently high energies, regardless of the details of the M and D components. In that respect it is a relatively model independent test of the two-component hypothesis. It should be interesting to watch how $p(z)$ changes when we extract it from multiplicity cross-sections at higher and higher energies. The present trend is encouraging (Fig. 45), but multiplicity measurements at the ISR should be quite interesting!

9. SPECULATIONS: POSSIBLE DEEPER APPLICATIONS OF THE STATISTICAL MECHANICS OF THE FEYNMAN GAS

As we have already remarked in the beginning of this Section, the discussion of the properties of the Feynman gas in terms of the pressure, surface tension, fugacity, etc. may help us to emphasize certain features and to consider them from a different angle but it does not yield any fundamental new information. In that respect the statistical mechanics of the Feyman gas is nothing but an amusing (and somewhat useful) mathematical toy. It is possible, however, that the statistical mechanics approach may yield much more profound results. It did not yet yield any such results, but we would like to make a few remarks concerning this possibility.

The "molecular potential" in the Feynman gas (as a function of distance) must somehow be related to the π–π interaction (as a function of energy and momentum transfer). The macroscopic features of the Feynman gas are related to properties of inclusive cross-sections, multiplicity distributions and similar "macroscopic" properties of multihadron production processes. In statistical mechanics we have relations between the intermolecular potential and the pressure, fugacity, etc. One such relation is given by:[26]

$$\frac{1}{z} = \int_0^\infty dy\, e^{-v(y) - py} \tag{161}$$

where y is the distance between two molecules (their rapidity difference), $v(y)$ is the two-body potential (related to the π–π interaction?), p the pressure and z the fugacity.

It is quite possible that relations of this kind may enable us to find interesting connections between the π–π interactions on one hand and the inclusive cross-sections and multiplicity distributions on the other hand. Such relations exist, of course, in specific models such as the AFS multiperipheral model, but such models are not realistic. The hope is, then, to correlate, through the statistical mechanics formalism, some features of the theoretically obscure physics of π–π interactions at low energies, to the equally obscure features of multihadron processes at high energies. Whether this can be done, we do not know, but it would be interesting to try.

V. Why Does the Pomeron Have a Unit Intercept?

1. INTRODUCTION

The phenomenological two component description of multihadron production provides us with a deep "two-component theoretical puzzle". The non-diffractive M-component is the part of the total cross-section through which the various non-diffractive mechanisms involving the known Reggeons "build up" the Pomeron via the unitarity relation. The puzzle posed to us by this component is: *How do the Reggeons build up the Pomeron and why does the Pomeron have a unit intercept*? The diffractive D-component is the part of the total cross-section in which the Pomeron contributes to the production of specific final states. In that case, the Pomeron is "building" itself and the puzzle here involves *the self-consistency of a "self-building" Pomeron, as well as the mechanisms by which the Froissart bound is respected in the presence of many Pomerons.*

We then have a "two-component puzzle"

(i) Why does the Pomeron have a unit intercept (or, how is it built by the Reggeons)?

(ii) How does the Pomeron respect the Froissart bound (or, how is it built by itself)?

Sections V and VI are, respectively, devoted to these two questions, assuming that the keys to the answers are contained, respectively, within the M- and D-components.

2. WHY IS $\alpha_P = 1$? THE "PROMOTION MECHANISM"

We assume that the Pomeranchuk singularity is at $\alpha_P(0) = 1$ and that, within the M-component of σ_{tot}, it is built by the non-Pomeron contributions to the n-particle production cross-section. How does it happen that $\alpha_p(0)$ is equal to one and not to 0·7 or 2·5, for example? The oldest mechanism suggested as an answer to this question is the AFS multiperipheral model[9] which assumes that the exchanged objects in the multiperipheral ladder are spin zero particles. Assuming no correlations, we then have:

$$\sigma_n = \frac{(g \log s)^n}{n!} s^{-2} \tag{162}$$

where g is the squared coupling constant of the ladder diagram. In Reggeistic terminology we would say that $\alpha_R = 0$. The resulting total cross-section is

$$\sigma_{tot} \propto s^{-2+g} \tag{163)}$$

Hence:

$$\alpha_P = g - 1 \tag{164}$$

Continuing our oversimplified analysis of this model, we would now say that if and only if $g = 2$, we have $\alpha_P = 1$. It is clear that the same model would yield a higher or lower value of α_P, if the interaction responsible for the production of particles would have been stronger or weaker, respectively.

In such models the value $\alpha_P(0) = 1$ seems like an "accident". The fundamental quantity is α_R which is independent of g, while α_P is an increasing function of g. In the $g \to 0$ limit we always have $\alpha_P = 2\alpha_R - 1$ and in the case of the AFS model: $\alpha_P(g = 0) = -1$.

Such models do provide us with a mechanism by which the power of s in σ_n (namely: $2\alpha_R - 2$) is "promoted" to the power of s in σ_{tot} (namely: $\alpha_P - 1$). However, the final value of α_P depends on the coupling constant and is not guaranteed in any way to be at 1 or below it. s-channel unitarity and the Froissart bound are not intrinsic ingredients of such models.

We then summarize:

$$\alpha_R\,(\text{any } g) = 0; \qquad \alpha_P(g) = -1 + f(g) \tag{165}$$

where $f(0) = 0$, and $f(g)$ is a monotonously increasing function of g.

3. AN ALTERNATIVE APPROACH: α_P IS ALWAYS AT ONE

An alternative philosophy with respect to α_P would be to assume that for some reason $\alpha_P = 1$ for any strength of the strong interactions. If that is the case, $\alpha_P = 1$ even when $g \to 0$. However we have seen that

$$(\alpha_P - 1) \underset{g \to 0}{\to} (2\alpha_R - 2) \tag{166}$$

Hence

$$\alpha_R \underset{g \to 0}{\to} 1 \tag{167}$$

This will happen in a theory which, in the $g \to 0$ limit, describes multihadron production in terms of vector meson exchanges. If we then increase g, α_P would tend to increase above 1. However, the Froissart bound forbids $\alpha_P > 1$. Consequently, some type of unitarization procedure will probably restore $\alpha_P = 1$. At the same time, it is likely that α_R, which was at one for $g = 0$, will be "pushed down" by the same mechanism that prevented α_P from increasing. In fact, in the absence of correlations we always have:

$$\alpha_P - 1 = 2\alpha_R - 2 + g \tag{168}$$

If $\alpha_P = 1$ for all g-values, we must then have:

$$\alpha_R = 1 - \tfrac{1}{2}g \tag{169}$$

The value of α_R *decreases* as g becomes larger! In such a model $\alpha_P = 1$ is the fundamental relation reflecting on one hand a vector meson theory in the $g \to 0$ limit and on the other hand the Froissart bound. The value of α_R depends on g and its physical value in the real world is determined by the strength of the strong interactions.

We then have:

$$\alpha_R(g) = 1 - h(g); \qquad \alpha_P\,(\text{any } g) = 1 \tag{170}$$

where $h(0) = 0$ and $h(g)$ is a monotonously increasing function of g.

We cannot perform experimental tests in order to decide which (if any) of the two alternatives described here corresponds to a correct theory of the strong interactions. Such tests would necessitate a change in the strength of the strong interactions, and are obviously in the domain of science fiction. However, from the theoretical point of view, the difference between the two approaches is both well defined and extremely important.

We believe[(27)] that the fixed-α_P type of approach, in which $\alpha_P = 1$ is a g-independent fundamental result and the value of α_R is "accidental", is much more appealing from the theoretical point of view. The challenge is, therefore, to construct a credible model for multihadron production which obeys the criteria defined in this section (eqn (170)).

Several such models have been proposed. We will devote the rest of this lecture to a brief discussion of thee such ideas. We then analyse what we consider to be the advantages and drawbacks of these models and conclude by suggesting directions for future search of such models.

4. A DUAL MODEL YIELDING $\alpha_P = 1$

The first model which we discuss here was proposed by Huan Lee.[(28)] He considered the various duality diagrams for producing n-hadrons. In the case of a two-hadron final state we have one planar diagram (Fig. 50a) and one non-planar diagram (Fig. 50b) in which the exchanged $q\bar{q}$ lines in the t-channel cross each other (and, in the dual model jargon the meson propagator in the t-channel is "twisted"). In the case of $2 \to 3$ hadrons we still have one planar diagram (Fig. 50c) but three non-planar ones (Fig. 50d) with one or more "twists". For $2 \to 4$ we have one planar (Fig. 50e) and 7 "twisted" (Fig. 50f) contributions. In general, every propagator can be either twisted or untwisted. An amplitude with an n-particle final state has $(n - 1)$ exchanged mesons in the multiperipheral ladder, leading to 2^{n-1} diagrams. Of these one and only one is planar (Fig. 50a, c, e) and $(2^{n-1} - 1)$ diagrams are non-planar.

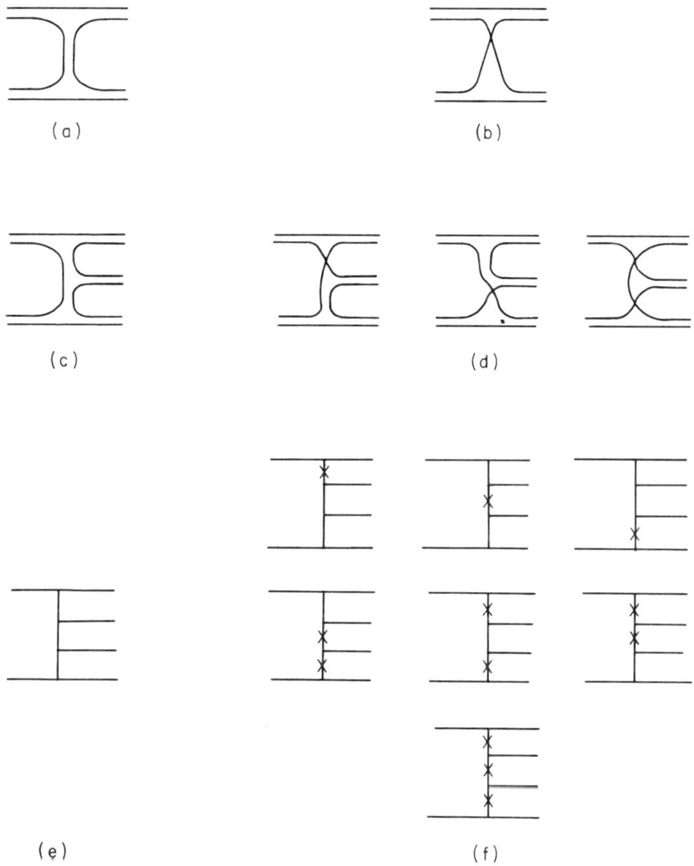

FIG. 50. "Untwisted" and "twisted" duality diagrams for 2, 3 and 4-particle final states.

If we assume exchange degeneracy, the contribution of any given diagram to the $2 \to n$ cross-section is independent of the number of "twists". If we further neglect all interference terms between different diagrams and neglect all correlations among the produced particles we will find:

$$\sigma_n \propto 2^{n-1} \frac{(g \log s)^n}{n!} s^{2\alpha_R - 2} \tag{171}$$

where 2^{n-1} is the number of different diagrams contributing to σ_n.

Next we consider the imaginary part of the forward elastic amplitude. It contains two components:

$$\operatorname{Im} A_{el} = R + P \tag{172}$$

where R contributes an s^{α_R-1} term to σ_{tot} and P contributes an s^{α_P-1} term. The duality diagrams representing R and P are shown in Fig. 51.

Note that the non-planar diagram (Fig. 50b) does not contribute to the *imaginary* part of the elastic amplitude (see also Section I.2). The optical theorem states that Im A_{el} is given by the sum of all multiplicity cross-sections σ_n. But those are given in terms of the diagrams of Fig. 50 and their contributions to Im A_{el} can be studied by considering the diagrams of Fig. 52. These

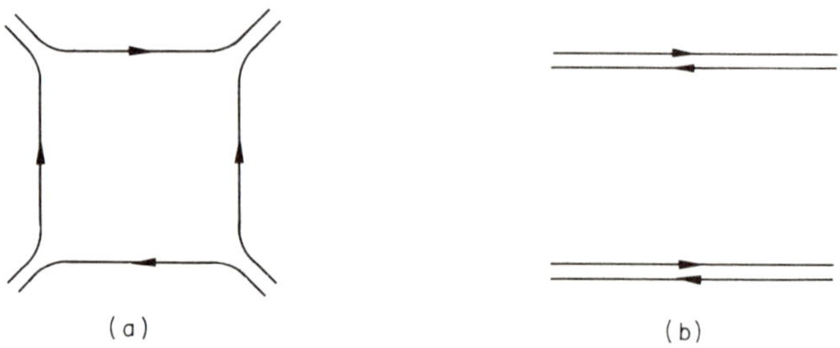

FIG. 51. Duality diagrams for the R and P components of Im A_{el}.

diagrams (for $n = 3$) correspond to the production mechanism of Fig. 50c and d. We have used here the no-interference assumption mentioned above (otherwise we should have considered diagrams such as Fig. 53).

Which diagrams in Fig. 52 contribute to which component of Im A_{el}? A quick comparison between Figs 51 and 52 immediately teaches us that the planar multihadron diagrams (such as 52a) are building up the R-component of Im A_{el} (Fig. 51a) while all the non-planar (twisted) multihadron diagrams (such as Figs 52b, c, d) build up the P-component of Im A_{el} (Fig. 51b). The quickest way to verify this is to eliminate all closed quark loops in the multihadron diagrams and to compare the topological structure of the remaining quark lines to that of the two components in Fig. 51.

If we now write:

$$R + P = \text{Im}\, A_{\text{el}} \propto \sigma_{\text{tot}} = \sum_n \sigma_n = \sum_n 2^{n-1} \frac{(g \log s)^n}{n!} s^{2\alpha_R - 2} \tag{173}$$

we immediately realize that of the 2^{n-1} terms in σ_n (eqn (171)), *one* term contributes to R while $(2^{n-1} - 1)$ terms contribute to P. Hence we find:

$$s^{\alpha_R - 1} \propto \sum_n^{\infty} \frac{(g \log s)^n}{n!} s^{2\alpha_R - 2} \tag{174}$$

$$s^{\alpha_P - 1} \propto \sum_n^{\infty} (2^{n-1} - 1) \frac{(g \log s)^n}{n!} s^{2\alpha_R - 2} \tag{175}$$

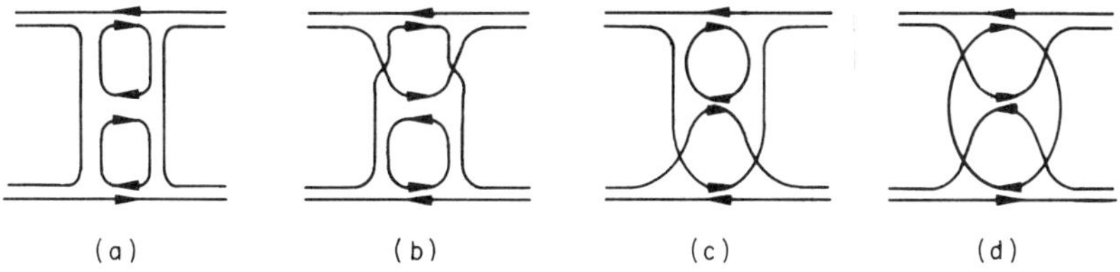

FIG. 52. Duality diagram description of the contributions of 3-particle states to the unitarity sum for Im A_{el}.

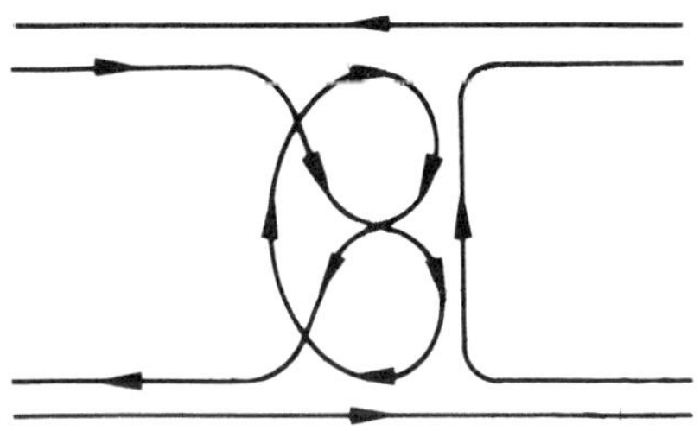

FIG. 53. An "interference" contribution. Such terms are neglected by H. Lee.

Comparing the asymptotic s-behaviour of the two sides of each of these equations, we get:

$$\alpha_R - 1 = g + 2\alpha_R - 2 \tag{176}$$

$$\alpha_P - 1 = 2g + 2\alpha_R - 2 \tag{177}$$

Equation (176) immediately yields:

$$\alpha_R = 1 - g \tag{178}$$

Hence, for $g \to 0$, $\alpha_R \to 1$ and for increasing values of g, α_R decreases. Substituting this result in eqn (177) we then find:

$$\alpha_P = 1 \text{ (independent of } g\text{!).} \tag{179}$$

The model therefore obeys the criteria suggested in section 3.

Let us review the important ingredients of this model:

(1) The number of diagrams for n-particle production contributing to the P component of Im A_{el} was larger (by a power of n) than the corresponding number for the R-component. Hence, while each diagram contributes $\sigma \propto s^{2\alpha_R - 2}$, the large number of diagrams for the P component "promotes" $2\alpha_R - 2$ to $\alpha_P - 1$ while the small number of diagrams for R promote $2\alpha_R - 2$ only to $\alpha_R - 1$.

(2) The identification of α_R within the multiperipheral ladder with the α_R of the R-component of Im A_{el} led to a unique relation between α_R and g. In the $g \to 0$ limit there is no "promotion". Hence $2\alpha_R - 2 = \alpha_R - 1$ or: $\alpha_R = 1$.

(3) The most interesting prediction: $\alpha_P = 1$ for any g, used both the above relation between α_R and g and the relation by which $2\alpha_R - 2$ is "promoted" to $\alpha_P - 1$. It was crucial that the ratio of the numbers of diagrams was essentially 2^n. Had it been a^n where $a \neq 2$, we would have found $\alpha_P \gtrless 1$ for $a \gtrless 2$.

(4) Two crucial technical assumptions were made here: the absence of correlations and the absence of interference terms. Both are probably not valid in nature. Unfortunately, any correlations might move α_P away from one. The result is, therefore, highly unstable, contrary to the philosophy developed in section 3 according to which $\alpha_P = 1$ should be a fundamental "stable" relation. The justification for neglecting the interference terms (such as Fig. 53) involves subtle technical questions related to the theory of dual resonance models. We will not enter such a discussion at this point.

(5) The most important drawback of the H. Lee model, in our opinion, is the fact that none of the physical ingredients which impose the Froissart bound, are present in the model. From this point of view, there is no guarantee whatsoever that $\alpha_P \leqslant 1$ in such a model. In fact, it is easy to get $\alpha_P > 1$ by relaxing the no-correlation assumption or the no-interference assumption. We believe that the only way to obtain $\alpha_P = 1$ for any g as a "stable" result (unchanged in the presence of small modifications of any sort) is to utilize the unitarity bound on the physical amplitude.

Nevertheless, we believe that this model has many interesting features and it might prove useful in various studies of the Pomeron intercept.

An interesting experimental test of the model[(28)] involves the n-dependence of the ratio between the σ_n contributions to the P and R components. The contribution to the R-component can be experimentally measured by studying the difference between particle and antiparticle cross-sections.

Hence, it is predicted that:

$$\frac{\sigma_n(K^-p) - \sigma_n(K^+p)}{\sigma_n(K^+p)} \sim 2^{1-n} \tag{180}$$

and similar relations for π^-p and π^+p, $\bar{p}p$ and pp, etc. These predictions are, however, sensitive to correlations, and to any type of correction to the simple model discussed here. *Note that throughout this lecture σ_n is really m_n—the non-diffractive part of σ_n!*

5. A MULTIPERIPHERAL MODEL WITH $I = 1$ EXCHANGES

We have seen that if we assume no correlations and if the ratio between the contributions of every σ_n to P and R is behaving like 2^n, we can extract the relations $\alpha_R = 1 - g$, $\alpha_P = 1$. In the Huan-Lee model these conditions were valid for a dual model without isospin or any other internal quantum numbers.

There is actually a much older, non dual, model which accomplishes the same result using assumptions which are different (and even less credible). This is a multiperipheral model in which, for some reason, all exchanged particles and all produced particles are assumed to have $I = 1$. Such a model was considered several years ago by Caneschi and Pignotti.[29] They studied a slightly different problem, but noticed that in such a model the ratio between the σ_n contributions to an $I = 1$ exchange in elastic scattering (necessarily an R component) and to an $I = 0$ exchange (which includes the Pomeron) is essentially 2^{-n}. The diagram-counting actually gives $\frac{1}{3}(2^n - 1)$ diagrams in P for every diagram in R. Consequently, all the equations of the Huan Lee model are reproduced, including the two final results:

$$\alpha_R = 1 - \tfrac{1}{2}g;\ \alpha_P = 1 \tag{181}$$

It is clear that there is no reason in the world to believe that $I = 1$ exchanges monopolize the multiperipheral ladder. In fact, it is practically clear that $I = 0$ exchanges are present, and are important. We mention this model here not because we believe in it, but rather to indicate that a wide class of models can yield the same result.

However, all of these models suffer from the main drawbacks mentioned at the end of Section V.4 (i.e. lack of unitarity constraints and no correlations).

6. MORE REALISTIC MODELS?

One may try more realistic models including exchange degenerate $I = 0$ and $I = 1$ tensor and vector exchanges as well as π-exchange in the multiperipheral ladder. Such models are very complicated, and can probably be studied only in the no-correlation approximation. Preliminary results indicate that in most cases one actually obtains $\alpha_P > 1$ in such models.[30] Furthermore the obtained α_P-values depend on α_R in a nontrivial way (through the phases of the signature factors). Another problem which has to be dealt with in such models is the determination of the correct phases of the amplitudes for n-particle production. These involve in addition to the usual signature factors of the exchanged trajectories, extra phase factors for each vertex.[31] As far as we know, a comprehensive study of such models has not yet been performed.

We suspect, however, that even more realistic multiperipheral models will fail to explain why $\alpha_P = 1$, as long as the unitarity bound is not manifestly obeyed by the model amplitudes.

7. A DUAL FEYNMAN GAS WITH TWO TYPES OF PARTICLES GIVING $\alpha_P = 1$

The last model discussed here is a scheme suggested by Veneziano[32] using Duality and the Feynman gas analogy. In many respects, this scheme is similar to the Huan-Lee model[28] (Section V.4), although the details are different.

Veneziano "untwists" the "twisted" duality diagrams contributing to n-particle production. Thus the twisted diagram of Fig. 54a is untwisted into 54b. In this way all duality diagrams become planar but the n produced particles are divided into two groups: n_R particles are emitted to the "right" and n_L particles to the "left", where $n = n_R + n_L$. Note that "right" and "left" here have no relation whatsoever to right and left directions in any real space. They only characterize the features of the duality diagrams.

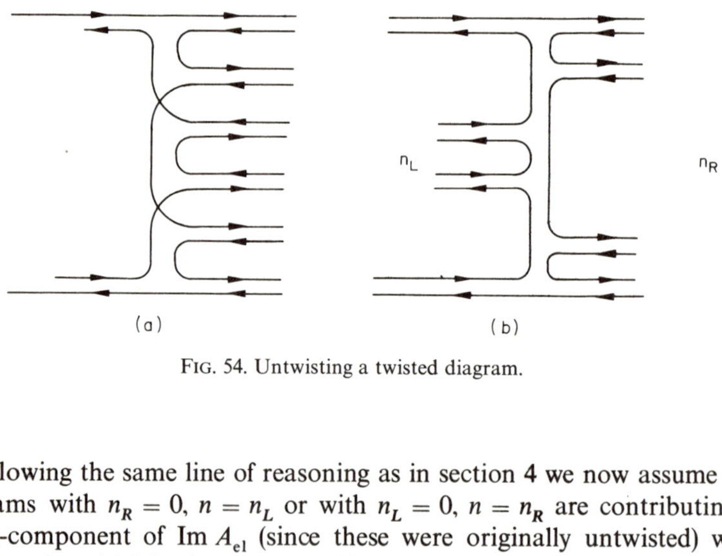

FIG. 54. Untwisting a twisted diagram.

Following the same line of reasoning as in section 4 we now assume that diagrams with $n_R = 0$, $n = n_L$ or with $n_L = 0$, $n = n_R$ are contributing to the R-component of Im A_{el} (since these were originally untwisted) while diagrams for which both $n_R, n_L \neq 0$ (diagrams which were originally twisted) contribute to the P-component.

Using the Feynman gas analogy we now consider a gas with two types of particles: "right" particles and "left" particles. We define an amplitude $T_{n_R n_L}$ such that:

$$T_n = \sum_{n_R + n_L = n} T_{n_R n_L} \tag{182}$$

and

$$\sigma_n = |T_n|^2 = \left| \sum_{n_R + n_L = n} T_{n_R n_L} \right|^2. \tag{183}$$

The generating function $Q(z_R, z_L)$ is:

$$Q(z_R, z_L) = \sum_n \left| \sum_{n_R + n_L = n} T_{n_R n_L} z_R^{n_R/2} z_L^{n_L/2} \right|^2 \tag{184}$$

Note that for $z = z_R = z_L$ we recover the usual generating function:

$$Q(z = z_R = z_L) = \sum_n z^n \sigma_n \tag{185}$$

We also have:

$$Q(z = z_R = z_L = 0) = \sigma_0 \propto s^{2\alpha_R - 2} \tag{186}$$

$$Q(z = z_R = z_L = 1) = \sigma_{\text{tot}} \propto s^{\alpha_P - 1} \tag{187}$$

When $z_R = 1, z_L = 0$ or $z_L = 1, z_R = 0$, the function $Q(z_R, z_L)$ represents the total contribution of all diagrams which were originally untwisted. This is the total contribution (summed over all multiplicities) to the R-component of Im A_{el}. We then have:

$$Q(z_R = 1, z_L = 0) = Q(z_R = 0, z_L = 1) \propto s^{\alpha_R - 1} \tag{188}$$

The pressure is defined as:

$$p(z_R, z_L) = \lim_{s \to \infty} \frac{\log Q(z_R, z_L)}{\log s} \tag{189}$$

Hence:

$$p(0, 0) = 2\alpha_R - 2 \tag{190}$$

$$p(1, 1) = \alpha_P - 1 \tag{191}$$

$$p(1, 0) = p(0, 1) = \alpha_R - 1. \tag{192}$$

Let us now assume that, in general, we can factorize the expression for $p(z_R, z_L)$ in the following way:

$$p(z_R, z_L) = F_1(z_R) + F_2(z_L) \tag{193}$$

In that case we have:

$$p(1, 1) + p(0, 0) = F_1(1) + F_1(0) + F_2(1) + F_2(0) \tag{194}$$

$$p(1, 0) + p(0, 1) = F_1(1) + F_1(0) + F_2(1) + F_2(0) \tag{195}$$

Hence:

$$p(1, 1) + p(0, 0) = p(1, 0) + p(0, 1) \tag{196}$$

$$\alpha_P - 1 + 2\alpha_R - 2 = \alpha_R - 1 + \alpha_R - 1 \tag{197}$$

Or:

$$\alpha_P = 1 \tag{198}$$

But why should we believe that eqn (193) is valid? If we neglect interference terms between different diagrams as well as correlations between "right" and "left" particles we will have no cross terms in $p(z_R, z_L)$ and eqn (193) will be valid. Note, however, that correlations among the "right" particles and/or among the "left" particles are allowed. In other words—the assumptions needed in order to get $\alpha_P = 1$ in the Veneziano approach are similar (but not identical) to the ones of Huan Lee (no interference terms, no correlations). The formulation is perhaps slightly more general, but the input and output of both models are essentially the same.

Consequently, our criticism of Section V.4 applies to the Veneziano approach as well. The $\alpha_P = 1$ result is highly sensitive to the assumptions made, and the Froissart bound is in no way guaranteed to be respected.

8. SUMMARY AND OUTLOOK

We have seen that it is theoretically attractive to postulate that $\alpha_P = 1$ is a fundamental feature of the strong interactions, and that it does not depend on the strength of the interaction. In such a case the leading non-Pomeron trajectory has a unit intercept when $g \to 0$, corresponding to a vector meson theory in the weak coupling limit. We then claim that this vector meson theory guarantees that $\alpha_P = 1$ for $g = 0$ and that $\alpha_P \geqslant 1$ for $g > 0$. On the other hand the Froissart bound guarantees that $\alpha_P \leqslant 1$ for all g. If $\alpha_P \geqslant 1$ and $\leqslant 1$, it must be exactly at one!

Several extremely interesting models actually obey this requirement. However, none of them includes among its input assumptions the physical ingredients that led to the Froissart bound. As a result, any slight violation of the assumptions of these models would "push" α_P away from one. The course we have to follow in order to overcome this difficulty is clear. We should try to construct a model, incorporating the unitarity bounds, and corresponding to a vector meson theory in the $g \to 0$ limit. Such a model might provide us with an interesting insight with respect to the Pomeron's intercept.

VI. The Truly Asymptotic Pomeron: Absorptive Effects in Multihadron Production

1. CONSISTENCY PROBLEMS

In the previous section we discussed the way in which the Pomeranchuk singularity in the forward elastic amplitude is produced by the non-diffractive contributions to multihadron final states. Regardless of how it is actually produced, and regardless of its exact nature, the Pomeron itself must "contribute" to the production mechanism of multihadron final states (the so-called diffractive component). When a sufficient number of such Pomerons appear in the mechanism for producing a given final state we usually find ourselves either in conflict with unitarity or in conflict with a self-consistent description of the Pomeron (or with both!).

These consistency problems are troublesome in the sense that they destroy almost any simple picture of the Pomeron. On the other hand, such a strong set of consistency requirement may guide us through many different ideas concerning the Pomeron, and help us to find its correct behaviour both at asymptotic energies and at present energies. This Section is devoted to a discussion of some of these problems, and of some of the solutions which have been offered for them.

Let us start by briefly reviewing two of the consistency problems:

(i) Suppose that the Pomeron is a simple pole at $\alpha = 1$ with zero slope. We then have $\sigma_{\text{tot}} \underset{s\to\infty}{\to} \text{const}$. Consider a multiperipheral diagram (Fig. 55) involving the exchange of n-Pomerons. The energy dependence of the cross-

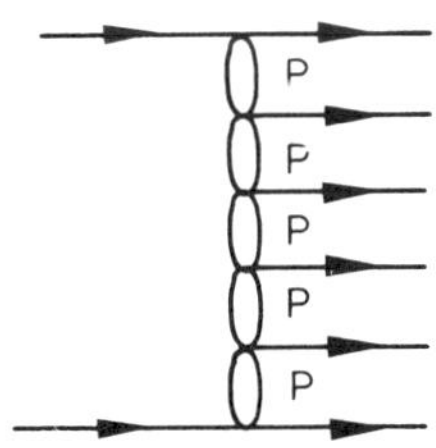

FIG. 55. A production mechanism involving many Pomerons.

section produced by such a mechanism will have a leading term behaving like $(\log s)^{n-1}$ as $s \to \infty$. This already contradicts our statement that $\sigma_{\text{tot}} \to \text{const}$. Moreover, for $n > 3$ this term alone violates the Froissart bound. What is even worse, is the fact that by summing over all values of n we may easily get $\sigma_{\text{tot}} \propto s^c$ where c is a positive power, thus violating the bound not only by logarithmic terms, but by a power of s. If the Pomeron has a slope, the cross-section for exchanging n-Pomerons increases as $s \to \infty$ only like $(\log \log s)^{n-1}$, but the sum over all multiplicities still increases like a power of s. This is the famous "Finkelstein–Kajantie disease"[33] which is the simplest example of the conflict both with unitarity and with self-consistency of the Pomeron pole.

(ii) Assume, again, that the Pomeron is a simple pole with $\alpha(0) = 1$, hence $\sigma_{tot} \to$ const. Consider the part of the total cross-section which is due to the "high mass diffraction excitation" or triple Pomeron vertex (Fig. 56) described

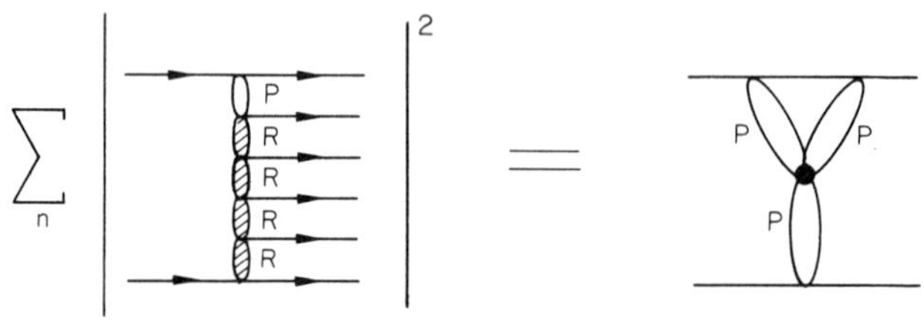

FIG. 56. The triple Pomeron term.

in Section III. This term contributes a cross-section which increases like $\log s$ (for $\alpha'_P = 0$) or $\log \log s$ (for $\alpha'_P \neq 0$) as $s \to \infty$, contradicting $\sigma_{tot} \to$const. Similar terms involving more Pomerons (Fig. 57) contribute terms which violate the Froissart bound. Again, a simple mechanism immediately leads to both types of contradictions.

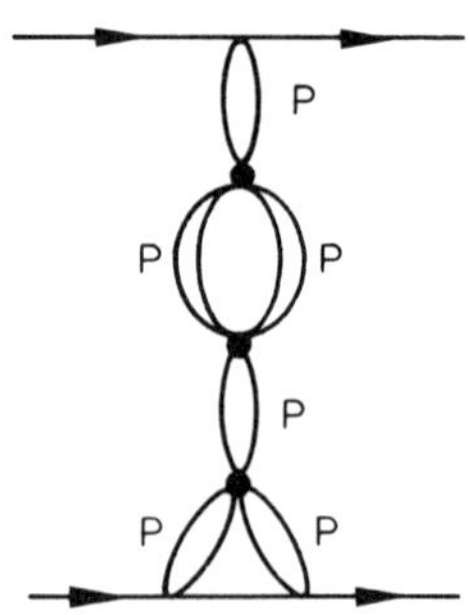

FIG. 57. A contribution to the total cross section involving three triple-Pomeron vertices and violating the Froissart bound.

This superficial discussion of two of the most serious inconsistencies serves to illustrate the kind of problem that we face. The Pomeron (pole?) which is produced by the non-diffractive production mechanisms may appear any number of times in diffractive production mechanisms. The question is, then, how to sum all of these diffractive contributions which include many Pomerons, and get a total cross-section which is again dominated by the Pomeron, in such a way that the Froissart bound is obeyed and self-consistency is preserved.

The problem is extremely complicated since it involves many different mechanisms involving various iterations of Pomeron exchange, while at the same time we do not really understand even the simplest Pomeron exchange mechanism.

2. THE SIGN OF THE CORRECTIONS TO SIMPLE POMERON EXCHANGE

Even at the simplest level of discussing the corrections due to two-Pomeron effects we face a difficulty. Suppose that we, again, assume that the Pomeron is an $\alpha = 1$ pole (and let us ignore its slope for simplicity). Assume, further, that this pole is generated by the non-diffractive component of the production cross-sections. The single Pomeron exchange term in forward elastic scattering then contributes a constant total cross-section. What will be the sign and energy dependence of the simplest correction to this term?

There are actually two "families" of such corrections. The first type of correction is the contribution of simple diffractive mechanisms to the total cross-section. If the elastic amplitude has a Pomeron exchange term, the elastic cross-section will include the squared contribution of this term. Since the elastic cross-section is part of the total cross-section, this squared Pomeron term will "correct" the original single Pomeron term. The correction is certainly positive, since the elastic cross-section is positive. Schematically we have the situation shown in Fig. 58: Im A_{el} includes a non-diffractive term

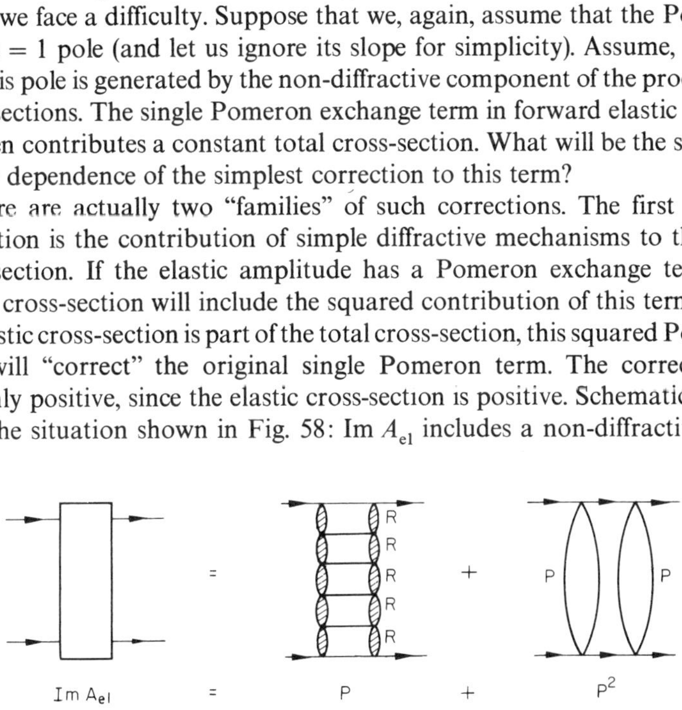

FIG. 58. If σ_{inel} is entirely non-diffractive and corresponds to a simple Pomeron pole term P, σ_{el} adds a positive P^2 term.

(σ_{inel}) which, by our assumption, produces the Pomeron pole, and a diffractive σ_{el} which produces a *positive* two-Pomeron cut correction:

$$\text{Im } A_{el} \propto \sigma_{inel} + \sigma_{el} = \sum_n T_n T_n^* + |A_{el}|^2 = P + P^2 \tag{199}$$

Actually we should replace σ_{el} by the total contribution of all diffractive production mechanisms. In that case the P^2 correction term will include the low mass as well as high mass diffraction excitations and it will increase with energy like a power of log *s*. We have therefore identified here a *positive* two-Pomeron correction to the basic Pomeron pole.

A second type of two Pomeron correction is offered to us by absorption. If the Pomeron pole is generated by the non-diffractive multiperipheral ladder, why shouldn't we include a correction due to initial and final state scattering of the two hadrons (Fig. 59)? The simplest way to do it is to follow the old

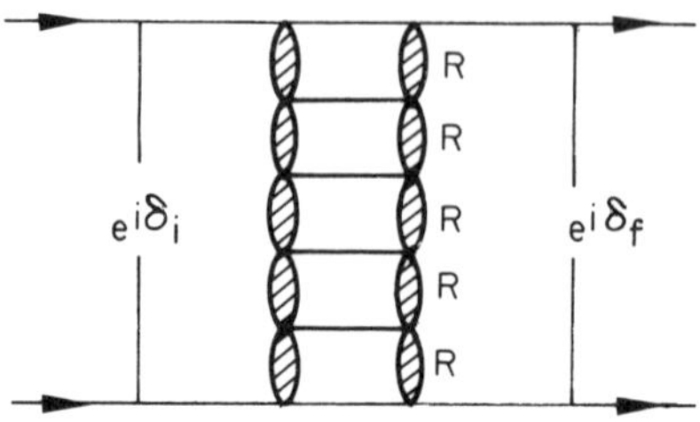

FIG. 59. A simple absorptive correction to the non-diffractive part.

method of Gottfried and Jackson[34] (borrowed from potential scattering) and to express the absorptive correction by factors of $e^{i\delta_i}$ and $e^{i\delta_f}$ where δ_i and δ_f are the (elastic) phase shifts of the initial and final states. Hence, if the uncorrected term was:

$$\text{Im}\, A_{el} = \sigma_{inel} = P \tag{200}$$

The corrected term will be:

$$\text{Im}\, A_{el} = e^{i\delta_i} P\, e^{i\delta_f} = S_i^{\frac{1}{2}} P S_f^{\frac{1}{2}} \tag{201}$$

Assuming that the initial and final elastic *S*-matrices are the same and defining

$$S_{el} = 1 + 2i\, A_{el};\ \text{Im}\, A_{el} = P \tag{202}$$

we then get:

$$\text{Im}\, A_{el} = S_{el} P = (1 - 2P)P = P - 2P^2 \tag{203}$$

Hence the absorptive two-Pomeron correction is negative.

By now we have an interesting situation:[35] We have two-Pomeron terms which are obviously positive and contribute to the total cross-section through the *D*-component (which includes σ_{el}). We also have two-Pomeron terms which are due to absorption. Those are obviously negative and they contribute to the non-diffractive *M*-component. There is no contradiction here.

The question is, however, which one of these terms is larger, and consequently, what is the energy dependence of the non-constant term in σ_{tot}.

If we consider only the elastic cross-section to be the positive P^2 term (eqn (199)), we immediately see that the simple absorption correction (eqn (203)) is twice as large[36] and the full correction is given by:

$$\sigma_{el} = P^2 \tag{204}$$

$$\sigma_{inel} = P - 2P^2 \tag{205}$$

$$\text{Im}\, A_{el} \propto \sigma_{tot} = P^2 + (P - 2P^2) = P - P^2 \tag{206}$$

In that case the two Pomeron cut correction is negative and for $\alpha'_P \neq 0$, it behaves like $(\log s)^{-1}$ as $s \to \infty$. We then have:

$$\sigma_{tot} \underset{s\to\infty}{\to} c_0 - \frac{c_1}{\log s} \tag{207}$$

where $c_0, c_1 > 0$. σ_{tot} is then predicted to rise.

On the other hand, we should remember that the elastic term is only a part of the total diffractive component. The full diffractive component may then be larger, and may "win" over the absorptive correction to the non-diffractive part. Moreover, the high mass diffraction excitation (triple Pomeron contribution) grows like $\log s$ (or $\log \log s$ for $\alpha'_P \neq 0$). Hence it may be the largest correction to the constant σ_{tot} and have we:

$$\sigma_{tot} \underset{s\to\infty}{\to} c_0 + c_2 \log s \tag{208}$$

where $c_0, c_2 > 0$. σ_{tot} is, again, rising!

At this point we may remark that present data cannot distinguish between these two functional forms (eqns (207), (208)) for σ_{tot}, although as $s \to \infty$ they are, of course, different.

However, equation (208) cannot be the final word in this story, since in addition to $\log s$ terms we have $(\log s)^n$ terms, due to diffractive contributions with many Pomerons (e.g. Figs 55 and 57). These will violate the Froissart bound unless cancelled by further negative absorptive corrections. But if such additional corrections are present (and they certainly *are* present) we are back where we started: We do not know whether the positive or the negative terms dominate.

The above discussion gives us only a taste of the problem that we face. It teaches us, however, one important lesson—the positive multi-Pomeron terms tend to explode, unless they are cancelled by absorptive corrections which restore unitarity and preserve the Froissart bound. We have already seen in Section I.4 that absorptive corrections corresponding to a logarithmically increasing absorption radius (in impact parameter space) can easily reduce the power of energy which appears in the asymptotic energy dependence of an amplitude. This is precisely what we have here. We have an amplitude which gives $\sigma_{tot} \propto s^c$ $(c > 0)$, and we apply an absorptive correction which (hopefully) would reduce c to zero (possibly leaving a $\log s$ or $\log^2 s$ dependence).

3. A SELF-CONSISTENT POMERON

Can we assume a certain form for the Pomeron exchange amplitude, allow this Pomeron to be exchanged any number of times in the amplitude for producing n-particles, sum the cross-sections over all multiplicities and find a total cross-section which is dominated by the same Pomeron with which we started, and is consistent with the Froissart bound?

This is the fundamental problem of "Pomeron physics". Several answers, none of which is entirely satisfactory, have been offered to this problem. We have no intention of reviewing all the various ideas proposed. Instead we would like to emphasize one approach which, in our opinion, has better chances of resembling the real solution:

The solution discussed here[37] actually answers a different question than the one asked above. The modified question includes absorptive corrections to the n-particle production amplitude. We can formulate it as follows:

1. Assume an explicit form for Pomeron exchange in the elastic amplitude.
2. Allow this Pomeron to be exchanged any number of times in the amplitude for producing n-particles.
3. Apply an absorptive correction to the n-particle production amplitude, using the ansatz of multiplying it by $\exp(i\delta_{el})$, where δ_{el} is the elastic phase shift corresponding to the *same* Pomeron term.
4. Compute the n-particle production cross-section and sum over all multiplicities to find the total cross-section.
5. Using the optical theorem determine the forward imaginary part of the elastic amplitude.
6. Demand that your input in step 1 is identical to the output of step 5.

This list of steps obviously defines a self-consistency equation for the Pomeron. Diagrammatically, this equation is shown in Fig. 60.

The equation itself is:

$$\operatorname{Im} A_{el} = \int |A_{el}|^2 \, d\phi_2 + (1 + 2i\,A_{el}) . \sum_n \int M_n M_n^* \, d\phi_n \tag{209}$$

where $1 + 2i\,A_e = S_{el} = \exp(2i\delta_{el})$, $d\phi_n$ is the phase space volume element and M_n is the amplitude for n-particle production obtained by exchanging n-Pomerons in a multiperipheral type mechanism, before any absorption

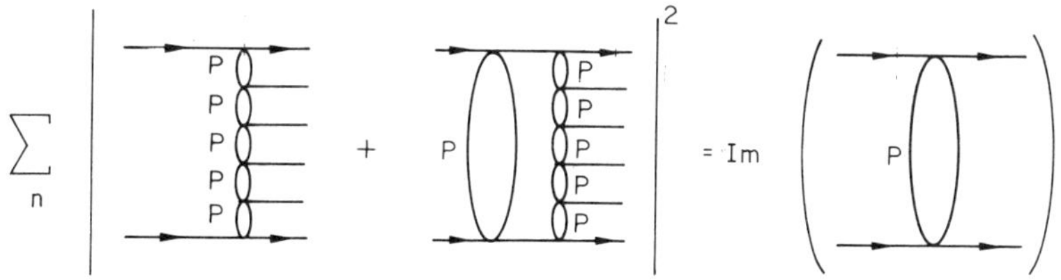

FIG. 60. Schematic diagrammatic presentation of the self consistency condition in an absorptive multi-Pomeron exchange model.

correction is applied. Clearly M_n is a function of A_{el}, and in that sense eqn (209) is an integral equation for A_{el}.

A solution to this equation was suggested by Finkelstein and Zachariasen[37] who observed that a completely absorbing disk with a logarithmically expanding radius could serve as a self-consistent Pomeron, where self-consistency is defined by the steps 1–6 outlined above. Such a Pomeron term would have:

$$\operatorname{Im} A_{el}(s, b) = \tfrac{1}{2} \text{ (for } b < R_0 \log s/s_0) \tag{210}$$

where b is the impact parameter and R_0, s_0 are free parameters. The region $b > R_0 \log s/s_0$ could have any kind of a "grey fringe", provided that Im A_{el} (s, b) falls sufficiently fast when $b \to \infty$. The elastic S-matrix, defined as:

$$S_{el} = 1 + 2i\, A_{el} \tag{211}$$

obeys:

$$S_{el}(s, b) = 0 \quad \text{(for } b < R_0 \log s/s_0) \tag{212}$$

This solution for the Pomeron obviously predicts:

$$\sigma_{tot} \underset{s \to \infty}{\to} \log^2 s \tag{213}$$

$$\sigma_{el} \underset{s \to \infty}{\to} \log^2 s \tag{214}$$

$$\frac{d}{dt}\left(\frac{d\sigma_{el}}{dt}\right) \underset{s \to \infty}{\to} \log^2 s \tag{215}$$

$$\frac{\sigma_{el}}{\sigma_{tot}} \underset{s \to \infty}{\to} \frac{1}{2} \tag{216}$$

The Froissart bound is obeyed. In fact, it is "saturated" by the model (as far as the energy dependence is concerned).

The Finkelstein–Zachariasen model is clearly not relevant to the data at present energies. We are not even sure that it represents the data anywhere.

However, it provides us with a natural (although probably oversimplified) way of observing the Froissart bound in the presence of multi-Pomeron exchanges. The absorptive mechanism invoked here may reflect only a part of the complicated corrections necessary in reality, but it is sufficient for restoring a reasonable energy dependence.

4. PROPERTIES OF THE "EXPANDING BLACK DISK" MODEL

Assuming that there is some grain of truth in this model we may ask for its predictions concerning quantities such as multiplicities, correlations, inclusive spectra, etc. These predictions, like the model itself, may have nothing to do with present or future data, but they may provide us with some insight into the general types of predictions of such absorptive models. A detailed analysis of the model was performed by Caneschi and Schwimmer.[38] Among their conclusions we mention the following:

1. The average multiplicity increases logarithmically with energy. However the integrated two particle correlation f_2 has a leading $\log^2 s$ term, reflecting long range correlations. This is obvious since the absorption mechanism (Fig. 60) operates between particles with widely different rapidities. We have:

$$\langle n \rangle \underset{s\to\infty}{\rightarrow} \log s \tag{217}$$

$$f_m \underset{s\to\infty}{\rightarrow} (\log s)^m \tag{218}$$

2. For every n, the multiplicity cross-section σ_n obeys:

$$\sigma_n \underset{s\to\infty}{\rightarrow} \text{const} \tag{219}$$

However, σ_{tot} increases like $\log^2 s$ as a result of the "opening up" of new multiplicity cross-sections at higher and higher energies.

3. The inclusive cross-section shows a central plateau which is independent of energy, when scaled by the total cross-section. In other words, the invariant inclusive cross-section for $y_{\text{cm}} = 0$ increases like $\log^2 s$.

4. The inclusive cross-section near $x = 1$ ("triple Pomeron region") obeys:

$$\frac{d\sigma}{dx} \underset{x\to 1}{\rightarrow} \frac{1}{1-x} \frac{1}{[\ln(1-x)]^3} \tag{220}$$

5. We can discuss the Feynman gas analogy of the model, and compute the generating function as well as the pressure of the gas. The obtained pressure is given by:

$$p(z) = \begin{cases} 0 & z < 1 \\ \frac{9}{2}c(z^{\frac{1}{3}} - 1) & z > 1 \end{cases} \tag{221}$$

where:

$$c = \lim_{s\to\infty} \frac{\langle n \rangle}{\log s} \tag{222}$$

and the pressure curve has a sharp break at $z = 1$.

It is interesting to note that, qualitatively, many of these features are similar to the ones obtained in our discussion of high mass diffraction excitation in Section III. Since the Finkelstein–Zachariasen model is essentially an "absorptive high mass diffraction excitation model" it is interesting to see that many of the properties of such a mechanism are not significantly changed by absorption.

5. AN ABSORPTIVE MULTIPERIPHERAL MODEL FOR INFINITE ENERGIES

The model discussed in the previous two sections involves the exchange of n-Pomerons in the n-particle production amplitude. This would be the most important mechanism when all the subenergies of neighbouring pairs of particles tend to infinity. However at any given finite energy range, we face a situation in which the subenergies for the n-particle production amplitude are substantially smaller than the total energy. In this case the explicit form of the two body amplitude which serves as the "building block" of the multi-peripheral amplitude M_n in eqn (209), may be quite different from the elastic amplitude which appears three times in that equation. In all of these three places, however, the relevant energy is the total energy and the same elastic amplitude should appear. It may, therefore, be useful to formulate an absorptive multiperipheral framework in which the "master equation" is:

$$A(s, b) = [A(s, b)]^2 + [1 - 2A(s, b)]\, M(s, b). \tag{223}$$

where $A(s, b) = \text{Im}\, A_{\text{el}}$, and M is the total inelastic cross-section (more precisely, its impact parameter projection) computed from an *unabsorbed* multi-peripheral model in which the basic two body amplitude is *not* necessarily the same as $A(s, b)$. The solution of eqn (223) is:

$$A(s, b) = \tfrac{1}{2}[2M(s, b) + 1 - \surd\{4[M(s, b)]^2 + 1\}]. \tag{224}$$

Assume now that at fixed b, $M(s, b) \underset{s\to\infty}{\to} s^{\beta(b)}$. If $\beta(b) < 0$, $M \underset{s\to\infty}{\to} 0$ and:

$$A(s, b) \underset{s\to\infty}{\to} M(s, b) + O(M^2) \to s^{\beta(b)} \tag{225}$$

The absorptive effects are negligible. If, on the other hand $\beta(b) > 0$, $M \underset{s\to\infty}{\rightarrow} \infty$ and:

$$A(s, b) \underset{s\to\infty}{\rightarrow} \frac{1}{2}\left[1 - O\left(\frac{1}{M^2}\right)\right] \rightarrow \text{const} \tag{226}$$

and the elastic amplitude is always bounded by the $A = \frac{1}{2}$ limit corresponding to complete absorption.

It would be interesting to speculate[39] on the hierarchy of applications of eqns (223) and (224). At "low" energies (presumably up to several hundreds BeV's) the basic two-body amplitude in the multiperipheral ladder is probably the Reggeon exchange amplitude. When we insert it into eqn (223) we get an elastic Pomeron (called the "bare" Pomeron by some authors). At higher energies it is more probable that several links in the multiperipheral chain will themselves involve this "bare Pomeron". If we insert such a term into eqn (223) we obtain an elastic amplitude corresponding to a logarithmically expanding black disk (sometimes called the "dressed" Pomeron). Finally, at still higher energies (by now, probably astronomical ones) we may have to allow these "dressed" Pomerons several times in the production amplitude. Then, and only then, we return to the situation of the Finkelstein–Zachariasen model in which the "expanding black disk" Pomeron reproduces itself through eqn (223).

This description[39] of the development of the Pomeron is clearly highly oversimplified. However, it focuses our attention on a few central ideas:

1. The "Pomeron" as seen at a certain range of energy may be quite different from the one observed at a significantly different energy.
2. Absorption corrections may become crucial only when $M \rightarrow \infty$, that is when the unitarity bounds are in danger.
3. Once the unitarity bound is threatened, the simplest absorptive correction can rectify the situation.
4. The phenomenology of the non-diffractive component of the total cross-section as well as the phenomenology of the single diffraction excitation term at present energies may be only remotely connected to the true asymptotic behaviour of the Pomeron.

All of these remarks leave us with no definite convincing solution for the puzzle of the Pomeron. However, we may already possess a starting point from which we should proceed.

6. SUMMARY AND OUTLOOK

We feel that absorption corrections are a necessary ingredient in any theoretical discussion of the asymptotic nature of the Pomeron. Even the simplest, almost childish, absorption scheme provides us with both self-consistency and consistency with unitarity. However, a much more mature formulation of these absorptive effects is necessary. In particular, there is no reason to believe that the only absorptive correction is given by the elastic phase shift in the initial state. It is clear that much more complicated effects should be taken into consideration, and that they should be treated in a more satisfactory way from either a field theoretical or an S-matrix point of view. Figure 61 displays some trivial examples of additional "Pomeron exchanges" which should be considered as a generalized initial (or final) state interaction. It is, therefore, clear that we face here a great theoretical challenge—to incorporate these unitarity effects into a strong interaction theory within a well defined theoretical framework.

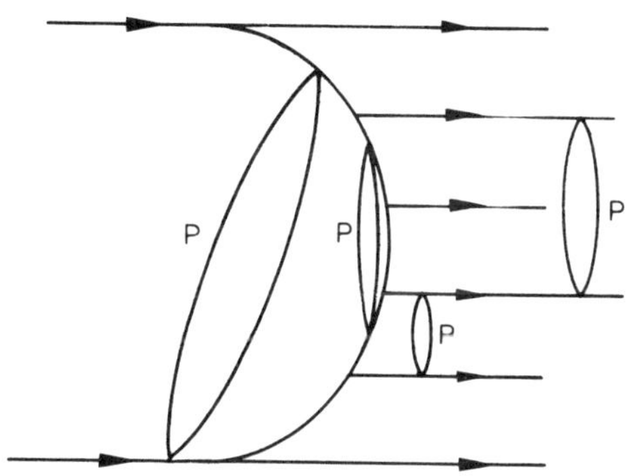

FIG. 61. Possible additional "absorptive" corrections.

At the same time we must tackle another question which is of immediate phenomenological importance. This is the problem of connecting the phenomenology of multihadron collisions at energies of, say, $s < 3000\ \text{BeV}^2$ to the true asymptotic properties of the Pomeron. Two extreme points of view can be advocated here:

(i) It is possible that we cannot handle any phenomenological discussion at present energies without finding a satisfactory consistent solution for the Pomeron puzzle.

(ii) Since all the conflicts of the Pomeron with unitarity stem from multi-Pomeron exchange mechanisms which are presumably negligible at present energies, and since the self-consistency problems of the Pomeron are avoided if we allow the Pomeron to be given by slightly different approximations for different energy variables, it is possible that all the difficult consistency problems discussed in this section are relevant only well above our present energy range. In that case, we may resort to a phenomenological discussion such as the one presented in Sections II and

III, while temporarily ignoring the true behaviour at $s \to \infty$. This possibility reminds one of the situation in quantum electrodynamics where one can safely compute electron–electron scattering amplitudes at, say, 100 MeV without worrying about the unitarity problems and the true asymptotic behaviour of *ee* scattering.

It will be very interesting to find out which of these opposing views is correct.

Acknowledgements

Much of the work reported in these lectures (especially in Sections II, III and IV) was performed in collaboration with E. Rabinovici. I would like to thank him for many fruitful discussions and a pleasant collaboration.

I also thank my colleagues at the Weizmann Institute and CERN for many illuminating discussions and helpful remarks.

Finally, I have enjoyed the delightful atmosphere at the Scottish universities summer school, and I am deeply thankful to all members of the organizing committee for their warm hospitality.

References

1. R. Dolen, D. Horn and C. Schmid. *Phys. Rev.* **166**, 1768 (1968).
2. H. Harari. *Phys. Rev. Letts* **22**, 562 (1969).
   J. L. Rosner. *Phys. Rev. Letts* **22**, 689 (1969).
3. G. Veneziano. *Nuovo Cimento* **57A**, 190 (1968).
4. H. Harari. *Phys. Rev. Letts* **20**, 1395 (1968); P. G. O. Freund, *Phys. Rev. Letts* **20**, 235 (1968).
5. G. F. Chew and S. Frautschi. *Phys. Rev. Letts* **7**, 394 (1961).
6. H. Harari. *Phys. Rev. Letts* **26**, 1079 (1971).
7. H. Harari and Y. Zarmi. *Phys. Rev.* **187**, 2230 (1969).
8. G. F. Chew and A. Pignotti. *Phys. Rev.* **176**, 2112 (1968).
9. D. Amati, S. Fubini and A. Stanghellini. *Nuovo Cimento* **26**, 896 (1962); K. G. Wilson, *Acta Phys. Austriaca* **17**, 37 (1963); R. P. Feynman. *Phys. Rev. Letts* **23**, 1415 (1969).
10. A. H. Mueller. *Phys. Rev.* **D4**, 150 (1971).
11. R. C. Hwa. *Phys. Rev. Letts* **26**, 1143 (1971); M. Jacob and R. Slansky. *Phys. Rev.* **D5**, 1847 (1972).
12. K. G. Wilson. Cornell preprint, CLNS-131 (1970) unpublished.
13. V. V. Ammosov *et al. Phys. Letts* **42B**, 519 (1972); J. Chapman *et al. Phys. Rev. Letts* **29**, 1686 (1972); G. Charlton *et al. Phys. Rev. Letts* **29**, 515 (1972); F. T. Dao *et al. Phys. Rev. Letts* **29**, 1627 (1972).
14. J. C. Sens. These proceedings, and references therein.
15. A. Bialas, K. Fialkowski and K. Zalewski. Krakow preprint (1972); J. Ellis, J. Finkelstein and R. D. Peccei. *Nuovo Cimento* **12A**, 763 (1972); K. Fialkowski. *Phys. Letts* **41B**, 379 (1972); M. LeBellac. Nice preprint (1972).
16. H. Harari and E. Rabinovici. *Phys. Letts* **43B**, 49 (1973).

17. W. R. Frazer, R. D. Peccei, S. S. Pinsky and Chung-I Tan. *Phys. Rev.* **D7**, 2647 (1973); K. Fialkowski and H. I. Miettinen. *Phys. Letts* **43B**, 61 (1973); J. D. Jackson and C. Quigg. NAL preprint (1972), unpublished.
18. M. G. Albrow *et al. Phys. Letts* **44B**, 207 (1973).
19. F. Sannes *et al. Phys. Rev. Letts* **30**, 766 (1973).
20. NAL–UCLA collaboration unpublished.
21. U. Amaldi, *et al. Phys. Letts*, **44B**, 112 (1973). S. R. Amendolia *et al. Phys. Letts* **44B**, 119 (1973).
22. W. R. Frazer and D. R. Snider. NAL preprint, 1973; D. Amati, L. Caneschi and M. Ciafalloni. CERN preprint, 1973; A. Capella, Min-Shih Chen, M. Kugler and R. D. Peccei. *Phys. Rev. Letts* **31**, 497 (1973); D. Horn and M. Moshe. NAL preprint, 1973; R. Roberts and D. P. Roy. Rutterford Laboratory preprint, 1973.
23. B. Alper *et al.* CERN preprint, 1973.
24. U. Amaldi. Private Communication.
25. J. D. Bjorken. Unpublished. M. Bander, *Phys. Rev.* **D7**, 2256 (1973); **D6**, 164 (1972); H. Harari and E. Rabinovici. Unpublished.
26. T. D. Lee *Phys. Rev.* **D6**, 3617 (1973).
27. H. Harari. *Phys. Rev. Letts* **29**, 1708 (1972).
28. H. Lee. *Phys. Rev. Letts* **30**, 719 (1973).
29. L. Caneschi and A. Pignotti. *Phys. Rev.* **180**, 1525 (1969).
30. H. Harari. Unpublished.
31. J. H. Weis. *Phys. Rev.* **D5**, 1043 (1972).
32. G. Veneziano. *Phys. Letts* **43B**, 413 (1973).
33. J. Finkelstein and K. Kajantie. *Phys. Letts* **26B**, 305 (1968).
34. K. Gottfried and J. D. Jackson. *Nuovo Cimento* **34**, 735 (1964).
35. J. Finkelstein and M. Jacob. *Nuovo Cimento* **56A**, 681 (1968).
36. L. Caneschi, *Phys. Rev. Letts* **23**, 254 (1969).
37. J. Finkelstein and F. Zachariasen. *Phys. Letts* **34B**, 631 (1971).
38. L. Caneschi and A. Schwimmer. *Nuclear Physics* **B44**, 631 (1972); **B48**, 519 (1972).
39. J. Finkelstein. Columbia University preprint (1973).

5. STATISTICAL BOOTSTRAP AND THERMODYNAMIC MODEL OF HIGH ENERGY STRONG INTERACTIONS

A. TOUNSI

Division de Physique Theorique, Institut de Physique Nucleaire and Laboratoire de Physique Theorique et Hautes Energies, Paris

I. Introduction

The thermodynamical model[1,2] belongs to that class of models in which multiparticle production of hadrons in high energy collisions is viewed as a two step process: objects made of highly excited hadron matter are created in the first step and then they decay in a more or less complicated way. Such objects are called clusters, novas or fireballs according to the different models.

In some of these models, such as the nova model[3] and the diffractive excitation model,[4] the production mechanism of fireballs is more or less well specified, but their decay is treated in a phenomenological way, in contrast to the thermodynamical model where fireball decay is completely determined by the statistical bootstrap hypothesis but where the production mechanism is not prescribed in a unique way.

The statistical bootstrap hypothesis[1] is indeed the general basis of the thermodynamical model which is, in turn, a particular expression of this hypothesis.

The statistical approach to multiparticle production was first proposed by Koppe[5] in 1948 and independently by Fermi[6] in 1952. In the so called Fermi model, the energy of collision concentrates in a small interaction volume V_0 where various kinds of particles, mainly pions, are created. A statistical equilibrium is assumed to be reached in the "gas" of these independent particles confined to the large normalization volume V.

The probability of an n-particle final state is then proportional to the density of final states given by phase space, multiplied by the probability $(V_0/V)^{n-1}$ that all n-particles will be simultaneously found inside V_0. This means that the S-matrix element for the transition between the initial state to the final n-particle state is taken to be a constant.

As we shall see in the following chapters, the thermodynamical model may be considered as a version of the Fermi model *deeply modified* by the introduction of the statistical bootstrap hypothesis.

II. The Statistical Bootstrap

In the original Fermi model[(6)] the effect of strong interactions between the particles inside the interaction volume is to establish some sort of statistical equilibrium so that each state which is accessible from the initial state will be excited to its average statistical strength. Quantitatively, strong interactions were taken into account just by their range, i.e. the radius of order 10^{-13} cm of the interaction volume. In the statistical bootstrap model where all hadrons are treated on the same footing, the interactions are taken more seriously. In the following section we show how they may be described at least from a global point of view by the knowledge of the hadronic mass spectrum.

1. MOTIVATION FOR THE STATISTICAL BOOTSTRAP

Consider the probability for an N-body final state in a collision:

$$P(E, N) = \int |\langle f|S|i\rangle|^2 \delta(E - \Sigma E_i)\delta^3(\Sigma \vec{p}_i) \prod_i^N \mathrm{d}^3\vec{p}_i \text{ ansatz,} \quad \text{(II.1)}$$

We wish to describe the interaction without making an explicit ansatz, like in the Fermi model, for the matrix element $|\langle f|S|i\rangle|^2$. We proceed in the following way. Among the N particles enclosed in a box of radius R, let us single out two particles 1 and 2 with relative distance and momentum r and p'. Because of the interaction between these two particles, the lth partial wave will behave asymptotically like:

$$\psi_l(r, p') \sim \frac{1}{r} \sin (p'r - \tfrac{1}{2}l\pi + \delta_l)$$

The boundary condition at R is

$$p'R - \tfrac{1}{2}\pi l + \delta_l(p') = n_l\pi$$

with $n_l = 0, 1, 2\ldots$, and the number of permitted partial wave states in $\{p', \mathrm{d}p'\}$ becomes:

$$\frac{\mathrm{d}n}{\mathrm{d}p'}\mathrm{d}p' = \left(\frac{R}{\pi} + \frac{1}{\pi}\frac{\mathrm{d}\delta_l}{\mathrm{d}p'}\right)\mathrm{d}p' \quad \text{(II.2)}$$

Thus the effect of the interaction between particles 1 and 2 is to add a term $\pi^{-1}\delta_l(p')/\mathrm{d}p'$ to the unperturbed level density $(R/\pi)\,\mathrm{d}p'$ due to the lth partial wave phase shift. Therefore, in (II.1), we can calculate phase space with distorted waves to eliminate from $|\langle f|S|i\rangle|^2$ the corresponding part of the interaction: the interaction manifests itself in a change of the level density. This is an old observation due to Beth and Uhlenbeck.(7)

Of course, in order to take into account the full S matrix, we have to apply the previous considerations to all phase shifts in all channels and to allow for creation and absorption of particles. Thus the effect of strong interactions might be represented by the total change of the level density in phase space.

One could find the preceding remarks without practical interest since nobody knows the set of phase shifts equivalent to the full S-matrix. Nevertheless the situation is far from being hopeless.

We know that strong interactions produce resonances. The existence of a resonance in the lth partial wave at $p' = p^*$ in the system of particles 1 and 2, causes the δ_l phase shift to increase by π when p' passes p^*. Assuming, for example, that the resonances have zero width we can write

$$\frac{1}{\pi}\frac{\delta_l(p')}{\mathrm{d}p'} = \delta(p' - p^*). \tag{II.3}$$

Then (II.2) inserted in (II.1) yields after summing over partial waves:

$$P(E, N) = \int |\langle S'\rangle|^2 \, \mathrm{d}R_N(E, m_1, m_2, \ldots, m_N) + \int |\langle S'\rangle|^2 \, \mathrm{d}R_{N-1}(E, m^*, m_3, \ldots, m_N), \tag{II.4}$$

where $\mathrm{d}R_N$ is an abbreviation for the N-particle phase space volume element. In the second integral, particles 1 and 2 are now replaced by a single particle of mass m^* equal to the invariant mass $M_{12}(p^*)$. In (II.4) we write $\langle S'\rangle$ instead of $\langle S\rangle$ to indicate that the part of the interaction responsible of the existence of a resonance between particle 1 and 2 is eliminated from $\langle S\rangle$. Thus, assuming the existence of a resonance, we are able to go from (II.1) to (II.4). This manipulation was first done by Belenky(8) but, until the work of Hagedorn, it was not pushed to its logical consequences: if we wish to shift all strong interactions to phase space we have to extend the Belenky procedure to all resonances, be they sharp or wide, and even to non resonant background. We thus arrive at the following expression for $P(E, N)$:

$$P(E, N) = \sum_l C_l \int \mathrm{d}R_l(E, m_1, \ldots m_l) \prod_i^l \rho(m_i)\, \mathrm{d}m_i, \tag{II.5}$$

where $\rho(m)$ is the complete hadronic mass spectrum. It counts the number of hadron states between m and $m + \mathrm{d}m$. In $\rho(m)$ is included the multiplicity factor $(2I + 1)(2J + 1)$ and a factor of 2 if the particle is different from its antiparticle. Not only resonant, but all phase shifts contribute to $\rho(m)$ by their derivatives.

The C_l in (II.5) are uninteresting constants left after the interactions have been shifted to phase space.

So far we have tried to make it plausible that the complete hadronic mass spectrum is equivalent to strong interactions. But how do we determine this still unknown function $\rho(m)$?

In 1965 Hagedorn[(1)] succeeded in deriving an asymptotic expression for $\rho(m)$ with the help of a statistical bootstrap hypothesis formulated in the canonical language of thermodynamics. Seven years later Frautschi reformulated[(9)] the hypothesis in microcanonical language by working directly in phase space, familiar to particle physicists, mainly for pedagogical reasons. Although, in Frautschi's formulation the words temperature and equilibrium are absent, one must bear in mind that phase space distribution means in fact the same as equilibrium. This formulation has been the starting point of several important works. Therefore we begin by Frautschi's derivation of the hadronic mass spectrum.

2. FRAUTSCHI'S FORMULATION OF THE STATISTICAL BOOTSTRAP

The first problem in Frautschi's formulation is the counting of states made of $n = 2, 3, \ldots, \infty$ non-interacting hadrons in a box of volume V_0. For n independent particles with total energy m, the density of states per unit interval of rest mass is given by the phase space of n particles:

$$P_n(m) = \left(\frac{V_0}{h^3}\right)^{n-1} \int \prod_{i=1}^{n} \mathrm{d}^3\vec{p}_i \, \delta\left(\sum_{i=1}^{n} E_i - m\right) \delta\left(\sum_{i=1}^{n} \vec{p}_i\right) \qquad \text{(II.6)}$$

This level density is expressed with the centre of mass at rest because we want to identify that density with the number of hadrons states per unit interval of rest mass; the motion states of the centre of mass are not to be counted as hadrons states.

Suppose for a while a world of mesons where each meson is a statistical n-plet ($n = 2, 3 \ldots$) of a unique meson m_0. Then the density of states, i.e. the mass spectrum of such mesons, would be

$$\rho(m) = \sum_{n=2}^{\infty} \left(\frac{V_0}{h^3}\right)^{n-1} \frac{1}{n!} \int \prod_{i=1}^{n} \mathrm{d}^3\vec{p}_i \, \delta\left(\sum_{i=1}^{n} E_i - m\right) \delta^3\left(\sum_{i=1}^{n} \vec{p}_i\right) \qquad \text{(II.7)}$$

with

$$E_i = \sqrt{(\vec{p}_i^{\,2} + m_0^2)}$$

The $1/n!$ takes care of the total symmetry of states of n bosons.

In a world where hadrons are made of real hadrons with mass spectrum

$\rho_{in}(m)$, the density of state $\rho_{out}(m)$ is

$$\rho_{out}(m) = \sum_{n=2}^{\infty} \left(\frac{V}{h^3}\right)^{n-1} \frac{1}{n!} \int \prod_{i=1}^{\infty} dm_i \, \rho_{in}(m_i) \int d^3 \vec{p}_i \delta\left(\sum_i E_i - m\right) \delta^3\left(\sum_i \vec{p}_i\right) \tag{II.8}$$

Here we sum not only over all possible momenta p_i but also over all possible masses m_i of the ith particle. Remember that $\rho_{in}(m)$ counts the number of states at mass m including spin, charge, baryon number, etc., and that, although the interactions are not explicitly taken into account, the bulk of them is included by counting all the resonant states. The factor $1/n!$ avoids double counting which would occur because of the summing over m_1, m_2, ..., m_n in all different orderings. Finally note that the Pauli principle has been ignored in (II.8), which results in an overestimate of $\rho_{out}(m)$. Nahm[10] has shown that this is not a large error. Furthermore, Hagedorn's formulation where the Pauli principle is explicitly taken into account leads to essentially the same hadronic mass spectrum as Frautschi's formulation.

Now how do we impose the bootstrap condition? In (II.8) $\rho_{out}(m)$ is the spectrum of hadronic states built up with hadron constituents whose spectrum is $\rho_{in}(m)$. The bootstrap approach requires all hadrons to be treated on the same footing. So $\rho_{out}(m)$ and $\rho_{in}(m)$ must in principle be the same function

$$\rho_{out}(m) = \rho_{in}(m) \quad \text{for all} \quad m > m_\pi \tag{II.9}$$

In practice, the present statistical approach cannot be expected to be very accurate at low masses and the bootstrap condition was imposed by Frautschi in the asymptotic form:

$$\frac{\rho_{in}(m)}{\rho_{out}(m)} \xrightarrow[m \to \infty]{} 1 \tag{II.10}$$

which is a stronger condition than the following one imposed by Hagedorn.[1]

$$\frac{\ln \rho_{in}(m)}{\ln \rho'_{out}(m)} \xrightarrow[m \to \infty]{} 1 \tag{II.11}$$

where $\rho'_{out}(m)$ as calculated by Hagedorn differs from $\rho_{out}(m)$ by including one particle states and the motion states of the centre of mass as allowed in a thermodynamical treatment. In such a treatment, condition (II.11) on entropies is natural. Besides, Hagedorn did not work directly on the level density but on its Laplace transform, the so-called partition function

$$Z(T, V) = \int_0^{\infty} \rho(m) \, e^{-m/T} \tag{II.12}$$

The partition function plays a central role in statistical thermodynamics because all expectation values of interesting quantities are calculable from $Z(V, T)$, as for instance the mean energy $\langle E \rangle$:

$$\langle E \rangle \equiv \langle m \rangle = T^2 \frac{\partial}{\partial T} \ln Z(V, T) \qquad \text{(II.13)}$$

After this rapid comparison between Frautschi's and Hagedorn's formulation, which makes it clear that they cannot be compared in every detail, let us go back to Frautschi's problem and solve eqn (II.8) under the condition (II.10).

We do not enter here into the details and we just give the method followed by Frautschi to get the solution. This method is similar in spirit to the one followed by Hagedorn.

Frautschi first showed that $\rho(m)$ in eqn (II.7), which corresponds to $\rho_{\text{in}}(m) = \delta(m - m_0)$, behaves for large m like

$$\rho(m) \sim \exp(b m^{\frac{3}{4}})$$

which is the familiar density of states of black-body radiation. So $\rho_{\text{out}}(m)$ in eqn (II.8) must rise at least as fast as $\exp(b m^{\frac{3}{4}})$. He then looked for self-consistent solutions in the form:

$$\rho_{\text{in}}(m) = c m^a \exp(b m^p) \qquad \text{(II.14)}$$

and he showed that:

(i) for $p < 1$, $\rho_{\text{out}}(m)$ grows exponentially faster than $\rho_{\text{in}}(m)$, so condition (II.10) is not fulfilled;

(ii) for $p > 1$, $\rho_{\text{out}}(m)$ grows exponentially slower than $\rho_{\text{in}}(m)$, so again condition (II.10) is not fulfilled;

(ii) for $p = 1$ there are solutions self-consistent not only in the exponent but also in the power a and in the numerical coefficient c, provided that

$$a < -\tfrac{5}{2}. \qquad \text{(II.15)}$$

Thus the asymptotic condition (II.10) is not strong enough to determine a unique self-consistent solution. Frautschi and Hamer[11] extended the bootstrap condition from asymptotic to finite masses in the form:

$$\frac{\rho_{\text{in}}(m)}{\rho_{\text{out}}(m)} = 1 \quad \text{for} \quad m \geqslant M \qquad \text{(II.15)}$$

where M is some finite mass. But, since in the statistical treatment the low mass states cannot be bootstrapped, Frautschi and Hamer[11] took

$$\rho_{\text{in}}(m) = \rho_{\text{out}}(m) + \rho(\text{low mass input}),$$

o

where ρ(low mass input) is a given known low mass input used to start generating higher mass states by a numerical iterative procedure. It turned out that this procedure leads to $a \approx -3$ and $1/b \simeq m_\pi$ independently of the low mass input. This result was confirmed by the important and very instructive work[10] of Nahm who succeeded in proving analytically that if

$$\frac{\rho_{\text{in}}(m)}{\rho_{\text{out}}(m)} = 1 + O(e^{-\lambda m}) \quad \text{for some } \lambda > 0$$

the value of a must be exactly -3.

In the work of Frautschi and Hamer, as well as in Nahm's, including or excluding exotic states has no effect on a and little on $1/b$.

So far we have stated a certain number of results without proof. Our aim in this chapter was to become acquainted with the philosophy and basic formulations of the statistical bootstrap and to exhibit the kind of solutions for the hadronic mass spectrum that can be expected from the bootstrap approach. In the next chapter we go into the details of a statistical bootstrap model with an explicit solution. Although this model is not the most realistic, it contains the main features of all statistical bootstrap models. Furthermore the technique for writing and solving the equation is rather general and has been more or less systematically employed by several authors.

III. Statistical Bootstrap Models with Explicit Solutions

There exist in the literature[12–15] several recent statistical bootstrap models with explicit solutions. I apologize for not being able to describe all of them in the present lecture notes. However I shall try to quote them at the right places.

1. COVARIANT BOOTSTRAP EQUATION

Yellin[13] writes Frautschi's bootstrap eqn (II.8) with the strong condition (II.15) in a covariant form:

$$\tau(M^2) = \delta_0(M^2 - m^2) + \sum_{l=2}^{\infty} \frac{B^{l-1}}{l!} \int dM_1^2 \ldots dM_l^2$$
$$\tau(M_1^2) \ldots \tau(M_l^2) \rho^{(l)}(M^2; M_1^2 \ldots M_l^2) \qquad \text{(III.1)}$$

where $\rho^{(l)}$ denotes the invariant l body phase space:

$$\rho^{(l)}(r^2, M_1^2, \ldots, M_l^2) = \prod_{i=1}^{l} \left\{ \int dr_i^4 \, \delta(r_i^2 - M_i^2)\theta(r_{i0}) \right\} \delta^4\left(r - \sum_{i=1}^{l} r_i\right) \qquad \text{(III.2)}$$

and where the inhomogeneous term $\delta(M^2 - m^2)$ represents the low mass input restricted for simplicity to one meson of mass m.

The constant B has dimension m^{-2} and plays the same role as the interaction volume V_0 divided by $(2\pi)^3$ does in the non-invariant formulation. (We now choose to work with units such that $\hbar = c = 1$.)

We take

$$B = B_0 \equiv \frac{\pi R^2}{\hbar^2} \quad \text{with} \quad R = m^{-1} \tag{III.3}$$

but other choices are possible.

2. EXPLICIT SOLUTION OF THE BOOTSTRAP EQUATION

We rewrite (III.1) in the form:

$$B\tau(r^2) = B\,\delta_0(r^2 - m^2) + \sum_{l=2}^{\infty} \frac{1}{l!} \int \delta^4\left(r - \sum_i r_i\right) \times \prod_{i=1}^{l} B\tau(r_i^2)\,\mathrm{d}^4 r_i. \tag{III.4}$$

In order to solve this equation we use a covariant Laplace transform.[15] For this purpose we define[16] a time like four vector b, a temperature T and its inverse β by

$$b = (b_0, \vec{b}); \qquad b^2 > 0, \qquad b_0 > 0$$

$$\beta = \frac{1}{T} \equiv \sqrt{b^2}$$

We introduce

$$Z(T, B) \equiv \int \mathrm{d}^4 r\, \mathrm{e}^{-b\cdot r}\, B\tau(r^2) \tag{III.5}$$

and:

$$z(T, m) \equiv \int \mathrm{d}^4 r\, \mathrm{e}^{-b\cdot r}\, B\delta_0(r^2 - m^2). \tag{III.6}$$

Then the Laplace transform of eqn (III.4) reads

$$Z(T, B) = z(T, m) + \sum_{l=2}^{\infty} \frac{1}{l!} \int \delta^4\left(r - \sum_i r_i\right) \mathrm{e}^{-b\cdot r}\, \mathrm{d}^4 r \prod_{i=1}^{l} B\tau(r_i^2)\,\mathrm{d}^4 r_i$$

The δ^4 function yields a factor $\Pi_{i=1}^{l}\, \mathrm{e}^{-b\cdot r_i}$ so that

$$Z(T, B) = z(T, m) + \sum_{l=2}^{\infty} \frac{1}{l!} Z^l(T, B), \tag{III.7}$$

or

$$Z = z + \exp(Z) - Z - 1 \tag{III.8}$$

which gives[10, 12]

$$z(T, m) = 2Z - \exp(-Z) + 1. \tag{III.9}$$

Now we assume that Z can be expressed as a power series of z:

$$Z = \sum_l g_l z^l, \tag{III.10}$$

where the coefficients g_l may be determined by identification after replacing z in (III.10) by the expression (III.9). Going back to the definitions (III.5) and (III.6) we transform (III.10) into

$$\int e^{-b \cdot r} B\tau(r^2)\, d^4r = \sum_{l=1}^{\infty} g_l \prod_{i=1}^{l} \int B\, \delta_0(r_i^2 - m^2)\, e^{-b \cdot r_i}\, d^4r_i$$

Using

$$\prod_{i=1}^{l} e^{-b \cdot r_i} = \int \delta^4(r - \Sigma\, r_i)\, e^{-b \cdot r}\, d^4r$$

we get

$$\int e^{-b \cdot r} B\tau(r^2)\, d^4r = \int d^4r\, e^{-b \cdot r} \sum_{l=1}^{\infty} g_l \int \delta^4(r - \Sigma r_i) \prod_{i=1}^{l} B\delta_0(r^2 - m^2)\, d^4r_i$$

for any b, which means that

$$B\tau(r^2) = \sum_{l=1}^{\infty} g_l \int \delta^4\left(r - \sum_i^l r_i\right) \prod_{i=1}^{l} B\delta_0(r_i^2 - m^2)\, d^4r_i \tag{III.11}$$

This solution was first obtained by Yellin.[12] The coefficients g_l are found to be[15]

$$g_1 = 1; \quad g_l = \frac{1}{2l!} \sum_k^{\infty} \frac{1}{k!} e^{-k/2} \left(\frac{k}{2}\right)^{k+l-1} \tag{III.12}$$

3. ASYMPTOTIC BEHAVIOUR OF $\tau(M^2)$

It is difficult to study the asymptotic behaviour directly using eqn (III.11). We investigate this behaviour in the following way.

First we notice that $z(T, m)$ as given by (III.9) is an elementary function of Z. When Z grows from 0 to ∞, $z(T, m)$ starts growing, reaches a maximum z_0 and then goes to $-\infty$. The maximum is

$$z_0 = 2 \ln 2 - 1 \tag{III.13}$$

and is reached at

$$Z_0 = \ln 2. \tag{III.14}$$

At the maximum, the second derivative of $z(T, m)$ is non-vanishing, which corresponds to a square root[10] singularity of Z at z_0:

$$Z \simeq Z_0 - \text{const.}\sqrt{(z - z_0)} \tag{III.15}$$

Let us show that this singularity determines the asymptotic behaviour of $\tau(M^2)$. We start from eqn (III.5), choose a reference frame where $(b = 1/T, 0, 0, 0)$ and use the notation $r = [\sqrt{(p^2 + M^2)}, \vec{p}]$. Then eqn (III.5) may be written:

$$\begin{aligned} Z(T, B) &= \int \mathrm{d}^4 r\, \mathrm{e}^{-b \cdot r} \int B\,\tau(r^2)\,\delta_0(M^2 - r^2)\,\mathrm{d}M^2 \\ &= \int_m^\infty \int \frac{\mathrm{d}^3\vec{p}}{2\sqrt{(p^2 + M^2)}} \exp\left(-\frac{\sqrt{(p^2 + M^2)}}{T}\right) B\tau(M^2)\,\mathrm{d}M^2 \\ &= \int_m^\infty B\tau(M^2)\,\mathrm{d}M^2 \int_0^\infty \exp\left(-\frac{\sqrt{(p^2 + M^2)}}{T}\right)\frac{2\pi\vec{p}^2\,\mathrm{d}p}{\sqrt{(p^2 + M^2)}} \end{aligned}$$

But:

$$\int_0^\infty \exp\left(-\frac{\sqrt{(p^2 + M^2)}}{T}\right)\frac{\vec{p}^2\,\mathrm{d}p}{\sqrt{(p^2 + M^2)}} \equiv MTK_1\left(\frac{M}{T}\right)$$

where $K_1(M/T)$ is the first modified Hankel function which behaves asymptotically for large M like $\sqrt{[\pi T/(2M)]}\,\mathrm{e}^{-M/T}$. Hence

$$Z(T, B) = \int_m^\infty B\tau(M^2) 2\pi M T K_1(M/T)\,\mathrm{d}M^2 \tag{III.16}$$

Similarly we can write

$$z(T, m) = \int_0^\infty B \exp[-\sqrt{(p^2 + m^2)}/T] \frac{2\pi\vec{p}^2\,\mathrm{d}p}{\sqrt{(p^2 + m^2)}} = 2\pi m T B K_1(m/T) \tag{III.17}$$

so that $Z(T, B)$ and $z(T, m)$ are related by

$$Z(T, B) = \int_m^\infty \tau(M^2) z(T, M)\,\mathrm{d}M^2 \tag{III.18}$$

Now $z(T, m)$ is a monotonically increasing function of T. There therefore exists a maximum value T_0 of T such that

$$z_0 = 2\ln 2 - 1 = z(T_0, m) = 2\pi m B T_0 K_1(m/T_0) \tag{III.19}$$

As a function of T, $Z(T, B)$ has a square root singularity at T_0 where it behaves like

$$Z(T, B) \underset{T \to T_0}{=} Z_0 - \text{const.} \sqrt{(T_0 - T)} \tag{III.20}$$

By inspection of (III.16) and because of the asymptotic behaviour of $K_1(M/T)$ for large M, one sees that $Z(T)$ can become singular only if $\tau(M^2)$ behaves asymptotically like

$$\tau(M^2) = \frac{c}{M^a} \exp(M/T_0) \tag{III.21}$$

Near $T = T_0$, $Z(T)$ may then be written

$$Z(T \to T_0, B) = F(T_0) + \text{const.} \int_m^\infty \exp\left[-M\frac{(T_0 - T)}{TT_0}\right] M^{(\frac{3}{2}-a)}\, \mathrm{d}M$$

giving

$$Z(T \to T_0, B) = F(T_0) + \text{const.} \left(\frac{TT_0}{T_0 - T}\right)^{a+\frac{5}{2}} \Gamma(a + \tfrac{5}{2}) \tag{III.22}$$

where we have made use of

$$\int_m^\infty \mathrm{e}^{-qx} x^{p-1}\, \mathrm{d}x = \frac{\Gamma(p)}{q^p}(1 + \ldots)$$

To get a square root singularity in (III.22) we need $a = -3$ which corresponds to the solution of Nahm and Yellin. One may object that the preceding proof is not quite rigorous. But Hagedorn and Montvay[15] have shown by inversion of K-transform (eqn (III.16)) that

$$\tau(M^2) = C_3 \frac{m}{M^3} \mathrm{e}^{M/T_0}\left(1 - C_4 \frac{T_0}{M} \cdots\right) \tag{III.23}$$

with:

$$C_3 = \frac{1}{2\pi}\left(\frac{m}{T_0}\right)\sqrt{[K_2(m/T_0)]}$$

$$C_4 = \frac{3}{2} - \frac{3}{8}\left(\frac{m}{T_0}\frac{K_1(m/T_0)}{K_2(m/T_0)} + \frac{1}{48} K_2(m/T_0)\right)$$

where $K_2(m/T)$ is the second modified Hankel function.

The maximum temperature T_0 as given by (III.19) comes out to be

$$T_0 = 1{\cdot}099\, m$$

in agreement with the value obtained by Frautschi and Hamer[11] and by Nahm:[10]

$$T_0 \simeq m \tag{III.24}$$

4. GENERALIZED BOOTSTRAP EQUATION

The bootstrap eqn (III.4) has been generalized in order to include not only the level density $\tau(M^2)$ but also particle and multiplicity distributions.

We start from the observation that if we divide the two sides of eqn (III.4) by $B\tau(r^2)$ we get 1 on the left-hand side and a sum of positive terms in the right-hand side. Thus we can define probabilities by

$$P_l(r, r_1, \ldots, r_l)\, \mathrm{d}^4 r_1 \ldots \mathrm{d}^4 r_l = \frac{1}{l!B\tau(r^2)}\, \delta^4(r - \Sigma r_i) \prod_{i=1}^{l} B\tau(r_i^2)\, \mathrm{d}^4 r_i \tag{III.25}$$

for $l \geqslant 2$. For $l = 1$ we take

$$P_1(r, r_1)\mathrm{d}^4 r_1 = \begin{cases} \delta^4(r - r_1)\, \mathrm{d}^4 r_1 & \text{if } r^2 = m^2; r_0 > 0 \\ 0 & \text{otherwise} \end{cases}$$

$P_l\ (r, r_1, \ldots, r_l)$ may be interpreted as the probability of finding a fireball with mass $M = \sqrt{r^2}$ in a state where it consists of l sub-fireballs $r_1 \ldots r_l$. Now if we assume that the decay of a fireball is the exact counterpart of its composition, then $P_l(r_1 \ldots r_l)$ is also the probability density that a fireball of four momentum r decays at the first step of decay into fireballs with momentum $r_1 \ldots r_l$. This assumption may be extended to all decay steps: the final decay products (pions of mass m in this model) correspond to the exact composition of the fireball. Such a hypothesis is called the "reciprocity principle" by Hagedorn and Montvay.[15] As observed by these authors this principle might be broken by final state interactions. Nevertheless we assume that it is exactly valid.

Let $V_l(r; k_1 \ldots k_l)\, \mathrm{d}^4 k_1 \ldots \mathrm{d}^4 k_l$ be the probability distribution that a fireball with mass $M = \sqrt{r^2}$ decays ultimately into (or consists of) l pions with four momentum $k_1 \ldots k_l$. The normalization condition gives

$$1 = \sum_{l=1}^{\infty} \int [\mathrm{d}^4 k_1 \ldots \mathrm{d}^4 k_l] V_l(r; k_1 \ldots k_l) \equiv \Sigma\, V_l(r^2) \tag{III.26}$$

where $V_l(r^2)$ is the total probability for l pion states and where V_1 is defined by:

$$V_1(r, k_1) = \begin{cases} \delta^4(r - k_1) & \text{if } r^2 = m^2;\ r_0 > 0 \\ 0 & \text{otherwise} \end{cases}$$

In order to write a general equation we introduce the generating functional[17]

$$V[r;\phi(k)] \equiv \sum_{l=1}^{\infty} \mathrm{d}^4k_1 \ldots \mathrm{d}^4k_l V_l(r;k_1 \ldots k_l)\phi(k_1)\ldots\phi(k_l) \quad \text{(III.27)}$$

Note that $V[r;\phi \equiv 1] = 1$.

From (III.27) one obtains[17] the exclusive probabilities $V_l\,(r;k_1 \ldots k_l)$ and the inclusive distributions $D_n(r;k_1 \ldots k_n)$:

$$V_l(r;k_1 \ \ldots k_l) = \frac{1}{l!}\frac{\delta^l}{\delta\phi(k_1)\ldots\delta\phi(k_l)} V[r,\phi(k)]_{\phi=0} \quad \text{(III.28)}$$

$$D_n(r;k_1 \ \ldots k_n) = \frac{\delta^n}{\delta\phi(k_1)\ldots\delta\phi(k_n)} V[r,\phi(k)]_{\phi=1} \quad \text{(III.29)}$$

For example the single particle distribution is

$$D_1(r,k)\,\mathrm{d}^4k \equiv D(r^2, r\cdot k)\delta_0(r^2 - m^2)\,\mathrm{d}^4k = D(r^2, r\cdot k)\frac{\mathrm{d}^3k}{2k_0}$$

The generating functional becomes[17] the generating function of the multiplicity distribution $\psi(r^2, k)$ when $\phi = h$:

$$\psi(r^2, h) \equiv V[r, \phi \equiv h] \quad \text{(III.30)}$$

and

$$V_l(r^2) \equiv \frac{1}{l!}\frac{\partial^l}{\partial h^l}\psi(r^2, \mathrm{h})_{h=0} \quad \text{(III.31)}$$

$$D_n(r^2) \equiv \frac{\partial^n}{\partial h^n}\psi(r^2, h)_{h=1} = \sum_{l=n}^{\infty} l(l-1)\ldots(l-n+1) \quad \text{(III.32)}$$

With this formalism, all distributions are known if we know $V[r,\phi(k)]$. We find it by the following argument.

Let a fireball consist of (or decay into) l sub-fireballs with four momenta r_i; each of these has the same internal (and decay) structure as the parent fireball, and decays *independently* into final states with distribution $V[r_i, \phi]$. Then the total $V[r, \phi]$ is the product of individual $V[r_i, \phi]$. This is true for each one of the possible configurations and hence

$$V[r,\phi] = \sum_{l=1}^{\infty} \int P_l(r, r_1, \ldots, r_l) \prod_{i=1}^{l} V[r_i, \phi]\,\mathrm{d}^4r_i \quad \text{(III.33)}$$

Replacing $P_l(r, r_1, \ldots, r_l)$ in (III.33) by the expression (III.25) yields

$$B\tau(r^2)V[r,\phi] = B\delta_0(r^2 - m^2)\phi(r) + \sum_{l=2}^{\infty}\frac{1}{l!}\int \delta^4(r - \Sigma r_i)\prod_{i=1}^{l} B\tau(r_i^2)V[r_i,\phi]\,\mathrm{d}^4r_i \quad \text{(III.34)}$$

This is the generalized bootstrap equation. It describes the composition of a fireball and its decay properties.

Obviously this equation reduces to the bootstrap equation (III.4) for the level density for $\phi \equiv 1$; when $\phi \equiv h$ we obtain the bootstrap equation for the multiplicity distribution. The different functional derivatives taken at $\phi = 0$ and $\phi = 1$ gives respectively the bootstrap equations for the different exclusive and inclusive distributions.

To solve eqn (III.34) we can proceed exactly in the same way as with the bootstrap equation (III.4) for the level density. By covariant Laplace transform, we obtain the same eqn as (III.8) but now Z is the Laplace transform of the generating function $V[r;\phi]$ and

$$z \equiv \int \mathrm{d}^4 r\, \mathrm{e}^{-b\cdot r} B\delta_0(r^2 - m^2)\phi(r)$$

As the relation between the functionals Z and z is the same (III.8) as before, it follows that

$$B\tau(r^2)\, V[r;\phi(k)] = \sum_{l=1}^{\infty} g_l \int \delta^4\left(r - \sum_{i=1}^{l} r_i\right) \left\{\prod_{i=1}^{l} \mathrm{d}^4 r_i B\delta_0(r_i^2 - m^2)\phi(r_i)\right\} \qquad \text{(III.35)}$$

which, as should be expected, brings back (III.11) when $\phi \equiv 1$.

From (III.35), all kinds of distributions can be obtained. For instance

$$V_l(r;k_1 \ldots k_l) = \frac{g_l}{B\tau(r^2)}\delta^4\left(r \quad \sum_{i=1}^{l} r_i\right)\prod_{i=1}^{l} B\delta_0(k_i^2 - m^2) \qquad \text{(III.36)}$$

which means that the pion distribution is proportional to phase space, the proportionality coefficient g_l being completely determined (III.12). For the study of the asymptotic behaviour of particle and multiplicity distributions, I refer the reader to the papers of Montvay[18] and of Montvay and Hagedorn.[15] They found the important result that the asymptotic forms are the same for a large class of models and in particular that the complete bootstrap and the linear bootstrap (see the following section) have nearly the same asymptotic limits.

A useful and intuitive result is the average multiplicity behaviour of the decay products of a mass M fireball for large M:

$$n(M) \simeq M/\bar{E} \qquad \text{(III.37)}$$

where

$$\bar{E} = \frac{mK_2(m/T_0)}{K_1(m/T_0)} \qquad \text{(III.38)}$$

5. LINEAR BOOTSTRAP

5.1. *The general case*

A special case of eqn (III.34) consists in the approximation that a fireball is made up of one other fireball plus any number of pions. Then in (III.34) all $B\tau(r_i^2)V[r_i, \phi]$ except one have to be replaced by $B\delta_0(k_i^2 - m^2)\phi(k_i)$, which leads to

$$B\tau(r^2)V[r, \phi(k)] = B\delta_0(r^2 - m^2)\phi(r)$$
$$+ \sum_{l=2}^{\infty} \frac{1}{(l-1)!} \int B\tau(r'^2)\{V[r', \phi(k)]\}$$
$$\times\, d^4r'\delta\left(r - r' - \sum_{i=2}^{l} k_i\right) \prod_{i=2}^{l} d^4k_i B\delta_0(k_i^2 - m^2)\phi(k_i) \quad \text{(III.39)}$$

This equation is *linear* in the unknown functional.

The covariant Laplace transform procedure applied to (III.39) yields the equation:

$$Z = z + Z[\exp(z) - 1] \quad \text{(III.40)}$$

or

$$Z = \frac{z}{2 - \exp(z)} \quad \text{(III.41)}$$

In the present approximation Z has a simple pole at

$$z_0 = \ln 2$$

and the maximum temperature T_0 is now determined by

$$2\pi m T_0 B K_1\left(\frac{m}{T_0}\right) = \ln 2$$

Here again one can show by K-transform inversion[15] that

$$\tau(M^2) \simeq \frac{C_p}{m} \frac{e^{m/T_0}}{M^{\frac{3}{2}}}\left(1 - \frac{3T_0}{8M} + \ldots\right) \quad \text{(III.42)}$$

with:

$$C_p = \frac{1}{\sqrt{2\pi}}\left(\frac{m}{T_0}\right)^{\frac{3}{2}} \frac{K_1(m/T_0)}{K_2(m/T_0)}$$

where the subscript p stands for pole approximation.

5.2. *The linear chain*

The simplest type of bootstrap results from the approximation that a fireball consists of a fireball plus one single pion (instead of an arbitrary number of them). In that case only the first term of the sum in eqn (III.39) remains and the Laplace transformation yields

$$Z = z + Zz \tag{III.43}$$

or

$$Z = \frac{z}{1 - z} = \sum_{l=1}^{\infty} z^l \tag{III.44}$$

so that all coefficients $g_l \equiv 1$. Comparison with eqn (III.11) shows that this corresponds to $\tau(r^2)$ being a sum over l-pion phase space with coefficients $g_l = 1$. In a Fermi statistical model the g_l would be $1/l!$ because the pions are indistinguishable. Since here this factor is missing, one might say that the linear chain bootstrap can be interpreted in two ways: either

a fireball consists of a fireball plus a pion

or a fireball consists of any number of *distinguishable* pions.

Even in the full bootstrap the coefficients g_l, as given by eqn (III.12) behave asymptotically[15] like:

$$g_l \underset{l\to\infty}{\Rightarrow} \text{const.}\ l^{-\frac{3}{2}} \tag{III.45}$$

which, compared to $1/l!$ is almost a constant. Thus even there the fireball may be considered to consist of "almost" distinguishable pions. This intimate relation between the bootstrap and distinguishable particles was already noticed in Hagedorn's first paper[1] and extensively discussed by Alexanian.[19] The maximum temperature in the linear chain case is given by the pole in eqn (III.44).

$$z = 1 = 2\pi m T_0 B K_1(m/T_0) \tag{III.46}$$

This was first derived (in a slightly different form) by Hagedorn[1] for $m \to 0$ and by Alexanian[19] for $m \neq 0$. Note that the simple pole in eqn (III.44) implies the same asymptotic behaviour (with different numerical constants) as in the general linear case, namely

$$\tau(M^2) \sim \text{const.} \frac{e^{M/T_0}}{M^{\frac{3}{2}}} \tag{III.47}$$

6. THE TRUNCATED BOOTSTRAP MODEL OF CHIU AND JOHNSEN

These authors[20] discuss the approximation that a fireball consists of two fireballs. Therefore the corresponding equation results if in our eqn (III.4) the sum on the r.h.s. is reduced to its first term $l = 2$. Laplace transforming it yields

$$Z = z + \tfrac{1}{2}Z^2 \tag{III.48}$$

or

$$Z = 1 - \sqrt{(1 - 2z)} \tag{III.49}$$

with a square root singularity at $z_0 = \frac{1}{2}$ and consequently with the same asymptotic behaviour (with different numerical constants) as the full bootstrap case.

We have presented this rapid review of different models in order to show the striking stability of the exponential mass spectrum against changes in the bootstrap assumptions:

full bootstrap: fireball = *any* number of fireballs

$$\tau(M^2) \sim \frac{\text{const.}}{M^3} \exp(M/T_0)$$

truncated: fireball = *two* fireballs

$$\tau(M^2) \sim \frac{\text{const.}}{M^{\frac{3}{2}}} \exp(M/T_0)$$

full linear: fireball = fireball + *any* number of pions

$$\tau(M^2) \sim \frac{\text{const.}}{M^{\frac{3}{2}}} \exp(M/T_0)$$

linear chain: fireball = fireball + *one* pion

$$\tau(M^2) \sim \frac{\text{const.}}{M^{\frac{3}{2}}} \exp(M/T_0)$$

The first case is the most general bootstrap one can write down and the last is the simplest which would deserve the name bootstrap. The other two are representatives of an infinite number of intermediate models. All of them have the exponential spectrum and the value of T_0 is in all cases of order m. See also the general discussion in Appendix B of Ref. (15).

IV. THE CANONICAL DESCRIPTION

In the preceding chapter we learned how the hadronic mass spectrum and decay properties of a mass M fireball can be derived from the covariant statistical bootstrap eqn (III.34). The solutions of this equation can be explicitly written down and their asymptotic behaviour for large M exhibits characteristic features of the bootstrap structure also found in the non-invariant formulation. However the explicit solutions are obtained in the form of an infinite series (see (III.35)) the general term of which involves $\rho^{(l)}$, the invariant l body phase space, but most of the applications to physical reactions use the framework of a canonical description. In such a description, a fireball is ascribed a fixed temperature instead of a fixed mass and its decay properties are easily computed with the help of statistical thermodynamics. But in view of the difficulties encountered in the statistical thermodynamics of real gases one may wonder whether such a description is possible. We have seen in the "bootstrap motivation" (Section II.1) that the interaction between particles may be pushed into phase space and hidden in the luxuriant level density $\rho(m)$ so that particles appear to be formally free. Therefore a thermodynamics of hadronic matter is not only possible but turns out to be, at least formally, no more difficult than the thermodynamics of ideal gases. However, in order to apply that thermodynamics to highly excited hadronic matter, one has to pay attention to the collective motions of hadronic matter generated by high energy collisions. These motions may be very complicated but we shall assume that they are produced only in the direction of the collision axis and do not give rise to turbulence. We shall see in Section V.1 how to separate the longitudinal collective motions from the *isotropic* thermal motion inherent in a statistical thermodynamical description. But how can we obtain from *isotropy* the observed anisotropic angular distributions of particles created in high energy collisions? We shall deal with this problem in Section V.1. So let us admit for the moment that collective motions have been separated off and that we are looking at a fireball at rest in thermal equilibrium.

1. THE PARTITION FUNCTION

In the canonical description, a fireball is considered as a thermodynamical system of volume V_0 in equilibrium at temperature T, but its energy is not fixed and fluctuates around a mean value $\bar{E}$. We know from statistical thermodynamics that all the information one can get about the system is contained in its partition function $Z(V_0, T)$:

$$Z(V_0, T) = \sum_i e^{-E_i/T}, \tag{IV.1}$$

where the E_i's are the energies corresponding to all possible states ψ_i of the system. In particular, all mean values can be obtained from $Z(V_0, T)$.

In each state ψ_i, the system (the fireball) is composed of non-interacting hadrons. Thus ψ_i is simply the direct product of single particle plane wave states. For a particle of mass m_γ only a discrete set $\{p_\alpha\}$ of plane waves is allowed in the box of volume V_0.

We denote by

$$\varepsilon_{\alpha\gamma} = \sqrt{(p_\alpha^2 + m_\gamma^2)} \tag{IV.2}$$

the energy of a particle of mass m_γ in a momentum state α and by $\nu_{\alpha\gamma}$ the occupation number which indicates how many particles of the kind γ are in a state α:

$$\nu_{\alpha\gamma} = \begin{cases} 0, 1 & \text{for fermions} \\ 0, 1, 2, \ldots & \text{for bosons} \end{cases} \tag{IV.3}$$

Thus the Pauli principle is easily taken into account.

With this notation, a state ψ_i may be represented by a matrix $\{\nu_{\alpha\gamma}\}_i \equiv \psi_i$ and we can write

$$E_i = \sum_{\alpha\gamma}^{\infty} \nu_{\alpha\gamma}\varepsilon_{\alpha\gamma} \tag{IV.4}$$

$$Z(V_0, T) = \sum_{\{\nu\}} \exp\left[-\frac{1}{T}\sum_{\alpha\gamma}^{\infty} \nu_{\alpha\gamma}\varepsilon_{\alpha\gamma}\right] \tag{IV.5}$$

or by interchanging product and summation:

$$Z(V_0, T) = \sum_{\{\nu\}} \prod_{\alpha\gamma} X_{\alpha\gamma}^{\nu_{\alpha\gamma}} = \prod_{\alpha\gamma}\left[\sum X_{\alpha\gamma}^{\nu_{\alpha\gamma}}\right] \tag{IV.6}$$

where

$$X_{\alpha\gamma} \equiv \exp(-\varepsilon_{\alpha\gamma}/T) \tag{IV.7}$$

and is always smaller than one.

Now we put

$$\gamma = \begin{cases} \phi & \text{for fermions} \\ \beta & \text{for mesons} \end{cases}$$

thus $\phi = 0, 1$ and $\beta = 0, 1, 2, \ldots$

Then (IV.6) becomes

$$Z(V_0, T) = \prod_{\alpha\beta} \frac{1}{1 - X_{\alpha\beta}} \prod_{\alpha\phi} (1 + X_{\alpha\phi})$$

and

$$\ln Z(V_0, T) = -\sum_{\alpha\beta} \ln(1 - X_{\alpha\beta}) + \sum \ln(1 + X_{\alpha\phi}). \tag{IV.8}$$

Now we go over to the continuum:

$$\sum_{\alpha} [\ldots] \Rightarrow \frac{V_0}{2\pi^2}\int_0^{\infty} [\ldots] p^2 \, dp \tag{IV.9}$$

$$\sum_{\beta\phi}[\ldots] \Rightarrow \int_0^\infty \rho_{B,F}(m)[\ldots]\, dm \tag{IV.10}$$

Thus

$$\ln Z = \frac{V_0}{2\pi^2}\int_0^\infty p^2\, dp\left[\int_0^\infty \rho_F(m)\ln(1 + X_{pm})\, dm - \int_0^\infty \rho_B(m)\ln(1 - X_{pm})\, dm\right] \tag{IV.11}$$

where

$$X_{pm} \equiv \exp\left(-\frac{1}{T}\sqrt{(p^2 + m^2)}\right) \tag{IV.12}$$

After expanding the logarithms we obtain

$$\ln Z = \frac{V_0}{2\pi^2}\sum_{n=1}^\infty \frac{1}{n}\int_0^\infty dm\, dp\, p^2\rho(m;n)X_{pm}^n \tag{IV.13}$$

where

$$\rho(m;n) \equiv \rho_B(m) - (-1)^n\rho_F(m) \tag{IV.14}$$

The integration over p yields

$$\int_0^\infty p^2 \exp\left\{-\frac{n}{T}\sqrt{(p^2 + m^2)}\right\} dp = m^2\frac{T}{n}K_2\left(\frac{nm}{T}\right) \tag{IV.15}$$

where K_2 is the second Hankel modified function. We have

$$K_2\left(\frac{nm}{T}\right) \xrightarrow[nm/T\to\infty]{} \sqrt{\left(\frac{\pi T}{2nm}\right)}\exp\left(-\frac{nm}{T}\right)$$

$$K_2\left(\frac{nm}{T}\right) \xrightarrow[nm/T\to 0]{} \frac{2T^2}{n^2m^2} \tag{IV.16}$$

Finally we obtain from (IV.13)

$$Z(V_0, T) = \exp\left\{\frac{V_0 T}{2\pi^2}\sum_{n=1}^\infty \frac{1}{n^2}\int_0^\infty \rho(m;n)\, m^2 K_2\left(\frac{nm}{T}\right) dm\right\} \tag{IV.17}$$

It can be shown[2, 21] that all terms in the sum may be neglected except the term $n = 1$. Thus within this approximation

$$Z(V_0, T) \simeq \exp\left\{\frac{V_0 T}{2\pi^2}\int \rho(m)m^2 K_2\left(\frac{m}{T}\right) dm\right\} \tag{IV.18}$$

where

$$\rho(m) = \rho_B(m) + \rho_F(m)$$

and is given by

$$\rho(m) \simeq cm^a \exp(m/T_0) \quad \text{(IV.19)}$$

Let us note here that the solutions first obtained by Hagedorn[1] using the weak bootstrap condition (II.11) correspond to

$$-\tfrac{7}{2} \leqslant a \leqslant -\tfrac{5}{2} \quad \text{(IV.20)}$$

He used the following argument.

Let $\rho(m)$ still be the unknown mass spectrum of strong interactions and call it $\rho_{\text{in}}(m)$ because it is the input in eqn (IV.18). Now we go back to eqn (IV.1), where we group together all terms of the sum, in which $E \leqslant E_i \leqslant E + \mathrm{d}E$. Let there be $\rho_{\text{out}}(E, V_0)\,\mathrm{d}E$ such terms (where $\rho_{\text{out}}(E)$ is the level density of the fireball at energy E), then the sum can be written:

$$Z(V_0, T) = \int_0^\infty \rho_{\text{out}}(E, V_0)\,\mathrm{e}^{-E/T}\,\mathrm{d}E \quad \text{(IV.1a)}$$

and this function Z must be identical with the one defined by eqn (IV.17). As momentum is not conserved in this description, one cannot demand exact equality of ρ_{out} and ρ_{in}. Requiring the asymptotic equality of $\log \rho_{\text{out}}$ and $\log \rho_{\text{in}}$ yields already the exponential spectrum (IV.19) with the condition (IV.20). This is other evidence for the incredible stability of the exponential solution against changes in the bootstrap condition (see also Sections III.5 and III.6).

We have seen that the solution $a = -\frac{5}{2}$ (first preferred by Hagedorn) is excluded in Frautschi's formulation (see (II.15)). The power of m in (IV.19) may look unimportant compared to the factor $\exp(m/T_0)$. It happens however that $a = -3$ and $a = -\frac{5}{2}$ correspond to completely different behaviour of the partition function for $T \to T_0$.

Indeed we already know (see (III.20)) that $Z(V_0T)$ is finite for $a = -3$ and has a square root singularity at $T = T_0$. On the other hand for $a = -\frac{5}{2}$ we get from (IV.18) and (IV.16)

$$Z(V_0, T) \simeq \exp\left\{V_0\left(\frac{T_0}{2\pi}\right)^{\frac{3}{2}} c \int_0^\infty \frac{\mathrm{d}m}{m^{\frac{5}{2}}}\, m^{\frac{3}{2}} \exp\left[-m(T_0 - T)/T_0T\right]\right\} + C(M_0, T_0) \quad \text{(IV.21)}$$

where M_0 is a fixed high mass beyond which we use the asymptotic forms (IV.19) and (IV.16) and where $C(M_0, T_0)$ is finite. It is easy to see that $Z(V_0, T)$

as given by (IV.21) behaves like

$$Z(V_0, T) \simeq \exp\left\{V_0\left(\frac{T_0}{2\pi}\right)^{\frac{3}{2}} c\ln\left(\frac{T_0^2}{M_0(T_0 - T)}\right)\right\} \tag{IV.21a}$$

Thus Z diverges for $a = -\frac{5}{2}$, when $T \to T_0$

2. PARTITION FUNCTION AND DISTRIBUTIONS

Not only thermodynamic functions can be derived from the partition function but also distribution functions.

2.1. *Momentum spectrum of freely created particles*

We look for the momentum spectrum of particles emitted by a fireball without restriction, "freely", due to conservation laws of quantum numbers. Since in the canonical description, a fireball is a thermodynamical system in equilibrium at a temperature T one may guess that the "hadronic gas" inside a fireball obeys a Planck law in analogy with black body radiation. This we are going to show explicitly.

Let us denote by $f_m(\vec{p}, T)$ the momentum spectrum of a particle of mass m emitted by a fireball at temperature T. By definition $f_m(\vec{p}, T)\,\mathrm{d}^3p$ is the average number of particles m with momentum between $\vec{p}$ and $\vec{p} + \mathrm{d}\vec{p}$ freely emitted without regard to the other particles: $f_m(\vec{p}, T)$ is thus an inclusive spectrum.

So we have to compute the mean number of particles of mass m and momentum $\vec{p}$. In the discrete notation this is the mean number of particles of kind m_k with momentum $\vec{p}_i$. Then $f_{m_k}(\vec{p}_i, T)$ is nothing but the mean value of the occupation number ν_{ik} defined by (IV.3):

$$f_{m_k}(\vec{p}_i, T) \equiv \bar{\nu}_{ik}. \tag{IV.22}$$

In order to compute $\bar{\nu}_{ik}$ we need to know the probability distribution of ν_{ik}. In (IV.6) one can see that the probability to find the system (our fireball!) in a state $\{\nu'\}$ is

$$W(\nu') = \frac{\Pi_{\alpha\gamma}\, X_{\alpha\gamma}^{\nu'_{\alpha\gamma}}}{Z}. \tag{IV.23}$$

Then the probability to find the system in any state belonging to a set S of states is given by

$$W(S) = \frac{1}{Z}\sum_{\{\nu\}}^{(S)}\prod_{\alpha\beta} X_{\alpha\gamma}^{\nu_{\alpha\gamma}}, \tag{IV.24}$$

where $\Sigma_{\{\nu\}}^{(S)}$ goes over all $\{\nu\} \in S$.

Now we need to know the probability to find ν_{ik} particles of momentum

$\vec{p}_i$ and kind k. This means that we are concerned with the particular set S of all states of which have ν_{ik} fixed while the other $\nu_{\alpha\gamma}$ assume all possible values. Thus in (IV.23) we have to sum over all $\nu_{\alpha\gamma}$ except ν_{ik} to find

$$W(S) \equiv W(\nu_{ik}) = \frac{1}{Z}\sum_{\{\nu\}}^{(S)}\prod_{\alpha\gamma} X_{\alpha\gamma}^{\alpha\gamma} \equiv \frac{1}{Z}\left\{\left(\sum_{11} X_{11}^{\nu_{11}}\right)\left(\sum_{12} X_{12}^{\nu_{12}}\right)\dots X_{ik}^{\nu_{ik}}\cdot\right\}$$

But the product in the curly brackets can be written

$$\{(\ )\dots X_{ik}^{\nu_{ik}}\dots(\ )\} = X_{ik}^{\nu_{ik}}\frac{Z}{\sum_\nu X_{ik}^\nu} \tag{IV.25}$$

where

$$\sum_\nu X_{ik}^\nu = \begin{cases} \dfrac{1}{1 - X_{ik}} & \text{when } k \text{ is a boson} \\ 1 + X_{ik} & \text{when } k \text{ is a fermion} \end{cases}$$

Thus we finally obtain

$$W(\nu_{ik}) = X_{ik}^{\nu_{ik}} \times \begin{cases} 1 - X_{ik} & \text{for bosons} \\ 1/(1 + X_{ik}) & \text{for fermions.} \end{cases} \tag{IV.26}$$

The mean value $\bar{\nu}_{ik}$ of the occupation number ν_{ik} is then

$$\bar{\nu}_{ik} = \sum \nu_{ik} W(\nu_{ik}) = \begin{cases} (1 - X_{ik})\sum_0^\infty \nu X_{ik}^\nu & \text{for bosons} \\ \sum_0^1 \nu X_{ik}^\nu/(1 + X_{ik}) & \text{for fermions} \end{cases} \tag{IV.27}$$

or

$$\bar{\nu}_{ik} = \frac{X_{ik}}{1 \mp X_{ik}} \qquad \begin{matrix} (-\text{ for bosons}) \\ (+\text{ for fermions}) \end{matrix} \tag{IV.28}$$

If we remember that $X_{ik} = \exp(-\sqrt{(p_i^2 + m_k^2)}/T)$ we can finally write (see eqn (IV.22))

$$f_{m_k}(\vec{p}_i, T) \equiv \bar{\nu}_{ik} = \left[\exp\left(\frac{\sqrt{(p_i^2 + m_k^2)}}{T}\right) \mp 1\right]^{-1}. \tag{IV.29}$$

Going over to the continuum we get

$$f_m(\vec{p}, T)\, \mathrm{d}^3p = \frac{V_0}{2\pi^2} p^2\, \mathrm{d}p\left[\exp\left(\frac{\sqrt{(p^2 + m^2)}}{T}\right) \pm 1\right]^{-1} \tag{IV.30}$$

which must be multiplied by the multiplicity factor $Z_m = (2J + 1)(2I + 1)$

when particle m has spin J and isospin I. Thus the momentum spectrum is, as guessed from the beginning, given by a Planck formula for massive particles. Let us add a few remarks about (IV.26).

Since $T < T_0 \simeq m_\pi$ we have for heavy hadrons

$$X_{ik} = \exp\left(-\frac{1}{T}\sqrt{(p_i^2 + m_k^2)}\right) \ll 1$$

Then

$$W(\nu_{ik}) \simeq X_{ik}^{\nu_{ik}} \simeq [W(\nu_{ik} = 1)]^{\nu_{ik}}$$

and (see IV.29):

$$\bar{\nu}_{ik} \simeq X_{ik} = W(\nu_{ik} = 1). \tag{IV.31}$$

This shows that for heavy particles we can neglect ± 1 in (IV.30) to get the same formula for fermions and mesons; furthermore the probability to find one particle k with momentum $\vec{p}_i$ is small and equal to the average number $\bar{\nu}_{ik}$. Also, the probability to find n such particles is the nth power of that for finding one.

2.2. *Multiplicity distribution of fireballs in a fireball: Composition of a fireball*

We first look for the probability that a fireball decays into n daughter fireballs in the first step decay (or first generation). This amounts to computing the probability $W(n)$ to find the fireball in any state characterized by $\Sigma_{\alpha\gamma}\nu_{\alpha\gamma} = n$. We have thus to apply eqn (IV.24) to the set S_n defined by

$$S_n = \left\{ \{\nu\}; \sum_{\alpha\beta} \nu_{\alpha\gamma} = n \right\} \tag{IV.32}$$

Hence:

$$W(n) = \frac{1}{Z} \sum_{\{\nu\}}^{(S_n)} \prod_{\alpha\gamma} X_{\alpha\gamma}^{\nu_{\alpha\gamma}} \tag{IV.33}$$

and all products of the sum are homogeneous of order $n = \Sigma\nu_{\alpha\gamma}$ in the variable $X_{\alpha\gamma}$.

In order to find $W(n)$ we introduce a generalized partition function defined by

$$Z(V, T, z) = \sum_{\{\nu\}} \prod_{\alpha\gamma} (zX_{\alpha\gamma})^{\nu_{\alpha\gamma}}, \tag{IV.34}$$

where z is an auxiliary variable. Now in all products that are homogeneous of order n we can factorize out z^n

$$\sum_{\{\nu\}}^{(S_n)} \prod_{\alpha\gamma} (zX_{\alpha\gamma})^{\nu_{\alpha\gamma}} = z^n \sum_{\{\nu\}}^{(S_n)} \prod_{\alpha\gamma} X_{\alpha\gamma}^{\nu_{\alpha\gamma}},$$

and thus

$$Z(V, T, z) = \sum_{n=0}^{\infty} z^n \left[\sum_{\{\nu\}}^{(S_n)} \prod_{\alpha\gamma} X_{\alpha\gamma}^{\nu_{\alpha\gamma}} \right]. \tag{IV.35}$$

Replacing the term in the brackets by its expression given by (IV.33) we find:

$$Z(V, T, z) = Z(V, T) \sum_{0}^{\infty} z^n W(n). \tag{IV.36}$$

Thus $Z(V, T)W(n)$ is the coefficient of z^n in the power expansion of $Z(V, T, z)$.

From (IV.13) when we replace X_{pm} by zX_{pm} we obtain via (IV.17) and (IV.18).

$$Z(V, T, z) \simeq \exp\left\{ z \frac{VT}{2\pi^2} \int_0^{\infty} \rho(m) m^2 K_2(m/T)\, \mathrm{d}m \right\}.$$

But to the same approximation

$$\frac{VT}{2\pi^2} \int_0^{\infty} \rho(m) m^2 K_2(m/T)\, \mathrm{d}m = \ln Z(V, T)$$

and thus

$$Z(V, T, z) = \exp\left[z \ln Z(V, T) \right] = \sum_{0}^{\infty} \frac{z^n (\ln Z)^n}{n!} \tag{IV.37}$$

Hence (see (IV.36))

$$Z(V, T) W(n) = \frac{(\ln Z)^n}{n!} \tag{IV.38}$$

or

$$W(n) = \frac{[\ln Z(V, T)]^n}{n!} \exp\left[-\ln Z(V, T) \right]. \tag{IV.39}$$

This is a Poisson distribution with mean value

$$N = \bar{n} = \sum n W(n) = \ln Z(V, T) \tag{IV.40}$$

We can arrive at (IV.40) in a more intuitive way. Let $\nu(m, T)$ be the total number of "particles" (fireballs) of mass m emitted (at the first generation) by a fireball at temperature T.

$$\nu(m, T) \equiv \int f_m(\vec{p}, T)\, \mathrm{d}^3\vec{p} = \frac{V z_m}{2\pi^2} \int p^2 \exp\left[-\sqrt{(p^2 + m^2)}/T \right] \mathrm{d}p$$

(Boltzmann statistics)

or

$$\nu(m, T) = \frac{Vz_m}{2\pi^2} m^2 TK_2(m/T). \tag{IV.41}$$

Let $n(m, T)$ be the number of "particles" emitted with mass between m and $m + \mathrm{d}m$:

$$n(m, T) \equiv \nu(m, T)\rho(m)\,\mathrm{d}m. \tag{IV.42}$$

The total number of particles with any mass is then

$$N(V, T) = \int_0^\infty n(m, T)\,\mathrm{d}m = \frac{VT}{2\pi^2}\int_0^\infty m^2\rho(m)TK_2\left(\frac{m}{T}\right) \tag{IV.43}$$

which gives again (approximation IV.18)

$$N(V, T) \simeq \ln Z(V, T) \tag{IV.44}$$

restricted to ordinary canonical description without momentum conservation and including one-particle states (see (IV.8)).

Thus a fireball consists of (or emits at the first step decay) an average number of sub-fireballs given by (IV.44).

From relations (III.20) and (IV.21) one can immediately see that for $T \to T_0$

$$N(V_0, T) = \text{const. if } \rho(m) \simeq cm^{-3}\exp(m/T_0) \tag{IV.45}$$

$$N(V_0, T) = \text{const.} \ln\left(\frac{T_0}{T_0 - T}\right) \text{if } \rho(m) \simeq cm^{-\frac{5}{2}}\exp(m/T_0) \tag{IV.46}$$

The same can be obtained from the microcanonical description. Indeed Frautschi[9] has shown that for high masses one can write $\rho_{\text{out}}(m)$ as a sum over n-particle phase space which takes the approximate form

$$\rho_{\text{out}}(m) = \sum_{n=2}^{\infty} \frac{x^{n-1}}{(n-1)!}\rho_{\text{in}}(m) \tag{IV.47}$$

The nth term of this sum represents the weight for decay into n sub-fireballs. As, however, the bootstrap condition requires

$$\rho_{\text{out}}(m) = \rho_{\text{in}}(m) \text{ for large } m$$

one has

$$\sum_{n=2}^{\infty} \frac{x^{n-1}}{(n-1)!} = 1$$

i.e.

$$x = \ln 2 \simeq 0{\cdot}69 \tag{IV.48}$$

Thus the probability that a fireball decays into n particles is

$$W(n) = \frac{(\ln 2)^{n-1}}{(n-1)!} \tag{IV.49}$$

which gives a mean value

$$\bar{n} = N = \sum_{n=2}^{\infty} nW(n) = 2{\cdot}4 \tag{IV.50}$$

This coincides with the result of Hagedorn and Montvay (Ref. (15), eqn (II.46)).

Note that numerically one finds:

$$W(2) = 0{\cdot}69; \qquad W(3) = 0{\cdot}24; \qquad W(4) = 0{\cdot}06$$

which means that 2 and 3 particle channels are by far the dominant ones. Furthermore,[9] for a 2 body channel a fireball decays into one other fireball and a light particle. (See Fig. 1.)

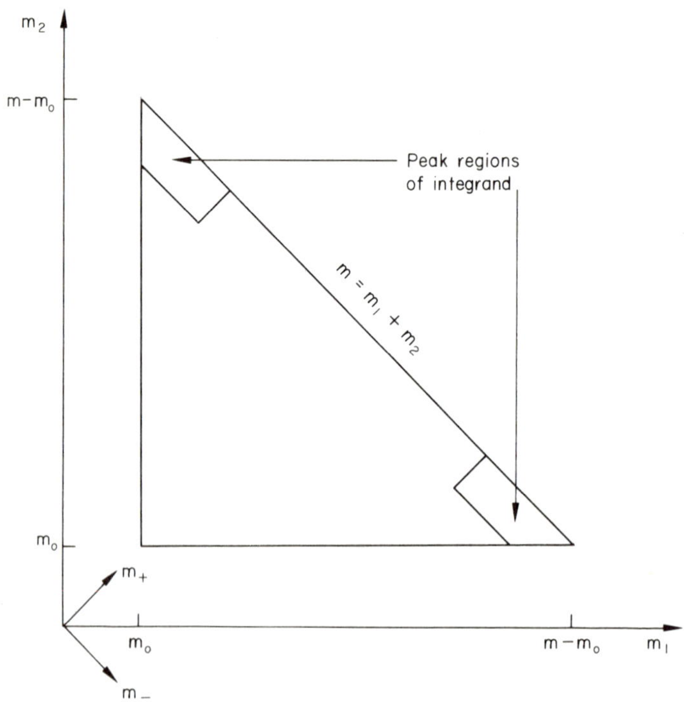

FIG. 1. Density of two particle states, $\rho_{\text{out}}^{n=2}(m)$ as a function of m_1 and m_2. With Frautschi's hadronic mass spectrum, a fireball of mass m decays preferentially into one heavy and one light particle. Figure taken from Ref. 9.

Finally, in view of the preceding results with respect to $N(V, T)$ for $T \to T_0$, we conclude that the decay of a fireball is different for $a = -\frac{5}{2}$ and $a = -3$. For $a = -\frac{5}{2}$ a fireball explodes into a large number of other fireballs, while for $a = -3$ (strong bootstrap) it decays preferentially into one other fireball and a light particle at each step. In the latter case we obtain the decay chain of Fig. 2.

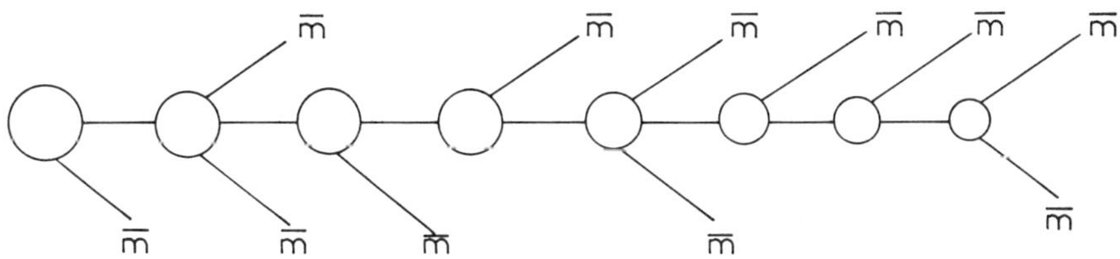

FIG. 2. Frautschi's linear decay chain. The average number of particles m_i is 1·4.

The number of steps in such a chain (we call it henceforth the length of the chain) can be easily approximated. The ith daughter fireball has a total energy $m_i + E_i$ where E_i is its (small) kinetic energy in the rest frame of the parent fireball M. Then the ith fireball decays into another fireball plus $1{\cdot}4 = (2{\cdot}4 - 1)$ light particles of average mass $\overline{m}$. Energy conservation relates to ith step to the next one by

$$(m_i + E_i) = (m_{i+1} + E_{i+1}) + 1{\cdot}4(\overline{m} + \tfrac{3}{2}T)$$

where $\frac{3}{2}T$ is the kinetic energy of the mass $\overline{m}$. If the decay chain ends after n steps we then obtain

$$1{\cdot}4n(\overline{m} + \tfrac{3}{2}T) \simeq M$$

or

$$n(M) \simeq \frac{M}{1{\cdot}4(\overline{m} + \frac{3}{2}T)} \qquad \text{(IV.51)}$$

Most of the light particles are pions so that $\overline{m} = 2m_\pi$ and since $T \sim T_0 \sim m_\pi$, we get

$$n(M) \simeq \frac{M}{5m_\pi} \qquad \text{(IV.52)}$$

Thus we reobtain directly a relation similar to (III.37).

3. PARTITION FUNCTION AND THERMODYNAMIC FUNCTIONS. THE LIMITING TEMPERATURE T_0

In any text-book on statistical thermodynamics one finds that the mean energy $\bar{E}$ see (eqn (II.13)) and the pressure P are related to the partition function by

$$\bar{E} = T^2 \frac{\partial}{\partial T} [\ln Z(V, T)] \tag{IV.53}$$

$$P = T \frac{\partial}{\partial V} [\ln Z(V, T)] \tag{IV.54}$$

For the same reasons as for $Z(V, T)$ these functions have different behaviour in the weak ($a = -\frac{5}{2}$) and in the strong bootstrap ($a = -3$) solutions:

$$\begin{aligned}
\bar{E}_{-\frac{5}{2}} &\xrightarrow[T \to T_0]{} C_1 \frac{T_0^2}{T_0 - T} \\
\bar{E}_{-3} &\xrightarrow[T \to T_0]{} C_2 \frac{T_0^2}{\sqrt{(T_0 - T)}} \\
P_{-\frac{5}{2}} &\xrightarrow[T \to T_0]{} C_3 \ln [T_0/(T_0 - T)] \\
P_{-3} &\xrightarrow[T \to T_0]{} C_4 + C_5 \sqrt{(T_0 - T)}
\end{aligned} \tag{IV.55}$$

where C_1, C_2, C_3, C_4, C_5 are constants, the value of which can be determined.[22]

We note that the mean energy diverges when $T \to T_0$ for the weak and for the strong bootstrap. Thus T_0 appears as a limiting temperature of hadronic matter in thermodynamical equilibrium. In fact T_0 is a universal limiting temperature for all matter in equilibrium.[22] In a collision problem the asymptotic constancy of T when $E \to \infty$ may be interpreted by saying that when we increase the collision energy it becomes easier to create new particles than to raise the kinetic energy of the produced ones.

The existence of a limiting temperature T_0 and the behaviour of the thermodynamical functions near T_0 have important consequences in astrophysics.[22, 23]

4. MEAN TRANSVERSE MOMENTA

If T tends to a limiting T_0, then the momentum distribution of particles emitted by highly excited fireballs must tend to the Planck distribution as $T \to T_0$. More precisely one finds that the limit of the mean value of the momentum of a particle of mass m is given by

$$\langle p(m, T) \rangle \equiv \frac{\int_0^\infty p^3 \, dp[\exp[\sqrt{(b^2 + m^2)}/T] \mp 1]^{-1}}{\int_0^\infty p^2 \, dp[\exp[\sqrt{(p^2 + m^2)}/T] \mp 1]^{-1}}$$

$$\langle p(m, T)\rangle \simeq 4\sqrt{\left(\frac{mT}{2\pi}\right)}K_{\frac{5}{2}}\left(\frac{m}{T}\right)\Big/K_2\left(\frac{m}{T}\right)$$
$$\Rightarrow 4\sqrt{\left(\frac{mT}{2\pi}\right)}$$

Thus we have for the mean transverse momentum $p_\perp$

$$\langle p_\perp(m, T)\rangle = \frac{\pi}{4}\langle p(m, T)\rangle \simeq \sqrt{\left(\frac{\pi m T}{2}\right)} \tag{IV.56}$$

which is limited when $T \to T_0$. Now if we *assume* that fireballs have only longitudinal motion, it follows from (IV.56) that the mean value of the transverse momenta of the observed particles is limited because of the existence of the limiting temperature T_0.

We feel that the recent data of CERN ISR[24] with respect to large transverse momenta do not contradict the statistical bootstrap: the slope of the momentum distribution of the bulk of the data is still governed by T_0, and one can probably account for the rest of the data by assuming that some fraction of the fireballs are produced with non-vanishing transverse momenta.

5. ENERGY MOMENTUM CONSERVATION

In the canonical description we forgot about energy-momentum conservation. This is a big drawback of that description when applied to particle collisions. So wc have to force energy momentum conservation on the final results. This is achieved[25] by cutting off the Planck momentum spectrum $f_m(p, T)$ at the kinematical limit p_{max}.

V. APPLICATIONS TO HIGH ENERGY COLLISIONS

1. VELOCITY DISTRIBUTION OF FIREBALLS

As emphasized in the beginning of Section IV, the canonical description applies to fireballs at rest after the collective longitudinal motions have been separated off. We give now a way of taking these motions into account. To be specific we consider first a proton–proton collision.[2]

The underlying idea is that a collision process consists of two steps. In the first step fireballs are produced and then decay, in the second step, with distributions determined by the statistical bootstrap.

Naturally fireballs are produced with different longitudinal velocities. Intuitively, one can, for example, imagine that the velocities depend on the impact parameter. Thus each fireball may have a longitudinal velocity ranging from zero to the initial velocity v_0 of the incoming proton. Instead of the velocity itself we use the variable

$$\lambda = \text{sign}\,(v)\frac{\gamma - 1}{\gamma_0 - 1} \tag{V.1}$$

with:

$$\gamma = \frac{1}{\sqrt{(1 - v^2)}}$$

obviously we have

$$-1 \leqslant \lambda \leqslant 1.$$

Another variable could be the rapidity η of the fireball in the C.M.S.;

$$\eta = \cosh^{-1}\gamma = \pm\cosh^{-1}[1 + \lambda(\gamma - 1)] \tag{V.2}$$

We keep for the moment the variable λ.

Let us denote by $F(\lambda)$ the velocity distribution of fireballs and let us consider a fireball with velocity λ. It emits particles with a Planck momentum distribution governed by its temperature T. This T is related to the mean energy of the fireball by eqn (IV.53). If we know the mean energy we can calculate T. As we shall see, the average energy of a fireball is, in the thermodynamical model, a function of the total energy of the collision and of the velocity λ of that fireball. Thus to a given λ will correspond a definite temperature $T(\lambda)$ and the momentum spectrum of a particle of mass m in a reference frame R (the C.M.S. or the lab. system) will be

$$W^{(R)}(\vec{p})\,\mathrm{d}^3\vec{p} = \int_{-1}^{+1} F(\lambda)\,L^{(R)}(\lambda, \gamma_0)\,\{f_m(p', T(\lambda, \gamma_0))\,\mathrm{d}^3p'\}\,\mathrm{d}\lambda \tag{V.3}$$

where $L^{(R)}(\lambda, \gamma_0)$ is the Lorentz operator which transforms the isotropic spectrum $f_m(p', T(\lambda, \gamma_0))\,\mathrm{d}^3p'$ seen in the fireball rest frame (we call it the λ-frame) into the spectrum seen in the R-frame.

In the original version[2] of the thermodynamical model $F(\lambda)$ was normalized in such a way that

$$\int_{-1}^{+1} F(\lambda)\,\mathrm{d}\lambda = 2. \tag{V.4}$$

Remembering that $F(\lambda)$ is the velocity distribution, this normalization amounts to saying that the average number of the fireballs (fireballs of the zero-th generation) just after the collision and before any decay is two.

Since $F(\lambda)$ is symmetric in a p–p collision the normalization condition (V.4) is equivalent to

$$\int_0^1 F(\lambda)\,\mathrm{d}\lambda = 1$$

and (V.5)

$$\int_{-1}^0 F(\lambda)\,\mathrm{d}\lambda = 1.$$

Thus $F(\lambda)$ may be interpreted as the probability for the incoming proton in the forward direction ($0 < \lambda < 1$) to be excited to a fireball state with velocity λ. The same interpretation is of course valid for the backward proton ($-1 < \lambda < 0$). One should note that if the average number of zero-th generation fireballs is n_F then one can either normalize $F(\lambda)$ to n_F or keep the normalization (V.4) and multiply the one particle momentum spectrum (V.3) by n_F. Of course the velocity distribution $F(\lambda)$ cannot be calculated from thermodynamics which applies to fireballs at rest and it has to be determined by other means[26] or to be considered as an empirical function of the model. Needless to say, this freedom on $F(\lambda)$ is a weak point of the model. However the abundance of high energy data allows its determination in a rather satisfactory way.

2. EXCITATION ENERGY. AVERAGE FIREBALL MASS

The link between the mean energy $\bar{E}$ and the temperature T is provided by relation (IV.53). But what do we choose for $\bar{E}$? In the original version of the thermodynamical model, $\bar{E}$ is defined with the help of an adiabatic hypothesis as follows.

Consider a piece of hadron matter belonging to the incoming proton. During the collision this piece is decelerated and its velocity decreases from v_0 (the velocity of the proton before the collision) to v. In the variable λ this means that $|\lambda|$ decreases from $|\lambda| = 1$ to $|\lambda| = (\gamma - 1)/(\gamma_0 - 1)$. During the deceleration process, the piece of matter is heated. The adiabatic hypothesis consists in postulating that all the kinetic energy of the incoming proton that has disappeared during the deceleration process is converted into heat (internal excitation).

Denote by ε_0 the average proton rest mass energy density.

$$\varepsilon_0 = m_p/V_0$$

with

$$V_0 = \tfrac{4}{3}\pi m_\pi^{-1}. \tag{V.6}$$

Then if the piece of matter has unit volume, its total energy is $E_0 = \varepsilon_0\gamma_0$. At the end of the deceleration process it has, in its rest frame, an energy density ε and a total energy $E = \varepsilon\gamma$. The adiabatic hypothesis states that

$$E = \varepsilon\gamma = \varepsilon_0\gamma_0 \tag{V.7}$$

or

$$\varepsilon = \frac{\varepsilon_0\gamma_0}{\gamma}. \tag{V.8}$$

Note that the same reasoning could have been applied to the incoming proton as a whole. This amounts to multiplying eqn (V.7) by V_0 to get

$$V_0\varepsilon_0\gamma_0 = V_0\varepsilon\gamma$$

or (V.9)

$$m_p\gamma_0 = m_F\gamma$$

Note that (V.9) means that the proton after the deceleration becomes a single fireball of volume V_0 and average rest mass $m_F > m_p$. Thus if all fireballs have volume V_0 and if one accepts the adiabatic hypothesis, each of the incoming protons cannot give rise to more than one fireball. This is consistent with the normalization (V.4).

The average mass m_F of a fireball depends on the total energy of the collision (through γ_0) and on the velocity λ (through γ and γ_0).

But other parametrizations of the fireball mass are possible. For example Kripfganz, G. Ranft and J. Ranft[27] use the following parametrizations:

(i) $m_F(\eta, s) = m_c\left(\dfrac{\sqrt{s}}{M_c}\right)^w \exp(-v\eta)$

with $0 \leqslant w \leqslant 1, 0 \leqslant v \leqslant 1$ (V.10)

(ii) $m_F(\eta, s) = m_c \ln\left[\dfrac{\sqrt{s}}{M_1}\exp(-v\eta)\right]$

where s is the total energy of the collision and η is the fireball rapidity.

The reason of these choices lies in the fact that the maximum kinematically allowed mass of a fireball moving with velocity λ is

$$m_F^{\max}(\lambda) = (\sqrt{s})[\gamma - \sqrt{(\gamma^2 - 1)}] \simeq (\sqrt{s})\,e^{-\eta} \equiv m^{\max}(s, \eta) \tag{V.11}$$

In case (i), $w = 1$ and $v = 1$ correspond to one fireball production as in the nova model[3] and $w = v = 0$ corresponds to $m_F = m_c =$ const., as in the multiperipheral model. The choices $0 < w < 1$ and $0 < v < 1$ interpolate between these limiting situations. The case (ii) corresponds to some sort of a general multiperipheral production of fireballs.

3. CONSERVATION LAWS

Equation (V.3) is the basic one for the calculation of inclusive momentum distributions. It gives the free production spectra, i.e. the spectra obtained when no conservation law is taken into account. It has to be modified in order to incorporate conservation laws. We deal here with additive quantum numbers like charge, baryon number, strangeness. These conservation laws are imposed on the partition function $Z(V_0, T)$ (see eqn (IV.6)) and are expressed with the help of the general probability rule (IV.24). I do not give the details of the derivation of the following formulae here and refer the reader to the papers[2, 28] of Hagedorn and Ranft.

3.1. *Conservation laws for pair and associate production*

If a particle a is constrained by a conservation law to be produced with a particle b belonging to a family B then the thermodynamical spectrum of a is given by

$$f_a(\vec{p}, T) = N_B(T) \cdot f_{m_a}(\vec{p}, T) \tag{V.12}$$

where

$$N_B(T) = \sum\nolimits_B^* \int f_{m_b^*}(\vec{p}, T)\, \mathrm{d}^3p \tag{V.13}$$

The sum Σ_B^* runs over all particles of the B family.

Example: if a is an antiproton, $\bar{p}$, then the sum Σ_B^* runs over the proton and all baryon resonances.

Similarly if a is a K^+, then the family B is composed of the K^- and all its resonances.

3.2. *Conservation of a quantum number of the incident particle*

If a particle a, say a proton, need not be newly created because it is one of the colliding particles, and if its quantum number is conserved (say the baryon quantum number) then its momentum spectrum is given by

$$f_a^{(c)}(\vec{p}, T) = \frac{1}{N_A(T)} f_{m_a}(\vec{p}, T) \tag{V.14}$$

where

$$N_A(T) = \sum\nolimits_A^* f_{m_a^*}(\vec{p}, T) \tag{V.15}$$

and where the superscript c stands for conserved.

The sum Σ_A^* runs over all resonances of the A family to which a belongs.

If a is a proton then A is the family of baryon resonances including a. One can check that (V.15) is the right formula since

$$\sum_A \int f_{m_a^*}(\vec{p}, T) = \frac{1}{N_A(T)} \sum_A f_{m_a^*}(\vec{p}, T) \tag{V.16}$$

which means that we have one out-going baryon, excited or not.

4. LOCALITY OF CONSERVATION LAWS AND PAIR AND ASSOCIATED PRODUCTION

Now we know how to modify the free spectrum $f_{m_a}(p, T)$ of a particle a when a conservation law requires that it must be produced with some particle b belonging to a family B. But *a priori*, particle a and particle b may be created in two different fireballs with velocities λ and λ', i.e. with temperatures $T(\lambda)$ and $T(\lambda')$, or in the same fireball.

λ-locality. If the particle a is produced in a fireball with velocity λ and particle b in a fireball or with velocity λ' then the differential contribution to the spectrum $W_a(p)$ of a (see eqn (V.3)) is

$$d^2 W_a(\vec{p}) = K(\lambda, \lambda')[N_B(T)F(\lambda')\, d\lambda']\, F(\lambda)\, d\lambda\, L(\lambda, \gamma_0)\{f_{m_a}(\vec{p}', T(\lambda))\} \tag{V.17}$$

where $K(\lambda, \lambda')$ is a correlation function.

This relation is obvious if one remembers that $F(\lambda)$ is the probability to find a fireball with velocity λ. Hagedorn and Ranft[2] have shown in studying the invariant $K\bar{K}$ mass distribution $f(M^2, T)$ (see Fig. 3) that the strangeness conservation law acts locally, i.e. $K(\lambda, \lambda')$ is essentially a $\delta(\lambda - \lambda')$ function:

$$K(\lambda, \lambda') = \delta(\lambda - \lambda') \tag{V.18}$$

This is the λ-locality of conservation laws. Then the momentum spectrum of the particle a produced in association with particle b is

$$W_a(\vec{p}) = \int_{-1}^{+1} [F(\lambda)]^2\, N_B(T(\lambda))\, L(\lambda, \gamma_0)\{f_{m_a}(\vec{p}', T(\lambda))\} \tag{V.19}$$

Strict locality: But one can assume[29] that particle a and particle b are created in the same fireball: this is the "strict locality" assumption. Then we have just to insert (V.12) into the curly brackets of (V.3) to obtain the spectrum of a produced in association with b:

$$W_a(\vec{p}) = \int_{-1}^{+1} F(\lambda)\, N_B(T)\, L(\lambda, \gamma_0)\{f_{m_a}(\vec{p}', T(\lambda))\} \tag{V.20}$$

Obviously λ-locality and strict locality are essentially assumptions about correlations. Strict locality is of course the only possibility in models where there is one fireball for a given λ (or for a given rapidity η).

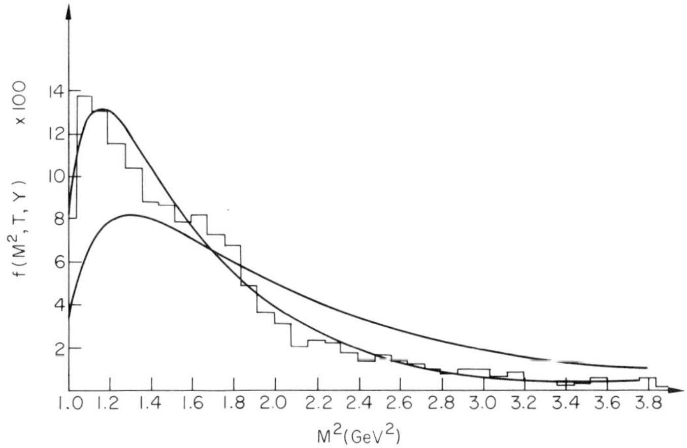

FIG. 3. Experimental invariant mass distribution $f(M^2)$ for $K\bar{K}$ from pp annihilation at 1·2 GeV/c compared to two thermodynamical distributions corresponding respectively to local and non-local strangeness conservation. Figure taken from Ref. 2.

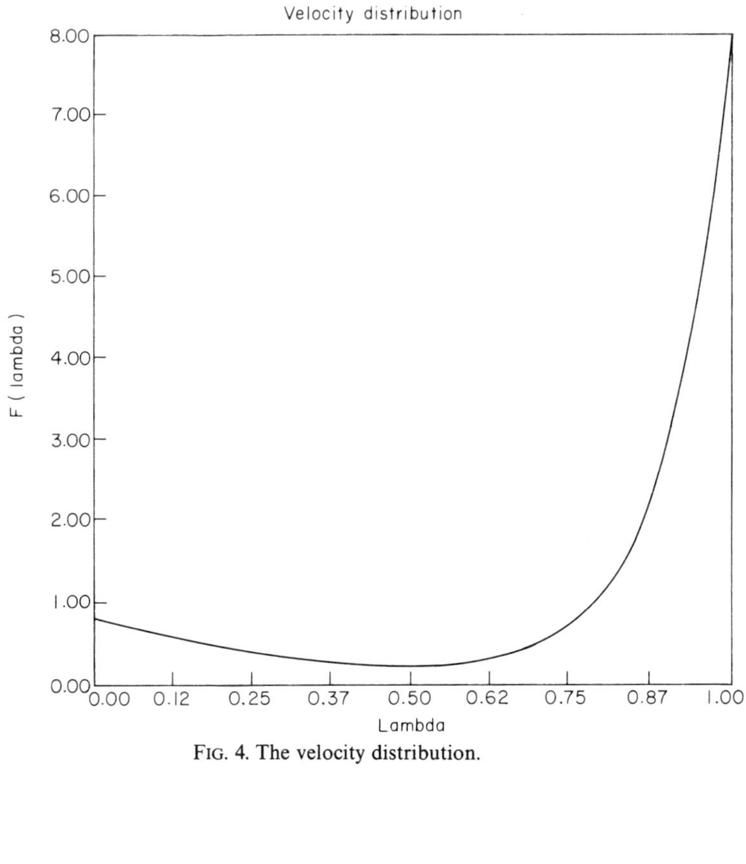

FIG. 4. The velocity distribution.

5. THROUGHGOING PARTICLE SPECTRUM

These are particles present in the initial state and "conserved" in the collisions (see Section V.3.1). This is the case, for example, of the incident proton. Its $W_p(\vec{p})$ spectrum is obtained by inserting (V.14) into the curly brackets of (V.3):

$$W_p(\vec{p}) = \int_{-1}^{+1} \frac{F(\lambda)}{N_A(T)} L(\lambda, \gamma_0)\{f_{m_p}(p', T(\lambda))\}\, d\lambda \tag{V.21}$$

In their original paper Hagedorn and Ranft used a special function $F_0(\lambda)$ in the calculation of the throughgoing particle spectrum. It has been shown[29, 30] that this is not necessary: a single velocity function is enough to account for all available data.

6. COMPLETE SPECTRA

The spectrum as given by (V.3), modified when necessary by conservation laws is in fact the spectrum of a particle of the first generation. As we already know, the production process is in fact more complicated, consisting of the decay chain of Fig. 2. Consider for example the π-spectrum. Each fireball m^* of the chain gives a contribution we denote by f^*_{π, m^*}. The total π-spectrum is the sum of contribution coming from all fireballs of all generations:

$$f_\pi(\vec{p}, T) = f_{m_\pi}(\vec{p}, T) + \sum_{m^*} f^*_{\pi, m^*}(\vec{p}, T) \tag{V.22}$$

where $f_{m_\pi}(\vec{p}, T)$ is the thermodynamical (Planck spectrum) of the first generation and where the Σ_{m^*} runs over all fireballs in the chain. For heavy fireballs the spectrum f^*_{π, m^*} is still a Planck spectrum. But for light fireballs which decay into two bodies f^*_{π, m^*} is a δ-function. Thus in (V.22) we group the thermodynamical contributions to obtain

$$f_\pi(\vec{p}, T) = Q^{(\pi)} f_{m_\pi}(\vec{p}, T) + \sum_{m^*}^{(2)} f^*_{\pi, m^*}(\vec{p}, T) \tag{V.23}$$

where now $\Sigma^{(2)}_{m^*}$ runs over the fireballs (resonances) which decay predominantly into two bodies.

In writing (V.23) we have tacitly assumed that all intermediate fireballs have the same temperature T, neglecting cooling and recoil effects. The factor $Q^{(\pi)}$ depends essentially on the length of the decay chain (see eqn (IV.52)) and thus it may be written.

$$Q^{(\pi)} = Q_\pi \frac{\varepsilon}{\varepsilon_0} \tag{V.24}$$

The f^*_{π, m^*} spectrum in (V.23) is the π-spectrum in the two body decay:

$$m^* \to \pi + x \tag{V.25}$$

where x is any particle allowed by the conservation laws. In the calculation of f^*_{π, m^*}, the resonance m^* is of course assumed to be produced with a thermodynamical spectrum.

In principle, all resonances which decay predominantly into two bodies have to be included in the sum $\Sigma^{(2)}_{m^*}$. But in most actual calculations,[30, 31] this sum was restricted to baryon resonances attached to the through-going particles. One then can write

$$\sum_{m^*}^{(2)} f^*_{\pi, m^*} \simeq \int \rho_B^{(2)}(m)\, f^*_{\pi, m^*}\, dm^* \tag{V.26}$$

where $\rho^{(2)}(m)$ counts only baryon resonances with not very high masses. The integration range in (V.26) is then finite and we can replace the integral by a finite sum:

$$\int \rho_B^{(2)}(m^*) f_{\pi, m^*} = \sum_{m^*} \rho_B^{(2)}(m^*) f^*_{\pi, m^*} \Delta m^* \equiv \sum_{m^*} A(m^*) f_{\pi, m^*} \tag{V.27}$$

which defines the weight factor $A(m^*)$. In practice[2] only a few resonances are included in the sum.

One has also to take into account isospin conservation in the production and in the decay of the m^* resonances. This may be achieved by a detailed Clebsch–Gordan analysis. But usually[31] this is treated in a global way by multiplying the sum $\Sigma^{(2)}_{m^*}$ in (V.23) by an "average" Clebsch–Gordan coefficient $C_{\pi^\pm}$ which appears as the total weight of all two-body isobar decay contributions to the $\pi^\pm$ momentum spectrum.

In this section we have explicitly discussed the complete spectrum of $\pi^\pm$. But this discussion applies of course to the complete spectrum of any particle of mass m. Then we denote by C_m the corresponding isospin total weight.

7. HAGEDORN–RANFT DIAGRAMS

These authors introduce a pictorial[2] representation of the various contributions the different complete spectra.

represents a fireball with velocity λ.

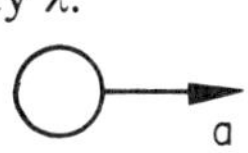

represents the free creation of a particle of mass m_a and corresponds to:

$$F(\lambda)L(\lambda, \gamma_0)\{f_{m_a}(\vec{p}', T(\lambda))\}$$

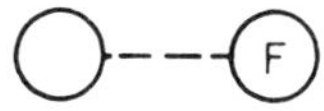

represents the creation of any particle of the F family and corresponds to the factor $N_F(T(\lambda))$ (see eqn (V.13)).

represents the absorption of an incident particle of the family F and corresponds to the factor $1/N_F(T(\lambda))$.

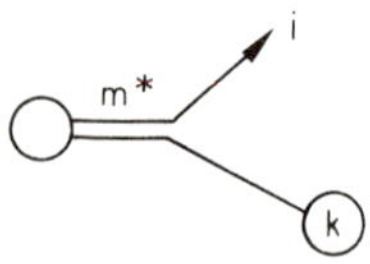

represents the spectrum of the particle i produced in the decay $m^* \to i + k$ and corresponds to $F(\lambda)L(\lambda, \gamma_0)\{f_{i,m^*}(\vec{p}, T(\lambda))\}$

By combining these symbols we obtain for example:

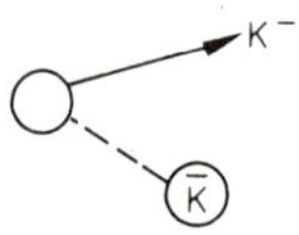

$$\equiv N_k(\lambda)F(\lambda)L(\lambda, \gamma_0)\{f_{m_K}(\vec{p}\,', T(\lambda))\}$$

for the K^- production (in the strict locality case)

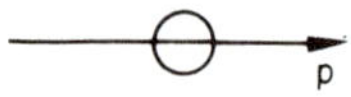

$$\equiv \frac{1}{N_B(T(\lambda))} F(\lambda)L(\lambda, \gamma_0)\{f_{m_p}(\vec{p}\,', T(\lambda))\}$$

for the through-going proton

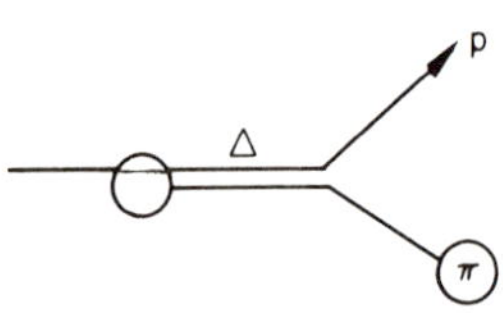

$$\equiv \frac{1}{N_B(T(\lambda))} F(\lambda)L(\lambda, \gamma_0)\{f^*_{p,m_\Delta}(\vec{p}\,', T(\lambda))\}$$

for the proton coming from an incident proton via a resonance Δ.

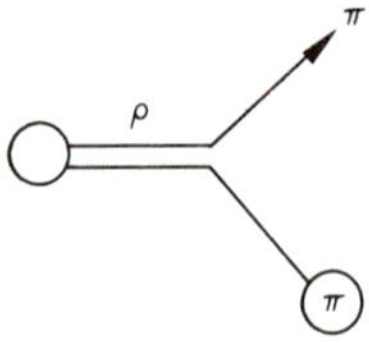

$$\equiv F(\lambda)L(\lambda, \gamma_0)\{f^*_{\pi,m_\rho}(\vec{p}\,', T(\lambda))\}$$

for a π created in ρ decay.

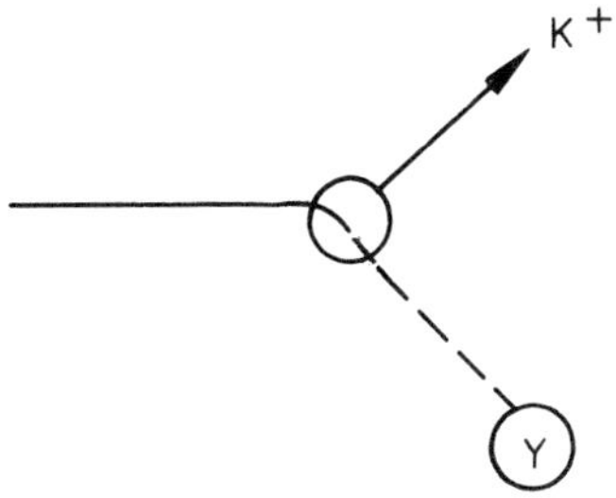

$$\equiv \frac{N_Y(\lambda)}{N_B(\lambda)} F(\lambda)L(\lambda, \gamma_0)\{f_{m_K}(\vec{p}', T(\lambda)\}$$

for the production of a K^+ associated with the transformation of the incident proton into a hyperon Y.

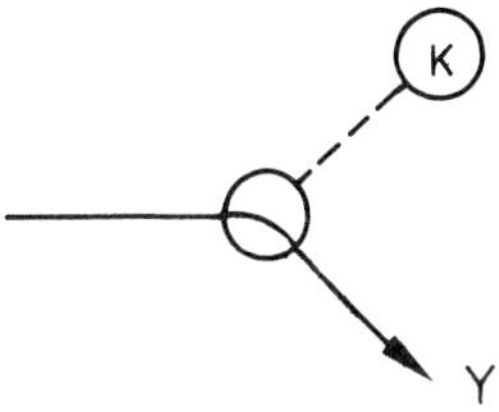

$$\equiv \frac{N_K(\lambda)}{N_B(\lambda)} F(\lambda)L(\lambda, \gamma_0)\{f_{m_Y}(\vec{p}', T(\lambda))\}$$

for the production of a hyperon Y in association with a K.

Tables I and II contain respectively the complete spectra of charged and neutral particles we have effectively used in our calculations.(32)

TABLE I.

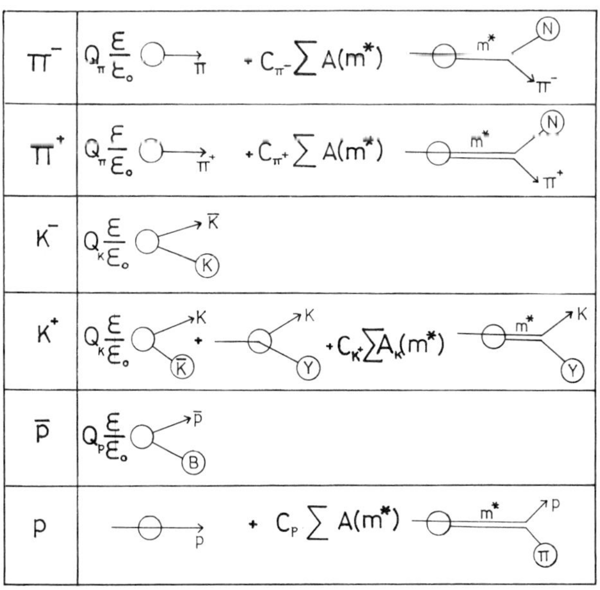

TABLE II.

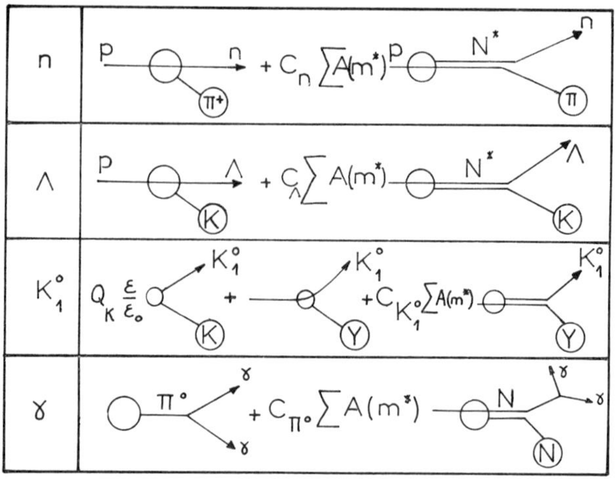

VI. SCALING, LIMITING FRAGMENTATION. COMPARISON WITH EXPERIMENTAL DATA

We first recall briefly a few definitions and some notation.

1. SCALING

Scaling is defined in the C.M.S. and we denote by an asterisk the quantities measured in this system.

The invariant differential cross-section $E^*(\mathrm{d}^3\sigma^*/\mathrm{d}^3\vec{p}^*)$ of the reaction $a + b \to c +$ anything has the scaling property if

$$E^* \frac{\mathrm{d}^3\sigma^*}{\mathrm{d}^3\vec{p}^*} \equiv f(p_{\parallel}^*; p_{\perp}, s) \to g(x, p_{\perp}) \qquad \text{(VI.1)}$$

for all x.

In (VI.1), E^*, $p_{\parallel}^*$ are the C.M. energy and longitudinal momentum of the observed particle c, $p_{\perp}$ is its transverse momentum and x is the Feynman variable defined by

$$x = 2p_{\parallel}^*/\sqrt{s} \qquad \text{(VI.2)}$$

One has

$$-\sqrt{[\alpha(s, m_c, m_R)]} \leqslant x \leqslant \sqrt{[\alpha(s, m_c, m_R)]} \qquad \text{(VI.3)}$$

where

$$\alpha(s, m_c, m_R) = \frac{1}{s^2}[s - (m_c + m_R)^2][s - (m_c - m_R)^2]$$

with m_R = minimum recoil mass required for production of particle c. Thus x is always in the interval $-1 \leqslant x \leqslant 1$ and the limits ± 1 are obtained when $s \to \infty$. Usually the function $g(x, p_\perp)$ is assumed to be continuous at $x = 0$. This assumption corresponds to scaling with a "central plateau" in the rapidity variable y^* of the observed particle.

$$y^* = \cosh^{-1}\left(\frac{E^*}{\mu}\right) = \sinh^{-1}\left(\frac{p_\parallel^*}{\mu}\right) = \ln\left(\frac{p_\parallel^* + \sqrt{(p_\parallel^{*2} + \mu^2)}}{\mu}\right) \quad \text{(VI.4)}$$

where

$$\mu = \sqrt{(p_\perp^2 + m_c^2)}$$

is the "longitudinal mass".

Since $p_\parallel^* = x(\sqrt{s})/2$ one can easily see that

$$y^* = \sinh^{-1}\left[\frac{x\sqrt{s}}{2\mu}\right] = \ln\left(\frac{\sqrt{s}}{2\mu}\right) + \ln\left[x + \sqrt{(x^2 + 4\mu^2/s)}\right]$$

$$x = \frac{2\mu}{\sqrt{s}} \sinh y^* \quad \text{(VI.6)}$$

$$dy^* = \frac{dp_\parallel^*}{E^*} = \frac{dx}{\sqrt{(x^2 + 4\mu^2/s)}}$$

and

$$E^* \frac{d^3\sigma^*}{d^3\vec{p}^*} = \frac{d^3\sigma^*}{dy^*\, d^2p_\perp} = \sqrt{(x^2 + 4\mu^2/s)} \frac{d^3\sigma^*}{dx\, d^2p_\perp} \quad \text{(VI.7)}$$

The rapidity variable has interesting properties:

(i) A constant interval $\xi \leqslant x \leqslant \xi$ in x corresponds to an increasing interval $-\ln[\xi(\sqrt{s})/\mu] \leqslant y^* \leqslant \ln[\xi(\sqrt{s})/\mu]$ in y^* as $x \to \infty$. In particular, a fixed interval around $x = 0$ corresponds to an interval in y^* which grows like $\ln[(\sqrt{s})/\mu]$. This is particularly interesting because all momenta $p_\parallel^*$ which grow less fast than $\sqrt{s}$ are mapped onto $x = 0$.

(ii) In a longitudinal Lorentz transformation characterized by $\gamma = (1 - \beta^2)^{-\frac{1}{2}}$ the rapidity y^* becomes

$$y = y^* + Y \quad \text{(VI.8)}$$

where

$$Y = \cosh^{-1} \gamma \quad \text{(VI.9)}$$

Thus the y^* distribution is merely translated by an amount $Y = \cosh^{-1}(\gamma)$. All these definitions and properties are well known.[34]

2. LIMITING FRAGMENTATIONS

Limiting fragmentation[35] is defined in the rest frame of either the target or the projectile. The invariant differential cross-section has the limiting fragmentation property if

$$E\frac{d^3\sigma}{d^3\vec{p}} \equiv F(\vec{p}, s) \to G(\vec{p}) \qquad \text{(VI.10)}$$

for fixed $|\vec{p}|$ where E and $\vec{p}$ are the energy and momentum of the observed particle (in the target or projectile frame).

One can demonstrate[34] that scaling and limiting fragmentation are equivalent for $|x| \gg 4\mu^2/s$ by showing that for any fixed value of x such that $x \ll -2\mu/\sqrt{s}$ (or $x \gg 2\mu/\sqrt{s}$) there corresponds a fixed value of $p_{\|}$ in the target rest frame (or in the projectile rest frame).

One can also show that $x = 0$ corresponds to $p_{\|} = \mu\sqrt{s}/2m_p$ which is finite neither in the target nor in the projectile rest frame. Consequently limiting fragmentation is not defined at the point $x = 0$.

3. LIMITING FRAGMENTATION AND SCALING IN THE THERMODYNAMICAL MODEL

The momentum spectra in the thermodynamical model has essentially the general form

$$W(\vec{p}) = \int_{-1}^{+1} G(\lambda)L(\lambda, \gamma_0)\{\phi(\vec{p}', \lambda, \gamma_0)\}\, d\lambda \qquad \text{(VI.11)}$$

where $G(\lambda)$ is some combination of $F(\lambda)$ and $N(T(\lambda))$ and $\phi(\vec{p}', \lambda, \gamma_0)$ is a combination of the Planck type momentum spectrum $f(p', T)$ with multiplicity factor $Q(\varepsilon)$ of the form given by (V.24).

If we assume $F(\lambda)$ is energy independent (or asymptotically energy independent), the energy dependence of $W(\vec{p})$ is contained

(i) in the factor $Q = Q_m \dfrac{\varepsilon(\lambda, \gamma_0)}{\varepsilon_0}$

(ii) in $T = T\,(\varepsilon(\lambda, \gamma_0))$

(iii) in the Lorentz operator $L(\lambda, \gamma_0)$.

The adiabatic hypothesis gives

$$\varepsilon(\lambda, \gamma_0) = \frac{\varepsilon_0\gamma_0}{\gamma} = \frac{\varepsilon_0\gamma_0}{1 + |\lambda|(\gamma_0 - 1)} \qquad \text{(VI.12)}$$

If $\gamma_0 \lambda_0 \gg 1$ for a fixed $\lambda_0 > 0$, i.e. if $\sqrt{s}/2m \gg 1/\lambda_0$ then (VI.12) becomes

$$\varepsilon(\lambda, \gamma_0) = \frac{\varepsilon_0}{\lambda} \tag{VI.13}$$

which means that for $|\lambda| > \lambda_0$, Q and T no longer depend on γ_0. Thus we are left with the energy dependence of the Lorentz operator $L(\lambda, \gamma_0)$. In order to show limiting fragmentation we write (VI.11) in the target frame in the form

$$\begin{aligned} W^{(t)}(\vec{p}) = & \int_{-1}^{-\lambda_0} G(\lambda) L^{(t)}(\lambda, \gamma_0) \{\phi(p', \lambda, \gamma_0)\} \,\mathrm{d}\lambda \\ & + \int_{-\lambda_0}^{\lambda_0} G(\lambda) L^{(t)}(\lambda, \gamma_0) \{\phi(p', \lambda, \gamma_0)\} \,\mathrm{d}\lambda \\ & + \int_{\lambda_0}^{-1} G(\lambda) L^{(t)}(\lambda, \gamma_0) \{\phi(p'. \lambda, \gamma_0)\} \,\mathrm{d}\lambda \end{aligned} \tag{VI.14}$$

where the superscript (t) stands for target.

We denote by

$$V_0 = (\gamma_0, \beta_0 \gamma_0)$$

the velocity four vector of the CM seen from the target.

$$V_{\mathrm{CM}} = (\gamma, \beta\gamma)$$

the velocity four vector for the fireball (λ) seen from the CM

$$V_{(\lambda)}^{(t)} = (\gamma_\lambda, \beta_\lambda \gamma_\lambda)$$

the velocity of the fireballs (λ) seen from the target.

Hence:

$$\begin{aligned} \gamma_\lambda &= \gamma_0(\gamma - \beta_0 \beta\gamma) = \gamma\gamma_0(1 - \beta\beta_0) \\ \beta_\lambda \gamma_\lambda &= \gamma_0(-\beta\gamma + \beta_0 \gamma) = \gamma\gamma_0(\beta_0 - \beta) \end{aligned} \tag{VI.15}$$

Under the condition (VI.13) one can easily show that

$$V^{(t)} \simeq \left(\frac{1 + \lambda^2}{2\lambda}, \frac{1 - \lambda^2}{2\lambda} \right) \tag{VI.16}$$

for $-1 \leqslant \lambda \leqslant -\lambda_0$.

Thus under that condition, $V^{(t)}(\lambda)$ is independent of γ_0. Since the Lorentz transformation

$$(\lambda) \rightarrow \text{target}$$

uses this energy independent $V^{(t)}(\lambda)$ we see that under condition (VI.13)

$L^{(t)}(\lambda, \gamma_0)$ does not depend on γ_0 for $-1 \leqslant \lambda \leqslant \lambda_0$. Thus under the same conditions

$$\underset{|\vec{p}|\ \text{fixed}}{W^{(t)}(\vec{p})} = \int_{-1}^{\lambda_0} G(\lambda) L^{(t)}(\lambda)\{\phi(p', \lambda)\}\, \mathrm{d}\lambda \tag{VI.17}$$

which does not depend on the primary energy.

A similar proof holds for the projectile. Hence the thermodynamical spectra satisfy limiting fragmentation if $F(\lambda)$ is independent (or asymptotically independent) of the primary energy of the collision.

Since scaling is equivalent to limiting fragmentation we have also scaling in the thermodynamical model for $|\lambda| > \lambda_0$. The central region $(-\lambda_0, \lambda_0)$ deserves particular attention. In this region

$$W^{(\text{centr.})}(\vec{p}) = \int_{-\lambda_0}^{\lambda_0} G(\lambda) L^{\text{CM}}(\lambda, \gamma_0)\{\phi(p', \lambda, \gamma_0)\}\, \mathrm{d}\lambda \tag{VI.18}$$

where

$$\phi(p', \lambda, \gamma_0) = \frac{\varepsilon}{\varepsilon_0}(\lambda, \gamma_0)\ \text{const.}\left[\exp\frac{E'}{T} \pm 1\right]^{-1}$$

with the $T \sim T_0$.

Because of the adiabatic hypothesis we can write

$$\left[E^* \frac{\mathrm{d}^3\sigma^*}{\mathrm{d}^3 p^*}\right]^{\text{centr.}} = \text{const.}\ \gamma_0 \int_{-\lambda_0}^{\lambda_0} E^* L^{\text{CM}}(\lambda, \gamma_0)\left[\frac{1}{\mathrm{e}^{E'/T} \mp 1}\right]\frac{\mathrm{d}\lambda}{\gamma} \tag{VI.19}$$

But

$$E^* L^{\text{CM}}(\lambda, \gamma_0)\left[\frac{1}{\mathrm{e}^{E'/T} \pm 1}\right] = \frac{E'}{\mathrm{e}^{E'/T} \pm 1} \tag{VI.20}$$

where the right-hand side is to be expressed in the CM variables:

$$E' = \mu \cosh(y^* - \eta) \tag{VI.21}$$

where y^* is the CM rapidity of the observed particle and $\eta = \cosh^{-1}\gamma$ is the rapidity of the fireball (λ) in the CMS. Then (VI.19) becomes

$$\left[\frac{\mathrm{d}^3\sigma^*}{\mathrm{d}y^*\, \mathrm{d}^2 p_\perp}\right]^{\text{centr.}} = \text{const.}\ \gamma_0 \int_{-\lambda_0}^{\lambda_0} \frac{\mathrm{d}\lambda}{\gamma}\left[\frac{\mu \cosh(y^* - \eta)}{\exp(\mu \cosh(y^* - \eta)/T) \pm 1}\right] \tag{VI.22}$$

We change the variable of integration and we obtain for $\gamma_0 \gg 1$

$$d\lambda = \pm \frac{d\gamma}{\gamma_0} = |\sinh \eta| . \frac{d\eta}{\gamma_0}$$

Thus (VI.22) becomes

$$\left[\frac{d^3\sigma^*}{dy^* \, d^2p_\perp}\right]^{\text{centr.}} = \text{const.} \int_{-\eta_0}^{\eta_0} d\eta \, |\tanh \eta| \left[\frac{\mu \cosh (y^* - \eta)}{\exp (\mu \cosh (y^* - \eta)/T) \pm 1}\right] \tag{VI.23}$$

It can be shown by a careful analysis[31] that (VI.23) leads to a central plateau of length $2\eta_0 \simeq 2 \ln (2\lambda_0\gamma_0)$ and with a small dip at $y^* \simeq 0$ of constant width. We do not enter into the details because this dip has not been observed experimentally. We simply wish to note that the dip is due to the factor $\tanh \eta$ in (VI.23).

In order to suppress this dip, we have to replace our $F(\lambda)$ by

$$\bar{F}(\lambda, \gamma_0) = \frac{F(\lambda)}{\tanh \eta} = F(\lambda) \frac{(\gamma_0 - 1) + 1/\lambda}{\sqrt{[(\gamma_0 - 1)^2 + 2(\gamma_0 - 1)/\lambda]}} \tag{VI.24}$$

It is easy to show that for $\gamma_0 \gg 1$

$$\bar{F}(\lambda, \gamma_0) \simeq F(\lambda) \qquad \text{for } \lambda \gg 1/\gamma_0$$

$$\bar{F}(\lambda, \gamma_0) = \frac{F(\lambda)}{\sqrt{(2\lambda\gamma_0)}} \qquad \text{for } \lambda \ll 1/\gamma_0 \tag{VI.25}$$

Thus nothing is changed in the fragmentation region $|\lambda| \gg 1/\gamma_0$ but the price of suppressing the dip is the introduction of a square root singularity at $\lambda = 0$ in $F(\lambda)$. In such a case the λ-locality (see (V.4)) which introduces $(F(\lambda))^2$ in the expression of the spectrum is ruled out.

4. FIT TO EXPERIMENTAL DATA

We do not give here the details of the fits.[30] We simply state that $F(\lambda)$ has been taken in the form

$$F(\lambda) = \frac{1}{N} [(1 - \lambda) e^{-a\lambda} + c\lambda e^{-b(1-\lambda)}] \tag{VI.26}$$

where N is determined by the normalization condition (V.5).

For the inclusive spectra of charged particles, the parameters C_m, and Q_m have been adjusted to experimental data at energies corresponding to proton laboratory momenta of 19 GeV/c and 30 GeV/c. The results at ISR energies are predictions.

TABLE III.

m	π^-	π^+	π^0	K^-	K^+	K^0	$\bar{p}$	p	n	Λ^0
Q_m	2·10			4·30			15·6	—		—
C_m	0·32	1·80	1·06	—	1·7	0·83	—	1·0	0·85	0·82
$F(\lambda)$	$a = 2{\cdot}05$			$b = 9{\cdot}33$				$c = 9{\cdot}93$		

The spectra of neutral particles are described with at most one free parameter (the C_m parameter) for each spectrum.[32] The values of our parameters are gathered in Table III, and some of our curves are displayed in Figs 5 to 13.

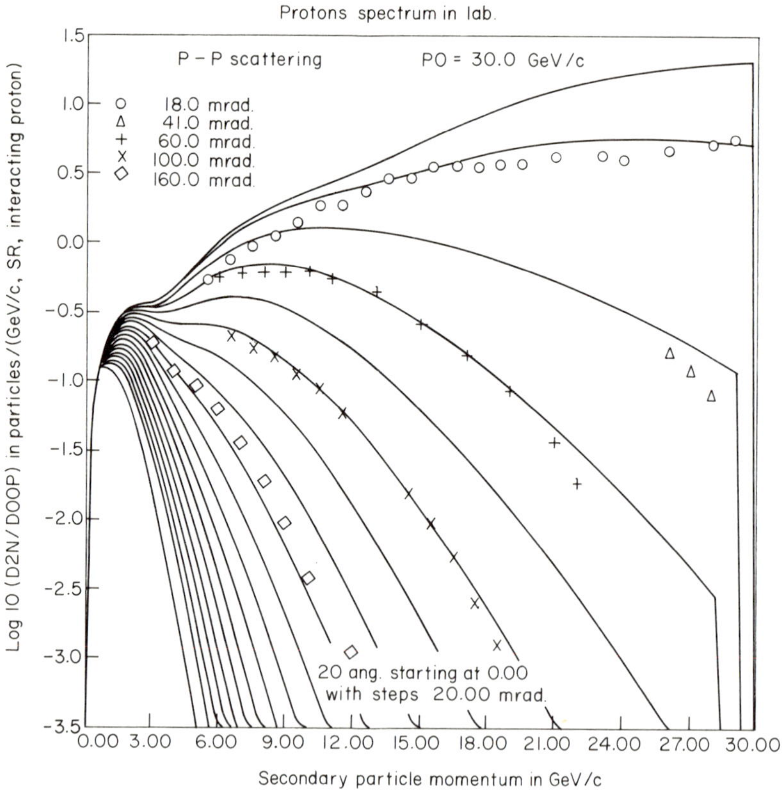

FIG. 5. Non-invariant inclusive spectrum of the proton for an incident momentum P_0 of 30 GeV/c.

VII. Correlations

To adopt the usual notation we put

$$f^{(1)}(\vec{p}^*) = E^* \frac{d^3N^*}{d^3\vec{p}^*} = \frac{E^*\, d^3\sigma^*}{\sigma_{\text{incl}}\, d^3\vec{p}^*} \tag{VII.1}$$

for the single particle distribution and

$$f^{(2)}(\vec{p}_1^*, \vec{p}_2^*) = E_1^* E_2^* \frac{d^6N^*}{d^3\vec{p}_1^*\, d^3\vec{p}_2^*} = \frac{E_1^* E_2^*}{\sigma_{\text{incl}}} \frac{d^6\sigma^*}{d^3\vec{p}_1^*\, d^3\vec{p}_2^*} \tag{VII.2}$$

for the two particle distribution.

Correlation functions are defined by

$$\rho^{(1)}(\vec{p}^*) = f^{(1)}(\vec{p}^*)$$
$$\rho^{(2)}(\vec{p}_1^*, \vec{p}_2^*) = f^{(2)}(\vec{p}_1^*, \vec{p}_2^*) - \rho^{(1)}(\vec{p}_1^*)\rho^{(1)}(\vec{p}_2^*) \tag{VII.3}$$

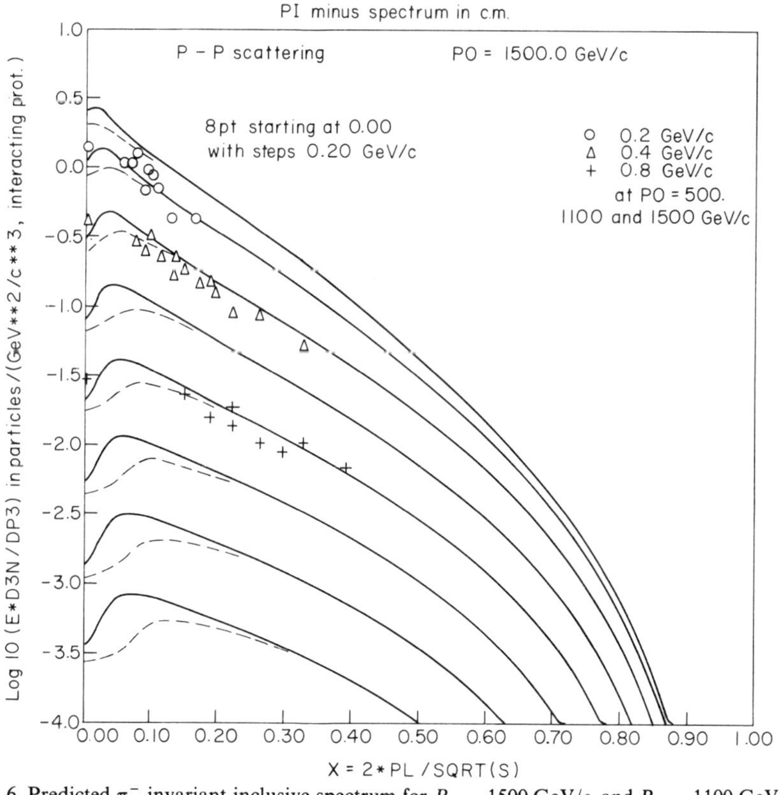

FIG. 6. Predicted π^- invariant inclusive spectrum for $P_0 = 1500$ GeV/c. and $P_0 = 1100$ GeV/c.

In the usual notation

$$R^{(1)} \equiv F^{(1)} \equiv \int \frac{d^3p^*}{E^*} f^{(1)}(\vec{p}^*) = \langle n \rangle$$

$$F^{(2)} \equiv \iint \frac{d^3\vec{p}_1^*}{E_1^*} \frac{d^3\vec{p}_2^*}{E_2^*} f^{(2)}(\vec{p}_1^*, \vec{p}_2^*) \tag{VII.4}$$

$$= \begin{cases} \langle n_i(n_i - 1)\rangle & \text{for identical particles 1 and 2.} \\ \langle n_i n_j \rangle & \text{for non-identical particles 1 and 2.} \end{cases}$$

The fully integrated two particle correlation is

$$R^{(2)} = \iint \frac{d^3\vec{p}_1^*}{E_1^*} \frac{d^3\vec{p}_2^*}{E_2^*} \rho^{(2)}(\vec{p}_1^*, \vec{p}_2^*) \tag{VII.5}$$

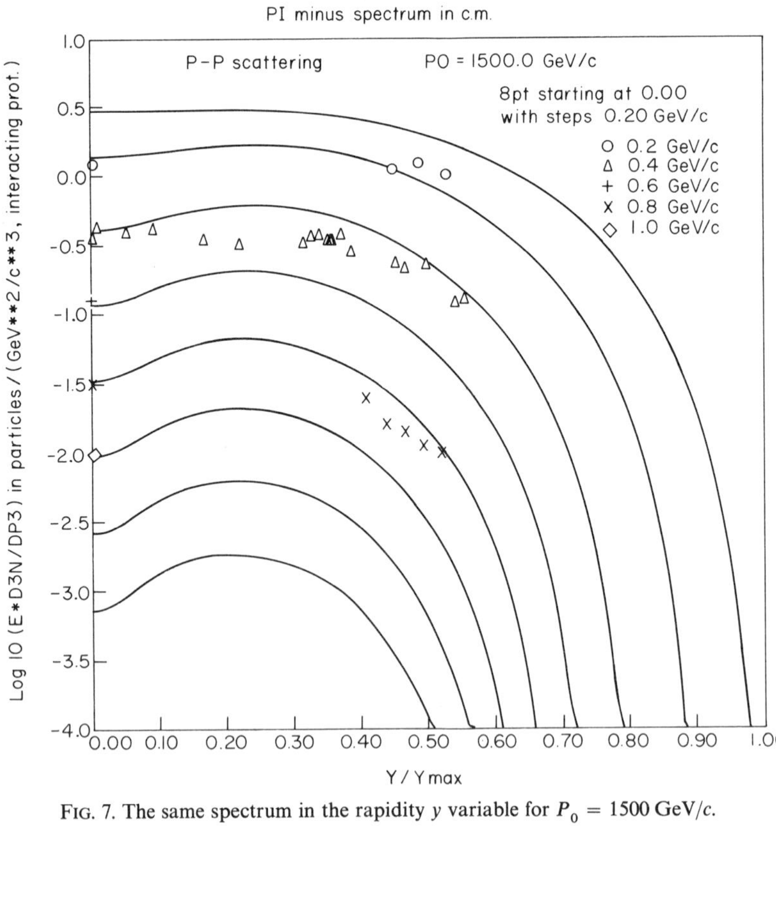

FIG. 7. The same spectrum in the rapidity y variable for $P_0 = 1500$ GeV/c.

$$= \begin{cases} \langle n_i(n_i - 1)\rangle - \langle n_i\rangle^2 & \text{for identical particles 1 and 2.} \\ \langle n_i n_j\rangle - \langle n_i\rangle\langle n_j\rangle & \text{for non-identical particles.} \end{cases} \tag{VII.6}$$

For identical particles, $R^{(2)}$ is related to the dispersion of the multiplicity distribution by

$$R^{(2)} = D^2 - \langle n\rangle \tag{VII.7}$$

Since

$$\frac{d^3\vec{p}^*}{E^*} = dy^*\, d^2p_\perp = \pi\, dy^*\, dp_\perp^2$$

the two particle rapidity correlation may be written as

$$C^{(2)}(y_1^*, y_2^*) = \iint d^2p_{\perp 1}\, d^2p_{\perp 2}\, \rho^{(2)}(\vec{p}_1^*, \vec{p}_2^*) \tag{VII.8}$$

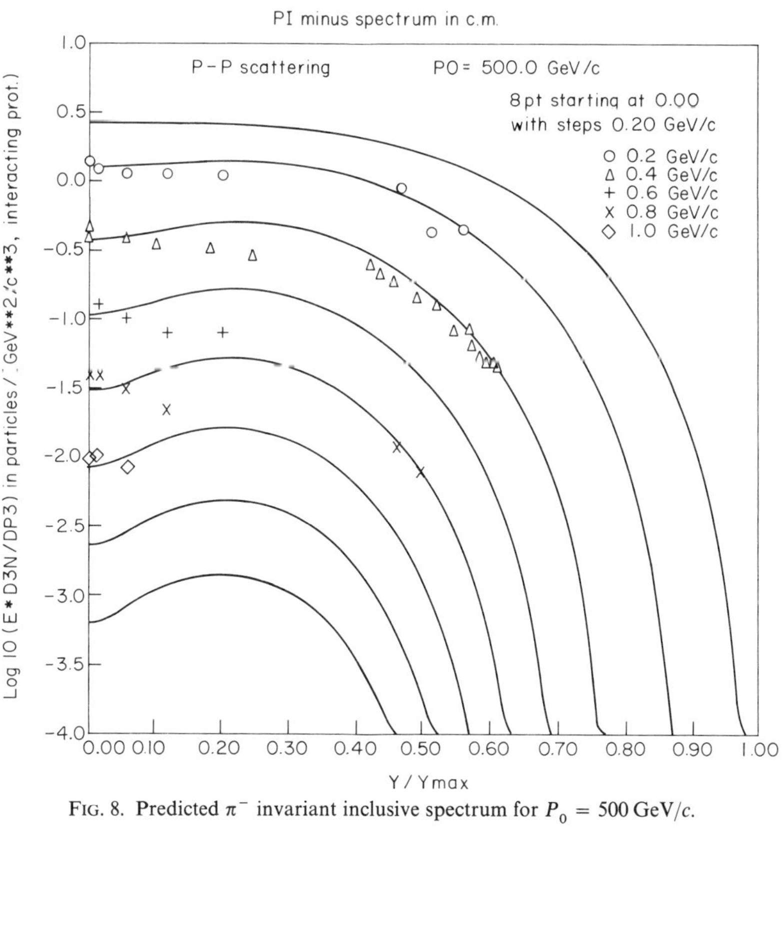

FIG. 8. Predicted π^- invariant inclusive spectrum for $P_0 = 500$ GeV/c.

I shall not enter here into the details of correlation calculations.[27, 36] I just indicate the method which is usually followed

1. ONE FIREBALL CONTRIBUTION

If the two particles 1 and 2 are emitted by the same fireball with velocity (λ) and temperature $T(\lambda)$, the two particle spectrum is

$$\frac{d^6N^*}{dy_1^*\, dy_2^*\, d^2p_{1\perp}\, d^2p_{2\perp}} = \int_{-1}^{+1} F(\lambda)\, Q^2(\lambda)\, [E_2'E_1' f_1(E_1', T(\lambda)) f_2(E_2', T(\lambda))]^{\mathrm{CM}}\, d\lambda$$

or (see eqn (6.21))

$$\frac{d^6N^*}{dy_1^*\, dy_2^*\, d^2p_{1\perp}\, d^2p_{2\perp}} = \int_{-1}^{+1} d\lambda\, F(\lambda) Q^2(\lambda) \mu_1 \cosh(y_1^* - \eta)$$

$$\times \exp\left[-\frac{\mu_1}{T}\cosh(y_1^* - \eta)\right] \mu_2 \cosh(y_2^* - \eta)\ \exp\left[-\frac{\mu^2}{T}\cosh(y_2^* - \eta)\right] \tag{VII.9}$$

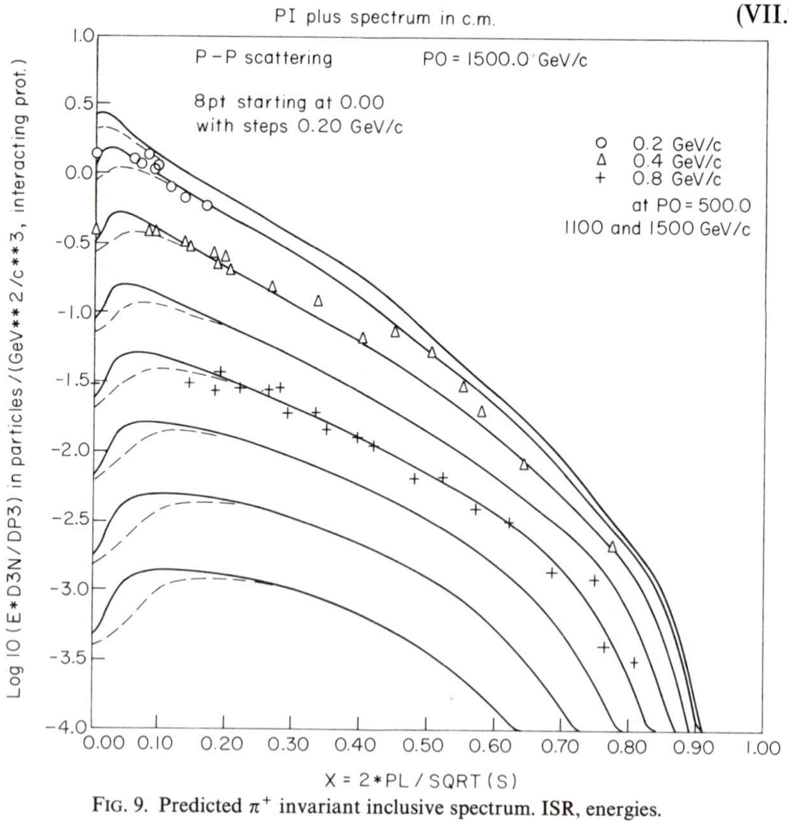

FIG. 9. Predicted π^+ invariant inclusive spectrum. ISR, energies.

where we have neglected the ± 1 in the Planck spectrum. Integration over the transverse momenta yields

$$\frac{d^2N^*}{dy_1^* \, dy_2^*} = \int_{-1}^{+1} d\lambda F(\lambda) Q^2(\lambda) \cosh(y_1^* - \eta) \cosh(y_2^* - \eta)$$

$$I\left(\frac{\cosh(y_1^* - \eta)}{T(\lambda)}\right) . \, I\left(\frac{\cosh(y_2^* - \eta)}{T(\lambda)}\right) \qquad \text{(VII.10)}$$

where:

$$I(A) = \int_0^\infty dp_{i\perp}^2 \, \mu_i \, e^{-\mu_i A}$$

The two particle rapidity correlation $C^{(2)}(y_1^*, y_2^*)$ is then obtained from (VII.2) and (VII.10). We show in Fig. 14, the result obtained by G. Ranft and J. Ranft for the reaction $K^+ + p \to \pi^-\pi^- +$ anything.

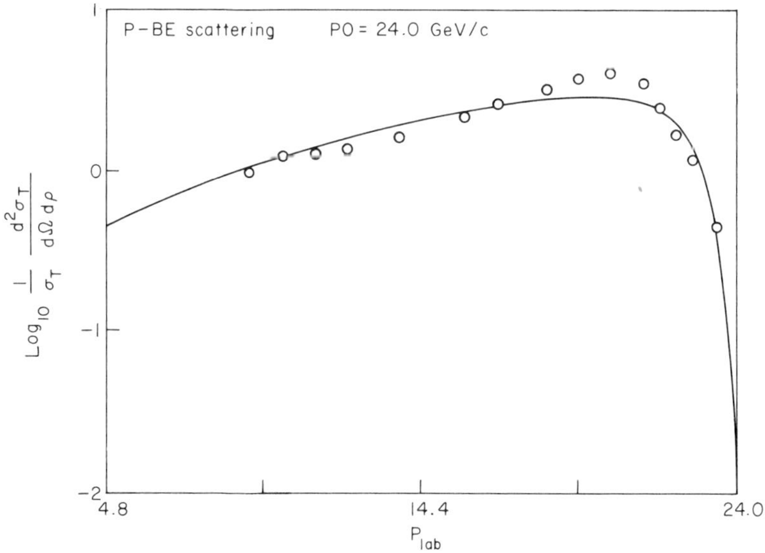

FIG. 10. Inclusive neutron spectrum from p–Be at 0° productioo angle for $P_0 = 24$ GeV/c.

2. TWO FIREBALL CONTRIBUTION

If particles 1 and 2 are emitted from two different fireballs of velocity λ_1 and λ_2 the two particle distribution in CMS is

$$\frac{d^6N^*}{dy_1^*\,dy_2^*\,d^2p_{1\perp}\,dp_{2\perp}} = \frac{\langle n_F(n_F-1)\rangle}{\langle n_F\rangle^2}\int_{-1}^{+1} d\lambda_1 \int_{-1}^{+1} d\lambda_2 F(\lambda_1)F(\lambda_2).$$
$$Q(\lambda_1)Q(\lambda_2)[E_1' f_1(E_1', T(\lambda_1)]^{CM}\,.\,[E_2' f_2(E_2', T(\lambda_2))]^{CM} \quad \text{(VII.11)}$$

where the factor $\langle n_F(n_F-1)\rangle/\langle n_F\rangle^2$ gives the probability of simultaneous production of two fireballs.

Similarly to (VII.10), the right hand side of (VII.11) can be expressed in terms of the rapidities η_1 and η_2 of the two fireballs and of the rapidities y_1^* and y_2^* of particles 1 and 2. Then from (VII.3) and (VII.11) one can obtain the two fireball contribution to the correlation function $C^{(2)}(y_1^*, y_2^*)$. The

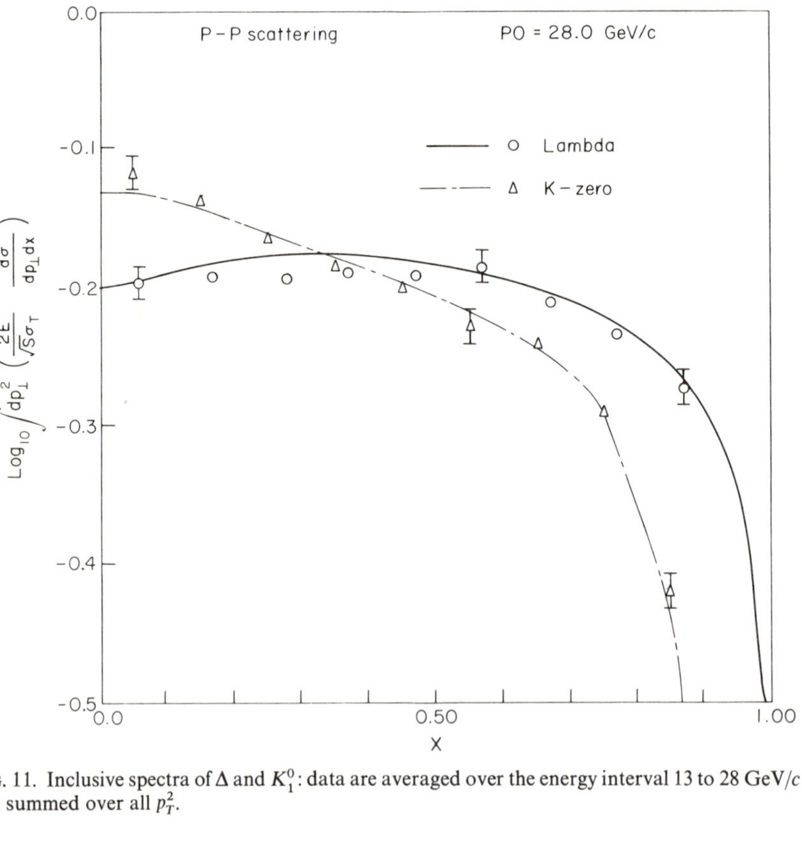

FIG. 11. Inclusive spectra of Δ and K_1^0: data are averaged over the energy interval 13 to 28 GeV/c and summed over all p_T^2.

preceding lines are a very brief introduction to correlation calculations in the thermodynamical model. For details I refer the reader to the papers of J. Ranft and his collaborators.[27, 36]

These authors use for the fireball mass the parametrizations given in Section V.2) by eqns (V.10) and they obtain for $F^{(1)}$ and $F^{(2)}$ the following asymptotic behaviour.

$F^{(1)} = \langle n \rangle \simeq \ln s$

$F^{(2)} = \langle n(n-1) \rangle \simeq s^{\frac{1}{2}w}$ in the case (i) with $w > 0$

$F^{(2)} = \langle n(n-1) \rangle \simeq A \ln s + B(\ln s)^2$ in the case (i) with $w = 0$

$F^{(2)} = \langle n(n-1) \rangle \simeq (\ln s)^2$ in the case (ii)

and for the dispersion of the multiplicity distribution

$$D \simeq \ln s \simeq \langle n \rangle \tag{VII.12}$$

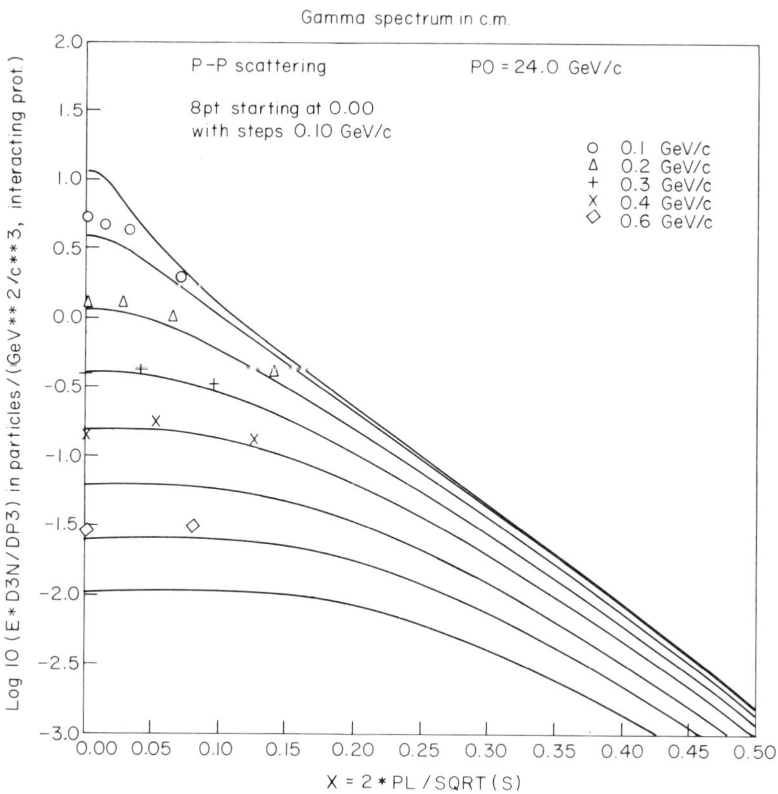

FIG. 12. Predicted γ inclusive spectra in p–p collisions at 24 GeV/c.

For the fully integrated correlation function $R^{(2)}$ they find

$R^{(2)} \simeq$ const. $\ln^2 s$ in the case (i) with $w = 0$

$R^{(2)} \simeq$ const. $s^{w/2}$ + const. $\ln^2 s$ in the case (i) with $0 < w \leqslant 1$

$R^{(2)} \simeq$ const. $\ln^2 s$ in the case (ii)

The asymptotic behaviour (VII.12) of the dispersion D of the multiplicity distribution is in agreement with the linear relation of Wroblewski.[37]

$$D = A \langle n_{ch} \rangle - B$$

and with the empirical regularity found by Czyzewski and Rybicki.[38] Therefore it seems that the multiperipheral production mechanism of fireballs dominates at high energy.[27]

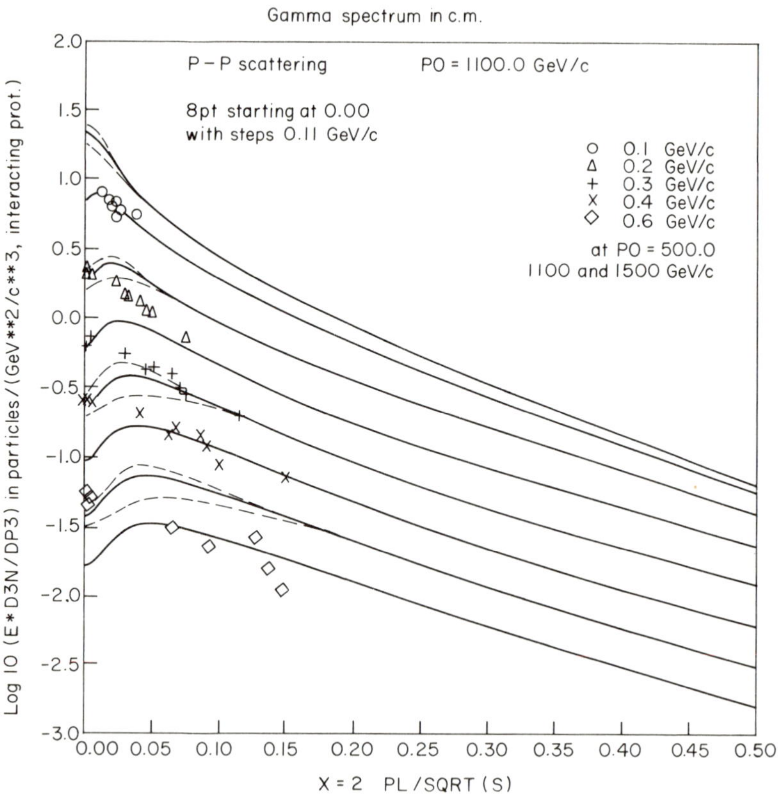

FIG. 13. Predicted γ inclusive spectra in p–p collisions at ISR energies. The dashed lines show the deviation from scaling predicted by the model for small x at $P_0 = 500$ and 1500 GeV/c.

However at the end of this very short summary on correlations it should be clear that in the thermodynamical model the decay of a fireball is well defined, but there is no definite prescription for their production mechanism. But obviously correlation studies can rule out certain assumptions about this mechanism.

VIII. Further Applications

I just give a list of more or less recent applications of the statistical bootstrap and of the thermodynamical model.

1. APPLICATIONS IN PARTICLE PHYSICS

(a) Collisions with non-identical particles in the initial state.

The formalism of Chapter 5 was slightly generalized[39] in order to be suitable for collisions with non-identical particles in the initial state such as $\pi^{\pm}p$, Kp.

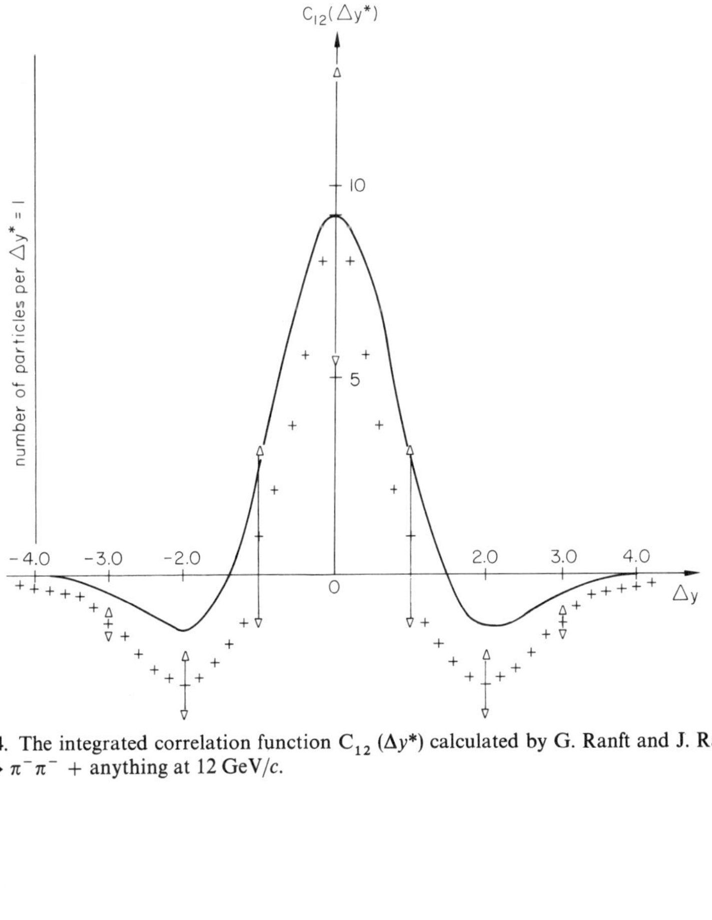

FIG. 14. The integrated correlation function C_{12} (Δy^*) calculated by G. Ranft and J. Ranft for $K^+p \rightarrow \pi^-\pi^-$ + anything at 12 GeV/c.

(b) Description of hadron production in $N\bar{N}$ and e^+e^- annihilation.[40, 41, 42]

(c) Production and decay of heavy vector mesons.[43]

(d) Ericson fluctuations in particle physics.[44]

This list is certainly not exhaustive.

2. APPLICATIONS IN ASTROPHYSICS

These are speculations[22, 23] about the "big bang", hadron star, etc.

ACKNOWLEDGEMENTS

I thank Dr R. Hagedorn for his generous help during the preparation of these notes, and J. Letessier and A. Bouquet for numerous and useful discussions.

REFERENCES

1. R. Hagedorn. *Suppl. Nuovo Cimento* **3**, 147 (1965).
2. R. Hagedorn and J. Ranft. *Suppl. Nuovo Cimento* **6**, 169 (1968).
3. M. Jacob and R. Slansky. *Phys. Rev.* **D5**, 1847 (1972).
4. R. C. Hwa. *Phys. Rev. Letts.* **26**, 1143 (1971).
5. H. Koppe. *Z.F. Naturforsch.* **3A**, 251 (1948).
6. E. Fermi. *Progr. Theor. Phys.* **5**, 570 (1950).
7. E. Beth and G. E. Uhlenbeck. *Physica* **4**, 915 (1937).
8. S. Z. Belenky. *Nucl. Phys.* **2**, 259 (1956).
9. S. Frautschi. *Phys. Rev.* **D3**, 2821 (1971).
10. W. Nahm. *Nucl. Phys.* **B45**, 525 (1972).
11. S. Frautschi and C. J. Hamer. *Phys. Rev.* **D4**, 2125 (1971).
12. J. Yellin. *Nucl. Phys.* **B52**, 583 (1973).
13. E. M. Ilgenfritz and J. Kripfganz. *Nucl. Phys.* **B56**, 241 (1973).
14. C. J. Hamer. *Phys. Rev.* **D8**, 3558 (1973).
15. R. Hagedorn and I. Montvay. *Nucl. Phys.* **B59**, 45 (1973).
16. B. Touckek. *Nuovo Cimento* **58**, 295 (1968).
17. A. H. Mueller. *Phys. Rev.* **D4**, 150 (1971). L. S. Brown. *Phys. Rev.* **D5**, 748 (1972).
18. I. Montvay. CERN preprint TH 1572 (1972); *Phys. Letts.* **42B**, 446 (1972).
19. M. Alexanian. *Phys. Rev.* **D4**, 2432 (1971), and **D5**, 922 (1972).
20. C. B. Chiu. "A Truncated Statistical Bootstrap Equation." University of Texas preprint, CPT-153 (1972); C. B. Chiu and J. Johnsen. *Phys. Letts.* **B42**, 475 (1972).
21. R. Hagedorn. "Thermodynamics of the Strong Interactions." Yellow Report CERN 71–12.
22. R. Hagedorn. *Astron. and Astrophys.* **5**, 184 (1970a).
23. K. Huang and S. Weinberg. *Phys. Rev. Letts.* **25**, 895 (1970).

24. M. Banner, G. Bassompierre, M. Croissiaux, J. Gresser, J. L. Hamel, R. Morand, J. P. Pansart, M. Riedinger, M. Schneegans, A. V. Stirling, J. Teiger, H. Zaccone and J. Zsembery. Contribution to the 16th Int. Conf. on High-Energy Physics, Batavia, 1972, paper no. 478; CERN-Columbia-Rockefeller Collaboration, Contribution to the 16th Int. Conf. on High-Energy Physics, Batavia, 1972 (no number assigned); B. J. Blumenfeld, F. W. Büsser, L. Camilleri, R. L. Cool, L. Di Lella, G. Gladding, L. M. Ledermann, L. Litt, A. Placci, B. G. Pope, S. L. Segler, A. M. Smith, J. K. Yoh and E. Zavattini (CERN-Columbia-Rockefeller Collaboration). Presented at the New York A.P.S. Meeting (29 Jan. 1973).
25. R. Hagedorn and J. Ranft. *Nucl. Phys.* **B48**, 191 (1972).
26. G. Ranft and J. Ranft. *Phys. Letts.* **32B**, 207 (1970).
27. J. Kripfganz, G. Ranft and J. Ranft. *Nucl. Phys.* **B56**, 205 (1973).
28. R. Hagedorn. *Suppl. Nuovo Cimento* **6**, 311 (1968b).
29. J. Letessier and A. Tounsi. *Nuovo Cimento* **11A**, 353 (1972).
30. J. Letessier and A. Tounsi, *Nuovo Cimento* **13A**, 557 (1973).
31. R. Hagedorn and J. Ranft. *Nucl. Phys.* **B48**, 159 (1972).
32. J. Letessier and A. Tounsi. *Nuovo Cimento* **15A**, 358 (1973).
33. R. P. Feynman. *Phys. Rev. Letts.* **23**, 1415 (1969), and in "High Energy Collisions." Proc. 3rd Conf., Stony Brook, September 1969.
34. W. R. Frazer, L. Ingber, C. H. Mehta, C. H. Poon, D. Silverman, K. Stowe, P. D. Ting and H. J. Yesian. *Rev. Mod. Phys.* **44**, 284 (1972).
35. J. Benecke, T. T. Chou, C. N. Yang and E. Yen. *Phys. Rev.* **188**, 2159 (1969).
36. J. Ranft and G. Ranft. *Nucl. Phys.* **B53**, 217 (1973).
37. A. Wroblenski. Warsaw University preprint IFD 72/2 (1972).
38. O. Czyzewski and K. Rybicki. *Nucl. Phys.* **B47**, 633 (1972).
39. Htun Than and J. Ranft. *Nuovo Cimento Letts.* **5**, 655 (1972).
40. C. J. Hamer. *Nuovo Cimento* **12A**, 162 (1972).
41. S. Rudaz. *Nuovo Cimento Letts.* **6**, 292 (1973).
42. J. Engels, K. Schilling and H. Satz. "Statistical and Thermodynamical Descriptions of Hadron Production in e^+e^- Annihilation." Preprint CERN TH 1674.
43. B. Margolis, W. J. Meggs and S. Rudaz. "Production and Decay of Heavy Vector Mesons in a Statistical Model." McGill University.
44. P. J. Carlson. "Ericson Fluctuations in High-Energy Hadron Physics" Report presented at the Fifth International Conference on High-Energy Physics and Nuclear Structure, Uppsala, 18–22 June 1973.

6. AN INTRODUCTION TO RENORMALIZABLE MODELS OF WEAK INTERACTIONS AND THEIR EXPERIMENTAL CONSEQUENCES

C. H. LLEWELLYN SMITH

CERN, Geneva

Introduction

There has been great excitement in weak interaction physics during the last two years stemming from the work of 't Hooft[1] who, with Veltman, B. W. Lee, Zinn-Justin, and others,[2] showed that certain models involving massive charged vector mesons are renormalizable. If we accept this (and think that renormalizability is a relevant criterion) the problem is then to construct models which might describe the real world and to confront them with experiment. This is the subject of these lectures.

Whether or not these models are "true", they have been very fruitful in focusing attention on the fact that many conventional hypotheses (prejudices) about weak interactions are very badly tested. The importance of the experiments which they suggest surely transcends the significance of the specific models proposed so far. For this reason, and also because there are by now many introductory accounts of the principles of building these models,[3–8] I have devoted a large part of these notes (Part II) to their experimental consequences.

In Part I, I have tried to motivate directly those ingredients of the models which have experimental implications. Rather than starting with the elegant notions of generalized gauge invariance and its spontaneous breakdown (which underlie all the models), and then using them to build models of weak interactions, we will be led to these ideas following an empirical path. We start with the difficulties of the current–current model and try to cure problems in a seemingly ad hoc manner as they arise; with the wisdom of hindsight, we can "derive" renormalizable models in this way. If the reader is hurried, or finds that the details obscure the trend of the argument in Part I, he is referred to Ref. 9 for an outline of the approach used here.

Part I

1. Problems with the phenomenological theory of weak interactions

In the "phenomenological theory", weak interactions are described by the Lagrangian

$$\mathscr{L} = \frac{G}{\sqrt{2}} J_\lambda^+(x) J_\lambda(x)$$

where

$$G = 10^{-5}/M_p^2$$
$$J_\lambda = J_\lambda^{\text{hadronic}} + J_\lambda^{\text{muonic}} + J_\lambda^{\text{electronic}}$$
$$J_\lambda^{\text{hadronic}} = \text{Cabibbo current}$$
$$J_\lambda^{\text{muonic}} = \bar{\mu}(x)\,\gamma_\lambda(1-\gamma_5)\nu_\mu(x)$$
$$J_\lambda^{\text{electronic}} = \bar{e}(x)\gamma_\lambda(1-\gamma_5)\nu_e(x)$$

In *lowest order* this Lagrangian gives an excellent description of all known leptonic and semileptonic processes[10] (apart from T-violation; there is also some trouble for non-leptonic processes, e.g. with the $\Delta I = \frac{1}{2}$ rule). In *next order* divergent integrals are encountered:

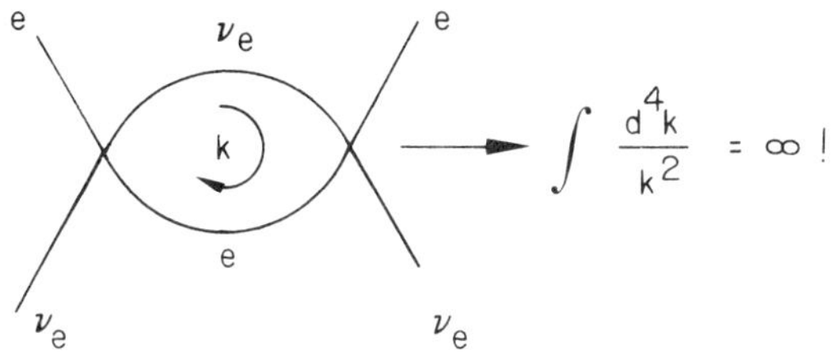

Unlike the case in QED, we cannot absorb these infinities simply by redefining the "bare" masses and coupling constants so as to reproduce the correct physical values in each order. To render all matrix elements finite, new arbitrary constants (which can, however, be determined by experiment) must be introduced in each order, i.e. the theory is non-renormalizable; such theories are considered further in the next section.

Sometimes *an effective Lagrangian philosophy* is adopted and higher orders are simply ignored. This must fail at high energies. In, for example, the amplitude

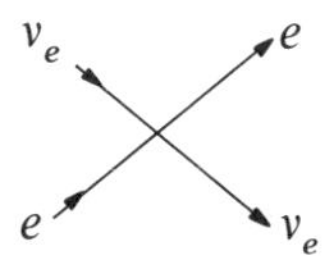

the electron mass can safely be neglected at high energy so that, on dimensional grounds,

$$\sigma \sim G^2 s, \qquad s = (p_e + p_{\nu_e})^2.$$

But this is an s-wave process (it takes place at a point with zero impact parameter); hence, by unitarity,

$$\sigma \leqslant \text{const.}/s$$

and the theory must fail at the "unitarity limit" $s \sim 1/G \sim 10^5\ \text{GeV}^2$. That this "bad high-energy behaviour" is connected with the non-renormalizability becomes clear when we consider calculating second-order processes in terms of lowest-order cross-sections using dispersion relations, e.g. in order G^2 the amplitude for forward νe scattering is given by[11]

$$A(\nu e \to \nu e)\Big|_{t=0} = \frac{1}{\pi}\int_0^\infty \left(\frac{\sigma^{\nu e}(s')}{s'-s} - \frac{\sigma^{\bar{\nu}e}(s')}{s'+s}\right) ds'.$$

Since $\sigma^{\nu e} \sim 3\sigma^{\bar{\nu}e} \sim G^2 s'$, we must make two subtractions:

$$A(\nu e \to \nu e)\big|_{t=0} = A(s=0, t=0) + sA^1(0,0) + \frac{s^2}{\pi}\int\left(\frac{\sigma^{\nu e}}{s'-s} - \frac{\sigma^{\bar{\nu}e}}{s'+s}\right)\frac{ds'}{s'^2}.$$

$A(0,0)$ and $A^1(0,0)$ are the first of the arbitrary constants alluded to above.

We therefore seek a theory with "good" high-energy behaviour which seems desirable

(i) because of the connection with renormalizability;
(ii) because perturbation theory must be totally misleading when it violates unitarity.

For the particular process considered above the situation is improved by *introducing a massive vector meson—W—*which plays a role analogous to that of the photon in QED:

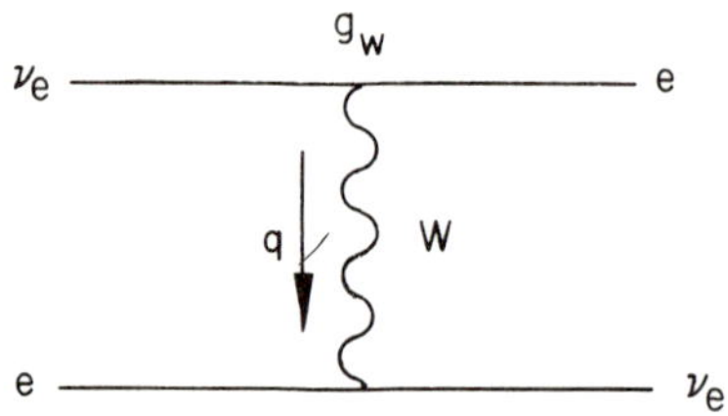

With $G/\sqrt{2} = g_W^2/M_W^2$, we obtain the current-current model as $q \to 0$. g_W is dimensionless and hence[12]

$$\frac{d\sigma^{\nu e \to \nu e}}{dq^2} \sim \frac{g_W^4}{(q^2 - M_W^2)^2}, \quad \sigma^{\nu e \to \nu e} \sim \frac{g_W^4}{M_W^2}$$

The partial wave amplitudes now only violate unitarity logarithmically (at energy $s \sim M_W^2\, e^{1/GM_W^2}$) and the forward dispersion relation needs only one subtraction (note that $\sigma^{\bar{\nu}e \to \bar{\nu}e} \sim g_W^4/s$).[13]

However, the situation is as bad as ever in the process:

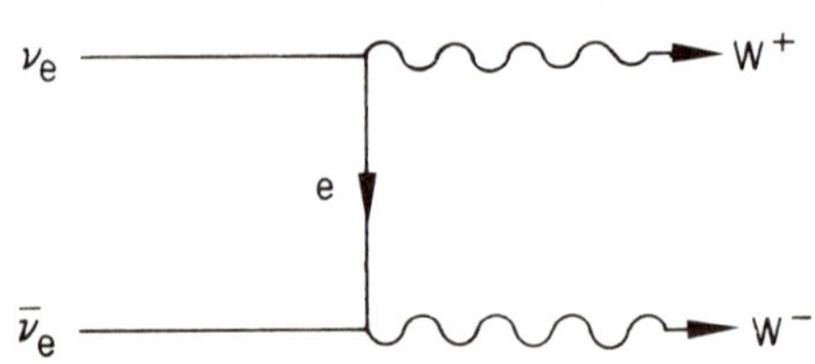

In its rest frame the possible polarization states of the W are described by the vectors

$$\begin{matrix} t \\ x \\ y \\ z \end{matrix} \underbrace{\begin{pmatrix} 0 \\ 1 \\ 0 \\ 0 \end{pmatrix}, \begin{pmatrix} 0 \\ 0 \\ 1 \\ 0 \end{pmatrix}}_{\varepsilon^T}, \underbrace{\begin{pmatrix} 0 \\ 0 \\ 0 \\ 1 \end{pmatrix}}_{\varepsilon^L}.$$

Under a Lorentz boost along $0Z$

$$\varepsilon^T \to \varepsilon^T$$

$$\varepsilon^L \to \left(\frac{|\vec{k}|}{M_W}, 0, 0, \frac{k_0}{M_W}\right) = \frac{k_\mu}{M_W} + 0\left(\frac{M_W}{k_0}\right)$$

Hence, dimensionally or by explicit calculation,

$$\sigma(\nu\bar{\nu} \to W_T W_T) \underset{s\to\infty}{\to} \text{const.}$$

but

$$\sigma(\nu\bar{\nu} \to W_L W_L) \underset{s\to\infty}{\to} \frac{g_W^4 s}{M_W^4}$$

which rapidly violates unitarity and makes two subtractions necessary in the forward dispersion relation for

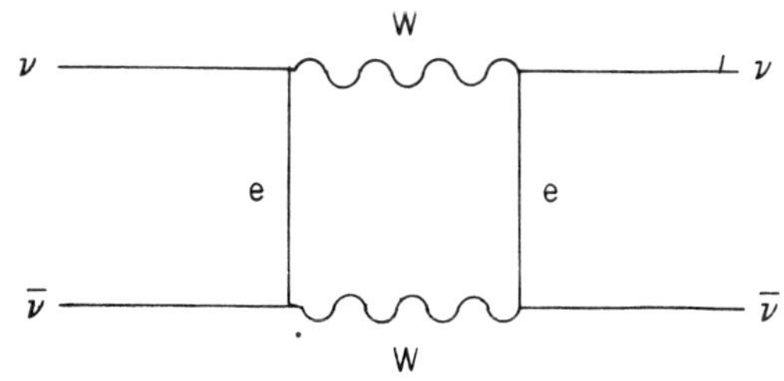

(the total cross-section in the crossed channel —$\sigma(\nu\nu)$—is zero in lowest order in the phenomenological model).

Alternatively, the W propagator is

$$\frac{\Sigma_{\text{pol.}}\varepsilon_\mu^*\varepsilon_\nu}{k^2 - M_W^2} = \frac{-g_{\mu\nu} + (k_\mu k_\nu/M_W^2)}{k^2 - M_W^2} \underset{k\to\infty}{\to} \text{const.}$$

(where the second expression is easily obtained from the first by using the vectors ε above and writing the result covariantly); the behaviour as $k \to \infty$ makes loop integrals diverge.

To *repeat*: There seems to be more hope of making a sensible theory when we introduce W's because g_W is dimensionless. However, serious trouble occurs with longitudinal W's because

$$\varepsilon_\mu^L = \frac{k_\mu}{M_W} + O\left(\frac{M_W}{k_0}\right)$$

at high energies. To keep the dimensions of cross-sections, etc., correct, the factor $1/M_W$ is accompanied by powers of E. What happens with longitudinal (virtual) γ's in QED? Do not similar problems occur? The answer is no, because of gauge invariance (this would be true even if the photon were massive, in which case the gauge invariance would be broken but the photon would still couple to an exactly conserved current). S-matrix elements are invariant under the replacement

$$\varepsilon_\mu^{\gamma_i} \to \varepsilon_\mu^{\gamma_i} + \lambda k_\mu^{\gamma_i}$$

in QED, which suggests that pieces of ε_μ proportional to k_μ are impotent. *It would obviously be desirable to generalize gauge invariance to massive charged vector fields*; we return to this later.

2. RELEVANCE OF RENORMALIZABILITY

It is important to keep in mind that there are many possible attitudes to the problems encountered above and that various cures have been proposed besides that espoused in these lectures (many of which have interesting experimental implications). We will not discuss these other attitudes here (they have been extensively reviewed in the literature),[14–17] but we will defend the point of view adopted here, that renormalizability is a relevant criterion.

If we start with a non-renormalizable Lagrangian $\mathscr{L}$ and proceed to calculate perturbatively, then (as mentioned above) we are gradually forced to introduce more and more new arbitrary constants; these constants characterize new terms which must be added to $\mathscr{L}$. Although we may not be forced to introduce a certain term until nth order, there is no principle which prevents it having arbitrary strength; therefore it should really be included in the original $\mathscr{L}$ which would then have an infinite number of terms! If weak interactions were described by a non-renormalizable $\mathscr{L}$, which could be treated perturbatively, the theory would involve an infinite number of parameters; there would seem to be no reason why the phenomena should be so simple or why "higher-order" processes should be so small (e.g. why $A(K_L^0 \to \mu^+\mu^-) < 2G^2M_K^4 m_\mu$ and $2\mathrm{Re}\langle K^0|R|\bar{K}^0\rangle \sim M_{K_L^0} - M_{K_S^0} \sim \frac{1}{2}\Gamma_{K_S^0}$), although this could be arranged. The success of renormalizable quantum electrodynamics (i.e. the theory of γ, e and μ with no intrinsic anomalous moments, in which $(g-2)_\mu$ has been successfully calculated to order e^6 in terms of e, m_e, and m_μ) suggests that it is a worthy ambition to strive to describe weak interactions with comparable economy. Note that if the photon coupled non-renormalizably to a charged vector meson, we would not be able to calculate its contribution to $(g - 2)_\mu$ and we would not understand why it did not spoil the success of the QED calculation.

It may be that weak interactions are described by a non-renormalizable $\mathscr{L}$ which cannot be treated perturbatively; in this case it would also be difficult to understand why the simple operator $\mathscr{L}_{\text{eff}}$ describes so much data so economically and successfully. However, it is interesting to note that a phenomenology of weak interactions can be developed outside the framework of perturbation theory.[18]

3. ELEMENTS OF RENORMALIZABLE MODELS

If the weak interactions are really mediated by a W, the badly behaved contribution to the amplitude for $\nu\bar{\nu} \to W^+W^-$ discussed above must exist; in a renormalizable model there must be other contributions which cancel it. They must occur in the same order in perturbation theory,[19] i.e. we need additional Born terms corresponding to new exchanges in the s, t, or u channels (we do not consider the possibility of contact interactions, which would necessarily be characterized by coupling constants with dimension M^{-n} with $n \geqslant 1$ and would therefore be expected to cause trouble in higher orders in which inverse powers of M may be compensated by powers of the energy). We now abstract certain lessons from an examination of some of the processes in which a cancellation must be invoked to cure the problems of the phenomenological model. (Later we will show that, to leading order, the desired cancellation can in fact be achieved in the ways suggested.)

(A) $\nu\bar{\nu} \to WW$

To cancel the offending part of the amplitude corresponding to

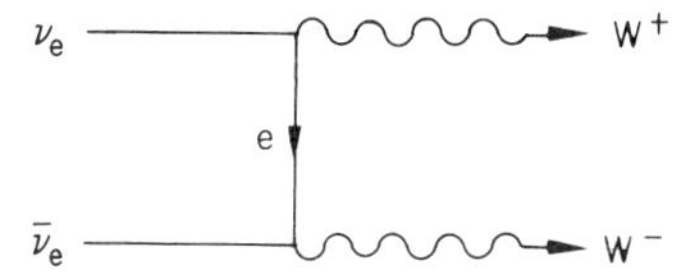

we must add the new exchanges:

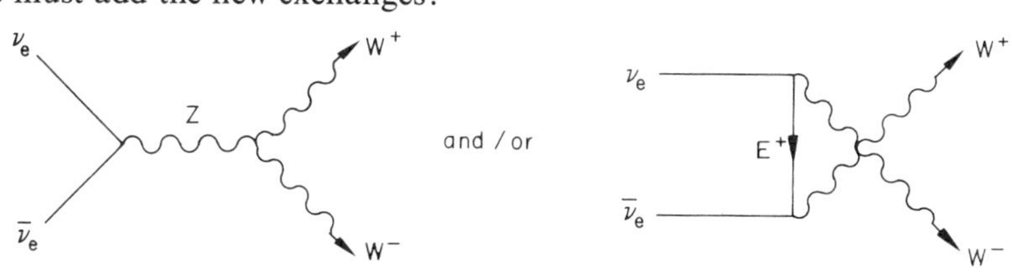

i.e. there must be neutral currents and/or "heavy" leptons. For reasons which will be spelled out in the next section, the new lepton will generally be in the u channel, i.e. *it will be an $E^+(M^+)$ with the same lepton number as $\nu_e, e^-(\nu_\mu, \mu^-)$.* (We assume the two-component neutrino theory and the usual lepton conservation laws here and below, so E^+ cannot be μ^+; small violations, in higher orders for example, would not effect the main line of argument and indeed occur in some renormalizable models which have been proposed.)

(B) $e^+e^- \to W^+W^-$

Trouble occurs in the amplitudes

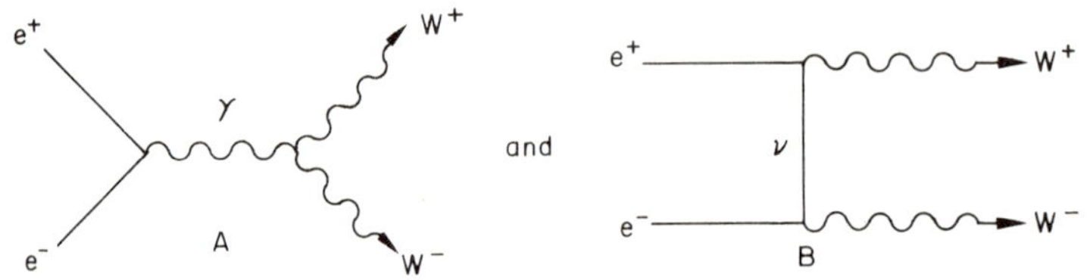

We cannot arrange to cancel the "bad" part of amplitude A with the "bad" part of amplitude B, since A conserves parity while B violates it maximally (we can polarize the electrons so that amplitude B vanishes in the approximation $m_e = 0$). Cancellation again requires new exchanges:

In principle, we could introduce one mechanism to cure diagram A (e.g. a neutral particle Z_1) and another to cure diagram B (e.g. another Z_2), but it is most economical to achieve both cures with the same new exchange [and use the same (related) new boson (heavy lepton) to cure $\nu\bar{\nu} \to WW$]. Since diagram A is electromagnetic and B is weak, *this leads us to unify weak and electromagnetic interactions. Note that in this case we expect $g \sim e$ and therefore*

$$M_W = \left[\frac{g^2\sqrt{2}}{G}\right]^{\frac{1}{2}} \sim \left[\frac{e^2\sqrt{2}}{G}\right]^{\frac{1}{2}} = 106 \text{ GeV}.$$

(C) $\bar{\nu}e \to W^- \gamma$

There are two diagrams which must occur

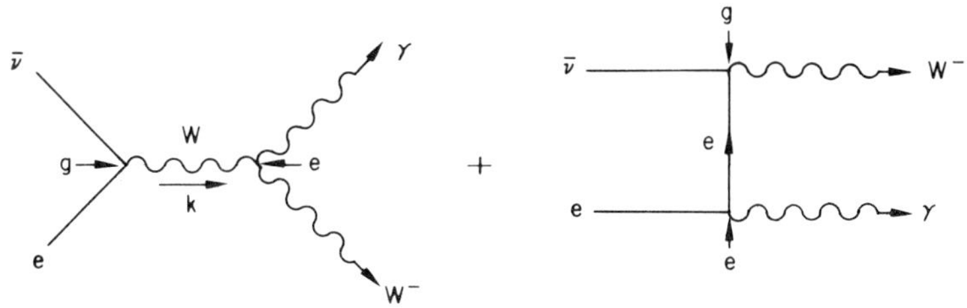

Both amplitudes violate parity maximally and have strength ge. There would clearly be a chance of arranging cancellation were it not for the fact that the first has a $k_\mu k_\nu / M_W^2$ term from the propagator which has no analogue in the second (so that, on dimensional grounds, the first diagram will be worse behaved than the second at high energies in general). Using the Dirac equation we find that this term gives a contribution

$$\sim \frac{m_e \bar{v}(1-\gamma_5)u}{M_W^2(s-M_W^2)} k^\mu J_\mu$$

where J_μ describes the $WW\gamma$ vertex. For cancellation to be possible, $k^\mu J_\mu$ must involve at least one power of the mass. Let us see what happens if we use the Lagrangian

$$\mathscr{L} = -\tfrac{1}{4}(\partial_\mu W_\nu - \partial_\nu W_\mu)^*(\partial_\mu W_\nu - \partial_\nu W_\mu)$$

and generate a vertex by the minimal substitution $\partial_\mu \to \partial_\mu - ieA_\mu$. With momenta labelled thus:

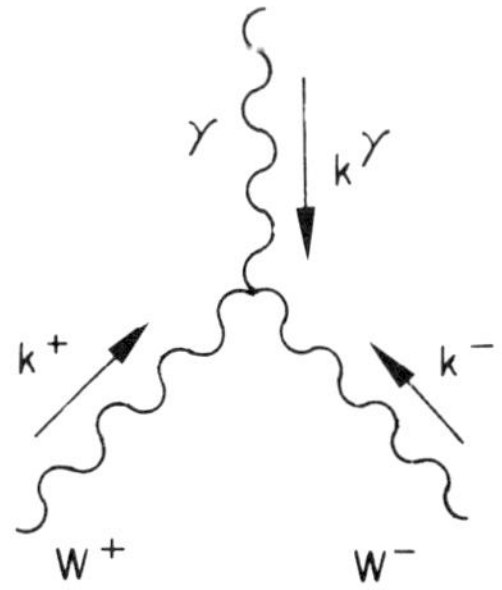

The vertex is proportional to

$$\begin{aligned}&\varepsilon^\gamma \cdot (k^+ - k^-)\varepsilon^+ \cdot \varepsilon^- \\ &+ \varepsilon^+ \cdot (k^- \qquad)\varepsilon^- \cdot \varepsilon^\gamma \\ &+ \varepsilon^- \cdot (\qquad -k^+)\varepsilon^\gamma \cdot \varepsilon^+\end{aligned}$$

in momentum space (where $\varepsilon^{+,-,\gamma}$ are the polarization vectors). The quantity $k^\mu J_\mu$ above is just this vertex with $\varepsilon^- \to k^-$; it is easy to check that it does have the desired property of being proportional to M_W. However, the vertex is gauge invariant by construction and therefore gives zero when we make the replacement $\varepsilon^\gamma \to k^\gamma$ (provided all other particles are on-mass shell and have physical polarizations such that $\varepsilon^i \cdot k^i = 0$). If, therefore, we antisymmetrize in the $+, -, \gamma$ labels, i.e. change the vertex to

$$\begin{aligned} &\varepsilon^\gamma \cdot (k^+ - k^-)\varepsilon^+ \cdot \varepsilon^- \\ &+ \varepsilon^+ \cdot (k^- - k^\gamma)\varepsilon^- \cdot \varepsilon^\gamma \\ &+ \varepsilon^- \cdot (k^\gamma - k^+)\varepsilon^\gamma \cdot \varepsilon^+ \end{aligned}$$

it must necessarily yield a power of M_W when we replace ε^- by k^-, since it would vanish if M_W were zero and there was complete $+, -, \gamma$ antisymmetry (it is actually proportional to M_W^2 so that the contribution of the $k_\mu k_\nu/M_W^2$ piece of the propagator is non-leading at high energies with this new vertex). The new vertex is still gauge invariant and corresponds to an extra piece

$$ieKW_\mu W_\nu(\partial_\mu A_\nu - \partial_\nu A_\mu)$$

in the Lagrangian with $K = 1$. With this vertex the W has a gyromagnetic ratio $\lambda = K + 1 = 2$ (according to the definition of Lee and Yang, K is called the "anomalous moment" but ambiguous moment would be a better term since, unlike the case in QED, the dimensions of the "anomalous" term and the original term are the same and there is no *a priori* reason to exclude either of them).

The γWW vertex we have "derived" is exactly that which is obtained automatically in a Yang–Mills theory, i.e. a theory with generalized gauge invariance (see Section I.6). It has long been known that the electrodynamics of spin-1 particles is best behaved at high energies if the vertices have the Yang-Mills form.[20] For example:

$$\begin{aligned} \sigma(e^+e^- \to W^+W^-) &\sim \frac{e^4 s}{M_W^4} \text{ for } \lambda \neq 2 \\ &\sim e^4/M_W^2 \text{ for } \lambda = 2 \\ \sigma(\gamma\gamma \to W^+W^-) &\sim \frac{e^4 s}{M_W^4} \text{ for } \lambda \neq 2 \\ &\sim e^4/M_W^2 \text{ for } \lambda = 2. \end{aligned}$$

Quite generally, in fact, the high-energy behaviour of the Born terms for a system of self-interacting spin-1 particles W_μ^i (of which one may be the photon) is radically better if the kinetic and interaction terms have the Yang–Mills form:

$$-\tfrac{1}{4}\sum_a |\partial_\mu W_\nu^a - \partial^\nu W_\mu^a + \sum_{bc} D_{abc} W_\mu^b W_\nu^c|^2$$

than in all other cases, provided D_{abc} is totally antisymmetric and satisfies the "Jacobi identity" which is given on p. 482 [assuming there are no coupling constants with inverse mass dimensions].(21, 22) *We conclude that if it is possible to construct a renormalizable model involving W's, the W–W interaction is likely to have the form appropriate to a Yang–Mills theory;* we assume this form below.

4. AN EXAMPLE: THE GEORGI–GLASHOW MODEL

We now show explicitly how the cancellation of the terms with the worst high-energy behaviour occurs in a "minimal" model with heavy leptons but no neutral currents [the Georgi–Glashow model].(23) since it is probably simpler than the other "minimal" model [due to Weinberg(24) and Salam(25)], with a neutral current but no new leptons, which we consider in Section I.11.

Consider first the offending amplitude corresponding to the diagram

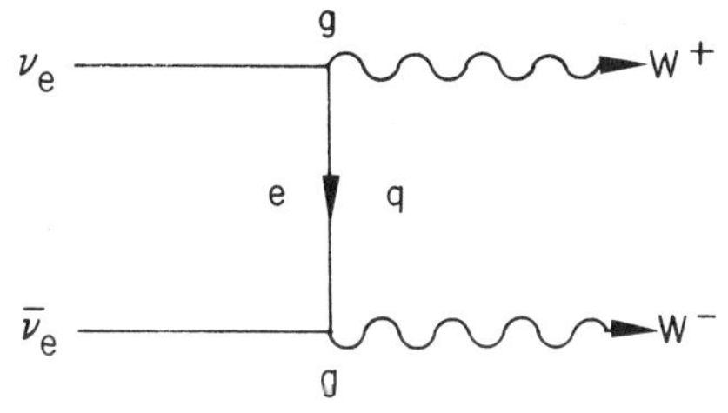

which is

$$M^{(1)} = g^2 \bar{v}(p_{\bar{\nu}}) \not\epsilon^-(1-\gamma_5)\frac{1}{\not q - m_e}\not\epsilon^+(1-\gamma_5)\,u(p_\nu)$$

To leading order we set $m_e = 0$ and, taking both W's to be longitudinally polarized, set

$$\varepsilon_\mu^+ \simeq \frac{k_\mu^+}{M_W}, \qquad \varepsilon_\mu^- \simeq \frac{k_\mu^-}{M_W}.$$

Rearranging and using the Dirac equation we get

$$M^{(1)} \simeq -\frac{2g^2}{M_W^2}\,\bar{v}\not q(1-\gamma_5)\,u.$$

It is not possible to cancel this with the diagram

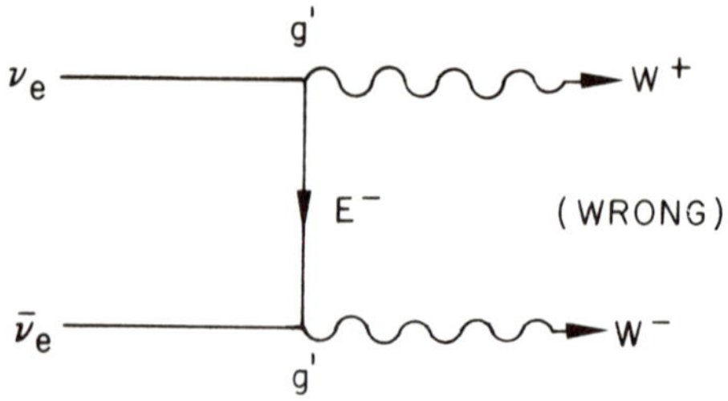

because, to leading order (in which we neglect M_E), the sum of the amplitudes would be proportional to $g^2 + g'^2$ which cannot be zero. Rather, as asserted in Section I.3, ν_e, e^- and E^+ must have the same lepton number and we must cancel with

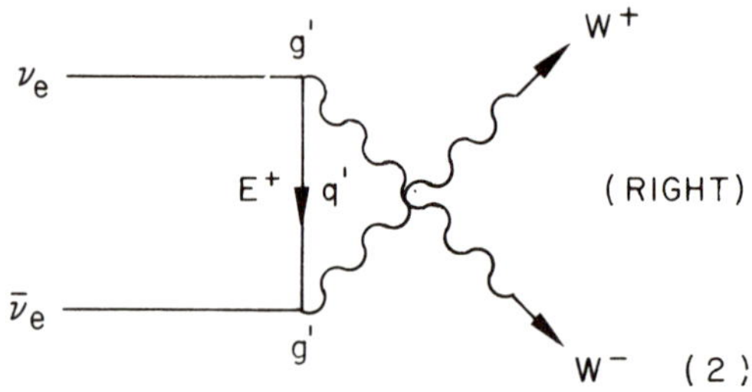

(2)

which obviously gives an amplitude

$$M^{(2)} \simeq -\frac{2g'^2}{M_W^2}\,\bar{v}\not{q}'(1-\gamma_5)\,u \simeq \frac{2g'^2}{M_W^2}\,\bar{v}\not{q}(1-\gamma_5)\,u$$

(using $q' = -q + p_\nu - p_{\bar{\nu}}$ and the Dirac equation) so that cancellation is achieved if

$$g^2 = g'^2.$$

Consider next the diagrams

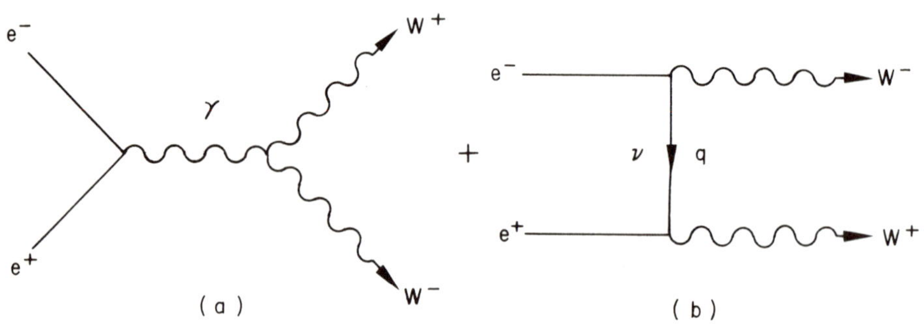

$M^{(b)}$ is obviously the same as $M^{(1)}$ to leading order, so that

$$M^{(b)} \simeq -\frac{2g^2}{M_W^2}\,\bar{v}\not{q}(1-\gamma_5)u.$$

With the γWW vertex "derived" in the previous section, another straightforward calculation gives:

$$M^{(a)} \simeq \frac{e^2}{M_W^2}\,\bar{v}\not{q}u = \frac{e^2}{2M_W^2}\,\bar{v}\not{q}[(1-\gamma_5)+(1+\gamma_5)]u.$$

The only way to cancel the leading part of both $M^{(a)}$ and $M^{(b)}$ simultaneously without a neutral current is by inventing a new t-channel exchange of a heavy neutral lepton ν':

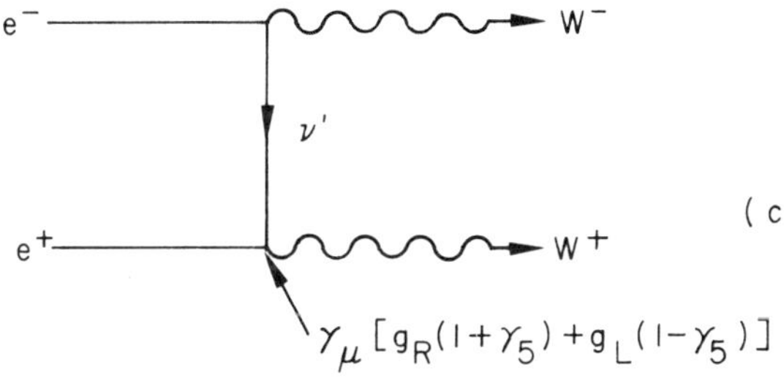

(c)

In direct analogy with the calculation of $M^{(1)}$ this gives:

$$M^{(c)} \simeq -\frac{2}{M_W^2}\,\bar{v}\not{q}[g_R^2(1+\gamma_5)+g_L^2(1-\gamma_5)]u$$

Setting $M^{(a)} + M^{(b)} + M^{(c)} \simeq 0$ we get

$$e^2 = 4g_R^2$$
$$e^2 = 4g^2 + 4g_L^2$$

(Note that if ν' was a doubly-charged object exchanged in the u-channel the sign of $M^{(c)}$ would be changed, making the cancellation of the coefficient of $1+\gamma_5$ impossible.) From the second equation it follows that

$$M_W = \left[\frac{g^2\sqrt{2}}{G}\right]^{\frac{1}{2}} < \frac{1}{2}\left[\frac{e^2\sqrt{2}}{G}\right]^{\frac{1}{2}} = 53\,\text{GeV}.$$

in this model.

Similarly, we can adjust the $E^+\nu'W$ coupling to obtain cancellations in $E^+E^- \to W^+W^-$ and $\nu'\bar{\nu}' \to W^+W^-$. The result of these labours can be summarized in the "isospin invariant" interaction

$$\mathscr{L}_{\text{int}} = e\vec{W}_\lambda.(\bar{\vec{\Psi}}\gamma_\lambda \times \vec{\Psi})$$

where the "isovector" $\vec{W}$ has components $W^{\pm}$ and γ and

$$\vec{\Psi} = \vec{\Psi}_R + \vec{\Psi}_L$$

where the isovectors $\vec{\Psi}_{R,L}$ have components:

$$\Psi_L = \frac{(1-\gamma_5)}{2}\begin{pmatrix} E^+ \\ \nu\cos\theta + \nu'\sin\theta \\ e^- \end{pmatrix}, \Psi_R = \frac{(1+\gamma_5)}{2}\begin{pmatrix} E^+ \\ \nu' \\ e^- \end{pmatrix}$$

with

$$g = \frac{e\cos\theta}{2}.$$

A similar "derivation" of the couplings in the Weinberg model would also have led to an "isospin invariant" interaction such as occurs in Yang–Mills theories, to which we shall turn after considering the general case.

5. CANCELLATION OF LEADING TERMS IN THE GENERAL CASE

Suppose we put all elementary fermions (F) (leptons and hadrons) in one big vector

$$\Psi = \begin{pmatrix} \nu_e \\ e \\ \nu_\mu \\ \mu \\ \\ ? \end{pmatrix}$$

(where ν_μ and μ^- could be replaced by $\bar{\nu}_\mu$ and μ^+ depending on what conservation laws we wish to embody in the theory). Let them interact with some vector fields $W_\mu^i (i = 1, 2, \ldots)$ so that the Born terms for $F_\beta \bar{F}_\alpha \to W^a W^b$ are:

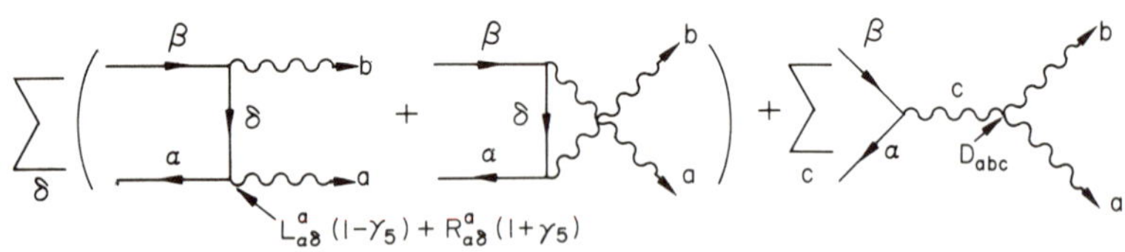

where D_{abc} is the $3W$ coupling of the form discussed in Section I.3. From the discussion of the Georgi–Glashow model, it should be clear that the sum of the worst behaved contributions from the first two diagrams is proportional to

$$\sum_\delta (L^a_{\alpha\delta}L^b_{\delta\beta} - L^b_{\alpha\delta}L^a_{\delta\beta})(1-\gamma_5) + (R^a_{\alpha\delta}R^b_{\delta\beta} - R^b_{\alpha\delta}R^a_{\delta\beta})(1+\gamma_5)$$

which must cancel a piece proportional to

$$\sum_c D_{abc}[L^c_{\alpha\beta}(1 - \gamma_5) + R^c_{\alpha\beta}(1 + \gamma_5)]$$

from the third diagram. In fact, with the standard convention for the $3W$ vertex, the cancellation conditions are

$$[L^a, L^b] = i \sum_c D_{abc} L^c$$

$$[R^a, R^b] = i \sum_c D_{abc} R^c.$$

We see that *an underlying group structure (characteristic of a Yang–Mills theory) is needed to achieve desirable high-energy behaviour to leading order* (the next to leading order "bad behaviour" must also be eliminated; we return to this below).

Although we began by trying to cure problems in an *ad hoc* way in lowest order, we might hope that the high degree of symmetry required to do this would lead to systematic cancellations in higher orders also.

Note that in quantum electrodynamics the analogue of the cancellation conditions is always satisfied. An element

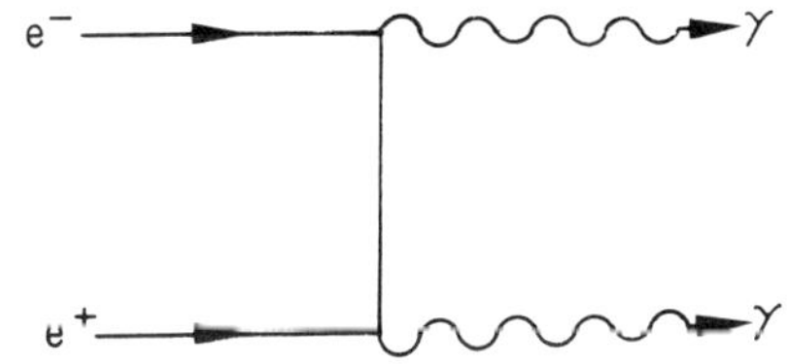

in any diagram is always accompanied by

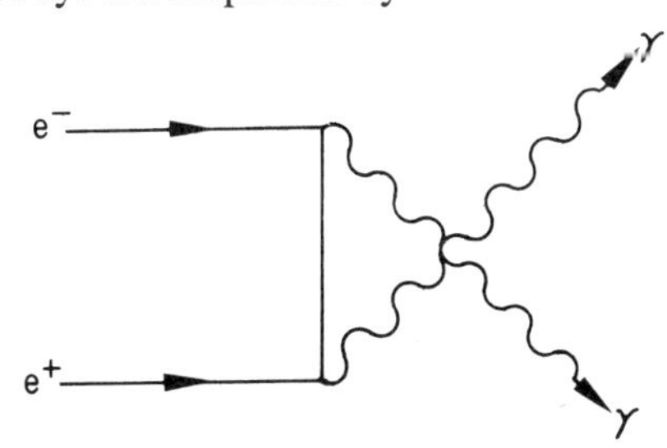

6. GENERALIZED GAUGE INVARIANCE

By imposing good high-energy behaviour, we have "derived" various properties which are probably necessary (but not sufficient) for renormalizability. We asserted that many of these properties are characteristic of Yang–Mills or generally gauge invariant theories.[26] We review these theories briefly in this section.

The main features of ordinary gauge invariance are listed in the left-hand column of Table I. Gauge transformations simply change the phases of all fields; the "gauge function" (phase) χ is neutral and so is the photon (which mixes with $\partial_\mu\chi$ under the transformation). We wish to generalize to charged vector (gauge) fields; we therefore require a charged-gauge function and hence, in place of charge conservation as a starting point, a global (constant over-all space time) transformation which mixes particles of different charge.

TABLE I

Electromagnetic gauge invariance	Generalized gauge invariance (isospin as an example)
Starting point: Formal expression of charge (isospin) invariance; $\mathscr{L}$ is invariant when:	
$\Psi_i \to \exp(ie_i\chi)\,\Psi_i$ (e_i is the charge of the particle i)	$\Psi^{(i)} \to \exp(ig\vec{T}_i\cdot\vec{\Lambda})\Psi^{(i)}$ ($\Psi^{(i)}$ are vectors representing isomultiplets labelled by i and $\vec{T}_i$ the appropriate isospin matrices)
Question: The choice of $\chi(\vec{\Lambda})$ is a convention; can it be chosen differently at different space-time points $[\chi = \chi(x), \vec{\Lambda} = \vec{\Lambda}(x)]$? *Answer:* Not without modification because kinematic terms in $\mathscr{L}$ are not invariant, for example:	
$\bar{\Psi}\gamma^\mu\partial_\mu\Psi \to \bar{\Psi}\gamma^\mu(\partial_\mu + ie\partial_\mu\chi)\,\Psi$	$\Psi\gamma^\mu\partial_\mu\Psi \to \Psi\gamma^\mu(\partial_\mu + ig\vec{T}\cdot\partial_\mu\vec{\Lambda})\Psi$
NECESSARY MODIFICATION to restore invariance:	
Introduce new field A_μ Define transf.: $A_\mu \to A_\mu + \partial_\mu\chi$	Introduce new isotriplet of fields $\vec{W}_\mu$. Define transf.: $\vec{W}_\mu \to \vec{W}_\mu + \partial_\mu\vec{\Lambda} + g\vec{\Lambda}\times\vec{W}_\mu + 0(g^2\Lambda^2)$ (g dependent terms occur because $\vec{W}_\mu$ is an isovector; it is "rotated" even if $\vec{\Lambda}$ = constant).
Replace ∂_μ by the "covariant derivative" $\partial_\mu - ieA_\mu$	Replace ∂_μ by the "covariant derivative" $\partial_\mu - ig\vec{T}\cdot\vec{W}_\mu$
This is a minimal requirement; functionals of $F_{\mu\nu} = \partial_\mu A_\nu - \partial_\nu A_\mu$ can also be added.	This is a minimal requirement; isospin invariant functionals of the isovector $\vec{G}_{\mu\nu} = \partial_\mu\vec{W}_\mu - \partial_\nu\vec{W}_\mu + g\vec{W}_\mu x\vec{W}_\nu$ can also be added.
Kinetic term is $-\frac{1}{4}F_{\mu\nu}F^{\mu\nu}$	Kinetic term is $-\frac{1}{4}\vec{G}_{\mu\nu}\cdot\vec{G}^{\mu\nu}$

TABLE I (*cont'd*)

Invariance requires $M_A = 0 (M_W = 0)$; a term $M_A^2 A_\mu A^\mu (M_W^2 \vec{W}_\mu \cdot \vec{W}^\mu)$ is not invariant.	
Theory is renormalizable (high degree of symmetry ⇒ systematic cancellation of divergences)	Theory may be renormalizable (high degree of symmetry ⇒ systematic cancellation of divergences).
Infra-red divergences tolerable. Still renormalizable with $M_A \neq 0$; A_μ couples to all charged particles and hence not to itself; the current to which A_μ couples is unchanged and remains conserved with $M_A \neq 0$.	Infra-red divergences intolerable. Not renormalizable with $M_W \neq 0$, without adding new ingredients; $\vec{W}_\mu$ couples to all particles with non-zero isospin and hence to itself; the current changes and is no longer conserved with $M_W \neq 0$ (without new ingredients).

An obvious example is isospin, but any unitary transformation which is not just a phase transformation (i.e. is "non-abelian") could do.

Choosing isospin as an example, the main features of generalized gauge invariance are shown on the right of Table I.

Note that the quantity

$$\vec{G}_{\mu\nu} = \partial_\mu \vec{W}_\nu - \partial_\nu \vec{W}_\mu + g \vec{W}_\mu \times \vec{W}_\nu$$

transforms as an isovector; kinetic and interaction terms in the Lagrangian will therefore be linked in a term

$$\vec{G}_{\mu\nu} \cdot \vec{G}_{\mu\nu}.$$

This form (written for an arbitrary unitary group) was already encountered at the end of Section I.3. Note further that the W-fermion couplings are proportional to the isospin matrices $\vec{T}_i$ which, of course, satisfy the commutation relations "derived" in Section I.5.

The ultra-violet divergences of Yang–Mills theory are supposed to be renormalizable[27] because of the high degree of symmetry, which leads to systematic cancellations order by order. However, this same symmetry forbids a W-mass term in the Lagrangian ($M_W = 0$ causes incurable infra-red divergences and is certainly incompatible with experiment); it also requires that either all lepton masses are zero or parity is conserved and the leptons form degenerate multiplets. This dilemma could perhaps be avoided if the symmetry were "spontaneously broken",[28, 29] i.e. if there were solutions in which the ground state did not enjoy the full symmetry of the Lagrangian, in which case the W could be massive and there could be fermion masses (differences) despite the original invariance. The classic example of spontaneous symmetry breaking is the ferromagnet; the Hamiltonian is rotationally invariant but the (infinite) number of ground states are not, since the spins are all aligned along some definite (but arbitrary) direction. Before exploring this idea, we continue in an empirical fashion to examine explicitly the difficulties which occur if the W's are massive and to try to cure them.

7. CANCELLATION OF RESIDUAL DIVERGENCES AND THE NECESSITY OF SCALAR (HIGGS) MESONS

In Section 4 we asserted that the high-energy behaviour of the Born terms for W–W scattering is especially gentle if the couplings have the Yang–Mills form. In the Georgi–Glashow model, for example, the sum of the diagrams

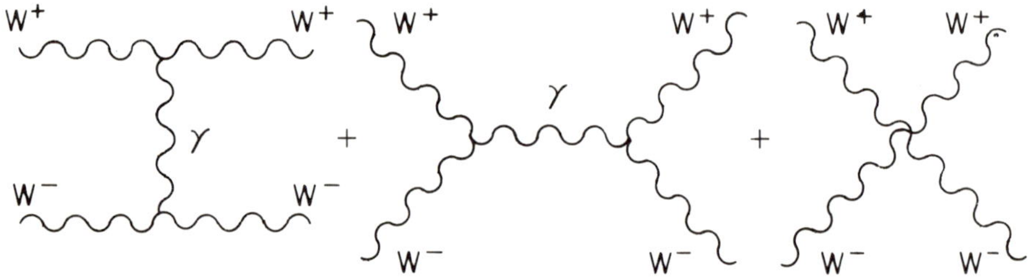

gives an amplitude which grows like $g^2 s/M_W^2$ as $s \to \infty$ when all four W's are longitudinal, whereas we might fear a $g^2 s^2/M_W^4$ growth, since each longitudinal W gives a factor $1/M_W$ (each term individually does grow like s^2). Although s is better than s^2, it still rapidly violates unitarity and is the source of divergence difficulties in higher orders (in models in which a massive neutral vector can be exchanged, giving a $k_\mu k_\nu/M_W^2$ term in the propagator, the amplitude could grow like s^6—but again the sum of diagrams actually grows at most like s if the couplings have the Yang–Mills form). We wish to cancel these residual s terms in the same order of perturbation theory. New exchanges are therefore required; we have already tried exchanging spin-1 and, since we wish to avoid the vicious problems associated with spins >1, they can only have spin-0. In the example considered above we need a neutral scalar:

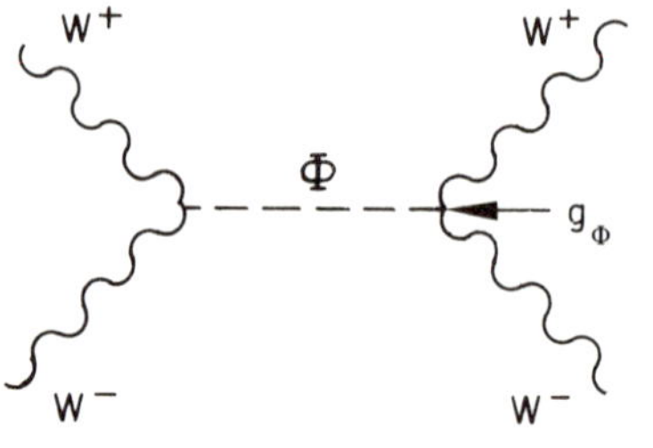

This gives a term $\sim g_\Phi^2 s/M_W^2$. Cancellation can be achieved and requires

$$g_\Phi \sim gM_W$$

Returning now to fermion–antifermion annihilation, explicit calculation shows that although (by construction) the conditions we imposed in, for example, $e^+e^- \to W^+W^-$ in the Georgi–Glashow model remove terms of order

$$\sim \frac{g^2 \bar{v} q u}{M_W^2},$$

pieces of order

$$\sim \frac{g^2 m_{\text{lepton}} \bar{v} u}{M_W^2}$$

remain. [Note that m_{lepton} is not necessarily m_e; in the Georgi–Glashow model it is $m_{\nu_e'} + O(m_e)$.] Such terms are unacceptable (note that $\bar{v}u \sim \sqrt{s}$ at large s); they rapidly violate unitarity and force us to introduce a subtraction in a dispersive calculation of the box diagram for $e^+e^- \to e^+e^-$, which is not allowed in a renormalizable theory. They are easily removed, however, using the scalar field Φ which we have already introduced. The diagram

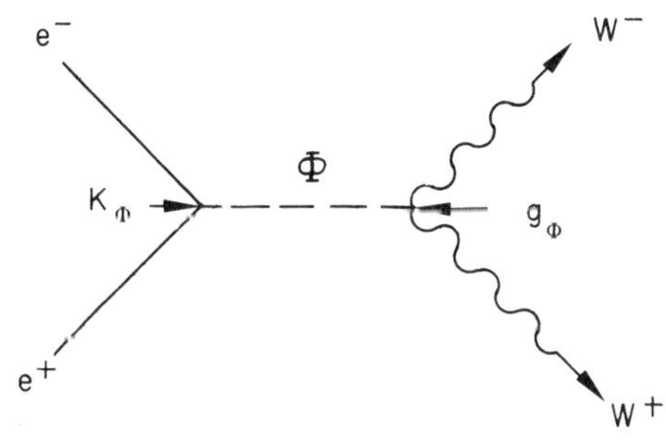

gives a contribution

$$\sim \frac{g_\Phi K_\Phi \bar{v} u}{M_W^2}$$

and cancellation is possible provided

$$K_\Phi \sim \frac{g^2 m_{\text{lepton}}}{g_\Phi} \sim \frac{g m_{\text{lepton}}}{M_W}.$$

The properties of the Φ's needed to make all WW[21, 22] and fermion anti-fermion[22] amplitudes well behaved can be worked out in the general case; the models discussed in subsequent sections can in fact be "derived" in this way. We will not go into this here, beyond noting that *scalar particles are necessary and must couple with strengths*

$$g_{\Phi} \sim gM_W$$

$$K_{\Phi} \sim g\frac{m_{\text{lepton}}}{M_W}$$

intimately connected with the masses.

8. THE HIGGS MODEL OF SPONTANEOUS MASS GENERATION

In Section I.6 we remarked that we would like to find a gauge theory in which the symmetry is "spontaneously broken" and hence masses of W's (and masses and/or mass differences of fermions) are generated spontaneously; in Section 7 we found that in well-behaved models there must be scalar particles intimately connected with the masses. These features are combined in the model invented by Higgs; we shall only sketch it here, referring to Higgs' classic papers,[29] reviews,[3,4] and other works[30] for more details.

Higgs studied the Lagrangian:

$$\mathscr{L} = -\tfrac{1}{4}F_{\mu\nu}F_{\mu\nu} - (\partial_\mu + ieA_\mu)\Phi^*(\partial_\mu - ieA_\mu)\Phi - M^2|\Phi|^2 - h|\Phi|^4,$$

where Φ is a complex field and, as usual,

$$F_{\mu\nu} = \partial_\mu A_\nu - \partial_\nu A_\mu.$$

$\mathscr{L}$ is obviously invariant under the gauge transformation:

$$\Phi \rightarrow e^{ie\chi}\Phi$$

$$A_\mu \rightarrow A_\mu + \partial_\mu\chi.$$

If $M^2 > 0$, this is just QED for a charged scalar particle. What if $M^2 < 0$? In this case (if $h > 0$) the classical potential $V(\Phi) = M^2|\Phi|^2 + h|\Phi|^4$ has the form

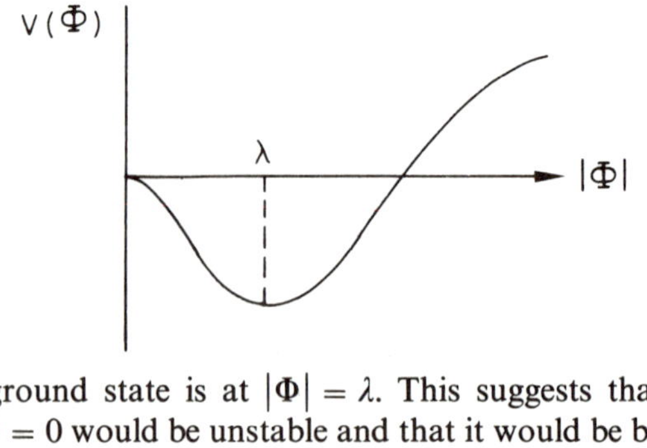

The classical ground state is at $|\Phi| = \lambda$. This suggests that perturbation theory about $\Phi = 0$ would be unstable and that it would be better to change

variables to $\Phi(x) = \lambda\, e^{i\xi} + \Phi'(x)$ with $\langle\Phi'(x)\rangle = 0$ in the ground state[31] (i.e. $\lambda = \sqrt{(-M^2/2h)}$). The phase ξ can be chosen at will; each choice gives one of an infinite number of possible ground states (in the theorists' jargon there is a "degenerate vacuum"). The symmetry is broken since these ground states are not invariant under gauge transformations, which turn one into another (note the analogy with the ferromagnetic example, introduced at the end of Section I.5, in which there is also an infinite number of non-rotationally invariant ground states which get turned into one another under rotations). Choosing $\xi = 0$, the term $e^2 A_\mu A_\mu |\Phi|^2$ becomes

$$e^2\lambda^2 A_\mu A_\mu + 2e^2\lambda A_\mu A_\mu \operatorname{Re}\Phi' + e^2 A_\mu A_\mu |\Phi'|^2$$

The vector field acquires a mass $M_W \sim e\lambda$ and we find a $A_\mu A^\mu \Phi'$ interaction with strength eM_W, such as we encountered in the previous section, where its necessary role was already discussed.

What happens to Φ which seemed to have an imaginary mass in the original Lagrangian? It is a simple exercise to show that if we rewrite Φ' in terms of two real fields Φ_1 and Φ_2 thus:

$$\Phi' = \Phi_1 + i\Phi_2$$

then Φ_1 has a positive real mass but Φ_2 is massless (the term linear in Φ vanishes); we seem to have traded one zero-mass particle for another. Φ_2 is the "Goldstone boson" which is supposed to occur when there is a "degenerate vacuum" (however, Goldstone's theorem does not apply to theories involving gauge fields in general and, as we shall see, the Goldstone boson is something of an illusion). Intuitively we can understand its presence since "different" ground states can only differ by containing different numbers of particles with zero energy and momentum. However, Φ_2 is spurious in some sense and it can be shown that there are no zero-mass poles in physical amplitudes. One way to gain a little understanding of this is to choose a different set of real fields χ, θ and B_μ which (it is hoped) yield the same S-matrix and set

$$\Phi = (\lambda + \chi)\, e^{i\theta}$$

$$A_\mu = B_\mu + \partial_\mu\theta$$

This is a gauge transformation with parameter θ (chosen to make Φ become real) followed by the change of variables $\Phi \to \lambda + \chi$ and $A_\mu \to B_\mu$; it therefore leaves the theory independent of θ as can easily be checked explicitly (χ is massive and the inapplicability of Goldstone's theorem is manifest). Thus one degree of freedom of the scalar field is spurious in this case. We can say that the original massless vector field, which would have two physical degrees of freedom, "eats up" part of the scalar field to become a massive particle with three degrees of freedom.

The new Lagrangian in this and more general cases is supposed to be renormalizable (like the original one with $M^2 > 0$). The proof is hard. The basic idea is outlined in the following *hand waving theoretical digression:*

If we try to quantize the Higgs Lagrangian using the variables Φ_1 and Φ_2 (or their generalization in more complicated cases) we meet the same problem as in QED; the momentum conjugate to A_0 is zero and the standard Hamiltonian quantization scheme breaks down. If we try to avoid the "creaking machinery of canonical field theory" (the phrase is J. S. Bell's) and define Feynman rules directly from the Lagrangian in the standard way, the difficulty occurs when we try to diagonalize the free propagator $\langle A_\mu A_\nu \rangle$; diagonalization would involve inverting the singular matrix $g_{\mu\nu}k^2 + k_\mu k_\nu$ which does not have a unique inverse. These problems are related to the gauge invariance and the consequent indeterminacy of the classical equations of motion, they would not occur if we fixed the gauge by using the variables χ and B_μ above and quantized them. In QED they can be avoided by fixing the gauge using Fermi's trick of adding $-\frac{1}{2}(\partial_\mu A_\mu)^2$ to the Lagrangian, which is not supposed to alter the physics because of the subsidiary conditions. Here, following 't Hooft, the gauge can be fixed by adding a term

$$-\tfrac{1}{2}\xi\left(\partial_\mu A_\mu + \frac{m\Phi_2}{\xi}\right)^2$$

[To ensure that the physics does not change, an additional piece must also be added in more complicated (non-abelian) models; it gives rise to the so-called Fadeev Popov ghost contributions in closed-loop diagrams[32].] The A_μ and Φ_2 propagators are

$$\langle A_\mu A_\nu \rangle = \frac{-g_{\mu\nu} + k_\mu k_\nu (1 - \xi)(m^2 - \xi k^2)^{-1}}{k^2 - m^2}$$

$$\langle \Phi_2 \Phi_2 \rangle = \frac{1}{-k^2 + (m^2/\xi)}.$$

For any $\xi \neq 0$ we have the so-called R-gauges; the propagators fall off as k^{-2} as $k \to \infty$ and, by power counting, the theory is renormalizable (R). However, the theory appears to be non-unitary. The pole residues are not positive definite (and it seems to be possible to produce particles with negative probability) but all such undesirable features depend on the parameter ξ. The proof of renormalizability uses the symmetry (expressed in the form of "Ward identities") to show that the S-matrix is independent of ξ and therefore has no undesirable features. In the limit $\xi \to 0$, the so-called U-gauge is obtained; manifest renormalizability is lost but the theory becomes manifestly unitary (U) and the "Goldstone ghost" Φ_2 acquires infinite mass. In the U-gauge (which corresponds to choosing the variables

χ, θ and B_μ above) the physical particle content of the theory is clear. (Note that for higher-order calculations it is dangerous to take the singular limit $\xi \to 0$ in the Feynman rules.) *End of digression.*

We now consider how to generate fermion masses spontaneously. If we add a fermion to the original Higgs model we can obviously give it any mass we like "by hand" without spoiling the gauge invariance. In more general models this is not so; consider, for example, a model based on some "isospin"-like symmetry in which the field $\vec{W}_\mu$ couples to

$$\bar{\Psi}\gamma_\mu(1 + \lambda\gamma_5)\vec{T}\Psi$$

Gauge invariance requires that this current is conserved and hence that in the original Lagrangian *either* the masses of all the fermions in the multiplet Ψ are zero, *or* they are all the same and $\lambda = 0$. At least some part of the fermion masses or mass differences must therefore be generated spontaneously in realistic models.[33] This can be done if we can write an invariant term in the Lagrangian with the structure (suppressing "isospin" indices) $K_\Phi \Phi \Psi_L \Psi_R$ + h.c. where $\Psi_{L,R}$ are left-handed and right-handed multiplets. When some components of Φ are "translated" to $\lambda + \Phi'$ we get (symbolically)

$$K_\Phi \Phi \bar{\Psi}_L \Psi_R + \text{h.c.} \to \underbrace{K_\Phi \lambda}_{m_s}(\bar{\Psi}\Psi) + \underbrace{K_\Phi}_{m_s/\lambda \,=\, g m_s/M_W} \Phi' \bar{\Psi}\Psi$$

where m_s is the "spontaneous" part of the mass. *We have recovered our empirical discovery of the last section that there will be scalar fields* (Φ') *coupled to fermions with strength* $\sim g(m_{\text{lepton}}/M_W)$.

The Higgs model is the prototype of the models discussed below in which there are, of course, more than one vector field. In this case the mechanics of mass generation, changing variables to the U-gauge (to get the particle content of the theory), etc., are much more complicated algebraically. Conceptually nothing very new is involved and we refer to the literature for details.[30] Practically, the most important result is that originally there must be (at least) one scalar field with the same charge as each vector which is to acquire mass. Each of these vectors "eats" one scalar in becoming massive (i.e. these scalars disappear in the U-gauge—cf. the field θ above); at least one massive scalar will remain. This and other results may be summarized in a recipe for model building.

9. RECIPE FOR MODEL BUILDING

1. Choose a (gauge) group; assign the vector fields to the "adjoint representation", so that their number is the same as the number of generators [e.g. 8 in $SU(3)$, 3 in $SU(2)$ or $O(3)$ etc.].

2. Choose a representation of the scalar (Higgs) fields and their charge assignments.
3. Choose representations for the left-handed and right-handed components of the fermions.
4. Couple the gauge fields invariantly to the Higgs fields and fermions.
5. Couple the Higgs fields invariantly and renormalizably to themselves.
6. Choose these couplings so that the potential of the Higgs fields is a minimum when neutral Higgs fields have non-vanishing vacuum expectation values.
7. Couple the Higgs fields invariantly to fermions.
8. Ensure that there are no "anomalies" (see Section I.10).

If we work with real vector fields (W^a_μ, $a = 1 \ldots m$) and real scalar fields (Φ_t, $t = 1 \ldots n$) this prescription gives the Lagrangian:

$$\mathscr{L} = \overline{\Psi}_i \left\{ i\delta_{ij}\partial\!\!\!/ - M_{ij} + gW\!\!\!\!/^{a}\left[\left(\frac{1-\gamma_5}{2}\right)L^a_{ij} + \left(\frac{1+\gamma_5}{2}\right)R^a_{ij}\right]\right.$$
$$\left. + \Phi^t\left[X^t_{ij}\left(\frac{1+\gamma_5}{2}\right) + X^{t+}_{ij}\left(\frac{1-\gamma_5}{2}\right)\right]\right\}\Psi_j$$
$$+\tfrac{1}{2}|\partial_\mu\Phi_t + P^a_{ts}W^a_\mu\Phi_s|^2$$
$$-\tfrac{1}{4}|\partial_\mu W^a_\nu - \partial_\nu W^a_\mu + D_{abc}W^b_\mu W^c_\nu|^2 + \mathscr{L}'(\Phi)$$

where repeated indices are summed and $\mathscr{L}'(\Phi)$ is an invariant containing Φ^3 and Φ^4 terms, the "group structure constants" $D_{abc} = D^*_{abc}$ satisfy the Jacobi identity:

$$[D^a, D^b] = D_{abc}D^c$$

where $D^a_{bc} = -D_{abs}$ and

$$[L^a, L^b] = iD_{abc}L^c$$
$$[R^a, R^b] = iD_{abc}R^c$$
$$[P^a, P^b] = D_{abc}P^c$$
$$L^cX^t - X^tR^c = iP^c_{st}X^t$$
$$L^aM = R^aM$$
$$L = L^+, R = R^+, M = M^+$$
$$P^a_{st} = P^{a*}_{st} = -P^{a*}_{ts}$$

In addition the "no anomalies" condition discussed in Section I.10 must be satisfied. $V(\Phi) = -\mathscr{L}'(\Phi)$ is supposed to have a minimum at $\Phi_t = \eta_t$ (η_t can only be non-zero if Φ_t is neutral). The recipe to obtain the U-gauge Lagrangian (in which the physical particle content is clear) is:

(a) If there are T non-zero n component vectors $P^a_{ts}\eta_s$, choose a basis such that $P^a_{ts}\eta_s = 0$ for all a for $t \leqslant n - T$.
(b) Set $\Phi_t \Rightarrow \eta_t + \Phi_t$ for $t \leqslant n - T$.
Set $\Phi_t \Rightarrow 0$ for $t > n - T$.
(c) Choose a basis in which the fermion mass matrix is diagonal and independent of γ_5 (this is always possible).
(d) Diagonalize the W and the Φ mass matrices.

10. ANOMALIES

There is one more condition which renormalizable models must satisfy: *there must be no [Bell–Jackiw–Adler type*[34]*] anomalies.* The simplest anomalies are associated with the triangle graphs:

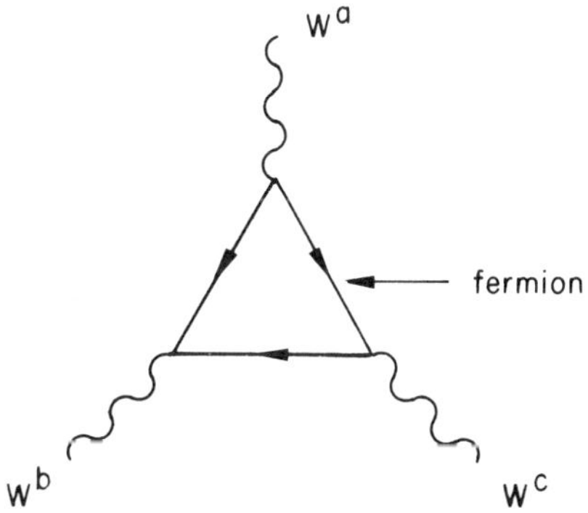

The amplitudes corresponding to these diagrams do not in general respect the symmetries of the theory embodied in the Feynman rules. This is caused by the divergent nature of the integrals because of which they need not have the full symmetry of the integrands (this can happen even when the integrals apparently converge because of cancellations between terms which are individually infinite). Furthermore, it can be shown that, in general, there is no way of regulating the integrals so that all the symmetries are respected. Such symmetry violations would spoil the renormalizability of gauge theories.[35] Fortunately it is possible to arrange that the sum of the anomalous contributions from all triangle diagrams in zero; this ensures the absence of all anomalies. The necessary and sufficient condition is:[36]

$$\mathrm{Tr}'\mathrm{Tr}\,[\gamma_5\{\Gamma^a, \Gamma^b\}\,\Gamma^c] = 0$$

for all a, b, and c, where $\{\,,\}$ denotes an anticommutator, the first trace acts on the γ matrices and the second on the indices i and j of the matrices

$$\Gamma^a_{ij} = L^a_{ij}(1 - \gamma_5) + R^a_{ij}(1 + \gamma_5).$$

If we want the theory to be renormalizable to all orders,[37] then this condition must be satisfied. It is a strong constraint on model building (although it can usually be satisfied in an *ad hoc* way by adding new multiplets of fundamental fields, which can be made heavy to avoid conflict with experiment). Note that there can be cancellations between the leptonic and hadronic contributions to anomalies.

11. ANOTHER EXAMPLE: THE WEINBERG–SALAM MODEL

It would be ridiculous to give another detailed account of this model in these notes since it has been so widely advertised. However, there are certain features which must be noted here since they will be needed in the discussion of neutral currents below. Any reader who is not familiar with all the other details of the model is urged to work them out as an exercise.

In this model there are no new leptons so there must be a massive neutral vector boson Z^0. $W^\pm$ connect left-handed electrons and ν_e which therefore belong to a doublet

$$\Psi_L = \frac{(1-\gamma_5)}{2}\begin{pmatrix}\nu_e \\ e\end{pmatrix}$$

while the right-handed electron $\Psi_R = [(1+\gamma_5)/2]e$ is a singlet (the model for muons is obtained by changing the label $e \to \mu$). From the available spin-1 particles $W^\pm$, Z^0 and γ we can form a triplet

$$\vec{W}_\mu = (W^\pm_\mu, Z_\mu \cos\theta_W + A_\mu \sin\theta_W)$$

and a singlet

$$B_\mu = -Z_\mu \sin\theta_W + A_\mu \cos\theta_W.$$

W_μ is the "isotriplet" gauge boson of an "isospin" group $[SU(2)]$ under which Ψ_L is a doublet and Ψ_R and B_μ are singlets. B_μ is the gauge boson of a group of phase transformations $[U(1)]$, like the gauge invariance of QED, under which $\vec{W}_\mu$ is a singlet ("charge" zero) and the leptons have "charges" which will be specified below [the invariance group is therefore $SU(2)\times U(1)$]. The reader is now invited to construct a scalar system which will make $W^\pm$ and Z^0 massive. Here we shall try to read the properties we shall need directly from the leptonic part of the Lagrangian which has the form:

$$\mathscr{L} = \overline{\Psi}_L\gamma^\mu(i\partial_\mu + g\vec{T}\cdot\vec{W}_\mu - g'B_\mu)\Psi_L + \overline{\Psi}_R\gamma^\mu(i\partial_\mu - g''B_\mu)\Psi_R$$

Observe that the mixture

$$gW^3_\mu - g'B_\mu$$

couples to ν_e; this combination must be orthogonal to the photon (i.e. be pure Z_μ) since the neutrino is neutral. Hence:

$$\tan\theta_W = g'/g.$$

The photon has the interaction:

$$-\frac{A_\mu}{\sqrt{(g^2+g'^2)}}\bar{\Psi}_e\gamma^\mu\left[2gg'\left(\frac{1-\gamma_5}{2}\right)+gg''\left(\frac{1+\gamma_5}{2}\right)\right]\Psi_e$$

Therefore, we set $g'' = 2g'$ to make electromagnetism conserve parity and keep the photon massless. If we did not want to give the electron a mass the current would be conserved for arbitrary g'' and this would not be necessary (we could make parity conserved by definition in this case). However, the industrious reader will find that, if he insists on the ordinary gauge invariance of QED, it is only possible to give the electron a mass if $g'' = 2g'$.

The electric and weak charges are obviously given by

$$e = \frac{-2gg'}{\sqrt{(g^2+g'^2)}} = -2g\sin\theta_W, \quad g_W = \frac{g}{\sqrt{2}}$$

Hence

$$M_W = \left[\frac{g^2\sqrt{2}}{2G}\right]^{\frac{1}{2}} = \frac{1}{2(\sqrt{2})|\sin\theta_W|}\left[\frac{e^2\sqrt{2}}{G}\right]^{\frac{1}{2}} = \frac{37\ \text{GeV}}{|\sin\theta_W|}$$

Expressing the physical particles Z_μ and A_μ, which are eigenstates of mass, in terms of W^3_μ and B_μ we get

$$M_Z^2 = M_W^2\cos^2\theta_W + M_B^2\sin^2\theta_W - 2\cos\theta_W\sin\theta_W M_{WB}^2$$

$$M_\gamma^2 = 0 = M_W^2\sin^2\theta_W + M_B^2\cos^2\theta_W + 2\cos\theta_W\sin\theta_W M_{WB}^2$$

$$M_{Z\gamma}^2 = 0 = (M_W^2 - M_B^2)\sin\theta_W\cos\theta_W + M_{WB}^2(\cos^2\theta_W - \sin^2\theta_W)$$

where M_W^2 is the common mass of the "isotriplet" $W^\pm$, W^3 (the whole Lagrangian being "isospin" invariant) and M_{WB} and $M_{Z\gamma}$ are off-diagonal mass terms. From these equations we obtain:

$$M_Z = \frac{M_W}{|\cos\theta_W|} = \frac{73\ \text{GeV}}{|\sin 2\theta_W|}$$

Note that θ_W is the only free parameter in the model.

Finally we record the form of the neutral current interaction which is:

$$\frac{e}{2\sin 2\theta_W}\,Z_\mu\,\bar{\nu}\gamma^\mu(1-\gamma_5)\nu + \frac{e\cot\theta_W}{4}\,Z_\mu\bar{e}\gamma_\mu[3\tan^2\theta_W - 1 + (1+\tan^2\theta_W)\gamma_5]e.$$

Note that in models (such as this) based on $SU(2) \times U(1)$ the condition that there be no anomalies takes the simple form[3]

$$\sum_i (2I_3^i)^2 Q_i = 0,$$

where the sum runs over all fermions, which have charge Q_i and third component of "weak isospin" I_3^i. In the Weinberg–Salam model, the e and μ systems contribute -2 to this sum and therefore hadrons (or additional leptons) must contribute $+2$.

12. MODELS; PROBLEMS AND HOPES

We have discussed the "minimal" models of leptons constructed by Weinberg and Salam and by Georgi and Glashow. It is straightforward to construct innumerable other models of leptons which do not contradict existing data.[38] Personally, not knowing why the muon exists, I have no prejudice about the lepton spectrum or its symmetries and can contemplate new multiplets of leptons comparatively serenely (although it would perhaps be nice if nature preferred a "minimal" model).

For hadrons, however, the situation is different. It is an established fact that low-lying hadronic states can be classified in $SU(3)$ multiplets and the eloquent advocacy of Gell-Mann has persuaded us that the weak and electromagnetic currents are components of octets of vector and axial currents which satisfy an $SU(3) \times SU(3)$ algebra. It is therefore natural to try to use quarks as the "fundamental hadronic fields", which must be specified if we wish to make a renormalizable model of semi-leptonic and non-leptonic weak interactions.[39]

However, it seems to be *impossible to make a model using three quarks with the conventional Cabibbo current*. Let us try; in this model there is a diagram

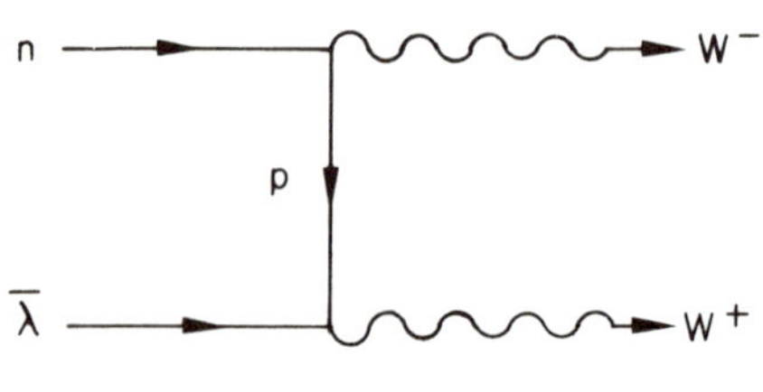

The corresponding amplitude grows like s at large energy. This cannot be cancelled by a new s-channel exchange in general, since such $\Delta S = 1$ neutral currents seem to be highly suppressed experimentally; this is true not only of neutral $\Delta S = 1$ currents coupled to leptons (which would contribute, e.g. to $K_L^0 \to \mu^+\mu^-$, $K^+ \to \pi^+ \nu\bar{\nu}$, or $K^+ \to \pi^+ e^+ e^-$), but also of self-conjugate $\Delta S = 1$ currents coupled to hadrons (which could mediate $\Delta S = 2$ non-leptonic processes in lowest order, in disagreement with the fact that

$$2\,\mathrm{Re}\,\langle \bar{K}^0|R|K^0\rangle \simeq M_{K_L^0} - M_{K_S^0} \simeq \tfrac{1}{2}\Gamma_{K_S^0} \approx 10^{-3}G^2M_K^5;$$

$$A(\Xi^0 \to p\pi^-) \gtrsim 3 \times 10^{-2} \times A(\Xi^0 \to \Lambda\pi^0)$$

In a renormalizable model this "bad" $n\bar{\lambda} \to W^+W^-$ amplitude must therefore be cancelled by new t or u-channel exchanges (unless a schizon scheme can be devised with a non-self-conjugate Z^0 not coupled to leptons); consequently there must be more than three "quarks" (or the Cabibbo structure must be changed). We consider three examples:

(1) In models based on $SU(2) \times U(1)$ the simplest possibility is to introduce a fourth "quark", denoted p', as first done by Glashow–Iliopoulos and Maiani.[40] In addition to the usual left-handed "Cabibbo doublet"

$$(1 - \gamma_5)\begin{pmatrix} p \\ n\cos\theta_c + \lambda\sin\theta_c \end{pmatrix}$$

there is a new doublet

$$(1 - \gamma_5)\begin{pmatrix} p' \\ -n\sin\theta_c + \lambda\cos\theta_c \end{pmatrix}.$$

The experimentally observed Cabibbo structure is explained by introducing a new quantum number called charm (C), which is supposed to be conserved by strong and electromagnetic interactions (it is analogous to strangeness), assigning $C = 0$ to p, n and λ but $C = 1$ to p'. Assuming that the known stable hadrons have $C = 0$, the $\Delta C = 1$ current contributed by the new doublet does not participate in their decays. (p' must be heavy to explain why charmed hadrons have not been seen; their properties are considered further in Section II.6.)

To leading order, in which masses are neglected, the sum of the amplitudes from the diagrams

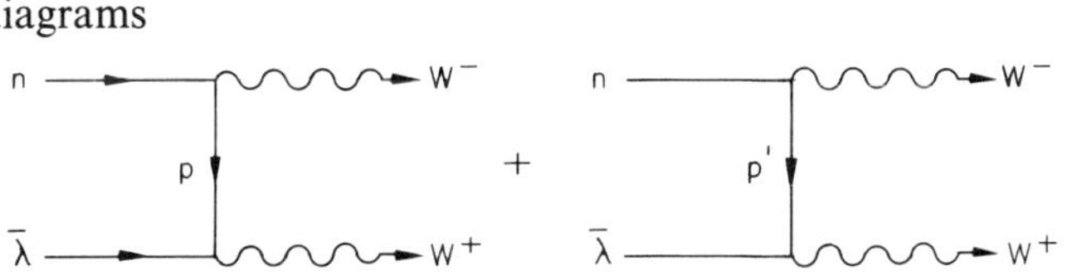

is zero and no contribution is needed from a ($\Delta S = 1$) neutral current [formally the Z couples to ($n\cos\theta_c + \lambda\sin\theta_c$) and ($-n\sin\theta_c + \lambda\cos\theta_c$) so the $\bar{n}\lambda Z$ coupling is zero].

This model can be married to the Weinberg–Salam model for leptons. Assigning charges $+1$ to p and p' and 0 to n and λ the resulting model is free of anomalies (provided the muon exists as well as the electron!); however, it is hard to build octets with these constituents. Alternatively, an anomaly-free model is possible with charges $+\frac{2}{3}$ for p and p' and $-\frac{1}{3}$ for n and λ provided we triple the number of hadronic fields by introducing two more for each one already present; these fields are supposed to be distinguished by some new "colour" quantum number but participate identically in the strong interactions.

Generalizations of the Glashow–Iliopoulos–Maiani trick can also be used to remove $\Delta S = 1$ neutral currents in more complicated models (see, for example, Ref. 41, Appendix B).

(2) The Han–Nambu model, in which there are nine integrally-charged quarks, was constructed to reproduce the results of the naïve low-energy quark model and of current algebra.[42] It turns out that it can be used as a basis for building renormalizable models of weak interactions. It has been married to the Georgi–Glashow [$O(3)$] model[43] and to the Weinberg–Salam [$SU(2) \times U(1)$] model[44] in ways which do not contradict experiment; however, the construction is rather arbitrary and the freedom has to be used in a totally *ad hoc* way to reproduce experimental facts (e.g. using the Georgi–Glashow scheme, the parameter θ, introduced in Section 4, which describes the leptons must be adjusted as a function of the hadronic parameters in order to obtain the observed universality of weak interactions). We refer to the literature for details.[43, 44]

(3) A model in which there are three quarks, but the Cabibbo current is changed, has been constructed by Georgi and Glashow.[45] They introduce left-handed and right-handed doublets:

$$\Psi_L = \frac{(1 - \gamma_5)}{2}\begin{pmatrix} p \\ n \end{pmatrix}, \Psi_R = \frac{(1 + \gamma_5)}{2}\begin{pmatrix} p \\ \lambda \end{pmatrix}$$

(the left-hand λ and the right-handed n being singlets).

The amplitude for $n\bar{\lambda} \to WW$, given by

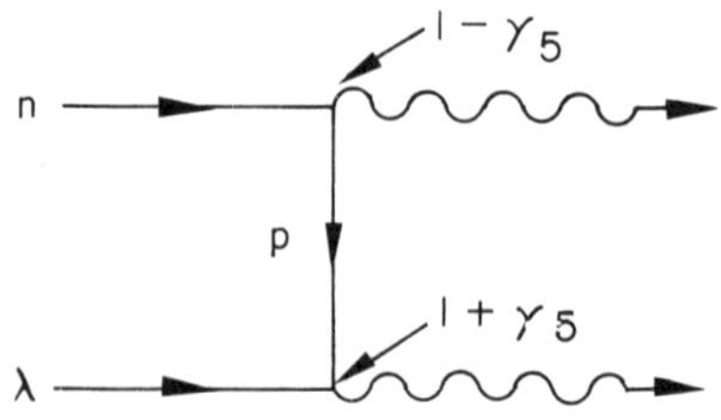

is zero if we neglect the masses and there is therefore no need for a fourth quark or a $\Delta S = 1$ neutral current.[46]

If this assignment is used with the $SU(2) \times U(1)$ model of Weinberg and Salam, it gives the unacceptable result that the Cabibbo angle is 45°. This is avoided by enlarging the group to $SU(2)_L \times SU(2)_R \times U(1)$ (so there are vectors $W_1^{\pm}$, $W_2^{\pm}$, Z_1, Z_2, γ). With double the usual number of neutrinos (ν_e^1, ν_e^2, ν_μ^1, ν_μ^2), which contribute incoherently to (semi) leptonic decays, and two new leptons (electron type E^+ and muon type M^+), a "pseudo Cabibbo structure emerges" which reproduces all the results of the Cabibbo theory except that the sign of g_A is reversed in $\Delta S = 1$ decays. Consequently the model is almost certainly excluded by the data.[47] We therefore give no details beyond noting one amusing implication of the fact that different (incoherent) mixtures of the two neutrinos are produced in π and K decay and therefore neutrino beams from π and K decay are different. Their properties may be summarized symbolically thus:

		Strength:
$\pi \to$ "ν_π" $+ \mu$ $\longrightarrow$	"ν_π" $+ A \to \mu^- + (S = 0)$:	$G \cos \theta_c$
	"ν_π" $+ A \to \mu^- + (S = 1)$:	$G \cos \theta_c \cos \xi$
$K \to$ "ν_K" $+ \mu$ $\longrightarrow$	"ν_K" $+ A \to \mu^- + (S = 0)$:	$G \cos \theta_c \cos \xi$
	"ν_K" $+ A \to \mu^- + (S = 1)$:	$G \sin \theta_c$

With $\xi = 0$ the results coincide with the standard model; with $\xi = \pi/2$ this is a little like the old "neutrino flip" hypothesis.[48] $\xi = \pi/2$ is certainly excluded by experiment but present limits on ξ are probably very weak. Despite the probable irrelevance of this model, we mention this result as an example of the culture shock produced by these new theories which cast doubt on hallowed prejudices.

There are many other models in the literature, most of which can be made to agree with the existing data. However, *to summarize the present situation*, all the models proposed so far seem to involve *ad hoc* arbitrary features; none can be taken very seriously, in my opinion. Consequently, in discussing possible experimental tests of renormalizable models, it is wise to abstract general features from several models. Thus we note that in a large class of renormalizable models, *either there are more than three quarks* (*and new hadronic degrees of freedom characterized by charm quantum numbers*) *or there are changes in the Cabibbo structure of both*†).

We close this section with a few miscellaneous remarks.

(1) In all these models the strength of the electromagnetic and weak interactions are of the same order; leptonic and semileptonic weak decays are "weak" because M_W is very large and the effective strength is of the order

† This is not true in the very interesting class of models devised by Bars, Yoshimura and Halpern[38, 50] and de Wit,[56] in which neither charmed particles nor changes in the Cabibbo current are needed (although the light cone and equal time space component algebras differ from the quark model algebras). We will not consider these models in these introductory notes, but urge the interested reader to consult the literature.

$$\frac{e^2}{q^2 - M_W^2} \sim \frac{-e^2}{M_W^2} (q^2 \ll M_W^2).$$

In non-leptonic weak processes, however, W's occur in closed loops and, in general, there might be non-leptonic $\Delta S = 1$ processes and/or $\Delta s = 0$ parity violating processes in order e^2, in gross contradiction with experiment. Weinberg has shown[49] that a sufficient (but not necessary) condition to avoid this is that the strong interactions are due to a neutral gluon field coupled to a conserved fermion current. In other models it is necessary to check that these unobserved effects are absent; this can easily be arranged in a large class of models by adjusting the counter terms available in the Lagrangian.[50, 51] However, even if we ensure that lowest-order effects have strength $G \sim e^2/M_W^2$, the second-order corrections will generally be of the order $G\alpha$. Care must therefore be taken to suppress the processes $K_L^0 \rightarrow \bar{\mu}\mu$ and $K^0 \leftrightarrow \bar{K}^0$ which are allowed in second order but are known to be of order $G\alpha^2$ (or less) and G^2, respectively. This can be done (for example, by using the Glashow–Iliopoulos–Maiani scheme and making $m_p \simeq m_{p'}$), but in general the necessity of suppressing these effects is a strong constraint on model building.[52]

(2) In models considered up to now, the universality of either the weak or the electromagnetic interactions has often been put in "by hand". In the discussion of the Han–Nambu model, we already remarked that weak universality has to be imposed artificially in the Georgi–Glashow scheme. In the Weinberg–Salam model it is possible to shift, for example, the charges of all hadrons by an arbitrary amount δ (although charge differences are universal). However, it is easy to conceive of models based on a single non-trivial (non-abelian) group in which there is one basic coupling constant and both weak and e.m. universality are automatic (note that, in contrast to QED, couplings can not be changed arbitrarily in gauge theories because the commutation relations are non-linear). Surely any "final" model must exploit this.

(3) In some models certain mass differences are necessarily finite and, in principle, calculable; this holds out the exciting prospect of understanding, for example, $m_p - m_n$ (it is even possible to make models in which $m_e \sim \alpha m_\mu$!).[53]

(4) These models may provide ways of understanding other phenomena such as the $\Delta I = \frac{1}{2}$[54] rule, CP-violation,[55] and the origin of the Cabibbo angle.[56]

A more detailed discussion of some of these topics and of other problems and hopes may be found in a paper by me to be published in the Proceedings of the 1973 Bonn Conference on Electron and Photon Interactions at High Energies.

PART II

1. EXPERIMENTAL CONSEQUENCES; GENERALITIES

Let us summarize the chain of argument in Part I which leads to the five results whose experimental consequences we will consider here in Part II.

(A) Weak interaction phenomenology → Vector currents → *vector bosons* (*1*) (with $g \sim e$, $M_W \sim 50$ GeV).

(B)

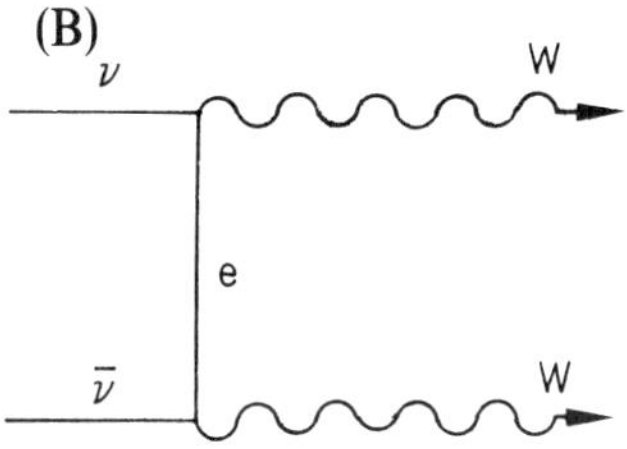

—amplitude grows too rapidly at high energy (violates unitarity and → non-renormalizable divergences).

In an acceptable perturbation theory there must be cancellations from other contributions of the same order, namely:

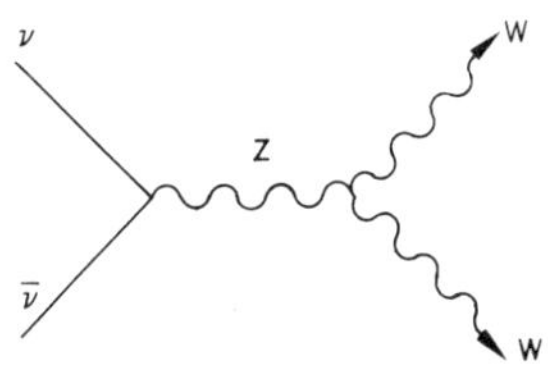

and/or

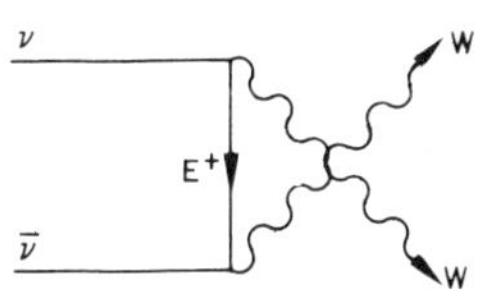

i.e. *neutral current* (2) and/or *heavy leptons* (3) [ν_e, e^-, E^+ (ν_μ, μ^-, M^+) generally have the same lepton number]. These cure the worst problems.

(C) Residual difficulties must be cancelled with a *scalar particle* (4)

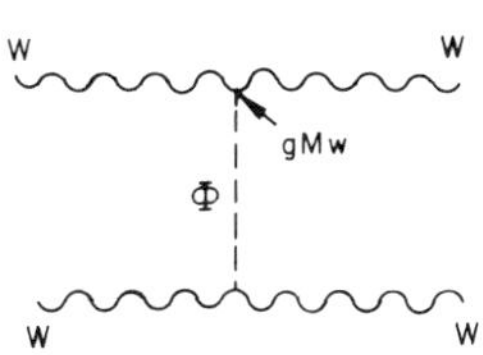

,

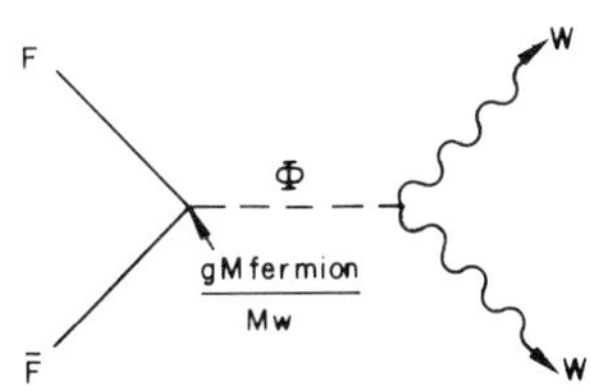

(D) 3 quarks + Cabibbo current seems to be impossible:

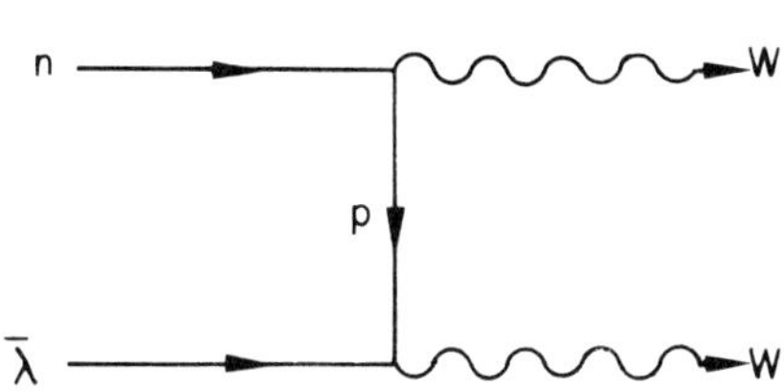

—cannot be cancelled by *s*-channel ($\Delta S = 1$) neutral current

∴ need

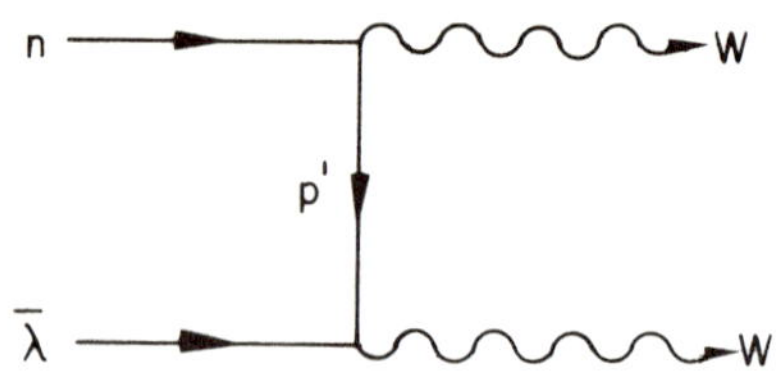

→ > 3 quarks

∴ need "*charmed*" *particles and/or other changes of Cabibbo theory* (5) in a large class of models.

2. CHARGED VECTOR BOSONS

2.1. *Decays and branching ratios*

The possible decays are

$$\begin{aligned} W &\to \mu\nu \\ &\to e\nu \\ &\to \textit{new leptons}\ (+\nu) \\ &\to \textit{hadrons.} \end{aligned}$$

Obviously

$$\Gamma > \Gamma(W \to \mu\nu) + \Gamma(W \to e\nu) = \frac{g^2 M_W}{3\pi} = \frac{G M_W^3}{(\sqrt{2})3\pi}$$

or

$$\tau < \frac{6 \times 10^{-19}}{(M_W \text{ in GeV})^3} \text{ secs.}$$

Since M_W is certainly $>1{\cdot}8$ GeV, τ is too short for W tracks to be visible.

There is obviously a close connection between hadronic W decay:

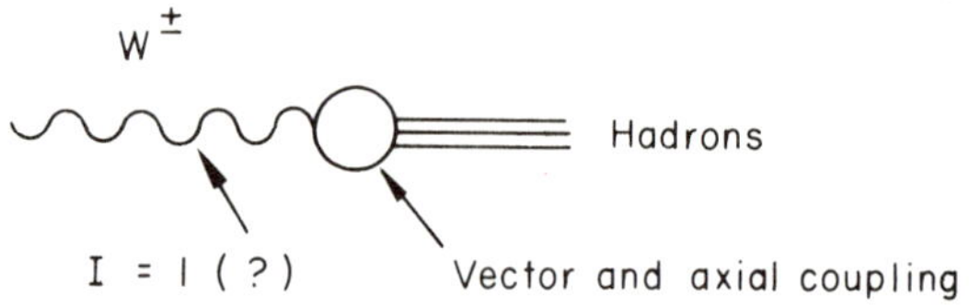

and "heavy photon" decay:

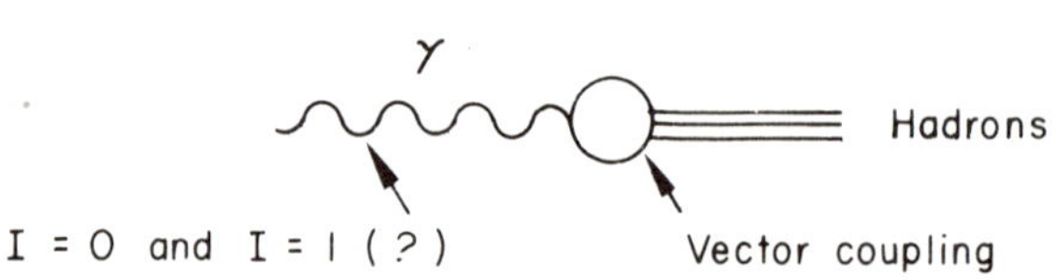

which can be observed in e^+e^- collisions (if it is not masked by Z, Φ, or other contributions). If we assume CVC for the vector $I = 1$ part and also that:

(i) the vector and axial couplings are the same at $s = M_W^2$ ("asymptotic chiral symmetry");

(ii) $$\left.\frac{\sigma(e^+e^- \to I = 1)}{\sigma(e^+e^- \to I = 0)}\right|_{s=M_W^2} = 3 \text{ ["asymptotic } SU(3)\text{"]}$$

we get

$$\frac{\Gamma(W \to \mu\nu) + \Gamma(W \to e\nu)}{\Gamma(W \to \text{Hadrons})} = \left.\frac{4\sigma(e^+e^- \to \mu^+\mu^-)}{3\,\sigma(e^+e^- \to \text{Hadrons})}\right|_{s=M_W^2}$$

(to check the coefficient, consider the free massless quark model in which the assumptions are all satisfied). The coefficient $\frac{4}{3}$ is likely to be wrong if some renormalizable model is correct, since there will be charm degrees of freedom beyond $SU(3)$, but the plausible result that the leptonic to hadronic branching ratios are of the same order for the electromagnetic and weak currents should be true in almost any "reasonable" model. Using the value of

$$R = \frac{\sigma(e^+e^- \to \text{Hadrons})}{\sigma(e^+e^- \to \mu^+\mu^-)} = 4{\cdot}7 \pm 1{\cdot}1$$

observed at $s = 16\,(\text{GeV})^2$ at CEA, the equation would give

$$\frac{\Gamma(W \to \mu\nu) + \Gamma(W \to e\nu)}{\Gamma(W \to \text{Hadrons})} \sim 22\%$$

If R has the CEA value at $s = M_W^2$ there will be sufficient probability of the W decaying into leptons to make searches based on these decays worth while. (However, comparison of the CEA data with lower-energy data from Frascati suggests that R may increase with s; in this case, which would be very surprising and interesting theoretically, the leptonic branching ratio might be very small if M_W is large.)

2.2. *Possible ways to discover W's*

(1) $e^+e^- \to W^+W^-$. The electromagnetic contribution to the cross-section can be calculated exactly[(20)] (given the ambiguous magnetic moment of the W, which we have in gauge theories) but, as we have seen, there may be other contributions which are of course model-dependent. The electromagnetic cross-section rises rapidly above threshold and there is a good signature from the leptonic decays. However, the energies of approved machines ($\leqslant 4{\cdot}5$ GeV/beam) are such that this process will not be able to compete with others in the near future.

(2)

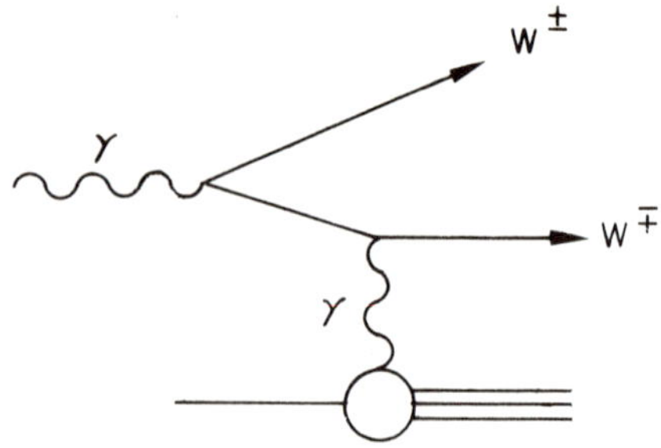

Again we can calculate the electromagnetic cross-section exactly,[57] but there can also be Z and Φ exchange contributions in renormalizable models (although they are likely to be small unless the masses are very small). The electromagnetic cross-section decreases rapidly with M_W and I guess that it is unlikely that W's with mass $\gtrsim 5$ GeV will be found in this way at NAL or CERN II.

(3)

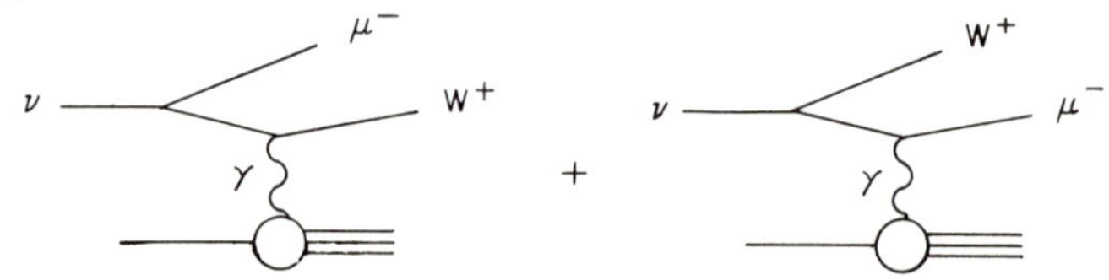

Again there can be (hopefully small) Φ and Z contributions in the same order. The photon exchange cross-section can be calculated exactly [see R. W. Brown *et al.*[58] and references therein]. The signature for leptonic decays is a μ^+ or e^+ with transverse momentum $p_T \sim M_W/2$, as well as a forward-going μ^-; such events should be easy to identify. When the W decays to hadrons, the event looks like ordinary inelastic scattering $\nu + A \to \mu^- + \ldots$ superficially. However, in contrast to genuine inelastic scattering, these events will be heavily concentrated in one corner of phase space, where they will appear with a characteristic (threshold) energy dependence which will make them easy to identify.[59] It is estimated that it should be possible to see W's of mass up to ~ 15 GeV in forthcoming ν experiments at NAL. The 1967 CERN HLBC experiment gives $M_W > 1{\cdot}8$ GeV (90% CL) independent of branching ratio;[60] a very early result from NAL[61] gives $M_W > 4{\cdot}2$ GeV (90% CL) if the branching ratio to $\mu\nu$ is $\geqslant 20\%$.

(4) $pp \to W + \ldots$. As in the discussion of W-branching ratios above we can estimate

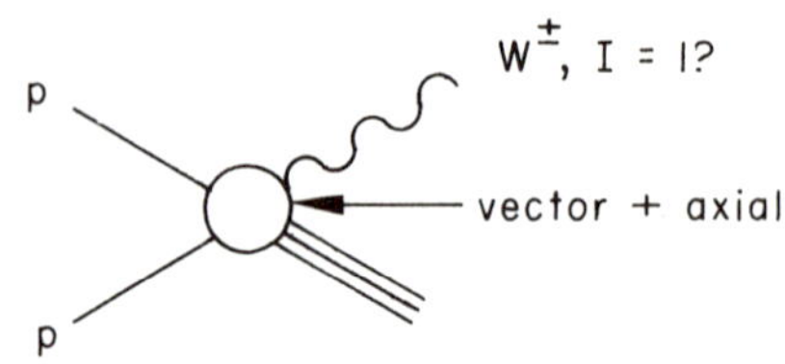

from

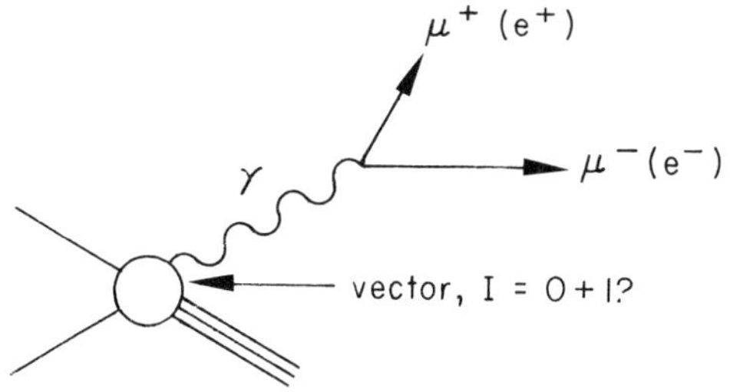

(if γ exchange is not masked by Z, Φ or other contributions), in a relatively model-independent way by making simple assumptions about the relative strengths of the vector and axial and the $I = 0$ and $I = 1$ contributions. Noting that there can be no V–A interference term in the total cross-section and averaging over $W^\pm$ and p and n targets (to eliminate interference terms between different isospins), CVC alone gives the inequality[62]

$$\begin{aligned}\sigma_W &\equiv \tfrac{1}{4}[\sigma_{W^+}(p) + \sigma_{W^-}(p) + \sigma_{W^+}(n) + \sigma_{W^-}(n)]\\ &= \sigma_W^{\text{vector}} + \sigma_W^{\text{axial}}\\ &\geqslant \sigma_W^{\text{axial}} = \frac{3GM_W^3}{(\sqrt{2}8)\alpha^2}\left(\frac{d\sigma_{\mu^+\mu^-}(p)}{dm_{\mu^+\mu^-}} + \frac{d\sigma_{\mu^+\mu^-}(n)}{dm_{\mu^+\mu^-}}\right)_{\text{isovector}}\end{aligned}$$

In most models $\sigma^{\text{vector}} \sim \sigma^{\text{axial}}$, but in photo and electroproduction $\sigma^{I=0} \ll \sigma^{I=1}$. It is therefore probably reasonable to assume that

$$\sigma_W \gtrsim \frac{3GM_W^3}{(\sqrt{2}8)\alpha^2}\frac{d\sigma_{\mu^+\mu^-}(p+n)}{dm_{\mu^+\mu^-}}$$

At present, the only data on $d\sigma_{\mu^+\mu^-}$ comes from an experiment performed at $s = 60\ \text{GeV}^2$ at BNL.[63] This can be used[62] to conclude that $M_W > 4{\cdot}5$ GeV if $B = \Gamma(W \to \mu\nu)/\Gamma(W \to \text{all}) > \frac{1}{4}$ on the basis of the limit $\sigma(pp \to W + \ldots)B < 6 \times 10^{-36}\ \text{cm}^2$ from another BNL experiment.[64] In order to interpret experiments at higher energies, however, a model is needed. Several are available but they give very different results. We consider, as an example, models based on the Drell–Yan mechanism[65] which predicts the appealing scaling law

$$\frac{d\sigma_{\mu^+\mu^-}}{dQ^2} = \frac{1}{Q^4}f(s/Q^2).$$

Unfortunately this cannot be confronted with the limit on $d\sigma(pp\to e^+e^- + \ldots)$ available from an ISR experiment[66] at $s = 2800\ \text{GeV}^2$ since the ISR and BNL experiments were sensitive to different ranges of Q^2/s.[67] Furthermore, in the absence of 4π detection, a model for the angular distribution is needed in order to interpret the experiments; this turns out to be model-dependent even in the framework of the Drell–Yan mechanism.[18] Hence, despite its widespread use, there is at present no evidence for the scaling law and, even

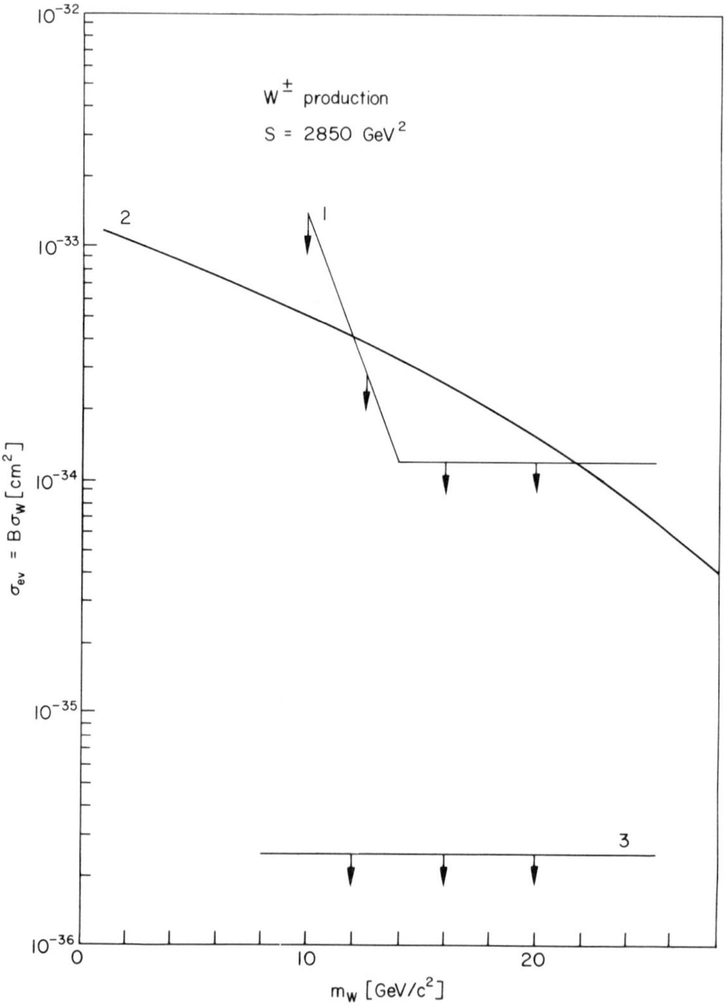

FIG. 1. *Curve* 1. Upper limit on $\sigma_{e\nu} = B\sigma_W$, where $B = \Gamma(W \to e\nu)/\Gamma(W \to \text{all})$, from the CERN-Columbia-Rockefeller experiment at the ISR. (There is some model dependence in calculating $B\sigma$ since the experiment did not have 4π detection efficiency.) *Curve* 2. Lower limit on $B\sigma_W$ assuming $B = \frac{1}{4}$ and using the lower bound on σ_W in the text with $d\sigma_{\mu\bar{\mu}}$ obtained by scaling the 30 GeV data according to the Drell-Yan scaling law. *Curve* 3. Upper limit which could be obtained from a proposed experiment. (This figure is taken from the CERN–Columbia–Rockefeller proposal for a future experiment[66].)

if there were, it should be used with caution in limited regions of phase space. *If* the scaling law is correct, the Brookhaven data can be scaled to give the W production cross-section at $\sqrt{s} = 55$ GeV depicted in Fig. 1; the experimental upper limit, which depends on a model for the angular distribution, would then exclude W's with mass between 12 and 21 GeV.

We conclude that *if* the Drell–Yan scaling law is correct, pp collisions are an excellent way to produce quite massive W's (as well as Z's, etc.). However, it may be that the scaling law is completely wrong and grossly overestimates the cross-section. Tests of the scaling law are therefore very important for this reason [and also because it provides a crucial test of the parton model[67]].

(5) Scaling violations. Consider the process

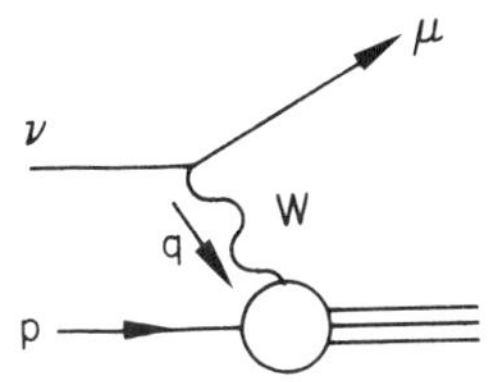

The cross-section depends on three structure functions F_i', which may be written

$$F_i' = \frac{F_i(x \equiv 2q\cdot p/(-q^2), q^2)}{(1 - q^2/M_W^2)^2}$$

which can be separated experimentally. It is logically impossible to show whether any q^2 variation of F_i' is due to F_i or to the propagator. However, analogous electron scattering experiments suggest that the F_i are approximately independent of q^2 or "scale" for large q^2 (there is indirect evidence for this in neutrino experiments); if, therefore, a plot of F_i' against q^2 at fixed x revealed a dipole distribution (superposed on some other weak q^2 dependence perhaps) it would be strong circumstantial evidence for the existence of a W. This interpretation would be very strongly supported if the ratio of the weak and electromagnetic structure functions $F^e_{2,1}$ and $F'^{\nu}_{2,1}$ had a dipole q^2 dependence (although even this would not prove that a W exists since the behaviour of the weak and the electromagnetic structure functions are *a priori* completely independent). A W propagator would of course affect all measurable quantities, some of which are more easily determined than the structure functions—for example, if scaling is valid, the cross-section would behave like s at large s for $M_W = \infty$ (or $\gg s$) with $\sigma^{\nu} \neq \sigma^{\bar{\nu}}$ in general, but like log s with $\sigma^{\nu} = \sigma^{\bar{\nu}}$ for $s > M_W^2$; however, such behaviour might well be due to scaling violation and not to a W.

Despite these cautious remarks, it is exciting to note that measurements of the double differential cross-section in possible future counter experiments at

NAL or CERN II might be able to demonstrate (circumstantially) the existence of W's with masses even up to the order of 70 GeV.

3. NEUTRAL CURRENTS

3.1. *Production and detection of Z's*

As in the case of W branching ratios, a plausible guess is that

$$\frac{\Gamma(Z \to \text{leptons})}{\Gamma(Z \to \text{hadrons})} \sim 1.$$

Perhaps the most auspicious reaction for producing high mass Z's is $pp \to Z + \ldots$. Like $d\sigma(pp \to W + \ldots)$, this can be estimated from $d\sigma(pp \to \gamma(\to \mu\bar{\mu}) + \ldots)$, although the connection is obviously more model dependent. As discussed above, it will not be clear whether pp collisions are a good source of W's, Z's, etc., until there are better limits on $d\sigma(pp \to \mu\bar{\mu}(e\bar{e}) + \ldots)$ at large s and Q^2. However, *if* the Drell–Yan mechanism and the quark parton model give the right order of magnitude this reaction will be a fruitful place to search for Z's; it might even be possible to see Z's with mass up to $\sim$37 GeV in future ISR experiments.[69] Jaffe and Primack[70] have shown that existing data already require that the Z needed in one model must have $M_W > 13$ GeV, if a Drell–Yan type model is valid.

Another possible reaction is $e^+e^- \to Z\gamma$. Z's might also be produced in high-energy, high-luminosity ep storage rings in the reaction:

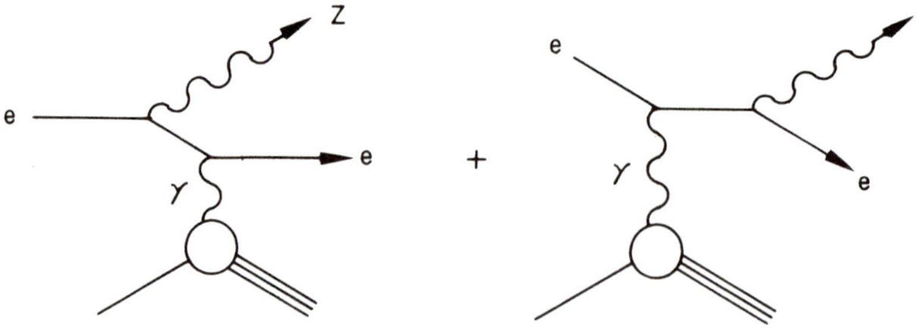

3.2. *Leptonic processes involving neutrinos*

If there are neutral currents they will obviously permit processes in lowest order which are forbidden in the conventional Feynman–Gell-Mann (FG) scheme, e.g.

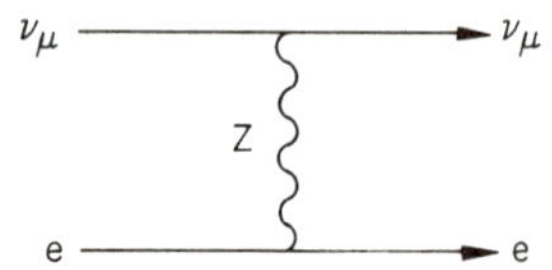

and change the cross-section of "allowed" processes, e.g.

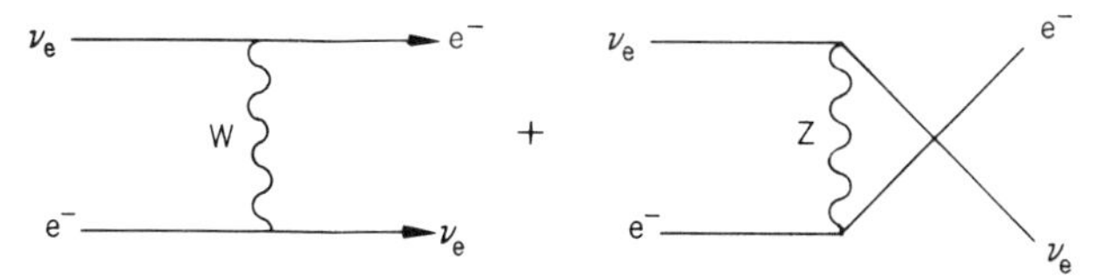

The cross-sections obtained in the Salam–Weinberg model are displayed in Fig. 2 as a function of the angle θ_W (see Part I, Section 11).

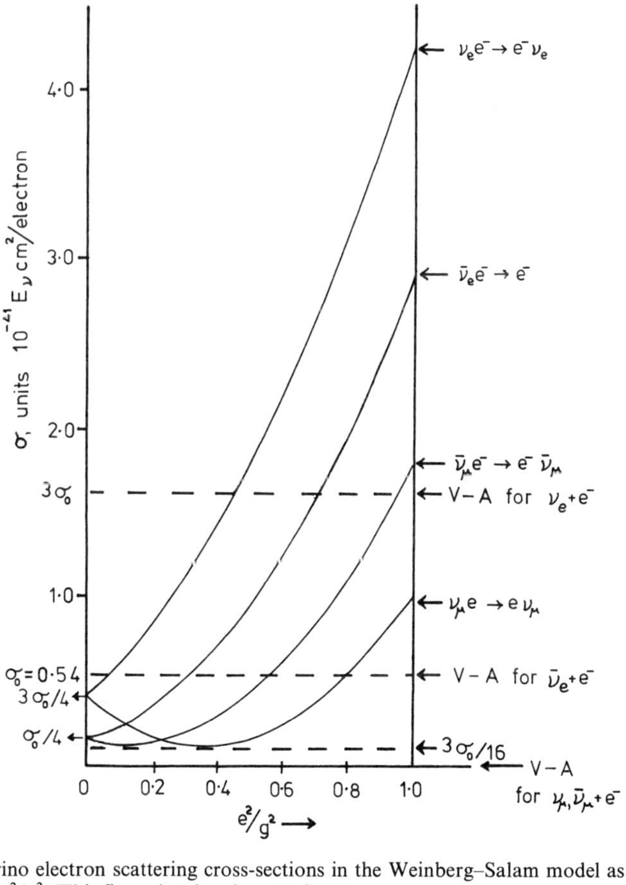

FIG. 2. Neutrino electron scattering cross-sections in the Weinberg–Salam model as a function of $\sin^2\theta_W = e^2/g^2$. (This figure is taken from Ref. 76.)

It may be possible to study the leptonic weak processes of "neutrino dissociation in the Coulomb field", e.g.

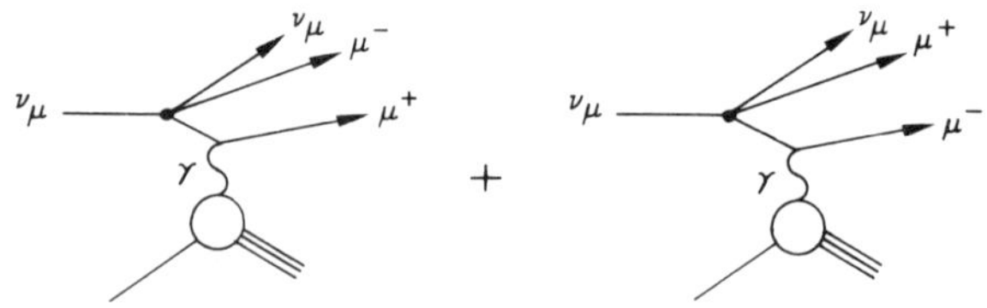

Cross-sections and angular distributions have been calculated[71] but, although seemingly reasonable numbers of events can be obtained, it may be very hard to distinguish them from the background.[72] Note that this experiment is complementary to the experiments discussed immediately below since it may be the only one in which it is possible to investigate "diagonal processes" such as $\nu_\mu \mu^- \to \mu^- \nu_\mu$, $\bar{\nu}_\mu \mu^+ \to \mu^+ \bar{\nu}_\mu$, $\nu_e e^- \to e^- \nu_e$, or $\bar{\nu}_e e^+ \to e^+ \nu_e$, which have never been observed although the cross-sections were predicted by Feynman and Gell-Mann 17 years ago (the last two are especially hard to investigate because of the relatively feeble ν_e and $\bar{\nu}_e$ fluxes); the importance of this investigation is independent of and equal to that of looking for purely neutral current processes.

Another possibility is to use atomic electrons as a target. This has been done by Gurr, Reines and Sobel[73] using an atomic reactor as a source of (low energy) $\bar{\nu}_e$'s. They give the result

$$\frac{\sigma(\bar{\nu}_e e \to \bar{\nu}_e e)}{\sigma_{\mathrm{FG}}} = 1{\cdot}1 \pm 1{\cdot}2$$

The limit 2·3 for the ratio corresponds to a partial cross-section limit of $6 \times 10^{-47}\,\mathrm{cm}^2$ in the experimental conditions! In view of the extreme difficulty of the experiment (the signal to noise ratio was of order 1) it is amazing that they were able to obtain any result at all. In the Weinberg–Salam model their result gives $\sin^2\theta < 0{\cdot}33$, allowing one standard deviation, or $\sin^2\theta < 0{\cdot}40$ at the 90% confidence level.[74]

In the CERN heavy liquid bubble chamber experiments there have also been searches for ν scattering on atomic electrons and one possible $\bar{\nu}_\mu + e^- \to \bar{\nu}_\mu + e^-$ event has been observed. The analysis of this event has very recently been reported.[75] In view of the potential significance of this event, it would be unwise to try to summarize the background analysis here; the reader is urged to study Ref. 75 himself. We eschew comment and quote the following table of numbers of single e^- events satisfying the experimental cuts $E_e > 300$ MeV and $\theta_e < 5°$:

Flux neutrinos/m^2	Weinberg predictions		Background	Observed
	Minimum	Maximum		
ν $1{\cdot}8 \times 10^{15}$	0·6	6·0	0·3 $\pm$ 0·2	0
$\bar{\nu}$ $1{\cdot}2 \times 10^{15}$	0·4	8·0	0·03 $\pm$ 0·02	1

(see Fig. 2 for a plot of the cross-sections as a function of the Weinberg angle). The authors state that if the event is interpreted in terms of the Weinberg theory, the 90% confidence limits on the Weinberg parameter are $0{\cdot}1 < \sin^2\theta_W < 0{\cdot}6$. The end of their paper reads "We conclude that the probability that the single event observed in the $\bar{\nu}$ film is due to non-neutral current background is less than 3%."

This event was obtained from 360,000 $\bar{\nu}$ pictures. "In the near future a further 10^6 $\bar{\nu}$ pictures will be taken at the PS, if possible with the booster, giving at least three times the intensity. Such a run will yield 8 events on the basis of the one event found so far or a minimum of 3 events if the cross-section is minimal".(76) This will allow us to draw much more definite conclusions.

3.3. *Semileptonic processes involving neutrinos*

The following experimental limits are given in the literature

$$\frac{\nu_\mu p \to \nu_\mu p}{\nu_\mu n \to \mu^- p} < 0{\cdot}21 \text{ (90\% C.L. CERN 1967; Ref. 77)}$$

$$\frac{\nu_\mu p \to \nu_\mu \Delta^+}{\nu_\mu p \to \mu^- \Delta^{++}} < 0{\cdot}31 \text{ (90\% C.L. ANL[p}\pi^\circ\text{] 1973; Ref. 78)}$$

$$\text{(90\% CERN[n}\pi^+\text{] 1967; Ref. 77)}$$

$$\frac{(\nu_\mu n \to \nu_\mu n\pi^0) + (\nu_\mu p \to \nu_\mu p\pi^0)}{2(\nu_\mu n \to \mu p\pi^0)} < 0{\cdot}14 \text{ (90\% C.L. Columbia; Ref. 79)}$$

$$< 0{\cdot}21 \text{ (90\% C.L. CERN, Ref. 76)}$$

(These two limits ignore charge exchange and all nuclear effects)

A few days before these lectures were given, the CERN Gargamelle Group announced the discovery of an excess of muonless events in their neutrino experiments which cannot be accounted for on the basis of conventional hypotheses (F. J. Hassert *et al. Phys. Letts.* **46B**, 138 (1973).) If this excess is attributed to neutral currents it gives

$$\left.\frac{\sigma(\nu_\mu A \to \nu_\mu + \ldots)}{\sigma(\nu_\mu A \to \mu^- + \ldots)}\right|_{E(\text{vis}) > 1 \text{ GeV}} = 0{\cdot}23 \pm 0{\cdot}03$$

$$\left.\frac{\sigma(\bar{\nu}_\mu A \to \bar{\nu}_\mu + \ldots)}{\sigma(\bar{\nu}_\mu A \to \mu^+ + \ldots)}\right|_{E(\text{vis}) > 1 \text{ GeV}} = 0{\cdot}46 \pm 0{\cdot}09,$$

where the errors are purely statistical. Subsequently the Harvard, Pennsylvania, Wisconsin group working at NAL have announced a non-zero result which is in good agreement with the CERN data. The experimental situation at the end of August 1973 is reviewed by G. Myatt in the Proceedings of the

Bonn International Symposium on Electron and Photon Interactions at High Energies. In the light of this exciting and important discovery, a considerable extension of the discussion of neutral currents in these lecture notes would be desirable. However, I have not extended the discussion in order to avoid redundancy since there will doubtless be much further study of the implications of neutral currents in the coming months, during which further experimental information may also be forthcoming.

These results are very hard to obtain because of the background of events initiated by neutrons (or K_L^0) which come around the shielding or from neutrino reactions in the shielding, coils, magnets, etc., Mann and Primakoff[80] have suggested a way of eliminating the neutron background in a hydrogen bubble chamber experiment. They note that because of charge symmetry

$$\sigma(np \to \pi^+ + \ldots) = \sigma(pn \to \pi^- + \ldots)$$

up to corrections of order α. Hence any substantial difference in the number of π^+'s and π^-'s cannot be due to neutron background. Taking, for example, events with the final states

$$\mu^- p\pi^+ + \text{neutrals}$$

$$pp\pi^- + \text{neutrals}$$

$$\pi^+ + \text{neutrals}$$

the neutral current cross-section is given in terms of the number of events by

$$\frac{\sigma(\nu p \to \nu n\pi^+ + \text{neutrals})}{\sigma(\nu p \to \mu^- p\pi^+ + \text{neutrals})} = \frac{N(\pi^+) - N(pp\pi^-)}{N(\mu^- p\pi^+)} + O(\alpha)$$

if neutrons are the only background (in order that the correction is really of order α or less it is necessary that the neutron flux $\phi(n)$ is less than or of order $\phi(\nu)\sigma(\nu)/\sigma(n)$, which is presumably true for energetic neutrons in equilibrium with the neutrino beam which produces them in the environment of the experiment).

We now consider some of *the theoretical bounds on semileptonic neutral current cross-sections* which can be obtained in models of the Weinberg–Salam type[81–84] (by this we mean models based on $SU(2) \times U(1)$ in which there are no new leptons). In such models the hadron–vector boson coupling is

$$\mathscr{L}_{\text{int}} = g\vec{J}_\mu \cdot \vec{W}_\mu + \tilde{g}J_\mu^s B_\mu$$

in the notation of Section 11 where $\vec{J}$ is the "weak isospin" current of the hadrons and J^s a "weak isoscalar" hadronic current. The decomposition of B and W_3 in terms of Z and γ gives:

$$eJ_\mu^{\text{e.m.}} = gJ_\mu^3 \sin\theta_W + \tilde{g}J_\mu^s \cos\theta_W$$

from which we obtain the effective Lagrangian for small $q^2(\ll M_W^2, M_0^2)$ processes:

$$\mathscr{L}_{\text{eff.}} = \frac{G}{\sqrt{2}}[\bar{\mu}\gamma_\lambda(1-\gamma_5)\nu_\mu J_\lambda^+ + \text{h.c.} + \bar{\nu}_\mu\gamma_\lambda(1-\gamma_5)\nu_\mu(J_\lambda^3 - 2\sin^2\theta_W J_\lambda^{\text{e.m.}})] + (\mu \to e).$$

Without specifying further the model for hadrons we cannot say anything about the matrix elements of J^3 in terms of those of J^+ and $J^{\text{e.m.}}$ and *there are no bounds on neutral current processes.* If, however, we are willing to make some identification between "weak" and strong isospin (which are *not* the same in general) the matrix elements may be connected by Clebsch–Gordan coefficients and progress is possible. In the 4 "quark" model discussed in Section 12 of Part I, e.g. the $\Delta S = 0$ charm-conserving ($\Delta C = 0$) parts of $J_{\text{weak}}^\pm$ are components of a strong isovector $\vec{J}_{\text{strong}}$ + strong isoscalar. Furthermore the electromagnetic current differs from the usual Gell-Mann–Nishijima form only by the addition of an extra isoscalar piece. The effective Lagrangian is therefore:

$$\mathscr{L}_{\text{eff.}}^{\substack{\Delta C=0\\ \Delta S=0}} = \frac{G\cos\theta_c}{\sqrt{2}}[\bar{\mu}\gamma_\lambda(1-\gamma_5)\nu_\mu\vec{J}_\lambda^+ + \text{h.c.} + \bar{\nu}_\mu\gamma_\lambda(1-\gamma_5) \times \nu_\mu(\vec{A}_\lambda^3 + (1-2\sin^2\theta_W)\vec{V}_\lambda^3 + \vec{J}_\lambda^s)] + (\mu \to e),$$

where $\vec{J}_\mu$ is the sum of the (vector and axial vector) strong isovectors $\vec{V}_\mu$ and $\vec{A}_\mu$ and J_s is a strong isoscalar.

The minimal assumptions about the hadronic current which are necessary to derive this effective Lagrangian are reasonable and we would probably try to incorporate them in any model. It must be stressed, however, that all the bounds in the literature use this $\mathscr{L}_{\text{eff.}}$ (or make stronger assumptions) and this could be wrong even if the Weinberg–Salam model of leptons is correct. We indicate how some of these bounds are obtained. Consider[82]

$$R = \frac{d\sigma(\nu p \to \nu x_1) + d\sigma(\nu n \to \nu x_2)}{d\sigma(\nu p \to \mu^- x_3) + d\sigma(\nu n \to \mu^- x_4)},$$

where x is some set of final states (e.g. $N\pi$ or $N\pi + N\pi\pi$ or all states perhaps) summed over all possible charge states. The neutron-proton average (or alternately the use of an $I = 0$ target) removes $I = 0$ and $I = 1$ interference terms. Using the fact that the $I = 0$ contribution is positive, $\mathscr{L}_{\text{eff}}$ and one C-G coefficient give

$$R \geqslant \frac{|A|^2 + x^2|V|^2 + xI}{2(|A|^2 + |V|^2 + I)}$$

where

$$x = 1 - 2\sin^2\theta w$$

and the muon mass has been neglected. The interference term I satisfies the Schwarz inequality:

$$2|A||V| \geqslant I$$

which, together with the fact that

$$|V|^2 \leqslant \frac{G^2}{\pi}\frac{Q^4}{4\pi\alpha}\left(\frac{\mathrm{d}\sigma(ep \to ex_1) + \mathrm{d}\sigma(en \to ex_2)}{2}\right) \equiv |V_{\text{e.m.}}|^2$$

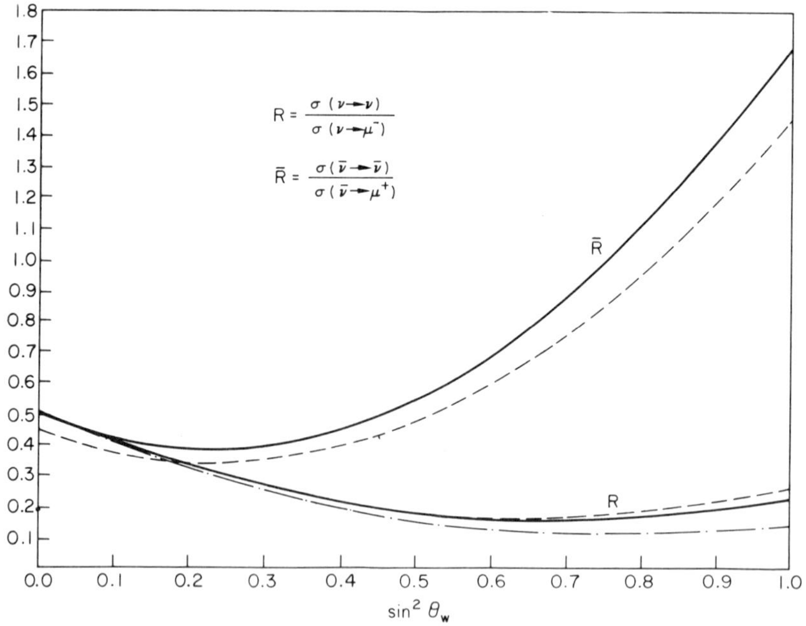

FIG. 3. Paschos–Wolfenstein lower bounds for inclusive neutral current cross-sections and model results obtained by Sehgal[85] in a quark parton model. –·–·–·–, lower bound for R (assuming scaling); ————, model in which nucleon = 3 quarks $[\sigma(\bar{\nu} \to \bar{\mu})/\sigma(\nu \to \mu) = \frac{1}{3}]$; – – – – – –, model in which nucleon = 3 quarks + sea $[\sigma(\bar{\nu} \to \bar{\mu})/\sigma(\nu \to \mu) = 0{\cdot}38]$
(The approximate lower bound for $\bar{R}$ is very close indeed to the model calculation.)

gives[81]

$$R \geqslant \frac{1}{2}\left[1 - 2\sin^2\theta_W \sqrt{\left(\frac{|V_{\text{e.m.}}|^2}{d\sigma_-}\right)}\right]^2$$

where

$$d\sigma_- = \frac{d\sigma(\nu p \to \mu^- x^3) + d\sigma(\nu n \to \mu^- x_4)}{2}$$

Somewhat more stringent bounds may be obtained using the detailed positivity properties of the structure functions. Thus, assuming scaling and using available electromagnetic and charged current data, Paschos and Wolfenstein obtained the bounds:[82]

$$R = \frac{\sigma(\nu p \to \nu + \ldots) + \sigma(\nu n \to \nu + \ldots)}{\sigma(\nu p \to \mu^- + \ldots) + \sigma(\nu n \to \mu^- + \ldots)} \geqslant \frac{1}{2}\left[\frac{2}{3} + \frac{x}{3} - (1 - x^2)\,0{\cdot}3\right]$$

$$\gtrsim 0{\cdot}23 \text{ with } \sin^2\theta_W < 0{\cdot}33$$

$$\bar{R} = \frac{\sigma(\bar{\nu} p \to \bar{\nu} + \ldots) + \sigma(\bar{\nu} n \to \bar{\nu} + \ldots)}{\sigma(\bar{\nu} p \to \mu^+ + \ldots) + \sigma(\bar{\nu} n \to \mu^+ + \ldots)} \gtrsim \frac{1}{2}(1 - x + x^2)$$

$$\gtrsim 0{\cdot}39 \text{ with } \sin^2\theta_W < 0{\cdot}33$$

These bounds for the total cross-sections are plotted in Fig. 3; they cannot be compared directly with the experimental results above, in which there is an energy cut. After the bounds have been modified for the cut, as has been done by G. Myatt (to be published in the Proceedings of the 1973 Bonn International Symposium on Electron and Photon Interactions at High Energies), comparison with the neutrino data gives

$$\sin^2\theta_W > 0{\cdot}27\ (90\%\ \text{C.L.})$$

and the antineutrino data gives

$$\sin^2\theta_W < 0{\cdot}49\ (90\%\ \text{C.L.}).$$

There is not a great deal of room for manoeuvre but, if these models are correct, we might expect the values of R and $\bar{R}$ to be close to the lower bounds. They would be saturated if the vector and axial amplitudes had equal strength and interfered maximally and if also the contribution of the isoscalar part of the neutral current were negligible. The first condition is probably almost satisfied for the charged current according to the CERN neutrino data, while the second is satisfied for the electromagnetic current according to both photoproduction data and a comparison of deep inelastic ν and e data. We might therefore expect these conditions to hold for the neutral current also.

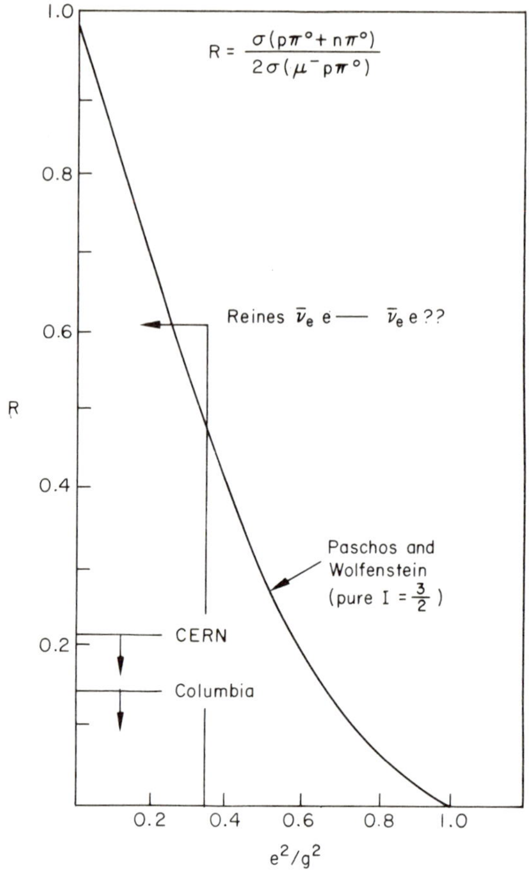

FIG. 4. The Paschos–Wolfenstein bound on

$$\frac{\sigma(\nu n \to \nu n \pi^0) + \sigma(\nu p \to \nu p \pi^0)}{2\sigma(\nu n \to \mu^- p \pi^0)}$$

in the Weinberg–Salam model, assuming $I = \frac{3}{2}$ dominance, compared to the experimental limits on R (neglecting nuclear charge-exchange effects) and the limit on $\sin^2 \theta_W$ from the experiment of Sobel, Gurr and Reines. (This figure is taken from Ref. 76.) The effect of weakening the assumptions is discussed in the text.

This expectation is borne out in the only available model which can easily account for the deep inelastic *ep*, *en* and νA data which is the (Gell-Mann-Zweig) quark parton model (or equivalently the quark light-cone algebra). As discussed in Section II.12, we must add a charmed quark p' in order to construct a renormalizable model (and also introduce colour triplets but they do not affect the discussion here). In first approximation the total ν

and $\bar{\nu}$ cross-sections can be accounted for by imagining that nucleons are made of three quarks from which the current scatters incoherently (the data imply that although additional $p\bar{p}$, $n\bar{n}$, $\lambda\bar{\lambda}$, or $\bar{p}'p'$ pairs may be important at small $x = Q^2/2\nu$ they make a small contribution to the total cross-section). An elementary calculation then gives the result that for energies below the charmed threshold

$$R \simeq \tfrac{1}{2} - \sin^2\theta_W + \tfrac{20}{27}\sin^4\theta_W$$
$$\bar{R} \simeq \tfrac{1}{2} - \sin^2\theta_W + \tfrac{20}{9}\sin^4\theta_W,$$

in the approximation $\theta_{\text{Cabibbo}} = 0$. This result was obtained by Sehgal[85] who also considered the corrections due to $Q\bar{Q}$ pairs; the results are plotted in Fig. 4. Myatt (*loc. cit.*) has plotted the predictions for the CERN experiment taking into account the energy cut. The neutrino data gives

$$0{\cdot}31 < \sin^2\theta_W < 0{\cdot}44,$$

according to this model, while the antineutrino data gives

$$\sin^2\theta_W < 0{\cdot}32.$$

It is easy to understand why $\bar{R} > R$ except for $\theta_W \simeq 0$ in this model and why this may be true more generally. According to the data $\sigma(\bar{\nu}A \to \mu^+ + \ldots)/\sigma(\nu A \to \mu^- + \ldots)$ is close to the lower limit of $\frac{1}{3}$ allowed by scaling (and the neglect of $\Delta S = 1$ and $\Delta C = 1$ processes or any other sources of violation of charge symmetry); this limit can only be reached when the vector and axial amplitudes have equal magnitude and interfere maximally. In our $\mathscr{L}_{\text{eff.}}$ the intrinsic strengths of the isovector parts of the V and A terms are different, except at $\theta_W = 0$, and using parton/light cone ideas we therefore expect

$$\frac{\sigma(\nu A \to \bar{\nu} + \ldots)}{\sigma(\nu A \to \nu + \ldots)} > \tfrac{1}{3}$$

for $\theta_W \neq 0$.

Turning to other processes, we shall only quote results which can already be confronted with experiment (more sophisticated bounds could be obtained for many processes if more charged-current and electromagnetic data were available[82–84]). Making some approximations, Weinberg[86] obtained

$$\frac{\sigma(\nu p \to \nu p)}{\sigma(\nu n \to \mu^- p)} \simeq \frac{1}{4}\frac{[(1.2)^2 + (1 - 4\sin^2\theta_W)^2]}{[(1.2)^2 + 1]}$$

(model dependent)

$$\frac{\sigma(\nu n \to \nu n)}{\sigma(\nu n \to \mu^- p)} \simeq \frac{1}{4}$$

for high energies *in a model in which there are no strange or charmed particles; this model cannot be correct in general*, although the result might represent a reasonable first guess for many models (e.g. in the 4 quark models it could be argued that there are few p' and λ quarks in the nucleon). With the more general form of $\mathscr{L}_{\text{eff.}}$ above the best we can do is

$$\frac{\sigma(\nu p \to \nu p) + \sigma(\nu n \to \nu n)}{2\sigma(\nu n \to \mu^- p)} \geqslant \frac{1}{4}\frac{[(1{\cdot}2)^2 + (1 - 2\sin^2\theta_W)^2]}{[(1{\cdot}2)^2 + 1]}$$

making the same approximations,[87] which have some validity for $E_\nu > 1\,\text{GeV}$.

Wolfenstein and Paschos,[32] using $\mathscr{L}_{\text{eff.}}$ above, obtained the bound:

$$\frac{\sigma(\nu p \to \nu\Delta^+)}{\sigma(\nu p \to \mu^-\Delta^{++})} \geqslant \tfrac{1}{3}[1 - 2\sin^2\theta_W Q]^2$$

where Q depends on other data and it is found that $Q^2 = 0{\cdot}20 \pm 0{\cdot}05$ for CERN energies. This bound is quite compatible with the experimental limit quoted above. However, assuming $I = \frac{3}{2}$ *dominance* it gives:

$$\frac{\sigma(\nu p \to \nu p\pi^0) + \sigma(\nu n \to \nu n\pi^0)}{2\sigma(\nu n \to \mu^- p\pi^0)} = \frac{3\sigma(\nu p \to \nu\Delta^+)}{\sigma(\nu p \to \mu^-\Delta^{++})} \geqslant 0{\cdot}50 \text{ for } \sin^2\theta_W < 0{\cdot}33$$

in apparent contradiction with experiment. The situation is displayed in Fig. 4. We make the following comments.

(1) B. W. Lee[88] has made a static model calculation which suggests that the actual value of the $I = \frac{3}{2}$ cross-section is not much above the Paschos–Wolfenstein lower bound (with his favoured choice of his one parameter, the model result is almost the same as the bound and with other apparently reasonable choices it can even be less!).

(2) $I = \frac{3}{2}$ dominance might be a bad approximation. With $\sin^2\theta_W < 0{\cdot}40$ and assuming that the $I = \frac{1}{2}$ and $I = \frac{3}{2}$ amplitudes are incoherent with $I = \frac{1}{2}$ contributing 30% to the cross-section, the bound is reduced to 0·19.

(3) In interpreting the experiments it was assumed that there were no charge-exchange effects in the nucleus; these may be important.[68] With an $I = 0$ target, bounds on inclusive processes can be obtained which do not depend on nuclear effects.[84] They can be applied to the Columbia and CERN experiments if it is a reasonable approximation to treat the targets (aluminium and freon, respectively) as consisting of $I = 0$ nuclei. The relevant bound depends sensitively on

$$V = \frac{\sigma(\nu A \to \mu^-\pi^+ + \ldots) + \sigma(\nu A \to \mu^-\pi^- + \ldots)}{2\sigma(\nu A \to \mu^-\pi^0 + \ldots)}$$

thus:

$$V \quad 5, \quad 4, \quad 3, \quad 2,$$

$$\frac{\sigma(\nu A \to \nu \pi^0 + \ldots)}{\sigma(\nu A \to \mu^- \pi^0 + \ldots)} \geqslant 0{\cdot}44, 0{\cdot}32, 0{\cdot}19, 0{\cdot}07,$$

for $\sin^2 \theta_W < 0{\cdot}4$. $r = 5$ corresponds to $I = \frac{3}{2}$ dominance. The Gargamelle experiment gives[89] $r \sim 2$.

Because of the uncertainties we cannot say that $\mathscr{L}_{\text{eff.}}$ is excluded.

3.4. *Neutral currents without neutrinos*

It is equally interesting to look for neutral current effects in processes not involving neutrinos (note that there are models in which there are Z's coupled to $\bar{e}e$ and $\bar{\mu}\mu$ but not to $\bar{\nu}\nu$). We consider three possibilities:

(1)

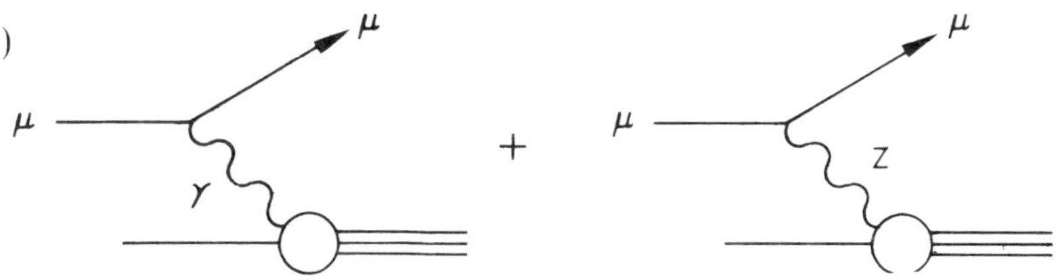

We expect that typically

$$\frac{\text{Int. term}}{\gamma \text{ exchange}} \sim \frac{g_Z^2}{e^2} \frac{q^2}{M_Z^2} \frac{1}{(1 - q^2/M_Z^2)}$$

and we would therefore guess that there might be effects of order

$$\frac{G}{4\pi\alpha} \frac{q^2}{1 - q^2/M_Z^2} \sim 10^{-4} \frac{q^2}{M_{\text{proton}}^2} \frac{1}{(1 - q^2/M_Z^2)}$$

We consider three ways in which a Z might manifest itself, referring to papers by Gatto and Preparata,[90] Love, Nanopoulos and Ross,[91] Derman,[92] and Petronzio[93] for detailed calculations:

(a) Apparent scaling violations; as in the case of the W it would be hard to prove the existence of a Z in this way.

(b) $\sigma_{\mu^+} \neq \sigma_{\mu^-}$, $\sigma_{e^+} \neq \sigma_{e^-}$ (even for spin averaged cross-sections). We guess that[94]

$$\frac{\dfrac{d^2\sigma_{\mu^+}}{d\omega dy} - \dfrac{d^2\sigma_{\mu^-}}{d\omega dy}}{\dfrac{d^2\sigma_{\mu^+}}{d\omega dy} + \dfrac{d^2\sigma_{\mu^-}}{d\omega dy}} \sim \frac{10^{-4} q^2}{M_p^2} \frac{1}{(1 - q^2/M_Z^2)} \frac{4y}{3 + \sqrt{\omega}}$$

(where $\omega = 2\nu/-q^2$, $y = 2\nu/(s - M^2)$. Detailed model calculations may be found in Refs 91–93. This asymmetry occurs because the $Z\mu\bar{\mu}$ (or $Ze\bar{e}$) vertex violates C as well as P (CPT and T being conserved). On its own, an asymmetry would not prove the existence of a neutral current since it could also be caused by two photon/one photon exchange interference (which would violate the Rosenbluth formula more violently however).

(c) Parity violation, which would be an incontrovertible signature for some non-electromagnetic phenomenon. With right- and left-handed polarized muons (electrons), or a target polarized parallel/antiparallel to the beam, we guess that the asymmetry will be

$$\frac{d\sigma_L - d\sigma_R}{d\sigma_L + d\sigma_R} \sim \frac{10^{-4} q^2}{M_p^2} \frac{1}{(1 - q^2/M_Z^2)}$$

Model calculations may again be found in Refs 91–93. With $|q^2| \to M_{\mu\bar{\mu}}^2$ similar estimates would also apply for the cross-section asymmetries with right/left-handed polarized photon beams in:

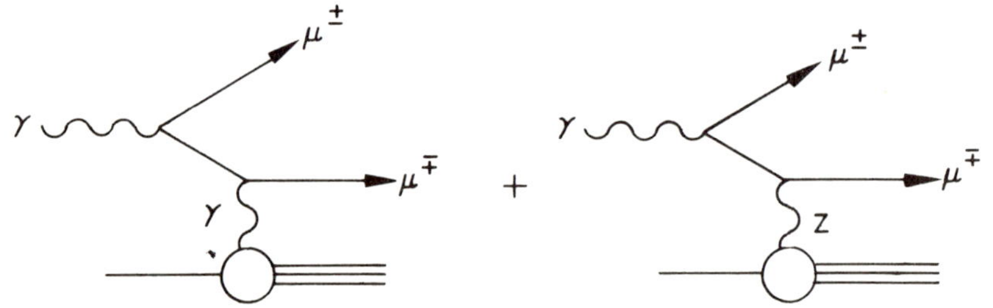

(2) It might be possible[95] to devise laser experiments which would be sensitive enough to see parity-violating effects in atoms of the order predicted by gauge theories.

(3) $e^+e^- \to \mu^+\mu^-$ (in principle similar effects could be seen in $e^+e^- \to \pi^+\pi^-$, $\Lambda\bar{\Lambda}$ etc. or more complicated hadronic channels; however, the cross-sections are probably too small for the effects to be visible, except perhaps in the inclusive processes, and we will not consider hadronic channels here[90]). Z exchange would give rise to three effects which might be observed:

(a) Change in the total rate.

(b) Longitudinal muon polarization.

(c) Asymmetry when $\theta_{\mu e} \to \theta_{\mu e} + \pi$ (caused by the fact that the $Z\mu\bar{\mu}$ and $Ze\bar{e}$ vertices violate C, but also by $2\gamma/1\gamma$ interference).

These effects have been considered by several authors.[90, 96–101] The value of effects (a) and (c) depends on a proper understanding of higher-order (radiative) corrections and beam polarization[99] (which can enhance the effects of interest). Bearing this in mind, we show in Fig. 5 results obtained for $s = (28\,\text{GeV})^2$ by Budney[100] in the Weinberg–Salam model considering only one photon and Z exchange (Budney makes a critical comparison between previous calculations). It seems that very high-energy, high-luminosity beams might well be able to detect neutral current effects. Since the effects are roughly proportional to $s = Q^2$, it is also clear from the figure (or from our order of magnitude guess, which is borne out by Fig. 5) that the prospects

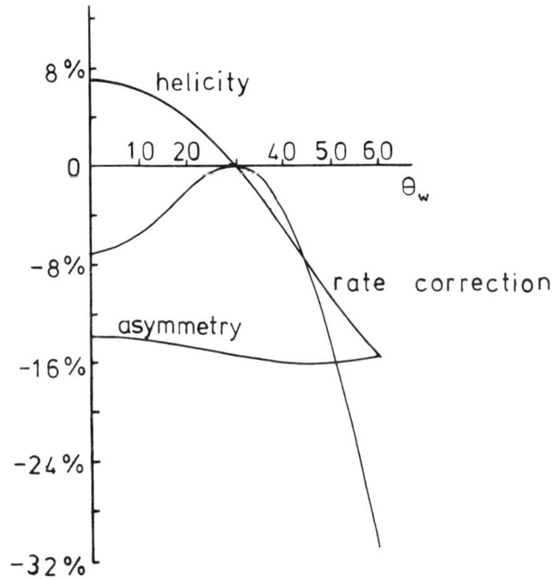

FIG. 5. Percentage change in the rate and amount of asymmetry and helicity in $e^+e^- \to \mu^+\mu^-$ due to Z exchange at $s = (28\ \mathrm{GeV})^2$, assuming the beams are polarized along the field of the bending magnets (taken from Ref. 100).

are dim for discovering the neutral currents predicted by this model with machines which exist or are under construction (although there may well be much larger effects unexpected by theorists).

4. HEAVY LEPTONS

We consider explicitly heavy muons (M) but the same remarks apply *mutatis mutandis* to heavy electrons (E) if muon and electron number are separately conserved. Some of the properties of the heavy leptons required in gauge theories are reviewed in Ref. 41, to which we refer for more details and references to the original literature. The experimental situation, as of one year ago, and more general ideas about possibly heavy leptons have been reviewed by Perl.[102]

4.1. *Charged leptons*

If an M^+ had mass

$$M_{M^+} > M_W$$

it would decay rapidly into $W^+\nu_\mu$ almost 100% of the time; we assume that this is not the case [note that in most models there are no theoretical limits on heavy lepton masses(103)]. The other possible decays are:

$$\begin{aligned} M^+ &\to \nu_\mu \mu^+ \nu_\mu \\ &\to \nu_\mu e^+ \nu_e \\ &\to \nu_\mu + \text{hadrons} \\ &\to M^0 \mu^+ \nu_\mu \\ &\to M^0 e^+ \nu_e \\ &\to M^0 + \text{hadrons} \end{aligned}$$

At least in simple models, $g_{M^+W\nu} = g_{\mu^-W\nu}$ (c/f Section 4 of Part I) in which case the $\nu\mu\nu$ and $\nu e\nu$ channels give a lower bound

$$\Gamma \geqslant 3\Gamma_\mu \left(\frac{M_{M^+}}{M_\mu}\right),$$

where the origin of the factor 3 is explained below, or

$$\tau \leqslant 0{\cdot}7\left(\frac{M_\mu}{M_{M^+}}\right)^5 \times 10^{-6} \text{ secs.}$$

Since experimentally(104) $M_{M^+} > 1$ GeV, it is clear that searches must be based on the decay modes and not on the identification of tracks. The calculation of branching ratios for the modes $M \to \nu_\mu + \ldots$ depends on the (virtual) W branching ratios in

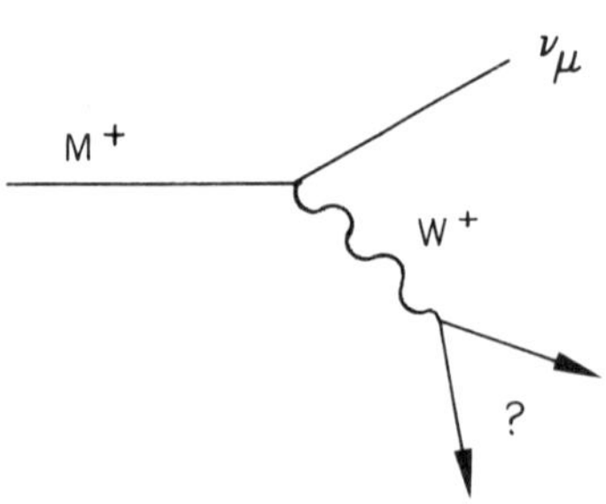

For the case $W \to$ leptons we obtain

$$\frac{\Gamma(M^+ \to \nu_\mu \mu^+ \nu_\mu)}{\Gamma(M^+ \to \nu_\mu e^+ \nu_e)} = 2,$$

for $M_{M^+} \ll M_W$ (but $M_{M^+} \gg M_\mu$). This factor of two occurs because there are two identical particles in the final state in one case (there are two diagrams which have the same amplitude, by a Fierz transformation, for $M_W \to \infty$; this gives a factor 2 in the amplitude and 4 in Γ but there is a factor $\frac{1}{2}$ in the phase space). Note that (neglecting M_μ) *this ratio would be one if M^+ had its own lepton number and hence its own neutrino*; therefore if heavy leptons are produced (in e^+e^- collisions e.g.) *it will be possible to identify the sort of neutrino into which they decay by studying the leptonic branching ratio.* (If $M^+ \to \nu_\mu + \ldots$, the leptonic branching ratio provides a measurement of M_{M^+}/M_{W}.)

Using the guesses of Section II.2, we obtain

$$\frac{\Gamma(M^+ \to \nu_\mu \mu^+ \nu_\mu) + \Gamma(M^+ \to \nu_\mu e^+ \nu_e)}{\Gamma(M^+ \to \nu_\mu + \text{hadrons})} \sim 2\left.\frac{\sigma(e^+e^- \to \mu^+\mu^-)}{\sigma(e^+e^- \to \text{hadrons})}\right|_{s \sim \frac{1}{3}M_{M^+}^2}$$

where the coefficient may be checked by considering the free quark model [in guessing that the right-hand side should be evaluated at $s \sim \frac{1}{3}M_{M^+}^2$ we have assumed that $M_W \gg M_{M^+}$ (see Fig. 7 of Ref. 41)].

If $M_{M^+} > M_{M^0}$ the same arguments would give

$$\frac{\Gamma(M^+ \to M^0 \mu^+ \nu_\mu)}{\Gamma(M^+ \to M^0 e^+ \nu_e)} = 1$$

$$\frac{\Gamma(M^+ \to M^0 \mu^+ \nu_\mu) + \Gamma(M^+ \to M^0 e^+ \nu_e)}{\Gamma(M^+ \to M^0 + \text{hadrons})}$$

$$\sim \left.\frac{4}{3}\frac{\sigma(e^+e^- \to \mu^+\mu^-)}{\sigma(e^+e^- \to \text{hadrons})}\right|_{s \sim \frac{1}{3}(M_{M^+} - M_{M^0})^2}$$

A guess is that

$$\frac{\Gamma(M^+ \to M^0 + \ldots)}{\Gamma(M^+ \to \nu_\mu + \ldots)} \sim (1 - Z^4)(Z^4 - 8Z^2 + 1) + 24\, Z^4 \ln(1/Z),$$

where

$$Z = \frac{M_{M^0}}{M_{M^+}},$$

but this is very model-dependent since it is sensitive to the vector/axial vector structure of the M^+M^0W vertex and, of course, to its strength (here we assumed that it is purely right or purely left-handed and equal in strength to the $M^+\nu W$ vertex); the complete formula is given in eqn (2.3) of Ref. 41.

We conclude that according to current conventional wisdom, an M^+ would have a sufficient branching ratio into leptons to make searches based on the leptonic modes worthwhile. We now consider various production processes:

(a)

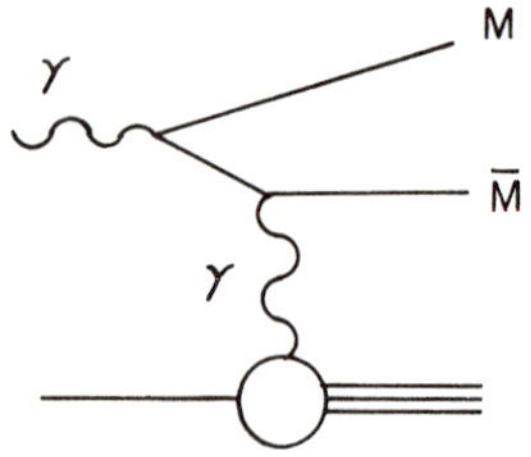

The electromagnetic cross-section can be calculated exactly but it turns out that its value is such that leptons with mass $\gtrsim 3{\cdot}5$ GeV are unlikely to be discovered in this way at NAL or CERN II. (In this reaction, and in (b) and (c) below, we ignore Z and Φ exchange out of ignorance; this is probably a good approximation in case (a) unless the masses are very small.)

(b) $e^+e^- \rightarrow M\bar{M}$. The cross-section rises very rapidly above threshold and decays such as

$$M^+ \rightarrow \mu^+ \nu_\mu \nu_\mu$$
$$M^- \rightarrow e^- \bar{\nu}_e \bar{\nu}_\mu$$

provide a clear signature. It should be possible to discover leptons with masses close to the beam energy in this way with high-luminosity machines. The present experimental limit obtained thus is $M_{M^+} > 1$ GeV (95% confidence).[104]

(c)

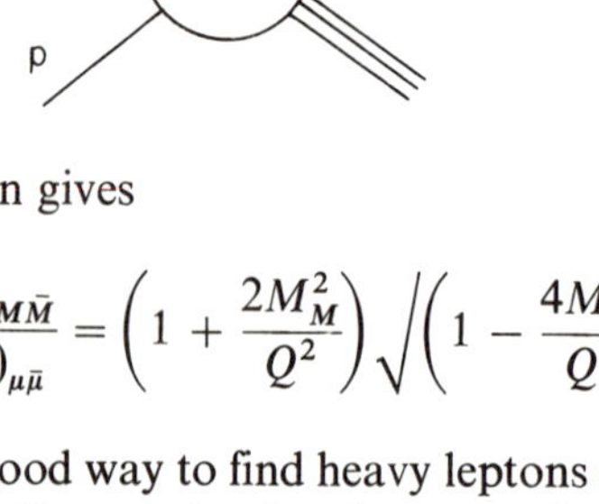

An elementary calculation gives

$$\frac{(\mathrm{d}\sigma/\mathrm{d}Q^2)_{M\bar{M}}}{(\mathrm{d}\sigma/\mathrm{d}Q^2)_{\mu\bar{\mu}}} = \left(1 + \frac{2M_M^2}{Q^2}\right)\sqrt{\left(1 - \frac{4M_M^2}{Q^2}\right)}.$$

Whether or not this is a good way to find heavy leptons depends on $\mathrm{d}\sigma_{\mu\bar{\mu}}$ (see Section II.2). *If* the Drell–Yan mechanism is correct, it should be possible to observe leptons with mass up to about 7 GeV in this way in conceivable experiments at the ISR.[105]

(d) $\nu_\mu + A \to M^+ + \ldots$

$\to \nu_\mu \mu^+ \nu_\mu$

$\to \nu_\mu e^+ \nu_e$

$\to \nu_\mu +$ hadrons

$\to M^0 + \ldots.$

Since the conventional process is $\nu_\mu + A \to \mu^- + \ldots$, this reaction has a spectacular signature. The cross-section

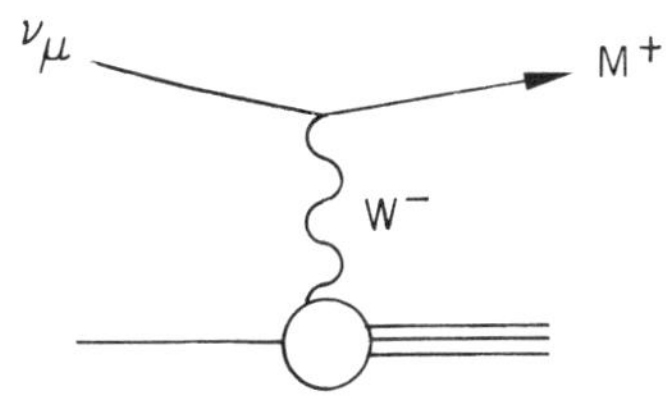

could easily be calculated from

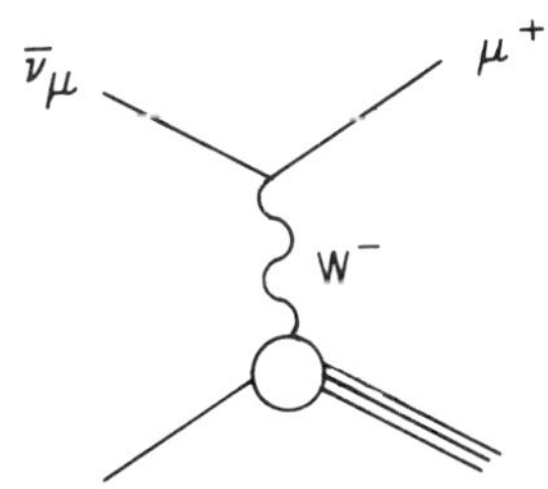

except for contributions which depend on M_{M^+}; luckily these terms can be bounded from below.[106] The results of a model calculation (which assumes scaling, charge symmetry—which may be wrong in many renormalizable models—and $g_{\nu MW} = g_{\nu_\mu W}$) is shown in Fig. 6 (the lower bound is realized in this model). It has been estimated that leptons with mass even as large as 18 GeV might be found in ν-counter experiments at CERN II or NAL in this way. Early results from the NAL monoenergetic neutrino experiment[107] already give the 90% C.L. limits $M_{M^+} > 2$ GeV, assuming $g_{WM^+\nu} = g_{W\mu\nu}$, $\Gamma(M \to \nu_\mu \mu \nu_\mu)/\Gamma(M \to \text{all}) \simeq 0{\cdot}3$ (which is a well founded guess for masses of order 2 GeV) and the cross-section calculation of Ref. 41.

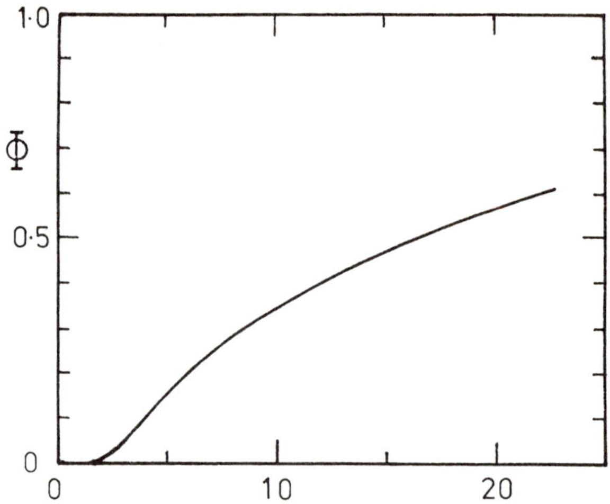

FIG. 6. Approximate model-dependent calculation of

$$\Phi = \frac{\sigma(\nu A \to M^+ + \ldots)}{\sigma(\nu A \to \mu^- + \ldots)}$$

(taken from Ref. 41). This model calculation coincides, within the errors of the approximation, with a much more general lower bound,[106] provided scaling and the charge-symmetry condition are valid.

4.2. *Neutral leptons*

We shall assume that $M_{M^0} < M_W$, otherwise the overwhelming decay modes are $M_{M^0} \to \mu^- W^+$, $e^- W^+$ (by our definition M^0 is a lepton and $\bar{M}^0$ an antilepton). The discussion of the decays $M^+ \to \nu_\mu + \ldots$ probably applies to $M^0 \to \mu^- + \ldots$ *mutatis mutandis* (except for a factor of $\frac{1}{2}$ in going from $M^+ \to \nu_\mu \mu^+ \nu_\mu$ to $M^0 \to \mu^- \mu^+ \nu_\mu$ because there are no longer identical particles) although the results are slightly more model-dependent because M^0 can couple to a left- or a right-handed μ^-. The discussion of $M^+ \to M^0 + X$ applies to $M^0 \to M^+ + \bar{X}$ if $M_{M^0} > M_{M^+}$.

We consider some possible production mechanisms

(a) $e^+e^- \to \bar{E}^0 \nu_e$ by W exchange

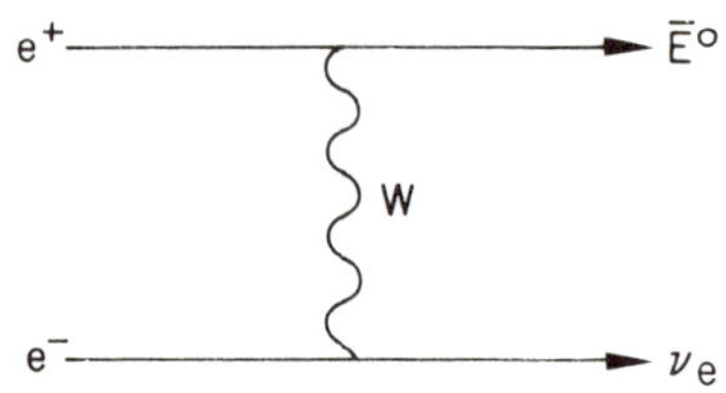

and/or Z exchange

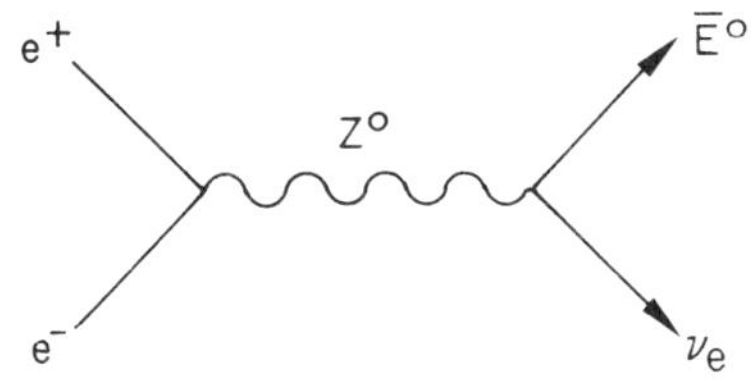

(in the models considered in Ref. 41, the $Z^0\nu_e\overline{E}^0$ coupling is zero but this is not a general result). W exchange alone gives the nominal result

$$\frac{\sigma(e^+e^- \to \overline{E}^0\nu_e)}{\sigma(e^+e^- \to \mu^+\mu^-)} \sim \frac{G^2 s^2}{\alpha^2\pi^2}\left(1 - \frac{M_{E^0}^2}{s.}\right)^2$$

while Z exchange alone gives:

$$\frac{\sigma(e^+e^- \to \overline{E}^0\nu_e)}{\sigma(e^+e^- \to \mu^+\mu^-)} \sim \left(\frac{G}{4\pi\alpha}\right)^2 \frac{s^2}{[1-(s/M_Z^2)]^2}\left(1 - \frac{M_{E^0}^2}{s}\right)^2\left(1 + \frac{2M_{E^0}^2}{s}\right)$$

although both cross-sections could be much larger or smaller in some models (see Fig. 11 in Ref. 41). It is unlikely that leptons with mass more than a few GeV could be discovered this way with the next generation of machines.

(b) $\nu_\mu A \to M^0 + \ldots$. The order of magnitude might be the same as for $\nu_\mu A \to M^+ + \ldots$ but it is obviously much more model-dependent.

(c) $\mu^- + A \to M^0 + \ldots$. Again we expect the same order of magnitude as in $\nu_\mu A \to M^+ + \ldots$. It has been estimated[108] that M^0's of up to about 10 GeV might be seen in this way in an experiment proposed at NAL.

5. SCALAR (HIGGS) MESONS

The Higgs mesons are almost the only ingredient of the new theories which had not been invented independently by ingenious (or underemployed) theorists. It would therefore be exceedingly interesting if they were discovered, but it turns out that they are very elusive and their properties extremely model-dependent.

M_Φ is completely free in all models [although in some cases it may be possible to use existing data to put bounds on it[52]] so single Φ exchange effects can be suppressed at will by making M_Φ large. Although $g_{\Phi WW} \sim gM_W$ in all models, the coupling to fermions is very model-dependent since it is proportional to the "spontaneous" part of the fermion mass—e.g. in the Weinberg–Salam model

$$g_{\Phi\mu\bar{\mu}} \sim g\frac{M_\mu}{M_W}, \qquad g_{\Phi e\bar{e}} \sim g\frac{M_e}{M_W},$$

but in the Georgi–Glashow model[109]

$$g_{\Phi\mu\bar{\mu}} \sim g\frac{M_M}{M_W},$$

$$g_{\Phi e\bar{e}} \sim g\frac{M_E}{M_W},$$

(this should be clear to the reader who has understood the discussion in Section I.7, of the role of the Φ in cancelling terms proportional to the fermion mass in $e^+e^- \to W^+W^-$ if he recalls the diagrams which contribute in each case). Presumably (?) "spontaneous" quark masses are much larger than M_μ so that in one class of models probably

$$\frac{\Gamma(\Phi \to \text{"light leptons"})}{\Gamma(\Phi \to \text{hadrons})} \ll 1$$

although possibly

$$\frac{\Gamma(\Phi \to \text{"heavy leptons"})}{\Gamma(\Phi \to \text{hadrons})} \sim 1$$

In other classes of models, however, all branching ratios might be of the same order (but, the decay being electromagnetic, we expect a lifetime less than or of the order 10^{-17} sec in all cases, so Φ tracks would not be visible).

We have been unable to think of any manifestation of the Φ's which is common to all models. Therefore, beyond making obvious remarks such as that a Φ^0 might make an appreciable contribution to $e^+e^- \to \mu^+\mu^-$ in some models and will break μ–e universality in general or that Φ's are available to account for unexpected phenomena[110] (e.g. in $K^+ \to \pi^0 e^+ \nu_e$, to which a Φ^+ can contribute in some models), we leave further discussion until when or if we have a really credible model.

6. CHARMED PARTICLES, CHANGES OF THE CABIBBO CURRENT, ETC.

The word charm has been used in many ways in the literature. By models with charm, we mean here any model in which there are new hadronic degrees of freedom beyond $SU(3)$ which are introduced (or exploited) to construct renormalizable models of weak interactions. Therefore in these charming models there are (by definition) changes from the conventional Cabibbo structure of the weak current. Furthermore, if charm is conserved by strong and electromagnetic interactions there will be new quasi-stable charmed particles with only weak decays. Consider two examples:

(a) Models with a fourth "quark" to which we attribute an additive quantum number $C = 1$, which is supposed to be exactly conserved by strong and electromagnetic interactions, the known hadrons having $C = 0$. In these models charm is exactly analogous to strangeness. (These ideas can of course be extended to models with arbitrary numbers of "quarks" with $C = 0, \pm 1 \ldots$.)

(b) Models in which there is a more complicated new symmetry group, under which known hadrons are supposed to be singlets. For example, in the Han Nambu model there is a symmetry $SU(3) \times SU(3)'$. The known hadrons are supposed to be $SU(3)'$ singlets.[111] There are additive quantum numbers Y' and I'_3 which are supposed to be conserved by strong and electromagnetic interactions. We assume here that $SU(3)'$ is conserved by strong interactions (although this is not true in all versions of the model). Charmed states are those which are not $SU(3)'$ singlets (by our definition); they can only be produced singly by weak interactions unless $Y' = I'_3 = 0$ in which case single electromagnetic production is also possible.

We consider *the decays of charmed particles*. Presumably heavy charmed particles decay into lighter ones by pion emission. If the ground state of mass M_c then decays purely by weak interactions, we might guess that the leptonic modes would give a lower bound to the width

$$\Gamma \gtrsim \Gamma(\Sigma \to N + \text{leptons})\left(\frac{M_c - M_{\text{proton}}}{M_\Sigma - M_{\text{proton}}}\right)^5$$

or

$$T \lesssim 10^{-10}\left(\frac{M_p}{M_c - M_p}\right)^5 \text{secs.}$$

(if it can decay electromagnetically τ would of course be very much smaller). This guessed bound is based on one channel; since the number of channels increases very rapidly with M_c, the larger M_c the better it is likely to be satisfied. Interesting estimates have been obtained by Snow[112] by summing all channels. The conclusion is that unless M_c is of order 2 GeV or less, charmed particles are unlikely to leave visible tracks. However, M_c may be less than 2 GeV and it is certainly worth looking for such tracks in any case.

The discovery of very short-lived particles must rely on identifiable decay modes. In models which use the Glashow–Iliopoulos–Maiani trick, decays (both leptonic and non-leptonic) in which there are strange particles in the final state are expected to dominate non-strange channels by a factor $\cot^2 \theta_{\text{Cabibbo}}$. One possible signature is therefore an increase of strange particle production above threshold. A better signature might be provided by the leptonic decays. An optimistic guess would be

$$\frac{\Gamma(C \to \text{leptons} + \text{hadrons})}{\Gamma(C \to \text{hadrons})} \sim 1 \text{ for } M_c \lesssim 2\text{ GeV}$$

which is much larger than the corresponding ratio for strange particle decays because of the increased phase space available for leptonic decays, which have at least 3 particles in the final state; however, this guess might easily be too large by one or even two orders of magnitude.

There can be associated production:

$$e^+e^-, p\bar{p}, pp, \text{etc.} \rightarrow C\bar{C}' + \ldots$$

$\quad\quad \rightarrow$ strange particle excess?

$\quad\quad \rightarrow$ strangeness non-zero state + leptons?

Snow has made some estimates of the cross-sections;[111] the conclusion is that it will probably be very hard to observe C's in this way. In the Han–Nambu model $SU(3)'$ non-singlets can be produced singly:

$$\gamma p \rightarrow C(I'_3 = Y' = 0) + \ldots$$
$$\quad\quad \rightarrow \gamma + \ldots.$$

This could give rise to pronounced threshold effects in photo and electro-production.

In all the models considered here:

$$\nu A \rightarrow \mu^- + C + \ldots$$

(Note that in models incorporating the Glashow–Iliopoulos–Maiani trick the cross-section for $\Delta C = 1$, $\Delta S = 0$ events will be suppressed by $\sin^2\theta_c$ generally, but if there are abundant $\Lambda\bar{\Lambda}$ pairs in the nucleon there may be frequent $\Delta S = 1$, $\Delta C = 1$ events which have an intrinsic strength $\cos^2\theta_c$.) As discussed below, the non-leptonic decays of C may give rise to apparent violations of selection rules in these reactions. The threshold effect, leptonic decays, and strange particle excess might also provide signatures.

We now consider some implications of changes in the conventional (Cabibbo) structure of the weak current.

(a) Improved tests of the Cabibbo predictions for hyperon decays are obviously desirable (presumably they will be available from experiments using hyperon beams at the new accelerators).

(b) Improved tests of the symmetry properties usually assumed for the weak current are very desirable (this is true independently of new theories since many of these properties are very badly tested). In most models the charge symmetry condition, the conventional isospin properties, the $\Delta S = \Delta Q$ rule, etc., etc., may fail, e.g. in the four-quark model the $\Delta C = +1$, $\Delta Q = +1$ current has a $\Delta S = 0$, $\Delta I = \frac{1}{2}$ piece (which is not related to the corresponding part of the $\Delta C = -1$, $\Delta Q = -1$ current by a charge reflection) and a $\Delta S = +1$, $\Delta I = 0$ piece. If a charmed particle is produced by a neutrino and then decays non-leptonically, there can be an apparent violation of the $\Delta S = \Delta Q$ rule, e.g.

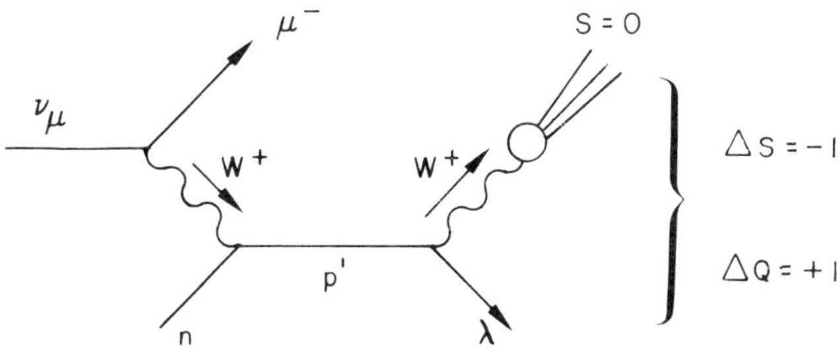

Tests of weak selection rules have been discussed repeatedly in the literature.[113] Some of these questions have recently been re-examined;[114, 115] we refer to the papers of De Rujúla and Glashow[115] for a catalogue of effects which can occur in various models as well as a discussion of some experimental tests.

(c) The Adler sum rule may obviously be changed if the structure of the weak current changes. This is true *a fortiori* of light cone/parton model sum rules and relations which depend sensitively on the choice of constituents as well as the symmetries of the currents (however, compared to the Adler sum rule, the derivations are on a very weak footing). The changes which occur have been considered by many authors[116] who find large and striking effects in many models. A particularly amusing example was pointed out by Love *et al.*[116] who considered a version of the Han–Nambu model in which, as well as left-handed currents, there are $SU(3)'$ octet right-handed currents; they come into play above the charmed threshold with the result that, according to light cone/parton ideas, the V–A interference term W_3 vanishes for inclusive processes—i.e. parity violating effects in inclusive processes are a low-energy phenomenon in this model! Catalogues of the effects which occur in various models may be found in the papers[116] by De Rujúla and Glashow, and Beg and Zee.

(d) All conventional hypotheses (prejudices) should be re-examined (cf. the discussion in Section I.12, of a model in which ν's from π and K decay are different).

ACKNOWLEDGEMENTS

I am very grateful to Jacques Prentki for his incisive comments on the whole manuscript and also to Cecilia Jarlskog and Noel Cottingham for critical remarks on Part I.

REFERENCES AND FOOTNOTES

1. G. 't Hooft. *Nucl. Phys.* **B35**, 167 (1971).
2. G. 't Hooft and M. Veltman. *Nucl. Phys.* **B44**, 189 (1972); and **B50**, 318 (1972). B. W. Lee and J. Zinn-Justin. *Phys. Rev.* **D5**, 3137 (1972); **D5**, 3121 (1972); **D5** 3150 (1972); **D7**, 1049 (1972). D. A. Ross and J. C. Taylor, *Nucl. Phys.* **B51**, 125 (1973). For reviews and further references see M. Veltman—Invited paper presented at the Bonn Conference on Electron and Photon Interactions, August 1973, to be published by North Holland Publishing Co. (this paper contains an interesting historical survey) and E. S. Abers and B. W. Lee, *Phys. Reports* **9C**, No 1 (1973).
3. B. Zumino. Lectures at the Cargese Summer Institute, July 1972, CERN TH 1550.
4. J. C. Taylor. "An Introduction to Gauge Fields and Renormalizable Theories of Charged Vector Mesons", Rutherford Lab. preprint, RPP/T/29 (1972).
5. L. M. Sehgal. "Unified Theories of Weak and Electromagnetic Interactions; An Elementary Introduction." Aachen preprint (1973).
6. A. De Rujúla. "An Elementary Introduction to the Unified Theories of Weak and Electromagnetic Interactions." Harvard preprint (1973).
7. Riazuddin. "An Introduction to Spontaneously Broken Gauge Theories of Weak and Electromagnetic Interactions." Daresbury preprint, DNPL-R25 (1973).
8. For a survey of developments up to September 1972, see articles by B. W. Lee, J. D. Bjorken, J. Primack and B. Zumino. In "Proc. XVI Int. Conf. on High Energy Physics." NAL (1972).
9. C. H. Llewellyn Smith. In "Links between weak and electromagnetic interactions." Rutherford Lab. Report RL-73-018 (1973).
10. See, for example, R. E. Marshak, Riazuddin and C. P. Ryan. "Theory of Weak Interactions in Particle Physics." Wiley–Interscience, New York, 1969.
11. If we neglect m_e, each spinor contains an explicit factor $S^{\frac{1}{4}}$ in the centre of mass; the amplitude therefore vanishes like s as $s \to 0$ and we can subtract the standard forward dispersion relation at $s = 0$ without introducing new parameters. This gives the equation in the text for the invariant amplitude; this equation also holds for the appropriate amplitude for $m_e \neq 0$.
12. The $k_\mu k_\nu/M_W^2$ term in the W propagator does not spoil this result since each k gives a factor m_e by virtue of the Dirac equation.
13. In a renormalizable theory we must ensure that to lowest order *either* $\sigma_{\text{tot}}^{\bar{\nu}e}$ and $\sigma_{\text{tot}}^{\nu e}$ vanish asymptotically *or*, if they tend to constants, $\sigma_{\text{tot}}^{\nu e}/\sigma_{\text{tot}}^{\bar{\nu}e} \to 1$ in order that the forward dispersion relation converges. The latter possibility is realized in the models discussed below. It is not hard to see that this requires the existence of neutral bosons (Z) and/or heavy leptons. We will follow a slightly different approach in motivating the existence of these objects and "deriving" their properties, but it is instructive to think about the role they play in the process considered here.
14. G. Segre. In "Springer Tracts in Modern Physics 52." Springer Verlag, Berlin, 1970.
15. E. P. Shabalin, Sovremennoe Sostoyanie Teorii Slabogo Vzaimodeistviya. Inst. of Th. and Expt. Studies, Moscow, report ITEP 724 (1969).
16. C. H. Llewellyn Smith. *Physics Reports* **3C**, No. 5 (1972).
17. C. H. Llewellyn Smith. CERN/ECFA/72/4 Vol. II.

18. T. Appelquist and J. D. Bjorken. *Phys. Rev.* **D4**, 3726 (1971). G. Preparata, *Phys. Letts.* **40B**, 253 (1972). (This paper works in the framework of non-renormalizable perturbation theory.) A. D. Dolgov, V. E. Zakarov and L. B. Okun. *Yadernaya Fizica* **14**, 1044 and 1247 (1971); *Phys. Letts.* **27B**, 298 (1971); *Nucl. Phys.* **B37**, 493 (1971). (This work is reviewed in the *Slabie Vzaimodeistviya Pri Visokikh Energiyakh*; lectures given at the International School of High-Energy Physics, Yerevan, 1971.) T. Appelquist, J. D. Bjorken and M. Chanowitz, SLAC PUB 1149 (1972).
19. Cancellations between different orders are impossible if perturbation theory is valid since the coupling constant can be varied arbitrarily (inside the radius of convergence).
20. Some references in which this point is made are: N. Cabibbo and R. Gatto. *Phys. Rev.* **124**, 1577 (1961). Y. S. Tsai and A. C. Hearn, *Phys. Rev.* **140B**, 721 (1965). S. Weinberg. In "Elementary Particles and Quantum Field Theory," Vol. 1 ed. S. Deser *et al.*, (M.I.T. Press, Cambridge, Mass., 1970;) K. J. Kim and Y. S. Tsai. *Phys. Rev.* **D7**, 1310 (1973).
21. J. M. Cornwall, D. N. Levin and G. Tiktopoulos, *Phys. Rev. Letts.* **30**, 1268 (1973).
22. C. H. Llewellyn Smith. CERN TH 1665 (1973) (to be published in *Physics Letters*).
23. H. Georgi and S. L. Glashow. *Phys. Rev. Letts.* **28**, 1494 (1972).
24. S. Weinberg. *Phys. Rev. Letts.* **19**, 1264 (1967); **27**, 1688 (1971).
25. A. Salam in Elementary Particle Physics eds. N. Svartholm, Almqvist and Wiksells p. 367. (Stockholm) 1968.
26. C. N. Yang and R. L. Mills. *Phys. Rev.* **96**, 191 (1954); R. Shaw. Ph.D. Thesis, Cambridge. Unpublished (1954). For a review see Ref. 4.
27. B. De Witt. *Phys. Rev. Letts.* **12**, 742 (1964); *Phys. Rev.* **162**, 1195 (1967) and **162**, 1239 (1967) derived the Feynman rules which give an ultra-violet renormalizable theory by power counting.
28. F. Englert and R. Brout, *Phys. Rev. Letts.* **13**, 321 (1964). F. Englert, R. Brout and M. F. Thiry. *Nuovo Cimento* **43A**, 244 (1966).
29. P. W. Higgs. *Phys. Letts.* **12**, 132 (1964); *Phys. Rev. Letts.* **13**, 508 (1964); *Phys. Rev.* **145**, 1156 (1966).
30. G. S. Guralnick, C. R. Hagen and T. W. B. Kibble. *Phys. Rev. Letts.* **13**, 585 (1964). T. W. B. Kibble. *Phys. Rev.* **155**, 1554 (1967). G. S. Guralnick, C. R. Hagen and T. W. B. Kibble, In "Advances in Particle Physics." Vol. 2 eds. R. L. Cool and R. E. Marshak. (Interscience), 1968.
31. The arguments here are all classical; they apply to the set of all tree diagrams but not necessarily to diagrams with closed loops. They should therefore be regarded only as suggestions which lead to a new Lagrangian (written in terms of the translated fields) which should then be quantized.
32. A particularly instructive discussion is given in G. 't Hooft and M. Veltman's "Diagrammer" (CERN yellow report 73–9). They consider, as an example, a gauge fixing function other than Fermi's with which Fadeev–Popov ghosts are necessary even in the algebraically uncomplicated case of QED.
33. More formally, each field Ψ must be split into a left-handed component $[\Psi_L = (1 - \gamma_5)/2 \cdot \Psi]$ and a right-handed component $[\Psi_R = (1 + \gamma_5)/2 \cdot \Psi]$ from which left-handed and right-handed multiplets are built (see Section I.5, where we show that the couplings of left-handed and right-handed fermions separately satisfy simple commutator conditions—"represent a Lie algebra"—from which it follows that left-handed and right-handed particles form multiplets

separately). A mass term can be written as $m\overline{\Psi}\Psi \equiv m(\overline{\Psi}_L\Psi_R + \overline{\Psi}_R\Psi_L)$; for this to be invariant, there must be left-handed and right-handed multiplets of the same dimension. In this case $L^i \equiv R^i$ and the interaction may be written $W^i_\mu(\overline{\Psi}_L L^i\gamma_\mu\Psi_L + \overline{\Psi}_R R^i\gamma_\mu\Psi_R) = W^i_\mu\overline{\Psi}L^i\gamma_\mu\Psi$, where $\Psi = \Psi_R + \Psi_L$ (cf. the discussion of the Georgi Glashow model in Section I.4). Parity is conserved by this interaction (but it can be violated by mixing induced by the spontaneous symmetry breaking).

34. J. S. Bell and R. Jackiw. *Nuovo Cimento* **60**, 47 (1967). S. L. Adler. *Phys. Rev.* **177**, 2426 (1969). See S. L. Adler, in "Lectures in Particles and Quantum Field Theory." Vol. 1, (M.I.T. Press, 1970) for a review.
35. M. Veltman, private communication. C. Bouchiat, J. Iliopoulos and P. Meyer. *Phys. Letts.* **38B**, 519 (1972). D. J. Gross and R. Jackiw. *Phys. Rev.* **D6**, 477 (1972).
36. H. Georgi and S. L. Glashow. *Phys. Rev.* **D6**, 429 (1972). This object gives the strength of the anomaly; it is proportional to the coefficient of the mass-independent part of the sum of triangles with an odd number of axial vertices and vectors W^a, W^b, W^c at the corners.
37. In processes which it might be practical to observe, anomalous contributions first occur in order e^6; Bjorken has argued that anomalous models should therefore not necessarily be rejected since the perturbation expansion may only be an asymptotic expansion and may break down in this order (however, according to the usual rules, an infinite number of arbitrary constants are needed to specify an anomalous theory completely and we would therefore reject it). See the remarks on p. 273 of Vol. II "Proc. XVI Int. Conf. on High-Energy Physics." NAL 1972.
38. For some models, see Refs. 23, 24, 25, 41, 43, 44, 45, 54, 55, 56, and B. E. Lee. *Phys. Rev.* **D6**, 1188 (1972). J. Prentki and B. Zumino. *Nucl. Phys.* **B47**, 99 (1972). S. Weinberg. *Phys. Rev.* **D5**, 1962 (1972). I. Bars, M. B. Halpern and M. Yoshimura. *Phys. Rev. Letts.* **29**, 969 (1972); *Phys. Rev.* **D7**, 1233 (1973). A. Salam and J. C. Pati, *Phys. Rev.* **D8**, 1460 (1973).
39. It could be argued that, since hadron dynamics can presumably not be treated perturbatively (this is true *a fortiori* in models such as the quark model in which hadrons are bound states), there may not be much sense in considering bare elementary hadron fields and demanding that the weak current couples to them renormalizably; strong interaction effects could soften the high-energy behaviour of amplitudes such as

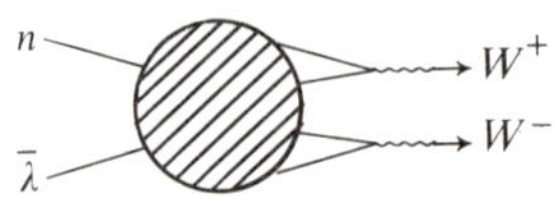

Current algebra/parton/light cone arguments suggest that this is not the case but this point of view is as hard to reject as it is to establish; we will not adopt it here.
40. S. L. Glashow, J. Iliopoulos and L. Maiani. *Phys. Rev.* **D2**, 1285 (1970).
41. J. D. Bjorken and C. H. Llewellyn Smith. *Phys. Rev.* **D7**, 887 (1973).
42. M. Y. Han and Y. Nambu. *Phys. Rev.* **139B**, 1006 (1965).
43. Y. Achiman. *Nucl. Phys.* **B56**, 635 (1973). M. A. B. Beg and A. Zee. *Phys. Rev. Letts* **30**, 675 (1973), and Rockefeller preprint COO-2232B-20 (1973). J. A. Rawls and L. Yu. *Phys. Rev.* **D7**, 2463 (1973).
44. H. J. Lipkin. *Phys. Rev.* **D7**, 1850 (1937). H. Georgi and S. L. Glashow. *Phys.*

Rev. **D7**, 561 (1973). R. N. Mohapatra. Maryland preprint 73-022 (1972). A. Love, D. V. Nanopoulos and G. G. Ross. *Nucl. Phys.* **B55**, 33 (1973).
45. H. Georgi and S. L. Glashow. *Phys. Letts.* **B44**, 191 (1973).
46. To remove the residual term, however, there must be a neutral Φ with coupling $\sim g(m_p/M_W)$. $\bar{n}\lambda$; this will give rise to unobserved $\Delta S = 2$ processes unless this neutral Φ is not self-conjugate (or $m_\Phi^2 \ll M_W^2$ and $m_p^2 \ll M_\Phi^2$). The simplest way to achieve this seems to involve a scheme in which no fewer than 10 physical Φ's remain (I am grateful to J. Prentki and B. W. Lee for a conversation on this point). Another possible criticism of this model, due to W. Alles and J. Julve Perez *Lett. al Nuovo Cimento* **8**, 123 (1973)) is that in conjunction with the usual approximate soft pion theorems it leads to results which disagree with successful predictions of the $\Delta I = \frac{1}{2}$ rule.
47. R. Oehme, E. C. Swallow and R. Winston. *Phys. Rev.* **D7**, 2124 (1973).
48. G. Feinberg, F. Gursey and A. Pais. *Phys. Rev. Letts* **7**, 208 (1961). (See E. C. M. Young, CERN yellow report 67-12, 1967, for a discussion of the experimental situation in 1967).
49. S. Weinberg. *Phys. Rev.* **D8**, 605 (1973) and *Phys. Rev. Letts.* **31**, 494 (1973).
50. I. Bars. "Parity Violation and Comparison Between Quark and *M* Schemes in Unified Gauge Theories." Berkeley preprint (1973).
51. There is abundant celebration of the beauty of the idea of unifying weak and electromagnetic interactions in the literature. In a sense, however, it is a very unnatural thing to do since parity is conserved in one case, while it is violated maximally in the other. In many models it is indeed hard to indulge in this unnatural practice with impunity.
52. B. W. Lee, J. R. Primack and S. B. Treiman. *Phys. Rev.* **D7**, 510 (1973).
53. See H. Georgi and S. L. Glashow. *Phys. Rev.* **D7**, 2457 (1973) for a review. (The first observation that mass differences can be calculated is due to 't Hooft, Ref. 1.)
54. B. W. Lee and S. B. Treiman. *Phys. Rev.* **D7**, 1211 (1973). A. Pais. *Phys. Rev.* **D8**, 625 (1973). M. A. B. Beg, *Phys. Rev.* **D8**, 664 (1973). T. P. Cheng. *Phys. Rev.* **D8**, 496 (1973).
55. R. N. Mohapatra. *Phys. Rev.* **D6**, 2023 (1972). A. Pais. *Phys. Rev. Letts.* **29**, 1912 (1973), and *Phys. Rev.* **D8**, 625 (1973) (Ref. 54 above). T. D. Lee. *Phys. Rev.* **D8**, 1226 (1973).
56. B. de Wit. *Nucl. Phys.* **B51**, 237 (1973).
57. K. J. Kim and Y. S. Tsai. *Phys. Letts.* **40B**, 665 (1972), and SLAC PUB 1168 (1973).
58. R. W. Brown, R. H. Hobbs and J. Smith. *Phys. Rev.* **D4**, 794 (1971). R. W. Brown and J. Smith. *Phys. Rev.* **D3**, 207 (1971) and references therein.
59. D. Cline, A. K. Mann and C. Rubbia. *Phys. Rev. Letts* **25**, 1309 (1970).
60. D. H. Perkins in Proc. 1969 CERN Conf. on Weak Interactions (CERN 69-7).
61. B. C. Barish *et al. Phys. Rev. Letts.* **34**, 180 (1973).
62. L. M. Lederman and B. G. Pope. *Phys. Rev. Letts.* **27**, 765 (1971).
63. J. Christenson *et al. Phys. Rev. Letts.* **25**, 1523 (1970).
64. P. Wanderer *et al. Phys. Rev. Letts.* **23**, 729 (1969).
65. S. D. Drell and T. M. Yan. *Phys. Rev. Letts.* **25**, 316 (1970).
66. F. W. Buesser *et al. Phys. Letts.* **B48**, 377 (1974), and proposal for an ISR experiment.
67. This point is made strongly by R. L. Jaffe in "Deep Inelastic Processes with Hadronic Initial States". M.I.T. Center for Th. Physics Pub. 344, 1973, who reviews the theoretical interest in testing the scaling law.
68. P. V. Landshoff and J. C. Polkinghorne. *Phys. Letts.* **40B**, 463 (1972) give a parton

formula for $d^2\sigma/dQ^2d\theta$ (which they evaluate in a specific model) where θ is the angle between Q and the collision axis. This quantity is very sensitive to the P_T distribution of the partons (if they all have zero transverse momentum it is proportional to $\delta(\theta)$). Experimentally, a more relevant quantity may be $d\sigma^2/dQ^2dQ_3$, which is not sensitive to $\langle P_T \rangle$ at very high energies provided $\langle P_T \rangle$ is small, and can easily be used to obtain the spectrum of the observed leptons. Assuming $\langle |\vec{P}_T| \rangle = 0$ (and hence $\langle |\vec{Q}_T| \rangle = 0$) and spin $\frac{1}{2}$ partons the model gives

$$\frac{d^2\sigma_{\mu^+\mu^-}}{dQ^2dQ_3} = \frac{4\pi\alpha^2}{3Q^4 \cdot Q_0} \sum_a \frac{F_{2a}(x_1)F_{2\bar{a}}(x_2)}{\lambda_a^2}$$

where

$$x_1 = \frac{Q_0 + Q_3}{\sqrt{s}}$$

$$x_2 = \frac{Q_0 - Q_3}{\sqrt{s}}$$

$$\lambda_a = \text{parton charge}$$

$$F_{2a}(F_{2\bar{a}}) = \text{(anti)parton contribution to } \nu W_2$$

$$\left(= \sum_{a,\bar{a}} F_{2a}(x).\right)$$

If partons are quarks, the fact that $\sigma^{\bar{\nu}A}/\sigma^{\nu A} \sim 0{\cdot}37$ tells us that the antipartons are concentrated at very small x. The dimuon cross-section is obviously very sensitive to the precise antiparton distribution assumed. If, e.g. $F_{2\bar{a}} = 0$ for $x > x_0$ then

$$\frac{d^2\sigma}{dQ^2dQ_3} = 0 \qquad \text{for} \qquad |Q_3| < \frac{Q^2 - x_0^2 s}{2\sqrt{s}}.$$

69. U. J. Becker *et al.* (MIT) "A Proposal to Search for the Heavy Neutral Boson Z^0 at the ISR".
70. R. L. Jaffe and J. Primack. MIT preprint 349 (1973).
71. A recent paper from which the literature may be traced is: R. W. Brown, R. H. Hobbs, J. Smith and N. Stanko. *Phys. Rev.* **D6**, 3273 (1972).
72. See J. Løvseth and M. Radomski. *Phys. Rev.* **D3**, 2686 (1971).
73. H. S. Gurr, F. Reines and H. W. Sobel. *Phys. Rev. Letts.* **28**, 1406 (1972).
74. H. H. Chen and B. W. Lee. *Phys. Rev.* **D5**, 1874 (1972).
75. F. J. Hassert *et al. Phys. Letts.* **46B**, 121 (1973).
76. D. C. Cundy in "Links Between Weak and Electromagnetic Interactions". Rutherford Lab. report RL-73-081.
77. D. C. Cundy *et al. Phys. Letts.* **31B**, 479 (1970).
78. W. Y. Lee. *Phys. Letts.* **40B**, 423 (1972).
80. A. K. Mann and H. Primakoff. *Phys. Rev.* **D8**, 196 (1973).
81. A. Pais and S. B. Treiman. *Phys. Rev.* **D6**, 2700 (1972).
82. E. A. Paschos and L. Wolfenstein. *Phys. Rev.* **D7**, 91 (1973).
83. C. H. Albright, B. W. Lee and E. A. Paschos. NAL THY 86 (1972) and footnote 18 of Ref. 82.
84. C. H. Albright, B. W. Lee, E. A. Paschos and L. Wolfenstein. *Phys. Rev.* **D7**, 2220 (1973).

85. L. M. Sehgal. *Nucl. Phys.* **B65**, 14 (1973) and *Phys. Letts.* **B48**, 133 (1974). See also R. B. Palmer. *Phys. Letts.* **46B**, 240 (1973) and C. A. Albright. *Phys. Rev.* **D8**, 3162 (1973).
86. S. Weinberg. *Phys. Rev.* **D5**, 1412 (1972).
87. We refer to Weinberg's paper for the details, but note that the approximation which includes neglecting the V–A interference term, is not valid at CERN energies. Myatt (quoted in Refs 76 and 89) has integrated Weinberg's differential cross-section for $\nu p \to \nu p$ exactly for the conditions of the CERN experiment; the exact value of our lower bound for $\sigma(\nu p \to \nu p) + \sigma(\nu n \to \nu n)$ for these conditions may be obtained from Myatt's curve (given in Refs 76 and 89) by changing $\sin^2 \theta_W \to \frac{1}{2} \sin^2 \theta_W$ in the scale.
88. B. W. Lee. *Phys. Letts.* **40B**, 420 (1972).
89. D. H. Perkins. In "Proc. XVI International Conference on High-Energy Physics". NAL 1972.
90. R. Gatto and G. Preparata. Nota Interna 479, University of Rome (1973).
91. A. Love, G. G. Ross and D. V. Nanopoulos. *Nucl. Phys.* **B49**, 513 (1972).
92. E. Derman. *Phys. Rev.* **D7**, 2755 (1973).
93. D. Petronzio. Roma preprint 411 (1972).
94. This formula was incorrectly stated to be a guess for the parity-violating asymmetry (considered below) in my paper "Neutrino Physics" (CERN/ECFA/72/4, Vol. II).
95. M. A. Bouchiat and C. C. Bouchiat. *Phys. Letts.* **B48**, 111 (1974).
96. J. Godine and A. Hankey. *Phys. Rev.* **D6**, 3301 (1972).
97. A. Love. *Nuovo Cimento Letts.* **5**, 113 (1972).
98. V. K. Chung, A. K. Mann and E. A. Paschos. *Phys. Letts* **41B**, 335 (1972).
99. R. W. Brown, Vu Khac Chung, K. O. Mikaelian and E. A. Paschos. NAL preprint NAL-THY-97 (1972).
100. R. Budny in "Links between Weak and Electromagnetic Interactions", Rutherford report RL-73-018 and *Phys. Letts.* **45B**, 340 (1973).
101. M. Kuroda and Y. Yamaguchi. *Prog. Th. Phys.* **49**, 2160 (1973).
102. M. Perl. SLAC PUB 1062 (1972).
103. An exception is the Georgi–Glashow model in which J. Primack and H. R. Quinn (*Phys. Rev.* **D6**, 3171, 1972) have shown that $(g-2)_\mu$ puts a bound on M_M as a function of M_W and M_Φ.
104. M. Bernadini *et al. Nuovo. Cimento* **17A**, 383 (1973). This limit assumes the branching ratio in Ref. 41, which for this mass depends only on assuming CVC for $\rho \to e\nu/\mu\nu$.
105. B. G. Duff in "Links Between Weak and Electromagnetic Interactions". Rutherford report RL-73-018.
106. C. H. Llewellyn Smith. *Nucl. Phys.* **B56**, 325 (1973).
107. B. C. Barish *et al. Phys. Rev. Letts.* **31**, 410 (1973).
108. R. Cester *et al.* NAL proposal "Feasible search for heavy neutral muons predicted by gauge theories", Princeton preprint (1973).
109. Models with these large $\Phi_{e\bar{e}}(\Phi_{\mu\bar{\mu}})$ couplings have a certain implausibility since they require that

$$M_{e(\mu)} = M^0 - M^s$$
$$M_{E(M)} = M^0 + M^s$$

where $M^0(M^s)$ is the bare (spontaneous) part of the mass so that the smallness of $M_e(M_\mu)$ seems to be a surprising accident.

110. They are also a source of potential troubles for model builders—see footnote 46 and Ref. 52.
111. This ensures that the "naïve quark model" results are the same as with Gell-Mann–Zweig quarks.
112. G. Snow. *Nucl. Phys.* **B55**, 445 (1972).
113. For a review, see Ref. 16.
114. P. Musset, CERN TC-L/Int. 73–1.
115. A. De Rujula and S. L. Glashow. *Phys. Letts.* **46B**, 381 (1973) and "Tests of Charge Symmetry and Scaling in Neutrino Physics", Harvard Preprint (1973).
116. R. Budney and P. N. Scharbach. *Phys. Rev.* **D6**, 3631 (1972). Riazuddin and Fayyazuddin. *Phys. Rev.* **D6**, 2032 (1972). S. P. de Alwis. *Phys. Letts.* **45B**, 253, 1973. A. Love, D. V. Nanopoulos and G. G. Ross. *Nucl. Phys.* **B55**, 33 (1973). A. De Rujula and S. L. Glashow. *Phys. Letts.* **46B**, 377 (1973). M. A. B. Beg and A. Zee. *Phys. Rev.* **D8**, 2334 (1973). Beg and Zee refer to a paper by S. Tamura and K. Fujii (*Prog. Theor. Phys.* **49**, 995 (1973)) and a Berkeley preprint by K. Lane and M. Suzuki which I have been unable to obtain.

7. SPONTANEOUS SYMMETRY BREAKING

P. W. HIGGS
Department of Physics,
University of Edinburgh, Edinburgh, Scotland.

I. INTRODUCTION

In his lectures Chris Llewellyn Smith has subjected you to some propaganda in favour of currently fashionable renormalizable models of weak and electromagnetic interactions, which are based on spontaneously broken local symmetries. His approach has been to show that, if you want the sum of tree graphs for any weak process to have a decent high energy behaviour, you need not only intermediate vector bosons but also a number of scalar bosons, with vertices obeying relations which are characteristic of this type of broken symmetry. This is an admirable way of converting sceptical phenomenologists but it is not the route by which Weinberg and Salam[1] were first led to invent such models. My task in these seminars is to tell you about the theoretical developments which led to their invention: since these developments were essentially complete as long ago as 1966, this will appear as an exercise in ancient history from the perspective of a particle physicist.

The invention of the first dynamically non-trivial relativistic model[2] to exhibit a spontaneously broken local symmetry was not motivated by any expectation of its relevance to leptonic weak interactions; it was conceived rather as a way—probably the only way—to evade the difficulty of the prediction of massless "Goldstone bosons" which was plaguing current attempts[3] to account for the breaking of hadron symmetries such as $SU(2)$ and $SU(3)$ in terms of spontaneous, rather than explicit symmetry breaking. These attempts had been initiated by Nambu,[4] who applied to particle physics some of the lessons he had learned from his contributions[5] to the theory of superconductivity.[6]

Thus, if one traces back the pedigree of the renormalizable models of the weak interactions, one ends up in the theory of various forms of condensed matter: there, as far back as 1928, one finds that spontaneous breaking of a symmetry is a feature of the theory of ferromagnetism;[7] at an intermediate date[8] it occurs again (in a form which, as we shall see, contains hints of later developments) in the theory of superfluidity. Since the interplay of symmetry and dynamics is easier to comprehend in some of these non-relativistic problems than in relativistic quantum field theories, I shall devote most of my first talk to them.

II. Symmetries—Exact and Broken

First we must state what is meant by spontaneous breaking of a symmetry. We shall be concerned here with the symmetry (exact or broken) of a quantum field theory/many-body theory under some compact Lie group G, whose generators $Q_1 \ldots Q_n$ satisfy the corresponding Lie algebra

$$[Q_a, Q_b] = if_{abc}Q_c.$$

To place spontaneous breaking of the symmetry in its setting, we shall distinguish four possible situations.

1. Exact symmetry

This is characterized by two conditions:

(a) The Lagrangian density $\mathscr{L}$ is invariant under the transformations of the fields which realize G.

$$\delta \mathscr{L} = 0. \tag{1a}$$

(b) The (unique) vacuum state is invariant under the unitary transformations on the state space which realize G.

$$Q_a|0\rangle = 0, \text{ all } a. \tag{1b}$$

Under these conditions one easily proves the invariance of the Hamiltonian,

$$[Q_a, H] = 0, \text{ all } a,$$

and uses Wigner's theorem to deduce the mass degeneracy of particle multiplets.

2. EXPLICITLY BROKEN SYMMETRY

The Lagrangian density consists of an invariant term plus symmetry breaking interactions.

$$\mathscr{L} = \mathscr{L}_0 + c\mathscr{L}_1,$$

where under G

$$\delta \mathscr{L}_0 = 0, \qquad \delta \mathscr{L}_1 \neq 0. \tag{2a}$$

($\mathscr{L}_1$ may be invariant under a subgroup $G' \subset G$.) Consequently the vacuum state is not invariant under G, but it would become invariant if the symmetry breaking interactions were switched off.

$$Q_a|0\rangle \to 0 \quad \text{as} \quad c \to 0. \tag{2b}$$

The mass degeneracy in particle multiplets is removed, the splitting being a function of c which tends to zero as $c \to 0$. This is the traditional view of the broken symmetries, $SU(2)$ and $SU(3)$, of hadron physics.

3. SPONTANEOUSLY BROKEN SYMMETRY

The Lagrangian density is formally invariant,

$$\delta \mathscr{L} = 0 \tag{3a}$$

but, as a consequence of the dynamics determined by $\mathscr{L}$, the vacuum state is non-unique. A complete space of many-particle states may be built on any one of these vacua and each one of them is non-invariant under G,

$$Q_a|0\rangle \neq 0 \text{ for some } a. \tag{3b}$$

(It may be invariant under a subgroup.) The Hamiltonian which acts on such a space of physical states is also non-invariant (but all the "different"

Hamiltonians describe the same physics!), so the particle multiplets are non-degenerate in mass.

4. SPONTANEOUSLY AND EXPLICITLY BROKEN SYMMETRY

Finally, one may have a situation where there is explicit symmetry breaking,

$$\mathscr{L} = \mathscr{L}_0 + c\mathscr{L}_1,$$

$$\delta\mathscr{L}_0 = 0, \qquad \delta\mathscr{L}_1 \neq 0, \tag{4a}$$

and, as a consequence of the dynamics determined by $\mathscr{L}_0$, one does *not* recover a unique invariant vacuum when the symmetry breaking interaction is switched off,

$$Q_a|0\rangle \not\to 0 \quad \text{as} \quad c \to 0. \tag{4b}$$

Similarly, the mass splitting within multiplets does not disappear as $c \to 0$. This is the situation encountered in some current theories of the broken chiral symmetries of hadrons.

All that has been said about the vacuum (i.e. the temporally and spatially homogeneous state of a *spatially infinite* quantized field/many-body system) in these four situations may be generalized to the finite temperature ensemble in equilibrium statistical mechanics. The advantage of considering this more general problem is that there are well known forms of condensed matter which behave in each of the four ways just described according to the values taken by the temperature and some external field. The best known of these is the infinite ferromagnet, for which the symmetry group G is that of spatial rotations $SO(3)$ on the spin vectors associated with each lattice point. Its behaviour at various temperatures T and in various applied magnetic fields $\mathscr{H}$ may be categorized as follows.

Paramagnetic phase, $T > T_c$ $\begin{cases} \mathscr{H} = 0\text{: exact symmetry;} \\ \mathscr{H} \neq 0\text{: explicit symmetry breaking;} \end{cases}$

Ferromagnetic phase, $T < T_c$ $\begin{cases} \mathscr{H} = 0\text{: spontaneous symmetry breaking;} \\ \mathscr{H} \neq 0\text{: spontaneous and explicit symmetry breaking.} \end{cases}$

(T_c is the Curie temperature.)

The dynamical reason for the spontaneous breaking of rotational symmetry which occurs below the Curie temperature is well understood: it arises from an effective interaction between neighbouring spins which favours their parallel, as against antiparallel alignment.[7] The symptom of symmetry breaking is the non-vanishing expectation value of a dynamic variable, here the spin density or magnetization, which transforms non-trivially under the symmetry group.

Note that for spontaneous breaking to occur it is essential that the ferromagnet be infinite in extent. An unconstrained finite ferromagnet would be free to rotate and, for $\mathscr{H} = 0$, its rotational ground state (and its finite temperature ensemble of rotational levels) would be $SO(3)$ symmetric. But, as the size tends to infinity, so does the moment of inertia: the rotational fine structure collapses, and it becomes physically impossible to turn the system from whatever direction its spontaneous magnetization has taken. Corresponding to this "physicist's" account of the way in which vacuum degeneracy sets in as the volume becomes infinite, one finds in a more mathematical treatment of the problem that the matrix elements of measurable quantities between states built upon different directions of spontaneous magnetization all tend to zero in the infinite volume limit. The symmetry remains only in the sense that all measurable properties of the infinite ferromagnet are independent of its direction of spontaneous magnetization.[(9)]

At first sight spontaneous symmetry breaking appears to conflict with what one learns in elementary Lagrangian mechanics about the connection between symmetries and conservation laws. One is taught that the invariance of a Lagrangian under a Lie group implies that the infinitesimal generators of the group are conserved. But for a spatially infinite field theory there is a subtlety in the derivation of the conservation laws which is often overlooked. It is straightforward in classical field theory to show that the invariance of a Lagrangian density $\mathscr{L}$ under an infinitesimal transformation, $\delta\phi = \varepsilon f(\phi)$, of the fields $\phi(x)$ implies, modulo the Euler–Lagrange equations, the *differential* conservation law

$$\partial_\mu j^\mu = 0, \tag{5}$$

where

$$j^\mu = \frac{\partial \mathscr{L}}{\partial(\partial_\mu \phi)} f(\phi).$$

(In quantum field theory there are circumstances in which one runs into trouble—the Schwinger–Adler–Bell–Jackiw anomalies[(10)]—even at this stage if one relies on manipulations which are valid classically.) But the step from equation (5) to the conservation of a generator involves an integration over a temporally thin but spatially unbounded slice of space–time, and this is where the subtlety appears. One finds that the generator, which is given by

$$Q(t) = \int d^3x\, j^0(\mathbf{x}, t),$$

obeys the equation of motion

$$\dot{Q}(t) = - \lim_{r \to \infty} \int d\Omega \, r^2 j_r(\mathbf{x}, t) \tag{6}$$

where

$$j_r = \mathbf{j} \cdot (\mathbf{x}/r).$$

The right-hand side of eqn 6 vanishes only if

$$j_r = o(r^{-2}) \text{ as } r \to \infty. \tag{7}$$

This condition may not be satisfied, and indeed is not satisfied in situations where there is spontaneous symmetry breaking.

III. The Bogoliubov Model of a Superfluid

Ferromagnetism is the example of spontaneous symmetry breaking which is easiest to picture, because the symmetry which is broken is geometrical. But the non-relativistic system which has the most features in common with the simplest relativistic field theories which behave in this way is the model of a superfluid which Bogoliubov discussed in 1947,[8] the weakly non-ideal Bose gas.

A system of N identical spinless non-relativistic particles, which interact via two-body forces, is described by the Hamiltonian

$$H = (2m)^{-1} \sum_{i=1}^{N} \mathbf{p}_i^2 + \tfrac{1}{2} \sum_{i=1}^{N} \sum_{j=1}^{N} v(|\mathbf{x}_i - \mathbf{x}_j|).$$

As is well known, in the absence of interaction the ground state of this system would be that in which all N particles occupy the zero momentum one-particle state—the phenomenon known as Bose condensation. The question is, what happens in the presence of a weak repulsive potential v? It is convenient to introduce a second quantized formalism by defining a field operator

$$\psi(\mathbf{x}) = V^{-\frac{1}{2}} \sum_{\mathbf{p}} a_{\mathbf{p}} \, e^{i\mathbf{p}\cdot\mathbf{x}}, \tag{8}$$

where V is the volume of the gas, $a_{\mathbf{p}}$ is an annihilation operator for a particle of momentum $\mathbf{p}$ and periodic boundary conditions are imposed so that the spectrum of $\mathbf{p}$ includes $\mathbf{p} = 0$ even for finite V. Then the Fock space Hamiltonian may be written

$$\hat{H} = (2m)^{-1} \int_V d^3x \, \nabla\psi^\dagger \cdot \nabla\psi$$

$$+ \tfrac{1}{2} \iint_V \mathrm{d}^3x \, \mathrm{d}^3y \, v(|\mathbf{x} - \mathbf{y}|) \psi^\dagger(\mathbf{x}) \psi^\dagger(\mathbf{y}) \psi(\mathbf{x}) \psi(\mathbf{y})$$

$$- \mu \int_V \mathrm{d}^3x \, \psi^\dagger \psi, \tag{9}$$

into which, for the sake of describing an ensemble which exchanges particles as well as energy with its surroundings, we have introduced the term $-\mu Q$, where μ is the chemical potential and Q, defined by

$$Q = \int_V \mathrm{d}^3x \, \psi^\dagger \psi, \tag{10}$$

is the particle number operator.

Bogoliubov argued that, for $V \to \infty$ and $N \to \infty$ but $N/V \to v$ (a finite density), the effect of v upon the original ground state would be to induce a few excitations of particles to non-zero momentum states, leaving the zero momentum state on average *macroscopically* occupied: that is to say, the expectation value of $a_0^\dagger a_0$ would be of order N. It would therefore be legitimate to neglect the commutator, $[a_0, a_0^\dagger] = 1$, compared with $a_0^\dagger a_0$ in the limit $N \to \infty$ and to treat a_0 and $a_0^\dagger$ as c-numbers. This argument can be made rigorous.

The conclusion is equivalent to the statement that the usual expansion of $\psi(\mathbf{x})$ (the infinite volume counterpart of eqn (8)) is inappropriate and should be replaced by

$$\psi(\mathbf{x}) = \mathrm{e}^{i\alpha} \sqrt{v_0} + \chi(\mathbf{x}), \tag{11}$$

where α is an arbitrary phase, v_0 is the density of the Bose condensate and

$$\chi(\mathbf{x}) = (2\pi)^{-\frac{3}{2}} \int \mathrm{d}^3p \, a(\mathbf{p}) \, \mathrm{e}^{i\mathbf{p}\cdot\mathbf{x}}, \tag{12}$$

the point $\mathbf{p} = 0$ being excluded from the support of $a(\mathbf{p})$ by the condition

$$\int \mathrm{d}^3x \, \chi(\mathbf{x}) = 0.$$

The Hamiltonian $\hat{H}$ may now be split into a "free field" term $\hat{H}_0$, which contains everything of degree up to 2 in χ and $\chi^\dagger$, and an interaction Hamiltonian $\hat{H}_1$ which contains the cubic and quartic terms. In the approximation in which $\hat{H}_1$ is neglected, the excitation spectrum is then found by diagonalizing $\hat{H}_0$: this is done by means of a Bogoliubov transformation which forms "quasiparticle" annihilation and creation operators out of linear combinations of χ and $\chi^\dagger$. One finds for the energy ω of a quasiparticle of momentum $\boldsymbol{k}$ the expression

$$\omega(\mathbf{k}) = \{(v_0/m)k^2\tilde{v}(k) + \tfrac{1}{4}k^4/m^2\}^{\frac{1}{2}}, \tag{13}$$

where

$$\tilde{v}(k) = \int d^3x\, v(|\mathbf{x}|)\, e^{-i\mathbf{k}\cdot\mathbf{x}}.$$

The condensate density v_0 may be determined, given the chemical potential μ, by minimizing the ground state energy with respect to v_0.[11] It is given by

$$v_0\tilde{v}(0) = \mu. \tag{14}$$

You may be wondering what all this has to do with spontaneous symmetry breaking, since there has so far been no mention of a symmetry. The symmetry here is a somewhat abstract one, like the internal symmetries of particle physics: the Hamiltonian $\hat{H}$ is invariant under the transformation

$$\psi(\mathbf{x}) \to e^{-i\theta}\,\psi(\mathbf{x}) \tag{15}$$

where θ is an arbitrary real constant. The generator of an infinitesimal transformation is just the particle number operator Q which was defined in eqn (10):

$$\begin{aligned}[][Q, \psi(\mathbf{x})] &= -\psi(\mathbf{x}),\\ [Q, \psi^\dagger(\mathbf{x})] &= \psi^\dagger(\mathbf{x}).\end{aligned} \tag{16}$$

But the annihilation operator representation (11) is clearly not invariant under the transformation (15): it is taken into a mathematically inequivalent representation with α replaced by $\alpha - \theta$. To each value of α there corresponds a *mathematically* distinct ground state. Nevertheless, just as for the infinite ferromagnet, the *physical* content of the model, which (in the approximation where the effects of $\hat{H}_1$ are neglected) is given by equations (13) and (14), is independent of α, which affects only the details of the Bogoliubov transformation. The analogue of spontaneous magnetization is the non-zero value of $\langle 0|\psi(\mathbf{x})|0\rangle$ which, if $\hat{H}_1$ is neglected, is just $e^{i\alpha}\sqrt{v_0}$.

In this model the non-conservation of Q in the infinite volume limit, as described by equation (6), is easily understood. The Heisenberg equation of motion for $\psi(\mathbf{x}, t)$ implies the differential conservation law (5), in which

$$j^0 = \psi^\dagger\psi, \qquad \mathbf{j} = (2im)^{-1}(\psi^\dagger\nabla\psi - \psi\nabla\psi^\dagger). \tag{17}$$

Consider first the usual expansion

$$\psi(\mathbf{x}) = (2\pi)^{-\frac{3}{2}}\int d^3p\, a(\mathbf{p})\, e^{i\mathbf{p}\cdot\mathbf{x}}.$$

If ψ were a classical field, it would be possible to form normalizable wave packets (solutions for which $\int d^3x|\psi|^2$ exists) of finite momentum and energy: such wave packets must have $|\psi|^2 = o(r^{-3})$ as $r \to \infty$ and the dependence of the phase factor on r must be no worse than $e^{\pm ikr}$ for some k, whence condition (7) is satisfied by the current (17). Applying this reasoning to matrix elements of the quantum mechanical operators j^μ between states built from single-particle wave packets of finite momentum and energy, one proves that Q is conserved.

Consider now the broken symmetry condition (11). It is now $\chi(\mathbf{x})$ which contains the annihilation operators for single particle wave packets and should therefore have, in matrix elements, the asymptotic behaviour normally ascribed to ψ. One sees (i) that j_r contains a term

$$(2im)^{-1}\sqrt{v_0}\,(e^{-i\alpha}\,\partial\chi/\partial r - e^{i\alpha}\,\partial\chi^\dagger/\partial r)$$

which is $o(r^{-\frac{3}{2}})$ as $r \to \infty$ and therefore results in non-conservation of Q, and (ii) that the presence in j^0 of terms such as v_0 leads to the divergence of the integral which defines Q. (More precisely, some but not all of the matrix elements of Q are infinite, so in models containing a non-trivial Lie group one may still use the Lie algebra in a class of physically interesting matrix elements for the purpose of current algebra calculations.) One may describe what is happening here in a more physical way by saying that the Bose condensate forms an infinite reservoir of particles with which the observable quasi-particle excitations can exchange particle number.

IV. Phonon-like and Plasmon-like Excitations

The feature of the Bogoliubov superfluid which has the most relevance to the behaviour of relativistic models is its excitation spectrum (13). In particular, the energy-momentum relation at small momenta is "phonon-like" for all interparticle repulsions which are of short range, that is

$$\lim_{k\to 0}\frac{\omega(k)}{k} = \sqrt{\left(\frac{v_0\tilde{v}(0)}{m}\right)} \tag{18}$$

provided that $\tilde{v}(0)$ exists: the right-hand side of (18) is the phonon velocity.[12] In relativistic models similar behaviour would be indicative of the existence of massless "Goldstone bosons".

However, if for small k

$$\tilde{v}(k) = O(k^{-2}),$$

we see that $\omega(0)$ does not vanish, but is finite: this is characteristic of a long-range repulsion

$$v(r) = O(r^{-1}) \text{ as } r \to \infty.$$

Indeed the Coulomb potential itself,

$$v(r) = q^2/4\pi r$$

where q is the charge on each particle, has the Fourier transform

$$\tilde{v}(k) = q^2/k^2,$$

which leads to the result

$$\omega(0) = |q| \sqrt{(v_0/m)}. \tag{19}$$

The right-hand side of (19) is the plasma frequency and the excitations of such a charged gas are known as plasmons.[13]

Although a glance at eqn (14) will show that there are difficulties in the theory of a simple charged Bose gas—unless v_0 is zero, the chemical potential is infinite!—there are more complex models involving a gas of charged bosons which do not suffer from these difficulties. Such a model is the BCS superconductor,[6] where bosons of charge $-2e$ are formed by the binding of electron pairs in the neighbourhood of the Fermi surface : in the presence of the positive ions Bose condensation occurs at a finite value of the chemical potential, but the excitation spectrum at low k is dominated by the Coulomb repulsion between the bosons and is therefore plasmon-like with $|q| = 2e$. The suggestion that plasmon-like excitations might occur in relativistic systems too was first made by Anderson.[14]

One may prove that, for short range forces, the property $\omega(k) \to 0$ as $k \to 0$ is a consequence of the broken symmetry condition

$$\langle 0|\psi|0\rangle \neq 0,$$

rather than an artefact of the approximations used in deriving (13), by examining certain spectral functions of the model. Consider the quantities

$$L^\mu(k, t - t') = \int d^3x\, e^{-i\mathbf{k}\cdot(\mathbf{x}-\mathbf{x}')} \langle 0|[j^\mu(\mathbf{x}, t), \psi(\mathbf{x}', t')]|0\rangle$$

and the associated spectral functions

$$\rho^\mu(k, \omega) = \int dt\, e^{i\omega(t-t')} L^\mu(k, t - t').$$

From the differential conservation law (5) we deduce that

$$\lim_{k\to 0} \frac{\partial}{\partial t} L^0(k, t) = -\lim_{r\to\infty} \int d\Omega\, r^2 \langle 0|[j_r(\mathbf{x}, t), \psi(\mathbf{x}', 0)]|0\rangle. \tag{20}$$

From the equal time commutator

$$[j^0(\mathbf{x}, t), \psi(\mathbf{x}', t)] = -\psi(\mathbf{x}, t)\, \delta(\mathbf{x} - \mathbf{x}') \tag{21}$$

we deduce that

$$\lim_{k\to 0} L^0(k,0) = -\langle 0|\psi|0\rangle. \tag{22}$$

If the right-hand side of eqn (20) vanishes, then (20) and (21) combine to give the result

$$\lim_{k\to 0} \rho^0(k,\omega) = -2\pi\langle 0|\psi|0\rangle\, \delta(\omega), \tag{23}$$

which shows that the excitation spectrum contains a branch such that $\omega(k) \to 0$ as $k \to 0$.

Now the right-hand side of (20) may be calculated as a series expansion in t by evaluating multiple commutators of $[j_r(\mathbf{x},0), \psi(\mathbf{x}',0)]$ with $\hat{H}$: one finds that their asymptotic behaviour is governed by that of the potential v and that for short range forces they are $O(r^{-2})$, so the result (23) follows. But a Coulomb potential $v(r) = O(r^{-1})$ gives rise to terms proportional to $\mathrm{d}v/\mathrm{d}r = O(r^{-2})$ in the commutators, and one finds that the right-hand side of (20) is not zero but finite. In the approximation where $\hat{H}_1$ is neglected one obtains the explicit result that in eqn (23) $\delta(\omega)$ is replaced by a combination of $\delta(\omega \pm \omega(0))$, where $\omega(0)$ is the plasma frequency.[15]

V. Relativistic Models: The Goldstone Theorem

Let us now look at how the proof of the relation between spontaneous symmetry breaking and phonon-like excitations is modified in relativistic models. Suppose we have a relativistic field theory with a Lagrangian density which is invariant under the transformation (15), where now $\psi(x)$ is replaced by a non-hermitean scalar field $\phi(x)$. The definition of the spectral functions may be written in the manifestly Lorentz covariant form

$$\rho^\mu(k^0,\mathbf{k}) = \int \mathrm{d}^4x\, \mathrm{e}^{ik_\nu x^\nu} \langle 0|[j^\mu(x), \phi(0)]|0\rangle. \tag{24}$$

(The metric is "West Coast": $+ - - -$.)

The differential conservation law (5) implies that

$$k_\mu \rho^\mu = 0. \tag{25}$$

But the Lorentz invariance of the theory (ϕ and j^μ being respectively scalar and vector) implies that ρ^μ must be a vector function of k^μ and the sign of k^0, spacelike vectors being excluded from its support:

$$\rho^\mu = k^\mu\{\sigma_1(k^2)\,\mathrm{sign}\, k^0 + \sigma_2(k^2)\}. \tag{26}$$

Combining (25) and (26) we get

$$\sigma_i(k^2) = C_i\delta(k^2). \tag{27}$$

The equal time commutator (21) implies the sum rule

$$\int_{-\infty}^{+\infty} \mathrm{d}k^0\, \rho^0(k^0, \mathbf{0}) = -\,2\pi\langle 0|\phi|0\rangle, \tag{28}$$

which determines C_1:

$$C_1 = -\,2\pi\langle 0|\phi|0\rangle. \tag{29}$$

Thus we have arrived at the Goldstone theorem:[16] in a manifestly covariant local quantum field theory spontaneous symmetry breaking gives rise to massless scalar bosons.[17] We have obtained this result without any of the discussion of the asymptotics of forces and of matrix elements of commutators containing j_r which was necessary in the non-relativistic case! We shall see when we look at specific models that our assumptions have in fact implicitly excluded long range forces.

VI. The Goldstone Model

The first[18] relativistic models to incorporate spontaneous symmetry breaking were those of Nambu and Jona–Lasinio.[4] They were designed to break a hadronic chiral symmetry by means of a fermion pairing mechanism analogous to that which occurs in a superconductor. The Lagrangian density in these models has the form[19]

$$\mathscr{L} = \bar{\psi} i\gamma^\mu \partial_\mu \psi + G\{(\bar{\psi}\psi)^2 + (\bar{\psi} i\gamma_5 \psi)^2\}, \tag{30}$$

in which the chiral invariant Fermi interaction was chosen so as to be attractive for low energy fermion–antifermion pairs of zero total linear and angular momentum. A massless bound state with the quantum numbers of the pion[20] was predicted. However, since the Lagrangian (30) is non-renormalizable, it was not clear at the time whether this phenomenon was genuine or merely an artefact of the approximations which had been used.

An important contribution towards the formulation of more credible models was made in the same year by Goldstone.[21] In order to have a model incorporating a spontaneously broken chiral symmetry which might be renormalizable, he replaced the Fermi interaction (30) by Yukawa interactions of the fermions with scalar and pseudoscalar fields, and induced spontaneous symmetry breaking by means of a quartic self interaction of these new fields.[22] The spin zero part of Goldstone's model, in its simplest form in which the broken symmetry group is an internal $U(1)$, is nothing but a relativistic version of the Bogoliubov superfluid, as the following considerations will show.

In order to satisfy all the requirements of relativity, the Bogoliubov Hamiltonian (9) must be modified in three respects:

(i) To take account of pair creation, which prevents us from having a conserved particle number, we must introduce distinct antiparticles and interpret the generator Q of a phase transformation (15) on a local non-hermitean scalar field $\phi(x)$ as the difference between particle and anti-particle number operators;

(ii) In the kinetic energy term the non-relativistic relation $E = p^2/2m$ must be replaced by $E = (p^2 + m^2)^{\frac{1}{2}}$;

(iii) To satisfy the requirement of causality either (a) the interaction must have a zero range,

$$v(|\mathbf{x}|) = \tilde{v}(0)\delta(\mathbf{x}),$$

or (b) we must introduce extra fields to mediate the interaction (e.g. the Maxwell field to mediate the Coulomb interaction). If we choose alternative (a) these three modifications lead[23] to a Lagrangian density of the form

$$\mathscr{L} = \partial_\mu \phi^\dagger \partial^\mu \phi + a\phi^\dagger \phi - b(\phi^\dagger \phi)^2, \tag{31}$$

where a and b are positive constants. This is Goldstone's simplest model.

Besides being renormalizable,[24] the Goldstone model has the merit of providing an example of spontaneous symmetry breaking which is easy to visualize and for which consequently the Goldstone theorem is easy to grasp intuitively. It is convenient to write the Lagrangian (31) in terms of hermitean fields ϕ_1 and ϕ_2, defined by

$$\phi = 2^{-\frac{1}{2}}(\phi_1 - i\phi_2), \; \phi^\dagger = 2^{-\frac{1}{2}}(\phi_1 + i\phi_2),$$

in a two-dimensional vector notation as

$$\mathscr{L} = \tfrac{1}{2}\partial_\mu \boldsymbol{\phi} \cdot \partial^\mu \boldsymbol{\phi} - \mathscr{V}(\boldsymbol{\phi}), \tag{32}$$

where the potential energy density $\mathscr{V}$ is given by

$$\mathscr{V}(\boldsymbol{\phi}) = -\tfrac{1}{2}a\boldsymbol{\phi} \cdot \boldsymbol{\phi} + \tfrac{1}{4}b(\boldsymbol{\phi} \cdot \boldsymbol{\phi})^2. \tag{33}$$

Now consider what this Lagrangian describes when the coordinates x are restricted to one dimension of space and one of time: one may imagine an infinite uniform string,[25] which can undergo transverse displacements $\boldsymbol{\phi}(x)$ relative to the axis $\boldsymbol{\phi} = 0$, to be placed in the axially symmetric potential (33). Since this potential corresponds to repulsion from the axis for small $\boldsymbol{\phi}$ and attraction towards it for large $\boldsymbol{\phi}$, the string as a classical system has its configuration of stable equilibrium where $\mathscr{V}$ is stationary and of non-negative curvature with respect to $\boldsymbol{\phi}$: that is anywhere in the bottom of the "trough" of $\mathscr{V}(\boldsymbol{\phi})$,

$$\boldsymbol{\phi} \cdot \boldsymbol{\phi} = a/b.$$

This condition defines a one-parameter family of stable equilibria, given by

$$\boldsymbol{\phi} = \lambda(\cos\alpha, \sin\alpha), \qquad \lambda^2 = a/b. \tag{34}$$

Although the matrix $\partial^2 \mathscr{V}/\partial\phi_i\partial\phi_j$ has a zero eigenvalue at these values of $\boldsymbol{\phi}$ on account of the axial symmetry, the equilibrium is *effectively* stable rather than neutral with respect to *uniform* displacements because the system at equilibrium has an infinite moment of inertia about the axis and therefore cannot be excited as a rigid rotator.(27) However, with respect to non-uniform displacements (to be more precise, those which tend to zero in both directions of the axis) the equilibrium *is* neutral in the sense that a wave of small amplitude, polarized in the direction orthogonal to the actual equilibrium configuration (34) (i.e. parallel to the bottom of the potential trough) and containing only small wavenumbers, can be excited at negligible cost in terms of potential energy, whether in the form of $\mathscr{V}$ or as elastic energy. For such an excitation one therefore expects

$$\lim_{k\to 0} \omega(k) = 0.$$

This is the Goldstone phenomenon.

The foregoing description was intended to provide some insight into another aspect—the classical field theoretic aspect—of spontaneous symmetry breaking. More formally, one finds the Goldstone excitation by examining the normal modes of small oscillation of the system about an equilibrium, chosen from the family (34). If one picks, say, $\alpha = 0$, then one has to linearize the field equations in terms of the displacements

$$\phi_1'(x) = \phi_1(x) - \lambda, \qquad \phi_2'(x) = \phi_2(x).$$

The linearized equations are

$$(\Box + m_0^2)\phi_1' = 0, \qquad m_0^2 = 2a, \tag{35}$$

and

$$\Box\phi_2' = 0. \tag{36}$$

The Fourier transform of the latter equation yields the dispersion relation of the Goldstone excitation. These results are, of course, independent of the number of coordinates x.

In the corresponding quantized field theory the classical treatment provides the tree approximation to various physically interesting quantities.(28) Thus, in this approximation, the vacuum expectation value $\langle 0|\boldsymbol{\phi}(x)|0\rangle$ is the classical equilibrium value, given by equation (34), and equations (35) and (36) are the relativistic wave equations for the particles which are the quanta of the corresponding small oscillations. In particular, eqn (36) describes a "Goldstone boson" of zero mass and spin. This particle does not acquire mass in higher orders of perturbation theory; the proof of the Goldstone theorem given earlier is applicable to the exact Lagrangian, j_μ being given by

$$j_\mu = i(\phi^\dagger \partial_\mu \phi - \phi \partial_\mu \phi^\dagger). \tag{37}$$

VII. How to Evade the Goldstone Theorem: Gauge Theories

Following the work of Goldstone, a number of attempts[3] were made to explain the breaking of hadronic $SU(2)$ and $SU(3)$ symmetries as spontaneous breaking induced by the self interactions of suitably chosen multiplets of scalar fields: all these models suffered from the same defect, the prediction of strongly interacting massless scalar particles. Meanwhile, the Goldstone theorem,[16] which showed that this was to be expected, was proved with greater and greater rigour: eventually it became a showpiece of axiomatic quantum field theory.[29, 30] It seemed puzzling to many people[31] at this time that there should be such a qualitative difference between the dynamical consequences of spontaneous symmetry breaking in relativistic models and in non-relativistic systems, where one example, the superconductor, was known not to have phonon-like excitations. The resolution of this puzzle came when it was pointed out[32] that the axioms upon which the Goldstone theorem was based were so restrictive as to be inapplicable to an important class of relativistic theories, namely gauge theories: since the only quantum field theory which could be judged successful at that date, quantum electrodynamics, is a gauge theory, this was indeed a serious restriction!

It is simple to give a formal explanation of why the Goldstone theorem is irrelevant to gauge theories. In such a theory each conserved current j_μ is coupled via the Maxwell equation

$$\partial_\mu F^{\mu\nu} = ej^\nu, \tag{38}$$

where e is the coupling constant, to an antisymmetric field tensor $F_{\mu\nu}$ which satisfies the other Maxwell equation

$$\partial_\lambda F_{\mu\nu} + \partial_\mu F_{\nu\lambda} + \partial_\nu F_{\lambda\mu} = 0. \tag{39}$$

The Lagrangian formulation of the theory, upon which the quantized version is based, uses as dynamic variables the potentials A_μ, defined by

$$F_{\mu\nu} = \partial_\mu A_\nu - \partial_\nu A_\mu \tag{40}$$

for which eqn (39) is the integrability condition. But these potentials are defined by eqn (40) only up to a gauge transformation

$$A_\mu \to A_\mu + \partial_\mu \chi$$

and a choice of gauge must be made in order to specify the corresponding quantum mechanical operators. As anyone who is acquainted with quantum electrodynamics knows, one can preserve manifest Lorentz covariance, by choosing a Lorentz gauge, only at the cost of including among the states unphysical "longitudinal and scalar photon" states: this is a high price to pay if one is primarily interested in the possible existence of a Goldstone boson.[33] If one wishes to have only physical degrees of freedom in the formalism, one is forced to use a gauge, such as a radiation gauge, in which the operators depend explicitly on a distinguished direction in space–time (a timelike direction for radiation gauge). But once manifest covariance is lost the spectral functions defined in eqn (24) need not have the form (26); instead in radiation gauge the Maxwell equations (38) and (40), the sum rule (28) and the Lorentz invariance of the spectrum together lead to the representation[34]

$$\rho_\mu = (n\,.\,k)\left\{\frac{(n\,.\,k)k_\mu - k^2 n_\mu}{(n\,.\,k)^2 - k^2}\right\}\sigma(k^2, \operatorname{sign} k_0), \tag{41}$$

where n_μ is a timelike unit vector, $n\,.\,k = n_\mu k^\mu$ and

$$\int_{-\infty}^{+\infty} \mathrm{d}k_0 k_0 \sigma(k_0^2, \operatorname{sign} k_0) = -2\pi\langle 0|\phi|0\rangle. \tag{42}$$

The existence of a phonon-like excitation (massless boson) is now just as dependent upon the dynamics by which σ is determined as was the case in non-relativistic theories. If, as is natural, we choose $n_\mu = (1, \mathbf{0})$, then it becomes obvious that the spatial components of ρ_μ have a $\mathbf{k}(\mathbf{k}^2)^{-1}$ singularity which ensures that the right-hand side of eqn (20) does not vanish, unless the coefficient of this singularity is zero as a consequence of σ having no support other than $k^2 = 0$. Thus there is no escape from the necessity of tackling the dynamics. Proofs of the Goldstone theorem which abandon manifest covariance but argue that relativistic causality supplies the needed asymptotics in eqn (20)[29] fail because in radiation gauge A_0 propagates instantaneously without disturbing the causal propagation of anything observable: this is the origin of the denominator in eqn (41).

At this point we should abandon formalism and consider what is involved from a more physical point of view. We saw earlier that it is the Coulomb repulsion which in non-relativistic systems turns the phonon-like excitation into a plasmon: that this is so in a superconductor was first stressed by Anderson,[14] who speculated that a similar phenomenon might occur in relativistic models. Gauge invariance is merely the price which one must pay for (or should I say the bonus which one gains from?) the electrodynamic extras which have to accompany the Coulomb force in a relativistic model.

VIII. Gauge Invariance and Mass

In parallel with the development of relativistic models involving spontaneous symmetry breaking there was for many years an independent line of development, consisting of attempts to build gauge models of strong, electromagnetic and weak interactions.[35] The starting point for this programme is the observation that the electromagnetic interactions of charged particles may be "generated" by the requirement that the Lagrangian density of the non-hermitean fields ϕ (or the corresponding spinor fields if the spin is not zero) be invariant under the *local* generalization of the transformation (15), in which θ is allowed to be a function of the space–time coordinates x. In order to maintain the invariance of derivative terms, the ordinary derivative $\partial_\mu \phi$ has to be replaced by the "covariant derivative", defined by

$$\nabla_\mu \phi = \partial_\mu \phi + ie\, A_\mu \phi \tag{43}$$

in which the potentials A_μ undergo, as the counterpart of (15), the gauge transformation

$$A_\mu \to A_\mu + e^{-1} \partial_\mu \theta. \tag{44}$$

The Lagrangian density, thus modified, must be completed by the addition of the gauge invariant expression $\frac{1}{4} F_{\mu\nu} F^{\mu\nu}$ which provides a kinematic term for the Maxwell field: no other $F_{\mu\nu}$-dependent term appears to be compatible with renormalizability. If, in place of (15), the transformations which are to be made coordinate-dependent form a general compact Lie group, then the requirement of local invariance necessitates the introduction of a multiplet of potentials $A_{\mu\alpha}$ belonging to the adjoint representation of the group.[36]

The principal obstacle to the success of this programme too was the apparent connection between gauge invariance and the zero mass of the quanta of the vector fields $A_{\mu\alpha}$. The vanishing of the renormalized photon mass in quantum electrodynamics had long been known to be a consequence of gauge invariance[37] and it seemed clear that the same arguments could be applied to every one-parameter subgroup of a general gauge group. However, it was shown by Schwinger[38] that gauge invariance alone does not prevent the spectral representation of the A_μ propagator from containing a delta function at $m \neq 0$; he speculated that such a massive spin one state might develop for sufficiently large values of the coupling constant e. At about the same time Boulware and Gilbert[39] constructed a model in which the gauge transformation (44) compensated the effect, not of the phase transformation (15), but of the inhomogeneous transformation

$$\phi(x) \to \phi(x) + \lambda\theta(x) \tag{45}$$

of a hermitean scalar field. It was nothing other than a Stueckelberg decomposition of a free massive vector field and its significant feature was the proportionality of the mass to the ϕ–A_μ coupling, which proved the feasibility of generating mass even with weak coupling, at least in a dynamically trivial model.

The credit for the first demonstration that spontaneous symmetry breaking, in the form of non-vanishing vacuum expectation values of a multiplet of scalar fields, provides the mass-generating mechanism in gauge theories belongs to Englert and Brout,[(40)] who calculated the $(\text{mass})^2$ matrix for the potentials $A_{\mu a}$ in the lowest order (e^2) of perturbation theory in which it occurs. Almost simultaneously, and independently, the realization that gauge theories provide the escape route from the Goldstone theorem led me to investigate the gauge invariant extension of models of the Goldstone type and to discover the same mechanism in a classical description of the fields.[(41)] Also independently and almost simultaneously, Guralnik, Hagen and Kibble[(42)] analysed the generation of mass in the Boulware–Gilbert model in terms of spontaneous breaking of symmetry under the transformation (45). Thus the two hitherto parallel lines of theoretical development, spontaneous symmetry breaking and gauge invariance, suddenly converged in a mutually beneficial union.

IX. Spontaneous Breaking of a Local Symmetry: The Gauge Invariant Extension of the Goldstone Model

The simplest dynamically non-trivial relativistic model[(40, 41)] which exhibits spontaneous breaking of a local symmetry is got from the Goldstone model by replacing $\partial_\mu \phi$ and $\partial_\mu \phi^\dagger$ respectively in the Lagrangian density (31) by the covariant derivatives

$$\nabla_\mu \phi = \partial_\mu \phi + ieA_\mu \phi,$$
$$\nabla_\mu \phi^\dagger = \partial_\mu \phi^\dagger - ieA_\mu \phi^\dagger$$

and adding the kinetic term $-\frac{1}{4}F_{\mu\nu}F^{\mu\nu}$. In the two-dimensional vector notation, one replaces (32) by

$$\mathscr{L} = \tfrac{1}{2}\nabla_\mu \boldsymbol{\phi} \cdot \nabla^\mu \boldsymbol{\phi} - \mathscr{V}(\boldsymbol{\phi}) - \tfrac{1}{4}F_{\mu\nu}F^{\mu\nu} \tag{46}$$

where

$$(\nabla_\mu \phi)_1 = \partial_\mu \phi_1 + eA_\mu \phi_2,$$
$$(\nabla_\mu \phi)_2 = \partial_\mu \phi_2 - eA_\mu \phi_1.$$

The Lagrangian density (46) is invariant under the local internal "rotation"

$$\begin{aligned}\phi_1(x) &\to \phi_1(x)\cos\theta(x) - \phi_2(x)\sin\theta(x),\\ \phi_2(x) &\to \phi_1(x)\sin\theta(x) + \phi_2(x)\cos\theta(x),\\ A_\mu(x) &\to A_\mu(x) + e^{-1}\partial_\mu\theta(x).\end{aligned} \tag{47}$$

The field equations of the model are

$$\begin{aligned}(\nabla_\mu\nabla^\mu\phi)_i + \partial\mathscr{V}/\partial\phi_i = 0, \qquad i = 1, 2,\\ \partial_\mu F^{\mu\nu} = ej^\nu = e(\phi_1\nabla^\nu\phi_2 - \phi_2\nabla^\nu\phi_1).\end{aligned} \tag{48}$$

These equations have a coordinate-independent solution (the equilibrium solution in the classical interpretation) which is the same as that of the Goldstone model, which for the potential (33) is given by eqn (34), together with

$$A_\mu = 0. \tag{49}$$

Once again, one may pick a particular direction, say $\alpha = 0$, for the equilibrium and study small oscillations of the system about this equilibrium by linearizing the eqns (48) in the displacements

$$\phi_1' = \phi_1 - \lambda, \qquad \phi_2' = \phi_2, \qquad A_\mu' = A_\mu \tag{50}$$

relative to the solution (34), (49). One finds that the massive scalar excitation (35) is unchanged, but that the Goldstone excitation (36) is replaced by

$$\partial_\mu(\partial^\mu\phi_2 - e\lambda A^\mu) = 0 \tag{51}$$

and that the linearized version of the Maxwell equation is

$$\partial_\mu F^{\mu\nu} = e\lambda(\partial^\nu\phi_2 - e\lambda A^\nu). \tag{52}$$

Equation (52), which is nothing other than a relativistic analogue of the phenomenological description of electromagnetic wave propagation in a superconductor, becomes immediately recognizable if we define a new vector field

$$B_\mu = A_\mu - (e\lambda)^{-1}\partial_\mu\phi_2. \tag{53}$$

It is the Proca equation[43]

$$\begin{aligned}\partial_\mu F^{\mu\nu} + m_1^2 B^\nu = 0,\\ F_{\mu\nu} = \partial_\mu B_\nu - \partial_\nu B_\mu,\end{aligned} \tag{54}$$

which on quantization describes a particle of spin 1 and mass

$$m_1 = e\lambda. \tag{55}$$

Equation (51) becomes the transversality condition

$$\partial_\mu B^\mu = 0.$$

In the quantized theory, eqns (34) and (49) determine vacuum expectation values and eqns (35), (51) and (52) determine propagators of the fields in the tree approximation. Because of the gauge invariance of the model, propagators are not fully determined until a choice of gauge (such as radiation gauge) has been made. Then one may proceed to calculate, in the same approximation (i.e. no closed loops in Feynman diagrams), the amplitudes for various physical processes (decay, scattering, etc.) involving the massive spin 0 and spin 1 particles described by eqns (34) and (54). One finds[(2)] that, although individual Feynman diagrams are gauge dependent, all gauge dependence disappears when one sums all the diagrams which contribute to a given process in the tree approximation. The resulting amplitudes are just those which one would calculate from the Lagrangian density

$$\mathscr{L}' = \tfrac{1}{2}\partial_\mu\phi\partial^\mu\phi - \mathscr{V}(\phi) - \tfrac{1}{4}F_{\mu\nu}F^{\mu\nu} + \tfrac{1}{2}e^2\phi^2 B_\mu B^\mu, \tag{56}$$

where $\phi(x)$ is now a Hermitian scalar field and

$$\mathscr{V}(\phi) = -\tfrac{1}{2}a\phi^2 + \tfrac{1}{4}b\phi^4,$$

in the tree approximation about the stable vacuum, given by

$$\langle 0|\phi(x)|0\rangle = \lambda, \qquad \langle 0|B_\mu(x)|0\rangle = 0.$$

But the Lagrangians (46) and (56) are equivalent, in classical field theory at least, in the sense that $\mathscr{L}'$ is got from $\mathscr{L}$ by the change of variables

$$(\phi_1(x), \phi_2(x)) = \phi(x)(\cos\alpha(x), \sin\alpha(x)),$$
$$A_\mu(x) = B_\mu(x) + e^{-1}\partial_\mu\alpha(x). \tag{57}$$

This transformation puts all the gauge dependence of the original Lagrangian $\mathscr{L}$ into the variable $\alpha(x)$,[(44)] which does not appear in the equivalent Lagrangian $\mathscr{L}'$.

Authors of subsequent papers on this model have preferred to use manifestly covariant gauges.[(33)] Its renormalizability was proved, in a family of such gauges, by B. W. Lee.[(45)] In these covariant gauges, the proof of renormalizability is fairly straightforward (one has to verify that, as in the σ-model, the renormalizability which one has for $a < 0$ in (33) is not spoiled by spontaneous symmetry breaking for $a > 0$, but to verify unitarity one has to check the cancellation of diagrams containing "scalar photons" with those containing "Goldstone ghosts" in intermediate states. Hence these gauges have become known as *renormalizable gauges*. If, on the other hand, one uses the equivalent Lagrangian (56), unitarity is manifest, but renormalizability is the consequence of intricate cancellations of most of the divergences to which the massive vector field propagator $(g_{\mu\nu} - m_1^{-2}k_\mu k_\nu)/(k^2 - m_1^2)$ normally gives rise. Hence this formalism has become known as the *unitary gauge*.

Note that in the unitary gauge, which corresponds to the phenomenological Lagrangian $\mathscr{L}'$ which one would write down if one knew the decay and scattering amplitudes, the original symmetry is quite unrecognizable since the fields $\phi(x)$ and $B_\mu(x)$ are invariants. Nevertheless, certain features of the structure of $\mathscr{L}'$ which are responsible for its renormalizability arise from the local symmetry of $\mathscr{L}$. The moral of the story for phenomenologists is perhaps that the most obvious formulation of a model of particle interactions may sometimes conceal its most important dynamical features.

X. Summary and Outlook

In Table I, I have summarized first the principal features of the two nonrelativistic systems whose dynamics I have discussed and second (below the double line) those of the relativistic models which developed out of them.

Nambu and Jona-Lasinio based their model on the BCS superconductor to the extent of spontaneously breaking their symmetries with *composite* scalar fields—this pairing mechanism is what is now called dynamic spontaneous symmetry breaking; but they failed to appreciate the importance of introducing Coulomb forces, and they used a nonrenormalizable interaction.

TABLE I

Model	Fields with $\langle 0\|\phi_i\|0\rangle \neq 0$	Coulomb force/ local symmetry	Excitation spectrum	Renormalizability in perturbation theory
Bogoliubov superfluid	Elementary	No	Phonons	—
BCS superconductor	Composite	Yes	Plasmons	—
Nambu and Jona-Lasinio	Composite	No	Goldstone bosons	Nonrenormalizable
Goldstone	Elementary	No	Goldstone bosons	Renormalizable
Higgs	Elementary	Yes	Massive spin 1 particles	Renormalizable
X	Composite	Yes	Massive spin 1 particles	Renormalizable

Goldstone introduced a considerable conceptual simplification by formulating renormalizable models of self-interacting scalar fields which resemble the Bogoliubov superfluid. My own contribution (based on the remark of Anderson), and that of Englert and Brout, was to recognize that the introduction of a Coulomb force, under the formal guise of a local symmetry, would turn the massless Goldstone bosons of the earlier models into the longitudinal states of massive vector particles.

Despite the fact that the models which contain elementary scalar fields were first introduced primarily for pedagogical reasons and were recognized[(46)] as providing probably only a phenomenological description of symmetry breaking in particle physics, they have been very fruitful lately as a basis for unifying weak, electromagnetic and even strong interactions.[(47)] In fact they have been too fruitful: the rules of the game are not strict enough and it is becoming recognized that, for reasons of theoretical economy alone, it may be desirable to restrict the group representations to which the scalars belong to those which can be found among the products $\bar{\psi}\psi$ of some elementary fermion fields. Dynamical spontaneous breaking of a local symmetry in a renormalizable model—the Brand X of the last entry of my table—thus appears as the latest objective for model builders.[(48)] Additional motivation for this new phase of the programme comes from the recent discovery that elementary scalar fields pose a threat to the asymptotic freedom which has recently been shown to be a feature only of non-Abelian gauge theories[(49)] and which provides a possible theoretical basis for Bjorken scaling.[(50)]

References

1. S. Weinberg. *Phys. Rev. Letts.* **19**, 1264 (1967); A. Salam. "Elementary Particle Physics." Ed. N. Svartholm p. 367. (Almquist & Wiksells, Stockholm) 1968.
2. For a detailed account of the simplest such model see P. W. Higgs. *Phys. Rev.* **145**, 1156 (1966). For a preliminary account see F. Englert and R. Brout. *Phys. Rev. Letts.* **13**, 321 (1964); P. W. Higgs. *Phys. Rev. Letts.* **13**, 508 (1964). For a more detailed account of the work of the former authors see F. Englert, R. Brout and M. F. Thiry. *Nuovo Cimento* **43A**, 244 (1966).
3. M. Baker and S. L. Glashow. *Phys. Rev.* **128**, 2462 (1962); S. L. Glashow. *Phys. Rev.* **130**, 2132 (1962); and several other authors.
4. Y. Nambu. Tenth International Conference on High Energy Physics, Rochester, 1960; Y. Nambu and G. Jona-Lasinio. *Phys. Rev.* **122**, 345 (1961); **124**, 246 (1961).
5. Y. Nambu, *Phys. Rev.* **117**, 648 (1960).
6. J. Bardeen, L. N. Cooper and J. R. Schrieffer. *Phys. Rev.* **106**, 162 (1957).
7. W. Heisenberg. *Z. Physik* **49**, 619 (1928). Spontaneous symmetry breaking was already a feature of the phenomenological theory of P. Weiss. *J. de Phys.* **6**, 667 (1907).

8. N. Bogoliubov. *J. Phys. USSR* **11**, 23 (1947).
9. I refer to a theoretical model which neglects the anisotropy arising from spin-lattice interactions in a real crystal.
10. J. Schwinger. *Phys. Rev.* **82**, 664 (1951); S. Adler. *Phys. Rev.* **177**, 2426 (1969); J. S. Bell and R. Jackiw. *Nuovo Cimento* **60**, 47 (1969).
11. This is not how it is done in Bogoliubov's paper, where it is v_0 which is fixed and μ which is varied, but the methods are easily shown to be equivalent.
12. The importance of this result for the theory of superfluidity is that, if a fluid has an excitation spectrum for which the minimum value of $\omega(k)/k$ is not zero, at flow velocities below this value the conversion of kinetic energy of flow into excitations, which at higher velocities gives rise to viscous damping, cannot occur, as was first shown by L. D. Landau. *J. Phys. USSR* **5**, 71 (1941).
13. Note that superfluidity can still occur, since the minimum of $\omega(k)/k$ is again non-zero.
14. P. W. Anderson. *Phys. Rev.* **130**, 439 (1963).
15. The arguments described here are of very general applicability, since they use only the differential conservation law (5) and the equal time commutator (21) which is the local expression of the transformation (15). They were first presented in detail in the context of ferromagnetism by R. V. Lange. *Phys. Rev. Letts.* **14**, 3 (1965); *Phys. Rev.* **146**, 301 (1966).
16. J. Goldstone, A. Salam and S. Weinberg. *Phys. Rev.* **127**, 965 (1962).
17. The definition (24) shows that they occur in intermediate states which are created from the vacuum by the scalar field $\phi(x)$.
18. I refer to spontaneous breaking of a symmetry under a Lie group. Suggestions that parity violation may be the result of spontaneous breaking of space reflection symmetry occur as early as 1958 in some of the papers of Salam and of Lee and Yang.
19. $SU(2)$ indices have been suppressed here.
20. An earlier model in which the pion is generated as a nucleon-antinucleon bound state is that of E. Fermi and C. N. Yang. *Phys. Rev.* **76**, 1739 (1949).
21. J. Goldstone. *Nuovo Cimento* **19**, 154 (1961).
22. The model is a version of the linear σ model: J. Schwinger. *Ann. Phys.* (*N.Y.*) **2**, 407 (1957); M. Gell-Mann and M. Lévy, *Nuovo Cimento* **16**, 705 (1960).
23. Details which have been hand-waved out of sight here include the choice of the conventional normalization for relativistic wave functions and the absorption of rest energies into the chemical potential.
24. The renormalizability of the linear σ model was proved by B. W. Lee. *Nucl. Phys.* **B9**, 649 (1969).
25. "Natural" units here are such as to give the value unity to the mass per unit length and the tension, and hence to the speed of wave propagation when $\mathscr{V} = 0$.
26. Such a situation would be realized if the string were uniformly charged and placed in a cylindrically symmetric static electric field, provided that the interaction between different parts of the string could be neglected.
27. Recall the infinite ferromagnet!
28. For a discussion of the relation between a classical field theory and the tree approximation to the corresponding quantized theory see D. G. Boulware and L. S. Brown. *Phys. Rev.* **172**, 1628 (1968).
29. R. F. Streater, *Proc. Roy. Soc.* (*London*) **A287**, 510 (1965).
30. D. Kastler, D. W. Robinson and A. Swieca, *Commun. Math. Phys.* **2**, 108 (1966).

31. A. Klein and B. W. Lee. *Phys. Rev. Letts.* **12**, 266 (1964). For the ensuing discussion see W. Gilbert. *Phys. Rev. Letts.* **12**, 713 (1964) and Ref. 32.
32. P. W. Higgs. *Phys. Letts.* **12**, 132 (1964).
33. That in a Lorentz gauge, where the Goldstone theorem is valid, the Goldstone boson is unphysical was first shown, in the model of Ref. 41, by T. W. B. Kibble, *Phys. Rev.* **155**, 1554 (1967).
32. A partial derivation of this representation is given in Ref. 32; an explicit example of it is discussed in the first paper of Ref. 2.
35. The following papers are primarily about strong interactions: C. N. Yang and R. L. Mills. *Phys. Rev.* **96**, 191 (1954); R. Shaw. Dissertation, Cambridge University (1954, unpublished); R. Utiyama. *Phys. Rev.* **101**, 1597 (1956); A. Salam and J. C. Ward. *Nuovo Cimento* **11**, 568 (1959); J. J. Sakurai. *Ann. Phys.* (*N.Y.*) **11**, 1 (1960); A. Salam and J. C. Ward. *Phys. Rev.* **136**, B763 (1964). The following are on weak interactions: S. A. Bludman. *Phys. Rev.* **100**, 372 (1955); *Nuovo Cimento* **9**, 433 (1958); S. L. Glashow. *Nucl. Phys.* **10**, 107 (1959); **22**, 579 (1961); A. Salam and J. C. Ward. *Phys. Letts.* **13**, 168 (1964). The last paper is a model of unified weak and electromagnetic interactions which contains many of the features of the later Weinberg–Salam model, Ref. 1.
36. M. Gell-Mann and S. L. Glashow. *Ann. Phys.* (*N.Y.*) **15**, 437 (1961).
37. J. Schwinger. *Phys. Rev.* **75**, 651 (1949); K. Johnson. *Nucl. Phys.* **25**, 435 (1961).
38. J. Schwinger. *Phys. Rev.* **125**, 397 (1962); **128**, 2425 (1962).
39. D. G. Boulware and W. Gilbert. *Phys. Rev.* **126**, 1563 (1962).
40. F. Englert and R. Brout. *Phys. Rev. Letts.* **13**, 321 (1964).
41. P. W. Higgs. *Phys. Rev. Letts.* **13**, 508 (1964).
42. G. S. Guralnik, C. R. Hagen and T. W. B. Kibble. *Phys. Rev. Letts.* **13**, 585 (1964).
43. This part of the linearized model is just the system studied in Refs 39 and 42.
44. The gauge transformation of the second kind becomes simply $\alpha(x) \rightarrow \alpha(x) - \theta(x)$.
45. B. W. Lee. *Phys. Rev.* **D5**, 823 (1972).
46. Englert and Brout, Ref. 40, stress this and give an example of how vector particle masses are generated in a model where the scalars are composite.
47. See Ref. 1 and the references following C. H. Llewellyn-Smith's contribution to this volume.
48. Among the first crop of detailed models of this type are those of J. M. Cornwall and R. E. Norton, *Phys. Rev.* **D8**, 3338 (1973); R. Jackiw and K. Johnson, *Phys. Rev.* **D8**, 2386 (1973).
49. G. 't Hooft, remark at the Marseille Seminar on Yang–Mills Theories, 1972; H. D. Politzer. *Phys. Rev. Letts* **30**, 1346 (1973); D. J. Gross and F. Wilczek. *Phys. Rev. Letts.* **30**, 1343 (1973); *Phys. Rev.* **D8**, 3633 (1973); S. Coleman and D. J. Gross. *Phys. Rev. Letts.* **31**, 851 (1973).
50. C. G. Callan and D. J. Gross, *Phys. Rev.* **D8**, 4383 (1973).

8. ALGEBRAIC PROPERTIES OF ELECTROMAGNETIC AND WEAK CURRENTS AND THEIR REPRESENTATION ON HADRON STATES

FREDERICK J. GILMAN

Stanford Linear Accelerator Center,
Stanford University, Stanford, California

I. INTRODUCTION

For almost a decade quarks have been used as the basis of a successful description of hadron spectroscopy. As originally introduced by Gell-Mann[1] and Zweig,[2] meson and baryon states are to be constructed from "constituent quark" building blocks as quark–antiquark and three quark states, respectively.

With internal angular momentum between the quarks, a simple non-relativistic prescription for building the hadron states gives both the $SU(3)$ and spin–parity quantum numbers to be expected. Particularly for baryons, where the physical states which exist are best determined, such a scheme is very successful.(3) While there are a number of meson states which are expected in such a quark model but have yet to be found, the relevant experiments are sufficiently difficult that these states could well exist and not have been observed as yet.(4)

At the same time, quarks have been used in another distinct way as the basis of constructing algebras from current components. Here one writes the relevant currents in terms of quark fields and the appropriate $SU(3)$ and Dirac matrices, commutes them using the naïve free field rules, and then throws away the quarks as a set of basis states—abstracting just the algebraic properties of the current commutators themselves.

In this way one obtains, for example, the chiral $SU(3) \times SU(3)$ algebra of the vector and axial-vector currents at equal times proposed by Gell-Mann.(5) This "algebra of currents" has also been very successful, being the basis of the Adler–Weisberger sum rule(6) and many other calculations.(7)

When extended from equal time commutators to those of current densities whose coordinates are on the light cone with respect to one another, one generates the quark light cone algebra.(8, 9) A powerful description of the scaling behaviour of the structure functions measured in deep inelastic scattering follows, with local relations as well as global relations or sum rules between different structure functions.

However, these two uses of quarks are not identical. In fact, identifying the three quarks which "constitute" a nucleon with the "current quark" fields which generate the vector and axial-vector currents at $q^2 = 0$ yields results like $g_A = \frac{5}{3}$ and $\mu_A(N) = 0$, in clear contradiction with experiment. Similarly, as $q^2 \to \infty$ in deep inelastic scattering a description of the current–nucleon interaction in terms of just three quarks with a symmetrical wave function, as in a naïve parton model, is known to fail.

The relation between these two uses(10) of quarks will be the main subject of the first several lectures. We will formulate the relation in terms of a transformation between basis states of the algebra of currents, going from a set of irreducible representations to those complicated mixtures of representations characteristic of the physical hadron under the action of the algebra of vector and axial-vector charges. The theoretical framework we will discuss will enable us to treat as a first step, vector and axial-vector transitions between hadron states at $q^2 = 0$. With the use of PCAC, all real photon and pion transitions fall within this domain.

Up to the present, the main evidence for the constituent quark model has been in the observed hadron spectrum and its mass pattern—the existence of identifiable $SU(6)$ supermultiplets.[3] More convincing evidence of the constituent quark makeup of the observed hadron spectrum would follow if we could demonstrate that the *algebraic relations* among the state vectors in the quark model also held true in the real world. Unfortunately the state vectors themselves are inaccessible to us—one only measures matrix elements or amplitudes. Therefore we must know the algebraic properties of the transition operators acting between the states, as well as the algebraic relations among the states, if one is to extract any information of this type. In other words, we have to solve two problems at once: what is the spectrum of hadron states and their relationship to one another; and what is the character of the operators which induce transitions between them.

The new development that we discuss in the first several sections is that we now have a hypothesis, abstracted from the free quark model, for the algebraic properties of the photon and pion transition operators.[11] Following work of Gilman, Kugler, and Meshkov,[12, 13] we shall conduct a fairly detailed analysis, exploring the basic algebraic structure and seeing how it is translated into matrix elements and widths for photon and pion transitions at $q^2 = 0$. We shall systematically go through meson and baryon decays, reviewing the present situation with respect to the comparison of theory and experiment. In general we find a very encouraging situation, and shall see that there now is good evidence for both the algebraic properties of the transition operators which we abstract and for the algebraic relations between hadron state vectors predicted by the constituent quark model.

In the last few sections we turn from $q^2 = 0$ to the opposite régime of $q^2 \to \infty$ and deep inelastic scattering, extending the current algebra to that of the vector and axial-vector current densities on the light cone.[8, 9] The resulting quark light cone algebra, and its concrete realization in the quark parton model, yields a concise description of the scaling behaviour of the structure functions in inelastic electron–, neutrino–, and antineutrino–nucleon scattering. We shall review the experimental situation in this regard, with emphasis on recent data from SLAC, NAL, and CEA and its implications for the quark light cone algebra or quark parton model.

Finally, with the coming advent of experiments with polarized lepton beams and targets, we devote the last section to this subject. Again with emphasis on the application of quark light cone algebra or of parton model ideas, the scaling behaviour of two additional (spin dependent) structure functions and the sum rules they satisfy are discussed in some detail.

II. Constituent Quarks and Hadron States

In order that we have the same hadron states in mind when we go to apply the theory of current matrix elements to actual transitions between hadrons, let us briefly review the constituent quark model and its comparison with the observed spectroscopy. In such a model, mesons are simply constructed as quark–antiquark ($q\bar{q}$) states and baryons as three-quark (qqq) states.

Since the quark is by assumption in the **3** representation of $SU(3)$, in the constituent quark model all mesons will be in representations contained in $\mathbf{3} \times \bar{\mathbf{3}} = \mathbf{1} + \mathbf{8}$. Similarly baryons must lie in representations contained in $\mathbf{3} \times \mathbf{3} \times \mathbf{3} = (\bar{\mathbf{3}} + \mathbf{6}) \times \mathbf{3} = \mathbf{1} + \mathbf{8} + \mathbf{8} + \mathbf{10}$ when constructed as qqq. By definition, states of either mesons or baryons which can not be obtained in this way are called exotic. Up to now, probably the strongest spectroscopic evidence for the constituent quark model lies not so much in the filling of every slot by an established particle, but in the continued absence of confirmed exotic states. While there are a couple of candidates for exotic Z^* baryons, there are literally hundreds of established or candidate states which are non-exotic.[14]

The rules for constructing the constituent quark model states involve taking $q\bar{q}$ for mesons and qqq for baryons and proceeding in an essentially non-relativistic manner by adding a total quark angular momentum, L, to the net quark spin, S, to form the total J corresponding to a given state. At this stage we assume exact $SU(3)$ and "forces" between the quarks which do not depend on $\vec{L} \cdot \vec{S}$ or $\vec{S} \cdot \vec{S}$ or couple quark spin and $SU(3)$, e.g. purely harmonic forces between quark and quark or quark and antiquark. As a result we may classify the resulting states in terms of the combined $SU(3)$ and $SU(2)$ (spin) representations of the quarks and the net $O(3)$ (internal angular momentum, L) representation. The states are therefore discussed in terms of representations of $SU(6) \times O(3)$. As usual, one treats the quarks as far as statistics is concerned as if they were parafermions of order three.[15] Alternately, if one deals with coloured quarks[16] they are fermions, but with all low lying states as singlets of the colour $SU(3)$.

In Table I the baryon states expected[15, 17] in the constituent quark model are given in terms of $SU(6)$ supermultiplets of increasing values of L (and increasing mass). Each $SU(6)$ supermultiplet is broken down into its $SU(3)$ and quark spin, S, components. Combination of S with L gives the various J^P states expected. A candidate for the non-strange baryon member of each $SU(3)$ and J^P multiplet is listed in each case using the standard notation: (πN partial wave)$_{2I,\,2J}$(mass in MeV). In several cases possible mixing between different states with the same J^P and isospin (I) in a given supermultiplet is indicated. In these cases the candidate member of a multiplet is chosen as the physical state which likely has a dominant component coming from a given quark model state. Also expected at the same mass as the **56** $L = 2$ level in a model with *purely* harmonic forces are **70** $L = 2$, **56** $L = 0$, **70** $L = 0$, and **20** $L = 1$ positive parity states.[15, 17]

As is immediately seen, good candidates exist for all the non-strange

TABLE I. Some baryon states expected in the constituent quark model and candidates for the $SU(3)$ singlets and for the nonstrange member of each other $SU(3)$ and J^P multiplet (see text).

$SU(6) \times O(3)$ multiplet	$SU(3)$ multiplet	Total quark spin, S	J^P	Candidate members of the multiplet
56 $L = 0$	**8**	$\frac{1}{2}$	$\frac{1}{2}^+$	$P_{11}(940), \ldots$
	10	$\frac{3}{2}$	$\frac{3}{2}^+$	$P_{33}(1236), \ldots$
70 $L = 1$ (possible mixing)	**1**	$\frac{1}{2}$	$\frac{1}{2}^-$	$Y_0^*(1405)$
	1	$\frac{1}{2}$	$\frac{3}{2}^-$	$Y_1^*(1520)$
	8	$\frac{1}{2}$	$\frac{1}{2}^-$	$S_{11}(1550), \ldots$
	8	$\frac{1}{2}$	$\frac{3}{2}^-$	$D_{13}(1520), \ldots$
	10	$\frac{1}{2}$	$\frac{1}{2}^-$	$S_{31}(1640), \ldots$
	10	$\frac{1}{2}$	$\frac{3}{2}^-$	$D_{33}(1690), \ldots$
	8	$\frac{3}{2}$	$\frac{1}{2}^-$	$S_{11}(1715), \ldots$
	8	$\frac{3}{2}$	$\frac{3}{2}^-$	$D_{13}(1700), \ldots$
	8	$\frac{3}{2}$	$\frac{5}{2}^-$	$D_{15}(1670), \ldots$
56 $L = 2$	**8**	$\frac{1}{2}$	$\frac{5}{2}^+$	$F_{15}(1688), \ldots$
	8	$\frac{1}{2}$	$\frac{3}{2}^+$	$P_{13}(1860), \ldots$
	10	$\frac{3}{2}$	$\frac{7}{2}^+$	$F_{37}(1950), \ldots$
	10	$\frac{3}{2}$	$\frac{5}{2}^+$	$F_{35}(1890), \ldots$
	10	$\frac{3}{2}$	$\frac{3}{2}^+$	$P_{33}(?), \ldots$
	10	$\frac{3}{2}$	$\frac{1}{2}^+$	$P_{31}(1860), \ldots$

baryon states in the **56** $L = 0$, **70** $L = 1$, and **56** $L = 2$, with the possible exception of a P_{33} state to complete the **56** $L = 2$. Furthermore, all non-strange baryon states known below 2·0 GeV in mass are thereby classified, except for P_{11} states at 1470 and 1780 MeV. At least one of these could well be assigned to an expected **56** $L = 0$, although it requires finding still another P_{33} state as its non-strange partner! The other P_{11} might possibly be assigned to a **70** $L = 0$ or **70** $L = 2$. Evidence for the existence of the latter multiplet has been recently summarized by Faiman, Rosner and Weyers.(18)

The meson states, which lie in either the **35** or **1** representations of $SU(6)$, are summarized in Table II. Again the $SU(3)$, quark spin S, and J^P for states which make up a given $SU(6) \times O(3)$ multiplet are listed, as well as non-strange mesons which are candidates among the physically observed mesons for identification with the quark model states.(14) In the case of mesons there is no trouble with extra states—many "slots" remain to be filled. The lack of isoscalar companions to the now established B meson, a decent A_1 candidate, and the isoscalar members of the $L = 2$ supermultiplet stand out as particularly needed discoveries.(4)

TABLE II. Some meson states expected in the constituent quark model and possible candidates for the isospin one and zero members of each $SU(3)$ and J^P multiplet.

$SU(6) \times O(3)$ multiplet		$SU(3)$ multiplet	Total quark spin, S	J^{PC}	Candidate members of the multiplet
35 $L = 0$	possible mixing	**8 + 1**	1	1^{--}	$\rho, \omega, \phi, \ldots$
		8	0	0^{-+}	$\pi, \eta, \ldots$
1 $L = 0$		**1**	0	0^{-+}	η'
35 $L = 1$		**8 + 1**	1	2^{++}	$A_2, f, f', \ldots$
		8 + 1	1	1^{++}	$A_1?, D, \ldots$
		8 + 1	1	0^{++}	$\delta, \sigma?, \ldots$
		8	0	1^{+-}	$B, ?, \ldots$
35 $L = 2$		**8 + 1**	1	3^{--}	$g, \omega_3, \ldots$
		8 + 1	1	2^{--}	$F_1?, ?, \ldots$
		8 + 1	1	1^{--}	$\rho'?, ?, \ldots$
		8	0	2^{-+}	$A_3?, ?, \ldots$

However, particularly for baryons where detailed phase shift analysis has permitted the establishment of states with dominantly inelastic decay modes, the fact remains that the expected quark model states exist and even fall in very well *identifiable supermultiplets*. Even for mesons, the $I = 1$ mesons with $L = 0$, 1, and 2 are almost (except for the A_1 and A_3??) all observed and are found in easily identified supermultiplets. Although we clearly are not dealing with a true symmetry and the supermultiplets are not even nearly degenerate, very few meson or baryon resonances do not fit into the clear pattern of $SU(6) \times O(3)$ supermultiplets.

III. The Transformation from Current to Constituent Quarks

Examination of the algebraic properties of hadron states and of the operators for current induced transitions between them naturally involves classifying transformation properties in a group-theoretical manner, i.e. in terms of irreducible representations of an appropriate Lie algebra or Lie group. For this purpose, consider the algebra formed by the 16 vector and axial-vector charges, $Q^\alpha(t)$ and $Q_5^\alpha(t)$, which are simply integrals over all space of the time components of the corresponding currents measurable in weak and electromagnetic interactions:

$$Q^\alpha(t) = \int d^3x V_0^\alpha(\vec{x}, t) \tag{1a}$$

$$Q_5^\alpha(t) = \int d^3x A_0^\alpha(\vec{x}, t). \tag{1b}$$

Here α is an $SU(3)$ index which runs from 1 to 8. At equal times these charges commute to form the algebra proposed by Gell-Mann,[5]

$$[Q^\alpha(t), Q^\beta(t)] = if^{\alpha\beta\gamma}Q^\gamma(t) \tag{2a}$$

$$[Q^\alpha(t), Q_5^\beta(t)] = if^{\alpha\beta\gamma}Q_5^\gamma(t) \tag{2b}$$

$$[Q_5^\alpha(t), Q_5^\beta(t)] = if^{\alpha\beta\gamma}Q^\gamma(t). \tag{2c}$$

This is the algebra of chiral $SU(3) \times SU(3)$, for it can be easily shown that eqns (2) are equivalent to the statement that the right-handed charges, $Q^\alpha + Q_5^\alpha$, and the left-handed charges, $Q^\alpha - Q_5^\alpha$, each form an $SU(3)$, and that they commute with each other—hence, chiral $SU(3) \times SU(3)$. For $\alpha = 1, 2, 3$ the Q^α's are the generators of isospin rotations: for $\alpha = 1, \ldots, 8$, they are the generators of $SU(3)$. The last of eqns (2), sandwiched between nucleon states moving at infinite momentum in the z direction, yields the Adler–Weisberger sum rule.[6]

Taken between states at infinite momentum[19], the Q^α's and Q_5^α's are "good" operators, i.e. they have finite (generally non-vanishing) values as $p_z \to \infty$. These values are the same as those of space integrals over the z components of the respective currents. If we adjoin to the integrals of the time component of the vector currents and the z-component of the axial-vector currents[21], integrals over certain "good" tensor current densities, the $SU(3) \times SU(3)$ algebra between states at infinite momentum can be enlarged still further. Letting the index α correspond to an $SU(3)$ singlet when $\alpha = 0$, we have a set of 36 charges which act like the operator $\lambda^0/2$ (proportional to the identity) plus an $SU(6)_W$ algebra of 35 generators whose elements commute like the products of $SU(3)$ and Dirac matrices: $\lambda^\alpha/2$, $(\lambda^\alpha/2)\beta\sigma_x$, $(\lambda^\alpha/2)\beta\sigma_y$, and $(\lambda^\alpha/2)\sigma_z$. We refer to this algebra, introduced by Dashen and Gell-Mann[20] in 1965 as the $SU(6)_W$ of currents. We denote these generators collectively by F^i, and use them to label the transformation properties of our states and operators. Note that $\beta\sigma_x$, $\beta\sigma_y$ and σ_z, which commute with z boosts and are "good" operators, are *not* the same as the spin components σ_x, σ_y and σ_z. The appropriate algebra to use is that of $SU(6)_W$ and not $SU(6)$. For quarks, $\beta = +1$ and quark spin and "W-spin" are the same, but for antiquarks, $\beta = -1$, and we have $-\sigma_x$, $-\sigma_y$ and σ_z instead of the antiquark spin components σ_x, σ_y, and σ_z.

In what follows we will label states or operators by their transformation properties under this $SU(6)_W$ algebra of currents. For this purpose we shall often use just the $SU(3) \times SU(3)$ subalgebra of the whole $SU(6)_W$ algebra of currents, as this subalgebra has elements which are directly measurable in weak and electromagnetic interactions. The overall $SU(6)_W$ representation will either be obvious or be made explicit. We will write

$$\{(A, B)_{S_z}, L_z\},$$

where A is the $SU(3)$ representation under $Q^\alpha + Q_5^\alpha$, B the representation under $Q^\alpha - Q_5^\alpha$, and S_z is the eigenvalue of Q_5^0, the singlet axial-vector charge [21]. The quantity L_z is then *defined* in terms of the z component of the total angular momentum J, as $L_z = J_z - S_z$. The "ordinary" $SU(3)$ content (under Q^α) of such a representation is just that of the direct product $A \times B$.

With such a labelling it is clear that, for example, the generator

$$Q_5^\alpha = \frac{Q^\alpha + Q_5^\alpha}{2} - \frac{Q^\alpha - Q_5^\alpha}{2}$$

transforms as $\{(8, 1)_0, 0\} - \{(1, 8)_0, 0\}$, while Q^α transforms as $\{(8, 1)_0, 0\} + \{(1, 8)_0, 0\}$. Being generators, they are in a **35** of the full $SU(6)_W$ of currents.

Representations of $SU(3) \times SU(3)$ can be built up from $(3, 1)_{\frac{1}{2}}$, $(1, 3)_{-\frac{1}{2}}$, $(1, \bar{3})_{\frac{1}{2}}$, and $(\bar{3}, 1)_{-\frac{1}{2}}$ which we *define* to be the *current quark* and *current antiquark* states with spin projection $\pm\frac{1}{2}$ in the z direction. The quarks form a **6** and the antiquarks a $\bar{\mathbf{6}}$ in $SU(6)_W$. As an example, consider the states of $q\bar{q}$ which can be formed with $L_z = 0$. The three states with quark spin $S = 1$ and $S_z = J_z = 1, 0$, and -1 are respectively labelled

$$S = 1, S_z = 1: \quad \{(3, \bar{3})_1, 0\} \tag{3a}$$

$$S = 1, S_z = 0: \quad \{(8, 1)_0, 0\} + \{(1, 8)_0, 0\} \text{ and } \{(1, 1)_0, 0\} + \{(1, 1)_0, 0\} \tag{3b}$$

$$S = 1, S_z = -1: \{(\bar{3}, 3)_{-1}, 0\}, \tag{3c}$$

while the $S = 0, S_z = J_z = 0$ state is written:

$$S = 0, S_z = 0: \{(8, 1)_0, 0\} - \{(1, 8)_0, 0\} \text{ and } \{(1, 1)_0, 0\} - \{(1, 1)_0, 0\}. \tag{4}$$

All these states lie in a **35** plus a **1** representation of the full $SU(6)_W$ of currents.

As another example, consider combining three current quarks to form a baryon. If we again take $L_z = 0$ and a symmetrical quark wave function, then we find the states with $S = \frac{1}{2}$ and $S = \frac{3}{2}$ transform as

$$\begin{aligned}
S = \tfrac{3}{2}, S_z &= \tfrac{3}{2}: \quad \{(10, 1)_{\frac{3}{2}}, 0\} \\
S = \tfrac{3}{2}, S_z &= \tfrac{1}{2}: \quad \{(6, 3)_{\frac{1}{2}}, 0\} \\
S = \tfrac{1}{2}, S_z &= \tfrac{1}{2}: \quad \{(6, 3)_{\frac{1}{2}}, 0\} \\
S = \tfrac{1}{2}, S_z &= -\tfrac{1}{2}: \{3, 6)_{-\frac{1}{2}}, 0\} \\
S = \tfrac{3}{2}, S_z &= -\tfrac{1}{2}: \{(3, 6)_{-\frac{1}{2}}, 0\} \\
S = \tfrac{3}{2}, S_z &= -\tfrac{3}{2}: \{(1, 10)_{-\frac{3}{2}}, 0\},
\end{aligned} \tag{5}$$

and they all lie in a **56** of the full $SU(6)_W$ of currents. In particular, if a nucleon at infinite momentum with $J_z = \frac{1}{2}$ acted under the algebra of currents as if it were simply composed of two current quarks with $S_z = \frac{1}{2}$ and one quark with $S_z = -\frac{1}{2}$ in a symmetrical wave function, we would have

$$|N\rangle = |\mathbf{56}, \{(6, 3)_{\frac{1}{2}}, 0\}\rangle \tag{6}$$

However, the $SU(3)$ content of $(6, 3)_{\frac{1}{2}}$ is just that of an octet (including the nucleon) and a decuplet (including the $\Delta(1236)$). Since Q_5^α is a generator of $SU(3) \times SU(3)$, it can only connect this representation to itself, i.e. for $\alpha = 1, 2, 3$ it can only connect the nucleon to the nucleon or to the $\Delta(1236)$. Furthermore, such a classification of the nucleon gives $g_A = \frac{5}{3}$. Both these results are in glaring contradiction with experiment. The nucleon *cannot* be in such a simple representation. This is already apparent from the Adler–Weisberger sum rule[6] itself, for it shows that the nucleon is connected by a generator of the algebra, the axial-vector charge Q_5^α (in the form of the pion field through the use of PCAC), to many higher mass N^*'s. Thus the nucleon and these N^*'s must be in the same representation of $SU(3) \times SU(3)$. Conversely, the nucleon state must span many different representations of the $SU(3) \times SU(3)$ of currents.

An attempt to describe *approximately* the nucleon state in terms of a sum of irreducible representations of $SU(3) \times SU(3)$ yields[22]

$$\begin{aligned}|N\rangle = \cos\theta|\{(6, 3)_{\frac{1}{2}}, 0\}\rangle + \sin\theta\,\{\sin\phi|\{(\overline{3}, 3)_{\frac{1}{2}}, 0\}\rangle \\ + \cos\phi\,[\cos\psi|\{(8, 1)_{\frac{3}{2}}, -1\}\rangle + \sin\psi|\{(3, \overline{3})_{-\frac{1}{2}}, 1\}\rangle]\},\end{aligned} \tag{7}$$

where θ, ϕ, and ψ are parameters to be fitted phenomenologically. It is clear that parametrizing states in a manner resembling the complicated nucleon wave function in eqn (7) is not the way to proceed in order to understand systematically the classification of higher resonances. The number of phenomenological parameters would increase so as to render the approach essentially useless.

Instead, one may assume[10, 11, 23, 24] that there exists a unitary operator, V, which transforms an irreducible representation (I.R.) of the algebra of currents into the physical state:

$$|\text{hadron}\rangle = V|\text{I.R., currents}\rangle. \tag{8}$$

The state $|\text{I.R., currents}\rangle$ is chosen as that irreducible representation of the algebra of currents which corresponds to baryons being built from just three current quarks and mesons from quark–antiquark. Thus, for example, the complicated nucleon state in eqn (7) is rewritten as

$$|N\rangle = V|\mathbf{56}, \{(6, 3)_{\frac{1}{2}}, 0\}\rangle. \tag{9}$$

All the complicated mixing of the real hadron states has been subsumed in the operator V.

In the following we will be interested in evaluating the hadronic matrix elements of a charge or current, say Q_5^α. Using eqn (8) we have

$$\langle \text{hadron}'|Q_5^\alpha|\text{hadron}\rangle = \langle \text{I.R.}', \text{currents}|V^{-1}Q_5^\alpha V|\text{I.R., currents}\rangle \tag{10}$$

The complications of hadronic states under the algebra of currents have now been transferred to the effective operator $V^{-1}Q_5^\alpha V$ which may be studied as an independent object. Moreover, if the operator $V^{-1}Q_5^\alpha V$ has simple transformation properties under the algebra of currents, the way is now open to systematically evaluate the matrix elements of Q_5^α between any two hadronic states.

The operator V serves another useful purpose. It is easy to see that if we define a new set of generators

$$W^i = VF^iV^{-1}, \tag{11}$$

then the W^i also form an $SU(6)_W$ algebra and furthermore, from the definition of V in eqn (8), hadron states transform as irreducible representations under the W^i corresponding to the naïve constituent quark model of hadrons. We therefore call the basis states of this new $SU(6)_W$ "constituent quarks" and identify the algebra with that of the $SU(6)_W$ of strong interactions.[25] Equation (8) can therefore be rewritten as

$$|\text{hadron}\rangle = |\text{I.R., constituents}\rangle = V|\text{I.R., currents}\rangle, \tag{12}$$

while eqn (10) becomes

$$\begin{aligned}\langle \text{hadron}'|Q_5^\alpha|\text{hadron}\rangle &= \langle \text{I.R.}', \text{constituents}|Q_5^\alpha|\text{I.R., constituents}\rangle \\ &= \langle \text{I.R.}', \text{currents}|V^{-1}Q_5^\alpha V|\text{I.R., currents}\rangle. \end{aligned}\tag{13}$$

From this standpoint the operator V just takes one from one set of basis states to another, or alternately, from one set of generators to another.

In the free quark model, the $SU(6)_W$ of strong interactions would be identical with the $SU(6)_W$ of currents if the quarks were restricted to have momentum purely in the z direction ($p_\perp = 0$). It is the transverse momentum of quarks which is the reason for breaking the identity of the two algebras. This is intuitive if we keep in mind that the $SU(6)_W$ of strong interactions[25] was conceived of as a collinear "symmetry". As we will see shortly, it is *not* a symmetry respected in strong interaction transitions—its conservation is badly violated in pion and photon decays.

As we are interested at the moment in limiting our attention to transitions between hadrons at $q^2 = 0$, we will be primarily concerned with the algebraic properties of the transformed axial-vector charge, $V^{-1}Q_5^\alpha V$, where Q_5^α is defined in eqn (1b), and the transformed first moment of the vector current, $V^{-1}D^{\alpha}_{\pm}V$, where $D^{\alpha}_{\pm}$ is defined as

$$D^{\alpha}_{\pm} = i\int d^3x\left[\frac{\mp(x \pm iy)}{\sqrt{2}}\right]V^{\alpha}_{0}(\vec{x}, t). \tag{14}$$

Taken between states at infinite momentum, commutators of Q^{α}_{5} lead to Adler–Weisberger sum rules,(6) while commutators of $D^{\alpha}_{\pm}$ lead to Cabibbo–Radicati sum rules.(26) Matrix elements of these operators are proportional to pion and photon transition amplitudes respectively. Their properties under the algebra of currents are that

$$Q^{\alpha}_{5} \text{ transforms as } \{(8, 1)_0 - (1, 8)_0, 0\} \tag{15a}$$

$$D^{\alpha}_{\pm} \text{ transforms as } \{(8, 1)_0 + (1, 8)_0, \pm 1\}. \tag{15b}$$

For guidance on what might be the algebraic properties of $V^{-1}Q^{\alpha}_{5}V$ and $V^{-1}D^{\alpha}_{\pm}V$ we turn to the free quark model. There Melosh(11) has been able to construct an explicit form of the operator V as

$$V = \exp\left[\left(\frac{i}{2}\right)\int d^3x\, q^{+}(x)\arctan\left(\frac{\vec{\gamma}_{\perp}\cdot\vec{\partial}_{\perp}}{m}\right)q(x)\right], \tag{16}$$

where m is the quark mass. The transformation V bears a strong similarity to the Foldy–Wouthuysen transformation, only restricted to transverse directions. As expected, if there was no transverse motion of the quarks ($\vec{p}_{\perp} \propto \vec{\partial}_{\perp} = 0$), or if $m \to \infty$, $V \to 1$ and "current" and "constituent" quarks coincide.

In a free quark model at $p_z = \infty$, either $V^{-1}Q^{\alpha}_{5}V$ or $V^{-1}D^{\alpha}_{\pm}V$ must connect only single quark states to single quark states; they thus have the general form:

$$V^{-1}Q^{\alpha}_{5}V \text{ or } V^{-1}D^{\alpha}_{\pm}V = \int d^3x q^{+}(x)\mathcal{O}(\partial_{\perp}, \gamma_i)\frac{\lambda^{\alpha}}{2}q(x), \tag{17}$$

where $\mathcal{O}$ is some function of the transverse derivatives ($\partial_{\perp}$) and the gamma matrices (γ_i). An explicit form of $\mathcal{O}$ was determined by Melosh(11) using the transformation in eqn (16), while Eichten *et al.*(27) argue that a large class of such functions exist. Without having a detailed dynamical formalism we are unable to make use of an explicit form, even if it were given to us. What is important here is that the operator is a "single quark" operator; i.e. it depends only on the coordinates of a single quark and it does not create connected $q\bar{q}$ pairs.

It is this property that we abstract from the free quark model and assume to hold in nature. In general, we assume that: *the operators* $V^{-1}Q^{\alpha}_{5}V$ *and* $V^{-1}D^{\alpha}_{\pm}V$ *have the transformation properties of the most general linear combination of single quark operators consistent with* $SU(3)$ *and Lorentz invariance.*

This is verified in the explicit free quark model calculations.[11,27] As $SU(3)$ is assumed conserved there, we have $V^{-1}Q^{\alpha}V = Q^{\alpha}$. The operator $V^{-1}Q_5^{\alpha}V$, with $J_z = 0$, contains two terms which transform under $SU(3) \times SU(3)$ as $\{(8,1)_0 - (1,8)_0, 0\}$ and $\{(3,\bar{3})_1, -1\} - \{(\bar{3},3)_{-1}, 1\}$ and behave as components of **35**'s of the full $SU(6)_W$ of currents.

The operator $V^{-1}D_+^{\alpha}V$, with $J_z = 1$, is slightly more complicated.[28] In general, there are four possible terms:[29] $\{(8,1)_0 + (1,8)_0, 1\}$, $\{(3,\bar{3})_1, 0\}$, $\{(\bar{3},3)_{-1}, 2\}$, and $\{(8,1)_0 - (1,8)_0, 1\}$. It appears that all four occur in the operator $V^{-1}D_+^{\alpha}V$ in the free quark model.[11,27] However, the last term, which corresponds to $q\bar{q}$ in a net quark spin $S = 0$, unnatural spin-parity state in the non-relativistic model (see eqn (4)), has no analogue with any natural spin-parity (in particular, vector meson) state of the quark model. Moreover, under a generalized parity transformation, $P\,e^{-i\pi J_y}$, which takes[22] $\{(A,B)_{S_z}, L_z\} \to \{(B,A)_{-S_z}, -L_z\}$, the first three terms do not change sign while the last one does. For the longitudinal ($J_z = 0$) component of the current this would eliminate the possibility of such a term. Therefore the $\{(8,1)_0 - (1,8)_0, 1\}$ term in D_+^{α} has no correspondence with any natural spin-parity meson state and cannot occur in the longitudinal component of the vector current. We shall often drop it in the following discussion and assume that $V^{-1}D_+^{\alpha}V$ only has terms which transform as[12, 13] $\{(8,1)_0 + (1,8)_0, 1\}$, $\{(3,\bar{3})_1\ 0\}$, and $\{(\bar{3},3)_{-1}, 2\}$, again in **35**'s of the $SU(6)_W$ of currents. Thus, in spite of the enormous complication of V itself, we abstract these remarkably simple algebraic properties of $V^{-1}Q_5^{\alpha}V$ and $V^{-1}D_+^{\alpha}V$ from the free quark model and postulate them to hold in the real world. We now proceed to apply this hypothesis to transitions between hadrons.

IV. Calculation of Photon and Pion Decays of Hadrons

We have now set up the basic apparatus necessary to carry out the application of the algebraic structure of the transformed currents to hadron transitions. First we have described how to classify states and operators in terms of the $SU(6)_W$ of currents and its subalgebra, chiral $SU(3) \times SU(3)$. Having seen that physical hadron states are not in irreducible representations of this algebra of currents, we defined an operator V, assumed unitary, which carries that irreducible representation of the algebra of currents characteristic of the quark model for any given hadron into the physical hadron state. Applying this to matrix elements of the axial-vector charge, Q_5^{α}, or the first moment of the vector current, D_+^{α}, and moving the transformation V from the state to the operator, shows that one must study the properties of the operators $V^{-1}Q_5^{\alpha}V$ and $V^{-1}D_+^{\alpha}V$. Appealing to the free quark model and the work of Melosh,[11] we abstract the algebraic properties of these operators, namely, we assume that in Nature $V^{-1}Q_5^{\alpha}V$ transforms as $\{(8,1)_0 - (1,8)_0, 0\}$ and $\{(3,\bar{3})_1, -1\} - \{(\bar{3},3)_{-1}, 1\}$ while $V^{-1}D_+^{\alpha}V$ transforms as $\{(8,1)_0 + (1,8)_0, 1\}$, $\{(3,\bar{3})_1, 0\}$, and $\{(\bar{3},3)_{-1}, 2\}$, all components of **35**'s of the full $SU(6)_W$ of currents.

To carry through this application of the algebraic structure of transformed currents to decays of real hadrons, we need several additional physical assumptions. Unfortunately very few weak axial-vector transitions have been measured. Until such data exist, we must content ourselves with a comparison of the theory for $q^2 = 0$ axial-vector matrix elements with amplitudes for pion decays. To relate matrix elements of Q_5 between states at infinite momentum to matrix elements of the pion field we need the PCAC hypothesis. Explicitly, for $\alpha = 1, 2, 3$ we assume:[30]

$$\partial_{\mu}A_{\mu}^{\alpha}(x) = \frac{1}{\sqrt{2}} f_{\pi}\phi_{\pi}^{\alpha}(x), \tag{18}$$

where $A_{\mu}^{\alpha}(x)$ is the axial-vector current and $f_{\pi} \simeq 135\,\text{MeV}$ is a constant related to the charged pion decay rate. The decay rate for hadron′ → hadron + π^- can then be computed in narrow resonance approximation in terms of matrix elements of $2^{-\frac{1}{2}}(Q_5^1 - iQ_5^2)$ between states at infinite momentum as

$$\Gamma(\text{hadron}' \to \text{hadron} + \pi^-)$$
$$= \frac{1}{(4\pi f_{\pi}^2)} \frac{p_{\pi}}{2J'+1} \frac{(M'^2 - M^2)^2}{M'^2} \sum_{\lambda} |\langle \text{hadron}', \lambda | 2^{-\frac{1}{2}}(Q_5^1 - iQ_5^2) | \text{hadron}, \lambda\rangle|^2, \tag{19}$$

where p_{π} is the pion momentum and the sum extends over all the possible common helicities, λ of the hadrons. The total width, Γ(hadron′ → hadron + π), may be obtained from eqn (19) by adding the π^+ and π^0 widths, which are related by isospin Clebsch–Gordan coefficients. Equation (19) may also be obtained in a more clearly covariant way by considering the narrow resonance approximation to the hadron′ intermediate state contribution to the Adler–Weisberger sum rule[6] obtained by taking eqn (2c) between hadron states at infinite momentum. In either case, we see that the width for hadron′ → hadron + π is directly fixed by matrix elements of Q_5, up to the validity[31] of PCAC. As a result, there are *no arbitrary phase space factors* in the calculation.

For photon decays we need no additional assumption to relate the width to the matrix element of the D_+^{α} operator of eqn (14) taken between states at infinite momentum. We have directly that in narrow resonance approximation[28]

$$\Gamma(\text{hadron}' \to \text{hadron} + \gamma)$$
$$= \frac{e^2}{\pi} \frac{p_\gamma^3}{2J' + 1} \sum_\lambda |\langle \text{hadron}', \lambda | D_+^3 + 3^{-\frac{1}{2}} D_+^8 | \text{hadron}, \lambda - 1 \rangle|^2, \tag{20}$$

where e is the proton charge, p_γ the photon momentum, and the sum extends over all possible helicities λ. Note that although the definition of D_+^α in eqn (14) involves only a first moment of the current, between states at infinite momentum all multipole amplitudes consistent with the spin and parity of the states enter matrix elements of D_+^α. Equation (20) may also be obtained from consideration of the narrow resonance approximation to the hadron′ contribution to the Cabibbo–Radicati sum rule[26] on hadron states. Again we have no arbitrary phase space factors.

For the present we shall use the narrow resonance approximation expressions, eqns (20) and (21), for pion and photon decay widths in order to make a comparison of the theory with experiment. For broad resonances in the initial and/or final state, or for decays of resonances where the physically available phase space is small, such an approximation introduces non-negligible errors.[32] However, we view the present comparison as being sufficiently accurate as a first test of the theory, particularly in view of the experimental errors on values for pion or photon decay widths. When the situation warrants it, the values of

$$|\langle \text{hadron}' | Q_5^\alpha | \text{hadron} \rangle|^2 \text{ and } |\langle \text{hadron}' | D_+^\alpha | \text{hadron} \rangle|^2$$

should be determined irrespective of any approximation in terms of contributions to Adler–Weisberger and Cabbibo–Radicati sum rules, respectively.

Recalling that, for example,

$$\langle \text{I.R.}', \text{constituents} | Q_5^\alpha | \text{I.R., constituents} \rangle$$
$$= \langle \text{I.R.}', \text{currents} | V^{-1} Q_5^\alpha V | \text{I.R., currents} \rangle,$$

we see that with the assumed algebraic properties of $V^{-1}Q_5^\alpha V$ (as abstracted from the free quark model), we know the transformation properties under the $SU(6)_W$ of currents of all quantities in a given matrix element. To make contact with experiment we make a physical assumption. Namely, we assume that we can identify the *observed* (non-exotic) *hadrons with constituent quark states*. In other words, we assume that there is a portion of the physical Hilbert space which is well approximated by the single particle states of the constituent quark model. As we saw in Section II, for baryons, composed of qqq, we do have candidates which fit very well into the $SU(6) \times 0(3)$ representations **56** $L = 0$, **70** $L = 1$, and **56** $L = 2$. For mesons we have correspondingly the $q\bar{q}$ states **35** $L = 0$, **1** $L = 0$, **35** $L = 1$, etc. Moreover, we shall assume that states with different values of the quark spin as well as

L_z and S_z are related as in the constituent quark model, i.e. by the $SU(6)_W$ of strong interactions. This will allow us to relate different helicity states of a given hadron to each other.

We then know the algebraic properties (under the *algebra of currents*) of all terms of a transformed matrix element of the physically observed states. Therefore we may use the Wigner–Eckart theorem and tables of Clebsch–Gordan coefficients to carry out the calculation from this point onward. Note that $SU(6)_W$ *invariance* of the transition operator under either the algebra of currents or of strong interactions *is not assumed*—only the transformation properties of the various terms are needed in the calculation.

More explicitly, for a given matrix element of Q_5^α we write the initial and final hadron states with $J_z = \lambda$ in terms of states with definite quark L_z and S_z. This involves coupling internal L and S to form total J for each hadron. After transforming to an $SU(6)_W$ of currents basis, the matrix element of the[33] $(8, 1)_0 - (1, 8)_0$ or $(3, \bar{3})_1 - (\bar{3}, 3)_{-1}$ term in $V^{-1}Q_5^\alpha V$ can then be written, using the Wigner–Eckart theorem applied to representations of the $SU(6)_W$ of currents, as a reduced matrix element times the produce of quark angular momentum, $SU(6)_W$, $SU(3)$, and W-spin Clebsch–Gordan coefficients.[34, 35, 36] For example, suppose we were calculating the matrix element of the $(8, 1)_0 - (1, 8)_0$ piece of $V^{-1}Q_5^\alpha V$ between initial and final states with common helicity λ, total angular momenta J and J', internal quark angular momenta L and L', quark spins S and S', $SU(6)_W$ representations R and R', and $SU(3)$ representations A and A' respectively. Then we have that

$$\begin{aligned}
&\langle R', A', L', S', J', \lambda, \text{currents}|\{(8, 1)_0 - (1, 8)_0, 0\}|R, A, L, S, J, \lambda, \text{currents}\rangle \\
&= \sum_{S_z S'_z} \underbrace{(L'L'_zS'S'_z|J'\lambda)(LL_zSS_z|J\lambda)}_{\substack{\text{quark angular momentum} \\ \text{Clebsch–Gordan coefficient}}} \underbrace{(R'|\mathbf{35}|R)}_{\substack{SU(6)_W \text{ Clebsch-} \\ \text{Gordan coefficient}}} \\
&\quad \underbrace{(A'|\mathbf{8}|A)}_{\substack{SU(3) \text{ Clebsch–} \\ \text{Gordan coefficient}}} \underbrace{(10WW_z|W'W'_z)}_{\substack{W\text{-spin Clebsch–} \\ \text{Gordan coefficient}}} \underbrace{\langle R', L', L'_z\|(8, 1)_0 - (1, 8)_0\|R, L, L_z\rangle}_{\text{Reduced matrix element}}. \qquad (21)
\end{aligned}$$

The W-spin Clebsch–Gordan coefficient follows since the $(8, 1)_0 - (1, 8)_0$ operator has $W = 1$ and $W_z = 0$. For any state, $W_z = S_z$. For baryons, $\vec{W} = \vec{S}$, while for mesons we have the conventional correspondence[25] (W–S flip),

$$\begin{aligned}
|W = 1, W_z = 1\rangle &= |S = 1, S_z = 1\rangle \\
|W = 1, W_z = 0\rangle &= -|S = 0, S_z = 0\rangle \\
|W = 1, W_z = -1\rangle &= -|S = 1, S_z = -1\rangle \\
|W = 0, W_z = 0\rangle &= -|S = 1, S_z = 0\rangle.
\end{aligned} \qquad (22)$$

The signs which result from using eqn (22) to convert from quark spin to W-spin are understood to be included in eqn (21) in the $SU(6)_W$ Clebsch–Gordan coefficient.

The reduction of the $(3, \bar{3})_1 - (\bar{3}, 3)_{-1}$ piece of $V^{-1}Q_5^\alpha V$ proceeds just as above, except that from eqn (22) it transforms under W-spin as

$$|W = 1, W_z = 1\rangle + |W = 1, W_z = -1\rangle.$$

As a result, the sum in eqn (21) is replaced by two sums involving the W-spin Clebsch–Gordan coefficients $(11WW_z|W'W_z')$ and $(1-1WW_z|W'W_z')$. For photon decays we need only recall that $(8, 1)_0 + (1, 8)_0$ is a $W = 0$, $W_z = 0$ object, while $(3, \bar{3})_1$ and $(\bar{3}, 3)_{-1}$ transform as $|W = 1, W_z = 1\rangle$ and $-|W = 1, W_z = -1\rangle$. Since the net J_z initially and finally must be the same for either hadron′ → hadron + π or hadron′ → hadron + γ decays, and since the net value of $W_z = S_z$ must also be the same by the W-spin Clebsch–Gordan coefficient in eqn (21) and its analogues, it follows that $L_z = J_z - S_z$ must also be additively conserved between the initial and final state (including the pion or photon operator).

The general algebraic structure of the results is now apparent.(12, 13, 37) All the Q_5^α matrix elements taken between hadron states in two given $SU(6)$ multiplets with *given* L_z and L_z' are related to at most one non-zero independent $SU(6)_W$ reduced matrix element, corresponding to the $\{(8, 1)_0 + (1, 8)_0, 0\}$, $\{(3, \bar{3})_1, -1\}$, or $-\{(\bar{3}, 3)_{-1}, 1\}$ pieces of $V^{-1}Q_5^\alpha V$. Similarly, there is at most one independent $SU(6)_W$ reduced matrix element for photon decays between states in two given $SU(6)$ multiplets with *given* values of L_z and L_z'. If L is zero, as is the case in essentially all cases of physical interest at the present, then of course $L_z = 0$ and the L_z' dependence of the $SU(6)_W$ reduced matrix element becomes trivial (in particular, the $\{(3, \bar{3})_1, -1\}$ and $-\{(\bar{3}, 3)_{-1}, 1\}$ pieces of $V^{-1}Q_5^\alpha V$, with $L_z' = -1$ and $+1$ respectively, have the same reduced matrix element). In such a case ($L = 0$) there are at most *two independent reduced matrix elements of* Q_5^α,

$$\langle R', L' \| (8, 1)_0 - (1, 8)_0 \| R, 0\rangle \tag{23a}$$

and

$$\langle R', L' \| (3, \bar{3})_1 - (\bar{3}, 3)_{-1} \| R, 0\rangle, \tag{23b}$$

and *four independent reduced matrix elements of* D_+^α taken between two given $SU(6)$ multiplets,

$$\langle R', L' \| (8, 1)_0 + (1, 8)_0 \| R, 0\rangle \tag{24a}$$

$$\langle R', L' \| (3, \bar{3})_1 \| R, 0\rangle \tag{24b}$$

$$\langle R', L' \| (8, 1)_0 - (1, 8)_0 | R, 0\rangle \tag{24c}$$

and

$$\langle R', L' \| (\overline{3}, 3)_{-1} \| R, 0 \rangle, \tag{24d}$$

given our assumption on the algebraic properties of $V^{-1} Q_5^\alpha V$ and $V^{-1} D_+^\alpha V$.

The algebraic structure of the theory presented here has much in common with relativistic quark model calculations, such as those of Ref. 38. In fact, the results of Ref. 38 may be cast into a form which permits the complete identification of certain parameters there with the reduced matrix elements discussed here. However, the assumption of a "potential" in the quark model calculations yields definite predictions of the reduced matrix elements themselves as they depend on masses and other parameters of the model. This is something we do not obtain using purely the algebraic structure discussed in Section III. Also very similar in algebraic structure, at least for decays to $L = 0$ hadrons, are some broken $SU(6)_W$ schemes[39]. The relation of such schemes, and in particular l-broken $SU(6)_W$, to the present theory is discussed in detail in Ref. 37. The results of assuming $SU(6)_W$ conservation for pion transitions are reproduced in the present theory by retaining only the $\{(8,1)_0 - (1,8)_0, 0\}$ term in $V^{-1} Q_5^\alpha V$ and using PCAC. The assumption of $SU(6)_W$ conservation plus vector dominance is equivalent to keeping only the $\{(3,\overline{3})_1, 0\}$ term in $V^{-1} D_+^\alpha V$.

V. Experimental Tests of the Transformation between Current and Constituent Quarks

With the basic features and assumptions of the theory described in the previous sections, we are in a position to apply it. We begin with the pionic decays of mesons. Only non-strange meson decays will be discussed in detail as all the corresponding strange meson decay rates are related to those we calculate by $SU(3)$. At the present time they add little to the experimental tests of the theory.

Consider the pionic transitions from **35** $L' = 1 \to$ **35** $L = 0$. From the previous section, there are two independent reduced matrix elements, $\langle L' = 1 \| (8,1)_0 - (1,8)_0 \| L = 0 \rangle$ and $\langle L' = 1 \| (3,\overline{3})_1 - (\overline{3},3)_{-1} \| L = 0 \rangle$. Rather than try to make a best fit in terms of all the measured decays, we simply use two inputs[40]: $\Gamma(A_2 \to \pi\rho) = 77$ MeV and $\Gamma_{\lambda=0}\,(B \to \pi\omega) = 0$. This latter condition is in agreement with experiments[41] which see a dominantly transverse decay, and corresponds to setting $\langle L' = 1 \| (8,1)_0 - (1,8)_0 \| L = 0 \rangle = 0$. All amplitudes are then multiples of $\langle L' = 1 \| (3,\overline{3})_1 - (\overline{3},3)_{-1} \| L = 0 \rangle$. The results are gathered in Table III. Predictions of particular interest are:

(1) $\Gamma(B \to \pi\omega)$ agrees within errors with $\pi\omega$ being the dominant (and so far, only observed) mode out of a total B (width[40] of 100 ± 20 MeV.

TABLE III. Decays of nonstrange **35** $L = 1$ mesons into **35** $L = 0$ mesons by pion emission.[a, b] All decay rates are fixed in terms of $\Gamma_{\lambda=0}(B \to \pi\omega) = 0$ and $\Gamma(A_2 \to \pi\rho) = 77$ MeV.

	Γ (predicted) (MeV)	Γ (experimental)[40] (MeV)
$A_2(1310) \to \pi\rho$	77 (input)	77 ± 20
$B(1235) \to \pi\omega$ $\lambda = 0$	0 (input)	dominantly $\lambda = 1$
$B(1235) \to \pi\omega$ $\lambda = 1$	76	100 ± 20 total width
$A_1(1070) \to \pi\rho$ $\lambda = 0$	52	?
$A_1(1070) \to \pi\rho$ $\lambda = 1$	26	
$A_2(1310) \to \pi$	17	16 ± 4
$\sigma(975) \to \pi$	35	~60 total width
$f(1260) \to \pi\pi$	118	125 ± 25
$\sigma(760?) \to \pi\pi$	234	Broad?

[a] Decay rates of the corresponding K^* states are simply obtained using $SU(3)$ for the matrix elements of Q_5.

[b] The ω, f and σ mesons are taken as ideal mixtures of singlets and octets, so as to be purely constituted by nonstrange quarks. Zweig's rule[2] is used to relate decay amplitudes involving the $SU(6)_W$ **35** and **1** parts of the $\lambda = 0$ ω, σ and f.

(2) $\Gamma(f \to \pi\pi)$ is in excellent agreement with experiment. Use of a d-wave phase space factor instead of the PCAC dictated factor changes the prediction by more than a factor of 2, destroying the agreement.

(3) $\Gamma(A_2 \to \pi\eta)$ is in excellent agreement with experiment.

(4) We predict a relatively narrow $A_1 \to \pi\rho$ with a dominantly longitudinal character. This is obviously not the non-resonance observed[42] in $\pi^{\pm}p \to (3\pi)^{\pm}p$, and there is no established state with which to compare our prediction.

(5) $\Gamma(\delta \to \pi\eta)$ agrees with the roughly known[40] total width.

(6) We have somewhat arbitrarily assigned the σ a mass of 760 MeV. While it is gratifying that the resulting $\Gamma(\sigma \to \pi\pi)$ is broad, the uncertainties in identifying the non-strange quark state with a particular observed $J^{PC} = 0^{++}$ hadron are very large.

Overall we find that experiment and theory compare quite favourably for $L' = 1 \to L = 0$ pionic decays of mesons. What is known about $L' = 0 \to L = 0$ and $L' = 2 \to L = 0$ pionic matrix elements between meson states also is quite consistent with the theory,[12, 13] although present data on these transitions does not provide a very restrictive test.[43, 44]

Encouraged by this we turn to baryons. As in the meson case, we discuss mainly the non-strange baryon decays. Our choice of multiplets discussed is motivated by the fact that they are the only ones where a fairly complete experimental comparison can be made.

For transitions of the type **56** $L' = 0 \to$ **56** $L = 0$ only the

TABLE IV. Decays of **70** $L = 1$ and **56** $L = 2$ baryons into **56** $L = 0$ baryons by pion emission. All rates are fixed by the D_{13} and S_{11} decays to πN for the **70** $L = 1$ decays, and by the F_{15} and P_{31} decays to πN for the **56** $L = 2$ decays. For two states which may be mixed, a combination of widths which is independent of mixing is used and listed under Γ (predicted).

Decay	Γ (predicted) (MeV)	Γ (experimental)[40, 47] (MeV)
$D_{13}(1520) \rightarrow (\pi N)_d$ $D_{13}(1700) \rightarrow (\pi N)_d$	Γ(1520) + 0·50 Γ(1700) = 79 MeV (input)	79 ± 20
$D_{13}(1520) \rightarrow (\pi\Delta)_d$ $D_{13}(1700) \rightarrow (\pi\Delta)_d$	Γ(1520) + 0·243 Γ(1700) = 30 MeV	10 ± 6
$S_{11}(1535) \rightarrow (\pi\Delta)_d$ $S_{11}(1715) \rightarrow (\pi\Delta)_d$	Γ(1535) + 0·264 Γ(1715) = 35 MeV	not seen
$D_{15}(1670) \rightarrow (\pi N)_d$	21 MeV	56 ± 14
$D_{15}(1670) \rightarrow (\pi\Delta)_d$	82 MeV	84 ± 21
$S_{31}(1640) \rightarrow (\pi\Delta)_d$	81	52 ± 20
$D_{33}(1690) \rightarrow (\pi N)_d$	19	32 ± 9
$D_{33}(1690) \rightarrow (\pi\Delta)_d$	55	not seen
$S_{11}(1535) \rightarrow (\pi N)_s$ $S_{11}(1715) \rightarrow (\pi N)_s$	Γ(1535) + 0·505 Γ(1715) = 116 (input)	116 ± 55
$D_{13}(1520) \rightarrow (\pi\Delta)_s$ $D_{13}(1700) \rightarrow (\pi\Delta)_s$	Γ(1520) + 0·243 Γ(1700) = 46	19 ± 10
$S_{31}(1640) \rightarrow (\pi N)_s$	18	48 ± 9
$D_{33}(1690) \rightarrow (\pi\Delta)_s$	61	172 ± 60
$F_{15}(1688) \rightarrow (\pi N)_f$	84 (input)	84 ± 25
$F_{37}(1950) \rightarrow (\pi N)_f$	74	92 ± 20
$F_{37}(1950) \rightarrow (\pi\Delta)_f$	65	37 ± 18
$F_{35}(1880) \rightarrow (\pi N)_f$	14	36 ± 18
$F_{35}(1880) \rightarrow (\pi\Delta)_f$	77	16 ± 16
$P_{33}(\quad) \rightarrow (\pi\Delta)_f$		?
$F_{15}(1688) \rightarrow (\pi\Delta)_f$	12	not seen
$P_{13}(1860) \rightarrow (\pi\Delta)_f$	57	not seen
$P_{31}(1860) \rightarrow (\pi N)_p$	75 (input)	75 ± 25
$P_{31}(1860) \rightarrow (\pi\Delta)_p$	8	not seen
$P_{33}(\quad) \rightarrow (\pi N)_p$		?
$P_{33}(\quad) \rightarrow (\pi\Delta)_p$		?
$F_{35}(1880) \rightarrow (\pi\Delta)_p$	44	not seen
$P_{13}(1860) \rightarrow (\pi N)_p$	118	75 ± 25
$P_{13}(1860) \rightarrow (\pi\Delta)_p$	5	not seen
$F_{15}(1688) \rightarrow (\pi\Delta)_p$	15	22 ± 7

$\{(8,1)_0 - (1,8)_0, 0\}$ term in $V^{-1}Q_5^{\alpha}V$ contributes. The two predictions made by the theory are:

(1) $F/D = \frac{2}{3}$ for the baryon decays. This is tested directly by the axial-vector contribution to the weak leptonic decays of the baryon octet, without need for PCAC. This prediction agrees quite well with experiment.[40].

(2)

$$\langle \Delta^0, \lambda = \tfrac{1}{2}|2^{-\frac{1}{2}}(Q_5^1 - iQ_5^2)|p, \lambda = \tfrac{1}{2}\rangle = (\tfrac{5}{4}\surd 2) \times \langle n, \lambda = \tfrac{1}{2}|2^{-\frac{1}{2}}(Q_5^1 - iQ_5^2)|p, \lambda = \tfrac{1}{2}\rangle. \tag{25}$$

This prediction also agrees with experiment.[45]

For **70** $L' = 1 \to$ **56** $L = 0$ and **56** $L' = 2 \to$ **56** $L = 0$ pionic decays there are two reduced matrix elements for each set of decays of the form $N^* \to \pi N$ and $N^* \to \pi\Delta$ where the N^* is a resonance in the **70** $L' = 1$ or **56** $L' = 2$ (see Table I).

Table IV compares the experimental partial widths[40,46,47] of the **70** $L' = 1$ and **56** $L' = 2$ baryons with theory.[12,13] We have again chosen to fit the reduced matrix elements to certain decays rather than doing an overall least squares fit. We observe that the agreement of experiment with theory is only qualitative, and that large experimental errors in the matrix elements are involved. One of the strongest disagreements is in the decays of the $D_{15}(1670)$, which cannot be mixed within the **70** $L = 1$ multiplet. The disagreement is in fact sharper than is apparent in Table IV, since the errors quoted on the πN and $\pi\Delta$ widths are correlated by the reasonably well-known inelasticity. While the theory predicts that less than 20% of the width is due to the πN decay, experiment indicates a 40% branching ratio.[40]

We must emphasize at this point that a large experimental ambiguity exists in evaluating the partial widths of resonances even when phase shift analysis results are known. In the case of strongly inelastic resonances, different ways of extracting resonance couplings may be used such as: extrapolations to the pole, K-matrix fits, Breit–Wigner fits, etc. These give widely varying estimates of partial widths. For example, the width of the $D_{13}(1520)$ decay to $\pi\Delta$ changes from 24 to 53 MeV depending on whether one used coupling estimates from the Argand diagram or a T-matrix pole fit.[47] There is also a theoretical error in our use of the narrow resonance approximation. This is particularly so for $\pi\Delta$ decays, where some of the larger discrepancies between theory and experiment occur. With this in mind, the overall situation for pionic widths of non-strange baryons is not unreasonable.

For real photon transitions, all helicity amplitudes can be obtained by taking matrix elements of $D_+^3 + (1/\surd 3)D_+^8$ between states at infinite momentum.

Such matrix elements of D^α_+ are equal to the sum of those of the possible terms in $V^{-1}D^\alpha_+ V$ discussed in Section III. At the present time only non-strange baryon transitions have been sufficiently investigated experimentally so as to provide a test of the theory.

For **56** $L' = 0 \to$ **56** $L = 0$ photon transitions, only the term in $V^{-1}D^\alpha_+ V$ which transforms as $\{(3, 3)_1, 0\}$ can make a non-zero contribution, because $L_z = 0$ in both the initial and final state. All matrix elements are therefore proportional to the single reduced matrix element $\langle$**56** $L' = 0\|(3, \bar{3})_1\|$**56** $L = 0\rangle$. For transitions between two octet members of the **56**, this term is characterized by an F/D value of $\frac{2}{3}$.

Between $J = \frac{1}{2}$ baryon states the matrix elements of $V^{-1}D^\alpha_+ V$ are interpretable as proportional to the total magnetic moment of the baryon.[11] As a result the theory gives[11]

$$\mu_T(n)/\mu_T(p) = -\tfrac{2}{3}, \tag{26}$$

the old $SU(6)$ result[48] which is rather close to experiment.[40] Furthermore, the ratio of $\sqrt{3}$ between the $\lambda = \frac{3}{2}$ and $\frac{1}{2}$ amplitudes for $\Delta \to \gamma N$ corresponds to a pure magnetic dipole transition with

$$\mu^*/\mu_T(p) = \tfrac{2}{3}\sqrt{2}, \tag{27}$$

if we use the relation

$$\langle\Delta, \lambda = \tfrac{1}{2}|D_+|N, \lambda = -\tfrac{1}{2}\rangle = \mu^*/\sqrt{2}. \tag{28}$$

A phenomenological analysis[49] of the data for pion photoproduction gives a value for $\mu^*/\mu_T(p)$ which is 1·28 ± 0·03 times the right-hand side of eqn (27). However, this is the result of finding the residue at the Δ pole in $\gamma N \to \pi N$. In our approach one should evaluate μ^* by taking the Δ contribution to the Cabibbo–Radicati sum rule (see Section IV). This results in a value[50] of $\mu^*/\mu_T(p)$ which is 0·9 ± 0·1 times the right-hand side of eqn (27), i.e. in quite satisfactory agreement with the theory. Equations (26) and (27) are standard $SU(6)$ results, as is to be expected since the $(3, \bar{3})_1$ term in $V^{-1}D^\alpha_+ V$ has the same transformation properties as the magnetic moment operator[48] used in $SU(6)$.

For **70** $L' = 1 \to$ **56** $L = 0$ and **56** $L' = 2 \to$ **56** $L = 0$ photon decays[13, 29, 51, 52] the situation is more complicated because of the presence of more independent reduced matrix elements than in the **56** $L' = 0 \to$ **56** $L = 0$ case. However, if we disregard the $\{(3, 3)_{-1}, 2\}$ as well as the $\{(8, 1)_0 - (1, 8)_0, 1\}$ term in $V^{-1}D^\alpha_+ V$, only two terms are left and there is a one to one correspondence with the results of quark model calculations:[53] the $(8, 1)_0 + (1, 8)_0$ term in $V^{-1}D^\alpha_+ V$ corresponds to the photon interacting with the quark convection current, while the $(3, \bar{3})_1$ term corresponds to the interaction with the quark magnetic moment. Of course, explicit quark model calculations with, say, harmonic potentials give the reduced matrix elements as well, something we do not obtain at all with the theory under discussion. Since the more restrictive quark model predictions are roughly consistent[38] with the experimental data on photon decay amplitudes,[54] so is the theory discussed here.

A more crucial test of the theory comes in predicting the relative signs of amplitudes for inelastic scattering. For, even if some mixing occurs or the identification of a given hadron with a particular state of the constituent quark model is only approximate, resulting in say a 30 or 40% error in the magnitude of a matrix element and a factor of two error in a width, the sign of an amplitude should still be preserved if the mixing is not too large.

For the reaction $\pi N \to N^* \to \pi\Delta$ we can compare our predictions to recent isobar model phase shift analyses.[46, 47, 55] Table V lists the theoretically predicted phases[12, 13, 56, 57] coming from the $(8, 1)_0 - (1, 8)_0$ and

TABLE V. Signs of resonant amplitudes in $\pi N \to N^* \to \pi\Delta$ for N^*'s in the **70** $L = 1$ and **56** $L = 2$. The arbitrary overall phase is chosen so that the $DD_{15}(1670)$ amplitude is negative.

	Amplitude in $\pi N \to \pi\Delta$	Theoretical sign from $(8, 1)_0 - (1, 8)_0$	Theoretical sign from $(3, \bar{3})_1 - (\bar{3}, 3)_{-1}$	Experimental solution $A^{(46, 47)}$	Sign solution $B^{(55)}$
70 $L = 1$	$DS_{13}(1520)$	−	+	−	+
	$DD_{13}(1520)$	+	+	−	+
	$SD_{31}(1640)$	+	−	−	−
	$DS_{33}(1690)$	−	+	+	+
	$DD_{15}(1670)$	−	−	−	−
	$DS_{13}(1700)$	−	+	+	+
56 $L = 2$	$FP_{15}(1688)$	+	−	+	+
	$FF_{37}(1950)$	+	+	+	+
	$FF_{35}(1880)$	+	+	+	+

$(3, \bar{3})_1 - (\bar{3}, 3)_{-1}$ pieces of $V^{-1}Q_5^{\alpha}V$ and the experimental results.[58] The theoretical predictions are of two kinds. First are those involving amplitudes with the same (l) partial wave in both the incoming and outgoing channel and which are therefore proportional to squares of matrix elements. These have well-defined signs regardless of the relative magnitudes of the reduced matrix elements of the $(8, 1)_0 - (1, 8)_0$ and $(3, 3)_1 - (3, 3)_{-1}$ terms in $V^{-1}Q_5^{\alpha}V$. The second kind of sign prediction depends on this relative magnitude, and may help us in deducing which term is dominant for pion decays from one $SU(6)$ multiplet to another.

We see that the signs in the original solution, A, of the experimental analysis disagrees with the theory even for predictions of the first kind as seen in Table V. We note, however, that the only disagreement is between the $D_{13}(1520)$ coupling and all other couplings. This sign cannot be changed by mixing the two D_{13} states. If the signs of this resonance can be reversed, one would have complete agreement between theory and experiment. We note that the analysis on which we base our comparison suffers from a lack of data between 1540 and 1650 MeV, i.e. between the $D_{13}(1520)$ and the other resonances in the **70** $L = 1$ and **56** $L = 2$. The relative phases of amplitudes above and below the gap are determined by continuity and K matrix fits. While a complete K matrix fit has yet to be done, it appears[55] that a second solution, B, of the phase shifts exists with the signs of the $D_{13}(1520)$ reversed with respect to solution A. This removes the most serious experimental failure among the many predictions of the theory as reviewed at Purdue.[56]

In solution B, the signs of the amplitudes are such as to indicate dominance of the $(3, \bar{3})_1 - (\bar{3}, 3)_{-1}$ term in $V^{-1}Q_5^\alpha V$ for **70** $L' = 1 \rightarrow$ **56** $L = 0$ decays, but dominance of the $(8, 1)_0 - (1, 8)_0$ term in **56** $L' = 2 \rightarrow$ **56** $L = 0$ decays. An analysis[59] of the signs of amplitudes for $\gamma N \rightarrow N^* \rightarrow \pi N$ shows that the measured signs are consistent with the theory at *both* the pion and photon vertex for **70** $L = 1$ and **56** $L = 2$ intermediate N^* resonances if the $(3, \bar{3})_1 - (\bar{3}, 3)_{-1}$ term dominates in **70** $L' = 1 \rightarrow$ **56** $L = 0$ pion transitions. Up to this time the experimental analysis of $\gamma N \rightarrow \pi N$ is insufficient to give confirmatory evidence on the dominance of the $(8, 1)_0 - (1, 8)_0$ term in **56** $L' = 2 \rightarrow$ **56** $L = 0$ pion decays.

Note that both our analysis of signs for **70** $L' = 1 \rightarrow$ **56** $L = 0$ pionic transitions of baryons and of widths for **35** $L' = 1 \rightarrow$ **35** $L = 0$ meson decays show that the $(3, \bar{3})_1 - (\bar{3}, 3)_{-1}$ term in $V^{-1}Q_5^\alpha V$ is dominant. This is completely contradictory to the assumption of $SU(6)_W$ conservation, which only allows the $(8, 1)_0 - (1, 8)_0$ term to be present. Also for photon decays from **70** $L' = 1 \rightarrow$ **56** $L = 0$ and **56** $L' = 2 \rightarrow$ **56** $L = 0$, the experimental amplitudes[57] indicate the importance of both the $(8, 1)_0 + (1, 8)_0$ and $(3, \bar{3})_1$ terms in $V^{-1}D_+^\alpha V$, while $SU(6)_W$ conservation plus vector dominance would allow only $(3, \bar{3})_1$.

Taking solution B for $\pi N \rightarrow \pi\Delta$, there are more than 20 signs in $\gamma N \rightarrow \pi N$ and $\pi N \rightarrow \pi\Delta$ which agree with experiment. Further, the domination of the $(3, \bar{3})_1 - (\bar{3}, 3)_{-1}$ term in $V^{-1}Q_5 V$ for **70** $L' = 1 \rightarrow$ **56** $L = 0$ pion decays is obtained consistently in both reactions. With that many signs correct, one begins to have the feeling that there is something right about the theory we have been employing.

Such a theory can be regarded as one more step in a program of abstracting algebraic properties from the free quark model, without necessitating the reality of free quarks themselves. Abstraction from the free quark model assures us that the assumed algebraic properties could be exact, and are at least consistent with relativity, invariance principles, etc. They are presumably the least complicated that one might expect to hold in the real world.

As an elegant and beautiful theory of the algebraic structure of weak and electromagnetic current induced transitions between hadrons at $q^2 = 0$, the present theory greatly unifies the treatment of weak and electromagnetic transitions with the systematics of hadron spectroscopy. With the identification of the observed hadrons to good approximation with constituent quark states and the use of PCAC, a powerful approximate theory of all pion and photon decay results.

As a theory of pion transitions, the present theory has much in common as far as general algebraic structure is concerned with both previous relativistic quark model calculations(38) and certain broken $SU(6)_W$ schemes.(39) We regard this theory of current-induced transitions, when supplemented by PCAC and/or vector meson dominance, as in fact providing a method of constructing a phenomenology of purely hadronic vertices(60) and providing justification for some aspects of these other theoretical schemes.

An important aspect of the present theory is that the comparison with experiment is in terms of amplitudes which are related in a straightforward way by Clebsch–Gordan coefficients, and decay widths are in turn related to these amplitudes in a non-arbitrary, known way. The agreement found, particularly in regard to signs, provides stronger support for the theory as a valid description of *both* photon and pion transitions, as well as for the identification of the observed hadron states with those of the constituent quark model.

VI. Deep Inelastic Scattering and Annihilation

1. Introduction

We now turn from the study of current induced transitions between hadrons at $q^2 = 0$ to the opposite regime of $q^2 \to \infty$. While it would be very interesting to extend our treatment of the transformed vector and axial-vector currents to $q^2 \neq 0$, little experimental information exists on matrix elements between specific hadrons. Instead we sum over all final hadrons obtainable in the interaction of a current with a given hadron and obtain quantities measured in the experimental investigation of inelastic electron–nucleon, neutrino–nucleon, and antineutrino–nucleon scattering for (61) space-like $q^2 (q^2 > 0)$ and in electron–positron annihilation into hadrons for time-like $q^2 (q^2 < 0)$.

2. INELASTIC ELECTRON–NUCLEON SCATTERING

For the case of inelastic electron scattering, the diagram of interest[62] is indicated in Fig. 1, where k and k' are the initial and final electron four-momenta, q is the four-momentum transfer carried by the virtual photon and p is the target nucleon's four-momentum. The final hadronic state n

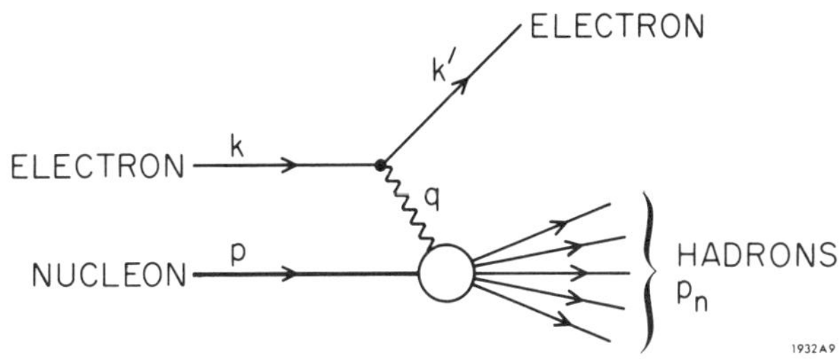

FIG. 1. Kinematics of inelastic electron–nucleon scattering.

then has four-momentum $p_n = p + q$ and invariant mass squared $W^2 = -(p + q)^2$. In the laboratory frame (initial nucleon at rest) with E and E' the energies of the initial and final electrons, the Lorentz scalar variable

$$\nu = -p \,.\, q/M_N = E - E' \tag{29}$$

is the virtual photon's energy, and the invariant momentum transfer squared is

$$q^2 = 4EE' \sin^2 \theta/2 \tag{30}$$

where θ is the scattering angle and the electron mass has been neglected compared to its energy. Knowing ν and q^2 from measuring the incident and scattered electron, the invariant mass W of the final hadrons is fixed by

$$W^2 = 2M_N\nu + M_N^2 - q^2. \tag{31}$$

The S-matrix element for the process in Fig. 1 may be written using the rules of quantum electrodynamics at the photon–electron vertex as

$$S_{fi} = \delta_{fi} + (2\pi)^4 i\delta^{(4)}(p_n + k' - p - k)\sqrt{\left(\frac{m_e^2}{EE'}\right)}(-e\bar{u}(k')i\gamma_\mu u(k)) \\ \times (\delta_{\mu\nu}/q^2)\langle p_n|J_\nu(0)|p\rangle, \tag{32}$$

where

$$J_\nu = e\left(V_\mu^3 + \frac{1}{\sqrt{3}}V_\mu^8\right) \tag{33}$$

is the (hadronic) electromagnetic current operator. Averaging over initial and summing over final electron and nucleon spins, we are led to an expression for the double differential cross-section in the laboratory for detection of only the final electron of the form

$$\frac{d^2\sigma}{d\Omega'\, dE'} = \frac{1}{(2\pi)^2}\left(\frac{E'}{E}\right)\left(\frac{e^2}{q^2}\right)^2 L_{\mu\nu} W_{\mu\nu}, \tag{34}$$

where the factor $L_{\mu\nu}$ arises from the trace of the gamma matrices due to the electron (neglecting the electron mass),

$$L_{\mu\nu} = \frac{1}{2}\left[k_\mu k'_\nu + k'_\mu k_\nu + \left(\frac{q^2}{2}\right)\delta_{\mu\nu}\right], \tag{35}$$

and the structure of the nucleon is summarized in

$$\begin{aligned} W_{\mu\nu} &= \frac{1}{2}\sum_{\substack{\text{nucleon}\\ \text{spin}}}\sum_n \left(\frac{1}{e^2}\right)\langle p|J_\mu(0)|n\rangle\langle n|J_\nu(0)|p\rangle (2\pi)^3\delta^{(4)}(p_n - p - q) \\ &= \frac{1}{2}\sum_{\substack{\text{nucleon}\\ \text{spin}}}\frac{1}{(2\pi e^2)}\int d^4x\, e^{-iq\cdot x}\langle p|[J_\mu(x), J_\nu(0)]|p\rangle \\ &= \frac{1}{2}\sum_{\substack{\text{nucleon}\\ \text{spin}}}\left(\frac{1}{2\pi}\right)\int d^4x\, e^{-iq\cdot x}\langle p|[V_\mu(x), V_\nu(0)]|p\rangle, \end{aligned} \tag{36}$$

where the second term in the commutator is zero by energy conservation for $\nu > 0$.

By Lorentz and gauge invariance the tensor $W_{\mu\nu}$ may be written as

$$\begin{aligned} W_{\mu\nu} = {} & W_1(\nu, q^2)(\delta_{\mu\nu} - q_\mu q_\nu/q^2) \\ & + W_2(\nu, q^2)(p_\mu - p\cdot q q_\mu/q^2)(p_\nu - p\cdot q q_\nu/q^2)/M_N^2. \end{aligned} \tag{37}$$

The quantity W_μ is just $(4\pi^2\alpha)^{-1}$ times the imaginary part of the Feynman amplitude for forward Compton scattering of virtual photons of mass^2 = $-q^2$. In terms of W_1 and W_2 the experimentally measured double differential cross-section resulting from combining eqns (34), (35) and (37) is

$$\frac{d^2\sigma}{d\Omega'\, dE'} = \frac{4\alpha^2 E'^2}{q^4}\left[2W_1(\nu, q^2)\sin^2\theta/2 + W_2(\nu, q^2)\cos^2\theta/2\right], \tag{38}$$

so that the structure functions W_1 and W_2, as they depend on ν and q^2, summarize the results of inelastic electron–nucleon scattering.

Now suppose that ν and q^2 are large, with ν/q^2 fixed. In the expression, eqn (36), for $W_{\mu\nu}$ in terms of the Fourier transform of a commutator of two currents the exponential is

$$\exp(-iq \cdot x) = \exp[-i(q_z z - q_0 t)] = \exp\{-i[\sqrt{(q^2 + \nu^2)}z - \nu t]\}$$
$$\simeq \exp[-i\nu(z - t)] \exp[-i(q^2/2\nu)z] \tag{39}$$

in this domain of ν and q^2 with $\vec{q}$ in the z direction. In order that the argument of the exponential should not become large and produce cancelling oscillations of the integrand, the region of integration in configuration space must satisfy

$$z - t \lesssim 0(1/\nu)$$
$$z \lesssim 0(2\nu/q^2). \tag{40}$$

Furthermore, since we want the commutator to be causal, it should vanish unless

$$x^2 = x_\perp^2 + z^2 - t^2 \leqslant 0, \tag{41}$$

which together with eqns (40) yields

$$x_\perp^2 \leqslant (t - z)(t + z) \lesssim 0(1/q^2). \tag{42}$$

Therefore, the important region of integration over the current commutator is $x^2 = x_\mu x_\mu \simeq 0$, i.e. along the light-cone.[63]

To gain theoretical insight into the commutator of two currents on the light-cone we again turn to the quark model and abstract certain properties, particularly algebraic ones, of such commutators from the free field case. In the free quark model one finds[8, 9]

$$\begin{aligned}[V_\mu^\alpha(x), V_\nu^\beta(0)] \underset{x^2 \to 0}{\simeq} \{ & if^{\alpha\beta\gamma}[(V_\nu^\gamma(x, 0) + V_\nu^\gamma(0, x))\delta_{\mu\lambda} \\ & + (V_\mu^\gamma(x, 0) + V_\mu^\gamma(0, x))\delta_{\nu\mu} - (V_\lambda^\gamma(x, 0) + V_\lambda^\gamma(0, x))\delta_{\mu\nu} \\ & + i\varepsilon_{\mu\nu\lambda\sigma}(A_\sigma^\gamma(x, 0) - A_\sigma^\gamma(0, x))] + d^{\alpha\beta\gamma}[(V_\nu^\gamma(x, 0) - V_\nu^\gamma(0, x))\delta_{\mu\lambda} \\ & + (V_\mu^\gamma(x, 0) - V_\mu^\gamma(0, x))\delta_{\nu\lambda} - (V_\lambda^\gamma(x, 0) - V_\lambda^\gamma(0, x))\delta_{\mu\nu} \\ & - i\varepsilon_{\mu\nu\rho\sigma}(A_\sigma^\gamma(x, 0) + A_\sigma^\gamma(0, x))]\} \frac{1}{4\pi} \partial_\lambda[\varepsilon(x_0)\delta(x^2)], \end{aligned} \tag{43}$$

where $V_\mu^\alpha(x)$ and $A_\mu^\alpha(x)$ are the vector and axial vector currents and

$$V_\mu^\alpha(x, 0) = :\bar{\psi}(x)(\lambda^\alpha/2)i\gamma_\mu\psi(0):$$

and

$$A_\mu(x, 0) = :\bar{\psi}(x)(\lambda^\alpha/2)i\gamma_\mu\gamma_5\psi(0): \tag{44}$$

are bilocal operators defined so that

$$V_\mu^\alpha(x, x) = V_\mu^\alpha(x)$$

and

$$A_\mu^\alpha(x, x) = A_\mu^\alpha(x). \tag{45}$$

All the $SU(3)$ properties are again summarized by the $SU(3)$ structure constants $f^{\alpha\beta\gamma}$ and $d^{\alpha\beta\gamma}$. In fact, if we specialize to the time components of the currents and go to the tip of the light cone ($x_0 = 0$) then we recover the old equal time algebra of Gell-Mann.[(5)]

An important property of eqn (43) is that it factorizes into a product of a c-number singularity which contains no masses or other dimensional parameters and a bilocal operator which carries the $SU(3)$ indices. When eqn (43), as the leading light-cone singularity in the limit, $\nu, q^2 \to \infty$, is inserted back in eqn (36) for $W_{\mu\nu}$, it is seen that Fourier transforms of matrix elements of the bilocal operator turn out to be the structure functions and that W_1 and νW_2 scale, i.e. are functions of $\omega = 2M_N\nu/q^2 = -2p \cdot q/q^2$. Because of the structure of the tensor indices in eqn (43), one also obtains the relation:

$$\frac{2M_N}{\omega} W_1(\omega) = \nu W_2(\omega), \tag{46}$$

which corresponds to the vanishing of the longitudinal relative to transverse photon–nucleon total cross-sections ($\sigma_L/\sigma_T = 0$) in the indicated limit.

Another, more mundane, way to see that it is νW_2 and W_1 which should scale if there is no dimensional parameter or scale in the virtual photon–nucleon interaction is to rewrite the double differential cross-section in eqn (38) in terms of the dimensionless variables

$$x = \frac{1}{\omega} = -q^2/2p \cdot q = q^2/2M_N\nu \tag{47}$$

and

$$y = \nu/E = p \cdot q/p \cdot k. \tag{48}$$

Then in the high energy régime where ν and q^2 are both large, we can write

$$\frac{d^2\sigma}{dx\,dy} = \left(\frac{4\pi\alpha^2}{q^4}\right)(2M_N E)[(1 - y)\nu W_2 + \tfrac{1}{2}y^2(2M_N x)W_1], \tag{49}$$

where terms of order M_N/ν and M_N/E have been neglected. The factor $(4\pi\alpha^2/q^4)$ is just the elastic electron scattering cross-section for a point particle. We see immediately that if the quantity in brackets, which involves the photon–nucleon interaction, is not to depend on some internal scale then νW_2 and $2M_N x W_1$ should only depend on a *dimensionless variable involving the photon–nucleon vertex*, i.e. they should depend on $x = -q^2/2p \cdot q = 1/\omega$.

The absence of any scale in the interaction of a virtual photon with a nucleon is realized explicitly in the parton model,[64, 65] which might also be regarded as a concrete representation of the light cone algebra. In the parton model one regards the nucleon as composed of *point constituents.* In an infinite momentum frame, each type (i) of parton, with charge Q_i (in units of e), is taken to have a distribution $f_i(x)$ in the fractional longitudinal momentum $x = p_z^{(\text{parton})}/p_z^{(\text{nucleon})}$. A straightforward calculation then shows that for spin $\frac{1}{2}$ partons[64, 65]

$$\nu W_2(x) = \sum_i Q_i^2 x f_i(x) = 2M_N x W_1(x), \tag{50}$$

where x is both the fractional longitudinal momentum of the struck parton and the value of the scaling variable $q^2/2M_N\nu$. Taking the partons to be quarks (and more generally, also antiquarks) one has a concrete representation of the quark light cone algebra. As with any particular representation of a given algebra, certain results may hold which do not follow necessarily in the general case.

The scaling behaviour exhibited by the data[66, 67] outside the region of prominent resonances for νW_2 and $2M_N W_1$ is shown in Fig. 2. There the values of the structure functions are plotted versus

$$\omega' = 1 + W^2/q^2 = \omega + M_N^2/q^2,$$

which is the same as ω is the limit ν, $q^2 \to \infty$. Clearly values of νW_2 and $2M_N W_1$ at the same ω' but different q^2 coincide, i.e. νW_2 and $2M_N W_1$ for the proton are functions of ω' to within the accuracy of the data for $q^2 > 1\ \text{GeV}^2$ and $1 < \omega' < 10$.

The validity of the relation $\nu W_2 = 2M_N x W_1$ as $\nu, q^2 \to \infty$ is more clearly examined in terms of the quantity $R = \sigma_L/\sigma_T$, which should then vanish as $\nu, q^2 \to \infty$ at fixed ω. A previous global average of R for the proton gave the value[67] $R_p = 0{\cdot}18 \pm 0{\cdot}10$. Newer data, but over essentially the same kinematic range, have recently been analysed and yield[68] a global average in agreement with this. More interestingly, there is some indication[68] that R_p is vanishing as $1/\nu$ for fixed values of $\omega \lesssim 5$. The analysis of the deuteron data taken in the same experiment shows[68] that $R_p = R_d = R_n$ to within the statistical errors of the measurement ($\pm 0{\cdot}04$).

The ratio of neutron to proton inelastic cross-sections, as extracted from deuterium data, shows consistency with the neutron structure functions scaling also.[69, 70] The ratio of n/p decreases from values near unity at small x (and large $\omega = 1/x$) to values[70] definitely below $\frac{1}{2}$ for $x > 0{\cdot}65$.

For values of x near one, where it appears that the n/p ratio is the smallest, one could hope to challenge the bounds from the quark light-cone algebra:[71]

$$\tfrac{1}{4} \leqslant n/p \leqslant 4. \tag{51}$$

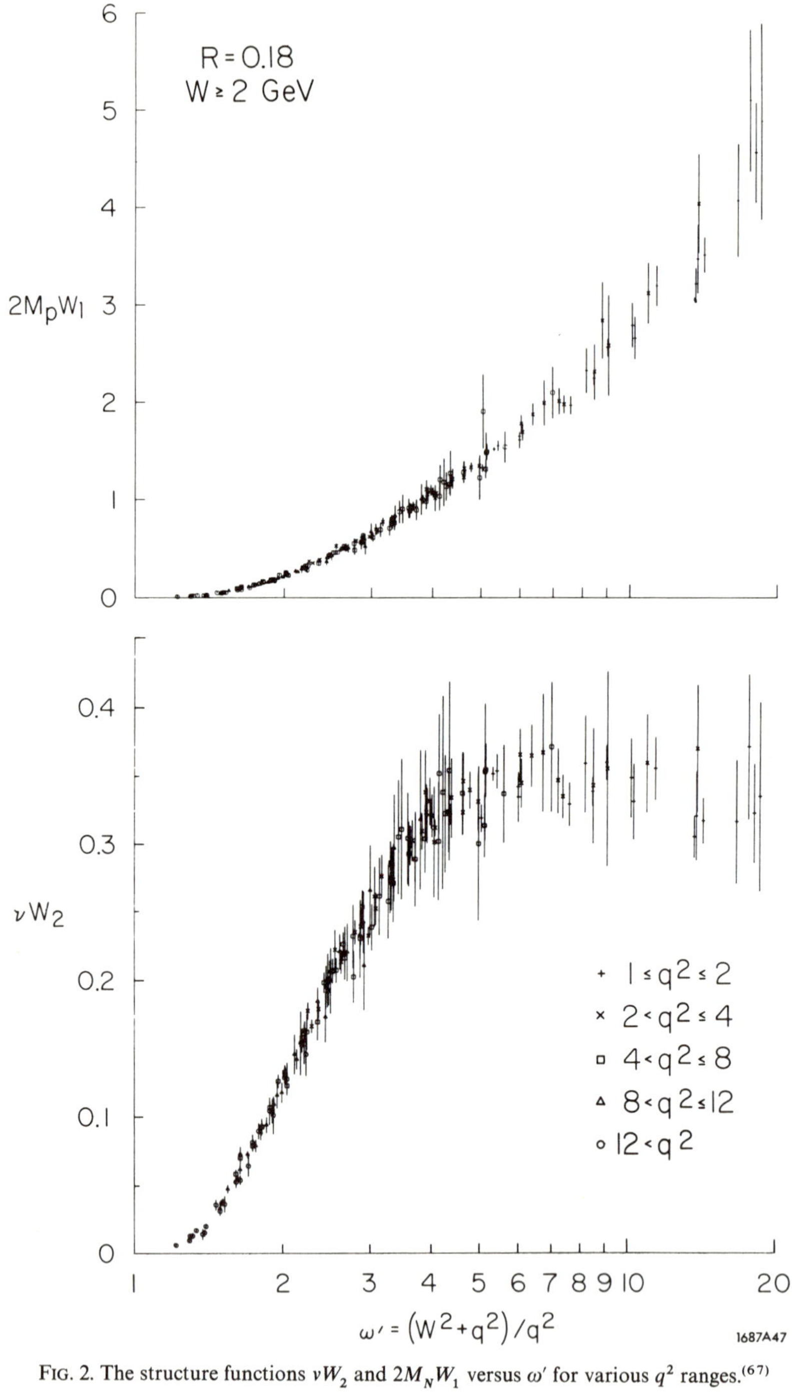

FIG. 2. The structure functions νW_2 and $2M_N W_1$ versus ω' for various q^2 ranges.[67]

Although present data extend to $x \simeq 0{\cdot}8$ and do not indicate any violation of the lower bound, experiments are under way to investigate the region near $x = 1$ in considerable detail.

Also of interest is the region of small x or large ω where the n/p ratio is expected to eventually approach unity on the basis of the dominance of the Pomeranchuk singularity in the photon–nucleon amplitude at large values of ω. Some recent data[72] on this ratio are shown in Fig. 3 where it is seen

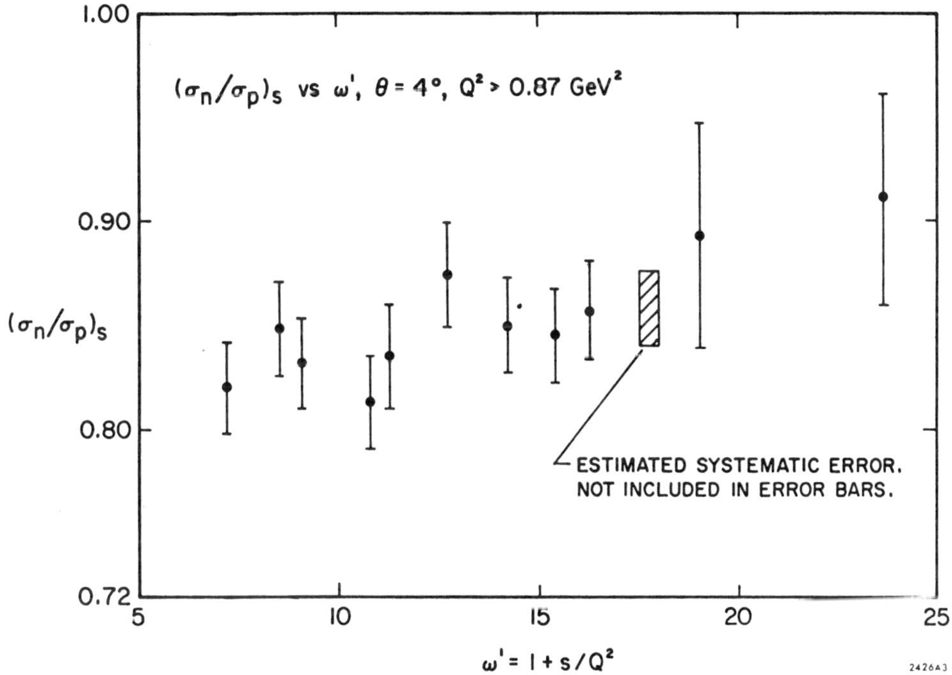

FIG. 3. The ratio of neutron to proton inelastic electron scattering cross-sections[72] for large values of ω'.

that even for values of ω between 10 and 20 the value of n/p is still only $\sim 0{\cdot}85$ and only slowly approaching unity. This bears directly on the convergence of the sum rule[73]

$$\int_1^\infty \frac{d\omega}{\omega}\left[\nu W_{2p}(\omega) - \nu W_{2n}(\omega)\right] = \tfrac{1}{3}. \tag{52}$$

This can be derived in parton models where the nucleon is composed of three "valence" quarks plus an isoscalar "sea" of $q\bar{q}$ pairs (plus neutrals), or it can be derived using exchange degeneracy arguments,[74] but it does not follow generally from the quark light cone algebra. Earlier evaluations of the sum rule using then existing data and Regge extrapolations for $\nu W_{2p}(\omega) - \nu W_{2n}(\omega)$ as $\omega \to \infty$ gave the estimate 0·19 ± 0·06 for the left hand side. The data shown in Fig. 3, however, gives[72]

$$\int_1^{20} \frac{d\omega}{\omega} [\nu W_{2p}(\omega) - \nu W_{2n}(\omega)] = 0{\cdot}18 \pm 0{\cdot}04, \tag{53}$$

and a rough estimate of the contribution from $\omega = 20$ to ∞ is 0·09. There is no longer an experimental basis for worrying about the sum rule's validity.

3. INELASTIC NEUTRINO– AND ANTINEUTRINO–NUCLEON SCATTERING

The extension of our discussion to inelastic neutrino and antineutrino scattering is easily made. Again neglecting lepton masses and averaging over nucleon spins, the double differential cross-section is[75]

$$\frac{d^2\sigma(\nu/\bar{\nu})}{d\Omega'\, dE'} = \frac{G^2 E'^2}{2\pi^2}\Big[2(\sin^2\theta/2)W_1^{(\nu/\bar{\nu})}(\nu, q^2) + (\cos^2\theta/2)W_2^{(\nu/\bar{\nu})}(\nu, q^2) \mp \frac{E + E'}{M_N}(\sin^2\theta/2)W_3^{(\nu/\bar{\nu})}(\nu, q^2)\Big], \tag{54}$$

with $G \simeq 1{\cdot}0 \times 10^{-5}/M_N$ the weak coupling constant. In the high energy limit, using the same variables x and y in eqns (47) and (48) as before, this becomes

$$\frac{d^2\sigma(\nu/\bar{\nu})}{dx\, dy} = \left(\frac{G^2 M_N E}{\pi}\right)\left[(1 - y)\nu W_2 + \frac{y^2}{2}(2M_N x)W_1 \mp y(1 - \tfrac{1}{2}y)x\nu W_3\right]. \tag{55}$$

The structure functions W_1 and W_2 now involve Fourier transforms of both vector–vector and axial vector–axial vector current commutators, while the new structure function W_3 involves only vector–axial vector commutators. Neglecting strangeness changing currents, the restriction of the weak, strangeness non-changing current to have isospin one implies that

$$W_i^{\nu p} = W_i^{\bar{\nu} n}$$
$$W_i^{\nu n} = W_i^{\bar{\nu} p}. \tag{56}$$

The quark light cone algebra in eqn (43) is also simply extended[8, 9] to include commutators of two axial-vector currents or a vector and axial-vector current. As an immediate consequence it follows that $2M_N x W_1$, νW_2, and νW_3 should scale, as can also be seen directly from eqn (55) using the argument that if the weak current–nucleon interaction doesn't involve parameters with dimensions, then those quantities in brackets on the right-hand side which characterize the current–nucleon vertex should only be dependent on the dimensionless quantity x. Scaling of $W_1 = F_1(x)$, $\nu W_2 = F_2(x)$, and $\nu W_3 = F_3(x)$ implies on integrating eqn (55) first over y and then over x that

$$\sigma_{\text{tot}}^{(\nu/\bar{\nu})} = \frac{G^2 M_N E}{\pi} \int_0^1 \mathrm{d}x \left[\frac{F_2(x)}{2} + \frac{2MxF_1(x)}{6} \mp \frac{xF_3(x)}{6} \right], \tag{57}$$

i.e. that the total neutrino or antineutrino cross-section rises linearly with the incident beam energy, E.

From the quark light cone algebra we have furthermore that[8, 9, 75]

$$2M_N x F_1(x) = F_2(x), \tag{58}$$

and the local relation between inelastic electron and neutrino scattering:

$$6[F_2^{ep}(x) - F_2^{en}(x)] = x[F_3^{\nu p}(x) - F_3^{\nu n}(x)]. \tag{59}$$

Various sum rules follow as well, including

$$\int_0^1 \frac{\mathrm{d}x}{x} [F_2^{\nu n}(x) - F_2^{\nu p}(x)] = 2, \tag{60}$$

the Adler sum rule,[76] which actually is supposed to hold for all q^2, and the sum rule[77]

$$\int_0^1 \mathrm{d}x [F_3^{\nu p}(x) + F_3^{\nu n}(x)] = -6. \tag{61}$$

In a parton model with only fermion constituents which interact with the current (no antifermions) one has as well that

$$F_2(x) = 2M_N x F_1(x) = -xF_3(x), \tag{62}$$

i.e. maximal V–A interference.

The simplest quantity with which to compare theory and experiment is the total cross-section summed over neutrino and antineutrino beams. From eqn (57) we find that the F_3 term cancels in the sum and using $F_2(x) = 2M_N x F_1(x)$ yields

$$\sigma_{\text{tot}}^{\nu N}(E) + \sigma_{\text{tot}}^{\bar{\nu} N}(E) = \frac{G^2 M_N E}{\pi} \frac{4}{3} \left[\frac{1}{2} \int_0^1 \mathrm{d}x (F_2^{\nu N}(x) + F_2^{\bar{\nu} N}(x)) \right]. \tag{63}$$

We write νN ($\bar{\nu} N$) to denote an average over neutrino (antineutrino) cross sections on protons and neutrons. Of course,

$$F_2^{\nu N}(x) = \tfrac{1}{2}(F_2^{\nu p}(x) + F_2^{\nu n}) = \tfrac{1}{2}(F_2^{\bar{\nu} p}(x) + F_2^{\bar{\nu} n}) = F_2^{\bar{\nu} N}(x). \tag{64}$$

The data from the Gargamelle experiment[78] are quite consistent with a linear rise with E of $\sigma_{tot}(E)$ for both neutrinos and antineutrinos. The coefficients of $G^2 M_N E/\pi$ yield[78]

$$\frac{1}{2}\int_0^1 dx(F_2^{\nu N}(x) + F_2^{\bar{\nu} N}(x)) = 0{\cdot}47 \pm 0{\cdot}07. \tag{65}$$

To relate this to electron scattering we must make some additional assumption beyond just the light cone algebra. We assume that in the x region which gives the most important contribution to $\int_0^1 dxF_2(x)$, one has only quark partons (no antiquarks). As we will see in a moment there is independent support for this from the ratio $\sigma_{tot}^{\bar{\nu}N}/\sigma_{tot}^{\nu N}$. With no antiquarks, one has the relation[79]

$$\tfrac{1}{2}(F_2^{\nu N}(x) + F_2^{\bar{\nu} N}(x)) = \tfrac{18}{5}(\tfrac{1}{2})(F_2^{ep}(x) + F_2^{en}(x)), \tag{66}$$

which together with the result from SLAC data (with a minor extrapolation),[66,67]

$$\frac{1}{2}\int_0^1 dx(F_2^{ep}(x) + F_2^{en}(x)) = 0{\cdot}15 \pm 0{\cdot}01, \tag{67}$$

predicts that

$$\frac{1}{2}\int dx(F_2^{\nu N}(x) + F_2^{\bar{\nu} N}(x)) = 0{\cdot}54 \pm 0{\cdot}04. \tag{68}$$

The agreement with the direct measurement, eqn (65), is obviously very good.[80]

Now let us return to $\sigma_{tot}(E)$ for neutrinos and antineutrinos separately. Rewriting eqn (57) we have

$$\sigma_{tot}^{(\nu/\bar{\nu})}(E) = \frac{G^2 ME}{\pi}\int_0^1 dx\, F_2(x)\left[\frac{1}{2} + \frac{1}{6} \pm \frac{B}{3}\right], \tag{69}$$

and therefore

$$\sigma_{tot}^{\bar{\nu}N}/\sigma_{tot}^{\nu N} = (2 - B)/(2 + B) \tag{70}$$

where

$$B = -\int_0^1 dx x F_3(x) \Big/ \int_0^1 dx F_2(x). \tag{71}$$

Purely from kinematic inequalities $|xF_3| \leqslant F_2(x)$ or $|B| \leqslant 1$. The extreme values of B correspond to maximal $V - A$ interference and are met for purely fermion partons ($B = +1$) or purely antifermion partons ($B = -1$). The experimental value of[78]

$$\sigma_{tot}^{\bar{\nu}N}/\sigma_{tot}^{\nu N} = 0{\cdot}38 \pm 0{\cdot}02 \tag{72}$$

gives[81]

$$B = 0{\cdot}90 \pm 0{\cdot}04, \tag{73}$$

and indicates almost purely fermion constituents in the region accessible to the Gargamelle experiment. Everything is quite consistent with the quark light cone algebra or the even more restrictive quark parton model with only a small component of antiquarks.

In the past few months the first data on inelastic neutrino scattering at NAL have been reported. The Caltech experiment[82] uses a "narrow band" beam with neutrinos of average energies of 50 and 145 GeV arising from decays of 160 GeV pions and kaons, respectively. While there are only 112 neutrino events from a steel target reported, they already allow some tentative conclusions.[82] If one assumes $2M_N x F_1(x) = F_2(x) = -xF_3(x)$, then eqn (55) becomes

$$\frac{\mathrm{d}^2\sigma^{\nu N}}{\mathrm{d}x\,\mathrm{d}y} = \frac{G^2 M_N E}{\pi} F_2^{\nu N}(x) \tag{74}$$

which is independent of y. Within large errors, the data are consistent with y independence. Although the flux is not known accurately, one can integrate over y and *compare the shape* of $\mathrm{d}\sigma^{\nu N}/\mathrm{d}x$ with what is seen at Gargamelle, or the better determined $F_2^{ep} + F_2^{en} \simeq F_2^{ed}$. This is shown in Fig. 4, with

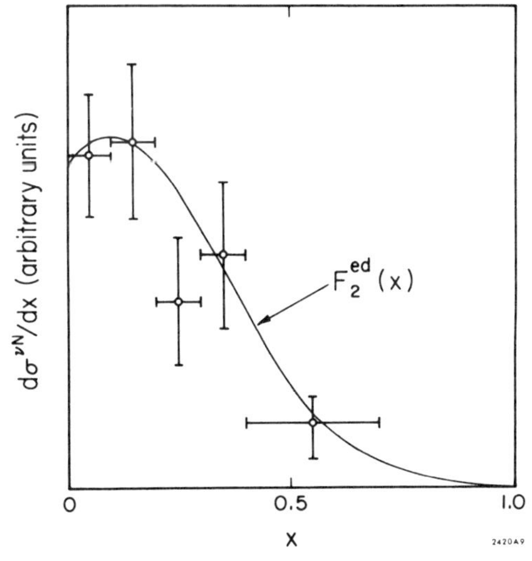

FIG. 4. The x distribution of inelastic neutrino scattering from the Caltech experiment[82] at NAL compared to $F_2^{ed}(x)$ measured at SLAC.

consistency seen between the x distribution measured with high energy neutrinos and that measured with electrons at SLAC.

A more stringent test of scaling is provided by the quantity

$$\langle q^2 \rangle = \langle xy2M_N E \rangle = 2M_N E \langle xy \rangle, \tag{75}$$

where

$$\langle f(x, y) \rangle = \frac{\int_0^1 \int_0^1 \mathrm{d}x\, \mathrm{d}y\, f(x, y)\, \mathrm{d}^2\sigma/\mathrm{d}x\,\mathrm{d}y}{\int_0^1 \int_0^1 \mathrm{d}x\, \mathrm{d}y\, \mathrm{d}^2\sigma/\mathrm{d}x\,\mathrm{d}y}. \tag{76}$$

$\langle q^2 \rangle$ should therefore rise linearly with E if there is scaling. The comparison with the results from the Caltech experiment for $\langle q^2 \rangle$ are shown in Fig. 5,

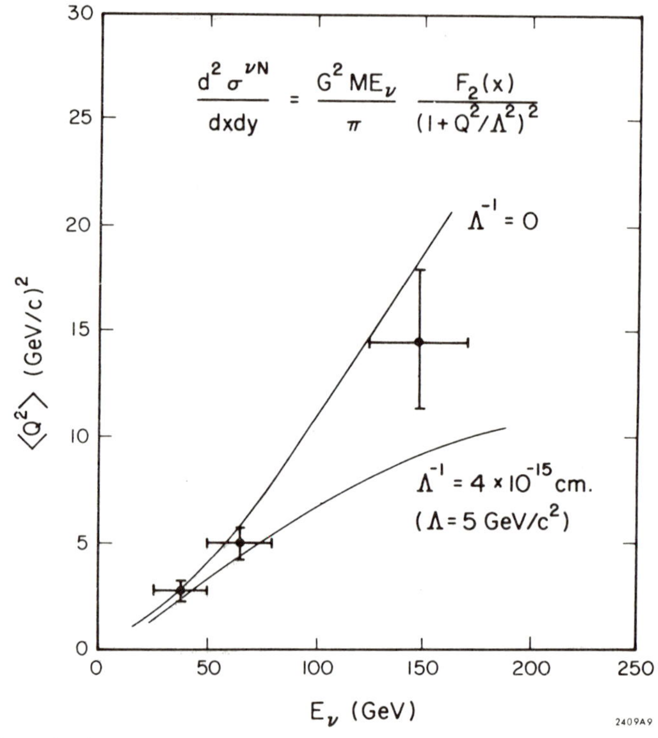

FIG. 5. Values of $\langle q^2 \rangle$ plotted versus the incident neutrino energy, E, from the Caltech experiment[82] at NAL.

where the curves are computed using the acceptance of the apparatus and assuming

$$\frac{\mathrm{d}^2\sigma}{\mathrm{d}x\mathrm{d}y} = \frac{G^2 M_N E}{\pi} \frac{F_2(x)}{(1 + q^2/\Lambda^2)^2} \tag{77}$$

Again the data are consistent with $\Lambda = \infty$, i.e. scaling. Note the large values of $\langle q^2 \rangle$ seen at NAL energies—in itself an indication of a "point-like" interaction.

A second experiment with a "broad band" beam of mean energy ~ 50 GeV has been carried out by a Harvard–Pennsylvania–Wisconsin collaboration.[83, 84] With about 300 neutrino and antineutrino events they are able to state that[84] $\sigma_{\text{tot}}^{\nu N} + \sigma_{\text{tot}}^{\bar{\nu} N}$ is very roughly ten times larger at a mean energy of 50 GeV than it is at 5 GeV in the Gargamelle experiment, just as expected from scaling (eqn (57)). The ratio $\sigma^{\bar{\nu} N}/\sigma^{\nu N}$ lies between $\frac{1}{3}$ and $\frac{1}{2}$ and is therefore consistent with the Gargamelle result of $0{\cdot}38 \pm 0{\cdot}02$. Noting that if

$$v = xy = 2E' \sin^2(\theta/2)/M_N, \tag{78}$$

then[85] $N^{-1}\, \mathrm{d}N/\mathrm{d}v$ is independent of flux and is a function of v alone if the structure functions scale. Analysis of their data shows that[84] both their neutrino and antineutrino data are consistent with scaling, and more particularly, with the $N^{-1}\, \mathrm{d}N/\mathrm{d}v$ curves calculated on the basis of the SLAC electron scattering data and the quark parton model.

Thus both high energy neutrino experiments seem to show that while we have increased the beam energies by an order of magnitude or better, nothing striking has changed from what was learned with Gargamelle. Everything seems remarkably consistent with the rather simple picture of scaling embodied in the quark light cone algebra and the quark parton model.

4. ELECTRON–POSITRON ANNIHILATION INTO HADRONS

There is one major indication of trouble with this simple picture: electron–positron annihilation into hadrons via one photon. The cross section for $e^+e^- \rightarrow$ hadrons is directly proportional to

$$\int \mathrm{d}^4x\, \mathrm{e}^{-iq\,.\,x} < 0|[J_\mu(x), J_\nu(0)]|0 >.$$

As a result, in the e^+e^- centre-of-mass where $\vec{q} = 0$ and $q_0 = \sqrt{(|q^2|)}$, the limit $q_0 \rightarrow \infty$ or $|q^2| \rightarrow \infty$ implies that $x_0 \approx 0$ and $\vec{x} \approx 0$, i.e. the tip of the light cone, is the important region of integration in that limit. As a result, assuming the quark model also gives the disconnected part of the commutator on the light cone, one finds

$$\sigma(e^+e^- \rightarrow \text{hadrons}) = \left(\frac{4\pi\alpha^2}{3|q^2|}\right) \sum_i Q_i^2, \tag{79}$$

where the sum is over the charged quark pairs creatable by the electromagnetic current. The quantity $4\pi\alpha^2/(3|q^2|)$ is just the cross-section for $e^+e^- \rightarrow \mu^+\mu^-$. Three quark constituents with charges $\frac{2}{3}$, $-\frac{1}{3}$, and $-\frac{1}{3}$ yield $\Sigma_i Q_i^2 = \frac{2}{3}$, while coloured quarks[16] give $\Sigma_i Q_i^2 = 2$, and the Han–

Nambu integrally charged quark scheme[(86)] has $\Sigma_i Q_i^2 = 4$. The present experimental situation is seen[(87)] in Fig. 6, including the most recent CEA results[(88)] for $\sigma(e^+e^- \to \text{hadrons})/\sigma(e^+e^- \to \mu^+\mu^-)$ of $4{\cdot}8 \pm 1{\cdot}1$ and $6{\cdot}3 \pm 1{\cdot}5$ at $|q^2| = 16$ and $25\ \text{GeV}^2$, respectively.

The seeming disagreement with our expectations leads one to ask whether

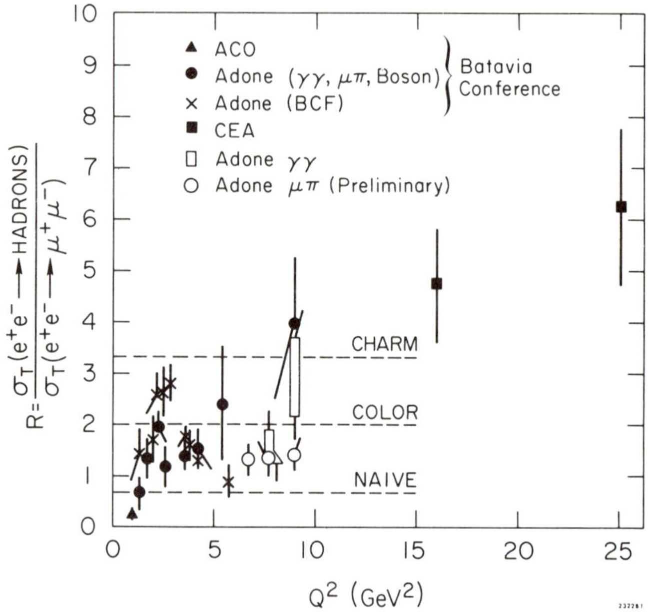

FIG. 6. Experimental results[(87)] for $\sigma(e^+e^- \to \text{hadrons})/\sigma(e^+e^- \to \mu^+\mu^-)$.

there is a breakdown in the scaling behaviour,[(89)] i.e. whether $\sigma(e^+e^- \to \text{hadrons})$ does not behave as $1/q^2$ as $|q^2| \to \infty$. A careful re-examination of the electron and neutrino data is also called for, particularly to look for the behaviour[(90)] of the moments $\int_0^1 dx\, x^n F_2(x)$ as powers of $(1/\ln q^2)$, as suggested in asymptotically free gauge theories.[(91)] Most important experimentally, one awaits the results from SPEAR and DORIS on the magnitude of $\sigma(e^+e^- \to \text{hadrons})$ at the same as well as still higher values of $|q^2|$.

VII. Inelastic Lepton Scattering with Polarized Beams and Targets

1. Introduction

In the previous lectures we discussed theoretical ideas and phenomenology which were mostly related to experiments already completed or in the process of being carried out. In this lecture we turn to a subject where no experiments have yet been accomplished, but where much of the previously discussed theoretical superstructure has immediate application.

In the near future the possibility of performing experiments with both polarized lepton beams and polarized targets will become a reality. The muon beams at BNL, CERN, or NAL automatically possess a longitudinal polarization because of their origin in the weak decays of pions. At SLAC a polarized electron source is being installed. The planned experiment[92] on deep inelastic scattering will be a technological *tour de force* involving production of a polarized lithium atomic beam, an ultraviolet flash lamp operating at 180 pulses/sec to knock off the electrons, injection into the accelerator and acceleration without loss of polarization, magnetic bending of the beam into the end station at which point $g - 2$ of the electron demands the energy be a multiple of 3·22 GeV so that the spin is rotated by a multiple of 180°, and finally the beam striking a longitudinally polarized proton target operating at 1°K in the field of superconducting magnet.

With this exciting background of approaching experiments, in this section I would like to discuss some of the theory and phenomenology related to inelastic lepton scattering with polarized beams and targets. As we shall see, a number of interesting questions having to do with amplitude analysis, scaling, parton models or light cone algebra, sum rules, quark models for resonance excitation, and even exotic J-plane singularity structure are subject to direct experimental answer with polarized beams and targets.

2. POLARIZED BEAMS

The scattering of a longitudinally polarized lepton beam on an unpolarized target should be independent of the lepton polarization if no hadrons are detected in the final state. This follows from parity conservation, and while this is well checked at small q^2 in the electromagnetic interactions of hadrons, an experiment[93] is planned at SLAC to verify this at large q^2 in deep inelastic scattering. Of course, if such a violation is found it is of enormous interest (due to weak interactions at an unexpected level?) and the theoretical questions discussed in the remainder of this talk becomes of secondary interest.

From this point on let us assume one photon exchange, together with the C, P and T invariance of electromagnetism. We use the kinematics and notation[61, 62] already given (see Fig. 1).

Suppose a final hadron (four-momentum p'_μ) is detected. Then $\vec{k}$ and $\vec{k}'$ define a plane (the lepton plane), as do $\vec{q}$ and $\vec{p}'$ (the hadron plane), with an angle ϕ between their normals.

Now recall for a moment the case of an unpolarized beam. There, to calculate a cross-section we must evaluate a lepton trace, $L_{\mu\nu}$, which is essentially the virtual photon density matrix. If we take the direction $\vec{q}$ as a z-axis, with x-axis in the lepton plane (so that the positive x direction is toward the leptons) then $L_{\mu\nu} = \frac{1}{2}(k_\mu k'_\nu + k'_\mu k_\nu + (q^2/2)\delta_{\mu\nu})$ can be written as

$$L_{\mu\nu} = \left(\frac{2}{1-\varepsilon}\right)\left(\frac{q^2}{4}\right)\left\{\begin{pmatrix} \sqrt{[\frac{1}{2}(1+\varepsilon)]} \\ 0 \\ -\sqrt{(q^2\varepsilon/\nu^2)} \\ 0 \end{pmatrix} \overbrace{\sqrt{[\tfrac{1}{2}(1+\varepsilon)]}\ 0\ -\sqrt{(q^2\varepsilon/\nu^2)}\ 0} + \begin{pmatrix} 0 \\ \sqrt{[\frac{1}{2}(1-\varepsilon)]} \\ 0 \\ 0 \end{pmatrix} \overbrace{0\ \sqrt{[\tfrac{1}{2}(1-\varepsilon)]}\ 0\ 0}\right\} \tag{80}$$

where

$$\varepsilon = \left[1 + 2\left(1 + \frac{\nu^2}{q^2}\right)\tan^2\theta/2\right]^{-1} \tag{81}$$

with θ = the lepton scattering angle in the laboratory. We have subtracted multiples[94] of q_μ from the virtual photon polarization vector so as to make $L_{\mu\nu} = 0$ when μ or $\nu = 4$. As is seen from eqn (80), the virtual photon is the incoherent sum of a piece with linear polarization in the x–z plane and a piece with linear polarization in the y-direction.

If $\varepsilon_\mu^{(\gamma)} f_\mu$ is the amplitude for

$$\text{“}\gamma(q)\text{”} + N(p) \to \text{hadron}'\,(p') + \ldots, \tag{82}$$

with the property

$$q_\mu f_\mu = 0, \tag{83}$$

then from eqn (80) we have that

$$\frac{d\sigma}{d^3p'} \propto |\sqrt{[\tfrac{1}{2}(1+\varepsilon)]}\, f_x - \sqrt{[q^2\varepsilon/\nu^2]} f_z|^2 + |\sqrt{[\tfrac{1}{2}(1-\varepsilon)]}\, f_y|^2$$

$$\propto \tfrac{1}{2}(|f_x|^2 + |f_y|^2) + \frac{\varepsilon}{2}(|f_x|^2 - |f_y|^2)$$

$$+ \varepsilon(q^2|f_z|^2/\nu^2) - \sqrt{[\tfrac{1}{2}\varepsilon(1+\varepsilon)]}\, 2\,\mathrm{Re}\,(f_x\sqrt{[q^2/\nu^2]}\, f_z^*). \tag{84}$$

The four terms in eqn (84) are in the form of the standard equation for, e.g. “γ” + $p \to \pi^+ + n$, where they are often labelled[95] $d\sigma_U/dt$, $d\sigma_P/dt$, $d\sigma_L/dt$, and $d\sigma_I/dt$ respectively, after their ϕ dependence is explicitly exhibited. This ϕ dependence is easily read off if f_x, f_y and f_z are rewritten in terms of helicity amplitudes (of the virtual photon). The four terms behave respectively as constant, cos 2ϕ, constant, and cos ϕ.

With this background, consider a longitudinally polarized beam where the leptons have helicity $\pm\frac{1}{2}$. The lepton trace now reads

$$L_{\mu\nu}^{(\pm)} = \tfrac{1}{2}(k_\mu k'_\nu + k'_\mu k_\nu + (q^2/2)\,\delta_{\mu\nu} \pm \varepsilon_{\mu\nu\lambda\sigma} k_\lambda k'_\sigma), \tag{85}$$

which can be rewritten with $\vec{q}$ as a z-axis, using the same manipulations as above, in the form[94]

$$L_{\mu\nu}^{(\pm)} = \left(\frac{2}{1-\varepsilon}\right)\left(\frac{q^2}{4}\right) \left\{ \begin{pmatrix} \sqrt{[\frac{1}{2}(1+\varepsilon)]} \\ \mp i\sqrt{[\frac{1}{2}(1-\varepsilon)]} \\ -\sqrt{[q^2\varepsilon/\nu^2]} \\ 0 \end{pmatrix} \underbrace{\overbrace{\sqrt{[\tfrac{1}{2}(1+\varepsilon)]} \;\; \pm i\sqrt{[\tfrac{1}{2}(1-\varepsilon)]} \;\; -\sqrt{[q^2\varepsilon/\nu^2]} \;\; 0}} \right\} \tag{86}$$

The virtual photon now can be represented in terms of a single elliptical polarization state. For incident leptons with helicity $+\frac{1}{2}$, the analogue of eqn (84) for detection of a final hadron is (summed over the helicities of all hadrons):

$$\frac{d\sigma}{d^3p'} \propto |\sqrt{[\tfrac{1}{2}(1+\varepsilon)]}\,f_x \pm i\sqrt{[\tfrac{1}{2}(1-\varepsilon)]}\,f_y - \sqrt{[q^2\varepsilon/\nu^2]}\,f_z|^2$$

$$\propto \tfrac{1}{2}(|f_x|^2 + |f_y|^2) + \tfrac{1}{2}\varepsilon(|f_x|^2 - |f_y|^2) + \varepsilon(q^2|f_z|^2/\nu^2)$$
$$- \sqrt{[\tfrac{1}{2}\varepsilon(1+\varepsilon)]}\,2\,\mathrm{Re}\,(f_x\sqrt{[q^2/\nu^2]}\,f_z^*)$$
$$\pm \sqrt{[\tfrac{1}{2}\varepsilon(1-\varepsilon)]}\,2\,\mathrm{Im}\,(f_y\sqrt{[q^2/\nu^2]}\,f_z^*), \tag{87}$$

where we have gained the last term when compared with eqn (84). An apparent term of the form $\sqrt{[\frac{1}{4}(1-\varepsilon^2)]}\,2\,\mathrm{Im}\,(f_x f_y^*)$ vanishes because of parity conservation, which becomes obvious when the amplitudes are rewritten in terms of (photon) helicity amplitudes.[96] In this form the ϕ dependence of the last term in eqn (87) is also easily seen to be that of sin ϕ.

Since the (complex) amplitudes f_x, f_y, and f_z are characterized by six real quantities, with one overall phase, there are five real numbers which give all possible physical information on the amplitude f_μ. As there are five independent quantities in eqn (87), we see that *all the physics contained in* f_μ *is obtainable with a polarized beam.*

Particular cases of interest in the literature are[97]

$$\text{“}\gamma\text{”} + N \to \pi + \text{anything}, \tag{88}$$

and[98, 99]

$$\text{“}\gamma\text{”} + N \to \rho^0 + N, \tag{89}$$

where one also has information on the ρ^0 density matrix coming from its two pion decay mode. It has recently been shown[98] explicitly that all possible joint density matrix elements of the virtual photon and rho meson are determinable through use of a polarized lepton beam and examination of the rho decay angular distribution. In short, the use of a polarized lepton beam alone allows one to perform complete amplitude analyses of the photon segment of virtual photon induced processes.

3. POLARIZED BEAM AND TARGET KINEMATICS

As in the unpolarized case, there are two equivalent ways to proceed, each particularly useful for certain purposes. One method is to define virtual photon–nucleon total cross sections; the other is to define structure functions. We begin with the definition of total cross-sections.

First we recall the unpolarized case,[62] where there are two independent photon–nucleon total cross-sections, σ_T and σ_L. By the optical theorem they are proportional to the imaginary part of the forward photon–nucleon scattering amplitude, $T_{\lambda'_\gamma\lambda'_N,\,\lambda_\gamma\lambda_N}(\nu, q^2, t = 0)$:

$$\sigma_T(\nu, q^2) \propto \tfrac{1}{2}(\mathrm{Im}\, T_{1\frac{1}{2},\,1\frac{1}{2}} + \mathrm{Im}\, T_{1-\frac{1}{2},\,1-\frac{1}{2}}) \tag{100a}$$

and

$$\sigma_L(\nu, q^2) \propto \mathrm{Im}\, T_{0\frac{1}{2},\,0\frac{1}{2}}. \tag{100b}$$

In fact, the usual tensor, $W_{\mu\nu}$, is also proportional to Im $T_{\mu\nu}$, and the $\mu, \nu = 1, 2, 3$ (z axis along $\vec{q}$) components are[100]

$$W_{ij} = \frac{K}{4\pi^2\alpha}\begin{pmatrix} \sigma_T & 0 & 0 \\ 0 & \sigma_T & 0 \\ 0 & 0 & \nu^2\sigma_L/q^2 \end{pmatrix}, \tag{101}$$

where $K = \nu - q^2/2M_N$ is the real photon (laboratory) energy necessary to produce the same centre-of-mass energy. The double differential cross-section for detecting a final lepton of energy E' if the incident beam has energy E is given by eqn (34) as

$$\frac{\mathrm{d}^2\sigma}{\mathrm{d}\Omega'\,\mathrm{d}E'} = \left(\frac{1}{2\pi}\right)^2\left(\frac{E'}{E}\right)\left(\frac{4\pi\alpha}{q^2}\right)^2 L_{\mu\nu}W_{\mu\nu},$$

so that using the lepton trace in eqn (80) we have

$$\frac{\mathrm{d}^2\sigma}{\mathrm{d}\Omega'\,\mathrm{d}E'} = \frac{\alpha}{4\pi^2}\frac{\mathrm{K}}{q^2}\left(\frac{E'}{E}\right)\left(\frac{2}{1-\varepsilon}\right)(\sigma_T + \varepsilon\sigma_L) = \Gamma(\sigma_T + \varepsilon\sigma_L), \tag{102}$$

which is the standard result.

In the polarized beam and target case, one is able to measure separately the total cross-sections $\sigma_{\frac{1}{2}}$ and $\sigma_{\frac{3}{2}}$ (net photon plus nucleon value of J_z equal to $\frac{1}{2}$ and $\frac{3}{2}$ respectively) and σ_{TL} defined by[101]

$$\begin{aligned} \sigma_{\frac{1}{2}}(\nu, q^2) &\propto \operatorname{Im} T_{1\frac{1}{2}, 1\frac{1}{2}}, \\ \sigma_{\frac{3}{2}}(\nu, q^2) &\propto \operatorname{Im} T_{1-\frac{1}{2}, 1-\frac{1}{2}}, \\ (\sqrt{2})\sigma_{TL}(\nu, q^2) &\propto \operatorname{Im} T_{1\frac{1}{2}, 0-\frac{1}{2}}, \end{aligned} \tag{103}$$

with the same proportionality constant as in eqn (100), so that

$$\sigma_T = \tfrac{1}{2}(\sigma_{\frac{1}{2}} + \sigma_{\frac{3}{2}}). \tag{104}$$

Therefore, there are two new measurable quantities, which we take to be $(\frac{1}{2})(\sigma_{\frac{1}{2}} - \sigma_{\frac{3}{2}})$ or the asymmetry,

$$A(\nu, q^2) = \frac{\sigma_{\frac{1}{2}} - \sigma_{\frac{3}{2}}}{\sigma_{\frac{1}{2}} + \sigma_{\frac{3}{2}}} = \frac{\frac{1}{2}(\sigma_{\frac{1}{2}} - \sigma_{\frac{3}{2}})}{\sigma_T}, \tag{105}$$

and σ_{TL}. For these quantities we have the bounds[102]

$$\left|\tfrac{1}{2}(\sigma_{\frac{1}{2}} - \sigma_{\frac{3}{2}})\right| \leqslant \sigma_T \text{ or } |A| \leqslant 1, \tag{106}$$

and

$$(\sqrt{2})\sigma_{TL} \leqslant \sqrt{(\sigma_{\frac{1}{2}}\sigma_L)} \text{ or } \sigma_{TL} \leqslant \sqrt{(\sigma_T\sigma_L)} = (\sqrt{R})\sigma_T. \tag{107}$$

It is not difficult to see from our definitions that if the nucleon target was polarized along $\pm\vec{q}$, only $\sigma_{\frac{1}{2}} - \sigma_{\frac{3}{2}}$ would contribute to the difference, while with polarization perpendicular to $\vec{q}$ (but still in the lepton plane), only σ_{TL} contributes to a difference between target spin orientations. However, in real life it is far easier to orient the target relative to the lepton beam, and not $\vec{q}$. For a nucleon target polarized antiparallel or parallel to the beam we find,[103]

$$\frac{d^2\sigma\uparrow\downarrow}{d\Omega'\,dE'} - \frac{d^2\sigma\uparrow\uparrow}{d\Omega'\,dE'} = 2\Gamma\left\{\left(\frac{E - E'\varepsilon}{E}\right)A\sigma_T + \frac{\sqrt{q^2}}{E}\varepsilon\sigma_{TL}\right\}, \tag{108}$$

to be compared with the unpolarized result,

$$\frac{d^2\sigma\uparrow\downarrow}{d\Omega'\,dE'} + \frac{d^2\sigma\uparrow\uparrow}{d\Omega'\,dE'} = 2\Gamma(\sigma_T + \varepsilon\sigma_L). \tag{109}$$

If we neglect σ_L and σ_{TL} as being small,[104] then we have that

$$\begin{aligned} \left(\frac{d^2\sigma\uparrow\downarrow}{d\Omega'\,dE'} - \frac{d^2\sigma\uparrow\uparrow}{d\Omega'\,dE'}\right)\left(\frac{d^2\sigma\uparrow\downarrow}{d\Omega'\,dE'} + \frac{d^2\sigma\uparrow\uparrow}{d\Omega'\,dE'}\right)^{-1} &= \left(\frac{E - E'\varepsilon}{E}\right)A(\nu, q^2) \\ &= \left(\frac{E - E'\varepsilon}{E}\right)\frac{\sigma_{\frac{1}{2}} - \sigma_{\frac{3}{2}}}{\sigma_{\frac{1}{2}} + \sigma_{\frac{3}{2}}}. \end{aligned} \tag{110}$$

This simple result shows that the virtual photon–nucleon asymmetry is "degraded" by the factor $(E - E'\varepsilon)/E$ in order to obtain the lepton beam–target asymmetry which is actually measured.

It is also possible to polarize the target perpendicular to the beam, but in the lepton plane. In this case we find[105]

$$\frac{d^2\sigma\uparrow\leftarrow}{d\Omega'\,dE'} - \frac{d^2\sigma\uparrow\rightarrow}{d\Omega'\,dE'} = 2\Gamma\left\{\sqrt{(\tfrac{1}{2})\varepsilon(1+\varepsilon)}\frac{\sqrt{q^2}}{E}A\sigma_T - \sqrt{\left(\frac{2\varepsilon}{1+\varepsilon}\right)}\frac{E - E'\varepsilon}{E}\sigma_{TL}\right\}. \tag{111}$$

Under presently contemplated experimental conditions, the $A\sigma_T$ term has an order of magnitude greater weight than the σ_{TL} term in eqn (108), while the situation is reversed in eqn (111). Thus, while it is possible in principle to separate $A\sigma_T$ and σ_{TL} by only running with the target polarized parallel or antiparallel to the beam at different values of E, E', and θ (but the same ν and q^2), in practice this is very difficult. The use of perpendicular polarization (eqn (111)) makes the task of separation much easier.[106, 107]

Isolation of σ_{TL} may be of some importance in obtaining information on $R = \sigma_L/\sigma_T$. As R is known to be small in deep inelastic scattering, it is difficult to measure accurately. Equation (107), however, shows that measurement of σ_T and σ_{TL} will give a *lower bound* on R. For the particular case of elastic scattering, where $\sigma_{TL} \propto G_M G_E$, measurement of σ_{TL} may provide the only way to study G_E at large q^2.

Just as there were two new cross sections to define in the spin dependent case, there are two additional structure functions besides the familiar W_1 and W_2. We write,[108]

$$\begin{aligned} W_{\mu\nu} &= W_1(\nu, q^2)(\delta_{\mu\nu} - q_\mu q_\nu/q^2) \\ &\quad + W_2(\nu, q^2)(p_\mu - p\cdot q\, q_\mu/q^2)(p_\nu - p\cdot q\, q_\nu/q^2)/M_N^2 \\ &\quad - d(\nu, q^2)\varepsilon_{\mu\nu\lambda\sigma}q_\lambda s_\sigma/4\pi M_N + g(\nu, q^2)s\cdot q\varepsilon_{\mu\nu\lambda\sigma}q_\lambda p_\sigma/4\pi M_N, \end{aligned} \tag{112}$$

where s_μ is a covariant spin four-vector associated with the nucleon target and possessing the properties

$$s\cdot p = 0, \qquad s\cdot s = +1. \tag{113}$$

In terms of the structure functions defined in eqn (112), the measured difference between double differential cross-sections with lepton and nucleon spins antiparallel and parallel is[103]

$$\begin{aligned} \frac{d^2\sigma\uparrow\downarrow}{d\Omega'\,dE'} - \frac{d^2\sigma\uparrow\uparrow}{d\Omega'\,dE'} &= \frac{4\alpha^2 E'}{q^2 E}\{(E + E'\cos\theta)\, d(\nu, q^2) \\ &\quad + (E - E'\cos\theta)(E + E')M_N g(\nu, q^2)\}\,(4\pi M_N)^{-1}. \end{aligned} \tag{114}$$

The catalogue is completed by giving the relationship of the structure functions to the previously defined total cross-sections:

$$\frac{1}{4\pi M_N}[\nu\, d(\nu, q^2) + M_N(\nu^2 + q^2)\, g(\nu, q^2)] = \frac{K}{4\pi^2\alpha}\frac{\sigma_{\frac{1}{2}} - \sigma_{\frac{3}{2}}}{2}$$

$$= \frac{K}{4\pi^2\alpha} A\sigma_T, \qquad (115a)$$

and

$$\frac{\sqrt{q^2}}{4\pi M_N} d(\nu, q^2) = \frac{K}{4\pi^2\alpha}\sigma_{TL}. \qquad (115b)$$

These allow one, for example, to go directly from eqns (108) to (114).

The particular choice of structure functions given in eqn (112) is of course not a unique one. In fact, each paper on this subject seems to define a new set, which can only be linear combinations of the ones given here. As will be seen in the next section, $d(\nu, q^2)$ and $g(\nu, q^2)$ are slightly more convenient when one wishes to use the inequalities for total cross-sections, eqns (106) and (107), to bound the integrands in certain sum rules.

4. SCALING AND SUM RULES

In Section VI we discussed the scaling behaviour[109] of the spin averaged structure functions: that as ν and q^2 become infinite, $W_1(\nu, q^2)$ and $\nu W_2(\nu, q^2)$ become nontrivial functions of $\omega = 2M_N\nu/q^2$. It comes as no surprise that a similar behaviour is expected on the same basis for the spin-dependent structure functions: one expects[110, 111] $\nu d(\nu, q^2)$ and $\nu^2 g(\nu, q^2)$ to scale. Neglecting q^2/ν^2 as ν and $q^2 \to \infty$ with ω fixed, $\nu d + M_N\nu^2 g \propto A(\nu, q^2) W_1$ from eqn (115a), and we see that $A(\nu, q^2)$ should also scale.

What magnitude and sign might we expect for these new structure functions in the scaling domain? For purposes of a very rough orientation and guide to our intuition, let us consider the most naïve parton model—something which cannot possibly be completely correct and which I will consequently call "dumb" quarks. Namely, let the nucleon consist of three point quarks: the proton is *ppn* and the neutron is *nnp*, i.e. just the three "valence" quarks with no $q\bar{q}$ sea, gluons, etc. Then in the unpolarized case, the usual parton manipulations yield that νW_2 and W_1 scale with

$$\nu W_2 = (2M_N/\omega) W_1 \text{ and } W_{1n}/W_{1p} = \tfrac{2}{3}.$$

The same calculations show, as expected, that νd and $\nu^2 g$ scale with

$$\nu^2 g = 0, \qquad (116)$$

for either the neutron or proton, and

$$\frac{\nu d + M_N \nu^2 g}{4\pi M_N} = \tfrac{5}{9} W_1 \quad \text{or} \quad A(\omega) = \tfrac{5}{9} \tag{117a}$$

for the proton and

$$\frac{\nu d + M_N \nu^2 g}{4\pi M_N} = 0 \quad \text{or} \quad A(\omega) = 0 \tag{117b}$$

for the neutron. Equation (116) is seen to follow in the detailed computation from the (assumed) Dirac point particle nature of the partons.

The positive, or at least nonnegative, values of A found in eqns (117) are to be expected on naïve grounds. For A in a parton model is just a measure of the extent to which the parton and nucleon spins are aligned, weighted by the parton charges squared. Therefore, since we don't expect the parton spins on the average to tend to line up opposite to the nucleon's spin, we expect $A \geqslant 0$.

In general, the use of a polarized target and beam could be thought of as weighting the individual parton contributions to the scattering by an additional factor: their spin alignment.[112] For example, in the naïve three quark model the p quark in the proton is weighted more heavily in $\sigma_{\frac{1}{2}}$ and less heavily in $\sigma_{\frac{3}{2}}$ than in the spin averaged cross section $\sigma_T = \frac{1}{2}(\sigma_{\frac{1}{2}} + \sigma_{\frac{3}{2}})$. If there is a connection of the parton charges to the excess[113] of positive hadrons projected forward in the hadronic final state at large q^2, then the excess should be different in $\sigma_{\frac{1}{2}}$ and $\sigma_{\frac{3}{2}}$.

Now let us turn from "dumb" quarks to what might be called "sophisticated" quarks. By this I mean the algebra of currents at equal time, and more generally the quark light cone algebra.[8] Here one abstracts certain algebraic properties from the free quark model; properties which might be exactly true in Nature. But one discards the quark themselves. These algebraic relations can often be converted into sum rules. For example, the equal time commutator of two space components of the vector current as abstracted from the quark model (see eqn (43)),

$$[V_i^\alpha(\vec{x}, t), V_j^\beta(\vec{y}, t)] = -i\varepsilon_{ijk}\, d^{\alpha\beta\gamma} A_k^\gamma(\vec{x}, t)\, \delta^{(3)}(\vec{x} - \vec{y}) + \text{terms symmetric in } i, j \tag{118}$$

leads to the sum rule[114]

$$\lim_{q^2 \to \infty} q^2 \int_0^\infty \frac{d\nu}{\nu} \left(\frac{\nu d + M_N \nu^2 g}{4\pi M_N} \right) = Z. \tag{119}$$

Rewriting this in the scaling limit gives[115]

$$\int_1^\infty \frac{d\omega}{\omega} A(\omega) \left(\frac{2M}{\omega} W_1(\omega) \right) = Z. \tag{120}$$

The quantity Z on the right-hand side of eqns (119) and (120) arises from the one nucleon matrix elements of the axial-vector current, A_k^γ, on the right-hand side of eqn (118). With the indices α and β chosen to correspond to the electromagnetic current, the index γ takes the values 3, 8, or 0, so that one has for protons and neutrons

$$Z_p = Z^{(3)} + Z^{(8)} + Z^{(0)}$$

and (121)

$$Z_n = -Z^{(3)} + Z^{(8)} + Z^{(0)}$$

$Z^{(3)} = g_A/6$ is directly measurable in weak interactions. Thus by taking the difference between eqn (120) for protons and for neutrons, one obtains a sum rule with the known quantity $Z_p - Z_n = g_A/3$ on the right-hand side. $Z^{(8)}$, arising from the same octet of currents as $Z^{(3)}$ is calculable if we know the F/D ratio of the axial-vector currents. If $(F/D)_{\text{axial}} = \frac{2}{3}$, then $Z^{(8)}/Z^{(3)} = \frac{1}{5}$.

It is instructive to examine the sum rule, eqn (120), in the naïve quark parton model discussed previously. There $A = \frac{5}{9}$ and

$$\int_1^\infty \frac{d\omega}{\omega}(2M_N/\omega)W_1(\omega) = 1$$

for the proton, while in the same naïve quark model $Z^{(3)} = \frac{5}{18}$ (since $g_A = \frac{5}{3}$), $Z^{(8)} = \frac{1}{18}$, and $Z^{(0)} = \frac{4}{18}$. Therefore $Z_p = (5 + 1 + 4)/18 = \frac{5}{9}$ and the left- and right-hand sides of eqn (120) are both $\frac{5}{9}$ for the proton. Similarly, since $A = 0$ and $Z_n = (-5 + 1 + 4)/18 = 0$, both sides are zero for the neutron. Therefore we might regard eqn (120) as being the correct "sophisticated" quark version of the "dumb" quark results in eqns (117).

Independent of any particular model recall that we expect $A \geqslant 0$. If this is to be correct for both proton and neutron,[116] $Z^{(8)} + Z^{(0)} > |Z^{(3)}|$. As a result $Z_p \geqslant 2Z^{(3)} = g_A/3 \simeq 0{\cdot}41$, which, as we will see in a moment, demands large asymmetries over a considerable ω range for the proton.

A second sum rule for the spin dependent structure functions,[117]

$$\int_0^\infty d\nu\, g(\nu, q^2) = 0, \tag{122}$$

or[115, 118]

$$\int_1^\infty d\omega\, \nu^2 g(\omega) = 0 \tag{123}$$

in the scaling limit, can be derived as a superconvergence relation or from the quark light cone algebra. Regge arguments indicate[117] that both eqns (122) and (123), where the amplitudes involved receive no contribution[119] from a Pomeranchuk pole, should converge on the basis of *known Regge pole intercepts.* In particular, one expects

$$A(\omega) \to 0 \text{ as } \omega \to \infty.$$

Can anything rigorous be said about the validity of these sum rules in the absence of actual data from polarized beam experiments? An examination of the left-hand side of eqn (120) shows that if we replace $A(\omega)$ by its upper bound (of 1), then we have

$$Z \leqslant \int_1^\infty \frac{d\omega}{\omega} \frac{2M_N}{\omega} W_1(\omega). \tag{124}$$

However, $(2M_N/\omega)W_1(\omega) \to$ constant as $\omega \to \infty$ if the usual assumption of dominance of the high ω behaviour by the Pomeranchuk singularity is made. The right-hand side of eqn (124) is then logarithmically divergent, and we have no rigorous bound which is useful in this case.

We can use bounds on $A(\omega)$ and presently existing data though to answer a lesser question. For example, suppose $A_p \leqslant 0{\cdot}5$ (remember that $A = 0{\cdot}5$ is a large asymmetry—it is equivalent to $\sigma_{\frac{1}{2}}/\sigma_{\frac{3}{2}} = 3$). How far in ω do we have to carry the integration in eqn (120) to get $Z = \frac{5}{9}$, as in the naïve parton model discussed above? Present data on $W_1(\omega)$ shows that the upper limit of integration must be $\gtrsim 50$. Even for $Z_p \geqslant g_A/3$, which we expect from the argument that A_p and A_n should be nonnegative, the upper limit of integration must be $\gtrsim 25$. Thus if Z_p obeys our expectations, *present data* on deep inelastic scattering *and the sum rule* in eqn (120) *indicate that we must have large asymmetries* over a very considerable range of ω values.[120] Put another way, if the sum rule in eqn (120) is correct, it converges slowly and receives important contributions from the large ω region. In this regard it is similar to the Adler sum rule,[121] and the parton model sum rule (discussed in Section VI)

$$\int_1^\infty \frac{d\omega}{\omega} [\nu W_{2p}(\omega) - \nu W_{2n}(\omega)] = \tfrac{1}{3}.$$

One might ask in this regard if the coefficient of A_k^γ in the commutation relation, eqn (118), and consequently the Z on the right-hand side of eqn (120), could be reduced by a factor K from the quark model value, making the sum rule easier to satisfy without large asymmetries. Here we are constrained by the Crewther relation[122]

$$3S = KR', \tag{125}$$

where K is defined as above ($K = 1$ in the quark model), S is the coefficient of the Adler anomaly,[123] and R' is the ratio

$$\sigma(e^+e^+ \rightarrow \gamma^{(3)} \rightarrow \text{hadrons})/\sigma(e^+e^- \rightarrow \mu^+\mu^-).$$

Since $S = \frac{1}{6}$ and $R' = \frac{1}{2}$ for ordinary quarks, and $S = \frac{1}{2}$ and $R' = \frac{3}{2}$ for coloured quarks, we see that the relation is satisfied in both these cases. Note that if S is fixed (say by PCAC and $\Gamma(\pi^0 \rightarrow \gamma\gamma)$) then large values of R' force K to be small and vice versa. One therefore has an interesting relation between the results of e^+e^- colliding beam experiments and those involving polarized lepton beams.

Finally, we note that if we subtract the second sum rule, eqn (122), from the first, eqn (119), and use eqns (107) and (115), we have

$$|Z| = \left| \lim_{q^2 \rightarrow \infty} q^2 \int_0^\infty \frac{\mathrm{d}\nu}{\nu} \frac{\nu d}{4M_N} \right| \leqslant \lim_{q^2 \rightarrow \infty} \int_0^\infty \frac{\mathrm{d}\nu}{\nu} \sqrt{(q^2 R)} W_1$$

$$\leqslant \lim_{q^2 \rightarrow \infty} \int_0^\infty \frac{\mathrm{d}\omega}{\omega} \left[\frac{2M_N}{\omega} W_1(\omega) \right] \sqrt{(\nu^2 R/q^2)}. \qquad (126)$$

Therefore, if $Z \neq 0$, R must not vanish faster than

$$q^2/\nu^2 = \frac{2M}{\omega} \frac{1}{\nu}$$

at fixed ω as $\nu, q^2 \rightarrow \infty$. Of course exactly the behaviour[124] $R = (q^2/\nu^2) f(\omega)$ is what is expected from the quark light cone algebra from which both sum rules were derived, so that everything is consistent in this regard.

5. DUALITY AND THE POLARIZATION ASYMMETRY IN DEEP INELASTIC SCATTERING

By now it is rather well accepted that there is a substantial non-diffractive component of the structure functions, at least for $1 \leqslant \omega \leqslant 10$, and that the distinguishable s-channel nucleon resonances exhibit a behaviour which is closely connected[125] with that of νW_2 near $\omega = 1$. Inasmuch as there is no Pomeranchuk pole contribution to the spin dependent structure functions, one might look here also for a connection of deep inelastic behaviour with that of s-channel resonances. To examine this question further, it is useful to construct a sum of resonances, in the spirit of the Veneziano model, which duplicates parton model results, and in particular the results of the naïve three quark model for the structure functions which we discussed earlier.

For this purpose, we consider the states generated by three quarks with harmonic potentials between them. We let the electromagnetic current interact with only one quark at a time, considering only interactions with the quark magnetic moments. This immediately leads to $R = \sigma_L/\sigma_T = 0$.

The ground and excited states of such a set of three quarks are classifiable in terms of $SU(6)$ representations, with the nucleon itself being in the ground state, the totally symmetric **56** representation. Since only one quark at a time is to be excited, only resonances in the totally symmetric **56** and mixed symmetry **70** representations can be produced. A little calculation[126] shows that summing over all possible excitations of resonances lying in a **56** at some given ν and q^2:

$$W_{1n}/W_{1p} = 12/17, \qquad A_p = 5/17, \qquad A_n = 0, \tag{127}$$

while for excitation of a **70**,

$$W_{1n}/W_{1p} = 3/5, \qquad A_p = 1, \qquad A_n = 0. \tag{128}$$

In particular there exists a combination of **56** and **70** s-channel nucleon resonances excited in photon–nucleon collisions which yield[126]

$$W_{1n}/W_{1p} = 2/3, \qquad A_p = 5/9, \qquad A_n = 0, \tag{129}$$

exactly the naïve three quark parton model results. Therefore it is possible to construct[126] a sum of s-channel resonances which for some, or for that matter all, ν and q^2 duplicates the naïve parton results. Varying the weights of different **56**'s and **70**'s with ν and q^2 then yields an infinity of models of the nondiffractive part of the structure functions in terms of an infinite sum of s-channel nucleon resonances.

Does this s-channel model make sense when compared to the behaviour of the known individual resonances which have been classified in **56**'s or **70**'s of $SU(6)$? In particular, we have the well known $P_{33}(1236)$ in a **56** (with the nucleon), the $D_{13}(1520)$ in a **70**, and the $F_{15}(1688)$ in another **56**.

Now in photoproduction (on a proton target) all these resonances have $\sigma_{\frac{3}{2}}(\text{resonance}) \geqslant \sigma_{\frac{1}{2}}(\text{resonance})$, i.e. $A_p(\text{resonance}) < 0$. Quark models[127] with harmonic forces[128] provide a neat explanation of this for the D_{13} and F_{15}. Both magnetic and convection current terms enter $\sigma_{\frac{1}{2}}$, where they cancel, while $\sigma_{\frac{3}{2}}$ comes from the convection current alone. Thus $\sigma_{\frac{3}{2}} \gg \sigma_{\frac{1}{2}}$ or $A \simeq -1$ for these two resonances. However, the same quark models also make a prediction. As $q^2 \to \infty$, the magnetic term dominates, just as we assumed in the model discussed above, and there is no longer any cancellation in $\sigma_{\frac{1}{2}}$ for these resonances. In fact $\sigma_{\frac{1}{2}} \gg \sigma_{\frac{3}{2}}$ or $A \simeq +1$ is predicted for these resonances as $q^2 \to \infty$. Detailed calculations[126] show that this change from $A(\text{resonance}) \simeq -1$ to $A(\text{resonance}) \simeq +1$ should occur rapidly as q^2 goes from zero to a few tenths of a GeV^2. These resonances would then have the same sign (positive) of A at large q^2 as is expected for deep inelastic scattering.

Experimentally there are no results from using polarized beams and targets, but the relevant information on the D_{13} and F_{15} is obtainable from observation of the angular distribution of their πN decays. Up to $q^2 \simeq 0{\cdot}6$ GeV^2, where large effects are expected, no appropriate change is observed in the $\pi^0 p$ angular distributions.[129] It appears that the quark model with harmonic forces predicts the wrong result, as the D_{13} and F_{15} still seem to have the opposite sign of A at $q^2 \simeq 0{\cdot}6$ GeV^2 from that expected in deep inelastic scattering and in the quark models of resonance excitation.

What then happens to duality for the spin dependent structure functions near $\omega = 1$? To discuss this one must first say what behaviour the deep inelastic scattering will have. On the basis of the sum rules we have discussed here and the ratio $\nu W_{2n}/\nu W_{2p}$ observed near $\omega = 1$, I expect[126] that both A_p and A_n will be $\simeq +1$ at $\omega = 1$ with A_n dropping rapidly to zero with increasing ω, but A_p staying large and going to zero only at very large values of ω. There are then two possibilities for the resonances: (1) all resonances eventually have $A_p > 0$ for large q^2 and one has a kind of "local duality" as in the case of νW_2, with each resonance mimicking the deep inelastic behaviour; or (2) only the nucleon (with $A = +1$), of the "prominent" resonances, has $A > 0$. A "global duality" situation holds, as in πN charge exchange at $t = 0$, and the average over many contributions of different sign must be taken to give the deep inelastic behaviour.[130] Experimentally, a few sweeps through the region of prominent nucleon resonances with a polarized beam and target should answer this. Perhaps next year at this time we will know the answer to whether possibility (1) or (2) is realized in Nature.

6. SUMMARY

We have now seen some of the many and varied theoretical or phenomenological questions which are susceptible to experimental answer with a polarized lepton beam and target. Among the possibilities we have examined are that:

(1) A detailed, and in principle complete, amplitude analysis of virtual photon induced processes can be carried out;

(2) The scaling behaviour of two new spin-dependent structure functions can be tested;

(3) The size of the asymmetry, A, which indicates in parton models the alignment of the charged constituent and nucleon spins can be measured. The various constituents are weighted in their contribution to the deep inelastic scattering in a different way from the spin averaged case;

(4) Rigorous sum rules exist which have right- and left-hand sides that

both can be directly measured, providing a test of the quark light cone algebra;

(5) The asymmetries of individual resonances can be measured, providing a very restrictive check of quark models of resonances and their electromagnetic excitation, as well as of the validity and form duality takes in spin-dependent scattering near $\omega = 1$;

(6) The large ν or ω behaviour of the structure functions is measurable, yielding additional information on Regge singularities which are essentially inaccessible in other processes, and in particular the position of the leading singularity of odd signature and even parity. An interesting period of the beginning of experimental answers to these and other questions lies immediately ahead.

References

1. M. Gell-Mann. *Phys. Letts.* **8**, 214 (1964).
2. G. Zweig. CERN preprints TH.401 and TH.412 (unpublished) (1964).
3. R. H. Dalitz. (*In* "Proceedings of the Second Hawaii Topical Conference in Particle Physics" (S. Pakvasa and S. F. Tuan, eds.) p. 325. University of Hawaii Press, Honolulu, Hawaii. (1968). Reviews the classification of hadron states according to the quark model.
4. F. J. Gilman. *In* "Experimental Meson Spectroscopy" (A. H. Rosenfeld and K. W. Lai, eds) AIP Conference Proceedings No. 8, p. 460. American Institute of Physics, New York (1972). Reviews the possible classification of the observed mesons in the quark model.
5. M. Gell-Mann. *Phys. Rev.* **125**, 1067 (1962).
6. S. L. Adler, *Phys. Rev. Letts.* **14**, 1051 (1965); W. I. Weisberger. *Phys. Rev. Letts.* **14**, 1047 (1965).
7. S. L. Adler and R. F. Dashen. *In* "Current Algebras". W. A. Benjamin, New York (1968). Reviews the applications of the algebra of currents.
8. H. Fritzsch and M. Gell-Mann. *In* "Scale Invariance and the Light Cone" Vol 2, p. 1. Gordon and Breach, New York (1971).
9. Y. Frishman. *In* "Proceedings of the XVI International Conference on High Energy Physics" (J. D. Jackson and A. Roberts, eds) p. 119. National Accelerator Laboratory, Batavia, Illinois (1972). Reviews the subject of light cone and short distance singularities. See references to previous work therein.
10. M. Gell-mann. *In* "Elementary Particle Physics" (P. Urban, ed.), p 733. Springer Verlag, New York (1972).
11. H. J. Melosh. Caltech thesis (unpublished) (1973).
12. F. J. Gilman, M. Kugler and S. Meshkov. *Phys. Letts.* **45B**, 481 (1973).
13. F. J. Gilman, M. Kugler and S. Meshkov. *Phys. Rev.* **D9**, 715 (1974).
14. J. Rosner (1973). Invited talk at the Berkeley Conference of the Division of Particles and Fields (August, unpublished). Reviews the recent developments in hadron spectroscopy.
15. O. W. Greenberg. *Phys. Rev. Letts.* **13**, 598 (1964).
16. W. A. Bardeen, H. Fritzsch and M. Gell-Mann. *In* "Proceedings of the Topical

Meeting on Conformal Invariance in Hadron Physics." Frascati (May 1972) (unpublished).
17. D. Faiman and A. Hendry. *Phys. Rev.* **173**, 1720 (1968).
18. D. Faiman, J. L. Rosner and J. Weyers. *Nucl. Phys.* **B57**, 45 (1973).
19. We always consider the classification of hadron states at infinite momentum, or, what is equivalent here, the classification of states at rest under the action of light-like charges.
20. R. Dashen and M. Gell-Mann. *Phys. Letts.* **17**, 142 (1965).
21. Recall the z and t components of the integrated axial-vector current are equal between states at infinite momentum.
22. H. Harari. *In* "Spectroscopic and Group Theoretical Methods in Physics" p. 363. North-Holland, Amsterdam (1968). Reviews the classification of states under chiral $SU(3) \times SU(3)$. See references therein.
23. R. F. Dashen and M. Gell-Mann. *Phys. Rev. Letts* **17**, 340 (1966) suggested a similar transformation in connection with saturating the local algebra.
24. F. Buccella *et al. Nuovo Cimento* **69A**, 133 (1970) and **9A**, 120 (1972) discuss a phenomenological scheme for transforming charges.
25. H. J. Lipkin and S. Meshkov. *Phys. Rev. Letts.* **14**, 670 (1965) and *Phys. Rev.* **143**, 1269 (1966); K. J. Barnes, P. Carruthers and F. von Hippel. *Phys. Rev. Letts.* **14**, 82 (1965).
26. N. Cabibbo and L. A. Radicati. *Phys. Letts* **19**, 697 (1966).
27. E. Eichten, J. Willemsen and F. Feinberg. *Phys. Rev.* **D8**, 1204 (1973). See also S. P. de Alwis, *Nucl. Phys.* **B55**, 427 (1973).
28. We need only consider D^α_+ with $J_z - +1$, since all matrix elements of D^α_- are related to those of D^α_+ by parity.
29. A. J. G. Hey and J. Weyers. CERN report TH. 1718 (unpublished) (1973).
30. M. Gell-Mann and M. Levy. *Nuovo Cimento* **16**, 705 (1960).
31. Intrinsic to the use of PCAC is a $\sim 10\%$ error for matrix elements. There is no contradiction between eqn (19) and the powers of momentum (p^{2l+1}) expected in an expression for a decay width from arguments based on threshold factors—the additional powers of momentum are implicitly contained in the matrix element of Q^α_5. Massive or massless final pions have momenta which typically differ by less than 10% for any given decay which we treat.
32. S. L. Adler and F. J. Gilman. *Phys. Rev.* **156**, 1598 (1967) discuss in detail the case of $\Delta \to \pi N$ and $\Delta \to \gamma N$.
33. When the context makes the meaning clear, we will often drop the L_z label for the transformation properties of operators.
34. J. C. Carter, J. J. Coyne and S. Meshkov. *Phys. Rev. Letts.* **14**, 523 (1965); C. L. Cook and G. Murtaza. *Nuovo Cimento* **39**, 531 (1965).
35. P. McNamee and F. Chilton. *Rev. Mod. Phys.* **36**, 1005 (1964).
36. W-spin Clebsch–Gordan coefficients are just those of $SU(2)$. We use the Condon–Shortley phase conventions with the W-spin state of the relevant piece of $V^{-1}Q^\alpha_5 V$ or $V^{-1}D^\alpha_+ V$ taken first.
37. A. J. G. Hey, J. L. Rosner and J. Weyers. *Nucl. Phys.* **B61**, 205 (1973) discuss the algebraic structure of pion decays and its relation to broken $SU(6)_W$ schemes. The algebraic structure of photon decays is also discussed in Ref. 29.
38. R. P. Feynman, M. Kislinger and F. Ravndal. *Phys. Rev.* **D3**, 2706 (1971).
39. The relevant broken $SU(6)_{W,\,\mathrm{strong}}$ schemes involve adding an $L = 1$ "spurion" in a **35**: see J. C. Carter and M. E. M. Head. *Phys. Rev.* **176**, 1808 (1968); D. Horn and Y. Ne'eman. *Phys. Rev.* **D1**, 2710 (1970); R. Carlitz and M. Kislinger.

Phys. Rev. **D2**, 336 (1970). Specific broken $SU(6)_W$ calculations schemes with the same algebraic structure as the theory considered here have been developed by L. Micu. *Nucl. Phys.* **B10**, 521 (1969); E. W. Colglazier and J. L. Rosner. *Nucl. Phys.* **B27**, 349 (1971); Petersen W. P. and J. L. Rosner. *Phys. Rev.* **D6**, 820 (1972).

40. P. Söding, *et al. Phys. Letts* **39B**, 1 (1972).
41. D. Cohen *et al. Phys. Rev.* **D8**, 23 (1973); M. Afzal *et al.* Nuovo Cimento, **15A**, 61 (1973); R. Ott. University of California Ph.D. Thesis and Lawrence Berkeley Laboratory preprint LBL-1547 (unpublished) (1972) and G. Lynch. Private communication. Y. Eisenberg. Private communication.
42. See, for example, R. Klanner. *In* "Experimental Meson Spectroscopy—1972" (A. H. Rosenfeld and K. W. Lai, eds.) AIP Conference Proceedings No. 8, p. 164 (1972). American Institute of Physics, New York.
43. F. J. Gilman and M. Kugler. *Phys. Rev. Letts.* **30**, 518 (1973) test particularly meson decays, using an additional assumption on the $\{(8, 1)_0 - (1, 8)_0, 0\}$ term in $V^{-1}Q_5^\alpha V$.
44. A. J. G. Hey and J. Weyers. *Phys. Letts.* **44B**, 263 (1973), use only the $SU(3) \times SU(3)$ subalgebra of $SU(6)_W$.
45. With the definitions $\langle n, \lambda = \frac{1}{2}|(2^{-\frac{1}{2}})(Q_5^1 - iQ_5^2)|p, \lambda = \frac{1}{2}\rangle = g_A/\sqrt{2}$ and $\langle \Delta^0, \lambda = \frac{1}{2}|2^{-\frac{1}{2}}(Q_5^1 - iQ_5^2)|p, \lambda = \frac{1}{2}\rangle = g^*/2$ the Adler–Weisberger sum rule reads $g_A^2 - g^{*2}$ + (higher mass contributions) = 1. The prediction then is that $g^* = \frac{4}{5}g_A \simeq 1{\cdot}0$, while the experimental value of g^* lies in the range 0·8 to 1·05 (see Ref. 22, particularly footnote 13).
46. D. Herndon *et al.* LBL preprint LBL-1065 (unpublished) (1972); R. Cashmore. Private communication. See also U. Mehtani *et al. Phys. Rev. Letts.* **29**, 1634 (1973).
47. R. Cashmore, *In* "Baryon Resonances—73", p 53, Purdue University, West Lafayette (1973).
48. M. A. B. Beg, B. W. Lee and A. Pais. *Phys. Rev. Letts.* **13**, 514 (1964).
49. R. H. Dalitz and D. G. Sutherland. *Phys. Rev.* **146**, 1180 (1966).
50. We use the results of F. J. Gilman and H. J. Schnitzer. *Phys. Rev.* **150**, 1362 (1966); S. L. Adler and F. J. Gilman (1967). Ref. 32, updated by recent photoproduction amplitude analysis to calculate μ^*.
51. F. J. Gilman and I. Karliner. Stanford Linear Accelerator Center, Report No. SLAC-PUB-1382, (unpublished) (1974).
52. A. Love and D. V. Nanopoulos. *Phys. Letts.* **45B**, 507 (1973).
53. L. A. Copley, G. Karl and E. Obryk. *Phys. Letts.* **29B**, 117 (1969) and *Nucl. Phys.* **B13**, 303 (1969); D. Faiman and A. W. Hendry. *Phys. Rev.* **180**, 1572 (1969); R. P. Feynman, M. Kislinger and F. Ravndal (1971). Ref. 38.
54. R. G. Moorhouse and H. Oberlack. *Phys. Letts.* **43B**, 44 (1973).
55. R. J. Cashmore (1973). Lectures at the Scottish Universities Summer School in Physics, this volume.
56. F. J. Gilman. *In* "Baryon Resonances—73", p 441. Purdue University, West Lafayette (1973).
57. The relative signs of resonant amplitudes in theories with generally similar algebraic structure have been considered by R. G. Moorhouse and N. H. Parsons. *Nucl. Phys.* **B62**, 109 (1973), in the quark model; D. Faiman and J. L. Rosner. *Phys. Letts.* **45B**, 357 (1973), in a broken $SU(6)_W$ scheme.
58. The forward, *s*-channel helicity partial wave amplitude for $\pi_1 N \to \pi_2 \Delta$, $T^J_{0\frac{1}{2}, 0\frac{1}{2}}$,

is related to the experimental partial wave amplitudes $T^J_{ll'}$ of Ref. 46 by

$$T^J_{0\frac{1}{2},0\frac{1}{2}} = (\text{positive number}) \times (l0\tfrac{1}{2}\tfrac{1}{2}|J\tfrac{1}{2})(l'0\tfrac{3}{2}\tfrac{1}{2}|J\tfrac{1}{2})(1I_z^{(\pi_1)}\tfrac{1}{2}I_z^{(N)}|I^{(N)}I_z^{(N)}) \times (\tfrac{3}{2}I_z^{(\Delta)}1I_z^{(\pi_2)}|I^{(N^*)}I_z^{(N^*)})T^J_{ll'}$$

where l and l' are the orbital angular momenta in the initial and final state respectively. Note the ordering in the isospin Clebsch–Gordan coefficients.

59. F. J. Gilman and I. Karliner. *Phys. Letts.* **46B**, 426 (1973).
60. I thank Professor M. Gell-Mann for several discussions suggesting this general framework for a phenomenology of strong interaction vertices.
61. We use a metric where $p \cdot p = \vec{p}^2 - p_0^2$, so that $q^2 > 0$ for space-like four-vectors q_μ. The fine structure constant $\alpha = e^2/4\pi \simeq 1/137$.
62. F. J. Gilman. *Phys. Reports* **4C**, 98 (1972), reviews the kinematics of inelastic scattering, scaling and the parton model. See the references to previous work therein.
63. R. L. Jaffe. *Phys. Rev.* **D6**, 716 (1972), treats the effect of singularities elsewhere in coordinate space.
64. R. P. Feynman. *Phys. Rev. Letts.* **23**, 1415 (1969).
65. J. D. Bjorken and E. A. Paschos. *Phys. Rev.* **185**, 1975 (1969).
66. M. Breidenbach *et al. Phys. Rev. Letts.* **23**, 935 (1969); E. D. Bloom *et al. Phys. Rev. Letts.* **23**, 930 (1969).
67. G. Miller *et al. Phys. Rev.* **D5**, 528 (1972).
68. E. M. Riordan. MIT thesis (unpublished) (1973).
69. E. D. Bloom *et al.* Stanford Linear Accelerator Center Report No. SLAC-PUB-796 presented to the XV International Conference on High Energy Physics, Kiev, USSR, 1970 (unpublished) (1973).
70. A. Bodek *et al. Phys. Rev. Letts.* **30**, 1087 (1973).
71. O. Nachtman. *Nucl. Phys.* **B38**, 397 (1972).
72. E. D. Bloom. Invited talk at the International Conference on New Results from Experiments on High Energy Particle Collisions, Vanderbilt University, Nashville, March 26–28, 1973 and Caltech preprint CALT-68-392 (unpublished) (1973).
73. K. Gottfried. *Phys. Rev. Letts.* **18**, 1174 (1967).
74. P. M. Fishbane and D. Z. Freedman. *Phys. Rev.* **D5**, 2582 (1972).
75. C. H. Llewellyn-Smith. *Physics Report* **3C**, 261 (1972), reviews the general subject of neutrino reactions. See references to previous work therein.
76. S. L. Adler. *Phys. Rev.* **143**, 1144 (1966).
77. D. J. Gross and C. H. Llewellyn-Smith. *Nucl. Phys.* **B14**, 337 (1969).
78. D. H. Perkins. *In* "Proceedings of the XVI International Conference on High Energy Physics" (J. D. Jackson and A. Roberts, eds.), Vol 4, p. 189. National Accelerator Laboratory, Batavia (1972). Reviews the data and related phenomenology on the basis of the Gargamelle experiment. See references therein.
79. The same result is also obtainable by assuming that the isoscalar electromagnetic current is responsible for 10% of $F_2^{ep}(x) + F_2^{en}(x)$ and there are equal vector and axial-vector contributions to $F_2^{\nu N}(x)$ or $F_2^{\bar{\nu} N}(x)$. See D. H. Perkins (1972), Ref. 78.
80. A more careful treatment makes (small) corrections for the non-zero value of the Cabibbo angle and for the excess of target neutrons over protons in nuclei. See D. H. Perkins (1972), Ref. 78.
81. Correction for the excess of target neutrons over protons in nuclei yields $B = 0{\cdot}86 \pm 0{\cdot}04$. See D. H. Perkins (1972), Ref. 78.

82. B. C. Barish *et al.* *Phys. Rev. Letts.* **31**, 565 (1973).
83. A. Benvenuti *et al.* *Phys. Rev. Letts.* **30**, 1084 (1973).
84. D. Cline. Invited talk at the Stanford Linear Accelerator Center Summer Institute in Particle Physics, July 1973 (unpublished) (1973), and references therein.
85. J. D. Bjorken, D. Cline and A. K. Mann. Stanford Linear Accelerator Center Report No. SLAC-PUB-1244 (unpublished) (1973).
86. M. Han and Y. Nambu. *Phys. Rev.* **139**, 1006 (1965).
87. R. F. Schwitters. *In* "Proceedings of the Summer Institute on Particle Physics", Stanford Linear Accelerator Center Report No. 167, Vol. I, p. 313 (1973) and references therein.
88. H. B. Newman. *In* "Proceedings of the Summer Institute on Particle Physics", Stanford Linear Accelerator Center Report No. 167, Vol. I, p. 255 (1973) and references therein.
89. M. S. Chanowitz and S. D. Drell. *Phys. Rev. Letts* **30**, 807 (1973).
90. E. D. Bloom. Invited talk at the VI International Conference on Electron and Photon Interactions at High Energy, Bonn (unpublished) (1973).
91. D. J. Gross. *In* "Proceedings of the Summer Institute on Particle Physics", Stanford Linear Accelerator Center Report No. 167, Vol. II, p. 251 (1973) and references therein.
92. V. W. Hughes *et al.* (1971). Yale-SLAC proposal E-80.
93. C. Y. Prescott *et al.* (1972). SLAC Group A proposal E-95.
94. We may subtract multiples of q_μ since the virtual photon is coupled to a conserved electromagnetic current of the hadrons. A similar manipulation of the lepton trace and discussion of the photon polarization is presented by J. Kuti. MIT Center for Theoretical Physics publication No. 234 (unpublished) (1971).
95. See, for example, the talk of K. Berkelman. *In* "Proceedings of the XVI International Conference on High Energy Physics" (J. D. Jackson and A. Roberts, eds) Vol 4, p. 41 (1973). National Accelerator Laboratory, Batavia, Illinois and references therein.
96. I thank Dr. R. Heimann for calling this to my attention. This term does not vanish if the helicities of all other particles are not summed over (e.g. with a polarized target).
97. T. P. Cheng and A. Zee. *Phys. Rev.* **D6**, 885 (1972).
98. K. Schilling and G. Wolf. *Nucl. Phys.* **B61**, 381 (1973).
99. The case of both a polarized beam and target has been considered by H. Fraas. University of Wurzburg preprint (unpublished) (1973).
100. We need only consider μ, $\nu = 1, 2, 3$ since we have previously put $L_{\mu\nu}$ in a form where $L_{\mu 4} = L_{4\nu} = 0$.
101. While $\sigma_{\frac{1}{2}}$, $\sigma_{\frac{3}{2}}$, σ_L, and σ_T are all sums of squares of amplitudes, and hence non-negative, σ_{TL} involves an interference between transverse and longitudinal amplitudes and has no positivity properties.
102. In this particular case, having assumed C, P, and T invariance, these inequalities follow simply from the positivity of $\sigma_{\frac{1}{2}}$ and $\sigma_{\frac{3}{2}}$ and the Schwartz inequality. The general case, which requires more powerful techniques, is discussed by M. G. Doncel and E. deRafael. *Nuovo Cimento* **4A**, 363 (1971).
103. We use the notation $\mathrm{d}^2\sigma\uparrow\downarrow/\mathrm{d}\Omega'\,\mathrm{d}E'(\mathrm{d}^2\sigma\uparrow\uparrow/\mathrm{d}\Omega'\,\mathrm{d}E')$ to denote the double differential cross-section for scattering a helicity $+\frac{1}{2}$ lepton on a nucleon target with spin direction antiparallel (parallel) to the beam. In all our equations we neglect the lepton mass.

104. Since σ_{TL} involves an interference between the transverse and longitudinal amplitudes, σ_{TL}/σ_T may not be small even though σ_L/σ_T is.
105. We denote by $d^2\sigma\uparrow\leftarrow/d\Omega'\,dE'$ ($d^2\sigma\uparrow\rightarrow/d\Omega'\,dE'$) the cross-section for helicity $+\frac{1}{2}$ leptons scattering on a nucleon target with spin direction perpendicular to the beam and toward (away from) the direction of the final lepton.
106. This was emphasized by C. E. Carlson and W. K. Tung. *Phys. Rev.* **D5**, 721 (1972), and also in Ref. 107.
107. A. J. G. Hey and J. E. Mandula. *Phys. Rev.* **D5**, 2610 (1972).
108. We use the structure functions defined by J. Kuti and V. F. Weisskopf. *Phys. Rev.* **D4**, 3418 (1971).
109. J. D. Bjorken. *Phys. Rev.* **179**, 1547 (1969).
110. L. Galfi *et al. Phys. Letts.* **31B**, 465 (1970); C. E. Carlson and W. K. Tung. (1972). Ref. 106; A. J. G. Hey and J. E. Mandula (1972). Ref. 107; C. Nash *Nucl. Phys.* **B31**, 419 (1971); M. Gourdin *Nucl. Phys.* **B38**, 418 (1972).
111. The general scaling behaviour of structure functions is treated by T. P. Cheng and A. Zee. *Phys. Rev.* **D6**, 3182 (1972).
112. More detailed results of parton models can be found in J. Kuti and V. F. Weisskopf. (1971). Ref. 108; M. Gourdin. (1972). Ref. 110; T. F. Walsh and P. Zerwas. DESY preprint DESY 72/36 (unpublished) (1972).
113. J. T. Dakin *et al. Phys. Rev. Letts.* **29**, 746 (1972); J. T. Dakin *et al. Phys. Rev. Letts.* **31**, 786 (1973).
114. J. D. Bjorken. *Phys. Rev.* **148**, 1467 (1966).
115. This sum rule has also been discussed in the scaling regime by C. E. Carlson and W. K. Tung. (1972), Ref. 106; A. J. G. Hey and J. E. Mandula. (1972), Ref. 107; C. Nash. (1971), Ref. 110; D. A. Dicus, R. Jackiw and V. L. Teplitz. *Phys. Rev.* **D4**, 1733 (1971); D. Wray. *Nuovo Cimento* **9A**, 463 (1972); B. Ward. Princeton preprint (unpublished) (1973), have also analyzed the sum rules for neutrinos.
116. J. Ellis and R. Jaffe (Stanford Linear Accelerator Center Report No. SLAC.—PUB-1288 (unpublished) (1973). have noted that the assumption that in a parton model of the nucleon the presence of λ and λ quarks is to be associated with the Pomeron, together with the absence of a Pomeron contribution to the spin dependent structure functions, allow one to relate $Z^{(0)}$ and $Z^{(8)}$. With an F/D value of 2/3 this implies $Z^{(3)} = Z^{(8)} + Z^{(0)}$, so that $Z_p = 2Z^{(3)}$ and $Z_n = 0$.
117. H. Burkhardt and W. N. Cottingham. *Ann. Phys.* (*N.Y.*) **56**, 453 (1971).
118. The sensitivity of this sum rule to model dependent assumptions in the light cone approach has been emphasized by T. P. Cheng and A. Zee. (1972), Ref. 111.
119. A detailed Regge analysis of the spin dependent structure functions is presented in Refs 94 and 117. The leading contributions correspond to an odd signature, positive parity (pole?) in one amplitude and an even signature, positive parity cut in the other.
120. Large asymmetries were previously argued on the basis of the sum rule in J. D. Bjorken. *Phys. Rev.* **D1**, 1376 (1970).
121. For recent discussions of the saturation of the Adler sum rule see E. A. Paschos. *In* "Proceedings of the XVI International Conference on High Energy Physics" (J. D. Jackson and A. Roberts, eds.) Vol 2, p. 166. National Accelerator Laboratory, Batavia (1973); S. F. Tuan. *Phys. Rev.* **D7**, 2092 (1973) and references therein.
122. R. Crewther. *Phys. Rev. Letts* **28**, 1421 (1969).
123. S. Adler. *Phys. Rev.* **177**, 2426 (1969).
124. This is spelled out explicitly in J. E. Mandula. *Phys. Rev.* **D8**, 328 (1973).

125. See F. J. Gilman (1972), Ref. 62 for a review of duality in electro-production and further references.

126. F. E. Close, F. J. Gilman and I. Karliner. *Phys. Rev.* **D6** 2533 (1972).

127. Photoproduction or electroproduction of the $P_{33}(1236)$ is predicted to be pure magnetic dipole in character by such quark models, which is equivalent to $\sigma_{\frac{3}{2}} = 3\sigma_{\frac{1}{2}}$ for this resonance. This agrees with the experimental data (see Ref. 129).

128. L. A. Copley, *et al.* (1969); Ref. 53; D. Faiman and A. W. Hendry (1969), Ref. 53; R. P. Feynman, *et al.* (1971), Ref. 38.

129. F. E. Close and F. J. Gilman. *Phys. Letts.* **38B**, 541 (1972), and references to experiments therein. Very recent data at $q^2 = 0{\cdot}45\ \text{GeV}^2$ on $\pi^+ n$ electro-production at forward angles, as presented by R. Meaburn, *et al.* Daresbury Nuclear Physics Laboratory Report DNPL/P163 (unpublished) (1973), do show an indication of a possible change in helicity structure of the $D_{13}(1520)$ and $F_{15}(1688)$. Further data and analysis is needed to clarify this situation.

130. See also the *s*-channel resonance structure and predictions for the asymmetry of M. Chaichian and S. Kitakado. *Nucl. Phys.* **B59**, 285 (1973), and the previous work of G. Domokos *et al. Phys. Rev.* **D3**, 1191 (1971); J. Cleymans. *Phys. Rev.* **D6**, 814 (1972).

9. AMPLITUDE ANALYSES IN THREE BODY FINAL STATES†

R. J. CASHMORE

*Stanford Linear Accelerator Center, Stanford University, Stanford, California**

† Work supported by the U.S. Atomic Energy Commission.
* Now at *Nuclear Physics Laboratory, Keble Road, Oxford, England.*

I. Introduction

In this Chapter I am going to attempt to review the status and methods of analysis of three body final states. In the last few years there has been a great deal of activity in the phenomenological analysis of these systems motivated by a variety of reasons. The three-body states considered have been obtained

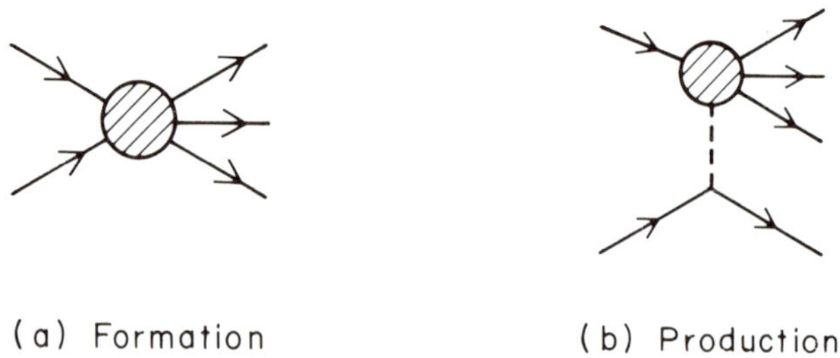

FIG. 1. The three body final states: (a) formation reactions; (b) production reactions.

in both formation and production experiments as indicated in Fig. 1 and the main emphasis has been in the "spectroscopic" aspects of the results.

In the formation experiments work has centred primarily around the reactions (Herndon *et al.*, 1972; Mast *et al.*, 1972)

$$\pi N \rightarrow \pi\pi N \tag{1}$$

$$K^- p \rightarrow \Lambda\pi\pi \tag{2}$$

Such reactions constitute a large part of the inelastic cross sections derived from the incident two body systems. Indeed for πN collisions with $E_{\text{c.m.}} < 2000$ MeV, the $\pi\pi N$ channel accounts almost completely for the inelastic cross section which itself is approximately 50% of the total. Thus if we are ever to understand the πN interaction it will be essential to have a description of these three body states. Furthermore, this low energy region is dominated by overlapping resonances ("the resonance region") many of which are highly inelastic. In our present attempts to understand the systematics of these resonances, the existence and branching fractions of each state are required. Of particular interest are the decays into $\pi\Delta$, $N\rho$, $\pi Y_1^*(1385)$, etc., which can be related to πN, $\bar{K}N$, $\Lambda\pi$, etc., in any symmetry scheme higher than $SU3$. The fact that the parent resonances are overlapping necessitates partial wave analyses which would be comparatively easy were the Δ, ρ, etc., stable particles (although of course $0^- + \frac{1}{2}^+ \rightarrow 0^- + \frac{3}{2}^+$ is necessarily more complicated than $0^- + \frac{1}{2}^+ \rightarrow 0^- + \frac{1}{2}^+$). Unfortunately they are not and their widths are sufficiently large that for much of the Dalitz plot their

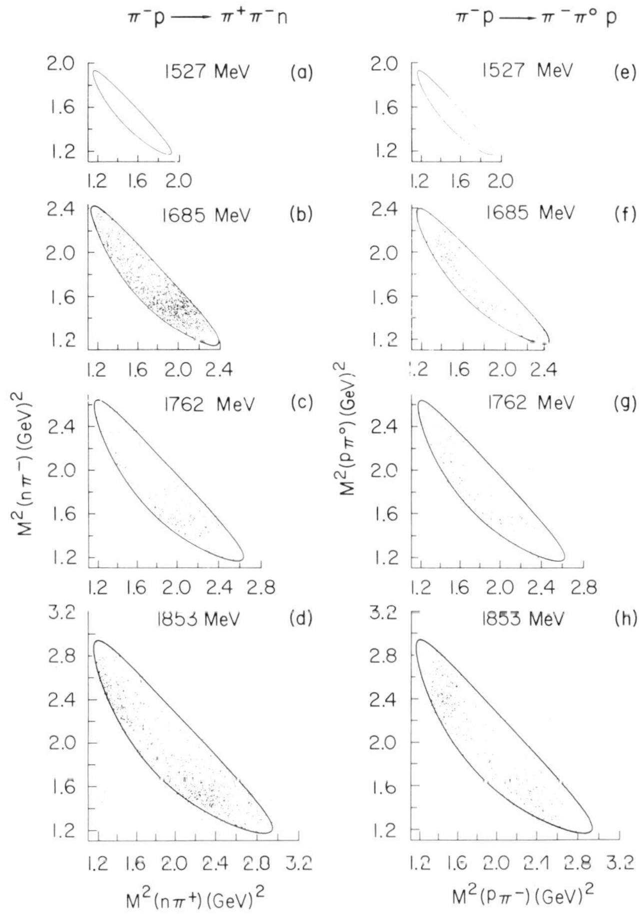

FIG. 2. Dalitz plot for $\pi^- p \to \pi^- \pi^+ n$.

amplitudes (Breit–Wigner) are large, as is clear from Fig. 2. In these circumstances it is difficult to select the data to produce a pure sample of e.g. $\pi N \to \pi\Delta$ and any detailed analysis of such selected data will be correspondingly suspect. Thus one must resort to other techniques which can take into account these overlapping resonance bands and perhaps even exploit their presence. Such methods and one in particular, the isobar model, together with their results will be the subject of these lectures. At this time the most important results have emerged from the analysis of reaction (1) but one should expect in the future valuable results from $\bar{K}N$ induced reactions, the Y^* situation being at least as complicated as the N^*.

The activity in production reactions is essential for a variety of reasons (Ascoli *et al.*, 1971). For meson systems parity conservation forbids the decay into two pseudoscalar mesons of any resonance belonging to the unnatural spin parity series 0^-, 1^+, 2^-. The first available state is then usually the three pseudoscalar meson system and thus we expect to observe such resonances there. The requirement of G-parity conservation means that for $G = -1$, $S = 0$ meson resonances the decay into two pions is forbidden. We then once again expect to see these states first in three pion systems, e.g. $A_2 \rightarrow 3\pi$. The whole discussion of the existence of many of these resonances is further

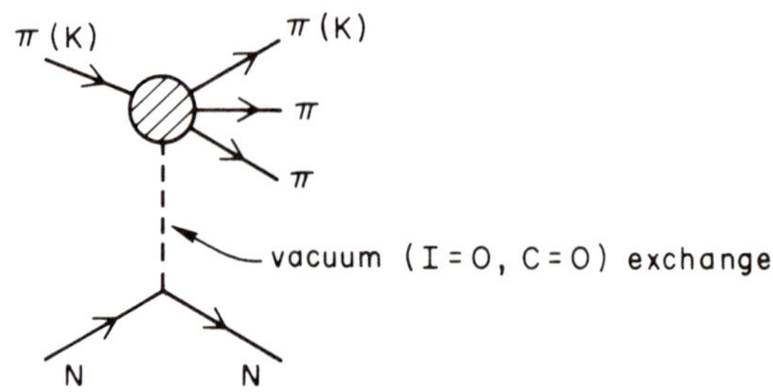

FIG. 3. Diffraction excitation.

complicated by dynamical processes which may occur. The diffractive excitation of a π or K (see Fig. 3) will automatically lead to a system of particles with unnatural spin parity if the following rule is obeyed

$$\Delta P = (-1)^{\Delta J}, \tag{3}$$

where ΔP is the change in parity and ΔJ the change in angular momentum. Hence we have long had the debate as to whether the low mass enhancement in the three pion system is due to a resonance or a dynamical effect (Deck mechanism). The resolution of this problem is vitally important to meson spectroscopy and awaits study of these three particle states in reactions in which diffraction dissociation is not possible, e.g. the charge exchange or hypercharge exchange reactions of Fig. 4

$$K^- n \rightarrow \pi^- \pi^+ \pi \Lambda, \tag{4}$$

$$\pi^+ n \rightarrow \pi^+ \pi^- \pi^0 p. \tag{5}$$

Of course we would also like a clear understanding of the process of diffraction excitation and as already noted this leads naturally to the consideration of three particle final states (these are in general the easiest to observe experimentally). Thus we will continue to study the reactions of

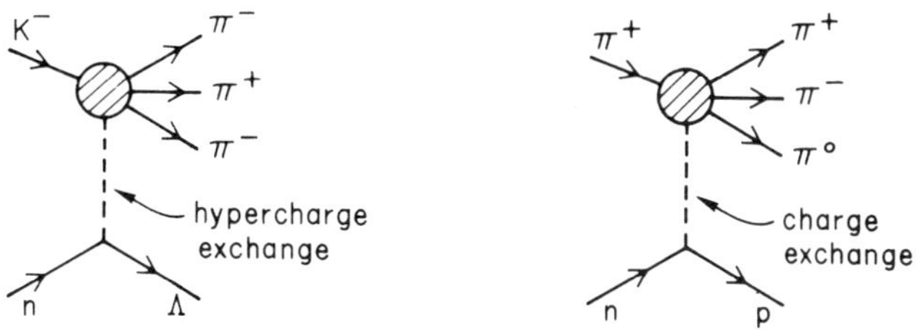

FIG. 4. Hypercharge and charge exchange reactions.

Fig. 3, but we will in addition direct our attention to the diffractive excitation of the nucleon as indicated in Fig. 5. Indeed the latter is possibly the best system in which to study this process as one already has detailed information on the baryon resonances spectrum. In general the analysis of these states in production reactions will always give information on the production dynamics.

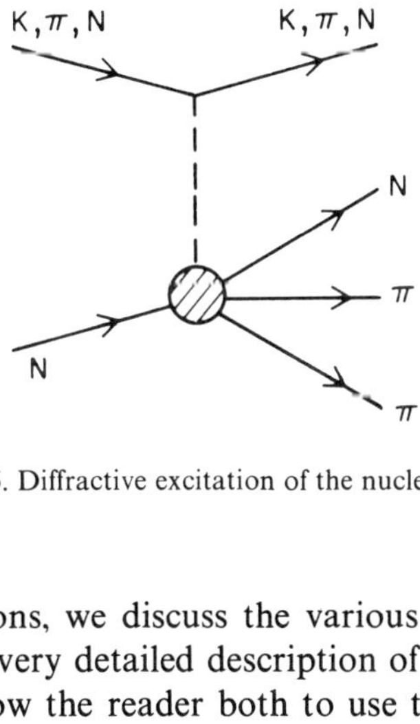

FIG. 5. Diffractive excitation of the nucleon.

In the following sections, we discuss the various techniques of analysis (Section II) leading to a very detailed description of the isobar model (Section III), in order to allow the reader both to use the method himself but also to be familiar with the approximations, definitions, etc., involved in any of the quoted results. Sections III.1 and III.2 will be concerned with the calculation of matrix elements while Sections III.3 and III.4 will contain the applications to formation and production reactions respectively. Sections IV and V describe and try to evaluate the results which have so far been obtained and finally in Section VI we summarize the situation and indicate in which directions we will proceed.

II. The Methods of Analysing Three-Body Final States

In the introduction I have already hinted at some of the methods one might apply and I would now like to discuss them in more detail. I will first deal with the approach in which one attempts to reduce the problem to one of the stable two body scattering by suitable cuts on the data. This will be followed by a discussion of the general methods of analysing three body final states. However the ambiguities, the variety of experimental data required, and the difficulties of interpretation render these approaches of little value at present making it necessary to use somewhat more model dependent methods. The most successful of these models is introduced and dealt with in detail in Section III.

1. Selection of Two Body Reactions

The oldest approach to analysing three-body final states is to attempt to isolate specific two-body reactions, e.g.

$$\pi^- p \to \pi^+ \Delta^- \tag{6}$$

$$\pi^+ p \to \pi^0 \Delta^{++}, \tag{7}$$

or

$$Q \to \pi K^*. \tag{8}$$

This is done by selecting events in which the invariant mass of a pair of particles lies in the required resonance band. As already indicated this can be a dangerous business. In general the resonances are wide and this means that at almost any point in the Dalitz plot their amplitudes are still quite large. Thus even though we believe we are selecting a clean sample of reaction (6) from the final state $\pi^+ \pi^- n$ the presence of the reaction (see Fig. 2)

$$\pi^- p \to \pi^- \Delta^+ \tag{9}$$

can produce appreciable interference effects within the Δ^- band. Since the helicity angle in the decay of the Δ^- is linearly related to the position along the Δ^- band, the decay distributions (and hence density matrix elements) are particularly susceptible to these interference effects. Of course the higher the centre of mass energy the smaller the proportion of resonance overlap and thus this technique becomes more reliable at higher energies.

The reason for attempting this isolation is that the calculations are comparatively simple (even if somewhat tedious). The complete expressions for the production differential cross section and decay density matrix in terms of partial wave amplitudes in the reaction

$$0^- + \tfrac{1}{2}^+ \to 0^- + \tfrac{3}{2}^+ \tag{10}$$

have been published (Brody and Kernan, 1969) and this has been the reaction most exhaustively studied in this manner. The limited data has necessitated energy dependent analyses, i.e. the partial wave amplitudes are parametrized in terms of Breit–Wigner resonances and energy dependent backgrounds and data at a variety of energies fitted simultaneously. This ensures continuity and hopefully reduces ambiguities which might exist by specifying the partial wave energy dependence.

The results from analyses of this type are probably satisfactory for the large dominant waves whereas the smaller partial waves are poorly determined. Such analyses do however provide a useful guide to and check on the results of more sophisticated methods.

The final objection to this type of analysis is that one does not obtain the relative phases of partial wave amplitudes for different channels, e.g.

$$\begin{aligned} \pi N &\rightarrow \pi\Delta \\ &\rightarrow N\rho \end{aligned} \qquad (11)$$

because the regions of the Dalitz plot which would allow this are explicitly removed from the analysis. As we shall see such quantities are of interest in higher symmetry schemes.

2. MODEL INDEPENDENT ANALYSES OF THE THREE BODY FINAL STATE

In general five variables are required to describe a three body final state. These variables are usually three Euler angles specifying the orientation of the three particle state with respect to some co-ordinate system together with two Dalitz plot variables (ω_1^2 and ω_2^2) giving effectively the energies of the three particles in their c.m. system. In the case of formation reactions using unpolarized targets one of these angles does not appear as it corresponds to arbitrary rotations about the beam direction. In order to use the data to maximum advantage it would be best to exploit all the correlations that exist. However the correlation between Dalitz plot populations and the angles requires specific dynamical assumptions, e.g. a resonance is produced in a particular angular momentum state.

In model independent analyses one selects regions of the Dalitz plot and attempts to find the J^P states associated with that area. The formalism for these types of analysis has been discussed exhaustively both for formation (Cashmore and Hey, 1972) and production reactions (Berman and Jacob, 1965). In both cases the angular orientation of the three-body plane is considered, the form of the distributions and the correlations present indicating the partial wave structure.

I will demonstrate some of the properties of this type of analysis by using the reaction

$$\pi N \rightarrow \pi\pi N \qquad (12)$$

as an example. In this case the differential cross section from an unpolarized target is written as (Cashmore and Hey, 1972)

$$\frac{d^4\sigma}{d\omega_1^2\, d\omega_2^2\, d\cos\theta\, d\phi} = \sum_{L,M} W_L^M(\omega_1^2, \omega_2^2)\, Y_L^M(\theta, \phi)\ldots, \tag{13}$$

where an integration has been performed over the angle about the incident beam direction, θ and ϕ the polar angles of the incident pion within a coordinate system defined by the final state particles. The object of any analysis

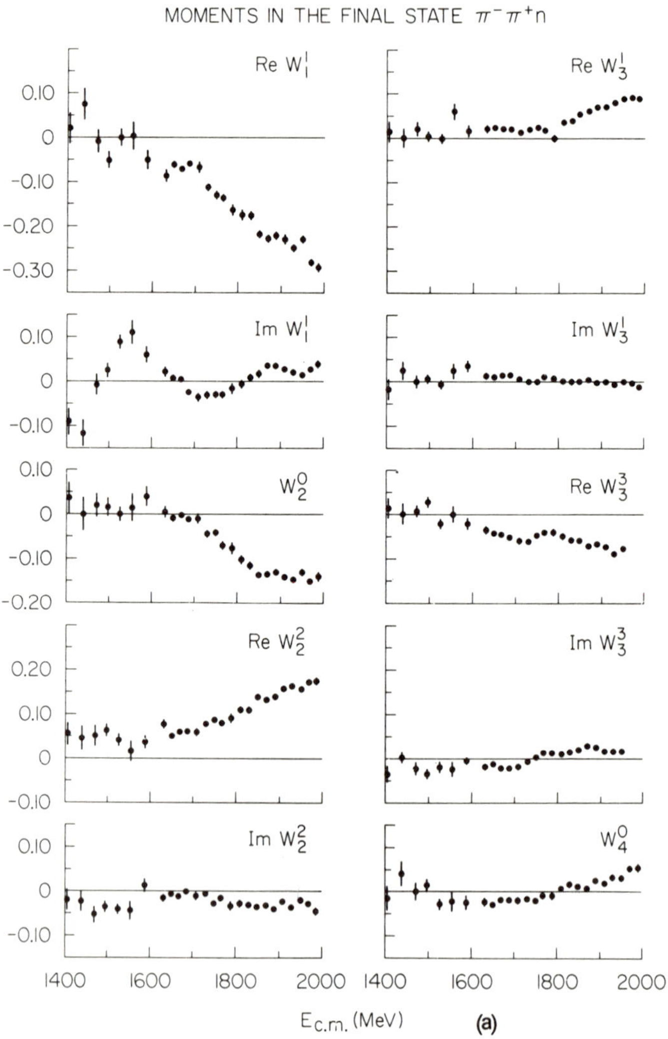

(a)

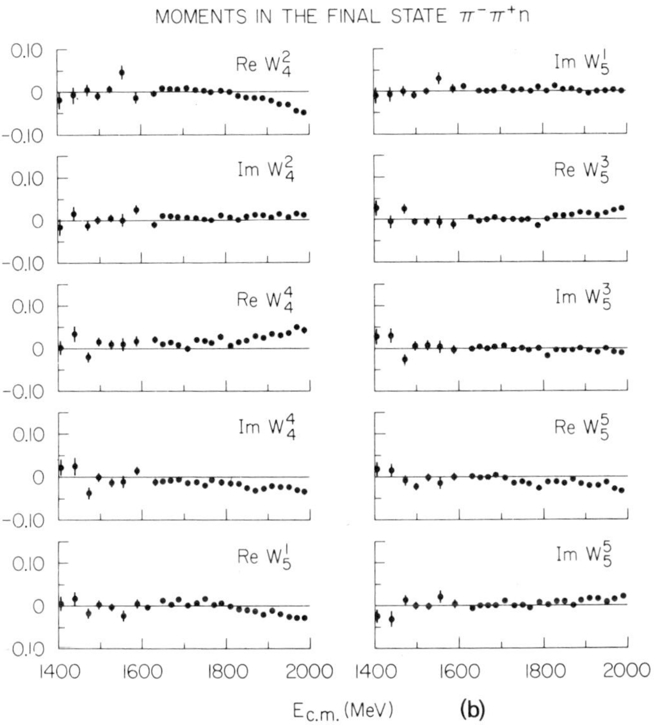

FIG. 6. The moments W_L^M as a function of energy in the final state $\pi^-\pi^+n$ normalized such that $W_0^0 = 1$. The x-axis is defined as $\vec{p}_N$ and the z-axis as $\vec{p}_{\pi^-} \wedge \vec{p}_{\pi^+}$.

of this type is then to determine the W_L^M as functions of Dalitz plot position. The W_L^M are then analogous to the expansion coefficients A and B of two body scattering differential cross sections and polarization distributions. These W_L^M possess many similar properties, e.g. (a) interference of waves with opposite (same) parity lead to terms with odd (even) values of L. (b) If J_{max} is the maximum angular momentum contributing to the reaction, and only one parity is present corresponding to this value, then $L_{max} = 2J_{max} - 1$ (for J half integral). However if waves of opposite parity are present with this value of J_{max} then $L_{max} = 2J_{max}$. (c) The W_L^M are bilinear products of the partial wave amplitudes $B_{J\Lambda}^{\mu\tau}(\omega_1^2, \omega_2^2)$ (Cashmore and Hey, 1972)

$$W_L^M \sim \sum_{\tau} \sum_{\substack{JJ' \\ \Lambda\Lambda'}} F(J, \Lambda, J', \Lambda', L)\, B_{J\Lambda}^{\mu\tau *} B_{J'\Lambda'}^{\mu\tau}\, \delta_{M\Lambda-\Lambda'} \tag{14}$$

A complete parity ambiguity exists unless the final state nucleon polarization is observed (Cashmore and Hey, 1972). This is equivalent to the Minami ambiguity of two body scattering.

In Figs 6a, b as an example, we can see the W_L^M averaged over the Dalitz plot for the final state $\pi^+\pi^- n$. The absence of moments with $L + M$ odd is required by parity conservation, while the large moments with L odd indicates the presence of waves of opposite parity. At a given energy the maximum L present does given an indication of the largest angular moment wave present. However one should not put too great a reliance on this since in the region of 1700 MeV, the $L = 4$ and 5 moments are small where we know the F15 and D15 resonances are important.

From the measured values of the $W_L^M(\omega_1^2, \omega_2^2)$ we would like to extract the partial wave amplitudes for this Dalitz plot position. It would then be possible to study the variation of these amplitudes with c.m. energy just as one does in conventional elastic scattering. Unfortunately there are many factors which reduce the value of this method.

(a) It is essential to make polarization measurements (preferably of the final baryon) to give enough W_L^M to allow the extraction of partial wave amplitudes.

(b) The partial wave amplitudes are functions of the Dalitz plot position, and ideally we would like to know them everywhere. The determination of the moments W_L^M requires $\sim$1000–2000 events and thus we can only hope to obtain these W_L^M integrated over regions of the Dalitz plot. At this time the lack of experimental data does not allow the use of even a coarse grid (characteristically there are $\sim$1000–10,000 events in a given channel at a given energy).

(c) The overall phase of the amplitudes is undetermined from one Dalitz plot position to the next.

(d) Optimistically one would hope to see variations of the partial wave amplitudes as a function of the Dalitz plot variables indicating the association of a J^P state with a particular decay decimal. However to extract couplings to these decay channels still requires a detailed model which predicts the variation of these amplitudes with Dalitz plot position.

However one must point out that the results of such an analysis are model independent and do represent a permanent record of the correlations which exist in the data. Furthermore these methods are useful when considering model dependent analyses in that they can provide an indication of waves which should be present (e.g. of opposite parity) and do sometimes limit the maximum angular momentum it is necessary to consider (although this can be misleading.

In the case of production reactions one usually considers the angular distribution in α, β and γ, the three angles of the problem (Berman and Jacob, 1965), for a given region of the Dalitz plot

$$\frac{d\sigma}{d\alpha\, d\cos\beta\, d\gamma} \sim \sum W_{JM\Lambda}(\omega_1^2, \omega_2^2)\, \mathscr{D}^J_{M\Lambda}(\alpha, \beta, \gamma)\ldots \tag{15}$$

and the $W_{JM\Lambda}$ are again products of the partial wave amplitudes. These results are published (Berman and Jacob, 1965) and we will see many of the general properties later in our discussion of model dependent analyses.

If the object of our analysis is finally to measure the couplings to decay channels then point (d) above suggests that it would be more sensible to make a specific model which will do this from the very beginning. This will immediately specify the correlations between the angular variables and the Dalitz plot position which we can exploit in our analysis. Of course the drawback is that it does become a model-dependent analysis. However the situation is somewhat analogous to the state of πN partial wave analyses of 10 years ago. Then one performed energy dependent partial wave analyses on small quantities of data. In this case we are performing mass-dependent (in the sense of Dalitz plot variables) partial wave analyses on similarly small quantities of data.

In the following section we introduce such a model dependent analysis and subsequently discuss it at length.

3. MODEL DEPENDENT ANALYSIS—THE ISOBAR MODEL

If we study Fig. 2 or Fig 7 we immediately notice the existence of strong resonance bands, the Δ and the ρ. Since we are not in a position fruitfully to follow a model independent discussion, this suggests a course in which wc insert as an integral part of our analysis the presence of these resonances. The method then consists of writing the transition amplitude for reaching a given final state as a coherent sum of two body processes as indicated in Fig. 8. The transition matrix is then written as (Deler and Valladas, 1966; Cashmore *et al.*, 1972)

$$T(W, \omega_1, \omega_2, \alpha, \beta, \gamma) = \sum_{\substack{JLSI\\ l}} A^{IJLSl}(W)\, C^I X^{JLSl}(W, \omega_1, \omega_2, \alpha, \beta, \gamma) \times B^L(\omega_1, \omega_2)\ldots, \tag{16}$$

where $\omega_1, \omega_2, \alpha, \beta, \gamma$ are the kinematical variables required to specify the reaction, C^I the product of all isospin Clebsch–Gordan coefficients, X^{JLSl} contains all factors related to angular momentum decompositions, $B^L(\omega_1, \omega_2)$ is the final state enhancement factor, e.g. a Breit–Wigner, where l is the orbital angular momentum in the decay of the isobar. The partial wave amplitudes $A^{IJLSl}(W)$ are then obtained in fits to the data.

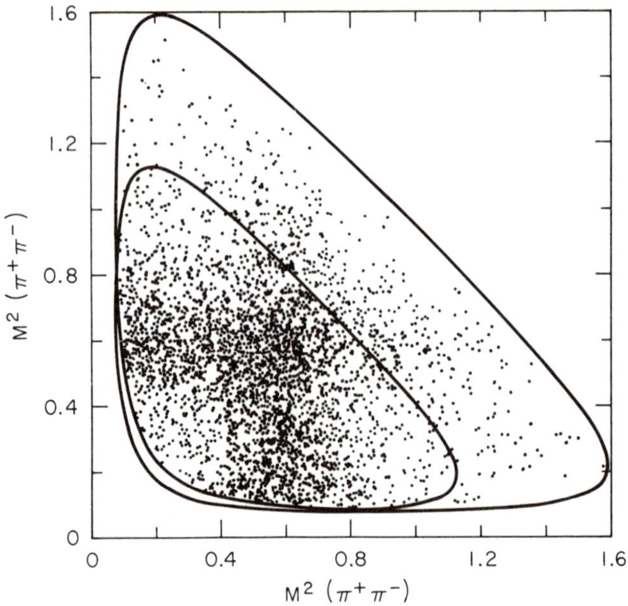

FIG. 7. $\pi^+\pi^-\pi^-$ Dalitz plot $M^2(\pi^+\pi_1^-)$ versus $M^2(\pi^+\pi_2^-)$ for events having 3π masses in the interval 1·0–1·4 GeV. The curves define the Dalitz plot boundaries for $M_{3\pi}$ = 1·20 and 1·40 GeV. The data is from 5 and 7·5 GeV/c π^-p collisions.

One considers many intermediate states in analysing the different reactions:

$$\pi N \to \pi\pi N \qquad \pi N \to \pi\Delta, N\rho, N\varepsilon$$

$$A \to 3\pi \qquad A \to \pi\varepsilon, \pi\rho, \pi f$$

$$Q, L \to K\pi\pi \qquad Q \to K\varepsilon, K\rho, \pi K^*, \pi(K\pi)_{\text{s wave}}, \pi K^*(1400)$$

and in order to unravel these, one would like to use all of the correlations present in the data.

However it is possible to obtain more limited results by analysis of the

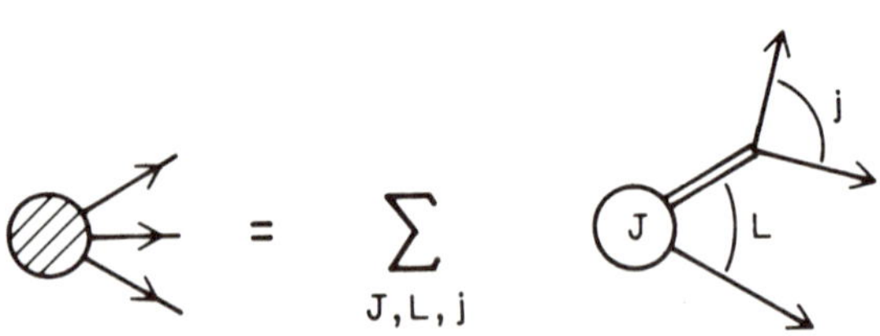

FIG. 8. The isobar model.

Dalitz plot distributions alone. In this case there are no longer terms corresponding to the interference of waves of different J or different parity P and moreover the different magnetic substates (J_z substates) in which a resonance may be formed all lead to the same Dalitz plot distribution. Thus it is now only possible to determine the contribution of each J, P state. Furthermore, ignoring the α, β and γ dependences of the cross-section will lead to a less reliable estimate of these partial wave amplitudes.

The following section is devoted to a detailed discussion of the calculations of these cross-section formulae and their properties while the results are reserved for Sections IV and V.

III. The Calculation of Transition Amplitudes to Three Particle States

In this Section we develop the formulae for the transition to a three particle state in formation reactions, e.g.

$$\pi N \to \pi\pi N \tag{17}$$

together with the similar results for the decay to three bodies of states obtained in production, e.g.

$$\begin{array}{l} \pi N \to A_2 + N \\ \qquad\quad \hookrightarrow \pi\pi\pi \end{array} \tag{18}$$

In Section III.1 we deal with that part of the calculation which is common to both results—the decay of a spin parity state $|J^P M\rangle$ which is defined in a co-ordinate system related to the final three particles. Section III.2 deals with the presence of two identical particles, e.g., two π^+'s and the isospin decompositions which occur. Section III.3 deals specifically with formation reactions while Section III.4 contains a similar discussion of production reactions. Finally Section III.5 remarks on the methods of applying this formalism in the analysis of experimental data.

Before beginning let us summarize the notation and symbols used in order to prevent their introduction in a random manner. It is not possible to do this completely but it is to be hoped that confusion will be reduced. The appendix deals with the definition of states, angular momentum projections, normalizations, phase spaces, properties of $\mathscr{D}$ functions and any detailed manipulations in an attempt to keep the text as clear as possible.

Notation

Final three particle system. Let j, k and l represent the final three particles. Let the diparticle be composed of particles k and l. All quantities pertaining to the diparticle are indexed by the subscript j. All quantities are defined in the three particle c.m. system S.

(a) Total three particle c.m. energy, total angular momentum and parity: W, J, P
(b) Z-component of J in system S: $-M$
(c) c.m. system four momenta: Q_j, Q_k, Q_l
(d) particle spins; $\sigma_j, \sigma_k, \sigma_l$
(e) c.m. system helicities; μ_j, μ_k, μ_l
(f) Intrinsic parities; η_j, η_k, η_l
(g) Mass of diparticle; ω_j
(h) Spin and c.m.s. helicity of the diparticle: j_j, λ_j
(i) Outgoing orbital angular momentum and total spin: L_j, S_j
In the diparticle rest frame we have the quantities
(j) Four momenta of decay particles: q_k, q_l,
(k) Helicities of decay particles: ν_k, ν_l
(l) Orbital angular momentum and total spin of the decay particles: l_j, s_j.

We then use LS coupling to give

$$\left.\begin{aligned} s_j &= \sigma_k + \sigma_l \\ j_j &= l_j + s_j \\ S_j &= \sigma_j + j_j \\ J &= L_j + S_j \end{aligned}\right\} \tag{19}$$

We assume that L_j and l_j are chosen to conserve parity

$$P = \eta_j \eta_k \eta_l (-1)^{L_j + l_j} \tag{20}$$

These definitions are summarized in Fig. 9.

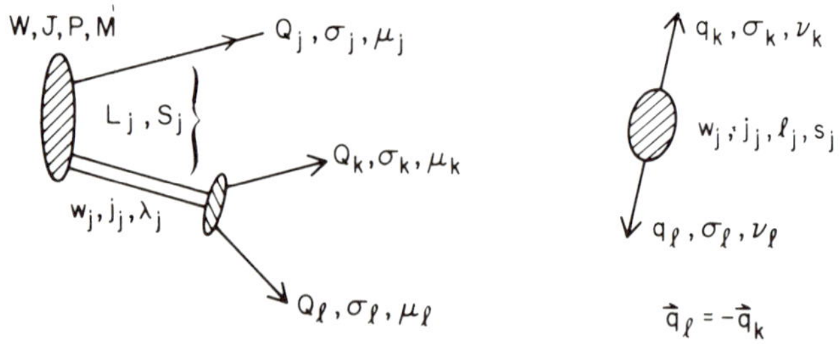

FIG. 9. Definitions in the three particle decay.

Incident two particle state in formation reactions

Let a and b represent the incident particles. In this case the c.m. system for the two particle state is the same as that for the three particle state.

(a) c.m.s. four momenta: p_a, p_b
(b) Particle spins: σ_a, σ_b
(c) c.m.s. helicities: μ_a, μ_b
(d) Intrinsic particles: η_a, η_b
(e) Incident orbital angular momentum and total spin: L, S.

Then

$$\left.\begin{aligned} S &= \sigma_a + \sigma_b \\ J &= L + S \end{aligned}\right\} \tag{21}$$

and we assume parity is conserved so that

$$P = \eta_a \eta_b (-1)^L \tag{22}$$

These are summarized in Fig. 10.

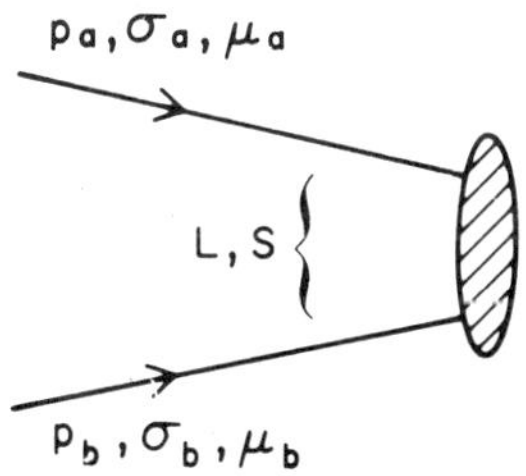

FIG. 10. Definitions in the incident two particle state in formation reactions.

Production reactions

In this case we have three other particles to describe besides j, k, and l of the three particle system we are considering. In this case the j, k, l c.m. system is not the same as the overall (ab) c.m. system. In general all quantities pertaining to a, b and c will be measured in the overall c.m. system.

(a) c.m.s. four momenta: p_a, p_b, p_c
(b) Particle spins: σ_a, σ_b, σ_c
(c) Helicities: μ_a, μ_b, μ_c
(d) Intrinsic parities: η_a, η_b, η_c.

These are summarized in Fig. 11.

For simplification in many of the following formulae n will be used to represent a set of quantities

Decay of three particle system: $n \equiv \{j, J, P, M; L_j, S_j; j_j, l_j, s_j\}$
Formation reactions: $n \equiv \{j, J, P, M; L, S; L_j, S_j; j_j, l_j, s_j\}$
Production reactions: $n \equiv \{j, J, P, M; L_j, S_j; j_j, l_j, s_j\}$

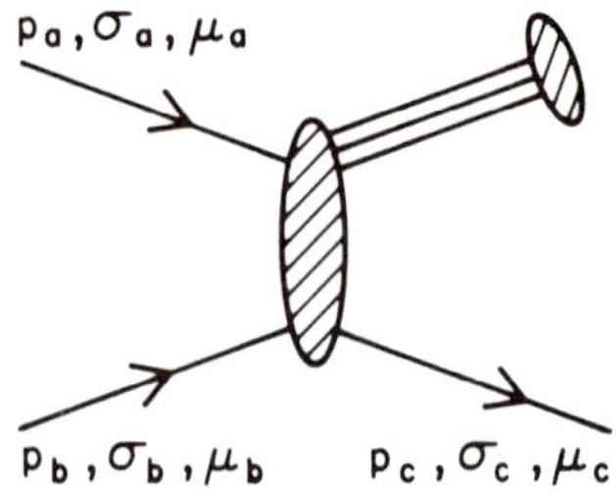

FIG. 11. Definitions in production reactions.

In some cases it may be necessary to display one particular quantity of this set, in which circumstances I will continue to use n to represent the remaining quantities.

1. THE DECAY OF AN INTERMEDIATE STATE INTO THREE FINAL PARTICLES

We first define a co-ordinate system S with respect to j, k and l as shown in Fig. 12. We now wish to consider the decay of a state $|nJM\rangle$ defined in this

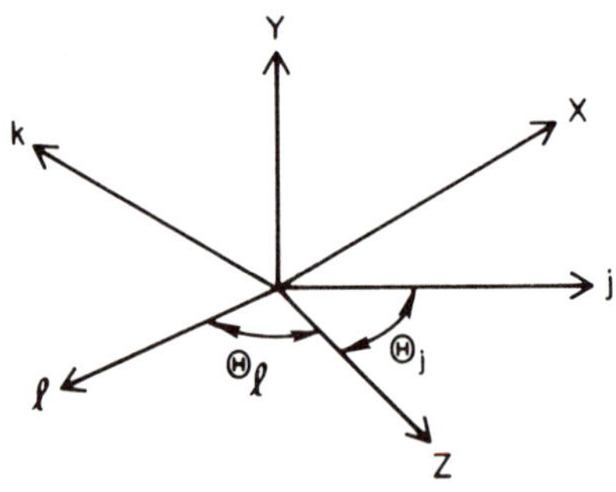

FIG. 12. The co-ordinate system S.

system to a state which is the product of three usual helicity states

$$|Q_j\mu_jQ_k\mu_kQ_l\mu_l\rangle = |Q_j\mu_j\rangle\,|Q_k\mu_k\rangle\,|Q_l\mu_l\rangle \tag{23}$$

i.e., we wish to calculate the transition

$$f_\mu^{nJM} = \langle\vec{Q}_j\mu_j\vec{Q}_k\mu_k\vec{Q}_l\mu_l|\,T\,|nJM\rangle, \tag{24}$$

where we envisage the reaction as first proceeding through a quasi-two-body state followed by the decay of one of these particles

$$f_\mu^{nJM} = \sum_{\mu_m,\,\mu_n}\int\frac{d^3Q_m}{2E_m}\frac{d^3Q_n}{2E_n}\langle Q_j\mu_jQ_k\mu_kQ_l\mu_l|\,T_2\,|Q_m\mu_mQ_n\mu_n\rangle \times \langle Q_m\mu_mQ_n\mu_n|\,T_1\,|nJM\rangle. \tag{25}$$

The isobar model then consists of writing

$$\langle Q_j\mu_j Q_k\mu_k Q_l\mu_l | T_2 | Q_m\mu_m Q_n\mu_n\rangle = \delta^3(Q_j - Q_m)\delta_{\mu_j\mu_m}\delta^3(Q_n - Q_k - Q_l) \times \langle Q_k\mu_k Q_1\mu_l | T_2 | Q_n\mu_n\rangle \tag{26}$$

so that

$$f_\mu^{nJM} = \sum_{\lambda_j} \langle Q_k\mu_k Q_l\mu_l | T_2 | -Q_j\lambda_j\rangle\langle Q_j\mu_j - Q_j\lambda_j | T_1 | JMn\rangle. \tag{27}$$

This is summarized in Fig. 13.

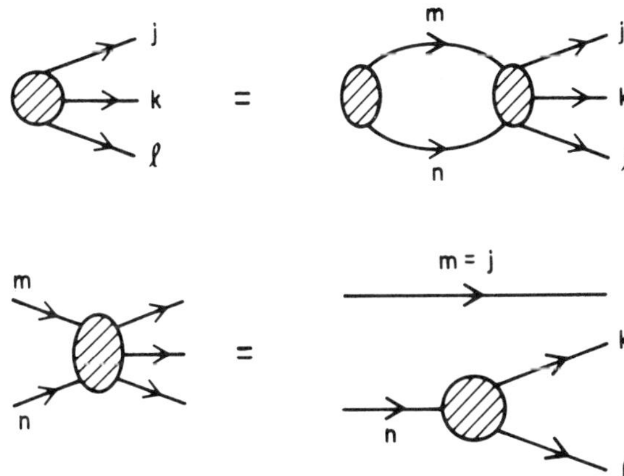

FIG. 13. The isobar model.

(1) *Primary decay* T_1

We write this using the decomposition into angular momentum states (see appendix)

$$\langle Q_j\mu_j - Q_j\lambda_j | T_1 | nJM\rangle = \sqrt{\left(\frac{2J+1}{4\pi}\right)}\left(\frac{4W}{Q_j}\right)^{\frac{1}{2}} \mathscr{D}^{J*}_{M,\mu_j-\lambda_j}(\Omega_j) \times \langle Q_j JM\mu_j | T_1 | nJM\rangle, \tag{28}$$

where the arguments of the $\mathscr{D}$ function are angles $\Omega_j = (\Phi_j, \Theta_j, -\Phi_j)$, the production angles of particle j in frame S. We can then perform a further partial wave decomposition converting from helicity states to LS states (see appendix)

$$= \left(\frac{4\pi}{Q_j}\right)^{\frac{1}{2}}\left(\frac{2L_j+1}{4\pi}\right)^{\frac{1}{2}} C(\sigma_j, j_j, S_j | \mu_j - \lambda_j)\, C(L_j, S_j, J | 0, \mu_j - \lambda_j) \times \mathscr{D}^{J*}_{M,\mu_j-\lambda_j}(j)\langle Q_j JML_j S_j | T_1 | nJM\rangle \tag{29}$$

The last partial wave amplitude is independent of M, due to rotational invariance and we will write it as

$$\langle Q_j JML_j S_j | T_1 | nJM \rangle = T^{nJ}_{1L_jS_j}(W, \omega_j) \tag{30}$$

(2) *Decay of isobar* T_2

We wish to evaluate the decay of the diparticle and this is most easily done in the isobar rest frame

$$\langle Q_k \mu_k Q_l \mu_l | T_2 | -Q_j \lambda_j \rangle = \sum_{\nu_k \nu_l} \mathscr{D}^{\sigma^*_k}_{\nu_k \mu_k}(\theta^k_j \hat{n}_k) \mathscr{D}^{\sigma^*_l}_{\nu_l \mu_l}(\theta^l_j \hat{n}_l) \langle q_k \nu_k q_l \nu_l | T_2 | j_j - \lambda_j \rangle \tag{31}$$

The $\mathscr{D}$ functions correspond to the rotations introduced by Lorentz transformations not along the directions of k and l (see appendix). The isobar when at rest is in a magnetic substate $|j_j \mp \lambda_j\rangle$ with respect to the direction Q_j since it was initially the χ state in the two particle helicity state (see appendix). We finally require the matrix element to a product of three helicity states as defined in eqn (23). Thus when we perform the angular momentum decomposition of the helicity matrix element in eqn (30) we must convert the second particle l from the conventional χ state to an ψ state (see appendix). Inserting the angular momentum decomposition gives

$$\langle q_k \nu_k q_l \nu_l | T_2 | j_j - \lambda_j \rangle = \left(\frac{4\omega_j}{q_k}\right)^{\frac{1}{2}} \left(\frac{2L_j + 1}{4}\right)^{\frac{1}{2}} C(\sigma_k, \sigma_l, s_j | \nu_k, -\nu_1)$$

$$C(l_j s_j j_j | 0, \nu_k - \nu_l) \mathscr{D}^{j_j^*}_{\lambda_j, \nu_k - \nu_l}(\Omega^d_k)(-1)^{\sigma_l - \nu_l} \langle q_k j_j - \lambda_j l_j s_j | T_2 | j_j - \lambda_j \rangle \tag{32}$$

where the arguments of the $\mathscr{D}$ function are $\Omega^d_k = (+\phi_j, \theta_j - \phi_j)$, the decay angles of isobar j in its rest frame defined using particle k. Due to rotational invariance we can then write

$$\langle q_k j_j - \lambda_j l_j s_j | T_2 | j_j - \lambda_j \rangle = \mathbf{D}^{j_j}_{l_j s_j}(\omega_j) \tag{33}$$

Thus the final amplitude is then written as

$$f^{nJM}_{\mu_j \mu_k \mu_l} = G^{nJM}_{\mu_j \mu_k \mu_l} T^{nJ}_{1L_jS_j}(W, \omega_j)\, \mathbf{D}^{j_j}_{l_j s_j}(\omega_j) \tag{34}$$

where

$$G^{nJM}_{\mu_j \mu_k \mu_l} = \frac{1}{4\pi}\left[\frac{4W}{Q_j}\frac{4\omega_j}{q_k}\right]^{\frac{1}{2}} \Bigg[\sum_{\lambda_j} C(\sigma_j j_j S_j | \mu_j - \lambda_j) C(L_j S_j J | 0 \mu_j - \lambda_j)$$

$$\mathscr{D}^{J^*}_{M, \mu_j - \lambda_j}(\Phi_j, \Theta_j, -\Phi_j) \Bigg\{ \sum_{\nu_k \nu_l} C(\sigma_k \sigma_l s_j | \nu_k - \nu_l) C(l_j s_j j_j | 0 \nu_k - \nu_l)$$

$$\mathscr{D}^{j_j^*}_{-\lambda_j, \nu_k - \nu_l}(\phi_j, \theta_j, -\phi_j)\, \mathscr{D}^{\sigma^*_k}_{\nu_k \mu_k}(\theta^k_j) \mathscr{D}^{\sigma^\dagger_l}_{\nu_l \mu_l}(\theta^l_j)(-1)^{\sigma_l - \nu_l} \Bigg\} \Bigg] \tag{35}$$

Note we do not include sums over L_j, S_j, l_j, s_j as these are specified in n, i.e. we are considering the transition by a very specific intermediate state.

The angles of the $\mathscr{D}$ functions are summarized in Figs 12 and 14.

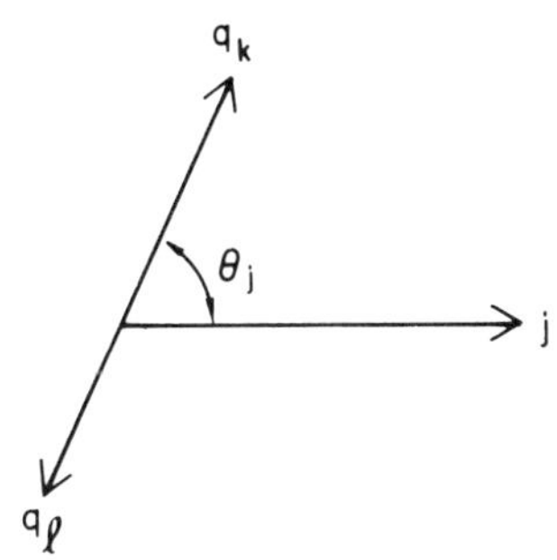

FIG. 14. The decay angles of the isobar.

(3) *The reduced matrix elements*

(a) *Initial decay matrix element.* $T_1^n(W, \omega_j)$. It is usual to extract from this the barrier penetration factors

$$\frac{Q_j^{L_j+\frac{1}{2}}}{\sqrt{(4W)}}$$

which have the correct threshold dependence.

The charge dependence may also be removed by including the isospin Clebsch–Gordan coefficients. Thus

$$T_1^n(W, \omega_j) = \frac{Q_j^{L_j+\frac{1}{2}}}{\sqrt{4W}} C(I^D, I^j, I \mid I_z^D I_z^j)\, \tau_1^n(W, \omega_j) \tag{36}$$

I^j, I_z^j isospin and z component of isospin for j
I^D, I_z^D isospin and z component of isospin for the isobar
I, $I_z = I_z^j + I_z^D$ total isospin and z-component of isospin.

It is often common to introduce further Q_j dependence through the Blatt and Weiskopf barriers $B_{L_j}(Q_j)$. These have the desirable property of damping the $Q_j^{L_j}$ factors at large Q_j. Other parametrizations may be used which possess this threshold behaviour, but in all cases we expect the $\tau_1^n(W, \omega_j)$ to be essentially independent of ω_j. All analyses assume this to be the case.

(b) *Decay matrix element.* $\mathbf{D}_{l_j s_j}^{j_j}(\omega_j)$. We can again remove the charge dependence from the decay term

$$\mathbf{D}_{l_j s_j}^{j_j}(\omega_j) = C(I^k, I^l, I^D \mid I_z^k I_z^l)\, D_{l_j s_j}^{j_j}(\omega_j) \tag{37}$$

where I^k, I^k_z isospin and z-component of isospin for k; I^l, I^l_z isospin and z-component of isospin for l.

To evaluate $D^{j_j}_{l_j s_j}(\omega_j)$ one uses either the Watson final state interaction theorem or a modified Breit–Wigner.

Watson:

$$D^{j_j}_{l_j s_j}(\omega_j) = \frac{e^{i\delta}\sin\delta}{q_k^{l_j+1}}\left(\frac{q_k}{4\omega_j}\right)^{\frac{1}{2}} \tag{38}$$

δ, elastic scattering phase shift at the energy ω_j

$\left(\frac{q_k}{4\omega_j}\right)^{\frac{1}{2}}$, is a factor added to ensure correct threshold behaviour.

Relativistic Breit–Wigner:

$$D^{j_j}_{l_j s_j}(\omega_j) = \frac{1}{\sqrt{\pi}}\frac{\sqrt{[\omega_0\Gamma_j(\omega_j)]}}{(\omega_0^2 - \omega_j^2) - i\omega_0\Gamma_j(\omega_j)} \tag{39}$$

$$\Gamma_j(\omega_j) = \Gamma_j(\omega_0)\left[\frac{q_k(\omega_j)}{q_k(\omega_0)}\right]^{2l_j}\frac{q_k(\omega_j)}{q_k(\omega_0)}\frac{\omega_0}{\omega_j} \tag{40}$$

and ω_0 is the resonance energy.

Non-relativistic Breit–Wigner:

$$D^{j_j}_{l_j s_j}(\omega_j) = (2\pi\omega_0)^{-\frac{1}{2}}\frac{\sqrt{[\Gamma_j(\omega_j)/2]}}{(\omega_0 - \omega_j) - i\Gamma_j(\omega_j)/2} \tag{41}$$

and $\Gamma_j(\omega_j)$ is defined as above.

Both of these forms are defined so that in the limit of zero width

$$\lim_{\Gamma_j \to 0}\int |D^{j_j}_{l_j s_j}(\omega_j)|^2\, d\omega_j^2 = \delta(\omega_0^2 - \omega_j^2). \tag{42}$$

In general the dependences of T_1 and D on the subenergy ω_j constitute the major assumptions and approximations of the isobar model.

(4) *Other isobars*

In the cases in which we are interested, each two-body subsystem may contain a number of isobars, e.g. in $\pi_1^+\pi_2^+\pi^-$ we can have $\rho(I = 1, J = 1)$ or $\varepsilon(I = 0, J = 0)$ states in either $(\pi_1^+\pi^-)$ or $(\pi_2^+\pi^-)$. The total amplitude for the decay of the state JM is then given by

$$f^{JM}_{\mu_j\mu_k\mu_l} = \sum_{n,j} G^{JMn}_{\mu_j\mu_k\mu_l}(j)T^{Jn}_{1L_jS_j}(W, \omega_j)\, D^{j_j}_{l_j s_j}(\omega_j)$$

i.e. we add all of the amplitudes together coherently. This does result in some slight double counting but this has been shown to be small and in practice is far less significant than the assumptions within the decay matrix elements.

In making this addition if identical particles are present care has to be made in making this addition in order to observe the correct symmetry. This is discussed in the next section.

2. THE PRESENCE OF TWO OR MORE ISOBARS, ISOSPIN DECOMPOSITION AND BOSE SYMMETRY

We are mainly interested in situations with two or more pions present in the final state and hence we require the final amplitude to be Bose-symmetric. Furthermore we will concentrate on the reactions

$$\pi N \to \pi\pi N \tag{43}$$

$$A \to \pi\pi\pi \tag{44}$$

and use these to demonstrate a method of calculation which always ensures the correct result. In the following discussion, we explicitly demonstrate the Clebsch–Gordan coefficients of eqns. (36) and (37) and derive the necessary factors of $\sqrt{2}$, etc., which must be inserted when we have identical particles present.

The isospin decomposition also allows the determination of the same partial wave amplitudes in different charge states where the interference phenomena are different. In practice this is a valuable point.

(1) $\pi N \to \pi\pi N$

Consider

$$\pi^+ p \to \pi^+\pi^0 p \tag{45}$$

$$\to \pi^+\pi^+ n \tag{46}$$

and decays through intermediate $\Delta\pi$ states.

$$|\tfrac{3}{2}\tfrac{3}{2}\rangle \to \frac{1}{\sqrt{2}}\{A_1|\Delta_1\pi_1\rangle + A_2|\Delta_2\pi_2\rangle\} \tag{47}$$

Here I use the pion not resonating to label the Δ state and the decay amplitude. The expression is clearly symmetric in π_1 and π_2. A_1 and A_2 will contain the kinematical information associated with the decay.

Now

$$|\tfrac{3}{2}\tfrac{3}{2}\rangle = \sqrt{\tfrac{3}{5}}\Delta^{++}\pi^0 - \sqrt{\tfrac{2}{5}}\Delta^+\pi^+ \tag{48}$$

We next consider the decay of the Δ and introduce a transition matrix element D which contains all the kinematics.

$$\left.\begin{aligned}\Delta^{++} &\to \pi^+ p\\ \Delta^{+} &\to \sqrt{\tfrac{1}{3}}\pi^+ n + \sqrt{\tfrac{2}{3}}\pi^0 p\end{aligned}\right\} \tag{49}$$

and if we write $G_i = A_i D_i$ we have

$$|\tfrac{3}{2}\tfrac{3}{2}\rangle \to \frac{1}{\sqrt{2}}[A_1\{\sqrt{\tfrac{3}{5}}\Delta_1^{++}\pi_1^0 - \sqrt{\tfrac{2}{5}}\Delta_1^+\pi_1^+\} + A_2\{\sqrt{\tfrac{3}{5}}\Delta_2^{++}\pi_2^0 - \sqrt{\tfrac{2}{5}}\Delta_2^+\pi_2^+\}]$$

$$= \frac{1}{\sqrt{2}}[A_1D_1\{\sqrt{\tfrac{3}{5}}\pi_2^+ p\pi_1^0 - \sqrt{\tfrac{2}{5}}(\sqrt{\tfrac{1}{3}}\pi_2^+ n\pi_1^+ + \sqrt{\tfrac{2}{3}}\pi_2^0 p\pi_1^+)\}$$

$$+ A_2D_2\{\sqrt{\tfrac{3}{5}}\pi_1^+ p\pi_2^0 - \sqrt{\tfrac{2}{3}}(\sqrt{\tfrac{1}{3}}\pi_1^+ n\pi_2^+ + \sqrt{\tfrac{2}{3}}\pi_1^0 p\pi_2^+)\}]$$

$$= \frac{1}{\sqrt{30}}\{[3G_1 - 2G_2]\pi_1^0\pi_2^+ p + [3G_2 - 2G_1]\pi_1^+\pi_2^0 p - \sqrt{2}[G_1 + G_2] \times \pi_1^+\pi_2^+ n\} \tag{50}$$

Thus the amplitude for obtaining a $\pi^0\pi^+p$ final state is then

$$\frac{\sqrt{2}}{\sqrt{30}}[3G_1 - 2G_2], \tag{51}$$

since there are two distinguishable ways of obtaining this and these must be added incoherently ($1 \equiv \pi^0, 2 \equiv \pi^+$). The amplitude for obtaining $\pi^+\pi^+n$ is

$$-\frac{\sqrt{2}}{\sqrt{30}}[G_1 + G_2]. \tag{52}$$

We now have the correct factors of $\sqrt{2}$ which account for the presence of the two identical π^+'s in the $\pi^+\pi^+n$ final state.

Similar calculations can be performed for all of the $\pi\pi N$ states obtained from $\pi^\pm p$ incident particles and these results are contained in Table I. This table includes the isospin decomposition of the incident state.

$$|\pi^+ p\rangle = |\tfrac{3}{2}\tfrac{3}{2}\rangle \tag{53a}$$

$$|\pi^- p\rangle = \sqrt{\tfrac{1}{3}}|\tfrac{3}{2} - \tfrac{1}{2}\rangle - \sqrt{\tfrac{2}{3}}|\tfrac{1}{2} - \tfrac{1}{2}\rangle \tag{53b}$$

(2) $I = 0, I = 1$ *meson decay into* 3π's

The other three particle final state so far analysed by the methods I describe is the 3π final state. Consider an A_1 or A_2 ($I = 1$) state decaying via the intermediate states $\pi\rho$ and $\pi\varepsilon$. Then we can write as above

$$|A\rangle \to \frac{1}{\sqrt{3}}A_\rho[\rho_1\pi_1 + \rho_2\pi_2 + \rho_3\pi_3] + \frac{1}{\sqrt{3}}A_\varepsilon[\varepsilon_1\pi_1 + \varepsilon_2\pi_2 + \varepsilon_3\pi_3] \tag{54}$$

Then for

$$\begin{aligned}|1\,1\rangle &= \sqrt{\tfrac{1}{2}}\rho^{+}\pi^{0} - \sqrt{\tfrac{1}{2}}\rho^{0}\pi^{+}\\ |11\rangle &= \varepsilon^{0}\pi^{+}\end{aligned} \tag{55}$$

and subsequent ρ, ε decay we obtain

$$|A\rangle \to \frac{1}{\sqrt{3}} A_{\rho}\{\sqrt{\tfrac{1}{2}}\pi_1^0[\sqrt{\tfrac{1}{2}}\pi_2^+\pi_3^0 - \sqrt{\tfrac{1}{2}}\pi_2^0\pi_3^+]B_{\rho}(2,3)$$

$$-\sqrt{\tfrac{1}{2}}\pi_1^+[\sqrt{\tfrac{1}{2}}\pi_2^+\pi_3^- - \sqrt{\tfrac{1}{2}}\pi_2^-\pi_3^+]B_{\rho}(2,3) + \text{similar terms for } \pi_2, \pi_3\}$$

$$+\sqrt{\tfrac{1}{3}}A_{\varepsilon}\left\{\frac{1}{\sqrt{3}}\pi_1^+[\pi_2^+\pi_3^- - \pi_2^0\pi_3^0 + \pi_2^-\pi_3^+]B_{\varepsilon}(2,3) + \text{similar terms for } \pi_2, \pi_3\right\} \tag{56}$$

Gathering terms together we eventually have

$$\begin{aligned}\tfrac{1}{6}[&\pi_1^+\pi_2^+\pi_3^-\{\sqrt{3}A_{\rho}(-B_{\rho}(2,3) + B_{\rho}(3,1)) + 2A_{\varepsilon}(B_{\varepsilon}(2,3) + B_{\varepsilon}(3,1))\}\\ &+ \pi_1^-\pi_2^+\pi_3^+\{\sqrt{3}A_{\rho}(-B_{\rho}(3,1) + B_{\rho}(1,2)) + 2A_{\varepsilon}(B_{\varepsilon}(3,1) + B_{\varepsilon}(1,2))\}\\ &+ \pi_1^+\pi_2^-\pi_3^+\{\sqrt{3}A_{\rho}(-B_{\rho}(1,2) + B_{\rho}(2,3)) + 2A_{\varepsilon}(B_{\varepsilon}(1,2) + B_{\varepsilon}(2,3))\}\\ &- \pi_1^+\pi_2^0\pi_3^0\{\sqrt{3}A_{\rho}(-B_{\rho}(1,2) + B_{\rho}(3,1)) + 2A_{\varepsilon}B_{\varepsilon}(2,3)\}\\ &- \pi_1^0\pi_2^+\pi_3^0\{\sqrt{3}A_{\rho}(-B_{\rho}(2,3) + B_{\rho}(1,2)) + 2A_{\varepsilon}B_{\varepsilon}(3,1)\}\\ &- \pi_1^0\pi_2^0\pi_3^+\{\sqrt{3}A_{\rho}(-B_{\rho}(3,1) + B_{\rho}(2,3)) + 2A_{\varepsilon}B_{\varepsilon}(1,2)\}]\end{aligned} \tag{57}$$

where the arguments of B represent the pions of the intermediate isobar, the first labelling the decay. This expression is symmetrical in any pair of pion labels because

$$\left.\begin{aligned}B_{\rho}(1,2) &= -B_{\rho}(2,1)\\ B_{\varepsilon}(1,2) &= B_{\varepsilon}(2,1)\end{aligned}\right\} \tag{58}$$

We see there are three ways of obtaining $\pi^+\pi^+\pi^-$ and thus the amplitude for obtaining this final state is just $\sqrt{3}$ times an individual expression. Similar calculations can be performed for the other charge states and also the $I = 0$ state. These results are summarized in Table II.

(3) *Bose symmetry*

If one now follows the same prescription for calculating the angular momentum factors, etc. in the transition matrix element (in the same overall coordinate system) then Bose symmetry will be automatically satisfied, i.e. we use the formalism described in Section III.1.

(4) *Summary*

In Tables I, II, and III which summarize the results we associate the product of the Clebsch–Gordan coefficients with the pair of particles resonating in the intermediate state. This is slightly different from the text where we used the particle not resonating as the label. However the change is obvious. Furthermore the subscripts 1, 2, etc., label specific charge states.

TABLE I. The Isospin Decomposition in $\pi N \to \pi\pi N$ Reactions

$I = \frac{3}{2}$

	$(\pi_1 N)_{I=\frac{3}{2}}$	$(\pi_2 N)_{I=\frac{3}{2}}$	$(\pi_1 N)_{I=\frac{1}{2}}$	$(\pi_2 N)_{I=\frac{1}{2}}$	$(\pi_1\pi_2)_{I=0}$	$(\pi_1\pi_2)_{I=1}$	$(\pi_1\pi_2)_{I=2}$
$\pi_1^+\pi_2^0 p$	$\frac{3}{\sqrt{15}}$	$\frac{-2}{\sqrt{15}}$	0	$-\sqrt{\frac{1}{3}}$	0	1	$-\sqrt{\frac{1}{5}}$
$\pi_1^+\pi_1^+ n$	$\frac{-1}{\sqrt{15}}$	$\frac{-1}{\sqrt{15}}$	$\sqrt{\frac{1}{3}}$	$\sqrt{\frac{1}{3}}$	0	0	$\sqrt{\frac{4}{5}}$
$\pi_1^-\pi_2^+ n$	$\frac{-3\sqrt{2}}{\sqrt{135}}$	$\frac{2\sqrt{2}}{\sqrt{135}}$	0	$\frac{\sqrt{2}}{\sqrt{27}}$	0	$-\sqrt{\frac{2}{9}}$	$\sqrt{\frac{2}{45}}$
$\pi_1^-\pi_2^0 p$	$\frac{-1}{\sqrt{135}}$	$\frac{4}{\sqrt{135}}$	$\frac{-2}{\sqrt{27}}$	$\frac{-1}{\sqrt{27}}$	0	$-\sqrt{\frac{1}{9}}$	$-\sqrt{\frac{3}{15}}$
$\pi_1^0\pi_2^0 n$	$\frac{-1}{\sqrt{135}}$	$\frac{-1}{\sqrt{135}}$	$\frac{1}{\sqrt{27}}$	$\frac{1}{\sqrt{27}}$	0	0	$\frac{2}{\sqrt{45}}$

$I = \frac{1}{2}$

	$(\pi_1 N)_{I=\frac{3}{2}}$	$(\pi_2 N)_{I=\frac{3}{2}}$	$(\pi_1 N)_{I=\frac{1}{2}}$	$(\pi_2 N)_{I=\frac{1}{2}}$	$(\pi_1\pi_2)_{I=0}$	$(\pi_1\pi_2)_{I=1}$	$(\pi_1\pi_2)_{I=2}$
$\pi_1^-\pi_2^+ n$	$\frac{-3}{\sqrt{27}}$	$\frac{-1}{\sqrt{27}}$	0	$\frac{-2\sqrt{2}}{\sqrt{27}}$	$-\frac{2}{3}$	$\frac{\sqrt{2}}{3}$	0
$\pi_1^-\pi_2^0 p$	$\frac{\sqrt{2}}{\sqrt{27}}$	$\frac{-\sqrt{2}}{\sqrt{27}}$	$\frac{-2}{\sqrt{27}}$	$\frac{+2}{\sqrt{27}}$	0	$-\frac{2}{3}$	0
$\pi_1^0\pi_2^0 n$	$\frac{\sqrt{2}}{\sqrt{27}}$	$\frac{\sqrt{2}}{\sqrt{27}}$	$\frac{+1}{\sqrt{27}}$	$\frac{+1}{\sqrt{27}}$	$\frac{\sqrt{2}}{3}$	0	0

TABLE II. Isospin Decomposition of $I = 0$ and $I = 1$ Decays to Three Pions

$I = 1$

	$(\pi_1\pi_2)_\rho$	$(\pi_2\pi_3)_\rho$	$(\pi_3\pi_1)_\rho$	$(\pi_1\pi_2)_\varepsilon$	$(\pi_2\pi_3)_\varepsilon$	$(\pi_3\pi_1)_\varepsilon$
$\pi_1^+\pi_2^+\pi_3^-$	0	$-\frac{1}{2}$	$+\frac{1}{2}$	0	$\frac{1}{\sqrt{3}}$	$\frac{1}{\sqrt{3}}$
$\pi_1^0\pi_2^0\pi_3^+$	0	$-\frac{1}{2}$	$+\frac{1}{2}$	$-\frac{1}{\sqrt{3}}$	0	0
$\pi_1^+\pi_2^-\pi_3^0$	0	$+\frac{1}{\sqrt{2}}$	$-\frac{1}{\sqrt{2}}$	$\sqrt{\frac{2}{3}}$	0	0
$\pi_1^0\pi_2^0\pi_3^0$	0	0	0	$-\frac{1}{3}$	$-\frac{1}{3}$	$-\frac{1}{3}$

$I = 0$

	$(\pi_1\pi_2)_\rho$	$(\pi_2\pi_3)_\rho$	$(\pi_3\pi_1)_\rho$	$(\pi_1\pi_2)_\varepsilon$	$(\pi_1\pi_3)_\varepsilon$	$(\pi_3\pi_1)_\varepsilon$
$\pi_1^+\pi_2^-\pi_3^0$	$-\frac{1}{\sqrt{3}}$	$-\frac{1}{\sqrt{3}}$	$-\frac{1}{\sqrt{3}}$	0	0	0

TABLE III. Isospin Decomposition of $I = \frac{1}{2}$ Decays to $K\pi\pi$

	$(\pi_1 K)_{I=\frac{1}{2}}$	$(\pi_2 K)_{I=\frac{1}{2}}$	$(\pi_1\pi_2)_{I=1}$	$(\pi_1\pi_2)_{I=0}$
$\pi_1^+\pi_2^0 K^0$	$+\frac{\sqrt{2}}{3}$	$-\frac{\sqrt{2}}{3}$	$+\sqrt{\frac{2}{3}}$	0
$\pi_1^+\pi_2^- K^+$	0	$+\frac{2}{3}$	$-\sqrt{\frac{1}{3}}$	$\sqrt{\frac{2}{3}}$
$\pi_1^0\pi_2^0 K^+$	$-\frac{1}{3\sqrt{2}}$	$-\frac{1}{3\sqrt{2}}$	0	$-\sqrt{\frac{1}{3}}$
$\pi_1^+\pi_2^- K^0$	$\frac{2}{3}$	0	$\sqrt{\frac{1}{3}}$	$\sqrt{\frac{2}{3}}$
$\pi_1^-\pi_2^0 K^+$	$+\frac{\sqrt{2}}{3}$	$-\frac{\sqrt{2}}{3}$	$\sqrt{\frac{2}{3}}$	0
$\pi_1^0\pi_2^0 K^0$	$-\frac{1}{3\sqrt{2}}$	$-\frac{1}{3\sqrt{2}}$	0	$-\sqrt{\frac{1}{3}}$

3. FORMATION REACTIONS

In this section we will develop the formalism to deal with reactions

$$a + b \rightarrow j + k + l \tag{59}$$

We begin with a calculation of the transition amplitudes, then derive formulae for all the observable quantities in reactions of type (17) and discuss the unpolarized cross-section in some detail. Finally we review the quantities of physical interest we wish to obtain using such an analysis.

(1) *Transition amplitude*

Suppose the initial beam has polar angles Θ, Φ in co-ordinate system S. We then consider the transition matrix element and make an LS partial wave decomposition

$$\begin{aligned} {}_S\langle nJM|T_p|\vec{p}_a\mu_a\vec{p}_b\mu_b\rangle &= \left(\frac{2J+1}{4\pi}\right)^{\frac{1}{2}}\left(\frac{4W}{p}\right)^{\frac{1}{2}} \mathscr{D}^J_{M\mu_a-\mu_b}(\Phi,\Theta,-\Phi)\cdot {}_S\langle nJM|T_p|pJM\mu_a\mu_b\rangle \\ &= \left(\frac{4W}{p}\right)^{\frac{1}{2}}\left(\frac{2L+1}{4\pi}\right)^{\frac{1}{2}} C(\sigma_a,\sigma_b,S|\mu_a,-\mu_b)C(LSJ|0\mu_a-\mu_b) \\ &\qquad \mathscr{D}^J_{M,\mu_a-\mu_b}(\Phi,\Theta,-\Phi)\ {}_S\langle nJM|T_p|pJMLS\rangle \end{aligned} \tag{60}$$

where rotational invariance implies that we can write

$$\langle nJM|T_p|pJMLS\rangle = T_p^{JLS}(W) \tag{61}$$

It is often convenient to again remove the charge dependence from this and also display explicitly some of its kinematical properties

$$T_p^{JLS}(W) = \frac{p^{L+\frac{1}{2}}}{\sqrt{4W}}\, C(I^aI^bI|I^a_zI^b_z)\tau_p^{JLS}(W), \tag{62}$$

These isospin Clebsch–Gordan coefficients have been explicitly included in Table I. Also note that eqn. (60) does not contain a sum over L, S as n implies a specific choice of these.

The final amplitude for the process (59) is then given by a coherent sum over all intermediate states

$$\begin{aligned} A_{\mu_a\mu_b\mu_j\mu_k\mu_l} &= \sum_n A^n_{\mu_a\mu_b\mu_j\mu_k\mu_l} \\ &= \sum_{n,J,M}\left(\frac{4W}{p}\right)^{\frac{1}{2}}\left(\frac{2L+1}{4\pi}\right)^{\frac{1}{2}} C(\sigma_a\sigma_bS|\mu_a\mu_b)C(LSJ|0\mu_a-\mu_b) \\ &\qquad G^{nJM}_{\mu_j\mu_k\mu_l}\mathscr{D}^J_{M,\mu_a-\mu_b}(\Phi,\Theta,-\Phi)T_p^{JLS}(W)T^{nJ}_{1L_jS_j}(W,\omega_j)\mathbf{D}^j_{lS_j}(\omega_j) \end{aligned} \tag{63}$$

where $G^{nJM}_{\mu_j\mu_k\mu_l}$ is given by (35). It is then convenient to define

$$T^n(W, \omega_j) = T_p^{JLS}(W)\, T^{nJ}_{1l_js_j}(W, \omega_j) \tag{64}$$

It is important to note that Σ_M may be removed since

$$\sum_M \mathscr{D}^{J*}_{M,\mu_j-\lambda_j}(\Phi_j, \Theta_j, -\Phi_j)\mathscr{D}^J_{M,\mu_a-\mu_b}(\Phi, \Theta, -\Phi) = \sum_M \mathscr{D}^J_{\mu_j-\lambda_j,M}(+\Phi_j, -\Theta_j -\Phi_j)$$

$$\mathscr{D}^J_{M\mu_a-\mu_b}(\Phi, \Theta, -\Phi) = \mathscr{D}^J_{\mu_j-\lambda_j,\mu_a-\mu_b}(\alpha, \beta, \gamma) \tag{65}$$

where

$$R(\alpha, \beta, \gamma) = R(\Phi_j, -\Theta_j, -\Phi_j)\, R(\Phi, \Theta, -\Phi) \tag{66}$$

(2) *Transition amplitude when using a polarized target or beam*

In this case we have an initial co-ordinate system defined when there is transverse polarization. If the polar angles of the beam are again Θ, Φ in system S, then the initial co-ordinate system will be related by an extra rotation α about the $0Z$ axis of S'(S' is the co-ordinate system obtained by rotation of S through the Euler angles Φ, Θ, $-\Phi$), i.e. we have

$$\left.\begin{array}{lll} 0XYZ \rightarrow 0X'Y'Z' & \text{Euler angles } \Phi, \Theta, -\Phi \\ 0X'Y'Z' \rightarrow 0xyz & \text{Euler angles } \alpha, 0, 0 \\ 0XYZ \rightarrow 0xyz & \text{Euler angles } \Phi, \Theta, \alpha-\Phi \end{array}\right\} \tag{67}$$

$0xyz$ is the frame in which the target polarization is defined. Then

$$A^{n'}_{\mu} = A^n_\mu\, \mathrm{e}^{-i(\mu_a-\mu_b)\alpha} \tag{68}$$

(3) *Transition amplitude when a final state particle decays*

If one of the final particles is a baryon which can undergo weak decay, e.g. $\Lambda \rightarrow p\pi^-$ then the decay angular distribution will give information on the parent baryon polarization. This dacay amplitude can then be directly introduced into the transition amplitude.

Suppose particle j of (59) undergoes weak decay. We express the decay matrix in terms of canonical spin states with respect to the helicity frame axes of particle j (see Fig. 15), i.e. the matrix element for

$$j \rightarrow 1 + 2 \tag{69}$$

is written as

$$\begin{aligned} B^D_{m_1m_2\mu_j} &= \langle \sigma_1 m_1 \sigma_2 m_2 | T_D | \sigma_j \mu_j \rangle \\ &= \sum_{L_dS_d} B^{L_dS_d} C(\sigma_1\sigma_2S_d|m_1m_2)C(L_dS_d\sigma_j|\mu_j - mm) Y^{\mu_j-m}_{L_d}(\theta_d, \phi_d) \end{aligned} \tag{70}$$

where

$$B^{L_dS_d} = \langle\sigma_1\sigma_2L_dS_d|T_D|\sigma_j\rangle \tag{71}$$

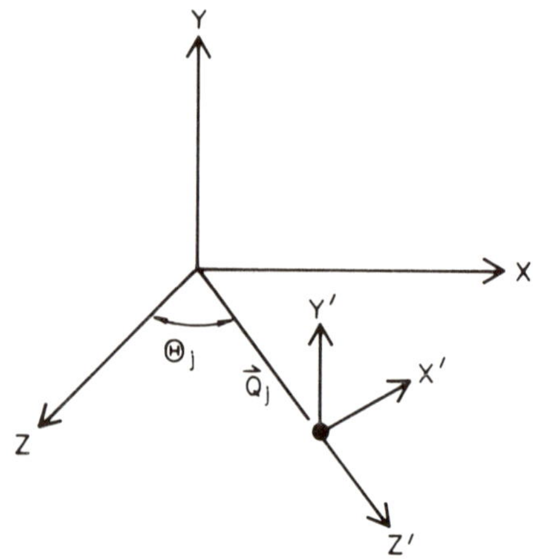

FIG. 15. The helicity frame axes $0x'y'z'$ for particle j. $0xyz$ are the axes of frame S.

Thus the final transition amplitude becomes

$$A^D_{\mu_a\mu_b m_1 m_2\mu_k\mu_l} = \sum_{\mu_j} B^D_{m_1m_2\mu_j}A_{\mu_a\mu_b\mu_j\mu_k\mu_l} \tag{72}$$

In the case of Λ decay obtained for instance in the reaction

$$K^-p \to \Lambda\pi^+\pi^- \tag{73}$$

many simplifications occur; $\sigma_2 = m_2 = 0$ and

$$B^D_{m_1m_2\mu_j} = \sum_{L_d} B^{L_d}C(l_d\sigma_1\sigma_j|\mu_j - m_1m_1)Y^{\mu_j-m_1}_{L_d}(\theta_d, \phi_d) \tag{74}$$

Further if we perform scattering from polarized targets the total transition amplitude will be (using (68))

$$A'^D_{\mu_a\mu_b m_1 m_2\mu_k\mu_l} = A^D_{\mu_a\mu_b m_1 m_2\mu_k\mu_l}\exp[-i(\mu_a - \mu_b)\alpha] \tag{75}$$

(4) *Experimental observables*

In order to simplify this discussion, we specialize to

$$M_1B_1 \to B_2M_2M_3 \tag{76}$$

where

$$B_1B_2 = \tfrac{1}{2}^+, M_1, M_2, M_3 = 0^-$$

e.g. reactions

$$\left.\begin{aligned} \pi N &\to N\pi\pi \\ \bar{K}N &\to \Lambda\pi\pi \end{aligned}\right\} . \tag{77}$$

We assume the initial polarization is specified in $0xyz$ and final baryon polarization in the helicity frame (see Fig. 15). The transition amplitudes are then given by eqn. (68).

(a) *Unpolarized cross-sections.* The initial density matrix is

$$\rho^i = \tfrac{1}{2} \tag{78}$$

and the differential cross-section is given by

$$I_0 = \mathrm{Tr}[A'\rho^i A'^{\dagger}] \Rightarrow \tfrac{1}{2}\sum_{\mu} |A'_{\mu}|^2 = \tfrac{1}{2}\sum_{\mu} |A_{\mu}|^2 \tag{79}$$

(b) *Asymmetry from a polarized target.* In this situation the initial density matrix is given by

$$\rho^i = \tfrac{1}{2}[1 + \mathbf{p}_b \cdot \boldsymbol{\sigma}_b] \tag{80}$$

and the differential cross section is

$$I_p = \mathrm{Tr}\,[A'\rho^i A'^{\dagger}] = I_0[1 + \vec{p}_b \cdot \vec{a}] \tag{81}$$

where

$$I_0 \vec{a} = \tfrac{1}{2}\,\mathrm{Tr}\,[A'\sigma_b A'^{\dagger}] \tag{82}$$

(c) *Final baryon polarization in scattering from an unpolarized target.* In this situation the final baryon density matrix is given by

$$I_0 \rho^f = \tfrac{1}{2} A' A'^{\dagger} \tag{83}$$

Then the polarization, P_L, of B_2 in a direction L is given by

$$I_0 P_L = \tfrac{1}{2}\,\mathrm{Tr}\,[A'A'^{\dagger}\sigma_L] \tag{84}$$

(d) *Depolarization tensor.* That is Final Polarization from a polarized target. If P_{Lj} is the polarization of particle j in direction L then

$$P_{Lj} = \mathrm{Tr}\,[\rho^f \sigma_{Lj}] \tag{85}$$

and we obtain

$$\begin{aligned} I_p P_{Lj} &= \mathrm{Tr}\,[A'\rho^i A'^{\dagger}\sigma_{Lj}] \\ &= I_0[P_{Lj} + \sum_i P_{ib} D_{ib,\,Lj}] \end{aligned} \tag{86}$$

where

$$I_0 D_{ib,Lj} = \tfrac{1}{2}\,\mathrm{Tr}\,[A'\sigma_{ib}A'^{\dagger}\sigma_{Lj}] \tag{87}$$

These results are summarized in Table IV.

TABLE IV. The Observable Quantities in $MB \rightarrow BMM$

I_0	$= \tfrac{1}{2}\{\lvert A_{++}\rvert^2 + \lvert A_{+-}\rvert^2 + \lvert A_{-+}\rvert^2 + \lvert A_{--}\rvert^2\}$
$I_0 A_x$	$= \mathrm{Re}\,[A_{++}A^*_{+-}\,\mathrm{e}^{i\alpha}] + \mathrm{Re}\,[A_{-+}A^*_{--}\,\mathrm{e}^{i\alpha}]$
$I_0 A_y$	$= \mathrm{Im}\,[A_{++}A^*_{+-}\,\mathrm{e}^{i\alpha}] + \mathrm{Im}\,[A_{-+}A^*_{--}\,\mathrm{e}^{i\alpha}]$
$I_0 A_z$	$= \tfrac{1}{2}\{\lvert A_{++}\rvert^2 + \lvert A_{-+}\rvert^2 - \lvert A_{+-}\rvert^2 - \lvert A_{--}\rvert^2\}$
$I_0 P_X^{(0)}$	$= \mathrm{Re}\,[A_{++}A^*_{-+}] + \mathrm{Re}\,[A_{+-}A^*_{--}]$
$I_0 P_Y^{(0)}$	$= -\mathrm{Im}\,[A_{++}A^*_{-+}] - \mathrm{Im}\,[A_{+-}A^*_{--}]$
$I_0 P_Z^{(0)}$	$= \tfrac{1}{2}\{\lvert A_{++}\rvert^2 + \lvert A_{+-}\rvert^2 - \lvert A_{-+}\rvert^2 - \lvert A_{--}\rvert^2\}$
$I_0 D_{xX}$	$= \mathrm{Re}\,[A_{+-}A^*_{-+}\,\mathrm{e}^{-i\alpha}] + \mathrm{Re}\,[A_{++}A^*_{--}\,\mathrm{e}^{i\alpha}]$
$I_0 D_{xY}$	$= -\mathrm{Im}\,[A_{+-}A^*_{-+}\,\mathrm{e}^{i\alpha}] - \mathrm{Im}\,[A_{++}A^*_{--}\,\mathrm{e}^{i\alpha}]$
$I_0 D_{xZ}$	$= \mathrm{Re}\,[A_{++}A^*_{+-}\,\mathrm{e}^{i\alpha}] - \mathrm{Re}\,[A_{-+}A^*_{--}\,\mathrm{e}^{i\alpha}]$
$I_0 D_{yX}$	$= -\mathrm{Im}\,[A_{+-}A^*_{-+}\,\mathrm{e}^{-i\alpha}] + \mathrm{Im}\,[A_{++}A^*_{--}\mathrm{e}^{i\alpha}]$
$I_0 D_{yY}$	$= \mathrm{Re}\,[A_{++}A^*_{-+}\,\mathrm{e}^{i\alpha}] - \mathrm{Re}\,[A_{+-}A^*_{-+}\,\mathrm{e}^{-i\alpha}]$
$I_0 D_{yZ}$	$= \mathrm{Im}\,[A_{++}A^*_{+-}\,\mathrm{e}^{i\alpha}] - \mathrm{Im}\,[A_{-+}A^*_{--}\,\mathrm{e}^{i\alpha}]$
$I_0 D_{zX}$	$= \mathrm{Re}\,[A_{++}A^*_{-+}] - \mathrm{Re}\,[A_{+-}A^*_{--}]$
$I_0 D_{zY}$	$= -\mathrm{Im}\,[A_{++}A^*_{-+}] + \mathrm{Im}\,[A_{+-}A^*_{--}]$
$I_0 D_{zZ}$	$= \tfrac{1}{2}\{\lvert A_{++}\rvert^2 + \lvert A_{--}\rvert^2 - \lvert A_{-+}\rvert^2 - \lvert A_{+-}\rvert^2\}$

The amplitudes $A_{\mu_f\mu_i}$ are written with $\mu_i, \mu_f = \pm\tfrac{1}{2}$. Also note that the order of the subscripts of A has been reversed.

We now turn to a detailed discussion of the scattering from an unpolarized incident state.

(5) *The differential cross-section from an unpolarized incident state*

In the appendix we give the phase space which corresponds to our normalization of states and this leads to the following differential cross section for reaction (59)

$$d\sigma = \frac{\pi^2}{Wp}\frac{1}{(2\sigma_a + 1)(2\sigma_b + 1)}\sum_{\mu}\lvert A_\mu\rvert^2\,d\rho \tag{88}$$

where A_μ is given by Eq. (63).

(a) *Conservation of Parity*. The variety of terms in this sum can be reduced by understanding the effects of parity. This is equivalent to setting $\mu \rightarrow -\mu$ in (63). We write

$$A^n_{\mu_a\mu_b\mu_j\mu_k\mu_l} = g^n_\mu T^n D^{j_j}_{l_j s_j} \tag{89}$$

and it can be shown (Cashmore *et al.*, 1972) that

$$g^n_{-\mu} = \eta_a\eta_b\eta_j\eta_k\eta_l(-1)^{\sigma_a+\mu_a+\sigma_b-\mu_b}(-1)^{\sigma_j+\mu_j+\sigma_k+\mu_k+\sigma_l+\mu_l}(g^n_\mu)^*. \tag{90}$$

For any specific problem this reduces the number of independent g^n_μ. In the case of reaction (76) we have

$$g^n_{-\mu} = (-1)^{\mu_f-\mu_i}(g^n_\mu)^* \tag{91}$$

where μ_i and μ_f are the helicities of the incident and final baryons. Since T^n is independent of μ we finally obtain in this case

$$A^n_{-\mu} = (-1)^{\mu_f-\mu_i} g^{n*}_\mu T_n D^{l_j}_{l_j s_j} \tag{92}$$

(b) *The cross-section due to isobars in the* (kl) *system and the relation to two-body scattering*. We first calculate the cross-section due to isobars in the (kl) system and then relate this to the equivalent expressions for stable two-body final states through zero width approximation for these isobars.

Using the fact that we can integrate over α (the angle of rotation about the incident beam) we can write

$$d\rho = \frac{\pi q_k Q_j}{8W\omega_j} d\omega_j^2 \, d\cos\theta_j \, d\cos\Theta \, d\Phi \tag{93}$$

The total cross-section is then

$$\sigma = \int \frac{\pi^2}{Wp} \overline{\sum_\mu} \sum_{nm} g^n_\mu g^{m*}_\mu X^n(W,\omega_j) X^{n*}(W,\omega_j) \frac{\pi q_k Q_j}{8W\omega_j} d\omega_j^2 \, d\cos\theta_j \, d\cos\Theta \, d\Phi \tag{94}$$

where

$$\overline{\sum_\mu} = \frac{1}{(2\sigma_a+1)(2\sigma_b+1)} \sum_\mu$$

and

$$X^n(W,\omega_j) = T^n(W,\omega_j) D^{j_j}_{l_j s_j}(\omega_j).$$

The above expression can then be reduced to give (Cashmore *et al.*, 1972)

$$\sigma = \frac{\pi}{p^2} \sum_n \frac{(2J+1)}{(2\sigma_a+1)(2\sigma_b+1)} \int |X^n(W,\omega_j)|^2 \, d\omega_j^2 \tag{95}$$

where we see that isobars of different quantum numbers in the (kl) system do not interfere.

If we now use Breit–Wigner form for $D^{j_j}_{l_j s_j}(\omega_j)$ and take the zero width limit the cross-section becomes

$$\sigma = \frac{\pi}{p^2}\sum_n \frac{2J+1}{(2\sigma_a+1)(2\sigma_b+1)}|T^{Jj_j}_{LSL_jS_j}(W,\omega_0)|^2 \tag{96}$$

which is the usual form in two-body scattering.

Finally we note that our forms of the kinematical factors (see eqns. (72) and (62)) mean that the total cross-section has the correct threshold dependence, i.e.

$$\sigma \propto \frac{1}{p^2}p^{2L+1}Q_j^{2Lj+1} \tag{97}$$

(6) *The quantities of physical interest*

The object of analysing reaction (59) is

(i) to measure the partial wave amplitudes τ^n where

$$T^n = T_p^{JLS}(W)T^{nJ}_{1L_jS_j}(W,\omega_j) \sim \frac{p^{L+\frac{1}{2}}}{\sqrt{4W}}\frac{Q_j^{L_j+\frac{1}{2}}}{\sqrt{4W}}\tau_p^{JLS}(W)\tau^{nJ}_{1L_jS_j}(W) \tag{98}$$

and we define

$$\tau^n = \tau_p^{JLS}(W)\tau^{nJ}_{1L_jS_j}(W) \tag{99}$$

The variation of this quantity will then indicate the presence of resonances (together with their decay channels) and its values can be compared with theoretical predictions.

(ii) to measure $D^{j_j}_{l_j}(\omega_j)$ in some specific cases. This has been used to try to determine the $I = 0\,(\pi\pi)$ phase shifts at low energies, a difficult thing to do by other techniques. Unfortunately the results are very dependent on the form of the model and there are some problems of interpretation making this a much less rewarding application.

In the rest of these lectures we concentrate on (i) as these are the most reliable results of this type of analysis and are providing the most excitement in this area of physics.

Finally in this section we list the isobar states that we have considered in our analysis of

$$\pi N \to \pi\pi N \tag{100}$$

for $E_{\text{c.m.}} < 2000$ MeV together with their partial waves. We have so far only considered the prominent Δ, ρ and ε ($I = 0\ \pi\pi$) final state interactions and their partial waves are listed in Table V, where the notation is $L_{\text{in}}, L_{\text{out}}, 2I, 2J$.

TABLE V. The Partial Waves Used in $\pi N \to \pi\pi N$

Isospin	Incident Wave	$\pi\Delta$	$N\rho_3$	$N\rho_1$	$N\sigma$
$I = \frac{1}{2}$	S11	SD11	SD11	SS11	SP11
	P11	PP11	PP11	PP11	PS11
	D13	DS13 DD13	DS13 DD13	DD13	DP13
	P13	PP13 PF13	PP13 PF13	PP13	PD13
	D15	DD15	DD15	DD15	DF15
	F15	FP15 FF15	FP15 FF15	FF15	FD15
	F17	FF17	FF17	FF17	
$I = \frac{3}{2}$	S31	SD31	SD31	SS31	
	P31	PP31	PP31	PP31	
	D33	DS33 DD33	DS33 DD33	DD33	
	P33	PP33 PF33	PP33 PF33	PP33	
	D35	DD35	DD35	DD35	
	F35	FP35 FF35	FP35 FF35	FF35	
	F37	FF37	FF37	FF37	

The 60 waves with angular momenta L, L', l each $\leqslant 3$. There are two nucleon-rho terms in the isobar model, indicated by ρ_3 and ρ_1, where the subscript indicates the coupling between the spin of the ρ ($l = d$) and the spin of the outgoing nucleon.

One can easily see how rapidly the number of partial waves grows. Furthermore certain aspects of the data already indicate that an improvement will be obtained by inclusion of P_{11}, D_{13} and F_{15} isobars particularly in the $\pi^+\pi^+n$ final state.

4. PRODUCTION REACTIONS

In these types of reactions we are concerned with three particle subsystems in final states containing four or more particles, i.e. reactions

$$a + b \to c + X \quad \llcorner\!\!\to j + k + 1 \tag{101}$$

of which there are many examples at present being studied, e.g.

$$\pi + p \to p + (A_1, A_2, A_3) \tag{102}$$

$$Kp \to p + (Q, L) \tag{103}$$

$$\pi, K, p + p \to \pi, K, p + N^* \tag{104}$$

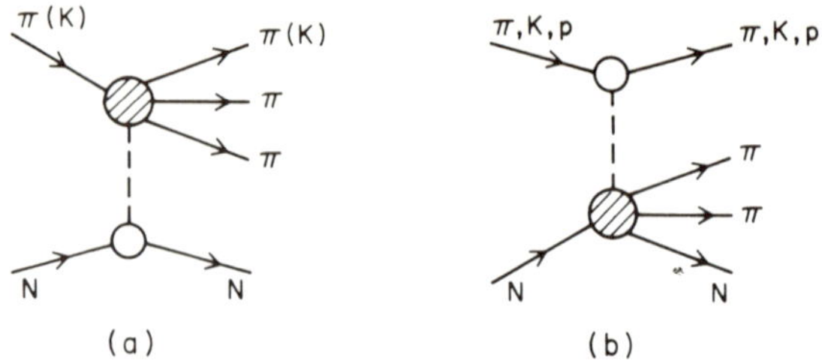

FIG. 16. Production reactions.

These are sketched in Fig. 16.

The variables we require to define the particles a, b, and c have already been given at the beginning of Section III. We do however have to introduce a new co-ordinate system S' in which to describe the production properties of the state X. S' will be related to the other particles a, b and c and will reflect our prejudices about the type of production process occurring, e.g.

(i) S' will be the Gottfried–Jackson system if we are interested in one particle exchange (see Fig. 17). $0Z'$ is defined as $\vec{p}\pi$ and $0Y'$ by $\vec{p}_{\text{in}} \wedge \vec{p}_{\text{out}}$ in the jkl rest system.

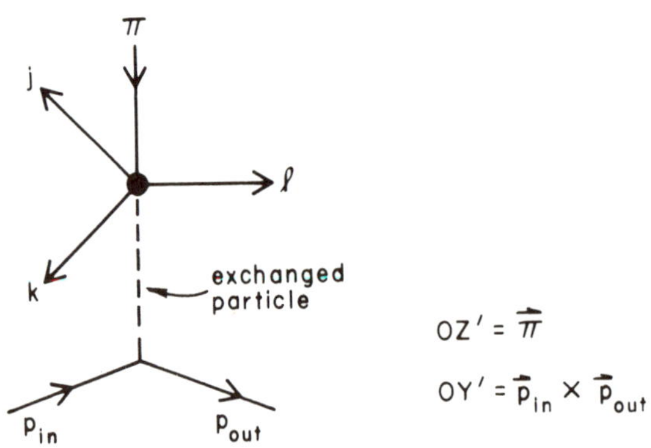

FIG. 17. The Gottfried–Jackson system.

(ii) S' will be the helicity frame if we are concerned with S-channel helicity conservation or absorption model predictions. In this case $0Z'$ is defined as $-\vec{p}_{\text{out}}$ in the jkl system. $0Y'$ has the previous definition.

(1) *Amplitude for production of X and its subsequent decay*

We assume that X is produced in an angular momentum substate $|nJM\rangle$

with respect to S'. Let the amplitude for the production of this state together with a final helicity μ_c of particle c from particles a and b be

$$R^{JpMn}_{\mu_a\mu_b\mu_c}(E, W, t) \tag{105}$$

where W is the mass of the three particle system, t is the four-momentum transfer to the three particle system and E is the total c.m. energy (i.e. in the ab system). We also include n in the state definition at this point to account for the fact that different channels of the same JPM may be produced with different helicity changes at the other vertex. In general, we will not include the total c.m. energy E in the arguments of R but it should always be understood to be there.

In order to describe the subsequent decay of X we require the following transition matrix element

$$A^{nJPm}_{\mu_j\mu_k\mu_l} = {}_S\langle nQ_j\mu_jQ_k\mu_kQ_l\mu_l|T|JPMn\rangle_{S'}$$

$$= \sum_m {}_S\langle Q_j\mu_jQ_k\mu_kQ_l\mu_l|T|JPmn\rangle_S \;{}_S\langle JPmn|JPMn\rangle_{S'} \tag{106}$$

$$= \sum_m {}_S\langle Q_j\mu_jQ_k\mu_kQ_l\mu_l|T|JPmn\rangle_S \;\mathscr{D}^J_{mM}(\alpha, \beta, \gamma) \tag{107}$$

where α, β, γ are the Euler angles defining the transformation from S to S'. We have already calculated the first factor in this sum—it is just the $f^{nJM}_{\mu_j\mu_k\mu_l}$ of Section III.1. Thus we can write

$$A^{nJPM}_{\mu_j\mu_k\mu_l} = \sum_m G^{nJm}_{\mu_j\mu_k\mu_l} T^{nJ}_{1L_jS_j}(W, \omega_j) D^{j_j}_{l_jS_j}(\omega_j)\, \mathscr{D}^J_{mM}(\alpha, \beta, \gamma) \tag{108}$$

$$= F^{nJM}_{\mu_j\mu_k\mu_l} T^{nJ}_{1L_jS_j}(W, \omega_j) D^{j_j}_{l_jS_j}(\omega_j) \tag{109}$$

Note that I have just written the amplitude for reaching the final state via one intermediate isobar state. The total amplitude is again obtained by coherently adding the terms due to other isobars ensuring that the correct symmetry properties are obtained. The forms of $T^{Jn}_{1L_jS_j}$ and $D^{j_j}_{l_jS_j}(\omega_j)$ have been discussed in Section III.1.

(2) *Properties of production and decay amplitudes under parity*

(a) *Production Amplitude:* $R^{nJM}_{\mu_a\mu_b\mu_c}$. Consider a co-ordinate system S' in which the Z axis is defined by a polar vector and the Y axis by an axial vector, e.g. the Gottfried–Jackson system as defined in Fig. 17. Then

$$R^{nJM}_{\mu_a\mu_b\mu_c} = \langle \phi_{JM}\psi_{\vec{q}_c\mu_c}|T|\psi_{\vec{q}_a\mu_a}\psi_{\vec{q}_b\mu_b}\rangle$$

$$= \langle \phi_{JM}\psi_{\vec{q}_c\mu_c}|P^{-1}PTP^{-1}P|\psi_{\vec{q}_a\mu_a}\psi_{\vec{q}_b\mu_b}\rangle$$

$$= \langle \phi_{JM}\psi_{-\vec{q}_c-\mu_c}|T|\psi_{-\vec{q}_a-\mu_a}\psi_{-\vec{q}_b-\mu_b}\rangle \eta_a\eta_b\eta_c(-1)^{\sigma_a-\mu_a+\sigma_b-\mu_b+\sigma_c-\mu_c}\eta_J$$

where $PTP^{-1} = T$ and $P\phi_{JM} = \eta_J\phi_{JM}$, i.e. η_J is the parity of state X.

If we now use an operator 0 which produces a rotation π about the Y axis we have

$$0\psi_{-\vec{q}-\mu} = \psi_{\vec{q}-\mu}$$
$$0\phi_{JM} = (-1)^{J-M}\phi_{J-M}$$

N.B. this assumes the Z-axis is in the production plane.

Thus eventually we have

$$R^{JM}_{\mu_a\mu_b\mu_c} = R^{J-M}_{-\mu_a-\mu_b-\mu_c}\eta_J\eta_a\eta_b\eta_c(-1)^{\sigma_a-\mu_a+\sigma_b-\mu_b+\sigma_c-\mu_c}(-1)^{J-M} \qquad (110)$$

(b) *Decay Amplitude:* $A^{nJM}_{\mu_j\mu_k\mu_l}$. In this case the properties under parity are most easily seen by setting $\mu \to -\mu$ and following a similar calculation to that for the formation reaction (Cashmore *et al.*, 1972). The result we obtain is that

$$A^{nJ-M}_{-\mu_j-\mu_k-\mu_l} = (-1)^{J-M}(-1)^{l_j+L_j}(-1)^{\sigma_j+\mu_j+\sigma_k+\mu_k+\sigma_l+\mu_l}$$
$$\times [F^{nJM}_{\mu_j\mu_k\mu_l}]^* T^{Jn}_{1L_jS_j}(X, \omega_j) D^{j_j}_{l_js_j}(\omega_j) \qquad (111)$$

(3) *The cross-section and its properties for the process $a + b \to c + j + k + l$*

The differential cross-section for the process is then given as

$$\frac{d^5\sigma(W, t, \mu_a\mu_b\mu_c\mu_j\mu_k\mu_l)}{d\alpha\, d\cos\beta\, d\gamma\, d\omega_1^2\, d\omega_2^2} \propto \left|\sum_{nJM} R^{nJM}_{\mu_a\mu_b\mu_c}(W, t)A^{nJM}_{\mu_j\mu_k\mu_l}(W, \alpha, \beta, \gamma, \omega_1^2, \omega_2^2)\right|^2 \qquad (112)$$

i.e. the production amplitude R is a function of W (the mass of the intermediate system) and t, while A is a well defined function of W, α, β, γ, ω_1^2 and ω_2^2, the parameters necessary to describe the final three particle state.

Note here we have five variables describing the decay whereas in the formation reaction (for unpolarized incident particles) we only require four. This is due to the fact that a, b and c define very specifically the initial coordinate frame (just as a polarized target would).

Thus if we consider the situation in which we do not observe the polarizations of the final particles in scattering from an unpolarized target we have

$$d\sigma \propto \sum_{\substack{\mu_a\mu_b\mu_c\\ \mu_j\mu_k\mu_l}} \left|\sum_{nJM} R^{nJM}_{\mu_a\mu_b\mu_c}(W, t)A^{nJM}_{\mu_j\mu_k\mu_l}(W, \alpha, \beta, \gamma, \omega_1^2, \omega_2^2)\right|^2 \qquad (113)$$

$$\propto \sum_{\substack{\mu_a\mu_b\mu_c\\ \mu_j\mu_k\mu_l}} \sum_{\substack{JMn\\ J'M'n'}} R^{nJM}_{\mu_a\mu_b\mu_c} R^{n'J'M'*}_{\mu_a\mu_b\mu_c} A^{nJM}_{\mu_j\mu_k\mu_l} A^{n'J'M'*}_{\mu_j\mu_k\mu_l} \qquad (114)$$

$$\propto \sum_{\mu_j\mu_k\mu_l} \sum_{\substack{nJM\\ n'J'M'}} \rho^{nn'}_{MM'} A^{nJM}_{\mu_j\mu_k\mu_l} A^{n'J'M'*}_{\mu_j\mu_k\mu_l} \qquad (115)$$

where we have defined an unnormalized density matrix

$$\rho_{MM'}^{nn'} = \sum_{\mu_a\mu_b\mu_c} R_{\mu_a\mu_b\mu_c}^{nJM} R_{\mu_a\mu_b\mu_c}^{n'J'M'*} \tag{116}$$

This density matrix possesses

(i) the usual property of hermiticity

$$\rho_{MM'}^{nn'} = \rho_{M'M}^{n'n*} \tag{117}$$

(ii) the property from conservation of parity of

$$\rho_{MM'}^{nn'} = \eta_n\eta_{n'}(-1)^{J+J'-M-M'}\rho_{-M-M'}^{nn'} \tag{118}$$

If $J = J'$ and $n \equiv n'$

$$\rho_{MM'}^{nn} = (-1)^{M-M'}\rho_{-M-M'}^{nn} \tag{119}$$

(4) *Structure of cross-section*

We can now demonstrate some aspects of the structure of these cross-sections which lead us to well-known results.

We first display explicitly the α, β, γ dependence of the distributions

$$d^5\sigma \propto \sum_{\substack{nn\\MM}} \rho_{MM'}^{nn'} \Big(\sum_{m,m'} \underbrace{\mathscr{D}_{mM}^{J}(\alpha,\beta,\gamma)\mathscr{D}_{m'M'}^{J'*}(\alpha,\beta,\gamma)}_{\substack{\text{The only terms containing}\\ \alpha,\ \beta,\ \gamma,\ M \text{ and } M'}} \ldots \Big) \tag{120}$$

If we first integrate over α we obtain the result that $m = m'$, i.e. the dependence of the differential cross-section on β and γ is given by

$$d^4\sigma \propto \sum_{\substack{nn'\\MM'}} \rho_{MM'}^{nn'} \left\{ \sum_m \mathscr{D}_{mM}^{J}(0,\beta,\gamma)\mathscr{D}_{mM'}^{J'*}(0,\beta,\gamma) \right\}$$

$$\propto \sum_{\substack{nn'\\MM'}} \rho_{MM'}^{nn'} \left\{ \sum_m \mathscr{D}_{mM}^{J}(0,\beta,\gamma)\mathscr{D}_{-m-M'}^{J'}(0,\beta,\gamma)(-1)^{m-M'} \right\} \tag{121}$$

We can now use the result that

$$\mathscr{D}_{mM}^{J}(0,\beta,\gamma)\mathscr{D}_{-m-M'}^{J'}(0,\beta,\gamma) = \sum_L C(J,J'L|m,-m)C(JJ'L|M,-M') \times \mathscr{D}_{0,M-M'}^{L}(0,\beta,\gamma)$$

$$= \sum_L C(J,J'L|m,-m)C(JJ'L|M,-M')\left(\frac{4\pi}{2L+1}\right)^{\frac{1}{2}} Y_L^{M'-M}(\beta,\gamma) \tag{122}$$

to display the (β, γ) dependence of the cross-section (121), i.e. the presence of spherical harmonics Y_L^m with $m \neq 0$ immediately implies the presence of off-diagonal terms in the density matrix and furthermore the maximum value of L can be a guide to the maximum J present ($L \leqslant 2J_{max}$).

If we now integrate over α, β and γ (i.e. we just look at the Dalitz plot population) we can use the orthogonality properties of the $\mathscr{D}$ function to write

$$\frac{d^2\sigma}{d\omega_1^2\, d\omega_2^2} \propto \sum_{\substack{n;n' \\ J=J' \\ M=M'}} \rho_{MM}^{nn'} \left(\sum_m \cdots \qquad\qquad \right) \tag{123}$$

Thus the Dalitz plot population depends only on ρ_{MM}^{JJ} and does not contain any terms corresponding to interferences between states which differ in J or M. Moreover since M does not appear within the parentheses we find that the Dalitz plot population is independent of M, the magnetic substate in which the state is initially produced. This result is expected since by integrating over α, β, and γ we have effectively destroyed any knowledge of the initial orientation.

The other result we would like to demonstrate is that waves of different parity do not interfere in the Dalitz plot distribution. To do this consider the interference of an isobar of type j with one of type k. We already know that we only have to consider states of the same J and M, and thus the general term is

$$\begin{aligned}
\frac{d^2\sigma}{d\omega_1^2\, d\omega_2^2} &\propto \int \sum_{\mu_j\mu_k\mu_l} \left[\sum_M \rho_{MM}^{nn'} A_{\mu_j\mu_k\mu_l}^{nJM}(j) A_{\mu_j\mu_k\mu_l}^{n'JM*}(k) \right] d\alpha\, d\cos\beta\, d\gamma \\
&\propto \int \sum_{\mu_j\mu_k\mu_l} \sum_{M\geqslant 0} \left(1 - \frac{\delta_{M,0}}{2}\right) \left[\rho_{MM}^{nn'} A_{\vec{\mu}}^{nJM}(j) A_{\vec{\mu}}^{n'JM*}(k) \right. \\
&\quad \left. + \rho_{-M-M}^{nn'} A_{\vec{\mu}}^{nJ-M} A_{\vec{\mu}}^{n'J-M*} \right] d\alpha\, d\cos\beta\, d\gamma \\
&\propto \int \sum_{\mu_j\mu_k\mu_l} \sum_{M\geqslant 0} \left(1 - \frac{\delta_{M,0}}{2}\right) \rho_{MM}^{nn'} \left[A_{\vec{\mu}}^{nJM}(j)\, A_{\vec{\mu}}^{n'JM*}(k) \right. \\
&\quad \left. + \eta\eta' A_{\vec{\mu}}^{nJ-M}(j)\, A_{\vec{\mu}}^{n'J-M*}(k) \right] d\alpha\, d\cos\beta\, d\gamma
\end{aligned} \tag{124}$$

But we know that

$$\sum_{\mu_j\mu_k\mu_l} \int A_{\vec{\mu}}^{nJM}(j)\, A_{\vec{\mu}}^{n'JM*}(k)\, d\alpha\, d\cos\beta\, d\gamma = \Delta^{Jnn'}(j, k) \tag{125}$$

i.e. independent of M

$$\frac{d^2\sigma}{d\omega_1^2\, d\omega_2^2} \propto \rho_{MM}^{nn'} \left[1 - \frac{\delta_{M,0}}{2}\right] \Delta^{Jnn'}(j, k)[1 + \eta\eta'] \tag{126}$$

Thus if $\eta = -\eta'$, i.e. waves of opposite parity

$$\frac{d^2\sigma}{d\omega_1^2\, d\omega_2^2} = 0. \tag{127}$$

Thus we have the well-known results that

(i) states of different J do not interfere in the Dalitz plot
(ii) states of different parity do not interfere in the Dalitz plot
(iii) the Dalitz plot density is independent of M, the magnetic substate in which a state is produced.

(5) *Application of analysis to* $(K)\pi + N \rightarrow (K)\pi + \pi + \pi + N$

Here we want specifically to consider the case where one of the incident particles has spin 0^- and is transformed into a three body state, i.e. production of the A_1, A_2, etc. We can write the cross-section

$$d\sigma \propto \sum_{\mu_i\mu_f} \left| \sum_{nJM} R^{nJM}_{\mu_i\mu_f}(W, t) A^{nJM}(W, \alpha, \beta, \gamma, \omega_1^2, \omega_2^2) \right|^2 \tag{128}$$

where μ_i and μ_f are the initial and final proton helicities and $\mu_j = \mu_k = \mu_l = 0$.
As before we can define a density matrix

$$\rho^{nn'}_{MM'} = \sum_{\mu_i\mu_f} R^{nJM}_{\mu_i\mu_f} R^{n'J'M'*}_{\mu_i\mu_f} \tag{129}$$

Clearly the object of any analysis is to determine the elements of the density matrix

$$d\sigma \propto \sum_{\substack{nJM\\ n'J'M'}} \rho^{nn'}_{MM'} A^{nJM} A^{n'J'M'*} \tag{130}$$

This equation will always hold even if we bin the data in W and t, i.e. we integrate over certain regions

$$\rho^{nn'}_{MM'}(W, \bar{t}) = \sum_{\mu_i\mu_f} \int_{t_1}^{t_2} R^{nJM}_{\mu_i\mu_f}(W, t)\, R^{n'J'M'*}_{\mu_i\mu_f}(W, t)\, dt \tag{131}$$

In Table VI we list all the waves one might use in studying A_1, A_2 decays considering only well known strong final state interactions

$$\begin{aligned} A_1, A_2 &\rightarrow \pi\varepsilon \\ &\rightarrow \pi\rho \end{aligned} \tag{132}$$

For the strange particles we have

$$\begin{aligned} Q, K^*(1400) &\rightarrow K\varepsilon, \qquad \pi(K\pi)_{S\text{-wave}} \\ &\rightarrow K\rho, \qquad \pi K^* \end{aligned} \tag{133}$$

TABLE VI. Partial Waves in Decay to Three Pseudoscalar Mesons for $l \leqslant 2$ (ε, ρ, f) and $L \leqslant 3$

J^P	$2J + 1$	L	l	Number of Amplitudes
0^-	1	0	0	1
		1	1	1
		2	2	1
1^-	3	1	1	3
		2	2	3
1^+	3	1	0	3
		0	1	3
		2	1	3
		1	2	3
		3	2	3
2^+	5	2	1	5
		1	2	5
		3	2	5
2^-	5	2	0	5
		1	1	5
		3	1	5
		0	2	5
		2	2	5
3^-	7	3	1	7
		2	2	7
3^+	7	3	0	7
		2	1	7
		1	2	7
		3	2	7
Total ($J \leqslant 2, l \leqslant 1$)				29

and there will be twice the number of amplitudes to consider. If one wishes to consider higher mass states, the A_3 and L, then we have to include amplitudes corresponding to decay into pseudoscalar and tensor mesons. Even if we restrict outselves to $J \leqslant 2$ we realize that we have an enormous density matrix to determine, e.g. a 29×29 density matrix. This will contain 841 complex elements which parity and hermiticity reduce to approximately

841 real numbers. These must also satisfy certain constraints, e.g.

$$|\rho_{ij}|^2 \leqslant \rho_{ii}\rho_{jj}$$

The imposition of these constraints is not entirely trivial and the task is one of certain difficulty! Instead one makes simplifying assumptions. These fall into two classes

(i) Resonance approximation—here one writes

$$\rho^{nn'}_{MM'} = \rho^{JJ'}_{MM'} C^n C^{n'*} \tag{134}$$

e.g.

$$\left.\begin{array}{l} 1^+ \to \pi\rho \\ \quad\;\; \to \pi\varepsilon \end{array}\right\} \begin{array}{l}\text{have the same density matrix just different}\\ \text{couplings to the final states.}\end{array}$$

Furthermore these couplings are assumed to be independent of M, i.e. $1^+\ M = 1$, $1^+\ M = 0$ decays have the same C. This immediately reduces the number of parameters to $(17 \times 17) + 9 \times 2 \sim 300$. This is the approach followed by the Illinois group.

(ii) One parametrizes the density matrix in terms of the production amplitudes

$$\rho^{nn'}_{MM'} = R^{nJM} R^{n'J'M'*} \tag{135}$$

and then attempts to determine the R^{nJM}. This has the advantage that we now only have 29 complex numbers to determine (58, if we allow both spin flip terms and spin-nonflip terms at the proton vertex) and the relations amongst density matrix elements are automatically ensured. However there is a disadvantage in that we have to assume that R^{nJM} are constant in any bin we consider or at least have the same t dependence. In reality

$$\rho^{nn'}_{MM'}(\overline{W}, \bar{t}) = \int\!\!\int R^{nJM} R^{n'J'M'*}\, \mathrm{d}W\, \mathrm{d}t \neq R^{nJM}(\overline{W}, \bar{t})\, R^{n'J'M'}(\overline{W}, \bar{t})\, \Delta W\, \Delta t \tag{136}$$

Even after these simplifications the number of parameters remains large and in order to make the problem tractable it is then essential to choose a subset of the partial waves. This is done in all analyses so far.

In using the last approach it is essential to realize that we really only determine the density matrix, the production amplitudes although having a physical foundation at this point only serve as a parametrization. Of course with more polarization experiments one will eventually be in a position to measure these.

(6) *Results of these analyses*

We finally list the results that one usually sees quoted from these analyses.

(i) The cross-section in a given J^P state is given by

$$\sigma^{J^P} = \int \sum_{\substack{MM' \\ nn'}} \rho^{nn'}_{MM'} A^{nJM} A^{n'JM'*} \, d\alpha \, d\cos\beta \, d\gamma \, d\omega_1^2 \, d\omega_2^2$$

$$= \int \sum_{\substack{M \\ n,n'}} \rho^{nn'}_{MM} A^{nJM} A^{n'JM*} \, d\alpha \, d\cos\beta \, d\gamma \, d\omega_1^2 \, d\omega_2^2 \tag{137}$$

Note that we retain n, n' even though $J = J'$, $P = P'$, $M = M'$ since these still contain the different possible decay channels, e.g.

$$\left.\begin{aligned} 1^+ &\rightarrow \pi\rho - s\text{-wave} \\ &\rightarrow \pi\varepsilon - p\text{-wave} \end{aligned}\right\}$$

(ii) The cross-section for a given J^P and a given decay channel

$$\sigma^{nJ^P} = \int \sum_M \rho^{nn}_{MM} |A^{nJM}|^2 \, d\alpha \, d\cos\beta \, d\gamma \tag{138}$$

Note $\sigma^{J^P} \neq \Sigma_n \sigma^{nJ^P}$ because of the presence of interference terms between different n, e.g.,

$$\left.\begin{aligned} 1^+ &\rightarrow \pi\rho - s\text{-wave} \\ &\quad\ \pi\rho - d\text{-wave} \end{aligned}\right\}$$

(iii) Normalized density matrix

$$\tilde{\rho}^{nn'}_{MM'} = \frac{\rho^{nn'}_{MM'}}{\sum_{n,M} \rho^{nn}_{MM}} \tag{139}$$

Of course within any J^P state $\Sigma_M \rho^{J^PJ^P}_{MM} \neq 1$ but these can be suitably renormalized.

(iv) The complex coupling constants C^n: As we will see the phases of these are important and from the C^n can be deduced the branching fractions for resonances.

(7) *Application to* $(K, p)\pi + N \rightarrow K(\pi) + \pi\pi N$

The next important application of this type of analysis will be to those types of reactions drawn schematically in Fig. 16b. We just want to remark at this point that for K, π induced reactions the problem is somewhat simpler. Since these particles are spinless we do not have to consider spin flip terms at this vertex. We do of course have to consider the transition from protons of different helicity to the various n, J^P, M states. However as we have seen these amplitudes are simply related by parity

$$f^{nJM} = f_+^{nJ-M} \eta(-1)(-1)^{J-M} \tag{140}$$

where $-$, $+$ are the initial helicities of the proton. For pp interactions this simplification clearly does not occur.

(8) *Application to polarized target experiments*

In this case it is most convenient to consider the helicity system. If the target proton is transversely polarized then this will define an initial co-ordinate system S''. The co-ordinate system S' will then be related to S'' by Euler angles 0, $-\theta_p$, $-\phi_p$ where θ_p and ϕ_p are production angles of the final

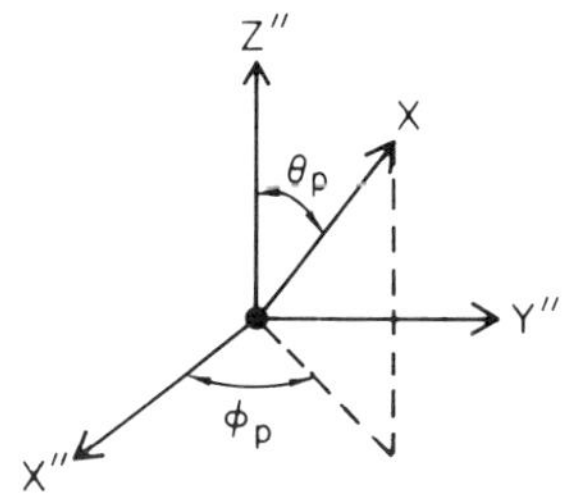

FIG. 18. Production angles of x in S'', a system defined by a polarized target.

multiparticle system with respect to S''. This is illustrated in Fig. 18. Note θ_p is defined for a given W and t. We then have amplitudes

$$R^{nJM}_{\mu_i\mu_f}(W, t, \phi_p) = R^{nJM}_{\mu_i\mu_f}(W, t)\, e^{-i\mu_i\phi_p} \tag{141}$$

where μ_i is the initial proton helicity.

We can then repeat the calculations of Section III.3 in order to obtain expressions for the asymmetrics from polarized targets, the polarization of the final recoil nucleon and the depolarization tensor of the nucleon. We will again obtain Table IV with the charge $\alpha \to -\phi_p$. It is also important to remember that the order of the proton helicities is reversed in the table.

(9) *Structure of the production amplitude* $R^{nJM}_{\mu_b\mu_c}(W, t)$

It is often convenient to extract the minimum angular momentum structure which must exist in this amplitude. Again the calculation is most easily performed in the helicity system in which we have

$$R^{nJM}_{\mu_b\mu_c}(W, t) = {}_{S'}\langle nJMq_c\mu_c|T|q_a\mu_a q_b\mu_b\rangle_{S''} \tag{142}$$

where S'' is a co-ordinate system defined with respect to the incident particles. If the production angle is θ_p and the initial state unpolarized we have

$$\begin{aligned} R^{nJM}_{\mu_b\mu_c}(W, t) &\sim \sum_r \mathscr{D}^r_{M-\mu_c,\mu_a-\mu_b}(0, -\theta_p, 0)\, R^{nJMr}_{\mu_b\mu_c} \\ &\sim \sum_r d^r_{M-\mu_c,-\mu_b}(0, -\theta_p, 0) R^{nJMr}_{\mu_b\mu_c} \end{aligned} \tag{143}$$

where r is the total angular momentum. For small θ_p we can write

$$d^r_{M-\mu_c,\,-\mu_b}(0, -\theta_p, 0) \sim \cos\left(\frac{\theta_p}{2}\right)^{|M-\mu_c-\mu_b|} \sin\left(\frac{\theta_p}{2}\right)^{|M-\mu_c+\mu_b|} \tag{144}$$

$$R^{nJM}_{\mu_b\mu_c}(W, t) \sim \left[\cos\frac{\theta_p}{2}\right]^{|M-\mu_c-\mu_b|} \left[\sin\frac{\theta_p}{2}\right]^{|M-\mu_c+\mu_b|} R'^{JMn}_{\mu_b\mu_c}(W, t) \tag{145}$$

If $M - \mu_c = -\mu_b$ (no helicity flip)

$$R^{nJM}_{\mu_b\mu_c} \sim \left(\cos\frac{\theta_p}{2}\right)^{|2\mu_b|} R'^{nJM}_{\mu_b\mu_c} \tag{146}$$

If $M - \mu_c = -\mu_b + 1$

$$R^{nJM}_{\mu_b\mu_c} \sim \left(\cos\frac{\theta_p}{2}\right)^{|-2\mu_b+1|} \left(\sin\frac{\theta_p}{2}\right) R'^{nJM}_{\mu_b\mu_c} \tag{147}$$

Now since

$$\begin{aligned} t &= (E_c - E_b)^2 - (\vec{p}_c - \vec{p}_b)^2 \\ &= m_c^2 + m_b^2 - 2E_bE_c + 2p_bp_c\cos\theta_p \end{aligned}$$

and

$$\begin{aligned} t_{\min} &= m_c^2 + m_b^2 - 2E_bE_c + 2p_bp_c \\ t - t_{\min} &= 2p_bp_c(\cos\theta_p - 1) \\ &= 2p_bp_c\,2\sin^2\theta_p/2 \end{aligned}$$

$$\sin\theta_p/2 \sim (t - t_{\min})^{\frac{1}{2}} \tag{148}$$

We have the well-known result that amplitudes with a net helicity flip of λ behave as $\sim(t - t_{\min})^{\lambda/2}$. This minimum t dependence can then be displayed explicitly and we then have to determine the $R'^{nJM}_{\mu_b\mu_c}$. If such a technique is employed then the approximations used in the parametrization of the density matrix (see eqn. (136)) can be improved.

Of course this will not necessarily be the total t-dependence of the amplitude but it is the minimum structure required by angular momentum considerations.

5. FITTING THE EXPERIMENTAL DATA

In fitting the experimental data, either from formation or production reactions, it is quite common to integrate over some of the variables, i.e. one fits projections of the data such as the Dalitz plot population. However, the best results will be obtained by making maximum use of the correlations that exist in the 4 (formation) or 5 (production) dimension variable spaces. This is most easily done by the maximum likelihood technique. One constructs a likelihood function

$$\mathscr{L} = \prod_{i=1}^{N} \sigma(\vec{\tau}, \vec{x}_i) \tag{149}$$

where $\vec{x}_i$ is the vector corresponding to the ith event in the multi-dimensional space, $\vec{\tau}$ is the vector of partial wave amplitudes and $\sigma(\vec{\tau}, \vec{x}_i)$ is the probability (cross-section) for such an event as calculated in Sections III.1, III.3, and III.4. Then log $\mathscr{L}$ is maximized by varying the parameters $\vec{\tau}$.

Finally a check is usually made that these sets of amplitudes $\vec{\tau}$ correspond to a good fit to the data by calculating the χ^2 for fits to the projections.

The results quoted in Sections IV and V are obtained using this technique.

IV. THE RESULTS OBTAINED FROM FORMATION EXPERIMENT

In this section I will attempt to summarize the situation that exists in the analysis of formation experiments. The results at this time bear mainly on the classification of resonance states and on understanding of their various decays within these classification schemes. The major results have come from analyses (Herndon *et al.*, 1972) of

$$\pi N \rightarrow \pi\pi N \tag{150}$$

but significant measurements have been made in the study of (Mast *et al.*, 1972)

$$K^- p \rightarrow \Lambda\pi\pi \tag{151}$$

and (Bland *et al.*, 1969)

$$K^+ p \rightarrow K\pi N \tag{152}$$

Before beginning detailed discussion of each set of results I would like briefly to review the motivations, some of which I have already noted in Section I, for pursuing such a study.

(a) *Identification of resonant states and classification in supermultiplets*

In the past few years the existence of many resonance states has been claimed from the partial wave analysis of two body reactions

$$\begin{matrix} \pi^+ p \rightarrow \pi^+ p \\ \pi^- p \rightarrow \pi^- p, \pi^0 n \end{matrix} \qquad I = \tfrac{1}{2} \text{ and } I = \tfrac{3}{2} N^* \tag{153}$$

$$\begin{matrix} K^- p \rightarrow K^- p, \bar{K}^0 n \\ \rightarrow \Lambda\pi, \Sigma\pi \end{matrix} \qquad I = 0, 1\ Y^* \tag{154}$$

Many of these states are very inelastic. Furthermore the identification of low spin resonances tends to be difficult in the presence of high spin states. For example in the resonance region of πN scattering at $E_{\text{c.m.}} \sim$ 1600–1700 MeV there are two spin $\frac{5}{2}$ resonances, the F15 and D15. However only after extended analysis were the lower spin states S31, S11, D33 finally identified. Both of these comments imply the need to study inelastic reactions. The first because this is where the resonances may be prolifically found and the second because momenta are on average lower and thus higher angular momentum states are expected to be suppressed ($L \sim qR$, with $R \sim 1$ fermi, or smaller). Indeed the Y^* situation would be in very poor shape were it not for the analyses of the inelastic two body final states $\Lambda\pi$, $\Sigma\pi$. Since the cross sections for $\pi\pi N$ and $\Lambda\pi\pi$, $\Sigma\pi\pi$, $\bar{K}\pi N$ are large one hopes for further improvements from study of these states.

With the identification of many of these states classification within supermultiplets becomes possible. We are all familiar with the $SU(3)$ classification of states into singlets, octets and decuplets. Of course there are many other irreducible representations of $SU(3)$ but these are the ones predicted by simple quark models (qqq for baryons and $q\bar{q}$ for mesons). For an N^* state with a given J^P all the other members of the octet must exist, just as for a Δ state all the other decuplet members must be present. At this point $SU(3)$ classification and the quark model work surprisingly well—many octets ($\frac{1}{2}^-, \frac{3}{2}^-, \frac{1}{2}^-, \frac{5}{2}^-, \frac{5}{2}^+$) and decuplets ($\frac{3}{2}^+, \frac{7}{2}^+$) are almost complete. The analysis of reactions

$$KN \to KN, \qquad K\pi N \tag{155}$$

if they demonstrated the existence of Z^*'s ($S = +1$ baryon resonances) could be fatal to simple quark models (requiring $qqqq\bar{q}$) and allow myriads of other resonant states. However it is possible that such states might be dynamical in origin thus allowing the quark model a reprieve.

Another highly successful symmetry scheme is that of $SU(6) \times 0(3)$ ($SU(3) \times SU(2) \times 0(3)$ of the quark states) (Dalitz, 1969; Greenberg, 1969). This attempts to group together known states of different J^P into yet larger multiplets. Of course fewer of these supermultiplets are identified as they contain more resonant states. Again the most popular model for baryons is the L excitation symmetric quark model of which the $[56, L = 0^+]$, $[70, L = 1^-]$, $[56, L = 2^+]$ multiplets appear on a sure footing.

However many states still need to be identified and the analysis of the inelastic reactions should provide the information with which to confirm or confound these schemes.

(b) *Couplings to decay channels*

The success of the classification schemes suggests that we might attempt to apply the same symmetry groups to the coupling constants. The application of $SU(3)$ to the decays

$$N^*, \quad \Delta \to N\pi$$
$$Y_0^*, \quad Y_1^* \to \Lambda\pi, \quad \Sigma\pi, \quad \bar{K}N$$
$$\Xi^* \to \Xi\pi$$

has already been discussed exhaustively by many authors (Levi Setti, 1969; Plane *et al.*, 1970). The undoubted success is summarized in Fig. 19 where the agreement is perfect (the one discrepancy has later been resolved (Plane *et al.*, 1970)).

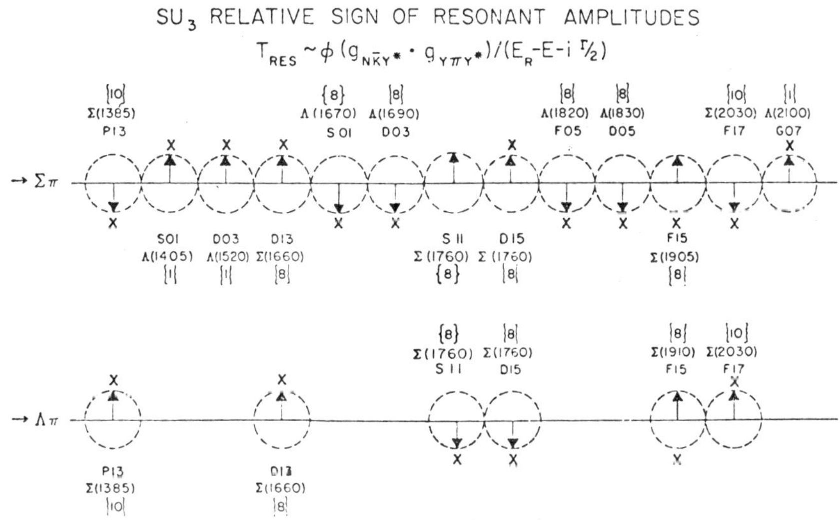

FIG. 19. Relative signs of Y^* couplings. The arrow is the prediction from $SU(3)$ and the X represents the experimentally observed coupling.

With the apparent success of $SU(6)$ classification schemes one would like to apply this symmetry group to the decays as well (Faiman, 1971; Faiman and Plane, 1972). This means that we can now relate

$$N^* \to N\pi \quad \text{and} \quad N^* \to \Delta\pi$$

(since the N and Δ belong to the same $SU(6)$ multiplet) as well as the decays of different J^P states. Furthermore since the ρ and π reside within the same **35** of $SU(6)$ we have relations for $N^* \to N\rho$ decays. It is immediately apparent that one needs to analyse the $\pi\pi N$ final states and with successful analysis of $\Lambda\pi\pi$, $\Sigma\pi\pi$, $\bar{K}\pi N$ one could again make the $SU(3)$ tests for

$$N^* \rightarrow \pi\Delta$$
$$Y_0^*, Y_1^* \rightarrow \pi Y_1^*(1385), \text{etc.}$$

The measurement of these decay couplings is clearly the next step in the development of our understanding of the resonance region.

(c) *Multichannel analyses*

If one can obtain a description of the inelastic states (i.e. the three body states for πN) for $E < 2000$ MeV we will essentially have all the information possible on the πN interaction. In any analysis we will be able to exploit the constraints of unitarity to the maximum effect. This should allow us to give not only resonance parameters but the complete analytic structure of the T-matrix. This will then be a sensitive test of any dynamical theory of resonances just as the coupling constants described in (b) are a fundamental test of the symmetry schemes.

(d) *Dynamical results*

With the advent of inelastic partial wave amplitudes it will be possible to construct finite energy sum rules relating the low and high energy behaviour of the scattering amplitudes, as was done in πN scattering. This should allow a new study of duality in a different set of reactions.

Clearly I have not been exhaustive in listing all the motivations but these few emphasize the great importance of this type of analysis.

1. $\pi N \rightarrow \pi\pi N$

This reaction has been extensively studied by the SLAC/LBL collaboration (Herndon *et al.*, 1972; Cashmore, 1973) and many of its results are confirmed by or consistent with other analyses (Bowler and Cashmore, 1970; Chinowsky *et al.*, 1970; DeBeer *et al.*, 1969a; 1969b; Brody *et al.*, 1971; Mehtani *et al.*, 1972). I will concentrate on the results of this analysis for a number of reasons.

(i) It spans the c.m. energy range $1300 < E < 2000$ except for a 100 MeV gap $1540 < E < 1650$ where the data is not yet available for analysis.

(ii) It utilizes the data in the most efficient manner making simultaneous maximum likelihood fits to the three major channels at each energy

$$\pi^- p \rightarrow \pi^+\pi^- n \qquad (156)$$

$$\pi^- p \rightarrow \pi^-\pi^0 p \qquad (157)$$

$$\pi^+ p \rightarrow \pi^+\pi^0 p \qquad (158)$$

TABLE VII. Number of Events for the Energy Bins Used in the Fits

C.M. Energy	Range (MeV)	$\pi^- p \to \pi^+ \pi^- n$	$\pi^- p \to \pi^- \pi^0 p$	$\pi^+ p \to \pi^+ \pi^0 p$
1310	1300–1330	1069	151	
1340	1330–1360	1664	11	
1370	1360–1380	2471	2	
1400	1380–1410	5049	964	78
1440	1430–1460	4918	1802	359
1470	1460–1480	3252	1629	175
1490	1480–1510	5555	3197	1523
1520	1510–1530	3241	2588	795
1540	1530–1560	3905	3285	1114
1650	1630–1670	6061	3757	2467
1690	1670–1710	5901	3689	1139
1730	1710–1750	3455	2630	4061
1770	1750–1790	3214	2352	2853
1810	1790–1830	2447	1541	3855
1850	1830–1870	3931	3183	6372
1890	1870–1910	5072	3170	12 690
1930	1910–1950	5817	4080	4298
1970	1950–1990	5277	3544	7744
Total	1300–1990	72 299	41 575	49 523

(iii) Excellent agreement with the inelastic cross-sections predicted by elastic phase shift analyses (EPSA).

(iv) Continuous solutions have been identified throughout the energy range from the independent solutions at each energy. This allows the presentation of reliable Argand diagrams for the first time.

(v) The analysis is of the type described in Section III.3 the intermediate states considered being

$$\pi N \to \pi\Delta$$

$$\to N\rho \quad S \text{ (the total } N\rho \text{ intrinsic spin)} = \tfrac{1}{2} \text{ or } \tfrac{3}{2} \text{ denoted as } \rho_1 \text{ or } \rho_3$$

$$\to N\varepsilon$$

The results are presented as partial wave amplitudes in an *LS* representation of the *T* matrix. The energies and the data considered are listed in Table VII where its lack of data from 1540 to 1650 is particularly conspicuous.

(vi) The fits to the experimental data are of good equality although they deteriorate at the higher energies. As can be seen in Fig. 20, a 4-*D* representation of the fit to $\pi^+\pi^-n$ at 1690 MeV, the enormous variations of structure

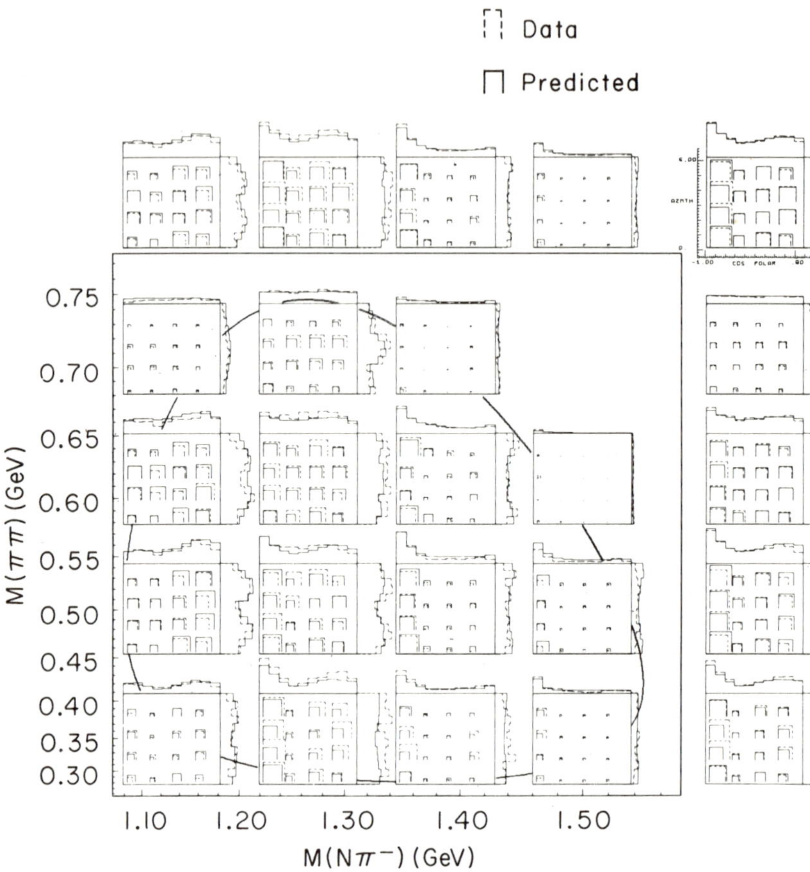

FIG. 20. Fits to the reaction $\pi^- p \rightarrow \pi^+\pi^- n$ at a c.m. energy of 1690 MeV. The figure contains $\cos\theta$ versus ϕ plots for individual regions of the Dalitz plot where $\cos\theta$ and ϕ are the polar angles of the incident pion in a co-ordinate system defined by the final state. The z-axis lies along $\vec{p}_N$ and the y-axis lies along $\vec{p}_{\pi^-} \times \vec{p}_{\pi^+}$. The plots outside the Dalitz plot are the sums of the corresponding plots within the boundary.

are extremely well-reproduced. Furthermore the partial wave amplitudes make excellent predictions of the $\pi^0\pi^0 n$ cross-section as demonstrated in Fig. 21.

(1). *Description of partial wave amplitudes*

Before showing the Argand diagrams of the T matrix elements leading to inelastic channels I have to emphasize one point. In analysing inelastic reactions at a given energy E the *relative phases* of the individual partial waves are only determined and not the absolute phase (a distinct difference

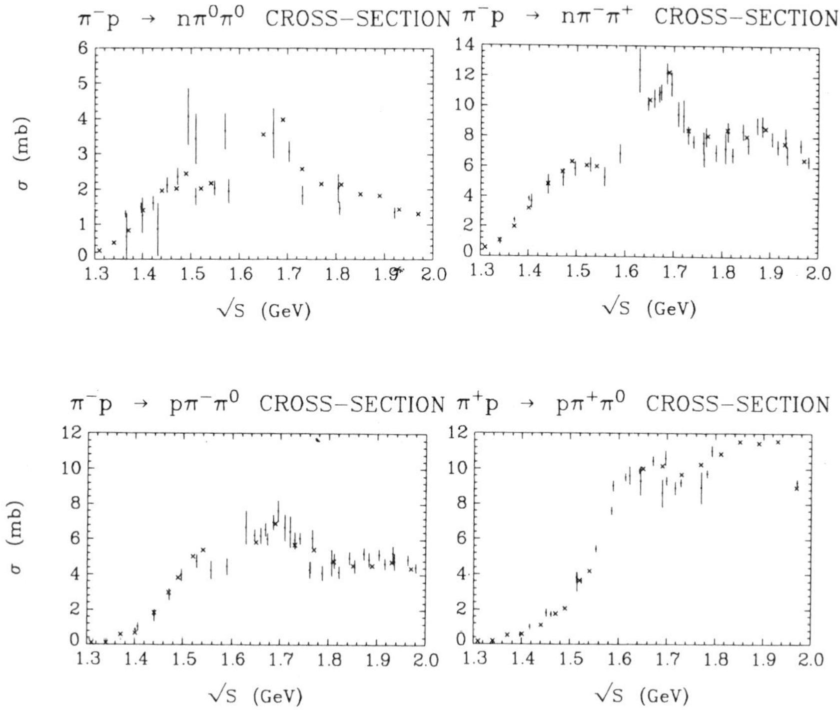

FIG. 21. Single pion production cross-sections. Data points are indicated by | and the predictions from our partial wave amplitudes by *x*.

from elastic phase shift analyses—there one essentially has the unscattered wave to define the absolute phase). Thus in order to present Argand diagrams it is necessary to specify the absolute phase at each energy (i.e. one free parameter for all waves, since their relative position is fixed). This has been done by using *K*-matrix fits to the *P*11, *D*15 and *F*15, *F*35 waves in the low, middle and high energy regions of the data. Now unfortunately the absence of data for $1540 < E < 1650$ makes it difficult to be certain of the correct continuity. Indeed the presence of two marginally different high energy solutions indicates that it would be possible to obtain radically different continuity through this region. I stress this point as it is of crucial significance in the comparison with theory at the present time. The Argand diagrams I will show correspond to one solution (A). Solution B differs qualitatively in *P*11 waves at the higher energies but at lower energies (<1540) it is possible that the partial wave amplitudes may be rotated by 180°. I will not discuss solution B as the work on it is far from complete, except to point out

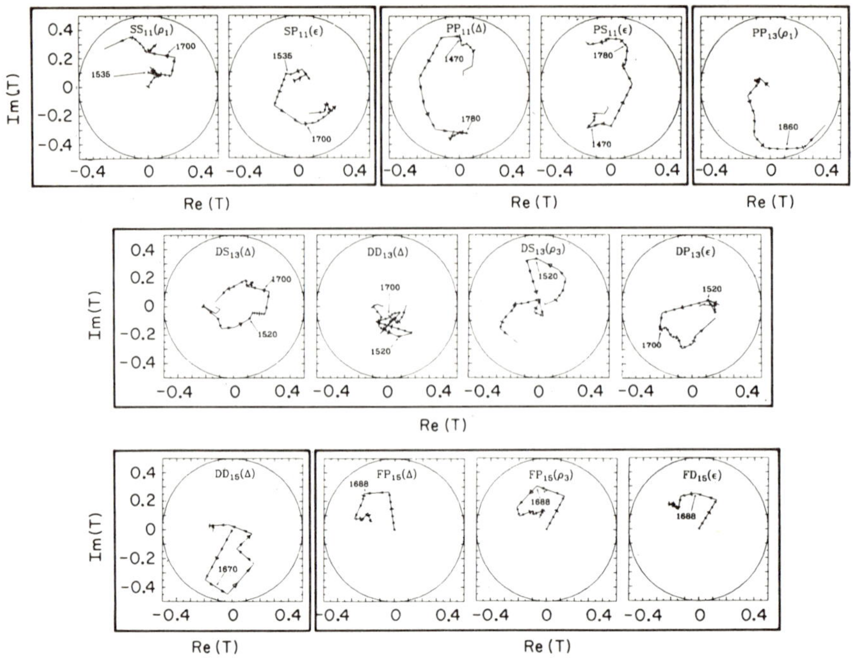

FIG. 22. $I = \frac{1}{2}$ partial wave amplitudes. Arrows are spaced every 20 MeV, with wide arrows every 100 MeV: base of wide arrows mark integral hundreds of MeV. Lower-l waves are plotted starting at $s = 1400$ MeV; higher-l waves only where they were first needed. Last arrowhead is always at 1940 MeV.

the changes of interpretation it would produce. Of course this would be unnecessary had the data in the "gap" been analysed!

(a) $I = \frac{1}{2}$ *states.* In Fig. 22 are the Argand diagrams of all the $I = \frac{1}{2}$ waves determined in the analysis. The figures contain a great deal of information and I will only point out the most important and striking features.

(i) The considerable motion in the Argand diagrams is not surprising, essentially all the structure is associated with the existence of known resonance states.

(ii) The following resonant states are clearly observed (the notation is L, L', $2I$, $2J$ where L is the incoming orbital angular momentum, L' the final state orbital angular momentum, I the isotopic spin, and J the total angular momentum)

$S11(1700)$ in $N\rho_1(SS11)$ and $N\varepsilon(SP11)$

$$\left.\begin{array}{l} P11(1470) \\ P11(\sim 1780) \end{array}\right\} \text{ in } \pi\Delta(PP11) \text{ and } N\varepsilon(PS11) \text{ (see Fig. 23)}$$

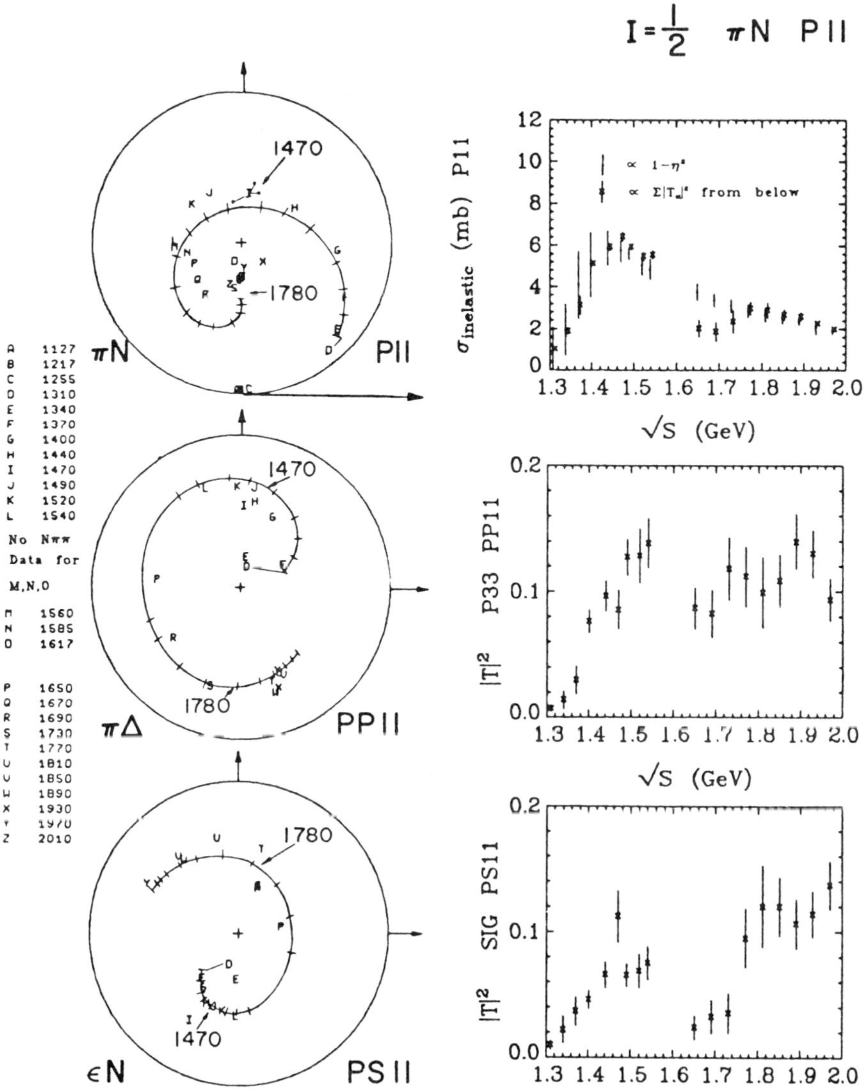

FIG. 23. Partial waves derived from the incident *P*11 wave.

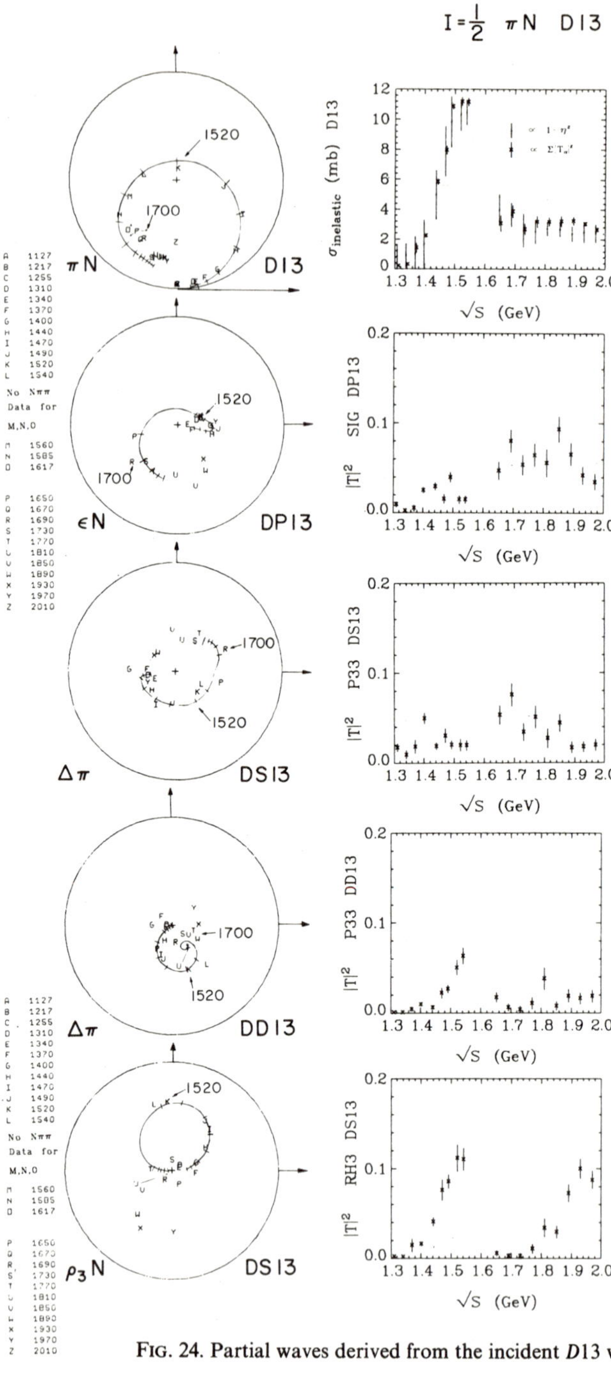

FIG. 24. Partial waves derived from the incident *D*13 wave.

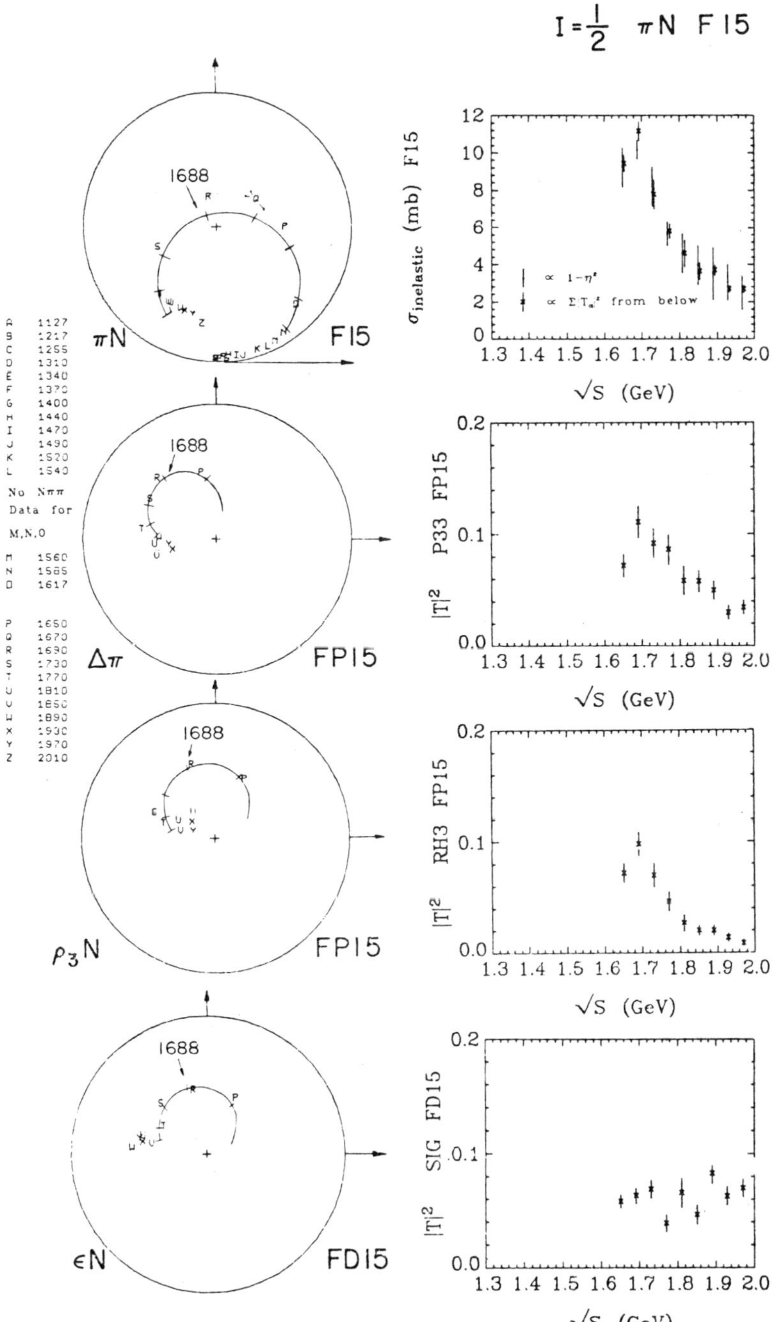

FIG. 25. The F_{15} partial wave.

$D13(1520)$	in $\pi\Delta(DS13$ and $DD13)$ and $N\rho_3(DS13)$ (see Fig. 24)
$D15(1680)$	in $\pi\Delta(DD15)$
$F15(1680)$	in $\pi\Delta(FP15)$, $N\rho_3(FP15)$, $N\varepsilon(FD15)$ (see Fig. 25)
$P13(\sim 1860)$	in $N\rho_1(PP13)$

(iii) $D13(1700)$: This state which has been hinted at in EPSA is definitely present in this analysis (see Fig. 24) decaying strongly into $\pi\Delta(DS13)$ and $N\varepsilon(DP13)$. This state has long been required to complete the $S = 0$ members of the $[70, 1^-]$ supermultiplet of negative baryon states.

(iv) Strong $N\rho$ couplings are observed for the $P13(1860)$, $D13(1520)$ and $F15(1690)$ resonances. This is not surprising as the last two resonances are strongly seen in photoproduction and application of VDM would imply this result.

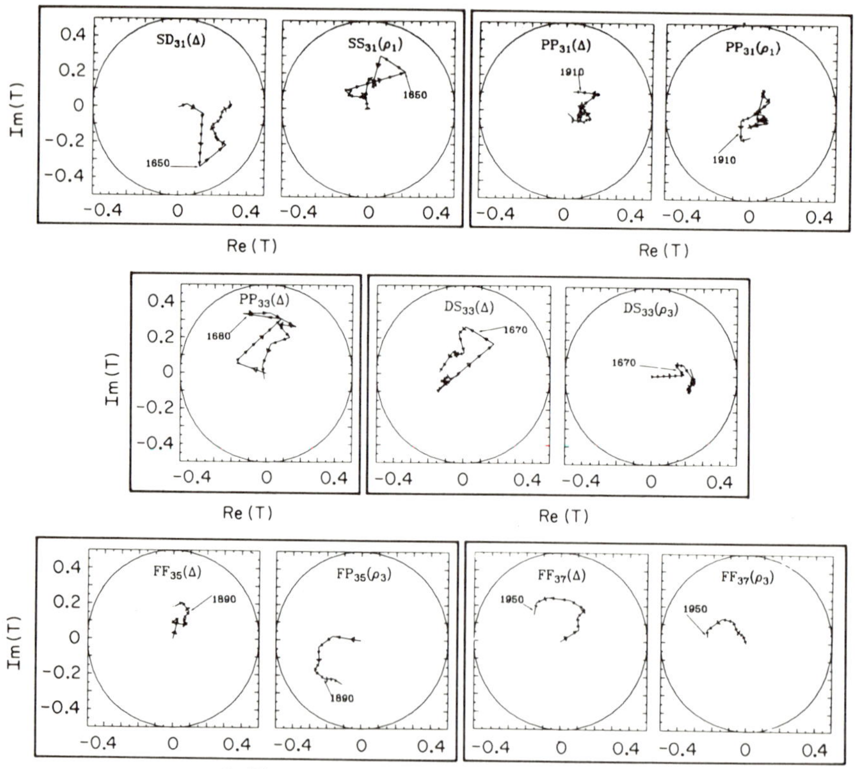

FIG. 26. $I = \frac{3}{2}$ partial wave amplitudes. Arrows are spaced every 20 MeV, with wide arrows every 100 MeV: base of wide arrows mark integral hundreds of MeV. Lower-l waves are plotted starting at $s = 1400$ MeV; higher-l waves only where they were first needed. Last arrowhead is always at 1940 MeV.

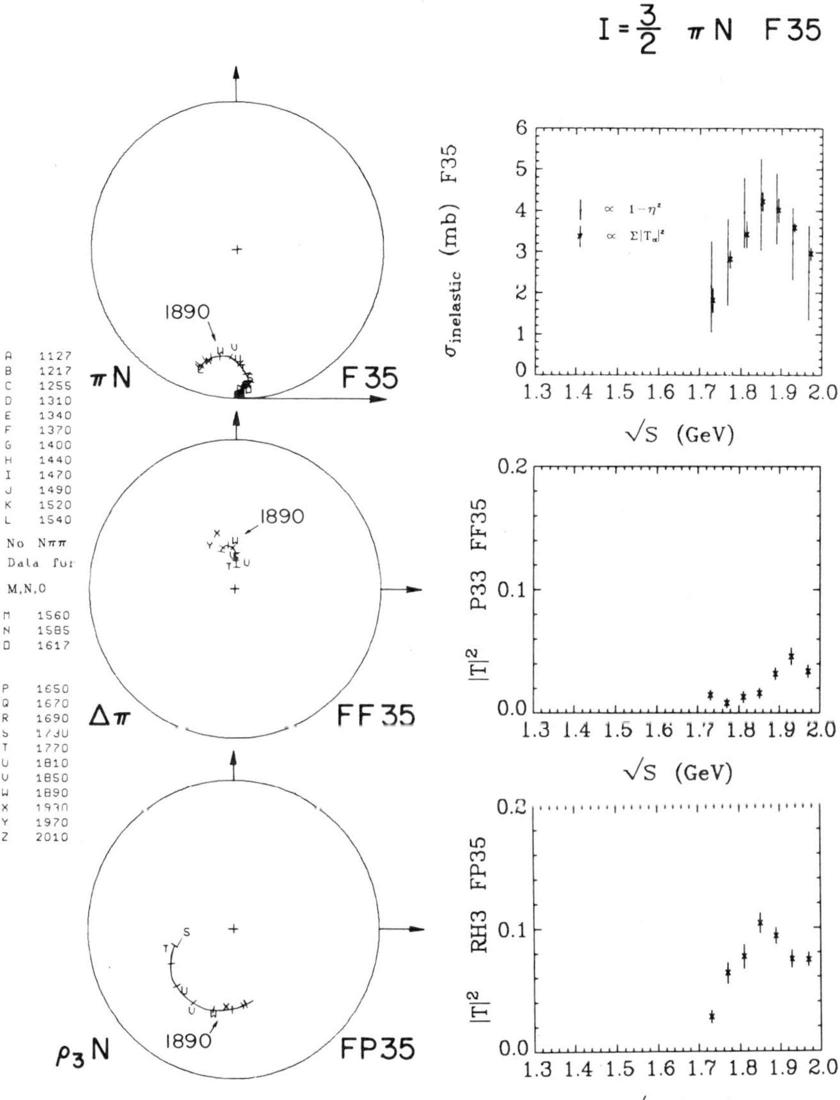

FIG. 27. Partial waves derived from the incident *F*35 wave.

(v) As can be seen in Figs. 23, 24 and 25 the agreement with the EPSA predictions is excellent.

(b) $I = \frac{3}{2}$ *states.* In Fig. 26 are the Argand diagrams of the $I = \frac{3}{2}$ waves.

(i) The following resonances can be clearly seen

*S*31(1600)	in $\pi\Delta(SD31)$ and $N\rho_1(SS31)$	Although only the upper part is observed (since $E_{c.m.} > 1650$ in our analysis)
*D*33(1670)	in $\pi\Delta(DS33)$	
*F*35(1890)	in $\pi\Delta(FF35)$ and $N\rho_3(FP35)$	(see Fig. 27)
*F*37(1930)	in $\pi\Delta(FF37)$ and $N\rho_3(FF37)$	

(ii) There is no evidence for the existence of a *P*33 resonant state within the energy range of the analysis (see Fig. 28).

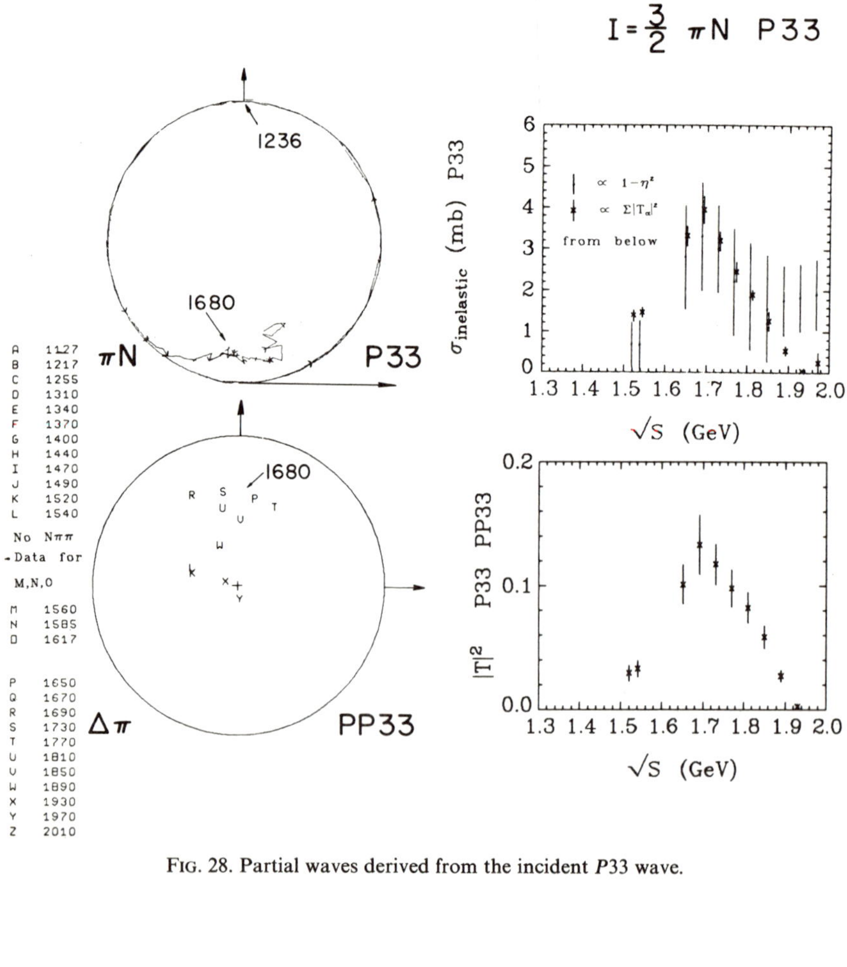

FIG. 28. Partial waves derived from the incident *P*33 wave.

(iii) Again the $N\rho$ couplings are not surprising as these resonances are strongly excited in photoproduction.

(c) *Further comments.* There are some problems in the analysis which can be used as a guide to the physics which may be present.

(i) The $\pi^+\pi^+n$ final state is badly fitted in the analysis. This is not surprising as the only intermediate states which may lead to this channel are $\pi\Delta$ states and from inspection of the π^+n mass projections it is clear that N^* isobars ($P11$, $D13$, $F15$) may be present. The absence of these states in the analysis probably also accounts for the inability to reach the EPSA predictions for the $P31$ and $D35$ waves ($\pi(F15)$ in an S-wave and $\pi(D13)$ in a P-wave would be derived from $D35$ whereas $\pi(P11)$ in an P-wave would come from $P31$).

(ii) The deterioration of the fits at higher energies is generally associated with being unable to fit the peripheral nucleon. This probably indicates the necessity of including π-exchange in the production of the $N\rho$ final state, so that higher partial waves are generated.

(2) *Comparison with theory*

(a) *Classifications.* The observation of the $D13$ resonance at 1700 MeV completes the $[70, 1^-]$ supermultiplet of baryon states. However the absence of any $P33$ states is embarrassing. One such state is required to complete the $[56, L = 2^+]$ while another is required as a partner of the Roper $P11(1470)$ resonance if this is classified within a radial excitation of the $[56, L = 0^+]$ (Dalitz, 1969; Greenberg, 1969; Feynman *et al.*, 1970).

(b) *Decays.* This area has seen most recent activity. In Fig. 29 I have drawn the arrows which correspond to the sign of the couplings of the resonances observed in solution A of the $\pi\pi N$ analysis. Should solution B be satisfactory it would imply changes of sign for the lower $P11$ and $D13$ states (as indicated by dashed arrows). To emphasize this problem I have also separated the low and high energy parts of the analysis. The predictions for these reactions have been derived from two directions.

Phenomenological analysis of baryon decays assuming a broken form of $SU(6)_W$ symmetry (Faiman, 1971; Faiman and Plane, 1972; Rosner, 1972; Faiman and Rosner, 1973)

Applications of results related to Melosh transformations (Melosh, 1973) assuming $SU(6)_W$ classification of the baryon states.

But both lead to essentially the same results. However the second approach does give some indication of why the broken form of $SU(6)_W$ appears.

These theories can be used to predict numerical results but one expects these to be less satisfactory then the prediction of signs, since they are more

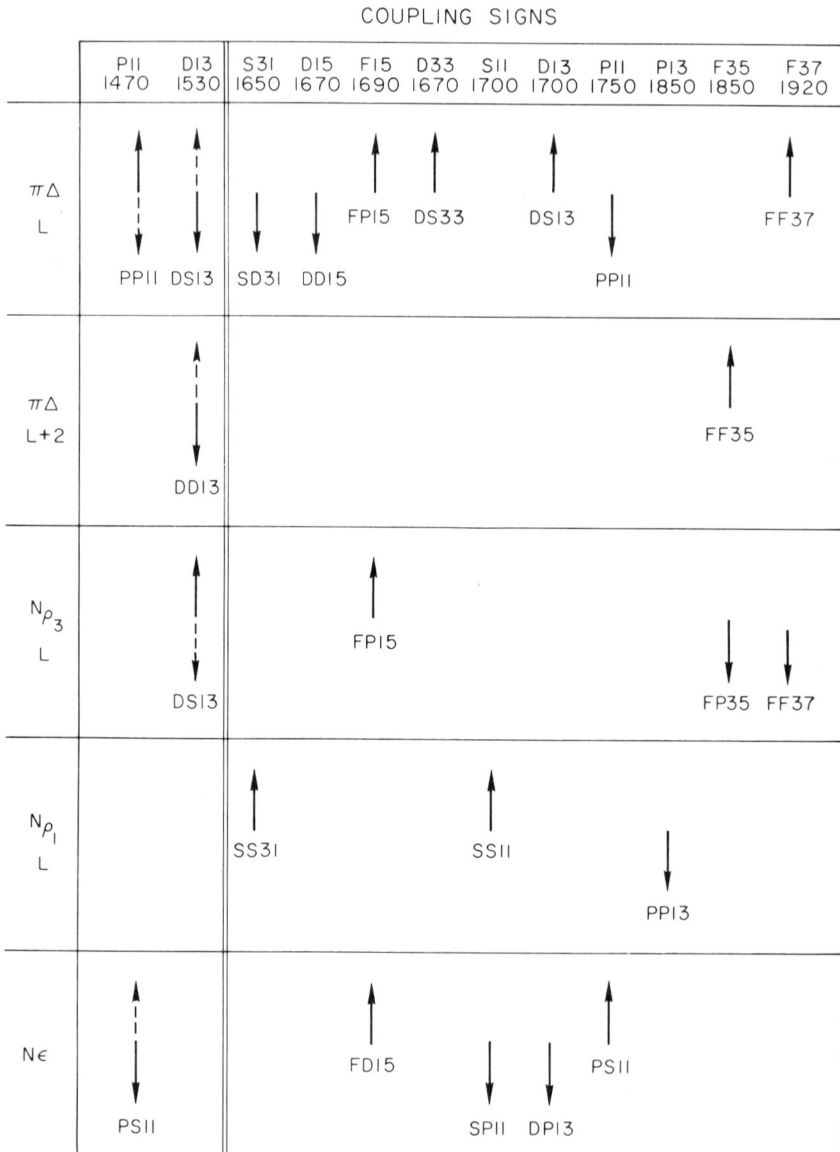

FIG. 29. Relative coupling signs of the resonances in all inelastic channels. The vertical double line separates the two regions, since the continuity is not unambiguously ascertained. The heavy arrows correspond to the present solution A. The dashed arrows will be the orientation if the second solution, B, is satisfactory.

susceptible to the details rather than the overall structure. Thus I will concentrate on the prediction and comparison of the signs.

(i) *Phenomenological analysis.* If one assumes $SU(6)_W$ invariance then the decays of all members of one multiplet into the members of another multiplet are related, i.e.

$$\begin{aligned}\langle H'\lambda'|T|H\lambda_H, M, \lambda_M\rangle \sim & [SU(6)_W \text{ Clebsch–Gordan coefficient}]\\ & \times [SU(3) \text{ Clebsch–Gordan coefficient}]\\ & \times [SU(2) \text{ isospin Clebsch–Gordan coefficient}]\\ & \times [SU(2)\ W \text{ spin Clebsch–Gordan coefficient}]\\ & \times \langle \| T_R \| \rangle \end{aligned} \tag{159}$$

where $\langle \| T_R \| \rangle$ is a reduced helicity matrix element. This is just the Wigner–Eckart theorem. Unfortunately it is already clear that this symmetry must be broken (e.g. for mesons it predicts that in $B \to \pi\omega$ the ω should have $\lambda = 0$ whereas in practice it clearly has $\lambda = 1$. In photoproduction of π's it predicts the helicity $\frac{3}{2}$ couplings of the $D13$ to be zero and these are the dominant couplings.)

As we see eqn (159) implies specific relations between different helicity matrix elements. As we have seen in Sections III.1 and III.3 this means specific relations must exist between the pair of waves allowed in decays to $\pi\Delta$. Thus the manner in which the symmetry is broken is to recast (159) into LS partial wave amplitudes and then assume that these are entirely decoupled, i.e. one uses all the Clebsch–Gordan coefficients of (159) for each partial wave but does not apply the relation between the two waves contained in (159) (Faiman and Plane), 1972; Faiman and Rosner, 1973). This is known as L-broken $SU(6)_W$.

Applying eqn (159) twice, once for the coupling to the initial state and once for the outgoing state, allows a calculation of the sign of the amplitude in, e.g.

$$\pi N \to \pi\Delta \tag{160}$$

In Fig. 30 I give clock diagrams (coupling signs) corresponding to the predictions (Faiman and Rosner, 1973) (for comparison with the SLAC/LBL (Herndon *et al.*, 1972) results all $I = \frac{3}{2}$ predictions should be multiplied by (-1) (Cashmore *et al.*, 1972)). One further point should be explained. $SU(6)_W$ sign means that the two orbital angular momentum states have the sign predicted by $SU(6)_W$ whereas "anti $SU(6)_W$" means that the two amplitudes have exactly the opposite relative sign compared to the $SU(6)_W$ prediction.

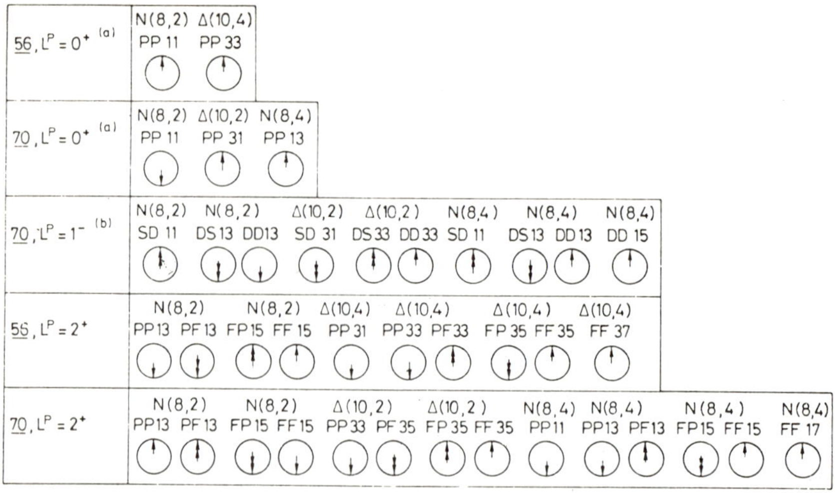

FIG. 30. Prediction of resonant phases in $\pi N \to N^* \to \pi\Delta$ corresponding to a choice of the "anti $SU(6)_W$" relative sign for the two partial waves in the decay. To obtain $SU(6)_W$ results, reverse all double handed clocks. Finally to compare with experimental results all $I = \frac{3}{2}$ waves must be multiplied by (-1) to give the same phase conventions.

In certain cases where incoming and outgoing waves are the same the sign of the amplitude is given just by Clebsch–Gordan coefficients. This immediately indicates a suspicious point of solution A clocks. The relative sign of the $DD15$ and $PP11(1470)$ $\pi\Delta$ decays implies that the $P11$ should belong to a [70] representation of $SU(6)_W$ whereas it is usually assigned to a [56]. Furthermore the relative signs of the $PP11(1470)$ and $DD13(1520)$ $\pi\Delta$ decays would imply that the $P11$ belongs to a [56] providing the $D13(1520)$ belongs to an $\{8\}^2$ of $SU(3) \times SU(2)_W$. Thus there would appear to be a problem in assigning the $P11$. However if solution B is valid these problems are immediately removed since the $P11(1470)$ and $D13(1520)$ amplitudes are changed in sign. However if solution A is the only tenable result we will immediately have to face the fact that $SU(6)_W$ symmetry in any of its broken forms can only be saved by configuration mixing. Not only is this unattractive it also implies the presence of other low lying multiplets. It is important to note that one has to consider $\pi\Delta$ decays to test this symmetry as relations amongst πN, $\Lambda\pi$, $\Sigma\pi$ only have one possible matrix element and we are thus only testing $SU(3)$ symmetry.

At the present moment it appears that the decays of the $[70, 1^-]$ favour anti-$SU(6)_W$ signs whereas the $[56, L = 2^+]$ decays indicate the $SU(6)_W$ sign. This is summarized in Table VIII.

(ii) *Melosh transformations.* The whole phenomenological analysis has

TABLE VIII. Signs of the Amplitudes for $\pi N \rightarrow N^* \rightarrow \pi\Delta$ for N^*'s in the $[70, L = 1]$ and $[56, L = 2]$

	Theory / Resonance	$SU(6)_W$ Sign $(8, 1)_0 - (1, 8)_0$	Anti-$SU(6)_W$ Sign $(3, \bar{3})_1 - (\bar{3}, 3)_{-1}$	Quark Model
$[70, L = 1]$ $[56, \vec{L} = 0]$	$DD13(1520)$	+* (−)	+* (−)	+* (−)
	$DS13(1520)$	− (+)	+ (−)	+ (−)
	$SD31(1640)$	+	−	
	$DS33(1690)$	+	−	
	$DS13(1700)$	+	−	
	$DD15(1670)$	−*	−*	−*
$[56, L = 2]$ $[56, \vec{L} = 0]$	$FP15(1688)$	−	+	+
	$FF35(1880)$	−*	−*	−*
	$FF37(1950)$	−*	−*	−*

Products of the experimental and theoretical signs (in various models) are presented. Signs which are independent of the model are denoted "*". Experiment and theory agree within the $[70, L = 1]$ or within the $[56, L = 2]$ if all the signs in any column are the same.

received important support in recent months for the work of Melosh (Melosh, 1973) and its developments (Gilman *et al.*, 1973; Gilman, 1973).

Many years ago Gell-Mann (Gell-Mann, 1962) proposed that the 16 vector and axial vector charges, $Q^\alpha(t)$ and $Q_5^\alpha(t)$ (integrals over space of the weak and electromagnetic current densities) commute at equal times to form an algebra

$$\begin{aligned} [Q^\alpha(t), Q^\beta(t)] &= if^{\alpha\beta\gamma} Q^\gamma(t) \\ [Q^\alpha(t), Q_5^\beta(t)] &= if^{\alpha\beta\gamma} Q_5^\gamma(t) \\ [Q_5^\alpha(t), Q_5^\beta(t)] &= if^{\alpha\beta\gamma} Q^\gamma(t) \end{aligned} \tag{161}$$

From these charges one can define $Q^\alpha(t) + Q_5^\alpha(t)$ and $Q^\alpha(t) - Q_5^\alpha(t)$ and these two charges commute with each other to give chiral $SU(3) \times SU(3)$. This group can be enlarged to form an $SU(6)_W$ algebra whose elements commute like $SU(3)$ and Dirac matrices: λ^α, $\lambda^\alpha\beta\sigma_x$, $\lambda^\alpha\beta\sigma_y$, $\lambda^\alpha\sigma_z$ [Note $\beta\sigma_x$, $\beta\sigma_y$ and σ_z are the generators of $SU(2)_W$ (Lipkin and Meshkov, 1965).] This algebra is referred to as the *$SU(6)_W$ of currents*. To label the various representations it is convenient to use the $SU(3) \times SU(3)$ subgroup.

$$(A, B)_{s_z} \tag{162}$$

where A is the representation of $Q^\alpha + Q_5^\alpha$, B the representation of $Q^\alpha - Q_5^\alpha$ and s_z is the value of the (current) quark spin component along the z-axis.

From previous comments it appears that the baryon states [and the meson states] lie in irreducible representations of $SU(6)_W \times 0(3)$. This $SU(6)_W$ algebra acts on constituent quarks, baryons being qqq and mesons $q\bar{q}$. Hadrons are simple in terms of this algebra, referred to as $SU(6)_W$ *strong*.

If these two algebras are identified as the same, many undesirable results follow, e.g. $g_A = \frac{5}{3}$, zero anomalous magnetic moment for the nucleon, no magnetic $N \to \Delta$ transition, etc.

It was proposed that there might be a unitary transformation between these two algebras

$$[SU(6)_{W,\,\text{strong}}] = V[SU(6)_{W,\,\text{currents}}]V^{-1} \tag{163}$$

The achievement of Melosh was to suggest a possible form for V motivated by the free quark model.

Instead of applying this transformation to the operators we can apply it to the states. Then

$$|\text{hadron}\rangle = |\text{I.R. constituents}\rangle = V|\text{I.R. currents}\rangle \tag{164}$$

i.e. the I.R. of constituents then becomes a complicated sum of irreducible representations of the $SU(6)_W$ of currents.

In order to apply these ideas to baryon decays we wish to consider hadron′ → hadron + π which is related through PCAC to

$$\langle\text{hadron}'|Q_5^\alpha|\text{hadron}\rangle.$$

Thus

$$\begin{aligned}\langle\text{hadron}'|Q_5^\alpha|\text{hadron}\rangle &= \langle\text{I.R.}'\text{ constituents}|Q_5^\alpha|\text{I.R. constituents}\rangle\\ &= \langle\text{I.R.}'\text{ currents}|V^{-1}Q_5^\alpha V|\text{I.R. currents}\rangle \end{aligned} \tag{165}$$

All we then have to know is $V^{-1}Q_5^\alpha V$ and thanks to Melosh we have a possible form for it. As we have seen Q_5^α itself transforms as $(8, 1)_0 - (1, 8)_0$ and Melosh found in the free quark model that

$$V^{-1}Q_5^\alpha V = R[(8, 1)_0 - (1, 8)_0] + S[(3, \bar{3})_1 - (\bar{3}, 3)_{-1}] \tag{166}$$

This algebraic property can now be extracted and exploited in calculating the decays of (165). Finally in order to complete the calculation we assume that the observed baryon resonances can be identified with I.R. of constituent quarks, and the different quark spin states are related by $SU(6)_W$ of strong interactions. The two terms of (166) can then be sandwiched between different states of the I.R. representations to give any transition matrix element just as in (159). We can immediately note the two parameters of this approach are then related to the different partial wave couplings of the broken $SU(6)_W$ predictions. Indeed the second term corresponds to that breaking.

In Table VIII I reproduce the comparison of signs from Gilman, Kugler and Meshkov (Gilman *et al.*, 1973) with the present solution A. The $[56, L = 2^+] \rightarrow [56, L = 0^+]$ decays have consistent signs and indicate the dominance of the $(8, 1)_0 - (1, 8)_0$ term (in the language of Rosner *et al.*—the $SU(6)_W$ like sign). However the $[70, 1^-] \rightarrow [56, L = 0^+]$ decays are inconsistent with both sets of relations. However if the $D13(1520)$ signs could be reversed there would be total agreement with the $(3, \bar{3})_1 - (\bar{3}, 3)_{-1}$ term (the anti $SU(6)_W$ of Rosner *et al.*). This is just the result that would occur should the solution B, at present under study, prove to be tenable. (The changes are recorded in brackets in Table VIII.

I can only stress that these comments would be unnecessary if data in the "gap" could be analysed. This is just the region which links the 1500 MeV and 1700 MeV groups of resonances.

(iii) *SU*(3) *comparisons.* With improved analyses of the $S = -1$ states it will eventually become important to relate

$$\left.\begin{array}{ll} N^* \rightarrow \pi\Delta(1235) & Y^* \rightarrow \pi\Sigma(1385) \\ & \quad\rightarrow \bar{K}\Delta(1235) \end{array}\right\} \tag{167}$$

$$\left.\begin{array}{ll} N^* \rightarrow N\rho & Y^* \rightarrow \Lambda\rho, \Sigma\rho \\ & \quad\rightarrow N\bar{K}^* \end{array}\right\} \tag{168}$$

and test the $SU(3)$ symmetry in these decays.

(iv) *Quark model comparisons.* At the present time quark model predictions for the photoproduction of π mesons from nucleons agree quite well with experiment (Moorhouse and Oberlack, 1973). The same dynamical arguments can be used to predict the $\pi\Delta$ decays of resonant states (Moorhouse and Parsons, 1973) (an essentially parameter free calculation).

Table VIII also contains the results of the comparison of these calculations with the results of the present experimental solution A. The same malady exists in that the $D13(1520)$ signs are again wrong compared to the others—of course this problem would be removed if the $D13(1520)$ signs could be reversed as we have noted before. However, one further discrepancy is present and that is in the sign of the $FP15$ $\pi N \rightarrow \pi\Delta$ transition and this indicates a problem with the detail calculations of the quark model interactions.

(c) *Multichannel analyses.* The ability to account for all of the πN inelasticity in many partial waves means that we are in a position to perform multichannel fits exploiting the constraints of unitarity to the fullest extent in attempting to understand the πN interaction (Cashmore, 1973).

The K-matrix formalism (Dalitz, 1962) is a convenient way of assuming unitarity in these analyses. However the K-matrix parameters obtained often have ridiculous values, i.e. they correspond to resonances having widths of ~500 MeV where one can see that this is not the case from inspection of the Dalitz plot. This problem probably results from the fact that one is not necessarily using the correct parametrizations for resonances and

TABLE IX. Pole Parameters and Γ_{partial} from Pole Residues

Wave	Pole	$\Gamma_{\pi N}$	$\Gamma_{\pi\Delta_L}$	$\Gamma_{\pi\Delta_L'}$	$\Gamma_{N\rho_3}$	$\Gamma_{N\rho_1}$	$\Gamma_{N\varepsilon}$	Other Channel	$\Gamma_{\text{tot}} = \Sigma\Gamma_i$
$S11$	1 $1498 - i\frac{66}{2}$	11				11	34	14 (ηN)	70
	2 $1648 - i\frac{103}{2}$	46				12	4	32 (ηN)	94
$P11$	1 $1383 - i\frac{200}{2}$	80	40				5		125
	2 $1724 - i\frac{290}{2}$	120	37				75		232
$P13$	1 $1728 - i\frac{159}{2}$	25				84			109
$D13$	1 $1515 - i\frac{143}{2}$	90	38	15	34		3		180
	2 $1646 - i\frac{114}{2}$	16	22	3	3		62		108
$D15$	$1666 - i\frac{159}{2}$	69	92						161
$F15$	$1672 - i\frac{155}{2}$	101	12		42		21		182
$S31$	$1605 - i\frac{59}{2}$	17	11			35			63
$D33$	$1650 - i\frac{150}{2}$	13	56		56				125
$F35$	$1824 - i\frac{282}{2}$	44		26	107				177
$F37$	$1866 - i\frac{255}{2}$	111	76		57			132 (junk)	380

background. However the *K*-matrix parametrization does give a "good" representation of the Argand diagram, i.e. of the *T*-matrix elements. In order to identify resonances and properties we now search the *T*-matrix for poles in the complex energy plane. The motivations for this procedure are

(i) pole positions and residues appear to be unique irrespective of the parametrization of the T-matrix, providing it is good (Lasinski and Barbaro-Galtieri, 1972; Ball *et al.*, 1972; Longacre, 1972)

(ii) we expect the pole position and residues to be closely related to the Breit-Wigner parameters. However the pole position does not equal (M_0, $i/2\Gamma_0$), the parameters of the Breit–Wigner in the physical region, and the residues are not necessarily equivalent to the partial widths. We expect these equalities to become very poor when we have either large backgrounds or wide resonances.

The results of these investigations are summarized in Table IX and correspond to the present solution A for $\pi N \to \pi\pi N$. One can observe that many resonance parameters are substantially different from the results of naive estimations and indeed this whole problem of resonance parameter extraction is a thorny one. However any detailed theory must predict T-matrices which should contain this pole structure.

Finally I might add a note of warning—before using any resonance parameters always review their origin. The estimates can be very different.

(d) *Dynamical calculations.* Attempts to understand the importance of local duality in the region 1500–1700 MeV for $\pi N \to \pi\Delta$ have been made (Kernan and Shepherd, 1969) but unfortunately that analysis used preliminary results of

$$\pi^- p \to \pi^+ \Delta^-$$

partial wave analysis which has been superseded. It is not clear that these results will still hold. When a final partial wave set is available that exercise will be repeated.

It will also be possible to attempt to correlate the predictions of π-exchange (a real amplitude) with the present partial waves and also study the relation of photoproduction amplitudes with the vector meson decays.

As one can see the results of such analyses are very rich in valuable data and clearly represent the next step in understanding strong interactions, at least in this energy region.

2. $K^- p \to \Lambda\pi\pi$

The only reliable results on this reaction have been obtained in the region of the $Y_0^*(1520)$ ($D03$) in an effort to make detailed measurements of its branching ratios (Mast *et al.*, 1972). This state is predominantly an $SU(3)$ singlet. If it were entirely such an object it would be forbidden to decay into $\pi Y_1^*(1385)$ since an $\mathbf{8} \times \mathbf{10}$ coupling does not contain a 1. This state is then a mixture of singlet and octet (there is a close by octet state, the $D03(1690)$)

$$|1520\rangle = \cos\theta|1\rangle + \sin\theta|8\rangle \tag{169}$$

$$|1690\rangle = -\sin\theta|1\rangle + \cos\theta|8\rangle \tag{170}$$

The value of this mixing angle can be estimated from the Gell-Mann–Okubo mass relation and from the two body decays $\bar{K}N$ and $\Sigma\pi$. The result is that $\theta = (21 \pm 5)^\circ$ (Plane *et al.*, 1970).

If we now consider the $\pi Y_1^*(1385)$ decays we see that the decay rates are simply related by

$$\frac{\Gamma(1520)}{\Gamma(1690)}\frac{PS(1690)}{PS(1520)} = \tan^2\theta \tag{171}$$

where PS represents the phase space in the decay. The value of θ from this relation should be the same as above but unfortunately the result indicates $\theta > 50^\circ$. We might expect this to be disastrous. However inspection of the $[70, 1^-]$ supermultiplet indicates yet another $D03$ Y_0^* state. If we now require mixing between all three states consistent results can be obtained (Faiman and Plane, 1972). This result then becomes an important factor in the description of the $[70, 1^-]$ decays, which were mentioned in the previous section.

At present it is important to stress the value of this type of analysis in K^-p reactions. Just as we are understanding the N^* situation with the advent of this new data, the Y^* region would be similarly helped and indeed our whole view of the symmetries of strong interactions.

3. $Kp \to K\pi N$

The obvious reason for studying this type of reaction is to identify exotic Z^*'s. The analyses so far performed have been of the "selection by cuts" type (see Section II.A). The reaction studied primarily is

$$K^+p \to K^0\Delta^{++} \tag{172}$$

and the results have given no indication of such resonant states with $I = 1$ (Bland *et al.*, 1969).

However at this point the elastic phase shift analyses of the $I = 0$ system are much more indicative of a resonant state at $E \sim 1800$ MeV—a Z_0^*. Again it would be desirable to study this possible state in the inelastic channels. Fortunately this is much simpler as the decay into $K\Delta$ is forbidden. At present this looks the most promising candidate for a Z_0^*.

As I mentioned briefly in the introduction to this section the existence of such states can be predicted in certain dynamical models. These consist of models in which there is strong π exchange in the t-channel leading to a strong K^*N cross-section. Under such circumstances this state can be a major factor in the intermediate states in the $KN \to KN$ amplitude (Aaron *et al.*, 1969; Aaron *et al.*, 1970) and the isospin crossing matrix then leads to much greater effects in the $I = 0$ KN system. Thus the observation of such a resonance would not be entirely disastrous to the quark model and moreover it would raise the question of just how frequently this type of situation occurs. Could many of the resonances, we now assume are elementary particles, have such an origin?

V. The Results from Production Reactions

As we have already discussed these isobar methods can be applied to the three body final states obtained in production reactions. The efforts have been almost exclusively limited to the three meson systems obtained in production reactions. These are grouped under the names $A_1(1^+)$, $A_2(2^+)$, $A_3(2^-)$, $Q(1^+)$, $L(2^-)$ and are central to the meson classification schemes but confused by the presence of diffraction dissociation. The analyses have been mainly of the type considering just Dalitz plot populations or moments ($\langle Y_L^M \rangle$) of the normal to the three particle plane. The truly complete isobar analysis has only been applied by the group at Illinois (Ascoli *et al.*, 1971; Ascoli, 1972; Illinois *et al.*, 1972) and more recently by the SLAC/LBL group (Lasinski *et al.*, 1973).

The Illinois group analysed almost exclusively the reactions

$$\pi^- p \to \pi^- \pi^- \pi^+ p \tag{173}$$

where the 3π system is isolated by considering only events in which the four momentum transfer from incident proton to outgoing proton is comparatively small ($t \lesssim 1{\cdot}0$) and none of the pions resonate with the proton to give a Δ state. The questions that one would like to answer in such an analysis are

(a) What are the spin parity states present as a function of mass?
(b) What are the production density matrix elements?
(c) What are the t-dependences of the processes and the production mechanisms of the states?
(d) What are the s-dependences of the production mechanisms of the states?

I will try to summarize the answers one has to these questions using mainly the results of the Illinois group (Ascoli *et al.*, 1971; Ascoli, 1972; Illinois *et al.*, 1972) (the SLAC/LBL results, although at an early stage in general support their conclusions). I will first tackle the 3π system and then comment briefly on the $K\pi\pi$ and $N\pi\pi$ situations.

1. THREE PION FINAL STATES $\pi\pi\pi$

In order to have sufficient statistics to perform the analyses the Illinois group and their collaborators have gathered together essentially three sets of data

(a) 5, 7, 7·5 GeV/c $\pi^- p \to \pi^-\pi^+\pi^- p$ ~30 000 events

(b) 11–25 GeV/c $\pi^- p \to \pi^-\pi^+\pi^- p$ ~15 000 events

(c) 40 GeV/c $\pi^- p \to \pi^-\pi^+\pi^- p$ ~13 000 events

(a) and (b) come from bubble chambers and (c) from the CERN boson spectrometer during its stay at Serpukhov. In discussing the results I will not always specify the exact source. An example of the 3π invariant mass distribution is given in Fig. 31.

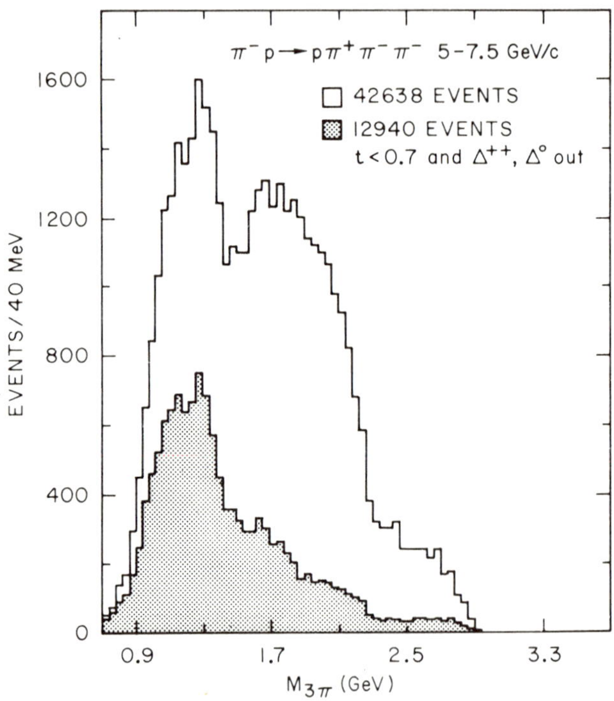

FIG. 31. The $\pi^+\pi^-\pi^-$ mass spectrum for the data at 5·0–7·5 GeV/c. The shaded histogram is for $t' < 0{\cdot}7$ GeV^2, $M_{p\pi^+} < 1{\cdot}4$ out, $1{\cdot}16 < M_{p\pi^-} < 1{\cdot}32$ out.

Unfortunately the data are not yet sufficient to allow analysis in both mass and t bins so that the variable of least importance is usually integrated over. Thus for spectroscopic studies one is mainly concerned with the mass dependence and hence t is integrated over, while for discussions of production mechanisms (dynamics) the most important variable is t and larger slices of mass are usually considered.

(1) *Spin parity states*

In Fig. 32 I show the intensities associated with the 0^-, 1^+, 2^+, 2^- and 3^+ waves. Three features are immediately obvious:

(a) the 1^+ enhancement around $\sim$1100 MeV – the A_1
(b) the 2^+ enhancement at $\sim$1300 MeV – the A_2
(c) the 2^- enhancement at $\sim$1700 MeV – the A_3

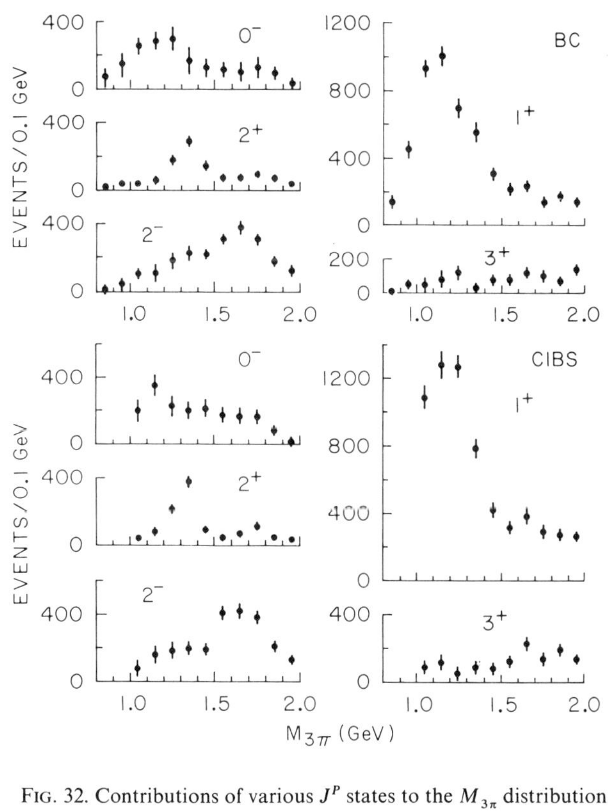

FIG. 32. Contributions of various J^P states to the $M_{3\pi}$ distribution.

Clearly it is important to know the different isobar (two body) states which contribute to this cross section. The 0^-, 1^+ and 2^- waves are shown in Figs. 33, 34, 35 and have contributions from the $\pi\rho$, $\pi\varepsilon$ (0^-, 1^+) and $\pi\rho$, $\pi f(2^-)$ states. The 2^+ wave only has contributions from the $\pi\rho$ system. The 1^+ wave is dominated by $(\pi\rho)$ in an s-wave whereas the 2^- is predominantly (πf) in an s-wave. This result has been known for some time as has the fact that Deck diagrams of the type of Fig. 36 give s-wave enhancements near threshold. One might hope to identify the 1^+, 2^- states as resonant by observing large rapid phase changes of these waves with respect to other partial waves. Unfortunately no such motion is apparent, e.g. in Fig. 37 we see the phase of the 2^- πf wave relative to other waves through the A_3 region. No resonant

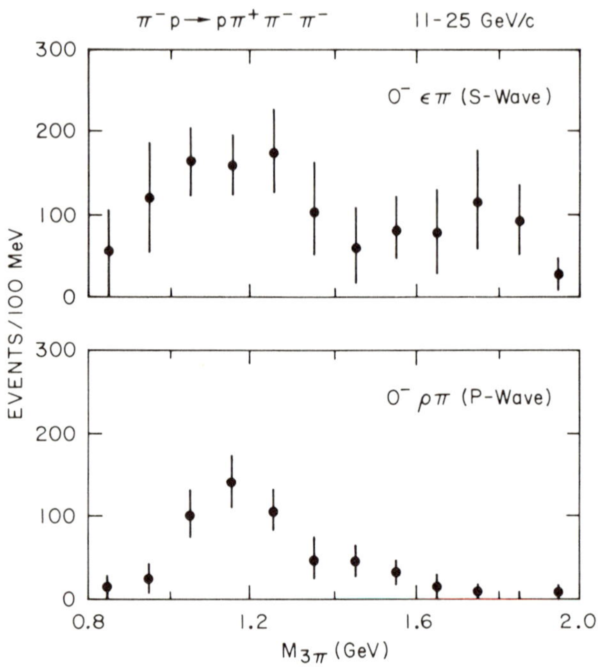

FIG. 33. Individual partial wave contributions to the 0^- state.

phase variation is observed. Figure 38 demonstrates the same result for the 1^+ $(\pi\rho)$ s-wave in the region of the A_1. These results thus give little encouragement for the identification of these states as resonances. However the presence of small resonant effects on a large background are not ruled out (this could produce little phase variation).

Of course it is always desirable to demonstrate when one can observe a "good" resonance. This is done for the A_2 in Fig. 39. There one sees the Breit–Wigner shape in the 2^+ $(\pi\rho)$ d-wave together with correct phase variation across the resonance.

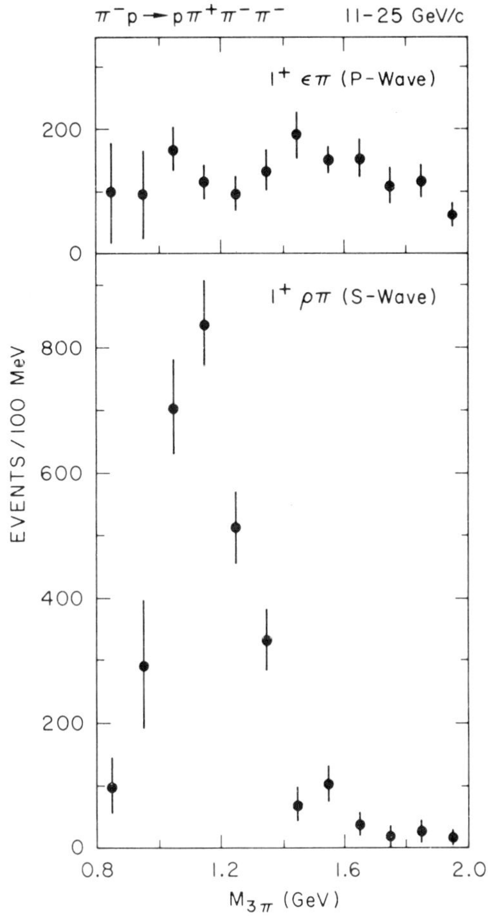

FIG. 34. Individual partial wave contributions to the 1^+ state.

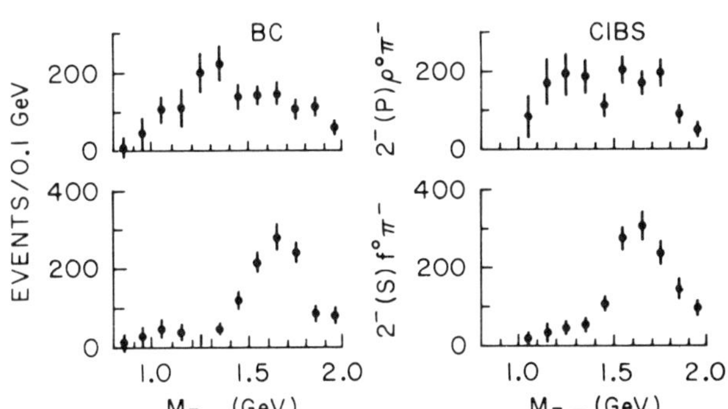

FIG. 35. Individual partial wave contributions to the 2^- state.

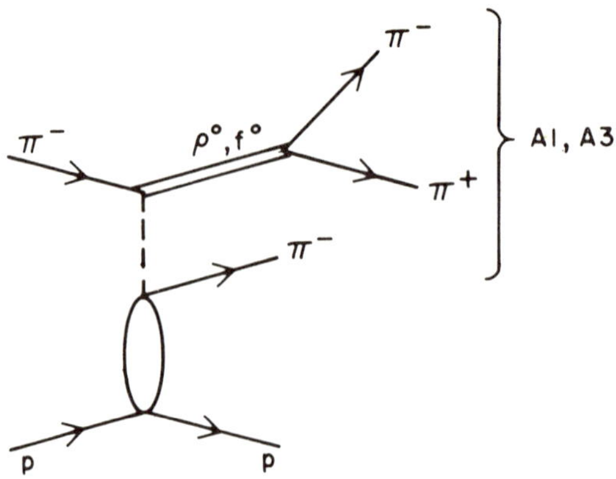

FIG. 36. The Deck mechanism for the A1 and A3.

Thus we can conclude that the only resonance positively identified is the 2^+ ($\pi\rho$) state, the A_2, and the 1^+ and 2^- waves although large and changing in intensity give little support for resonance interpretations.

(2) *Density matrix elements*

In the last two years the question of *s*-channel helicity conservation has

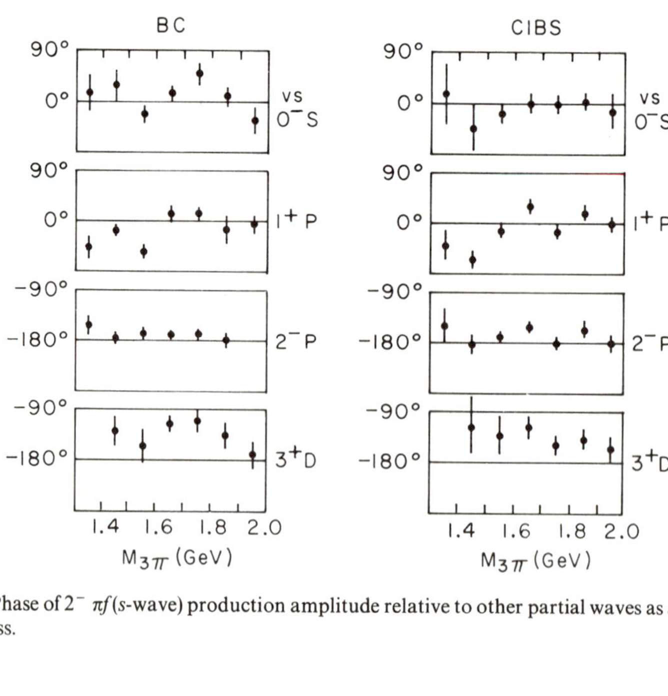

FIG. 37. Phase of 2^- πf (*s*-wave) production amplitude relative to other partial waves as a function of 3π mass.

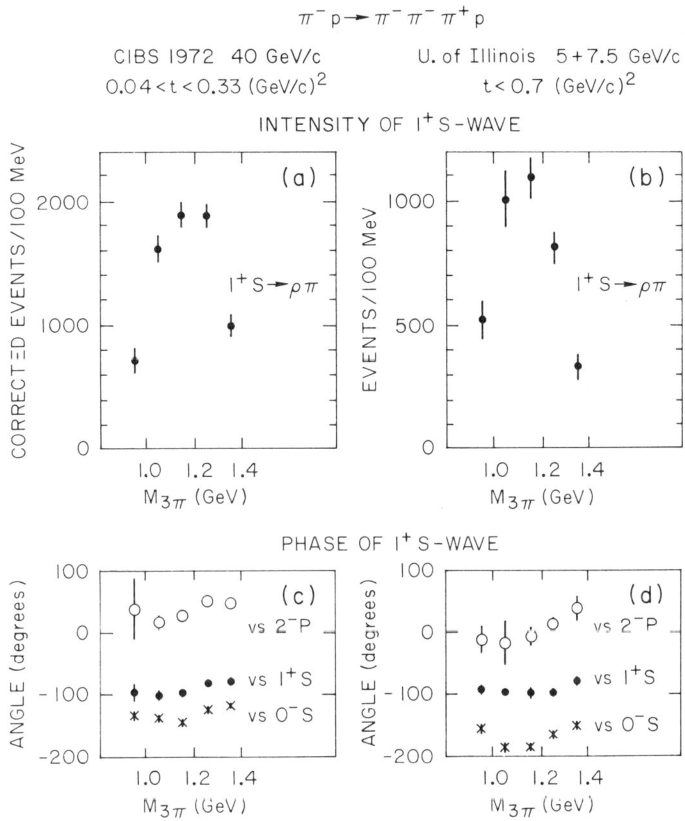

FIG. 38. The $1^+\pi\rho$ (s-wave). Its intensity and phase relative to other partial waves as a function of 3π mass.

received a great deal of attention. It appears to be essentially conserved for reactions (Gilman *et al.*, 1970)

$$\pi N \to \pi N \tag{174}$$

$$\gamma N \to \rho^\circ N \tag{175}$$

However these reactions do not involve spin changes whereas excitation of the A_1 and Q enhancements does ($0^- \to 1^+$). Thus attention has been focused in the density matrix elements of the 1^+ state measured both in the t-channel (Gottfried–Jackson) and s-channel helicity systems (see Section III.4 for definition of these). In Fig. 40 one sees ρ_{00} evaluated in the t-channel where it is ~0·9–1·0. This indicates predominantly t-channel helicity conservation although the existence of Re ρ_{01} indicates that some $M = \pm 1$ production occurs. However these density matrix elements evaluated in

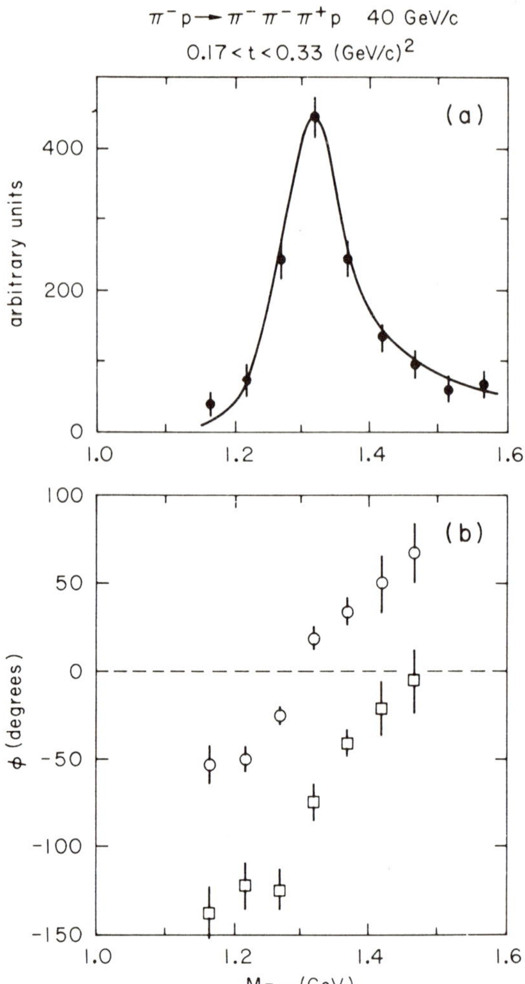

FIG. 39. The $2^+\pi\rho$ (d-wave). Its intensity and phase variation relative to the 1^+ waves as a function of 3π mass.

the s-channel system would give values of $\rho_{00} \sim 0{\cdot}4$, Re $\rho_{01} \sim 0{\cdot}3$ at the higher t value. Thus exact helicity conservation does not occur in either system but it is closer to being true in the t-channel.

It should also be noted that the 2^- system is also produced predominantly in the $M = 0$ substate.

In contrast the A_2 is produced almost entirely in the $M = \pm 1$ states.

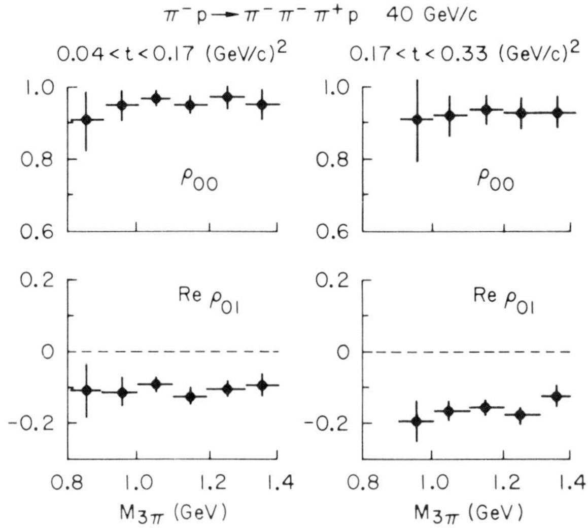

FIG. 40. The density matrix of the 1^+ state.

In fact $\rho_{11} \sim \rho_{1-1}$ indicating the production of a pure state $|21\rangle + |2-1\rangle$. The near equality of these two density matrix elements indicates that natural spin parity exchange dominates (Ader *et al.*, 1968) since

$$\sigma_{\text{natural}} \propto \rho_{11} + \rho_{1-1}$$
$$\sigma_{\text{unnatural}} \propto \rho_{11} - \rho_{1-1} \tag{176}$$

This result is expected if the reaction occurs through f^0 or ρ^0 exchange.

(3) *Production mechanisms and t-dependence*

The dominance of natural parity exchange in the production of A_2 suggests the presence of f^0 and/or ρ^0 exchange terms. However by comparing A_2 production in the reactions

$$\pi^- p \to A_2^- p$$
$$\pi^+ n \to A_2^0 p \tag{177}$$

it is apparent that f^0 exchange dominates. Further if the coupling is spin non-flip at the proton vertex (as we guess from elastic scattering) then there will be a net helicity flip leading to zero cross-section in the forward direction, i.e. at $t' \sim 0$. This is demonstrated in Fig. 41.

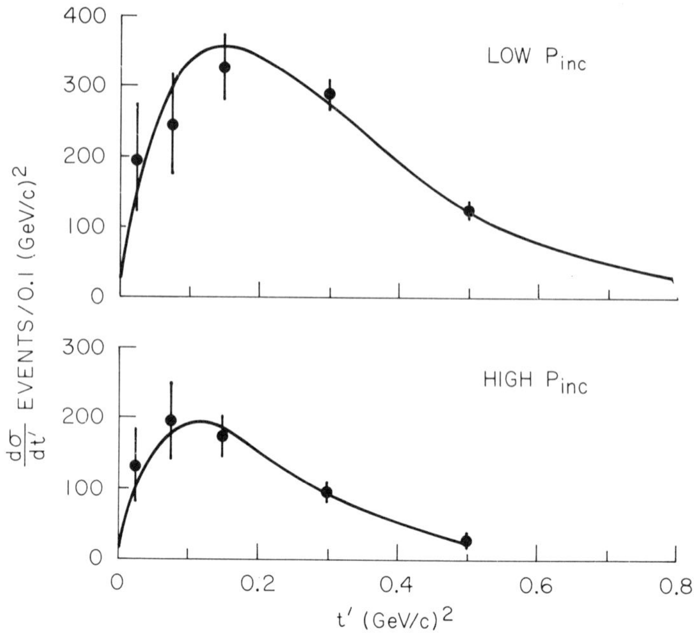

FIG. 41. Production angular distribution of spin parity 2^+ events. Curves are fits of the form $t' e^{-Bt'}$ with $B = 6{\cdot}41$ and $B = 9{\cdot}04$ for low (5–7·5) and high (11–25) incident π momenta.

(4) *s-Dependence of cross-sections*

The ratio of A_1 to A_2 to A_3 cross-sections is surprisingly energy independent even up to the 40 GeV/c. If these cross-sections are parametrized as

$$\sigma \propto p_{\text{lab}}^{-n} \, (\sim s^{-n}) \tag{178}$$

then the value of the exponent is

$$A_1: n \sim 0{\cdot}5 \pm 0{\cdot}2$$

$$A_2: n \sim 0{\cdot}57 \pm 0{\cdot}08$$

$$A_3: n \sim 0{\cdot}8 \pm 0{\cdot}3$$

Furthermore if one looks at the unnatural and natural parity exchange contributions to the A_2 cross-section, Fig. 42, one sees that it is the unnatural parity contribution that is falling extremely fast.

The fact that the A_2 cross-section does not appear to drop any more rapidly than the A_1 has caused speculation that it may be produced by Pomeron exchange in violation of the empirical rule

$$\Delta P = (-1)^{\Delta J} \tag{179}$$

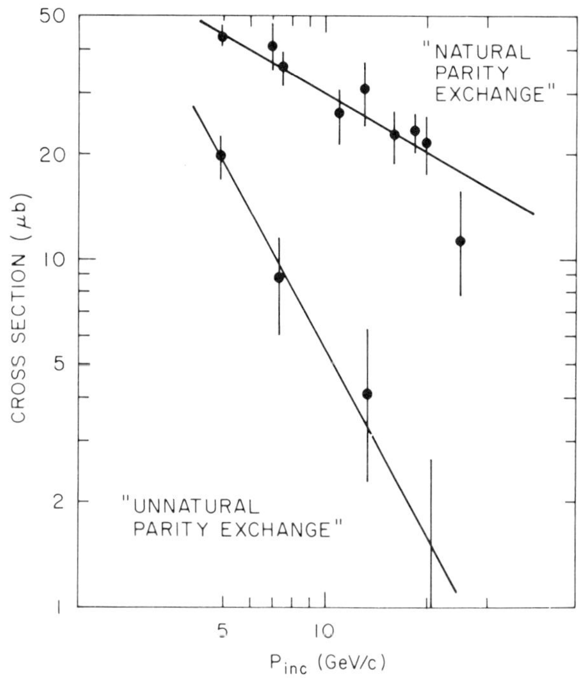

FIG. 42. Cross-sections as a function of incident momentum for production of 2^+ by "natural parity exchange" and "unnatural parity exchange".

At this time this can only be speculative and awaits more detailed measurements.

Finally one can attempt to extract information from the relative phases of the A_2 and A_1 waves. If we write

$$\frac{\rho_{A_2A_1}}{\sqrt{(\rho_{A_1A_1})}} \propto \frac{1}{m_{A_2}^2 - m_{3\pi}^2 - im_{A_2}\Gamma_{A_2}} \bar{\rho}_{A_2A_1} \tag{180}$$

then we might expect $\bar{\rho}_{A_2A_1}$ to be constant in phase and magnitude if the A_1 phase does not change (independent of $m_{3\pi}$). This will then represent the relative phase δ of the A_2 and A_1 production amplitudes ($\delta = \phi_{A_2} - \phi_{A_1} = \arg(\bar{\rho}_{A_2A_1})$). The results indicate a slight variation of δ with c.m. energy, $\delta \sim -46^\circ$ at 5 GeV/c and $\sim -75^\circ$ at 40 GeV/c but it does appear to be independent of t'. Now if one assumes the A_1 is produced through the Deck mechanism, Fig. 36, its production phase is that of πp diffraction scattering, i.e. pure imaginary, $\phi_{A_1} = 90°$. This then implies

$$\phi_{A_2} \sim 44° \text{ at } 5\,\text{GeV}$$
$$\sim 15° \text{ at } 40\,\text{GeV}.$$

These phases are clearly inconsistent with Regge model predictions

(a) pure f^0 exchange $\rightarrow \phi_{A_2} \sim 145°$ ($t' \sim 0{\cdot}15$)

(b) pure P exchange $\rightarrow \phi_{A_2} \sim 90°$

(c) mixture of f^0 + Pomeron exchange with Im (f^0 exchange) $> 0 \rightarrow \phi_{A_2} \sim 120°$

However if we have

(d) f^0 + Pomeron with Im (f^0 exchange) < 0

agreement is possible but the energy dependence of ϕ_{A_2} on p_{lab} in this latter case would be opposite to the observation.

We can summarize these results by the statement that we appear to have problems with our interpretation of the A_2 production mechanism.

(5) *Summary*

This will have given a feeling for the great quantity of information that has been extracted from the analysis of these three pion production reactions, both spectroscopic and dynamical (production mechanisms). However, I think it is true to say that the hoped for goal of proving a resonance contribution in the A_1 has not been reached. Clearly at least a large proportion of the 1^+ enhancement is dynamic in origin (Deck mechanism) and it is below the sensitivity of these experiments to detect a resonant contribution in this environment. The only hope is to consider reactions in which diffraction dissociation is impossible for the identification of this state, e.g.

$$K^- n \rightarrow \pi^- \pi^+ \pi^- \Lambda^0$$
$$\pi^+ n \rightarrow \pi^+ \pi^- \pi^0 p \tag{181}$$

Many of these comments apply equally well to the case of the 2^- A_3 enhancement.

These analyses have increased interest in the production mechanism of the A_2, the contribution of Pomeron exchange remaining a tantalizing possibility. Measurements of the $K\bar{K}$ decay of the A_2, if measured with a good absolute normalization, should help resolve this problem.

However it does appear as though further investigation of the 3π system in these reactions will not be terribly rewarding and new information on the resonant states will come from reactions such as (181).

2. $K\pi\pi$ FINAL STATES

Here the situation is more complicated than for 3π's. We expect (from the $q\bar{q}$ model) *two* 1^+ mesons, the Q's, the $K^*(1400)$ and a similar 2^- enhancement, the L. At this time there are no published results from an isobar model analysis. Analyses have concentrated around the Dalitz plot populations, πK^* decays obtained by cuts, and general angular momentum considerations (see Section II.2). The main results can be summarized as follows

(a) 1^+ waves are dominant in the mass region $1{\cdot}0 < M(K\pi\pi) < 1{\cdot}4$ with decays into both πK^* and $K\rho$. In fact the $K\rho$ decay is strongest near threshold.
(b) In this reaction t-channel helicity conservation is ruled out.
(c) Comparison of Q and $\bar{Q}$ states in

$$K_L^0 p \to K_S^0 \pi^+ \pi^- p \tag{182}$$

indicate the probable presence of other contributions to the production mechanism besides Pomeron exchange (diffraction dissociation).
(d) Varying production mechanisms of two interfering 1^+ states could explain the changes in $K\pi\pi$ spectrum that are observed.

The application of isobar model analyses to reactions of the type

$$\begin{aligned} K^\pm p &\to K^\pm \pi^+ \pi^- p \\ &\to \left.\begin{matrix} K^0 \\ \bar{K}^0 \end{matrix}\right\} \pi^\pm \pi^0 p \end{aligned} \tag{183}$$

will clearly result in advances similar to those in the 3π system. However the problem is more complicated because of the increased number of decay channels which must be considered.

3. $N\pi\pi$ FINAL STATES

There is even less to say about these states, the work being little and the results few. The nucleon having spin $\frac{1}{2}$ makes the problem somewhat more complicated and yet again increases the number of partial waves. Some Dalitz plot analyses have been performed and result in the statements that $J^P = \frac{1}{2}^+$ dominates in the region of 1200–1400 MeV, $J^P = \frac{3}{2}^-$ is present around 1500 MeV and $\frac{5}{2}^+$ large at $\sim$1700 MeV. As we have seen in the analysis of $\pi N \to \pi\pi N$ these $P11$, $D13$, $F15$ states are important at these masses and are the very states we expect in diffraction excitation if the $\Delta P = (-1)^{\Delta J}$ rule holds.

Application of the isobar model approach is clearly going to bring about a dramatic improvement. Furthermore these would seem to be the obvious states in which to study diffraction dissociation as the formation reactions give us a clear bench mark.

VI. Summary and Conclusions

The isobar model analysis gives an excellent phenomenological description of the data on three body final states both in formation and production reactions.

The major problems in the analysis of

$$\pi N \rightarrow \pi\pi N \tag{184}$$

is associated with the lack of results from the data that exist in the energy range $1540 < E < 1650$. The measurement of the $\pi^0\pi^0 n$ final state together with experimental results on single pion production from polarized targets will be sensitive tests of the present partial wave amplitudes. However, problems do already exist in the $\pi^+\pi^+ n$ final state and at the higher energies, and any future analysis will have to introduce both $I = \frac{1}{2}$ isobar intermediate final states and π-exchange contributions. The wealth of information one has already gained suggests that such an effort will be fruitful. This initial information has already been valuable in discussing higher symmetry schemes for the strong interaction in the resonance region. The existence of the $D13(1700)$ is vital to $SU(6)_W$ schemes but the absence of the $P33$ states is a nagging embarrassment. This is just the beginning of the theoretical activity that should now occur in an effort to reproduce the essentially complete T-matrices that we obtain.

In the future one might expect to see similar analyses of

$$\begin{aligned} \bar{K}N &\rightarrow \Lambda\pi\pi \\ &\rightarrow \Sigma\pi\pi \end{aligned} \tag{185}$$

where the weak decay of the final baryon gives a good description of its polarization. The Y^* situation needs new information to unravel the exceedingly complicated set of states which should be present (for every octet and decuplet there must be Y_1^* states while Y_0^* states are singlets and members of octets). The advent of this information will then allow tests of the $SU(3)$ content of any theory we have for these three body reactions.

The analyses of the positive strangeness KN induced reactions is going to be absolutely vital in testing the validity of quark model schemes. However the preliminary results from analyses of reactions

$$K^+p \rightarrow K^0\Delta^{++} \tag{186}$$

give little indication of $I = 1$ resonant structure and I do not believe any different conclusions will appear from the more sophisticated isobar analysis. The more intriguing situation is in the $I = 0$ state where there is not only a candidate for a Z_0^* at $\sim$1800 MeV but dynamical calculations, considering π exchange in the t-channel, do give poles in the s-channel. The Z_0^* system is more likely to contain resonances from this type of mechanism due primarily to the fact that the $s - t$ channel isospin crossing matrix predicts the strongest effects in the $I_S = 0$ state from isospin 1 exchange in the t channel (π) to make the K^*N intermediate state (Aaron *et al.*, 1969; Aaron *et al.*, 1970). Thus the detailed comparison of observed T-matrix elements with those predicted by theory will be a sensitive test of the validity of such calculations. Indeed this whole question is a somewhat sensitive one—every new resonant state that we find is automatically regarded as a new elementary particle and classified as such. However resonances may occur through an accident of the size of inter-particle forces and the proximity of thresholds, e.g. is the $Y_0^*(1405)$ a bound state of the $\bar{K}N$ system just as the deuteron is a bound state of the $^3S_1$ pn interaction? Such questions (if they have any meaning) will only be resolved by detailed analysis and careful thought.

Let me now turn to production reactions. The question of the existence of the A_1, A_3, Q's, L is still open and will eventually be resolved by analysis of the hypercharge or charge exchange reactions

$$\begin{aligned} K^-p &\rightarrow \Lambda\pi^+\pi^-\pi^0 \\ \pi^-p &\rightarrow \Lambda K^0\pi^+\pi^- \end{aligned} \tag{187}$$

There is still no reason to believe that they will be produced as pure J^P states (i.e. no other large partial waves present) in such reactions. I expect the isobar model type of analysis will still be required to determine the J^PM states present and their decay channels. As we have seen one automatically obtains an enormous amount of information on the production mechanism of such states and this has already led to a number of surprises, e.g. the energy dependence of the A_2. In attempts to understand the phenomenon of diffraction excitation both at meson and at nucleon vertices the reactions

$$\begin{aligned} K^{\pm}p &\rightarrow K^{\pm}\pi^+\pi^-p \\ \pi^{\pm}p &\rightarrow \pi^{\pm}\pi^+\pi^-p \end{aligned} \tag{188}$$

will continue to receive a lot of attention. Perhaps most results will emerge from the nucleon vertex studies (i.e. diffractive excitation of N^*'s) where one already has information on the spectrum of states from the analyses of the formation reactions. Finally the increase in the use of polarized targets will lead to an increased understanding of the production mechanisms of these states.

It is clear that we have only begun to extract all of the latent information in these three body states. The indications we have so far are that this information can be very rich indeed and will be a valuable tool in unravelling the mysteries of the strong interaction.

Appendix

In this appendix I gather together the definition of states, the angular momentum projections, normalizations, phase spaces, cross section formulae, etc. The conventions are those of Cashmore *et al.* (1972) and are essentially taken from Jacob and Wick (JW) (1959).

1. Particle states, normalizations, cross sections and angular momentum decomposition

We use the phase convention of JW but the normalization is different. If $\psi_{p\lambda}$ represents a state with momentum p along the Z axis and helicity λ, then the *general state* is defined by

$$|p\theta\phi\lambda\rangle = R(\phi, \theta, -\phi)\,\psi_{p\lambda} \tag{189}$$

and

$$\langle p'\theta'\phi'\lambda'|p\theta\phi\lambda\rangle = 2E\delta^3(\vec{p}' - \vec{p})\,\delta_{\lambda\lambda'} \tag{190}$$

we also define states $\chi_{p\lambda}$ by

$$\chi_{p\lambda} = (-1)^{S-\lambda}\,\mathrm{e}^{-i\pi J_y}\,\psi_{p\lambda} = (-1)^{S-\lambda}\,\psi_{-p\lambda} \tag{191}$$

The general χ state is then given by

$$|-p\theta\phi\lambda\rangle = R(\phi, \theta - \phi)\,\chi_{p\lambda} \tag{192}$$

These states are denoted by a minus sign on p. Thus

$$|-p\theta\phi\lambda\rangle = (-1)^{S-\lambda}|p\pi - \theta, \phi + \pi, \lambda\rangle \tag{193}$$

We also need to know how the states $|p\theta\phi\lambda\rangle$ transform under *Lorentz transformations*. Let the Lorentz transformation be l where $p' = lp$ and let $U(l)$ be the unitary operator corresponding to l. Wick (1962) has shown that

$$U(l)|p\theta\phi\lambda\rangle = \sum_{\nu} \mathscr{D}^S_{\nu\lambda}(\Omega\hat{n})|p'\theta'\phi'\nu\rangle \tag{194}$$

where $\hat{n}$ is a vector along $\vec{p} \wedge \vec{p}'$. We will require this result to transform from an isobar j rest frame to the overall c.m. system (see Fig. 43). Ω is then an angle of rotation about the normal to the three particle plane and is given

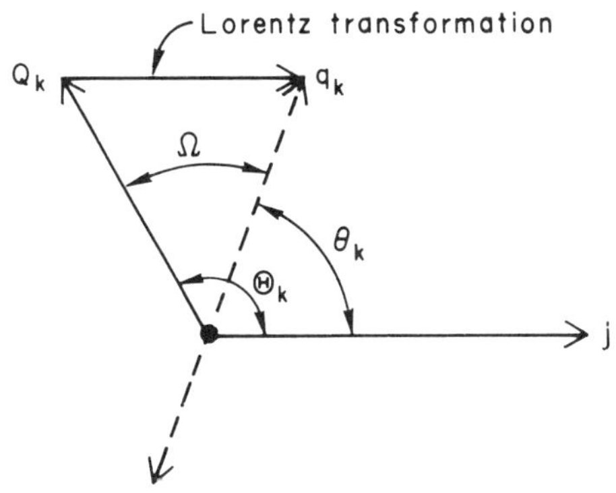

FIG. 43. Wigner angle—the rotations due to Lorentz transformations.

(for particle k) by

$$\cos \Omega = (\cosh \rho - \cosh \sigma_k \cosh \sigma_k')/\sinh \sigma_k \sinh \sigma_k' \tag{195}$$

where

$\tanh \rho = v_j =$ velocity of j in the c.m.s.

$\tanh \sigma_k = v_k =$ velocity of k in the c.m.s.

$\tanh \sigma_k' = v_k' =$ velocity of k in the isobar rest frame.

Ω can also be calculated as

$$\Omega = \Theta - \theta - \omega \tag{196}$$

where ω is the Stapp angle and is given by

$$\sin \omega = \sin \Theta \frac{\beta_i \gamma_i \beta_k \gamma_k [1 + \gamma_i + \gamma_k + \gamma_k^{(i)}]}{(1 + \gamma_i)(1 + \gamma_k)(1 + \gamma_k^{(i)})} \tag{197}$$

where

$\beta_i - \beta$ for the isobar in the c.m. system

$\beta_k - \beta$ for particle k in the c.m. system

$\beta_k^{(i)} - \beta$ for particle k in the isobar rest frame.

Multiparticle final states are defined as the direct product of one particle states. Thus

$$|\vec{p}_1 \lambda_1 \, |\vec{p}_2 \lambda_2 \cdots |\vec{p}_n \lambda_n\rangle = |p_1 \theta_1 \phi_1 \lambda_1\rangle \cdots |p_n \theta_n \phi_n \lambda_n\rangle \tag{198}$$

and

$$\langle \vec{p}_1\lambda'_1\vec{p}'_2\lambda'_2 \ldots \vec{p}'_n\lambda'_n | \vec{p}_1\lambda_1\vec{p}_2\lambda_2 \ldots \vec{p}_n\lambda_n \rangle = \prod_{i=1}^{n} (2E_i)\, \delta^3(\vec{p}'_i - \vec{p}_i)\, \delta_{\lambda_i \lambda'_i} \tag{199}$$

For two body states it is sometimes more convenient to use

$$\begin{aligned} \vec{P} &= \vec{p}_1 + \vec{p}_2 \\ \vec{p} &= (\vec{p}_1 - \vec{p}_2)/2 \end{aligned} \tag{200}$$

Letting (p, θ, ϕ) be the polar co-ordinates of p, we have

$$|\vec{P}p\theta\phi\lambda_1\lambda_2\rangle = |\vec{p}_1\lambda_1\rangle |\vec{p}_2\lambda_2\rangle \tag{201}$$

where the states on the right-hand side are either ψ or χ states. It can then be shown (Cashmore *et al.*, 1972) that in the c.m.s., $\vec{P} = 0$, the normalization is

$$\langle \vec{P}' = 0, p'\theta'\phi'\lambda'_1\lambda'_2 | \vec{P} = 0, p\theta\phi\lambda_1\lambda_2 \rangle$$

$$= \delta(W' - W)\, \delta^3(\vec{P}' - \vec{P})\, \delta^2(\omega' - \omega) \frac{4W}{p} \tag{202}$$

where $\omega = (\theta, \phi)$ and $W = E_1 + E_2$.

For the normalizations of (190) the number of particles of type i in a volume V is $2E_i V/(2\pi)^3$. In a volume V the total number of states available is $V \mathrm{d}^3 p_i/(2\pi)^3$, so that the density of final states per particle is $\mathrm{d}^3 p_i/2E_i$. Thus the number of three-particle final states available, $\mathrm{d}\rho_F$ is given by

$$\mathrm{d}\rho_F = \frac{\mathrm{d}^3 p_1}{2E_1} \frac{\mathrm{d}^3 p_2}{2E_2} \frac{\mathrm{d}^3 p_3}{2E_3}, \tag{203}$$

With the normalization of (190) the incident flux is

$$\frac{2E_a}{(2\pi)^3} \frac{2E_b}{(2\pi)^3} \left(\frac{p_a}{E_a} + \frac{p_b}{E_b} \right) = \frac{4}{(2\pi)^6} pW \tag{204}$$

The transition probability/unit volume/unit time is

$$\frac{\delta^4(p_{\text{out}} - p_{\text{in}})}{(2\pi)^4} |M|^2 \, \mathrm{d}\rho_F \tag{205}$$

where M corresponds to the transition matrix element with our normalization of states. Now Berman and Jacob (1965) have shown that

$$\begin{aligned} \mathrm{d}\rho &= \delta^4(p_{\text{out}} - p_{\text{in}})\, \mathrm{d}\rho_F \\ &= \tfrac{1}{8} \mathrm{d}E_1\, \mathrm{d}E_2\, \mathrm{d} \cos\Theta\, \mathrm{d}\Phi\, \mathrm{d}\alpha \end{aligned} \tag{206}$$

where Θ, Φ and α are the Euler angles specifying the orientation of the final three-particle state with respect to the incident system. Equation (206) may also be rewritten in many forms using the kinematical relations that exist

$$
\begin{aligned}
\mathrm{d}\rho &= \frac{1}{8}\frac{q_1 Q_1}{2W\omega_1}\,\mathrm{d}\omega_1^2\,\mathrm{d}\cos\theta_1\,\mathrm{d}\cos\Theta\,\mathrm{d}\Phi\,\mathrm{d}\alpha \\
&= \frac{1}{8}\frac{q_1 Q_1}{W}\,\mathrm{d}\omega_1\,\mathrm{d}\cos\theta_1\,\mathrm{d}\cos\Theta\,\mathrm{d}\Phi\,\mathrm{d}\alpha \qquad (207)\\
&= \frac{1}{8}\frac{1}{4W^2}\,\mathrm{d}\omega_1^2\,\mathrm{d}\omega_2^2\,\mathrm{d}\cos\Theta\,\mathrm{d}\Phi\,\mathrm{d}\alpha
\end{aligned}
$$

We can summarizc by writing

$$\mathrm{d}\sigma = \frac{\pi^2}{F}|M|^2\,\mathrm{d}\rho \qquad (208)$$

where $F = pW$.

We now turn to the decomposition of the two-particle states into angular momentum states. We work in the two-particle c.m. system and assume particle 2 is in a χ state, i.e.

$$|P = 0, p\theta\phi\lambda_1\lambda_2\rangle = |\vec{p}_1\lambda_1\rangle|-\vec{p}_1\lambda_2\rangle = R(\phi, \theta, -\phi)\,\psi_{p_1\lambda_1}\chi_{p_1\lambda_2} \qquad (209)$$

We now define the following state of total angular momentum J and z-component M by

$$|P = 0, pJM\lambda_1\lambda_2\rangle = N_J\int \mathscr{D}^J_{M\lambda}(\phi, \theta, -\phi)|P = 0, p\theta\phi\lambda_1\lambda_2\rangle\,\mathrm{d}^2\omega \qquad (210)$$

where $\mathrm{d}^2\omega = \mathrm{d}\cos\theta\,\mathrm{d}\phi$ and $\lambda = \lambda_1 - \lambda_2$. Choosing

$$N_J = \left(\frac{2J+1}{4\pi}\right)^{\frac{1}{2}}\left(\frac{p}{4W}\right)^{\frac{1}{2}} \qquad (211)$$

ensures that

$$
\begin{aligned}
&\langle P' = 0, p'J'M'\lambda'_1\lambda'_2|P = 0, pJM\lambda_1\lambda_2\rangle \\
&\qquad = \delta_{JJ'}\,\delta_{MM'}\,\delta_{\lambda_1\lambda'_1}\,\delta_{\lambda_2\lambda'_2}\delta^3(P' - P)\,\delta(W' - W) \qquad (212)
\end{aligned}
$$

and we then have

$$
\begin{aligned}
&\langle P = 0, p\theta\phi\lambda_1\lambda_2|P' = 0, p'JM\lambda'_1\lambda'_2\rangle \\
&= \left(\frac{2J+1}{4\pi}\right)^{\frac{1}{2}}\left(\frac{4W}{p}\right)^{\frac{1}{2}}\mathscr{D}^{J*}_{M\lambda}(\phi, \theta, -\phi)\,\delta(W' - W)\,\delta^3(P' - P)\,\delta_{\lambda_1\lambda'_1}\,\delta_{\lambda_2\lambda'_2} \qquad (213)
\end{aligned}
$$

In terms of the orbital and spin angular momenta, L and S, we have

$$|P = 0, pJMLS\rangle = \sum_{\mu_1 \mu_2} \left(\frac{2L + 1}{2J + 1}\right)^{\frac{1}{2}}$$

$$\times\, C(LSJ|0\,\mu_1 - \mu_2)\, C(S_1 S_2 S|\mu_1 - \mu_2)\,|P = 0, pJM\mu_1, \mu_2\rangle \quad (214)$$

with the normalization

$$\langle pJ'M'L'S'|pJMLS\rangle = \delta_{JJ'}\,\delta_{MM'}\,\delta_{LL'}\,\delta_{SS'} \quad (215)$$

2. $\mathscr{D}$ FUNCTIONS

We use the following definition of the rotation operator

$$R(\alpha, \beta, \gamma) = \mathrm{e}^{-i\alpha J_z}\,\mathrm{e}^{-i\beta J_y}\,\mathrm{e}^{-i\gamma J_z} \quad (216)$$

Since the product of two rotations is again a rotation, we have that

$$R(\alpha, \beta, \gamma) = R(\alpha'', \beta'', \gamma'')\, R(\alpha', \beta', \gamma') \quad (217)$$

The elements of the matrix corresponding to R are given by

$$\mathscr{D}^J_{m_1 m_2}(R) = \langle jm_1|R|jm_2\rangle \quad (218)$$

and in the terms of matrices (217) becomes

$$\mathscr{D}^J_{m_1 m_2}(\alpha, \beta, \gamma) = \sum_m \mathscr{D}^J_{m_1 m_2}(\alpha'', \beta'', \gamma'')\,\mathscr{D}^J_{m_1 m_2}(\alpha', \beta', \gamma') \quad (219)$$

The matrix elements can be simplified to

$$\mathscr{D}^J_{m_1 m_2}(\alpha, \beta, \gamma) = \mathrm{e}^{-im_1\alpha}\, d^j_{m_1 m_2}(\beta)\, \mathrm{e}^{-im_2\gamma} \quad (220)$$

where the $d^j_{m_1 m_2}(\beta)$ are real. These functions satisfy the general relations

$$\begin{aligned} d^j_{m_1 m_2}(\beta) &= (-1)^{m_1 - m_2}\, d^j_{m_2 m_1}(\beta) = (-1)^{m_1 - m_2}\, d^j_{-m_1 -m_2}(\beta) \\ d^j_{m_1 m_2}(\pi - \beta) &= (-1)^{j - m_2}\, d^j_{-m_1 m_2}(\beta) = (-1)^{j + m_1}\, d^j_{m_1 -m_2}(\beta) \\ d^j_{m_1 m_2}(-\beta) &= d^j_{m_2 m_1}(\beta), \qquad d^j_{m_1 m_2}(\beta + 2\pi) = (-1)^{2j}\, d^j_{m_1 m_2}(\beta) \end{aligned} \quad (221)$$

The matrix elements have the following normalizations

$$\int \mathscr{D}^j_{m_1 m_2}(\alpha, \beta, \gamma)\,\mathscr{D}^{j'*}_{m_1 m_2}(\alpha, \beta, \gamma)\,\mathrm{d}\alpha\,\mathrm{d}\cos\beta\,\mathrm{d}\gamma$$

$$= \frac{8\pi^2}{2j + 1}\,\delta_{jj'}\,\delta_{m_1 m_1}\,\delta_{m_2 m_2'} \quad (222)$$

and

$$\int d^j_{m_1 m_2}(\theta)\, d^{j'}_{m_1 m_2}(\theta)\,\mathrm{d}\cos\theta = \frac{2}{2j + 1}\,\delta_{jj'} \quad (223)$$

3. THE REACTION $a + b \rightarrow c + d$

It can be shown (Cashmore *et al.*, 1972) that the cross-section for this reaction, with these normalizations of states, is

$$\sigma = \frac{\pi}{p^2} \sum_J \frac{(2J + 1)}{(2\sigma_a + 1)(2\sigma_b + 1)} \sum_{LSL'S'} |\langle QJML'S'|T|pJMLS\rangle|^2 \tag{224}$$

For $\pi N \rightarrow \pi N$ this becomes

$$\sigma(J^P) = \pi\lambda^2(J + 1/2)|\langle J^P|T|J^P\rangle|^2. \tag{225}$$

References

Aaron, R., Teplitz, D. C., Amado, R. D., and Young, J. E. (1969). *Phys. Rev.* **187**, 2047.

Aaron, R., Amado, R. D. and Silbar, R. R. (1970). Los Alamos preprint LA-DC-11961.

Ader, J. P., Capdeville, M., Cohen-Tannoudji, G., Salin, Ph. (1968). *Nuovo Cimento* **56A**, 952.

Ascoli, G., Brockway, D. V., Eisenstein, L., Ioffredo, M. L., Kruse, U. E., Schultz, P. F., Caso, C., Tomasini, G., von Handel, P., Schilling, P., Costa, G., Ratti, S., Daronian, P., Mosca, L., Brenner, A. E., Harrison, W. C., Heyda, D., Johnson, Jr., W. H., Kim, J. K., Law, M. E., Mueller, J. E., Salzberg, B. M., Sisterson, L. K., Johnston, T. F., Prentice, J. D., Steenberg, N. R., Yoon, T. S., Carroll, J. T., Erwin, A. R., Morse, R., Oh, B. Y., Robertson, W., and Walker, W. D. (1971). *Phys. Rev. Letts* **26**, 929.

Ascoli, G. (1972). Contributed paper to the 1972 International Conference on High Energy Physics, 16th, National Accelerator Laboratory, Batavia, Illinois, September 6–13, 1972, No. 443.

Ball, J. S., Campbell, R. R., Lee, P. S., and Shaw, G. L. (1972). *Phys. Rev. Letts* **28**, 1143.

Berman, S., and Jacob, M. (1965). *Phys. Rev.* **B139**, 1023.

Bland, R. W., Bowler, M. G., Brown, J. L., Kadyk, J. A., Goldhaber, G., Goldhaber, S., Seeger, V. H., and Trilling, G. H. (1969). *Nucl. Phys.* **B13**, 595.

Bowler, M. G., and Cashmore, R. J. (1970). *Nucl. Phys.* **B17**, 331.

Brody, A. D., Kernan, A. (1969). *Phys. Rev* **182**, 1785.

Brody, A. D., Cashmore, R. J., Kernan, A., Leith, D. W. G. S., Levi, B. G., Shen, B. C., Herndon, D. J., Price, L. R., Rosenfeld, A. H., and Söding, P. (1971). *Phys. Letts* **34B**, 665.

Cashmore, R. J., and Hey, A. J. G. (1972). *Phys. Rev.* **D6**, 1303.

Cashmore, R. J., Herndon, D. J., and Söding, P. (1972). Lawrence Berkeley Laboratory LBL-543.

Cashmore, R. J. (1973). *In* Proceedings of the International Conference on Baryon Resonances, Purdue, West Lafayette, Indiana, April 20–21, 1973.

Chinowsky, W., Mulvey, J. H., and Saxon, D. H. (1970). *Phys. Rev.* **D2**, 1790.

Dalitz, R. (1962). "Strange Particles and Strong Interactions" Oxford University Press.

Dalitz, R. (1969). In "Pion-Nucleon Scattering" (G. L. Shaw and D. Y. Wong, eds). John Wiley and Sons, Inc., New York.

DeBeer, M., Deler, B., Dolbeau, J., Neveu, M., Diem, Nguyen Thuc, Smadja, G., and Valladas, G. (1969a). *Nucl. Phys.* **B12**, 599.

DeBeer, M., Deler, B., Bolbeau, J., Neveu, M., Diem, Nguyen Thuc, Smadja, G., and Valladas, G. (1969b). *Nucl. Phys.* **B12**, 617.

Deler, B., and Valladas, G. (1966). *Nuovo Cimento* **45A**, 559.
Faiman, D. (1971). *Nucl. Phys.* **B32**, 573.
Faiman, D., and Plane, D. E. (1972). *Nucl. Phys.* **B50**, 379.
Faiman, D., and Rosner, J. (1973). CERN preprint TH. 1636, Phases of resonant amplitudes: $\pi + N \rightarrow \pi + \Delta$ (revised, unpublished).
Feynman, R., Pakvasa, S., and Tuan, S. F. (1970). *Phys. Rev.* **D2**, 1267.
Gell-Mann, M. (1962). *Phys. Rev.* **125**, 1067.
Gilman, F., Pumplin, J., Schwimmer, A., and Stodolsky, L. (1970). *Phys. Letts* **31B**, 387.
Gilman, F., Kugler, M., and Meshkov, S. (1973). Stanford Linear Accelerator Center preprint SLAC-PUB-1235, Pionic transitions as tests of the connection between current and constituent quarks.
Gilman, F. (1973). *In* Proceedings of the International Conference on Baryon Resonances, Purdue, West Lafayette, Indiana, April 20–21, 1973.
Greenberg, O. (1969). *In* Proceedings of the International Conference on Elementary Particles, 5th, Lund, Sweden, 1969. Berlingska Boktryckeriet, Lund.
Herndon, D. J., Longacre, R., Miller, L. R., Rosenfeld, A. H., Smadja, G., Soding, P., Cashmore, R. J., and Leith, D. W. G. S. (1972). Lawrence Berkeley Laboratory and Stanford Linear Accelerator Center preprint LBL-1065/SLAC-PUB-1108. A partial wave analysis of the reaction $\pi N \rightarrow \pi\pi N$ in the c.m. energy range 1300–2000 MeV.
Illinois-GHMS-H-ABBCCH-ND-W Collaboration and CIBS Collaboration (1972). *In* Proceedings of the International Conference on Experimental Meson Spectroscopy, 3rd, Philadelphia, Pennsylvania, April 28–29, 1972 (Presented by G. Ascoli).
Jacob, M. and Wick, G. C. (1959). *Ann. Phys.* **7**, 404.
Kernan, A. and Shepherd, H. K. (1969). *Phys. Rev. Letts* **23**, 1314.
Lasinski, T. and Barbaro-Galtieri, L. (1972). *Phys. Letts* **39B**, 1.
Lasinski, T., Ronat, E. E., Rosenfeld, A. H., Tabak, M., and Cashmore, R. J. (1973). Preliminary studies of the 3π system in $\pi^+ p \rightarrow \pi^+\pi^+\pi^- p$ at 7 GeV/c.
Levi Setti, R. (1969). *In* Proceedings of the International Conference on Elementary Particles, 5th, Lund, Sweden, 1969. Berlingska Boktryckeriet, Lund.
Lipkin, H. and Meshkov, S. (1965). *Phys. Rev. Letts* **14**, 670.
Longacre, R. (1972). Ph.D. Thesis, Lawrence Berkeley Laboratory LBL-948 (unpublished).
Mast, T. S., Alston-Garnjost, M., Bangerter, R. O., Barbaro-Galtieri, A., Solmitz, F. T., and Tripp, R. D. (1972). *Phys. Rev. Letts* **28**, 1220.
Mehtani, U., Fung, S. Y., Kernan, A., Schalk, T. L., Williamson, Y., Birge, R. W., Kalmus, G. E., and Michael, W. (1972). *Phys. Rev. Letts* **29**, 1634
Melosh IV, H. J. (1973). Caltech Thesis (unpublished).
Moorhouse, G. and Oberlack, H. (1973). *Phys. Letts* **43B**, 44.
Moorhouse, G. and Parsons, N. (1973). Glasgow University preprint.
Plane, D. E., Baillon, P., Bricman, C., Ferro-Luzzi, M., Meyer, J., Pagiola, E., Schmitz, N., Burkhardt, E., Filfuth, H., Kluge, E., Oberlack, H., Barloutaud, R., Granet, P., Porte, J. P., and Prevost, J. (1970). *Nucl. Phys.* **B22**, 93.
Rosner, J. (1972). *In* 1972 International Conference on High Energy Physics, 16th, National Accelerator Laboratory, Batavia, Illinois, September 6–13, 1972, Appendix to Baryon resonances and related phenomenology
Wick, G. C. (1962). *Ann. Phys.* **18**, 65.

10. SOME EXPERIMENTS ON MULTIPLE PRODUCTION

KENNETH G. WILSON

Laboratory of Nuclear Studies, Cornell University, Ithaca, New York

INTRODUCTION

A program of experiments is described mainly on secondary particle spectra to test scaling hypotheses derived from the multiperipheral model. It is assumed that diffraction dissociation and multiperipheral processes are distinct effects, and the consequences of this for the scaling laws are explained. Feynman's analogy linking multiple production to the statistical mechanical distribution functions of a gas is outlined, and based on this analogy it is suggested that one looks for a correlation length in the two particle spectrum of secondaries.

High energy scattering cross-sections of pions and protons show a preponderance of multiple production processes. For example, the mean multiplicity of charged secondaries is about 4 for 30 GeV p–p interactions, and cosmic ray data suggest a mean multiplicity of order 10 at much higher

Editor's footnote
This previously unpublished paper (Cornell preprint CLNS-131, 1970) is referred to at various places in these proceedings and is therefore included for its relevance to current developments in the field.

energies.[1] Multiple production will be a dominant effect at the NAL accelerator and CERN storage rings, and there is currently much interest in developing an experimental programme to study it.

Actually, multiple production is already dominant at the energies of the present Brookhaven and CERN accelerators, but experimental studies of it have been sporadic and always subordinate to the study of elastic and quasi-elastic cross-sections.[2] A major problem preventing more systematic study of inelastic events has been the question of what variables to observe in highly multiple events. A programme to measure the complete differential cross-sections for a process with, for example, five secondary particles is prohibitive in both time and expense; one must select particular aspects of the cross section to measure.

In the case of elastic or quasi-elastic scattering, one must also be selective in any given experiment. One looks at a limited range of incident energy and scattering angle, and one selects particular final states. Usually the choice is dictated by a desire to test a particular theoretical model, a Regge pole fit, an optical model calculation, etc. However, as the total number of experiments becomes large, one collects a reservoir of data on elastic and quasi-elastic scattering covering most of the range of energies, angles and final states accessible to experiment; that is the total range covered by all data from all experiments is more limited by experimental than by theoretical considerations. The importance of this is that a future theory of strong interactions is likely to find its best test in different ranges of energy, angle, etc., than any particular model now under consideration so one would prefer not to have the range of data collected in the sum total of high energy experiments limited by currently popular model theories.

It is out of the question to use the same approach in experiments on multiple production; because of the number of variables involved even the sum total of all multiple production experiments will provide only a small fraction of the data that is experimentally accessible. Therefore it is important to think about the choice of experiments to be performed, in particular to try to maximize the possibility that the experimental data collected will remain useful despite continuing changes in theoretical fashions.

The purpose of this paper is to propose an experimental programme for studying multiple production, which takes into account the problems mentioned above. The programme is a set of specific and feasible experiments, whose immediate aim is to test some currently popular theoretical models.[3] However it will be argued that the results of these experiments will help to characterize the general features of multiple production independently of any model. As a basis for setting up the experiments and predicting their outcome, it will be supposed that there are three types of multiple production processes, namely: (a) multiperipheral events, (b) diffraction and diffraction dissociation, and (c) multi-Regge exchange events.[4] We shall give a simple but qualitative definition for each of the three types of processes; these definitions will *not* involve specific models requiring detailed calculation and parameter fits in order to compare with experimental data. The experiments to be proposed will test whether multiple production processes can be separated into these three categories.

Before discussing these ideas in detail, some general comments will be made about the purposes for doing high energy experiments.

There does not exist a real theory of strong interactions at present, and the models of high energy processes one studies at present are no substitute for such a theory. It is the ultimate aim of experiment and theory to try to obtain a real theory, so it is worth considering how particular experiments will affect the finding and testing of such a theory. First, though, one must say what one means by a "real theory of strong interactions". My view is that there are four essential requirements for a real theory:

A. It must be derived from a few fundamental principles comprehensible to both experimentalists and theorists.
B. Any free parameters in the theory must appear explicitly and obviously as a consequence of the fundamental principles (just as s and $\hbar$ are explicit and obvious in ordinary quantum mechanics, and c is explicit and obvious in relativity) and there should be no arbitrary functions in the theory.
C. The fundamental principles should imply a set of equations containing the fundamental parameters whose solution will describe all aspects of strong interactions including the complete S matrix (even the S matrix for n particles going to m particles for any n and m) and all matrix elements of the weak and electromagnetic currents. For a given set of values of the parameters the equations should have one and only one solution; if this cannot be proven there should at least be plausible physical arguments suggesting it.
D. One should be able to determine qualitative features of the solution of the equations from qualitative features of the equations, or better from qualitative statements of the fundamental principles without using the equations at all.

One has to be an idealist to believe that a theory will be found satisfying these four requirements. So the author is prepared to be flexible but would be very sceptical about any proposed theory that seriously violates any of these requirements.

At present there are two principles (in addition to the principles of quantum mechanics) which one hopes will be a part of a real theory when it is found, namely, locality and Gell-Mann's current commutators. Model-independent tests of these principles are of vital importance. Locality can be tested, at least partially, by checking forward dispersion relations and rigorous bounds on high energy cross-sections. However, these tests do not involve detailed experiments on multiple production processes, so they will not be discussed further here.[5] One has some ideas on what the fundamental parameters of strong interactions are, namely, the strengths of the $SU(3) \times SU(3)$ breaking terms in the Lagrangian are probably fundamental parameters. It is difficult to learn much about these parameters in a model-independent way from high energy cross-sections. We have not even a glimmering of an idea what the equations of strong interactions would be, let alone how to obtain qualitatively or quantitatively the solution of these equations.

It is likely that the equations of strong interactions will be as complicated as the many-body equations of non-relativistic quantum mechanics or classical physics. In consequence it will probably be very difficult to get detailed quantitative solutions of them; the best one can hope for is to determine the basic qualitative features of their solution. Because of this I think the experimental programme on multiple production should be aimed at finding clear-cut qualitative features of these processes rather than trying to have precise numerical data on particular processes just to test a particular model. Also, I think it is more important to understand processes with large cross-sections than to study processes with small cross-sections. For example, the elastic cross-sections at large angles where the cross-sections are $\sim 10^{-33}$ or less may well be useless for either finding or testing a real theory because there can be very many competing small effects which would become important in calculating a cross section of this size. Where a cross-section is large there is more hope that a few qualitative features of the theory will be sufficient to determine the behaviour of the cross-section.

There are already known a number of simple properties of high energy cross sections which are good examples of the "clear-cut qualitative features" that one should look for. We know that (at presently accessible large energies) total cross-sections are constant (at least roughly). Elastic or quasi-elastic cross-sections requiring exchange of internal quantum numbers (isospin, strangeness, etc.) fall with s roughly like $s^{(\alpha-1)}$ where α depends only on what is exchanged and not the particular process. In multiple production processes, we know that the transverse momentum of secondaries is bounded, having a mean value around 300 MeV independent of the incident energies. The probability of finding a secondary with transverse momentum $p_\perp$ much larger than 300 MeV falls rapidly as $p_\perp$ increases, perhaps exponentially. As new energy ranges open up it is important to check that these results continue to hold.

Now the three types of high energy events discussed in this paper will be defined. It will be assumed that total cross-sections are strictly constant at high energy (i.e. they do not increase with energy, even logarithmically).

First consider diffraction and diffraction dissociation.[6] These are the processes that would be described by single Pomeron exchange, if one can describe these processes by Regge theory. They include elastic and quasi-elastic scattering where no quantum numbers are exchanged. They also include processes in which the incident or target particle (or both) fragment into several particles where the several particles are not just the decay products of an N^* or ρ or etc.[7] It is assumed that the cross-sections for diffraction or diffraction dissociation to a fixed final state are constant at high energy. By a "fixed final state", I mean that the fragments of the target particle have fixed momenta in the lab system independent of the incident energy, while the fragments of the incident particle have longitudinal momenta which are fixed fractions of the incident energy (fixed x in the Feynman language) and fixed transverse momenta.

The assumption that total cross-sections are constant puts a strong restriction on the nature of diffraction dissociation. Consider the partial cross-section σ_n (diff) for producing n secondaries by diffraction dissociation at high energies. Because of the constant total cross-section, $\Sigma_n \sigma_n$(diff) must be finite. Since σ_2, σ_3, etc. are fixed with energy, this means σ_n (diff) $\rightarrow 0$ as $n \rightarrow \infty$. This means that diffraction dissociation will go predominantly to low multiplicity intermediate states. In other words the mean multiplicity in diffraction dissociation must be constant independent of energy.* If the average multiplicity of all inelastic events increases indefinitely with energy there must be other processes besides diffraction dissociation.

By the same argument the probability that a pion of longitudinal momentum k_z is emitted as a fragment of the target in diffraction dissociation to a final state of fixed multiplicity must decrease rapidly as k_z increases; to be precise the cross-section must fall faster than $1/k_z$, since the integral over k_z must converge.[8] To produce pions of large k_z is possible only if large multiplicities are allowed so that large k_z pions appear simultaneously with low k_z pions rather than in separate events. But by the previous argument large multiplicities are unimportant for diffraction dissociation.

So our picture of diffraction dissociation is that diffraction dissociation cross-sections are constant with energy for fixed final states, that the final states resulting from diffraction dissociation are predominantly of low multiplicity, and that final state particles will mostly be of low energy either in the lab system or the projectile system. There will be a tail of events not satisfying these criteria, but it will be assumed that this tail is negligible compared to multiperipheral cross-sections.

Secondly, consider double Regge pole exchange. A typical double Regge pole exchange process is shown in Fig. 1: $p + p \rightarrow p + p + \rho^0$ where the ρ^0 is at rest in the centre-of-mass system, with pions exchanged between the

* This is strictly true only for the median multiplicity.

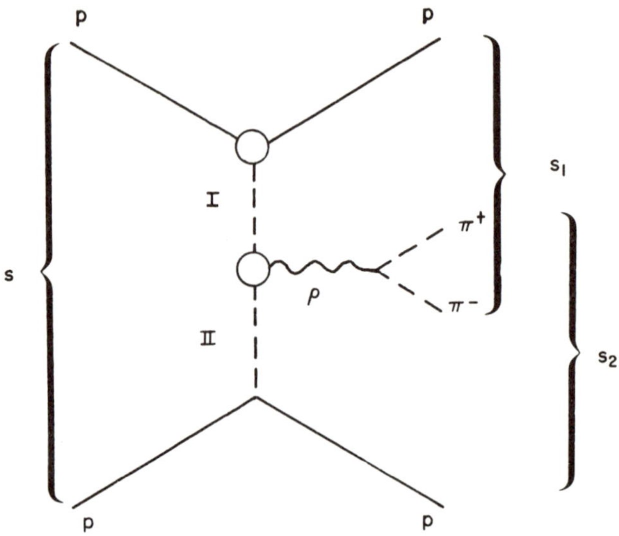

FIG. 1. Double Regge pole exchange graph for process $p + p \rightarrow p + p + \pi^+ + \pi^-$. The exchanges can be π exchange or any combination of Pomeron and ρ exchange. The invariants s_1 and s_2 are the invariant mass squared for the Regge poles I and II respectively.

ρ^0 and each proton. Double Regge pole exchange is expected to apply when the particles emitted at the intermediate vertex are highly relativistic with respect to both incident particles; in other words when the particle s across each Regge pole is large (s_1 and s_2 of Fig. 2 must be large). For any exchange other than double Pomeron exchange these double exchanges should fall rapidly with energy and hence be a negligible part of the total cross-section. If there is double Pomeron exchange these processes are important, and wreck the picture of this paper because with double Pomeron exchange one has a process with a fixed number of particles in the final state where secondaries can come out with any k_z. However, because diffraction may well not be an exchange process at all, and because multiple Pomeron exchange causes theoretical difficulties,(9) it will be assumed here that double Pomeron exchange is either non-existent or very small. One of the experiments described later is to test this assumption.

Finally, consider multiperipheral processes. The multiperipheral model of Amati, Fubini, and Stanghellini(10) is at first sight a special model based on multiple pion exchanges which one would not want to take very seriously. But as Amati *et al.* found, the multiperipheral model exhibits simple scaling laws for inelastic processes at high energies(10) (including the scaling law for the single particle spectrum rediscovered seven years later by Feynman(3)).

These scaling laws were shown by Amati *et al.* to be independent of the details of the model, such as the values of coupling constants; one can also have other particle exchanges besides pions. By multiperipheral processes I mean any processes satisfying the scaling laws predicted from the multiperipheral model. The author has stated elsewhere the rules for constructing such scaling laws.[10] The best way to introduce these scaling laws is, I think,

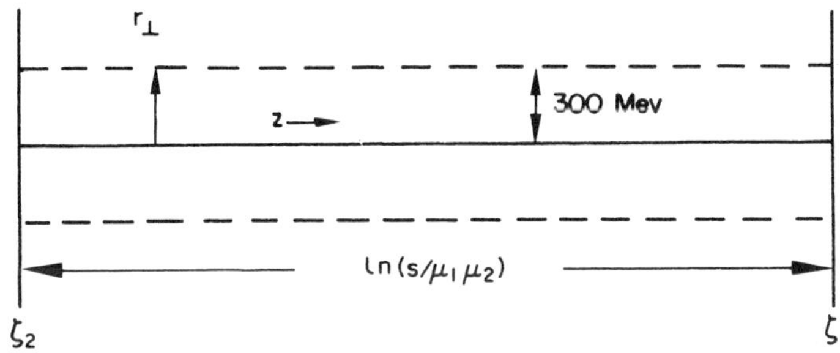

FIG. 2. Sketch of the Feynman gas showing walls at ζ_1 and ζ_2 and the bound on $|\mathbf{r}_\perp|$. There can be some leakage beyond the walls, as noted in the text. μ_1 and μ_2 are the incident particle masses.

to use an analogy invented by Feynman.[11] This analogy links multiparticle production cross-sections to the multiparticle distribution functions of a classical gas, with the total cross-section becoming the partition function of a gas. This analogy is very much on Feynman's mind when he discusses his parton model of high energy collisions, although it is not discussed in his papers.

In the Feynman gas analogy, a secondary particle with momentum $\mathbf{k}$ and energy k_0 corresponds to a gas particle at position $\mathbf{r}$; the components of $\mathbf{r}$ are

$$x = k_x \tag{1}$$

$$y = k_y \tag{2}$$

$$z = \ln\left[(k_z + k_0)/m_\perp\right] \tag{3}$$

and

$$m_\perp = (m^2 + k_x^2 + k_y^2)^{\frac{1}{2}} \tag{4}$$

Here m is the mass of the secondary and k_x and k_y are the transverse components of the momentum. The variable z is Feynman's "rapidity"[(3)] and is used because a Lorentz transformation in the z direction on k is equivalent to a translation of z.

It will be assumed that multiple production cross-sections are defined using the invariant form of phase space d^3k/k_0 for each secondary. When one changes variables from $\mathbf{k}$ to $\mathbf{r}$, invariant phase space becomes just d^3r. We also need the form of the energy-momentum conservation δ-functions written in terms of the $\mathbf{r}$ variables. One notes that

$$k_0 + k_z = m_\perp e^z \tag{5}$$

$$k_0 - k_z = \frac{m_\perp^2}{k_0 + k_z} = m_\perp e^{-z} \tag{6}$$

Let $\mathbf{s}_1$ and $\mathbf{s}_2$ be the $\mathbf{r}$ variables of the incident particles while $\mathbf{r}_1, \ldots, \mathbf{r}_n$ are the coordinates of the secondaries. Let the incident momenta be $\mathbf{p}_1$ and $\mathbf{p}_2$, the secondary momenta $\mathbf{k}_1 \ldots \mathbf{k}_n$. Let $p = p_1 + p_2 - k_1 - \ldots - k_n$; then the energy–momentum conservation δ functions are

$$\delta(p_0)\,\delta^3(p) = 2\delta(p_0 + p_z)\,\delta(p_0 - p_z)\,\delta^2(\mathbf{p}_\perp) \tag{7}$$

One now has

$$p_+ = p_0 + p_z = \mu_1 e^{\zeta_1} + \mu_2 e^{\zeta_2} - m_{1\perp} e^{z_1} - \ldots - m_{n\perp} e^{z_n} \tag{8}$$

$$p_- = p_0 - p_z = \mu_1 e^{-\zeta_1} + \mu_2 e^{-\zeta_2} - m_{1\perp} e^{-z_1} - \ldots - m_{n\perp} e^{-z_n} \tag{9}$$

$$\mathbf{p}_\perp = \mathbf{s}_{1\perp} + \mathbf{s}_{2\perp} - \mathbf{r}_{1\perp} - \ldots - \mathbf{r}_{n\perp} \tag{10}$$

where ζ_1 is the z component of $\mathbf{s}_1$ and ζ_2 the z component of $\mathbf{s}_2$; μ_1 and μ_2 are the masses of the incoming particles (which have no transverse momentum so $\mu_1 = \mu_{1\perp}, \mu_2 = \mu_{2\perp}$). The vector $p_\perp$ contains the transverse components (p_x, p_y) of p.

The total cross-section, written in $\mathbf{r}$ variable form, is a function

$$\sigma_T(\mathbf{s}_1, \mathbf{s}_2) = \sum_{n=2}^{\infty} \frac{1}{n!} \int d^3r_1 \ldots \int d^3r_n\, 2\delta(p_+)\,\delta(p_-)\,\delta^2(\mathbf{p}_\perp) \times \sigma_n(\mathbf{r} \ldots, \mathbf{r}_n, \mathbf{s}_1, \mathbf{s}_2) \tag{11}$$

where $\sigma_n(\mathbf{r}_1, \ldots, \mathbf{r}_n, \mathbf{s}_1, \mathbf{s}_2)$ is the invariant cross-section[(12)] for producing n secondaries with $\mathbf{r}$ variables $\mathbf{r}_1 \ldots \mathbf{r}_n$. One must also sum over particle species $(\pi^+, \pi^0, \pi^-, K^+$, etc.) but this will not be written explicitly.

In the Feynman analogy $\sigma_T(\mathbf{s}_1, \mathbf{s}_2)$, with $\mathbf{s}_1$ and $\mathbf{s}_2$ fixed, is a partition function of a gas, and the functions $\sigma_n(r_1, \ldots, \mathbf{r}_n, \mathbf{s}_1, \mathbf{s}_2)$ are the n-particle distribution functions for the gas. The gas has some strange features, namely the restriction on the positions of particles given by the δ-functions and the fact that transverse coordinates x and y are measured in GeV; these features will not upset the usefulness of the analogy.

The most important effect of the δ-functions is to bound the longitudinal positions z_i of the gas particles. Assume the invariant $s = (p_1 + p_2)^2$ is large, and that one is in the lab system. Then one finds

$$\zeta_1 = \ln [s/\mu_1\mu_2] \tag{12}$$

$$\zeta_2 = 0 \tag{13}$$

The δ function for p_+ puts an upper bound on the z_i, namely they cannot be much larger than ζ_1 (z_i can be larger than ζ_1 if $m_{i\perp}$ is less than μ_1, but the maximum value of $z_i - \zeta_1$ is $\ln(\mu_1/m_i)$ where m_i is the mass of the secondary, and this quantity is a constant independent of s). Similarly, the δ function for p_- puts a lower bound on z_i, namely it cannot be much less than $\zeta_2 = 0$ (z_i cannot be less than $-\ln(\mu_2/m_i)$. Hence the quantities ζ_1 and ζ_2 define boundaries in the z direction for the gas; one can imagine the gas being confined between walls at ζ_1 and ζ_2. The separation of the walls is proportional to $\ln s$ and goes to infinity as $s \to \infty$. The wall at ζ_1 will be called the "incident particle wall"; the wall at ζ_2 is the "target wall". The Feynman gas is illustrated in Fig. 2.

In the transverse direction there are also kinematic bounds but these are not noticeable except near the walls at ζ_1 and ζ_2; away from these walls the dynamic property of bounded transverse momenta keeps the gas largely confined to a tube of radius ~ 300 MeV.

In a real gas with over 10^{23} particles one does not even think of measuring the distribution functions for those 10^{23} particles. Instead one measures thermodynamic (statistical) average properties of the gas, such as the density. A density $\rho(\mathbf{r})$ is readily defined for the Feynman gas; to be precise it is a function $\rho(\mathbf{r}, \mathbf{s}_1, \mathbf{s}_2)$. The density–density correlation function is another statistical average which is readily defined for the Feynman gas; it is a function $g(\mathbf{r}_1, \mathbf{r}_2, \mathbf{s}_1, \mathbf{s}_2)$. In real gases a knowledge of the density and density correlation functions (as a function of temperature and pressure, say) determines all the properties of the gas of practical interest. By analogy a knowledge of the density and density–density correlations of the Feynman gas should be invaluable for characterizing its properties.

The density $\rho(\mathbf{r}, \mathbf{s}_1, \mathbf{s}_2)$ is simply the invariant single particle spectrum for the scattering problem. The definition of the density ρ is that $\rho(\mathbf{r}, \mathbf{s}_1, \mathbf{s}_2)\, d^3r$ is the average number of particles to be found in the volume d^3r. From the partition function of eqn (11) it follows that ρ is

$$\rho(r, s_1, s_2) = \frac{1}{\sigma_T(s_1, s_2)} \sum_{n=2}^{\infty} \frac{1}{(n-1)!} \int d^3r_1 \ldots \int d^3r_{n-1}$$
$$\times\, 2\delta(p_+)\, \delta(p_-)\, \delta^2(\mathbf{p}_\perp)\, \sigma_n(\mathbf{r}, \mathbf{r}_1, \ldots, \mathbf{r}_{n-1}, \mathbf{s}_1, \mathbf{s}_2) \tag{14}$$

This is just the probability density per collision to produce a secondary with position variable $\mathbf{r}$ [i.e. momentum $\mathbf{k}$ given by eqns (1)–(4)]. If one converts from position variables to momentum variables, ρ becomes a function $\rho(k, p_1, p_2)$. Since d^3r is d^3k/k_0, $\rho(k, p_1, p_2)\, d^3k/k_0$ is the probability per collision to produce a secondary of momentum k in a range d^3k. This ρ is an inclusive quantity in Feynman's language; all events are included which have a secondary in this range. In practice one defines separate densities for each particle species, π^+, π^0, π^-, K^+, etc. I have normalized the spectrum to the total cross-section instead of the total inelastic cross-section as is sometimes done; the question of normalization will be reconsidered later.

To define the two particle correlation function one first defines a joint probability density $P(\mathbf{r}, \mathbf{r}', \mathbf{s}_1, \mathbf{s}_2)$ for finding one particle at $\mathbf{r}$, another at $\mathbf{r}'$. P is given by

$$P(\mathbf{r}, \mathbf{r}', \mathbf{s}_1, \mathbf{s}_2) = \frac{1}{\sigma_T(s_1, s_2)} \sum_{n=2}^{\infty} \frac{1}{(n-2)!} \int d^3r_1 \ldots \int d^3r_{n-2}$$

$$\times\ 2\delta(p_+)\,\delta(p_-)\,\delta^2(\mathbf{p}_\perp)\,\sigma_n(\mathbf{r}, \mathbf{r}', \mathbf{r}_1, \ldots, \mathbf{r}_{n-2}, \mathbf{s}_1, \mathbf{s}_2) \qquad (15)$$

If the particles of the gas were uncorrelated the joint probability density $P(\mathbf{r}, \mathbf{r}', \mathbf{s}_1, \mathbf{s}_2)$ would be a product of two single particle densities. The correlation is defined by subtracting this product from P:

$$g(\mathbf{r}, \mathbf{r}', \mathbf{s}_1, \mathbf{s}_2) = P(\mathbf{r}, \mathbf{r}', \mathbf{s}_1, \mathbf{s}_2) - \rho(\mathbf{r}, \mathbf{s}_1, \mathbf{s}_2)\,\rho(\mathbf{r}', \mathbf{s}_1, \mathbf{s}_2). \qquad (16)$$

The most interesting property of the correlation function for a real gas is the correlation length ξ. Qualitatively ξ measures the maximum separation $|\mathbf{r} - \mathbf{r}'|$ for which $g(\mathbf{r}, \mathbf{r}')$ is appreciably different from zero. Quantitatively ξ is defined from the expectation that $g(\mathbf{r}, \mathbf{r}')$ falls off exponentially in the separation $|\mathbf{r} - \mathbf{r}'|$ when this separation is large. One then defines ξ by the asymptotic form

$$g(\mathbf{r}, \mathbf{r}') \sim \exp\{-|\mathbf{r} - \mathbf{r}'|/\xi\} \qquad (17)$$

apart from a power of $|\mathbf{r} - \mathbf{r}'|$.

By analogy with the real gas one should define a correlation length ξ for the Feynman gas. This makes sense only at high energies where large values of $|\mathbf{r} - \mathbf{r}'|$ are possible and one might be able to see experimentally the function $g(\mathbf{r}, \mathbf{r}')$ dropping to zero as $|\mathbf{r} - \mathbf{r}'|$ increases. Consider, to be specific, the π^+–π^- correlation function; let $\mathbf{r}$ refer to the π^+, $\mathbf{r}'$ to the π^- and let k and k' be the corresponding momenta. Let s_{+-} be the invariant mass of the π^+–π^- pair; in terms of the $\mathbf{r}$ variables, one has

$$s_{+-} = 2m_\pi^2 + 2m_\perp m_\perp' \cosh(z - z') - 2\mathbf{k}_\perp \cdot \mathbf{k}_\perp' \qquad (18)$$

(m_π is the π mass). The "perpendicular masses" $m_\perp$ and $m'_\perp$ are typically approximately equal to $k_\perp$ and $k'_\perp$, all being of order 300 MeV. So for s_{+-} to be large, $z - z'$ must be large. One expects strong correlation between the π^+ and π^- when $s_{+-} \simeq m_\rho^2$ (m_ρ is the ρ mass). The best one can hope for is that $g(\mathbf{r}, \mathbf{r}', \mathbf{s}_1, \mathbf{s}_2)$ becomes small when $s_{+-} \gg m_\rho^2$. To see this fall-off one has to let $z - z'$ become large. In principle one can also make s_{+-} large by going to large transverse momenta, but this means going well out into the tail of transverse momentum distribution. This involves looking at small cross-sections of the kind one would prefer to avoid.(13) So what one should look for is a correlation length in $g(\mathbf{r}, \mathbf{r}', \mathbf{s}_1, \mathbf{s}_2)$ as a function of the longitudinal variables z and z'; an experiment to measure this correlation function is included in the programme.

The Feynman gas analogy makes clear, I think, that any programme to study multiple production should include experiments on the single particle spectrum and two-particle correlation functions.

To motivate the general predictions of the multiperipheral model, imagine that the particles of the Feynman gas interact only through short range forces. In particular, assume the range of the forces stays fixed as s increases, so that for large s the range of forces is small compared to the separation of the walls of the gas. I shall not give a precise definition for the idea of "short range forces", as it is hard to formulate as a mathematical property of the distribution functions σ_n. Further discussion of this problem is given in the Appendix. The hypothesis suggests some properties of the density and correlation functions for large s. The part of the gas that is well away from the walls has no direct interactions with the walls, in fact being many interaction lengths away from them. One would then expect the properties of the gas in this region to be independent of the precise location of the walls. This means $\rho(\mathbf{r}, \mathbf{s}_1, \mathbf{s}_2)$ should be independent of $\mathbf{s}_1$ and $\mathbf{s}_2$ when $\zeta_1 - z$ and $z - \zeta_2$ are large: $\rho(\mathbf{r}, \mathbf{s}_1, \mathbf{s}_2) \to \rho(\mathbf{r})$ for $\zeta_1 - z$ and $z - \zeta_2 \to \infty$. By Lorentz invariance in the collision plane, which is translational invariance along the z axis for the Feynman gas, $\rho(\mathbf{r})$ cannot depend on z so $\rho(\mathbf{r})$ depends only on x and y, i.e. only on $k_\perp$.(14) The part of the gas near the incident particle wall ($z = \zeta_1$) is many interaction lengths away from the target wall; for this range of z, ρ should depend on $\mathbf{s}_1$ but not on $\mathbf{s}_2$: $\rho(\mathbf{r}, \mathbf{s}_1, \mathbf{s}_2) \to \rho(\mathbf{r}, \mathbf{s}_1)$ when $z - \zeta_2 \to \infty$ holding $\zeta_1 - z$ fixed. By translational invariance $\rho(\mathbf{r}, \mathbf{s}_1)$ can depend only on $k_\perp$ and $z - \zeta_1$. Translated into momentum variables, this means that $\rho(k, p_1, p_2)$ depends only on $k_\perp$ and k_0/p_{10} (Feynman's x variable(15)) when $p_{10} \to \infty$ with x held fixed. These are just the scaling predictions of Amati, Fubini, Stanghellini,(10) Feynman,(3) and Yang *et al.*(3) for the single particle spectrum.

Similarly, the correlation function $g(\mathbf{r}, \mathbf{r}', \mathbf{s}_1, \mathbf{s}_2)$ should be independent of $\mathbf{s}_2$ when z, z', and ζ_1 all go to ∞, and in this limit should depend on z, z', and

ζ_1 only in terms of the differences $z - z'$ and $z' - \zeta_1$. So if one holds $k_\perp$, $k'_\perp$, and $z' - \zeta_1$ fixed (i.e. fixed inelasticity x' for the particle with momentum r'), the correlation function should depend only on $z - z'$ and not on ζ_1 (i.e. not on s). Hence the correlation length ξ should also be independent of s, for fixed $k_\perp$, $k'_\perp$, and $z' - \zeta_1$. More generally, the correlation length cannot become large as s increases no matter what values one uses for $k_\perp$, $k'_\perp$, or z', since a dependence on s is possible only if $g(\mathbf{r}, \mathbf{r}', \mathbf{s}_1, \mathbf{s}_2)$ depends simultaneously on $\mathbf{s}_1$ and $\mathbf{s}_2$, and for sufficiently large s this is excluded by the short range force picture. (Once one has a function independent of $\mathbf{s}_1$ or $\mathbf{s}_2$, one can write a scaling law for that function from which one shows that ξ is bounded.)

If the density of the Feynman gas is independent of z except near the walls, the mean number $\bar{n}$ of particles in the Feynman gas is of order $\zeta_1 - \zeta_2 \sim \ln s$. There can be large fluctuations in the density over small ranges of z, but these are unlikely to be coherent over the entire volume of the gas so fluctuations in the number n of particles about $\bar{n}$ should be small relative to $\bar{n}$, presumably of order $(\bar{n})^{\frac{1}{2}}$.(16) In particular it is unlikely that n will be 2 or 3 or 4 when s is sufficiently large. But this is in clear contradiction with the assumption that diffractive cross-sections are constant as $s \to \infty$. This provides the motivation for trying to distinguish diffractive processes from multiperipheral processes; we imagine that it is only the multiperipheral part of the inelastic cross-sections which act like the distribution functions of a gas with short range forces. It would then only be the multiperipheral cross-sections which produce a multiplicity proportional to $\ln s$ on the average, with fluctuations of order $[\ln s]^{\frac{1}{2}}$. Since diffractive cross-sections are mostly of low multiplicity, while the multiplicity of multiperipheral processes increases indefinitely with s, they should be clearly separated at sufficiently large s. In particular, for very large s there should be values of the multiplicity n with $1 \ll n \ll \ln s$ for which both the diffractive and multiperipheral cross-sections should be very small. One of the experiments is designed to test whether multiperipheral processes can be distinguished by their multiplicity from diffractive processes.

We must now re-examine the problem of how to normalize the single particle spectrum. This normalization is important due to the definition of the two particle correlation function $g(\mathbf{r}, \mathbf{r}', \mathbf{s}_1, \mathbf{s}_2)$. In particular, whether $g(\mathbf{r}, \mathbf{r}', \mathbf{s}_1, \mathbf{s}_2)$ goes to zero as $|\mathbf{r} - \mathbf{r}'|$ becomes large depends on how $\rho(\mathbf{r}, \mathbf{s}_1, \mathbf{s}_2)$ and $P(\mathbf{r}, \mathbf{r}', \mathbf{s}_1, \mathbf{s}_2)$ are normalized [see eqn (16)]. If it is only the multiperipheral part of the inelastic cross-sections which behave like distribution functions of a gas with short range forces then it is desirable to define a partition function which is a sum only over multiperipheral cross-sections and to define the density and correlation functions using this partition function. What this means is that both P and ρ should be normalized using σ_M, the multiperipheral part of the cross-section, rather than either σ_T or the total inelastic cross-section. Let us denote by $\rho_M(\mathbf{r}, \mathbf{s}_1, \mathbf{s}_2)$ and $g_M(\mathbf{r}, \mathbf{r}', \mathbf{s}_1, \mathbf{s}_2)$ the density and correlation function for the multiperipheral cross-sections normalized to σ_M. That is, $\rho_M(\mathbf{r}, \mathbf{s}_1, \mathbf{s}_2)\, \mathrm{d}^3r$ is the probability per multiperipheral interaction that a secondary is found in a multiperipheral interaction with position variable $\mathbf{r}$ in a range d^3r. It will now be assumed that it is $\rho_M(\mathbf{r}, \mathbf{s}_1, \mathbf{s}_2)$ and $g_M(\mathbf{r}, \mathbf{r}', \mathbf{s}_1, \mathbf{s}_2)$ (rather than $\rho(\mathbf{r}, \mathbf{s}_1, \mathbf{s}_2)$ and $g(\mathbf{r}, \mathbf{r}', \mathbf{s}_1, \mathbf{s}_2)$ which satisfy the Amati–Fubini–Stanghellini scaling laws.

One now has the problem of determining whether ρ and g will also satisfy scaling laws or whether one must measure ρ_M and g_M separately. The answer is that some scaling laws apply also to ρ and g, some do not. For example, the scaling law $\rho_M(\mathbf{r}, \mathbf{s}_1, \mathbf{s}_2) \to \rho_M(\mathbf{r}, \mathbf{s}_1)$ when ζ_1 and $z \to \infty$ holds also for ρ. To see this one defines $\rho_D(\mathbf{r}, \mathbf{s}_1, \mathbf{s}_2)$ to be the single particle spectrum for diffractive events, normalized to the total diffractive cross-section σ_D. At high energies we assume the total cross-section σ_T is approximately equal to $\sigma_D + \sigma_M$; then

$$\rho(\mathbf{r}, \mathbf{s}_1, \mathbf{s}_2) = \frac{\sigma_D}{\sigma_D + \sigma_M} \rho_D(\mathbf{r}, \mathbf{s}_1, \mathbf{s}_2) + \frac{\sigma_M}{\sigma_D + \sigma_M} \rho_M(\mathbf{r}, \mathbf{s}_1, \mathbf{s}_2) \tag{19}$$

One further assumption: assume that σ_M approaches a constant as $s \to \infty$ (i.e. does not go to zero). Now it is easily seen that $\rho_D(\mathbf{r}, \mathbf{s}_1, \mathbf{s}_2)$ becomes independent of $\mathbf{s}_2$ when z and $\zeta_1 \to \infty$ due to diffractive cross-sections individually being constant for large s. Since σ_D and σ_M are constants for large s, the combined spectrum ρ is also independent of s_2, in the limit z and $\zeta_1 \to \infty$.

An example of a result holding for $g_M(\mathbf{r}, \mathbf{r}', \mathbf{s}_1, \mathbf{s}_2)$ which does *not* hold for the full correlation function $g(\mathbf{r}, \mathbf{r}', \mathbf{s}_1, \mathbf{s}_2)$ is the result that $g_M(\mathbf{r}, \mathbf{r}', \mathbf{s}_1, \mathbf{s}_2) \to 0$ when $|z - z'| \gg \xi$ where ξ is a fixed correlation length. Suppose for example that ζ_1 is very large, $z' \sim \zeta_1$ and z lies in the range $0 \ll z \ll \zeta_1$ (so that $\zeta_1 - z$ and $z - \zeta_2$ are both $\gg 1$). Then there are no diffractive events which give secondaries of position $\mathbf{r}$; as a result the correlation function $g(\mathbf{r}, \mathbf{r}', \mathbf{s}_1, \mathbf{s}_2)$ can be written

$$\begin{aligned} g(\mathbf{r}, \mathbf{r}', \mathbf{s}_1, \mathbf{s}_2) &= \frac{\sigma_M}{\sigma_M + \sigma_D} P_M(\mathbf{r}, \mathbf{r}', \mathbf{s}_1, \mathbf{s}_2) - \frac{\sigma_M}{\sigma_M + \sigma_D} \rho_M(\mathbf{r}, \mathbf{s}_1, \mathbf{s}_2) \\ &\times \frac{\sigma_M \rho_M(\mathbf{r}', \mathbf{s}_1, \mathbf{s}_2) + \sigma_D \rho_D(\mathbf{r}', \mathbf{s}_1, \mathbf{s}_2)}{\sigma_M + \sigma_D} = \frac{\sigma_M}{\sigma_M + \sigma_D} g_M(\mathbf{r}, \mathbf{r}', \mathbf{s}_1, \mathbf{s}_2) \\ &- \frac{\sigma_M \sigma_D}{(\sigma_M + \sigma_D)^2} \rho_M(\mathbf{r}, \mathbf{s}_1, \mathbf{s}_2)[\rho_D(\mathbf{r}', \mathbf{s}_1, \mathbf{s}_2) - \rho_M(\mathbf{r}', \mathbf{s}_1, \mathbf{s}_2)] \end{aligned} \tag{20}$$

It follows from this formula that the full correlation function $g(\mathbf{r}, \mathbf{r}', \mathbf{s}_1, \mathbf{s}_2)$ will not be zero for $|z - z'|$ large unless the diffractive spectrum ρ_D is equal to the multiperipheral spectrum ρ_M. This is impossible for all $\mathbf{r}'$ because $\rho_M(\mathbf{r}', \mathbf{s}_1, \mathbf{s}_2)$ is constant when $\zeta_2 \ll z' \ll \zeta_1$ while $\rho_D(\mathbf{r}', \mathbf{s}_1, \mathbf{s}_2)$ is zero in this range.

The problem of separating multiperipheral and diffractive events will be discussed in connection with individual experiments.

To conclude the discussion of multiperipheral events there are two miscellaneous observations to be made. First, there has always been some reluctance to measure secondary pion spectra on the grounds that most of them are probably decay products of ρ's, A_1's, etc., and so one would prefer to know the spectra of ρ's and A_1's. However, the scaling laws of Amati, Fubini, and Stanghellini for pion spectra are equally valid whether or not the π's are mostly decay products.[10] The intuitive reason for this is that the decay process is itself a short range effect in $\mathbf{r}$-space [for example a π^+ and π^- with an invariant mass of order m_ρ cannot have a large separation in z, from eqn (18)]. Hence if the cross-sections for producing ρ's and other resonances obey the multiperipheral scaling laws, then so will the cross-sections for the π's resulting from ρ decays. This means that to test the multiperipheral scaling laws it is pointless to distinguish π's from ρ decay from uncorrelated π's.

The second observation is this. In p–p collisions it is possible for pions to be emitted backwards in the *lab* system due to the low mass of the pion. The minimum z for the pion is $-\ln(m_p/m_\pi)$ where m_p is the proton mass (this was shown earlier). This value for z corresponds to a rather fast backward pion. It is not easy to find dynamic mechanisms that would produce such fast backward pions; they are too fast to be decay products of low-lying N^*'s. So there should not be many such pions. By symmetry there cannot be many pions going much faster than the incident proton either, for such pions would be going backwards in the rest system of the incident proton. However a pion having the same velocity as the incident proton and a transverse momentum of about 300 MeV has only $\frac{1}{3}$ the energy of the incident proton. Hence the spectrum of pion in p–p collisions should fall rapidly as the longitudinal momentum k_z of the pion increases once $k_z/E > \frac{1}{3}$ where E is the incident proton energy. This is the explanation for the exponential fall-off in k_z seen experimentally for $k_z/E > \frac{1}{4}$ in pion spectra in p–p collisions.[17] This exponential tail is useful for experimental purposes, as will be discussed later. No similar tail is expected for high energy π's in π–p interactions; in π–p interactions the π spectrum should be reasonably flat up to the kinematic limit $k_z/E \simeq 1$.

Now a specific programme of experiments will be described. They test the theoretical ideas already described. However, if the predictions are correct the experiments can be interpreted without making numerical fits to a particular model; rather the experiments will define non-trivial properties of multiple production which all future models would have to agree with. The experiments should all be feasible with present-day techniques. I have tried to define the simplest and most practical experiment to test each theoretical prediction; I hope that these experiments will not be replaced by more elaborate "improvements" of them. The individual experiments are not new; what is important is that they form a sensible programme.

EXPERIMENT 1: PARTIAL CROSS SECTIONS AS A FUNCTION OF MULTIPLICITY

In this experiment one considers p–p interactions at fixed but large s. One measures the partial cross-section $\sigma_{nc}(s)$ for producing n charged secondaries, as a function of n. No distinction is made between π^+, π^-, p, $\bar{P}$, etc.: all charged particles are counted, and all neutrals are ignored. What one is looking for is a dip in a plot of $\sigma_{nc}(s)$ versus n. Namely, at sufficiently high s the partial cross-section $\sigma_{nc}(s)$ should first decrease with n as one moves out of the diffraction dissociation region, and then increase again as n approaches the mean multiplicity $\bar{n}_c(s)$ of charged secondaries from multiperipheral processes. One expects $\bar{n}_c(s)$ to be proportional to $\ln s$, so the larger s is the more pronounced the dip should be.

There is no guarantee that a dip will occur at the energies of NAL or the CERN ISR, even if the theoretical picture of diffraction dissociation plus multiperipheralism is correct. So an absence of a dip, while disappointing, does not disprove the picture. The importance of the experiment is that a dip, if found, would be clear cut and model independent evidence for the existence of two separate processes in multiple production.

If one could detect neutrals as easily as charged particles one would measure $\sigma_n(s)$, the partial cross-section for producing n secondaries including neutrals instead of $\sigma_{nc}(s)$. Because the mean number $\bar{n}(s)$ of all secondaries in multiperipheral processes is larger than $\bar{n}_c(s)$, the dip is likely to appear at a lower energy in the function $\sigma_n(s)$ than in $\sigma_{nc}(s)$. But I believe the charged multiplicity remains the simplest experiment to do even if large bubble chambers can see γ-rays, and I urge that an experiment on charged multiplicities not be held up in hopes of doing a measurement including neutrals.

The next group of experiments concern the single particle spectrum. For purposes of this discussion I shall write ρ as a function of $k_\perp$, k_z, and E, where $k_\perp$ is the transverse momentum of the secondary, k_z is the longitudinal momentum in the *lab* system of the secondary, and E is the incident energy. Phase space is d^3k/k_0: $\rho(k_\perp, k_z, E)\, d^3k/k_0$ is the probability per collision to find a secondary of momentum $\mathbf{k}$ in a range d^3k. There are separate spectra for π^+, π^-, K^+, etc.; there are also according to theory separate spectra $\rho_D(k_\perp, k_z, E)$ and $\rho_M(k_\perp, k_z, E)$ for diffractive events and multiperipheral events. The spectrum ρ_M is normalized to the total multiperipheral cross-section $\sigma_M(s)$ instead of the total cross-section; ρ_D is normalized to the total diffractive cross-section. When it is necessary to separate ρ_M or ρ_D from ρ we will discuss how to do it, if possible.

I beg all experimentalists not to plot single particle spectra as angular distributions. It is $k_\perp$, not an angle, which has an average value around 300 MeV independent of k_z; this fact makes plots vs. $k_\perp$ and k_z much simpler to interpret than angular distributions. I personally prefer the use of the lab variable k_z or else the rapidity z to a centre-of-mass variable, since the multiperipheral scaling laws look simpler in terms of these variables.

Experiment 2: Beam Survey from a Hydrogen Target

In this experiment one measures the spectrum $\rho(k_\perp, k_z, E)$ of high energy secondary π^- mesons from p–p collisions. The measurement is made varying E holding $k_\perp$ and k_z/E fixed (fixed x in Feynman's language). The prediction to be tested is that $\rho(k_\perp, k_z, E)$ is independent of E when $k_\perp$ and k_z/E are held fixed. The range of E should be ~ 30 GeV and up; $k_\perp$ should be of order 300 MeV and $k_z/E \sim \frac{1}{3}$. The choice of $k_\perp$ and k_z is meant to ensure that one is in a region where the spectrum is large. The choice of π^- over other particles is for simplicity, since all but a few negative secondaries are π^-. A good experiment would be to hold $k_\perp$ fixed, measure the spectrum as a function of k_z and E and try to fit the data to an exponential depending only on k_z/E.

The experiment can also be done in π^+p or π^-p collisions but in this case one does not have the exponential behaviour in k_z which makes for a good experiment. On the other hand, one can get to higher values of k_z for the secondary π if the incident particle is a π, since the limitation $k_z/E \lesssim \frac{1}{3}$ no longer applies.

There is no need to separate diffractive from multiperipheral contributions since both are predicted to obey the Amati–Fubini–Stanghellini–Feynman scaling law.

Experiment 3: Factorization in the Single Particle Spectrum

In this experiment one measures the spectrum $\rho(k_\perp, k_z, E)$ for backwards π^- in the lab in both p–p and π–p collisions. The energy E is high and fixed; I suggest holding $k_\perp$ fixed and fitting to an exponential in k_z. What one is testing for is whether these spectra are equal for p–p and π–p collisions, that is whether $\rho(k_\perp, k_z, E)_{\pi p} = \rho(k_\perp, k_z, E)_{pp}$. According to the picture described earlier the multiperipheral spectrum $\rho_M(k_\perp, k_z, E)$ should be the same for high energy π^-p and pp collisions as long as one is looking at low energy secondaries; this is because a low energy particle is a particle near the target wall in the Feynman gas and cannot tell what particle was incident at the other wall which is many interaction lengths away. (This prediction is another general consequence of the multiperipheral model.[10]) It is not clear whether or not the diffractive component $\rho_D(k, k_z, E)$ should be the same for π–p and p–p collisions. If diffraction is described by a simple Regge pole then the factorization property of Regge poles requires that $\rho_D(k_\perp, k_z, E)$ be the same for π–p and p–p. If diffraction is something else than Regge exchange then $\rho_D(k_\perp, k_z, E)$ could very well be different (at least in magnitude, if not in shape) for π–p and p–p scattering. If this is the case then one will see a violation of factorization when one performs this experiment. I do not expect there will be large violations of factorization, so I think it is important to do this experiment carefully so that one can make accurate comparisons of πp and pp collisions at the same values of $\mathbf{k}$; also the experiment will have to be done at several incident energies E to see if any violations of factorization persist as E increases. One does not have to use incident π's and p's of the same energy; one might instead use π's and p's with the same velocity, which means the π would have $\frac{1}{7}$ of the energy of the proton. I would suggest a compromise: Let the incident π's have $\frac{1}{3}$ the energy of the incident protons.

There is no way in this experiment to separate ρ_M from ρ_D; a separate experiment will be proposed later to test factorization in multiperipheral processes separately.

This experiment tests whether factorization holds when ρ has both diffractive and multiperipheral contributions. Even if factorization breaks down, it is still possible for ρ_D and ρ_M separately to factorize; if $\rho_D \neq \rho_M$ and if the ratio σ_M/σ_{tot} is different for πp and pp collisions, then ρ will not factorize (see eqn (19)).

EXPERIMENT 4: THE dk_z/k_z LAW

In this experiment one measures the spectrum $\rho(k_\perp, k_z, E)$ of secondary π^- in p–p collisions. The energy E is held fixed and should be the highest energy available. The spectrum is measured as a function of k_z holding $k_\perp$ fixed. As usual $k_\perp$ should be of order 300 MeV, while k_z should be in the intermediate range, say 1 GeV $< k_z < E/4$. What one is testing for is whether $\rho(k_\perp, k_z, E)$ is independent of k_z in the range 1 GeV $\ll k_z \ll E/4$. This prediction is called the "dk_z/k_z" law because longitudinal phase space has the form dk_z/k_z in this region.

The problem with this experiment is that the prediction that ρ is independent of k_z is valid only for secondaries which in the Feynman gas analogy must be many interaction lengths away from both walls. In experiments 2 and 3 the only requirement is that the walls be separated by many interaction lengths. So if "many interaction lengths" turns out to be a distance

z_0, then the bound for s in experiments 2 and 3 is $\zeta_1 - \zeta_2 = \ln(s/\mu_1\mu_2) > z_0$ but the bound for this experiment is $\ln(s/\mu_1\mu_2) > 2z_0$. To double $\ln s$ means squaring s; for example, if experiments 2 and 3 require $s > 30\ \text{GeV}^2$, this experiment requires $s > 900\ \text{GeV}^2$! (μ_1 and μ_2 are of order 1 in GeV for p–p collisions and can be neglected in these units). One is further squeezed in this experiment because the pions must have z's considerably less than ζ_2 which means energies $\ll E/3$ instead of energies $\ll E$. This latter squeeze can be avoided by using πp instead of pp collisions; if one can get enough incident pions with more than $\frac{1}{3}$ the beam energy the πp-experiment is probably better.

The prediction is precisely that $\rho_M(k_\perp, k_z, E)$ is independent of k_z in the intermediate range; but according to the picture of diffraction described earlier $\rho_D(k_\perp, k_z, E)$ is negligible in this range so $\rho(k_\perp, k_z E)$ should also be independent of k_z in the intermediate range.

EXPERIMENT 5: SEARCH FOR DOUBLE POMERON EXCHANGE

In this experiment one measures the cross-section for the specific reaction $p + p \rightarrow p + p + \pi^+ + \pi^-$. One looks in particular at π^+–π^- pairs whose centre of mass is at rest in the centre-of-mass system of the incident particles. The cross-section is studied as a function of the incident energy E and the invariant mass squared s' of the pair. What one is looking for is (a) the energy dependence of the reaction, for fixed s', and (b) whether the ρ peak broadens as E increases.

The theory of this experiment is as follows. Double pomeron exchange (see Fig. 1), if it exists, can only produce $I = 0\ \pi$ pairs and cannot produce ρ's. Pomeron exchange with the incident particle plus ρ exchange with the target particle can produce ρ's. Double Pomeron exchange if it exists should become the dominant process compared to other exchanges at sufficiently high energies. There should not be a narrow resonance like the ρ in the $I = 0$ channel; if $I = 0$ pairs become important the ρ peak in the π^+–π^- mass distribution should appear to be broadened due to these pairs. The cross-section for producing ρ's should fall with energy; for example, if ρ's are produced by Pomeron exchange $+\rho$ exchange, one finds the cross-section should behave as $E^{-\frac{1}{2}}$, E being the incident energy. (To be precise the cross-section is the cross-section for producing π^+–π^- pairs with fixed windows in the centre of mass for the π^+ and π^- and no restrictions on the final state of the protons.) A naïve double Pomeron model predicts a constant cross-section for π^+–π^- pairs with $I = 0$; there are enough problems with double Pomeron exchange[(9)] that I am unwilling to make this or any other prediction for the $I = 0$ pair cross-section.

There is no sign of $I = 0$ production at 30 GeV energies, so it is unlikely to overwhelm ρ production at NAL or CERN energies; at best the $I = 0$ cross-section might become comparable to the ρ cross-section.

In summary one looks at the energy dependence of the ρ cross-section to see if it falls with E according to a Pomeron $+\rho$ exchange model or another exchange model; one looks for a broadening of the ρ peak to see if there is double Pomeron exchange.

EXPERIMENT 6: CORRELATION LENGTH EXPERIMENT

In this experiment one measures the two-particle spectrum $P(k_\perp, k_z, k'_\perp, k'_z, E)$ for production of a π^+ of momentum $\mathbf{k}$ and a π^- of momentum $\mathbf{k}'$ in $\pi^- p$ collisions. That is the probability of finding a π^+ or momentum $\mathbf{k}'$ and a π^- of momentum $\mathbf{k}'$ in ranges d^3k and d^3k' in the final state, per collision, is $P\,\mathrm{d}^3k\,\mathrm{d}^3k'/k_0k'_0$. This is an inclusive experiment in the Feynman terminology. For fixed and large E, one holds $k_\perp$, $k'_\perp$, and k'_z fixed, with $k_\perp$ and $k'_\perp$ of order 300 MeV and $k'_z \sim E$ (say $k_z = 2E/3$). P is then measured as a function of k_z. In other words one looks at a fast forward π^- of fixed momentum, while varying the longitudinal momentum of the π^+. One wants a wide range of k_z, say from 500 MeV up to the beam momentum. Along with P, one should also measure $\rho(k_\perp, k_z, E)$ and $\rho(k'_\perp, k'_z, E)$ for use in the analysis discussed below.

If one uses the naïve analogy to a gas then what one should do with the data is to compute the two particle correlation function $g(k_\perp, k_z, k'_\perp, k'_z, E)$:

$$g(k_\perp, k_z, k'_\perp, k'_z, E) = P(k_\perp, k_z, k'_\perp, k'_z, E) - \rho(k_\perp, k_z, E)\,\rho(k'_\perp, k'_z, E) \tag{21}$$

What is interesting is to see if this correlation function exhibits a correlation length, i.e. see if it goes to zero as k_z becomes small compared to k'_z and E. This is the first thing to investigate with the data. However, according to the theory of this paper it is the multiperipheral correlation function g_M (instead of g) that should have a correlation length. If g does not go to zero when k_z is small compared to k'_z and E, then it is worth looking for a correlation length in multiperipheral processes alone.

For k_z in the intermediate range (300 MeV $\ll k_z \ll E$), both P and $\rho(k_\perp, k_z, E)$ should be free of diffractive contributions and hence differ only in normalization from P_M and ρ_M. Hence, if g_M shows a correlation length, one should have

$$P(k_\perp, k_z, k'_\perp, k'_z, E)/\rho(k_\perp, k_z, E) \simeq P_M/\rho_M \simeq \rho_M(k'_\perp, k'_z, E) \tag{22}$$

There is no independent way to measure $\rho_M(k'_\perp, k'_z, E)$, so the content of this equation is that P/ρ should be independent of k_z when k_z is in the intermediate range. This suggests that one look for a correlation length by seeing if P/ρ does approach a constant for $k_z \ll E$. If so, the way to obtain a correlation length is to fit the departure of P/ρ from the constant to an exponential in the gas variable z, or else find the value of $\xi = |z - z'|$ above which P/ρ is constant.

It is quite possible that P/ρ will be constant for k_z in the intermediate range, but depart from this constant when k_z is small and P and ρ again have diffractive contributions.

For 25 GeV incident pions there is no intermediate range for k_z, because diffraction dissociation or isobar decays can produce pions at rest in the centre-of-mass system. This will not be possible at NAL energies, unless diffraction dissociation involves intermediate states of mass > 2 GeV (cf. Franzini[2]).

The main aim in choosing experimental parameters is to have an intermediate range for k_z not contaminated with diffraction dissociation products. It is important for this reason not to let $k_\perp$ be much smaller than 300 MeV. The reason for this is that the invariant mass squared of the π^+ and π^-, for k_z in the intermediate region, is

$$s_{+-} \simeq k_\perp^2 k_z'/k_z. \tag{23}$$

If $k_\perp^2$ is small then s_{+-} will be small even if the π^+ is transverse in the centre of mass (i.e. in the middle of the intermediate range). In practice one would like $s_{+-} > 4\ \text{GeV}^2$ or so when the π^+ is transverse in the centre of mass, to be well away from diffraction dissociation effects. One does not want to increase $k_\perp$ much beyond 300 MeV because then one gets into the tail of the transverse momentum distribution and out of the interesting region.[13]

Experiment 7: Test of Factorization in Multiperipheral Processes

In this experiment one measures $P(k_\perp, k_z, k_\perp', k_z', E)$ and $\rho(k_\perp', k_z', E)$ for very large fixed E in both πp and pp collisions. The momentum k' refers to a π^+ coming out at 90° in the centre of mass, its momentum is held fixed. The momentum k refers to a π^- which is backwards in the lab system (as in the previous factorization experiment). Define

$$\rho_{M(\exp)}(k_\perp, k_z, E) = P(k_\perp, k_z, k_\perp', k_z', E)/\rho(k_\perp', k_z', E) \tag{24}$$

The purpose of this experiment is to see whether $\rho_{M(\exp)}(k_\perp, k_z, E)$ is the same for πp and pp collisions.

The only difference experimentally between this experiment and Experiment 3 is that instead of measuring the backwards pion spectrum for all collisions one is now measuring the backwards pion spectrum only for those collisions which emit a transverse π^+ in the centre of mass as well as a backwards pion. In other words one is using the transverse pions as an event or beam monitor. This excludes diffractive events according to our theoretical picture; both P and ρ should be proportional to P_M and ρ_M and the ratios P/ρ should equal P_M/ρ_M. Furthermore the π^+ and π^- will be well separated in z so $P_M(k_\perp, k_z, k'_\perp, k'_z, E)$ should factor into $\rho_M(k_\perp, k_z, E)\rho_M(k'_\perp, k'_z, E)$. Hence one expects

$$\rho_{M(\exp)}(k_\perp, k_z, E) \simeq \rho_M(k_\perp, k_z, E) \tag{25}$$

Hence comparing $\rho_{M(\exp)}(k_\perp, k_z, E)$ for πp and pp collisions is a test of factorization for multiperipheral processes alone. As in Experiment 3 one should make this test at several energies E so one can see whether violations of factorization decrease as E increases.

These seven experiments constitute the experimental programme. If any of these experiments agrees well with the prediction, the result will be interesting whether or not one likes the diffractive plus multiperipheral picture of multiple production. Every one of the predictions cited can be wrong and it would be remarkable if they all were to be verified. If it turns out that diffractive events are not distinguishable from multiperipheral events, Experiment 7 may not be worth pursuing. There are many reasonable ways to modify the experiments proposed here but I hope that such modifications will be examined critically to determine if they achieve the objectives of these experiments as clearly and as simply as the specific experiments cited here.

ACKNOWLEDGEMENTS

When I first studied multiperipheralism I was much helped by conversations with Dr. A. Krzywicki. In the last year I have been helped and reinspired by several conversations with Professor R. P. Feynman. The discussion of two particle correlations described here resulted from conversations with Professor L. Brown and Drs. R. Peccei and D. Silverman at the Aspen Institute for Physics. I wish to thank also Professors S. Drell, J. Bjorken, W. Fraser, J. Peoples, and Dr. P. Roy for discussions.

APPENDIX. SHORT RANGE FORCES AND BOUNDED TRANSVERSE MOMENTUM

In the text the hypothesis of short range forces in the Feynman gas was used without specifying it precisely. It was also pointed out that it is hard to reconcile the observed bounded transverse momentum of secondaries with one's intuitive picture of the Feynman gas if it has short range forces. The multiperipheral model suggests a definition of short range forces from which one predicts bounded transverse momenta: the purpose of this Appendix is to describe this definition and show how it leads to bounded transverse momenta.

An intuitive picture of short range forces would be that the dependence of

$\sigma_n(\mathbf{r}_1, \mathbf{r}_2, \ldots, \mathbf{r}_n, \mathbf{s}_1, \mathbf{s}_2)$ on one of the $\mathbf{r}$'s, say $\mathbf{r}_1$, would be affected only by other $\mathbf{r}$'s which are near $\mathbf{r}_1$. A quantitative way of saying this is that if one changes $\mathbf{r}_1$ to $\mathbf{r}_1'$ the exact value of $\mathbf{r}_2$ does not matter if $|\mathbf{r}_1 - \mathbf{r}_2|$ is large, i.e.

$$\frac{\sigma_n(\mathbf{r}_1', \mathbf{r}_2, \ldots, \mathbf{r}_n, \mathbf{s}_1, \mathbf{s}_2)}{\sigma_n(\mathbf{r}_1, \mathbf{r}_2, \ldots, \mathbf{r}_n, \mathbf{s}_1, \mathbf{s}_2)} = \frac{\sigma_n(\mathbf{r}_1', \mathbf{r}_2', \ldots, \mathbf{r}_n, \mathbf{s}_1, \mathbf{s}_2)}{\sigma_n(\mathbf{r}_1, \mathbf{r}_2, \ldots, \mathbf{r}_n, \mathbf{s}_1, \mathbf{s}_2)} \tag{A.1}$$

in a limit in which $|\mathbf{r}_2 - \mathbf{r}_1| \to \infty$ while $|\mathbf{r}_1 - \mathbf{r}_1'|$ and $|\mathbf{r}_2 - \mathbf{r}_2'|$ are held fixed. But this is not the prediction of the multiperipheral model. In the multiperipheral model one must first introduce a set of momentum transfer variables. If $k_1, \ldots, k_n$ are the four-momenta of the secondaries, then one defines

$$\begin{aligned} q_1 &= p_1 - k_1 \\ q_2 &= q_1 - k_2 \\ q_{n-1} &= q_{n-2} - \mathrm{k}_{n-1} = \mathrm{k}_n - p_2 \end{aligned} \tag{A.2}$$

(see Fig. 3).

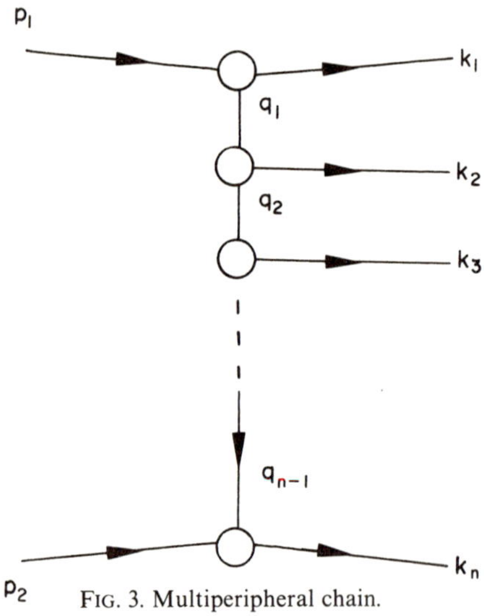

FIG. 3. Multiperipheral chain.

In the model of multiple meson exchanges, the k_i are the momenta of secondaries in order of emission from the multiperipheral chain (Fig. 3). If one writes the cross section for emitting n secondaries as a function $\sigma_n(q_1, \ldots, q_{n-1}, p_1, p_2)$ of the $n - 1$ momentum transfers, then the multiperipheral cross section factorizes into functions depending on neighbouring q_i (i.e. q_1 and q_2, or q_2 and q_3, etc.). The principal source of this dependence on neighbouring q's is the kinematic restriction that $(q_i - q_{i-1})^2 = m_i^2$ where m_i is the mass of the ith secondary; the propagators of the multiperipheral chain depend only on a single q_i. If the vertices of the chain are not point-like, they also depend on neighbouring q_i.

The first generalization from the multiperipheral model is that the short range force idea should be stated in terms of momentum transfers q_i instead of secondary momenta k_i. But there are still problems. The multiperipheral model predicts a factorization in terms of an ordering of the q_i rather than in terms of separation of q_i from q_j. That is, the dependence of σ_n on q_i in the multiperipheral model is independent of the values of q_{i+2}, q_{i-2}, etc., regardless of whether q_{i+2} or q_{i-2} is close to q_i or not, and dependent on $q_{i\pm 1}$ even if $(q_i - q_{i\pm 1})^2$ is large. In summary the σ_n's of the multiperipheral model look more like a random walk distribution function than a gas distribution with short range forces.

An analysis of the multiperipheral model shows that on the average the q_i tend to order themselves so that $q_{i0} > q_{i+10}$ and $k_{iz} > k_{i+1z}$; the highest energy transfers occur at the end of the chain associated with the incident particle, and the secondaries with the highest longitudinal momentum are emitted near this end of the chair. Also the quantities $(q_i - q_{i-j})^2$ tend to increase as j increases, so on the average the factorization of the multiperipheral is similar to a short range force hypothesis. This suggests that the way to generalize the multiperipheral model is as follows. First, one defines an ordering of the secondary momenta, namely, they should be numbered so that $k_{1z} \geqslant k_{2z} \geqslant \ldots \geqslant k_{nz}$. (An alternative is to order in terms of the z variables: $z_1 \geqslant z_2 \geqslant z_3 \ldots$). Then one defines $q_1, \ldots, q_{n-1}$ by eqns (A.2). Then one supposes that the dependence of $\sigma_n(q_1, \ldots, q_{n-1}, p_1, p_2)$ on q_i is determined only by those q_j for which $(q_j - q_i)^2$ is small. Since $s_{ij} = (q_i - q_j)^2$ is the total mass squared of the secondaries numbered $i + 1$ to j (or $j + 1$ to i if $j < i$), it is reasonable to suppose a dependence on s_{ij} when s_{ij} is in the resonance region; the crucial assumption is that there will be no appreciable dependence on s_{ij} when s_{ij} is beyond the region of the principal resonances.

There is another part to any short-range force hypothesis, namely, the dependence of $\sigma_n(q_1, \ldots, q_{n-1}, p_1, p_2)$ on q_i must not be changed if one changes the total number of secondaries but leaves unchanged the q_j near q_i (in a gas analogy this is saying the distribution function for particles in one region cannot be changed if one adds or subtracts particles in a far-away region). This is also true in the multiperipheral model and should be demanded for any generalization from it. A quantitative statement of this requirement will not be given.

As far as the properties of one and two particle spectra are concerned, a short range force hypothesis in terms of the q_i is as good as a short range force hypothesis in terms of the k_i themselves. The reason is that $k_i = q_i - q_{i-1}$ and q_i and q_{i-1} are close variables; if k_i and k_j are well separated the variables q_i and q_{i-1} tend also to be well separated from q_j and q_{j-1}, and as a result the particles of momentum k_i and k_j tend to be uncorrelated. One also assumes uncorrelated dependence on p_1 or

p_2 and q_j if $(p_1 - q_j)^2$ or $(p_2 + q_j)^2$ is large; this then leads to factorization of the dependence of spectra on the secondary momenta k and the incident momenta when the secondary is (in the gas analogy) well away from the walls. These results can be shown in detail for the multiperipheral model itself using the multiperipheral integral equation of Amati, Fubini, and Stanghellini;[10] to show that scaling laws hold for a generalized short range force picture quantitatively is more difficult but it seems reasonable that they should still hold.

Finally, using the momentum transfer variables q_i it is easy to bound the transverse momenta of secondaries. All that is necessary is to bound the momentum transfers q_i^2. The reason that this bounds the transverse components $k_{i\perp}$ of k_i is the following. First, one shows that the 2-vector (q_{i0}, q_{iz}) is spacelike (except possibly for i near 1 or n). The reason is simple. Consider the two-vectors (p_{10}, p_{1z}), (p_{20}, p_{2z}), (k_{10}, k_{1z}), $\ldots$, (k_{n0}, k_{nz}). These are all timelike two-vectors, with masses $\mu_1, \mu_2, m_{1\perp}, \ldots, m_{n\perp}$. If the two-vector momentum transfer from p_1 to $k_1 + \ldots + k_i$ is timelike also it acts like a final state particle in one of two ways: either in the sense of splitting p_1 into $q_i + k_1 + \ldots + k_i$ or in splitting p_2 into $-q_i + k_{i+1} + \ldots + k_n$. For the first splitting to be possible the mass μ_1 of p_1 must exceed the sum of the masses $m_{j\perp}$ of $k_1 \ldots k_i$ but since $m_{j\perp} \geqslant m_j$, this sum usually exceeds μ_1 (except for special cases when i is small and the outgoing particles have lower rest masses than the incident particle). Likewise it is unlikely for the sum of $m^j_\perp$ from $j = i + 1$ to $j = n$ to be less than μ_2. Hence (q_{i0}, q_{iz}) is spacelike. But this means $q_{i\perp}^2 < |q_i^2|$, so if $|q_i^2|$ is bounded then $q_{i\perp}^2$ is also; since $k_i = q_i - q_{i-1}$, $k_{i\perp}^2$ is bounded also.

This argument breaks down if the incident particles have large perpendicular momenta, for then the two-vectors (p_{10}, p_{1z}) and (p_{20}, p_{2z}) have masses $\mu_{1\perp}$ and $\mu_{2\perp}$ which are large and it is quite easy for (q_{i0}, q_{iz}) to be timelike. In summary, all links of the multiperipheral chain and its generalizations know the direction of the momentum of the incident particle; however the magnitude of this momentum is forgotten through a random walk effect as one goes many links away from the incident particle.

References

For more recent references see the other articles in this Volume.

1. See, e.g., O. Czyzewski in "Proceedings of the 14th International Conference on High Energy Physics, Vienna," p. 377.
2. For recent experimental studies of multiple production, see the talks of P. Franzini, M. Koshiba, W. D. Walker, and R. Panvini. *In* "High Energy Collisions". (C. N. Yang *et al.* eds) Gordon and Breach, New York, (1969). See also J. L. Day *et al. Phys. Rev. Letts* **23**, 1055 (1969); D. B. Smith *et al., ibid.* **23**, 1064 (1969).
3. For some current theoretical work on multiple production, see T. T. Chou and C. N. Yang. *Phys. Rev. Letts* **25**, 1072 (1970); J. Benecke, T. T. Chou, C. N. Yang, and E. Yen, *Phys. Rev.* **188**, 2159 (1970); R. P. Feynman. *In* "High Energy Collisions"

(Ref. 2); G. F. Chew, T. Rogers and D. R. Snider, UCRL-19457, preprint (1970); N. F. Bali, L. S. Brown, R. D. Peccei, and A. Pignotti, *Phys. Letts* **33B**,175 (1970).
4. It is part of the Feynman picture (Ref. 3) that one should distinguish multiperipheral events from diffractive events.
5. For recent work on bounds, see R. J. Eden. Invited talk at the Wisconsin Conference "Expectations for Particle Reactions at the New Accelerators" (April 1970).
6. I shall use the term "diffractive processes' to include both diffraction and diffraction dissociation. The original reference on diffraction dissociation is M. Good and W. Walker. *Phys. Rev.* **120**, 1857 (1960).
7. The ideas of this paper differ in spirit from the limiting fragmentation hypothesis of Benecke *et al.* (Ref. 3) in that we distinguish two types of processes (diffractive and multiperipheral) contributing strongly to the total cross-section. Some of the predictions discussed here agree with predictions of Benecke *et al.*
8. See also T. T. Chou and C. N. Yang (Ref. 3).
9. J. Finkelstein and K. Kajantie. *Phys. Letts* **26B**, 305 (1968); G. F. Chew and A. Pignotti. *Phys. Rev.* **176**, 2112 (1968).
10. D. Amati, A. Stanghellini, and S. Fubini. *Nuovo Cimento* **26**, 896 (1962). For a review see S. Fubini. *In* "Strong Interactions and High Energy Physics", (R. G. Moorhouse, ed.) Plenum Press, New York, (1964), p. 259. For another review of the scaling laws see K. Wilson. *Acta Phys. Austr.* **17**, 37 (1963). For an attempt to check the scaling laws against cosmic ray data see O. Czyzewski and A. Krzywicki. *Nuovo Cimento* **30**, 603 (1963).
11. R. P. Feynman, privatc communication.
12. The function σ_n has been normalized in such a way that no factors of 2π occur in eqn (11); also the incident particle is assumed to be highly relativistic so the velocity factor is a constant and included in the normalization.
13. As noted earlier, the trouble with looking at small cross-sections is that they may be affected by small effects which are negligible in large cross-sections.
14. It is hard to understand, given the hypothesis of "short range forces," how $\rho(\mathbf{r}, \mathbf{s}_1, \mathbf{s}_2)$ can depend even on $k_\perp$. This problem arises because the separation of $k_\perp$ into k_z and $k_\perp$ is defined in terms of the direction of the momentum of the incident particle, whereas the hypothesis of short range forces suggests that the density at $\mathbf{r}$ should be independent of the momentum of the incident particle. The resolution of this problem is that the definition of "short range forces" that is suggested by the multiperipheral model has some subtle features that permit $\rho(\mathbf{r}, \mathbf{s}_1, \mathbf{s}_2)$ to depend on the directions but not the magnitudes of ρ_1 and ρ_2. This is explained in the Appendix.
15. I apologize for having used x both as the x coordinate of $\mathbf{r}$ and as Feynman's inelasticity variable; from now on x will appear explicitly only as an inelasticity.
16. The distribution in n for multiperipheral processes has been predicted to be a Poisson distribution; see, e.g. G. F. Chew and A. Pignotti. *Phys. Rev. Letts* **20**, 1078 (1968).
17. This explanation is found in, for example, C. N. Yang in "High Energy Collisions" (Ref. 2). For an experimental curve see S. Drell, invited paper presented to Conference on "Expectations for Particle Reactions at the New Accelerators" (Madison, April 1970).
18. For a review of double Pomeron exchange see E. L. Berger, invited paper at the Irvine Regge Pole Conference (Dec. 1969). For a test of double Pomeron exchange see R. Lipes, G. Zweig, and W. Robertson, *Phys. Rev. Letts* **22**, 433 (1969).

AUTHOR INDEX

Numbers in parentheses are reference numbers and are included to assist in locating references in the text where the authors' names are not mentioned. Numbers in italics indicate the pages on which references are listed in full.

C

D

E

F

G

I

J

K

L

N

O

P

T

SUBJECT INDEX

Q

R

S